Introduction to

Biology and Biotechnology

Second Edition

Introduction to
Biology and Biotechnology
Second Edition

K. Vaidyanath
M.Sc, Ph.D
Professor of Genetics (Retd.),
Osmania University, Hyderabad (AP)

K. Pratap Reddy
M.Sc, Ph.D
Professor of Zoology,
Osmania University, Hyderabad (AP)

K. Satya Prasad
M.Sc, Ph.D
Professor of Botany,
Osmania University, Hyderabad (AP)

BSP BS Publications
An Imprint of **BSP Books Pvt. Ltd.**
4-4-309/316, Giriraj Lane,
Sultan Bazar, Hyderabad - 500 095

Introduction to Biology and Biotechnology, Second edition
by **Vaidyanath, K. Pratap Reddy, K. Satya Prasad**

1st Edition 2004 (*Introduction to Applied Biology and Biotechnology*)
2nd Edition, 2008

Third Reprint, 2023

Published by :

BSP **BS Publications**
An Imprint of **BSP Books Pvt. Ltd.**
4-4-309/316, Giriraj Lane, Sultan Bazar,
Hyderabad - 500 095
Phone : 040 - 23445688
e-mail: info@bspbooks.net
www.bspbooks.net

Printed at

Repro India Limited.

Price: Rs. 995.00

ISBN: 987-81-7800-152-4 (Paperback)

Dedicated to

Our parents, teachers

and

the young minds

Preface to Second Edition

The first edition of this book appeared in 2004. Since then much water flowed under the bridge of Biotechnology. Developments in the areas of biotechnology have radically altered our concept of biological applications from the point of view of sustainable development and human welfare. Some of the developments are breathtakingly fresh. Many genetically modified plants, animals and microbes including recombinant human gene products have flooded the market, which were well-nigh impossible some three decades ago. Such is the revolutionary pace of biotechnology, especially after the development in the fields of Genomics, Proteomics and Structural Biology, that it has opened up plethora of possibilities in all spheres of human activity especially, towards the development of welfare society. The next one hundred years and beyond, the basic Biology and Biotechnology are going to be dominant factors on the horizon of industrial production and new applications, making possible sustainable development of man, material and ecology.

This book has been conceived primarily to provide an overview of many of the fundamentals of biology and biotechnology for the benefit of undergraduate and postgraduate students of engineering and technology, besides basic sciences, specializing in biotechnology. It may also serve other students, pursuing Agriculture, Veterinary Science, Medicine.

The book is totally revised and designed to cover some of the new topics like Human Biology, Genetics, Molecular Biology and Photosynthesis etc., as per the requirements of changed syllabi of B.Tech I year of the Indian Universities.

The authors hope that this revised edition is immensely useful to the students.

-Authors

Preface to Second Edition

Preface to First Edition

Biotechnology – the art and science of marshalling living organisms, part of the tissues and cells to produce economically important products to better human lives and environment is catching up and developing rapidly. Efforts are on by the developed as well as by developing countries to join forces to make biotechnology the most dynamic growth engine of economy in 21st century. The research and developments in the areas of biotechnology is alive with excitements and many revolutionary advances have been made. Centre to these developments is Biology which widened ultimately embracing the whole gamut of biology especially the areas of Cell Biology, Tissue culture, Immunology, Developmental Biology, Genetics, Molecular Biology, Biochemistry, Microbiology, Medicine, Agriculture and Animal Husbandary. Consequently many Indian universities are drawing up syllabi for undergraduates as well as for postgraduates in the faculties of science, technology and engineering. One problem faced by the engineering students, specializing in biotechnology, is that their knowledge of biology is of school level and they need to be tutored about the basics of biology as well as applications leading to the development of biotechnology. The basic principles that the engineering students have to learn are few simple elementary facts of Biology.

This book is conceived primarily for undergraduate students of engineering and other basic sciences specializing in Biotechnology, nevertheless, it may be useful to students of Botany, Zoology, Agriculture, Veterinary science and Medicine, both at undergraduate and postgraduate levels. The book covers the syllabi of JNTU, Anna and other Universities for B.Tech course in Biotechnology.

The book is designed in such a way that not only enough description of accessary areas of basic biology, but also their applications, and the type of experiments from which the concepts have been developed, compelling the reader to collate the evidences for the concept and comprehend their plausible technological applications and limitation.

The authors have referred and taken material liberally from the standard books in presenting the concepts, the list of which is given under general bibliography at the end of the book. Several important research finding have also been incorporated in the book from various sources, research journals, review articles, etc., but neither the names of researchers nor the journals in which their work was reported have been listed in the text barring few exceptions, the way of style of this book warranted.

The authors do hope that this book will be of immense help to the students of Biotechnology, both engineering as well as basic science stream.

-Authors

Acknowledgements

Several people helped in conceiving this book. Our thanks goes to many of our colleagues for their comments upon draft of early chapters and also to anonymous reviewers for their helpful comments. We thank J. Madhavi for her skillful rendering of illustrations and patient editing of error-ridden draft of this book, especially chapter IV and VI. We also thank M. Maruthi Prasad for help in compiling the information on photosynthesis and endocrine system. We also thank our publisher (BS Publications) for their steady interest in bringing out the book.

Finally, our gratitude goes to our respective family members, without whose constant support, we could not have accomplished this task.

-Authors

Contents

PART - A

FUNDAMENTALS OF BIOLOGY

CHAPTER 1

INTRODUCTION TO MICROBIAL WORLD

CHAPTER 3

INTRODUCTION TO ANIMAL BIOLOGY

Chapter 4

Introduction to Human Biology

CHAPTER 6

INTRODUCTION TO GENETICS AND MOLECULAR BIOLOGY

PART - B

ELEMENTS OF BIOTECHNOLOGY

CHAPTER 7

INTRODUCTION TO BIOTECHNOLOGY

CHAPTER 8

MICROBIAL BIOTECHNOLOGY

Chapter 11

Molecular Medicine and Medical Biotechnology

Part - A
Fundamentals of Biology

1 Introduction to Microbial World

1.0 Introduction

The microbial and human world is interwined closely. The relationship is often helpful and positive in many ways but also harmful especially in terms of misery and devastation caused by microbes leading to untold morbidity and death. Despite microbes causing diseases, through the ages people have conquered and managed to exploit them beneficially for the human welfare in a variety of ways. Biotechnologists, through genetic engineering techniques manipulated them to benefit mankind. Some of the developments like Sewerage treatment, manufacturing of organic acids, enzymes used in industrial processes, biofuels, biofertilizers, biopesticides, biosensors, single cell proteins etc; are listed below :

1.1 Microorganisms : A General Account

Microorganisms are ubiquitous living forms first to evolve, and present most abundantly on wide ranging habitats on earth. With the advent of the microscope in the 17th century, Leeuwenhock observed the multitude of diversity of microorganisms. He provided the earliest descriptions of bacteria in the proceedings of the Royal Society. Until the end of 18th century, not much progress in microbiology was made after Leeuwenhock. Italian naturalist Lazzaro Spallanzani produced experimental evidence to disprove the view of spontaneous generation. Later Louis Pasteur demonstrated the presence of microorganisms in the air and their role in fermentation and spoilage, which demolished the theory of spontaneous generation once for all. Louis Pasteur and Robert Koch greatly influenced the microbiology until the middle of 20th century. They examined the role of microorganisms as causative agents of infectious diseases and concentrated on the prevention of such diseases through immunization and sanitation. Koch and his associates developed simple methods for the isolation and culturing of microorganisms on solid media. He has defined rules to prove the causative factors for disease, ways and means for the isolation and identification of the pathogen.

There are diverse group of microorganisms. In fact the range of variation is mind boggling. The biological diversity or ('*biodiversity*'), a well known concept, is defined as the variability among living organisms which includes diversity within the species (intraspecific), between species (inter specific) and of ecosystems, to which the organisms come to adapt and occupy. The basic unit however is species which is defined as a group of similar strains distinguished from other similar groups by genotypic, phenotypic and ecological characteristics.

Living organisms are distinguished into prokaryotes and eukaryotes based on their cell structure. Whittaker proposed a system of classification (Fig. 1.1) in which five kingdoms are invoked consisting of plantea, fungi, animalia, protista and monera encompassing all organisms depending on cell structure and energy yielding systems. Microorganisms are distributed in the kingdoms of monera, protista, fungi and partly in plants.

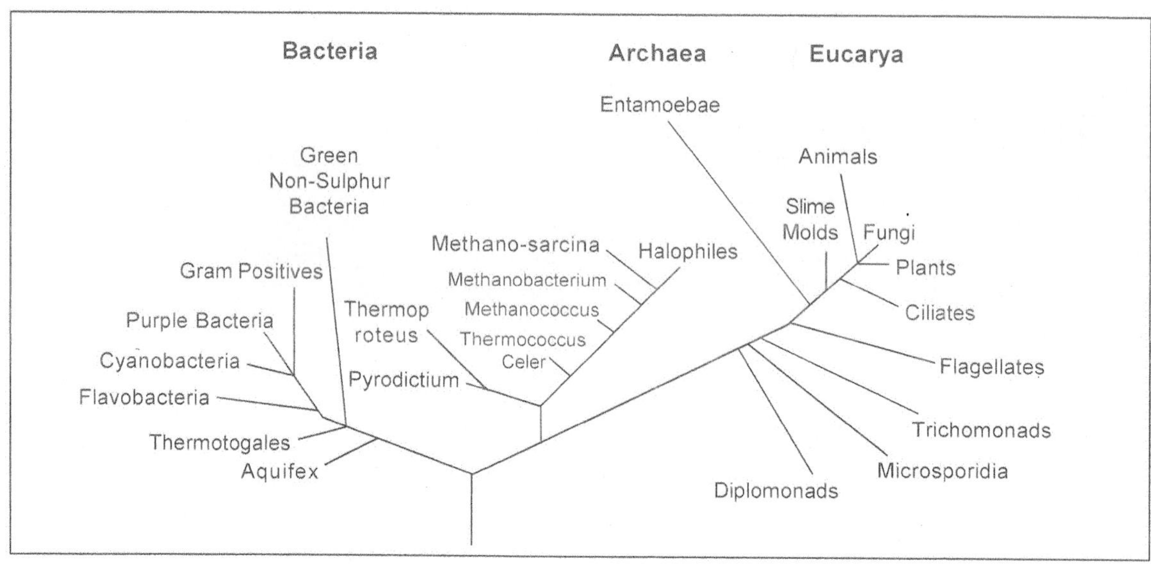

Fig. 1.1 Phylogenetic tree based on 16S and 18S rRNA homologies (Olsen and Woese, 1993)

Bacterial species includes strains with 70% or more DNA homology and 5% or less thermal stability. Recent studies of macro molecular (16S r RNA) homologies radically changed the classification of microbes into three domains, *Archaea*, *Bacteria* and *Eucarya*. The phylogenetic tree (Fig 1.1) based on 16S ribosomal RNA homologies reveal three evolutionary lines diverged from a common ancestral progenitor from Archaea, Bacteria and Eucarya (Woese *et al.*,1990). The divergence into three lines appears to have taken place early in the evolution, with Archaea diverging between Bacteria and Eucarya. The Eucarya (closer to the eukaryotes) evolved from a hypothetical ancestor "*Eurcaryote*" by acquiring cytoplasm and organelles from Bacteria and Archaea.

Mutations, genetic recombination and natural selection being the driving forces played a great role in the evolution of new microbial species. Variations in the genetic make up and natural selection favoured the proliferation of some kinds of microbes and eliminated others. Microbial diversity increased

with new kinds of microorganisms appearing as a result of evolution. Biodiversification of microorganisms has been a continuous process for over 3.85 billion years and the great biodiversity of microbes has yet to see the light of the day. Hawksworth (1991) and others estimated the total biological species including microorganisms and noted only 4 % of viruses, 5% of fungi and 12% of bacteria are known out of the total number of estimated species (Table 1.1).

Table 1.1 *Estimation of biological diversity (Hawksworth, 1991)*

Group	Species (Known)	Total Species (Estimated)	Percentage of Known species
Viruses	5000	1,30,000	4
Bacteria	4760	40,000	12
Fungi	69,000	15,00000	5
Algae	40,000	60,000	67
Bryophytes	17,000	25,000	68
Gymnosperms	750	-	-
Angiosperms	2,50,000	2,70,000	93
Protozoa	30,800	1,00,000	31

1.2 Bacteria

Anton van Leeuwenhock discovered tiny microbes called *animalcules* with his microscope in rain water, pond water and from most of the materials he examined. He described all the unicellular microorganisms such as algae, yeasts, protozoa and bacteria for the first time in 1676. The first description of bacteria has come from an important letter to the Royal Society of London in 1683 where he sketched the rod, sphere and spiral forms of bacteria and their movement. After Leeuvenhock, French microbiologist Louis Pasteur and German scientist Robert Koch established the role of bacteria in fermentations and as causal agents of diseases respectively. Robert Koch established his germ theory of disease in 1876 through a series of procedure known as *Koch's postulates*. These postulates became a guide for relating a specific microorganism to a single disease and for the development of pure culture technique. Fifty years period between 1860-1910 was considered as a golden age of microbiology with the expansion of our knowledge on bacteria that lead to the development of an independent discipline called *bacteriology*.

1.2.1 Distinguishing Characters

Bacteria are characteristically microscopic, unicellular organisms which reproduce by fission. Bacteria and blue-green algae are typically prokaryotic. But bacteria differ from blue-green algae and from all other organisms of plant kingdom in lacking a photosynthetic pigment chlorophyll '*a*'. However, some species are ubiquitous in their distribution inhabiting all the natural habitats like soil, water and air. Bacteria also occur in various foods, and food products such as milk, butter, cheese and milk beverages. They grow on various fruits, vegetables, staple foods like cereals, sugar and flour and perishable foods like meat and fish.

1.2.2 Classification

Bacteria are distinguished into different groups based on the following characters :

1. *Morphological characters :* Including cell shape and size, staining reactions, presence or absence of spore or reproductive forms and motility.

2. *Cultural characters :* Colony form on solid media and nutritional requirements for multiplication in liquid media.
3. *Biochemical characters :* Include metabolic end-products and presence or absence of a particular enzyme or pathway.
4. *Serological characters :* Concern the nature of the surface antigens as revealed by suitable specific antibodies.
5. *Molecular characters :* Include the base sequences of the DNA, GC ratios and nucleic acid-hybridisation.

Bacteria are classified into 33 sections according to Bergey system. The Bergey's Manual of systematic Bacteriology provide authentic descriptions of bacterial species. Gram negative bacteria are described in 1 – 11 sections while sections 12 – 17 deal with the Gram positive, phototrophic bacteria. Archaea bacteria are described in 18 – 25 sections. Actinomycetes and other filamentous bacteria are described in 8 sections from 26 – 33.

The rigid cell wall determines the shape of the bacterial cell and are highly variable in size and shape. Bacteria are distinguished by their cell shapes such as spheres in coccoid bacteria, rods, spirally grown cells, comma shaped cells, box or plate shaped and filamentous bacteria with stalks and branched systems. Basically three cell shapes are distinguished in bacteria Fig. 1.2.

1. The spheres (coccus),
2. The rod (bacillus) and
3. The curved rod (one slight curve-Vibrio or a spiral or helix).

Bacteria are usually unicellular but sometimes daughter cells do not separate after cell division resulting in a multicellular undifferentiated structure. The shape of the multicelled structure depends upon the planes of cell division. Spherical cells exhibit a variety of shapes depending on the plane of cell division (Fig. 1.2).

Diplococcus - Bacterial cells divide in one plane and remain in pairs.

Streptococcus - Cells remain attached to form chains.

Pediococcus - Cells divide in two planes and form plates.

Sarcina - Division of cells takes place regularly in three planes to produce a cube.

Staphylococcus - Irregular division of cells in three planes produce bunches of cocci.

The rods always divide only in one plane and if they remain attached to each other it results in the formation of a filament. Spiral bacteria are predominantly unattached and exhibit variation in the length and tightness of spiral in different species. Some bacteria possesses a stalk that helps in attachment to a solid substratum using a holdfast at the tip.

Some bacteria divide by budding as in yeasts (*Rhodomicrobium*) with assymmetric cell division and polarised growth. Motility in bacteria is facilitated by flagella, by gliding or by axial filaments.

1.2.3 Structure

Bacterial cell is protected by a definite cell wall. Some bacteria possess capsules one to many flagella (Fig. 1.3). Bacterial, cell contains circular DNA, 70S type of ribosomes and mesosomes. Cell structure of bacteria is typically prokaryotic and differs in shape and plane of arrangement. In prokaryotic cells, nucleic acid is distributed as discrete masses throughout the cytoplasm. Nuclear membrane and other cell organelles are absent.

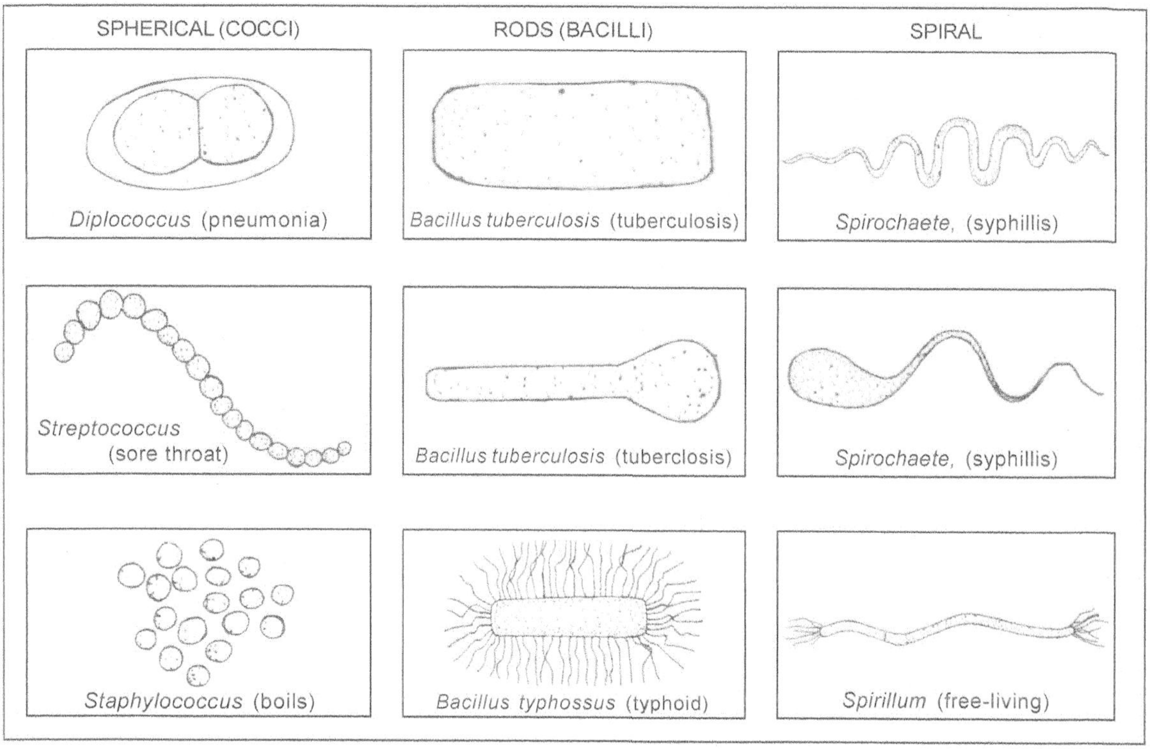

SPHERICAL (COCCI)
RODS (BACILLI)
SPIRAL
Diplococcus (pneumonia)
Bacillus tuberculosis (tuberculosis)
Spirochaete, (syphillis)
Streptococcus (sore throat)
Bacillus tuberculosis (tuberclosis)
Spirochaete, (syphillis)
Staphylococcus (boils)
Bacillus typhossus (typhoid)
Spirillum (free-living)

Fig. 1.2 Bacteria in different shapes (Cocci, Bacilli, Spirilli)

Flagella
Cytoplasm
70 s ribosomes
mesosome
Nucleoid (DNA)
Plasmamembrane
Capsule

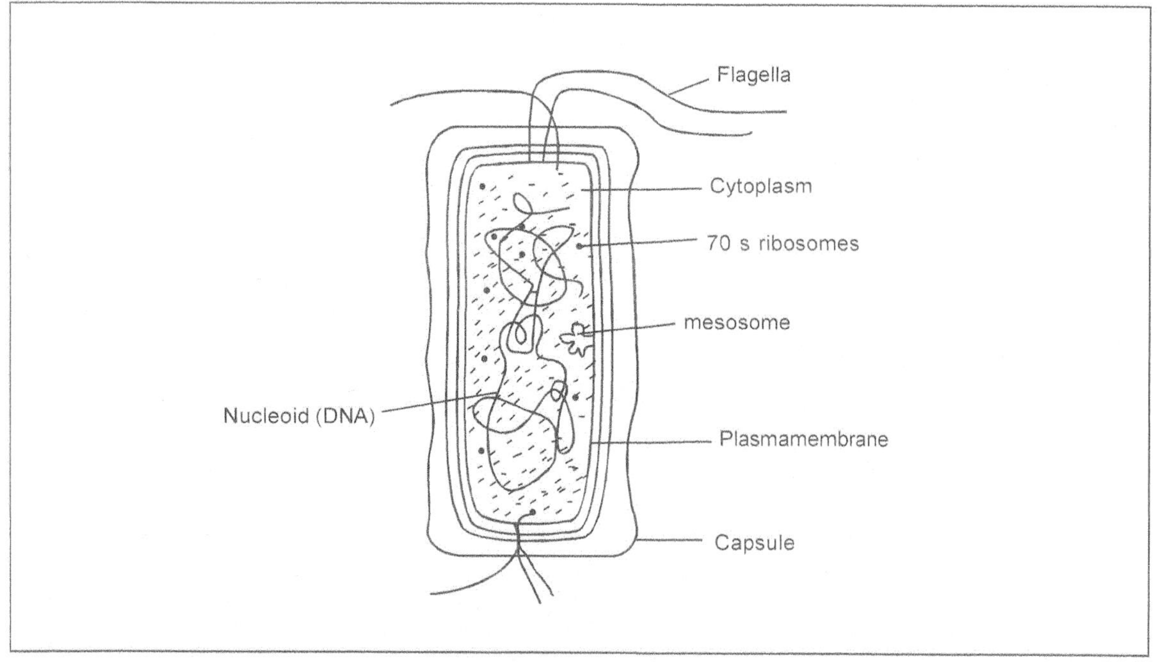

Fig. 1.3 Structure of a typical bacterial cell

The cytosol or cytoplasm is a concentrated solution containing a variety of enzymes, coenzymes and metabolites in the form of an aqueous fluid or semifluid or matrix. It provides a chemical environment for cellular activities and intermediary metabolism.

1.2.4 Capsule

The bacterial cell wall in some bacteria is surrounded by a gel layer called *Capsule*. It is an amorphous, shrunken layer synthesized by cell membrane and consists of a polysaccharide in water. The capsule provides protection against phagocytes and viruses. It prevents loss of water under dehydration and hydration conditions in many habitats, and aids in the uptake of essential cations.

The cell wall surrounds the cell membrane. The wall components are strongly antigenic. The rigidity and strength of cell wall is mainly due to strong fibres composed of heteropolymers known as *peptidoglycons* or *mucopeptides*. These fibres form a tough meshwork rather than solid structure. The fibres facilitate the movement of water, minerals and cell wastes through the cell wall. Bacterial cell wall is made up of network of peptidoglycon which consists of alternating units of N-acetyl-glucosamine and N-acetyl muramic acid with β-1-4 linkages. Peptidoglycan determines the shape of the cell and accounts for 40-80% of total dry weight. The cell wall of Gram-negative bacteria is comparatively thinner than the cell wall of Gram-positive bacteria. Differences between cell walls of Gram-positive and Gram-negative bacteria are summarised in Table 1.2.

Table 1.2 *Differences between cell walls of Gram-positive and Gram-negative bacteria*

	Gram positive	**Gram negative**
Gram staining	Retain crystal violet and appear dark violet	Looses crystal violet and - takes counter stain safranine to appear pink or red
Outer membrane	Absent	Present
Peptidoglycan	Thick (several layers)	Thin (single layer)
Lipids & lipoprotein	Low	High
Lipopolysaccharides	Absent	High
Teichoic acid	Present (mostly)	Absent
Periplasmic space	Absent	Present
Flagella	Possess 2 rings in basal body	Possess 4 rings in basal body
Strength, Resistance to desication & suscepti-bility to Penicillin	High	Low
Susceptibility to Strep-tomycin, Tetracycline & Chloremphenicol	Low	High

Gram-positive and Gram-negative bacteria vary in the susceptibility to various antibiotics. Gram-positive bacteria are highly susceptible to penicillin and less susceptibility to other antibiotics like streptomycin, tetracycline and chloremphenicol. Gram-positive bacteria mostly contain teichoic acid while it is totally absent in Gram-negative bacteria. Lipids and lipopolysaccharides are rich in Gram-negative bacteria. The cell envelope of Gram-negative bacteria is bilayered consisting of mainly lipoproteins, lipopolysaccharides and phospholipids.

Plasma membrane is situated just beneath the cell wall consisting of proteins, lipids, oligosaccharides and water. It is a continuous phospholipid bilayer embedding globular proteins. It provides permeability barrier and facilitates the transport of inorganic nutrients. It consists of enzymes of biosynthetic pathways and possess attachment sites for bacterial chromosome and plasmid DNA.

Mesosomes the inner plasma membrane invaginates to form mesosomes which are supposed to perform the respiratory activity but they are not analogous to mitochondria due to the absence of outer membrane. Mesosomes are also implicated as sites for synthesis of some of wall membranes.

The internal matrix of cell inside the plasma membrane is called *Cytoplasm* and it consists of water (80%), protein, carbohydrates, lipids, inorganic ions and low molecular weight compounds. The prokaryotic cytoplasm differs from eukaryotic cytoplasm in lacking cytoskeleton, cytoplasmic streaming and compartmentation of organelles.

Ribosomes : Ribosomes account for 30% of the total dry weight of the bacterial cell. Ribosomes exist freely in the cytoplasm and are smaller and less dense than eukaryotic ribosomes. Ribosomes are of 70S type unlike eukaryotes (80S type). 'S' is a Svedberg unit measuring the rate of sedimentation during ultracentrifugation which depends on size, shape and weight of the particles. They are made up on two subunits of sedimentation constants *viz.*,30S and 50S which combine to form the characteristic prokaryotic 70S ribosome. The 30S subunit contains one RNA molecule of 16S and the 50S subunit contains two RNA molecules of 5S and 23S.

Nuclear body is referred as nucleoid and is of primitive type. Nuclear body is an amorphous, globular mass of fibrillar, chromatic material which occupies about 10-20% of the cell volume. Double-stranded DNA forms a single circular chromosome in each nucleoid. Unlike eukaryotes, histones are absent in the nuclear material. No mitotic or meiotic phenomenon are seen in the nuclear body. Meselson and Stahl (1958) demonstrated chromosome replication by semiconservative method.

A number of inclusion bodies or storage granules are dispersed in the cytoplasm which may some times account for 50% of the dry weight. These include two major groups such as membrane enclosed inclusion bodies and inclusion bodies without membrane. Membrane (non-unit type membrane 2-4 nm thick) enclosed inclusion bodies include chlorosomes, carboxysomes, magnetosomes, gas vacuoles, polyhydroxybutyrate (PHB) granules and glycogen granules. Cyanophycin granules and phycobilisomes lack any membrane.

Flagella : Most motile bacteria possess long (up to 2mm) thin (20 nm dia) helical appendages known as *flagella*. Flagella of bacteria have no definite membrane like eukaryotes. Flagella consists of protein molecules called *flagellin*. The arrangement and number of flagella on a bacteria cell can be useful in identification and classification. There are five types of flagellation pattern in bacterial cells (Fig. 1.4) and bacteria without flagella are referred as atrichous.

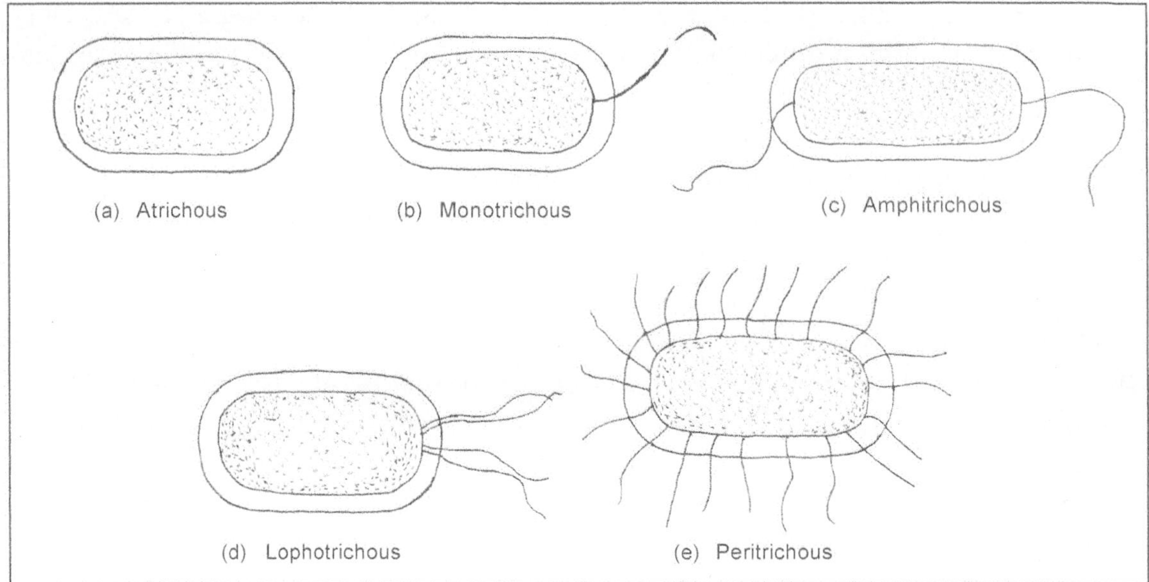

Fig. 1.4 Flagella in Bacteria

(a) Atrichous - No flagella are present.
(b) Monotrichous - A single flagellum at one pole of the cell.
(c) Amphitrichous - Flagella present at both poles of the cell
(d) Lophotrichous - Several or numerous flagella at one pole of the cell
(e) Peritrichous - Flagella present all over the surface of the cell

1.2.5 Pili, Fimbriae and Spinae

Pili and Fimbriae are non-flagellar appendages of cell. They have been observed mostly in Gram-negative bacteria and are made up of protein sub units of pilin. Pilin sub-units are arranged helically to form a filament. The filament is usually straight and shorter than a flagellum.

In some bacteria the longer pili on male cells act as a bridge between conjugating cells. The smaller pili may serve as anchoring tools *e.g.*, pathogens to their host. Both flagella and pili are antigens and can serve as specific sites of attachment for bacteriophages. Pili are plasmid coded and the plasmid codes for conjugation and resistance to antibiotics. Fimbriae are involved in adhesion to the Eukaryotes.

In some Gram-positive bacteria tubular pericellular appendages called spinae are found and they are made up of single protein moiety *spinin*, that are known to help adjust cells to environmental conditions such as salinity, p^H, temperature etc.

Nutrition in Bacteria

On the basis of mode of nutrition, bacteria can be distinguished into 2 major groups :

1. Heterotrophs and 2. Autotrophs.

In heterotrophs the sources of 'C' and energy are generally the same organic substances which may serve as building blocks as well as provide energy.

In autotrophs the 'C' source is CO_2 while energy is drawn by oxidation of inorganic compounds. Based on the mode of metabolism, source of 'C' and energy, bacteria are further recognised into 4 groups such as (i) Photo autotrophs (ii) Photo heterotrophs (iii) Chemo autotrophs and (iv) Chemo heterotrophs.

1.2.6 Reproduction

Bacteria multiply most commonly by asexual means such as fission and also produce endospores under unfavorable conditions. Endospores primarily serve as a means of survival rather than multiplication.

Binary fission: This is the most common method of reproduction in bacteria. In this, fission occurs at a rapid rate resulting in the division of cells. The cell splits transversely into two equal cells by Binary fission. The nuclear material undergoes replication and the two daughter chromosomes separate into each cell (Fig. 1.5).

Endospores : In certain genera like *Bacillus* and *Clostridium*, endospores are formed with in the cells. Spores are commonly formed in the centre of the cell.

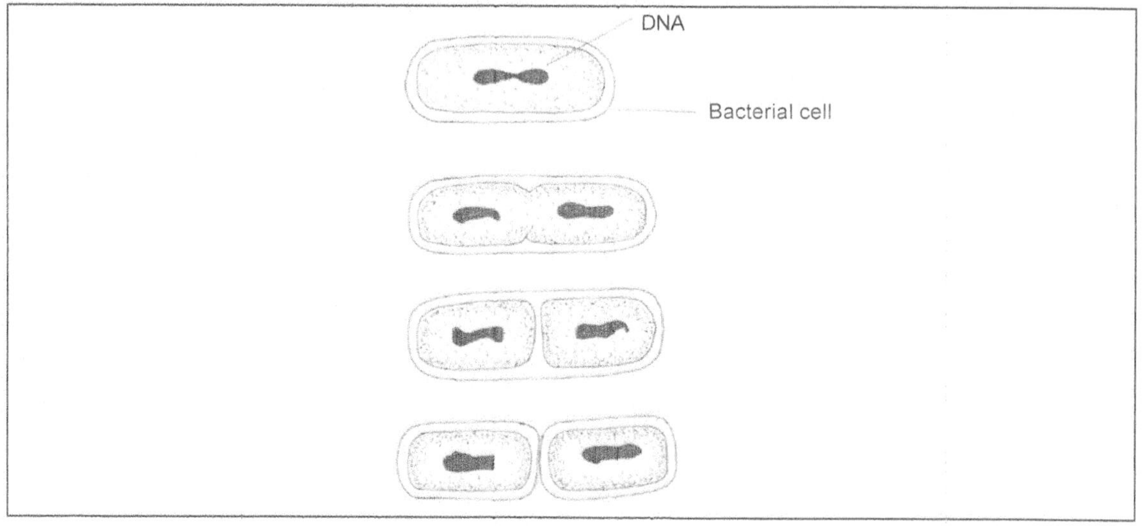

Fig. 1.5 Binary fission in Bacteria

1.2.7 Recombination Mechanisms

Transfer of genetic material in bacteria takes place by three mechanisms :

1. Conjugation 2. Transformation and 3. Transduction

1.2.7.1 Conjugation

Conjugation was first discovered by Joshua Lederberg and Edward Tatum (1946) in *Escherichia coli*. It involves transfer of genetic material from one cell to another during a period of physical living contact. In conjugation one cell acts as donor of DNA while others are recipients. Lederberg and Tatum observed the production of prototrophs when strains of *E. coli* auxotrohpic to different nutrients were grown together for some generations. An auxotroph is one that cannot synthesize an essential compound due to the lack of an important enzyme system. In this process the deficiencies are complimented due to genetic recombination and it is termed as "*conjugation*". For example, among two strains of *E. coli* one was unable to synthesize an essential compound 'A' and the other could not synthesize an essential compound 'B'. Neither strain was able to grow in a culture medium lacking both 'A' and 'B'. But when the two strains are grown together, these strains showed growth on the medium lacking both A and B. Apparently the genes for synthesis of compound A and B passed between the cells resulting in a recombined chromosome that could produce both the compounds.

Subsequently it was shown that some cells behave as donors (male) and others as recipients (female) and when these two are brought together, the F-factor (which is a plasmid) is quickly transferred to the recipient. The recipient becomes a male (donor). The donor forms a *sex pilus* which helps in attachment to the recipient. Thus conjugation requires physical contact Fig. 1.6. The donor is designated as F^+ and the recipient as F^-. Essentially conjugation is the process of gene transfer between cells of opposite mating types that are in physical contact with each other.

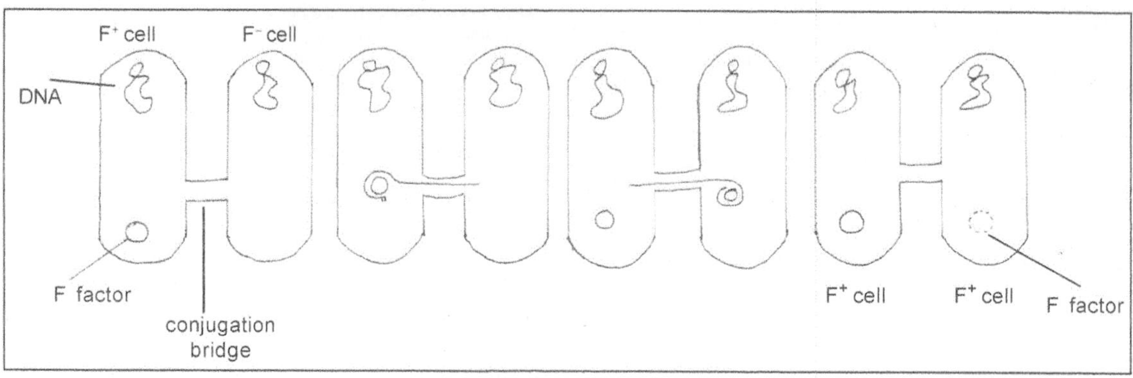

Fig. 1.6 Transfer of 'F' factor during conjugation

1.2.7.2 Transformation

Transformation involves the transfer of relatively small segment of naked DNA from a donar (male) cell to recipient (female) cell (Fig. 1.7).

Frederick Griffith (1928) discovered the phenomenon of transformation. He conducted a series of experiments with laboratory mice and two types of pneumonia pathogens, *Streptococcus pneumonia* (*Diplococcus pneumonia*). The pneumonia-causing bacterium has two strains, one is pathogenic (virulent) with smooth capsulated cells (S) and the other nonpathogenic (avirulent strain with rough) non-capsulated cells (R). S strain being virulent killed mice while the R strain (non-pathogenic) did not cause death of mice. The heat killed strain also could not affect the death of mice like R strain. However, when heat killed S strain mixed with R strain was injected the mice died (Fig. 1.7).

Some substance from the heat-killed cells transformed the rough colony producing R strain to smooth colony producing S strain thus making it virulent that killed mice. This phenomenon was called *transformation*. Griffith thought that the substance was a protein. Later, Avery *et al* (1944) showed that the transforming principle is a nucleic acid. Hershey and Chase (1952) obtained the evidence that DNA is the genetic material. Major steps in transformation are shown in Fig. 1.7.

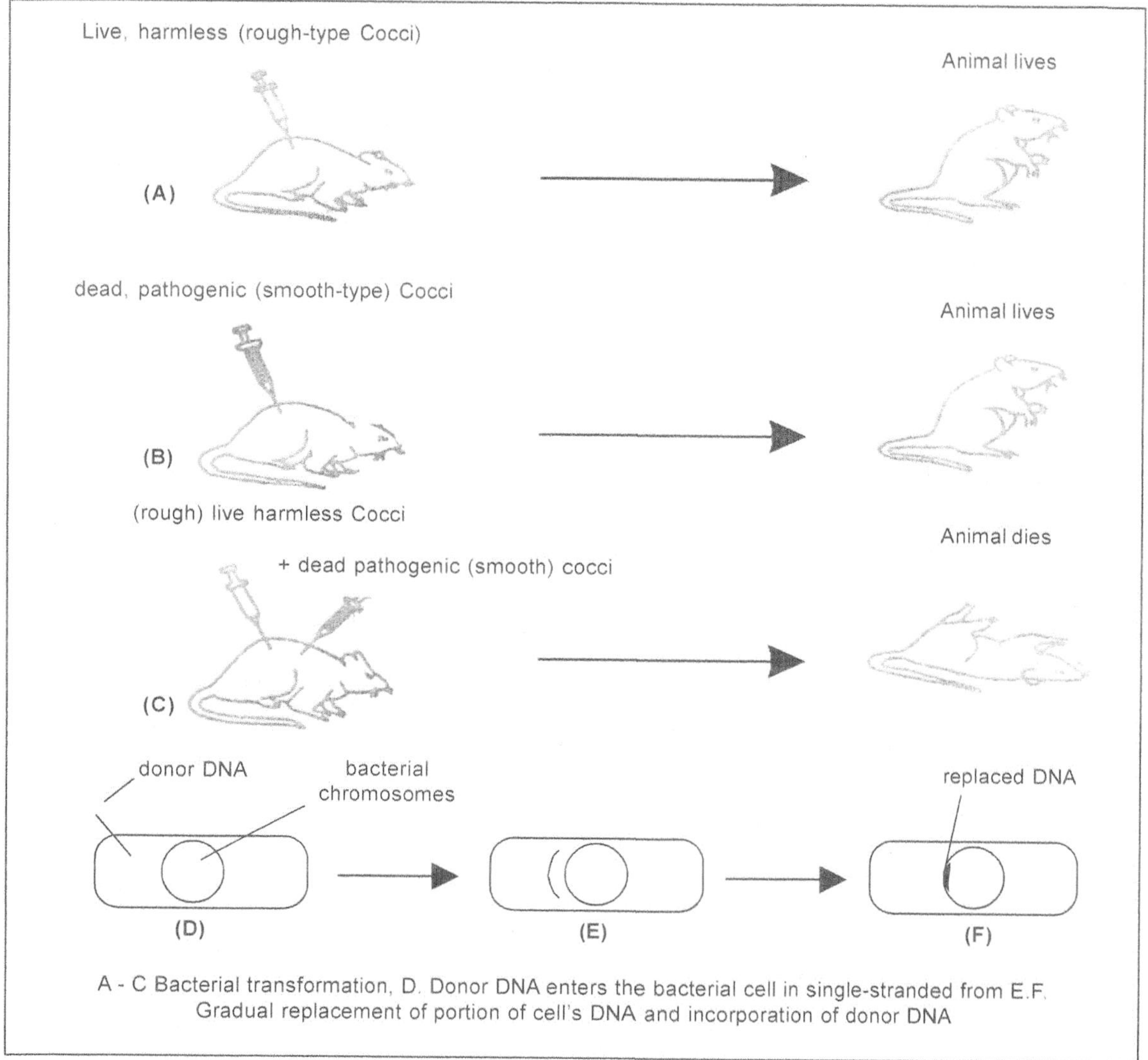

Fig. 1.7 Transformation in bacteria (Griffith's experiment on mice)

1.2.7.3 Transduction

Transfer of genetic material from one cell to another cell by a bacteriophage (virus) is called transduction. Zinder and Lederberg (1952) discovered the phenomenon of transduction in *Salmonella typhimurium*. Transduction involves infection by a bacteriophage, its replication in bacterial cell and finally lysis of the bacterial cell to release virus particles.

Infection by a bacteriophage takes place in several stages such as adsorption, penetration, replication, assembly, lysis and release. Bacteriophage particles attached to specific receptor site on bacterial cell wall - followed by the penetration of viral genetic material into bacterial cell. Viral DNA replicates independently by using cell machinery of the host. Along with the replication of phage DNA into multiple copies, phage proteins are also synthesized simultaneously. Consequently, assemblage of phage particles takes place. Lysis of the bacterial cell results in the release of virus particles.

Bacteriophage reproduction is of two types, the virulent phage and the temperate phage. Virulent phages reproduced by lytic cycle as they destroy the host bacterial cell (e.g. T-phages and Lambda (λ) phage).

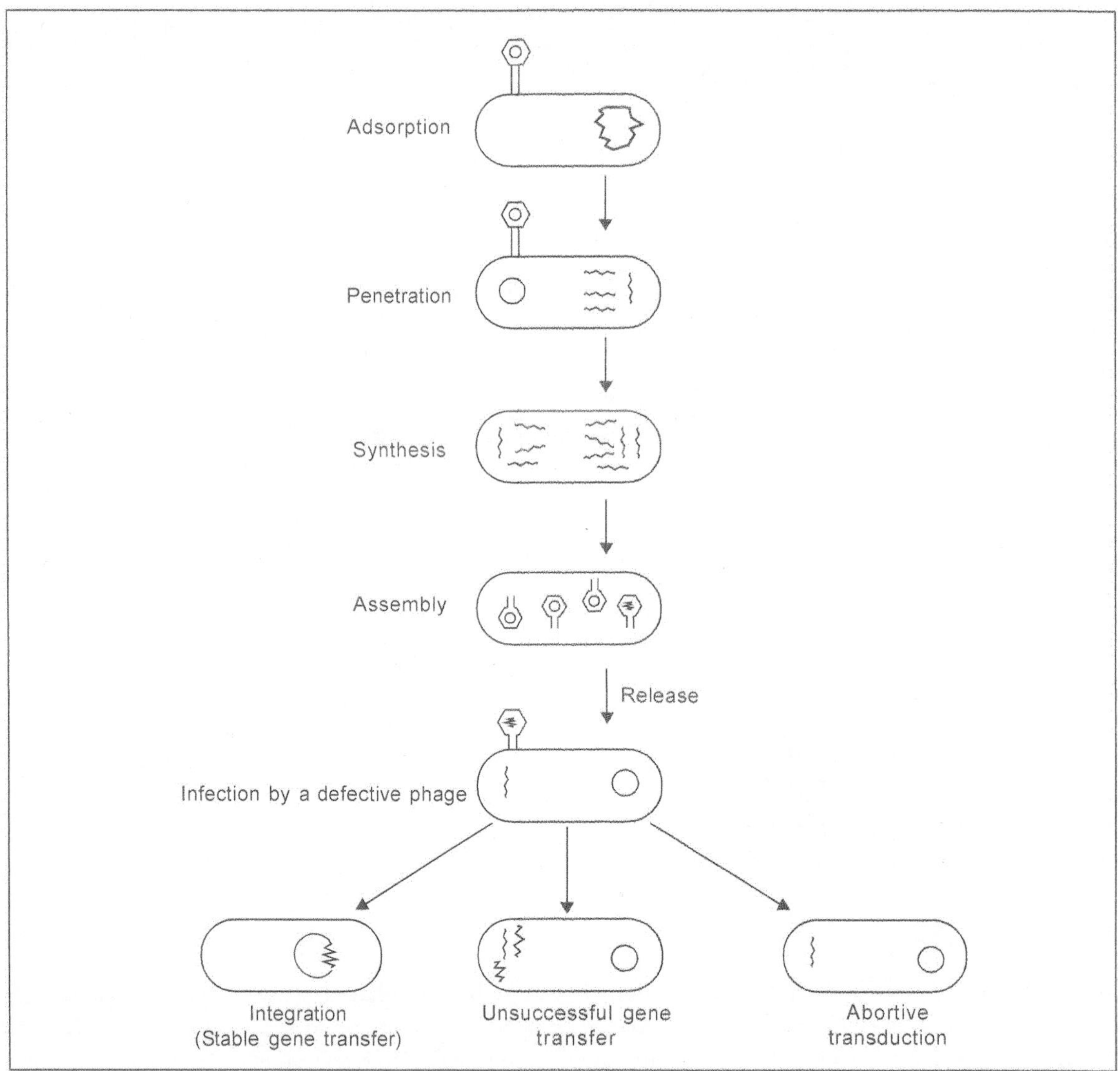

Fig. 1.8 Outlines of Generalised transduction by bacteriophages

In contrast, the temperate phages do not lyse the host cell. The viral genome behaves like 'F' factor and gets integrated with bacterial chromosome. This latent form of phage DNA that remains within the bacteria and integrates with chromosome is called a *Prophage*. Bacteria containing prophage are called *Lysogenic bacteria* and the relationship between the phage and the host is referred as lysogeny.

Transduction occurs usually between closely related species of the same genus (i.e intragenic). Intergeneric transduction also occurs between closely related enteric genera like *E. coli* and *Salmonella* species. Transduction is of two types, the generalised transduction and specialized transduction.

Generalised transduction occurs during the lytic cycle of virulent or temperate phages (Fig. 1.8). After the release of phage DNA into bacterial cell, it multiplies while the bacterial chromosome gets fragmented. During the assembly phase of the phage DNA and capsids, the bacterial chromosome fragments may also get packed by mistake. Since the capsid can take about 44kb of DNA, some or all viral DNA may be left behind in the process. Therefore the quantity of bacterial DNA carried by phage DNA depends mainly on the size of the capsid. Phage P22 carries about 1% of the bacterial chromosome while phage P1 of *E. coli* can carry about 2-25% of bacterial genome. This defective phage is referred as a generalized transducing particle. The genome is introduced into the host cell upon infection to another bacterium by the defective phage (Fig. 1.8). The transferred bacterial DNA (exogenete) may get integrated into the recipient bacterial chromosome (endogenete). However, about 70-90% of the transferred DNA may not be integrated with the recipient chromosome but survives and expresses itself.

1.2.7.4 Complete Transduction

When the exogenete introduced by a defective phage is integrated into the recipient bacterial chromosome, it is referred as complete transduction. In contrast, when the exogenote fails to integrate into the endogenote (recipient bacterial chromosome) it is called abortive transduction. The recipient bacteria that contain non-integrated transduced DNA and are partially diploid are called abortive transductants. Abortive transductants up on division produce only two daughter cells containing exogenote in each generation while other cells do not contain the exogenote.

1.2.7.5 Specialized (restricted) Transduction

Transfer of only a few restricted genes of the bacterial chromosome to the recipient bacterial cell in certain temperate phages is known as specialized or restricted transduction (Fig. 1.9(a-h)) when a phage genome is introduced in the bacterial cell, it gets integrated with bacterial cell, it gets integrated with bacterial chromosome as prophage (A). The DNA becomes free containing a small segment of bacterial chromosome (B). It multiplies and disintegrates the bacterial chromosome (C). Assembly of the phage DNA with bacterial chromosome takes place and subsequently the phages are released from the bacterial host (D). This defective phage upon infection to another bacterium introduces its DNA comprising a piece of bacterial chromosome (E). The genes of the phage can insert with homologous DNA of the infected bacterium (F). However, crossing over may occur between the homologous gene loci of the bacterial chromosome and the donar DNA attached with phage genome (G) resulting in the integration of the donar DNA with recipient DNA (H).

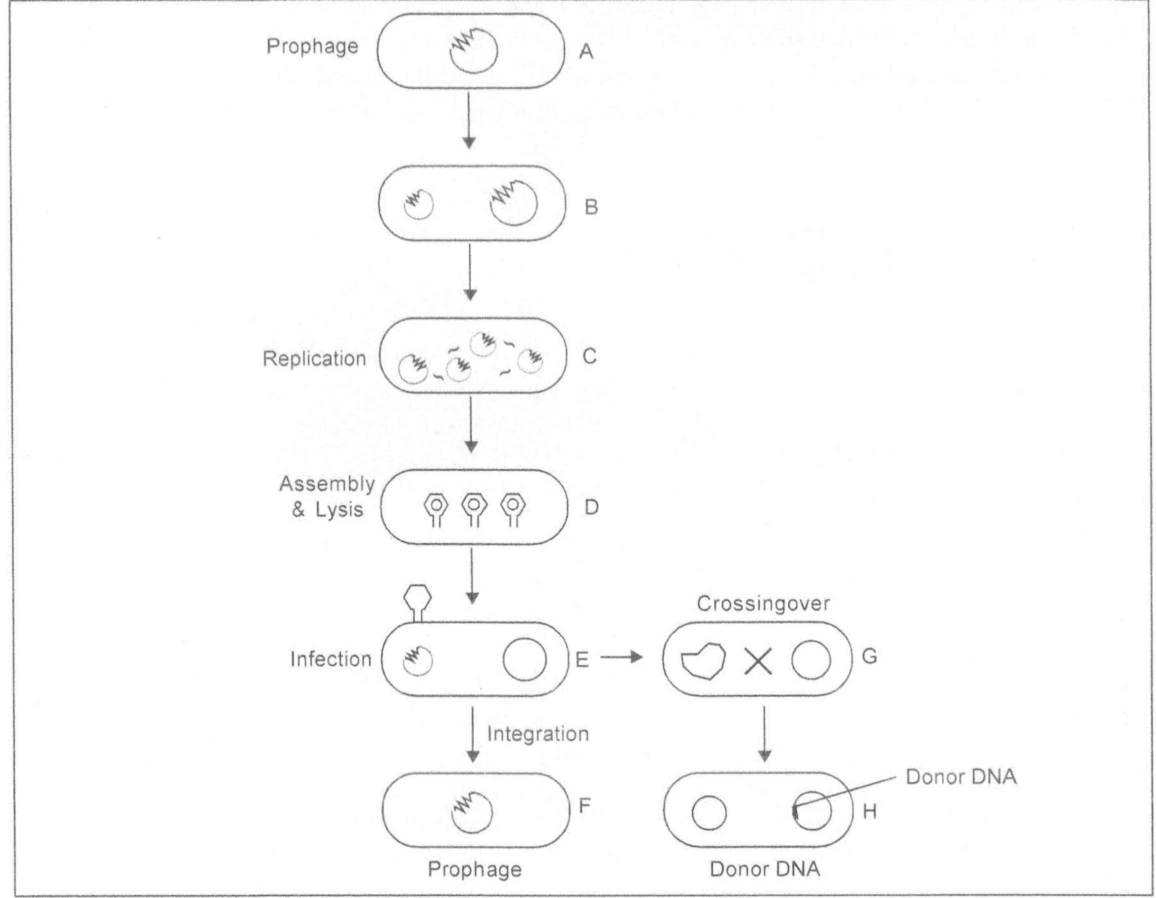

Fig. 1.9(a-h) Specialized transduction by a temperate bacteriophage

1.3 Algae

Algae are classified into 11 classes by *Fritch*, based on pigmentation, reserve food, flagellation and reproduction.

1. *Chlorophyceae* include green algae with chlorophyll a, b, xanthophyll and carotenoids in their chloroplasts. The cell wall is made up of cellulose and starch is the reserve food. Spores are ciliate and motile. Sexual reproduction is isogamous, anisogamous and oogamous.
   e. g : Chlamydomonas, Volvox, Chlorella, Spirogyra.
2. *Xanthophyceae :* Xanthophyll pigment along with chlorophyll '*e*' gives green – yellow colour to this group of algae. Pyrenoids are absent and fat is the food material. Sexual reproduction is by fusion of ciliate gametes and cilia are of different length.
   e. g : Vaucheria, Microspora.
3. *Chrysophyceae:* Besides chlorophyll, yellow–green phycocyanin is the main pigment. Algae are unicellular, multicellular or colonial . The cell wall exists in the form of two over lapping halves. Stored food is oil or insoluble carbohydrate, Leuosin.
   e .g : Chrysosphaeria.

4. *Bacillariophyceae* includes diatoms which contain diatomin in their chloroplasts. These are unicellular nonmotile algae with pyrenoids. Chlorophyll b is replaced by chlorophyll C
   e. g : Navicula, Pinnularia.

5. *Cryptophyceae:* These are red, green–blue, olive–green or green coloured algae with two large chloroplasts in each cell. Pyrenoids are present. These are fresh water and marine algae.
   e. g : Cryptomonas.

6. *Dinophyceae* : The algae are dark yellow or brown or red in colour. Stored food is starch or oil. Chromatophores are present in the nucleus.
   e. g : Peridinium.

7. *Chloromonadineae :* Algae are bright green or olive green in colour. Xanthophyll is in abundance. Reproduction takes place by longitudinal division
   e. g : Vacuolaria.

8. *Eugleninae :* Chlorophyll is present and have naked ciliated reproductive organs resembling microscopic animal
   e. g : Euglena.

9. *Phaeophyceae :* Yellow–brown marine algae with fucoxanthin as the main pigment. Chlorophyll '*a*' and carotenes are present. Chlorophyll 'c' replaces chlorophyll b. Laminarin, mannitol and fats are the storage material. Zoospores are biciliate and the cilia are of different sizes. Zygote has no resting period.
   e. g : Sargassum, Fucus.

10. *Rhodophyceae :* These are red algae due to the presence of phycoerythrin pigment besides phycocyanin, chlorophyll a, carotene and xanthophyll in small quantities. Floridean starch is the storage food. Sexual reproduction is oogamous.
    e. g : Polysiphonia.

11. *Myxophyceae (= Cyanophyceae) :* These are prokaryotes with phycocyanin as the main pigment. Other pigments like phycorythrin, chlorophyll b and carotenes are present in low quantity. Chlorophylls are found in thylacoids. Myxophycean starch is the storage food. No sexual reproduction and no motile stage in the algae. Algae are unicellular or filamentous.
    e. g : Nostoc, Anabaena, Lyngbya.

1.4 Fungi

Fungi show a wide range of activities as they depend on autotrophs like green plants and also utilize dead organic matter. If the food is obtained from the host the fungus is termed as a parasite and if it obtains food from dead organic material it is termed as a saprophyte. Fungi parasitizing plants often produce disease in them hence they may be referred as pathogens. Fungal pathogens are known to cause about 70% of the plant diseases. In contrast fungi relatively cause few diseases of man and other animals. In fact potato blight disease caused by *Phytophthora infestans* is the culprit of Irish famine that has taken lives of over a million people. Similarly, great Bengal famine (1943) was the result of severe outbreak of *Helminthosporium oryzae* on rice crop.

Fungi are capable of parasitzing themselves and insects. When fungi parasitize other fungi, they are known as mycoparasites and if they parasitize insects they are referred as entomogenous fungi. Mycoparasites and entomogenous fungi are of great significance as commercial disease control agents in biological control programmes. As saprophytes, fungi form an important component of biodegradation which is essential in the biosphere, ensuring the recycling of carbon and mineral nutrients for plant growth. On the negative side, fungi are involved in biodeterioration such as food spoilage, rot of wood, leather and fabrics. Food spoilage fungi produce mycotoxins (Aflatoxins) which are potent carcinogens.

1.4.1 Fungi in Biotechnology

Edible fungi or mushrooms are directly used as food and yeasts are useful in making beer, wine, bread and cheese. Several yeasts (*Candida utilis, Saccharomyces cervisiae*) are used as single cell protein. Fungi are used in several important, industrial processes such as citric acid production. The discovery of miracle drug penicillin produced by *Penicillium notatum* has changed the course of modern medicine. Fungi produce innumerable commercially important antibiotics including penicillins, cephalosporins, griseofulvin and fusidic acid. Fungal enzymes in commercial use include α-amylase, amyloglucosidase, pectinases and proteases.

1.4.2 Classification

Fungi have been traditionally considered as a subkingdom of plant kingdom (hence in chapter 1 of this book) and however Whittaker (1969) regarded them as a separate kingdom on par with plants and animals. Firstly fungi are distinguished into wall-less fungi (Myxomycota) and true walled fungi (Eumycota). Ainsworth (1973) divided Eumycota into five subdivisions *viz.*, Mastigomycotina, Zygomycotina, Ascomycotina, Basidiomycotina and Deuteromycotina. Zoosporic (flagellated asexual spores) fungi are included in Mastigomycotina which are distributed into two classes *Chytridiomycetes* (characteristically possess single posterior whiplash flagellum) and *Oomycetes* (biflagellate zoospores). Sporangiospore forming aseptate fungi are grouped into Zygomycotina which is further divided into saprophytic Zygomycetes and parasitic (in the guts of arthropods) Trichomycetes.

Ascomycotina members are typically septate mycelial or yeast like forms which characteristically produce sexual spores (ascospores) in an ascus. Members of the class *Hemiascomycetes* are yeasts or mycelial forms with naked ascocarp (i.e., not enclosed in a fruiting body). In *Euascomycetes*, the asci are typically enclosed in an ascocarp.

Sub-division Basidiomycotina consists of three classes : Teliomycetes, Hymenomycyetes and Gasteromycetes. Asexual spores are absent and the plant body is mycelial (mycellium septate) or yeast. Sexual spores are typically formed on a *basidium*. The parasitic (rusts and smuts) fungi are included in the class Teliomycetes. No special fruiting body (basidiocarp) is found in this class to enclose the basidia. Toadstools and bracket fungi are included in the class Hymenomycytes in which basidia are exposed. In Gasteromycetes, basidiocarps are various and encloses the basidia. Imperfect fungi or Deuteromycotina are characterised by the absence of sexual reproduction or it is unknown. Mycelium is septate or yeast like. It is divided into three classes, Blastomycetes (typically yeasts), Hyphomycetes and Coelomycetes. Hyphomycetes are mycelial and include conidial fungi which are formed on simple hyphae or hyphal branches (conidiophores). In Coelomycetes, conidia are formed from conidiophores in flask-shaped structures (pycnidium) or on a pad of saucer shaped tissue (acervulus).

1.5 Protozoa

Protozoans are microscopic animalcules. A protozoan is a complete unicellular eukaryotic organism. Over 50,000 species of protozoans are known till date. Locomotion is through pseudopodia, flagella or cilia. These are mostly heterotrophic in nutrition while some are autotrophs. Protozoans are considered under four subphyla based on the type of locomotory organ.

***Subphylum* 1 :** *Sarcomastigophora* - Organs of locomotion are pseudopodia or flagella or both. The organisms are uninucleate or multinucleate. In this subphyla, 3 classes are recognised. They are: Mastigophora or Flagellata; Sarcodina or Rhizopoda; and Opalinata.

***Subphylum* 2 :** *Sporozoa* – Includes endoparasites. Locomotory organelles and contractile vacuoles are absent but possess an apical complex of ring like, tubular, filamentous organelles at some or other stage of their life cycle . Spores are formed but with no polar filaments. The sporozoans are divided into two classes : (a) Telosporea (b) Piroplasmea.

***Subphylum* 3 :** *Cnidospora* (Microspora) – It includes intracellular endoparasites, especially on insects. No locomotory organelles and contractile vacuoles present. The spores differ from sporozoans in possessing one to four polar filaments. On the basis of the mode of spore formation cnidospora is divided into two classes : (a) Myxosporea (b) Microsporea.

***Subphylum* 4 :** *Ciliophora* – Includes largest group of animal–like, structurally complex protozoans with cilia as locomotory organelles. Cilia are replaced in the adult by sucking tentacles for feeding, A large macronucleus of trophic function and one to several micronuclei of reproductive function are present. Sexual reproduction is through conjugation and asexually reproduce by transverse fission. Ciliata is the only class in this subphylum.

1.6 Viruses

Viruses are microscopic, filterable and obligate intracellular parasites which require a living host for multiplication. They infect all types of organisms such as animals, plants, bacteria, algae, fungi and insects. Lwoff defined viruses as " *infectious, potentially pathogenic nucleoproteins with only one type of nucleic acid which reproduce from their genetic material, unable to grow and divide and devoid of enzymes*". According to Lwoff and Tournier (1971), enzymes for energy metabolism (Lipman system) and ribosomes are absent in viruses. Information for the production of enzymes, synthesis of ribosomal proteins, r RNA and soluble t RNA is absent.

Viruses exhibit following distinguishing characters unlike other microorganisms :

(a) They possess a single nucleic acid,

(b) They are potentially infectious,

(c) Inability to grow the genetic material only,

(d) Reproduction from the genetic material only,

(e) Absence of Lipman system (Lack of enzymes for energy metabolism),
(f) Lack of information for the production of enzymes in the energy cycle,
(g) Lack of ribosomes,
(h) Lack of information for the synthesis of ribosomal proteins and
(i) Absence of information for rRNA an tRNA synthesis.

1.6.1 History

Beijerinck observed differences between bacteria and filterable viruses and put forward the concept of *contagium vivum fluidum* (living infectious fluid). Electron microscopy and development of ultracentrifugation techniques in the 20th century revealed the nature, morphology and other details of viruses. Stanley, an American chemist crystallized the virus particles for the first time in 1935 from infected tobacco plants awarded Noble prize for his discovery in 1946. Bawden and Pirie analysed the crystallized particles and revealed their chemical nature as of protein and ribonucleic acid (RNA). Later on Fraenkel-Conrat confirmed the genetic material of tobacco mosaic virus (TMV) as the RNA.

In 1916, Frederick W. Twort in England and Felix d'Herelle of Pasteur Research Institute in France have independently observed the lysis of bacterial cultures and this lytic effect was noticed even after passing through the bacterial filters. Twort suggested the lysis may be due to virus and d'Herelle (1917) named it as a bacteriophage (bacteria eater). Hollings (1962) discovered mycoviruses in the cultivated white button mushroom. Cyanophages infecting blue-green algae were discovered by Safferman and Morris (1963). Satellite viruses were discovered by Kassanis (1966).

Diener and Raymer (1967) discovered viroids as smaller infectious agents from potato spindle tuber diseases. Properties of viroids differ basically from viruses in the following features. (a) the pathogen exists *in vivo* as encapsulated in the RNA (b) no virion like particles in the infected tissue (c) the infectious RNA is of low molecular weight (d) autonomous replication of the infections RNA in susceptible cell (i.e., no helper virus is required) and (e) the infectious RNA consists of one molecular species only.

1.6.2 Morphology

Viruses exhibit a variety of shapes such as spheroid or cuboid (adenoviruses), elongated (Potato viruses), flexuous or coiled (beet yellow), rod shaped (Tobacco mosaic), bullet shaped (rabies virus), filamentous (bacteriaophage M 13), pleomorphic (alfalfa mosaic) etc. Size of the viruses vary from 20 nm to 300 nm in diameter. They are smaller than bacteria and larger than protein and nucleic acid molecules. Virus, the complete assembly of the infectious particle is technically known as '*virion*'. Chemically virus is a nucleoprotein. The nucleic acid core is surrounded by a protein coat known as *capsid.* It consists of a large number of capsomer sub-units. Electron microscopy and crystallography studies reveal three morphological types such as,

1. Helical viruses e.g. T M V (naked capsids), Influenza virus (enveloped capsid).
2. Polyhedral (Icosahedral) viruses
   e.g., Adenovirus , poliovirus (Naked capsid)
   Herpes simplex virus (Enveloped capsid)
3. Complex viruses – e.g., Vaccinia virus (capsids not clearly identified);
   Bacteriaophages (some other structures are attached to capsids).

Certain plant, animal viruses and bacteriophages are surrounded by a thin membranous envelope of 10 – 15 nm thickness made up of proteins, lipids and carbohydrates. Lipids provide flexibility to the shape and size. Envelope proteins are of viral origin and lipid and carbohydrates may be derived from host membranes. Lipids of the viral envelope belong to four classes.

(i) Phospholipids,

(ii) Cholesterol,

(iii) Fatty acids and

(iv) Glycolipids.

Virus nucleic acid is either a single or double stranded DNA or RNA molecule. The nucleic acid may be linear or circular with plus or minus polarity. Viral proteins are of four types; (i) envelope proteins (ii) nucleocapsid protein (iii) core protein and (iv) viral enzymes. Carbohydrates like galactose, mannose, glucose, fucose, glucosamine, galactosamine are found in viral envelope. The carbohydrates are hexoses and hexoamine occur in the form of glycoprotein and for glycolipids (e.g. Influenza virus, Parainfluenza virus etc).

Viruses are generally classified depending on the host they attack and the disease they produce (e.g., Plant viruses, poliovirus etc). Lwoff, Horne and Tournier (1962) classified viruses based on (i) the nature of nucleic acid (DNA or RNA) (ii) symmetry of the viral particle (Helical, Icosahedral, cubic, cubic–tailed), (iii) presence or absence of envelope, (iv) diameter of the capsid and (v) capsomer number of the capsid. This system of classification is popularly known as LHT system and it divides phylum Vira into two subphyla, the Deoxyvira (DNA Viruses) and Ribovira (RNA Viruses). LHT system was adopted by the Provisional Committee on Nomenclature of Viruses (PCNV) formed by the International society of Microbiological society.

Plant viruses are mostly RNA viruses. Plant viruses are distinguished into 16 groups by the plant virus committee of P C N V. Some of the important plant viruses are tobacco mosaic virus (Tobacco virus group), potato virus (Potexvirus group), tobacco necrosis virus, tomato spotted wilt virus, cauliflower mosaic virus (a double stranded (ds) DNA virus) and cucumber mosaic virus etc.

Viral diseases are common in humans and animals such as small pox, influenza and common cold. Baltimore (1971) classified animal viruses into six groups depending on the relationship between virion, nucleic acid and m RNA transcription. Some of the important animal viruses include Papova virus (all ds DNA viruses), adenovirus, herpes virus, pox virus and Rhabdo virus, picorna virus, Reovirus, Retrovirus (all RNA viruses). Bacteriophages are of four types :

1. The ss DNA Bacteriophages e.g., X ∅ 174, M 13
2. The ds DNA phages – e.g., T_2 T_4
3. The ss RNA phages – e.g., P1 HP1
4. The ds RNA – e.g., ∅ 6

All the phages possess a nucleic acid enclosed in a capsid which is made up of capsomers. The capsomers inturn contains protomers or protein subunits. Six morphological types are distinguished among bacteriophages such as Type A–∅ ds RNA T–even (T_2, T_4, T_6) phages; Type B–ds DNA phages (T_1, T_5); Type C–ds DNA phages (T_3, T_7); Type D–ss DNA phages (∅174); Type E–ss RNA phages (F_2, MS 2); and Type F–ss DNA phages (fd, f1).

T-phages are the virulent double stranded DNA phages infecting the bacterium, *E. coli*. There are seven coliphages of T-series; three even and four odd number phages. The T_2 is including a tadpole shaped structure consisting of five important sub-structures such as the head, head-tail, connector, tail plate and fibers (Fig. 1.10). The virion is naked, icosahedral tailed. The head is an elongated hexagonal like prism consisting of 10-faced equatorial bands measuring 95 × 15 nm with about 2000 capsomers.

The phage consists of long helical tail. The tail is connected to the head with a connector having a collar with attached whiskers (Fig. 1.10). Tail consists of inner tubular core surrounded by a contractile sheath. The sheath connects the head at one end and the base plate at the other end of the tail. Base plate is hexagonal with six spikes or tail fibers at its six corners. Tail fibers help in recognizing the specific receptor sites on the surface of the bacterial cell wall.

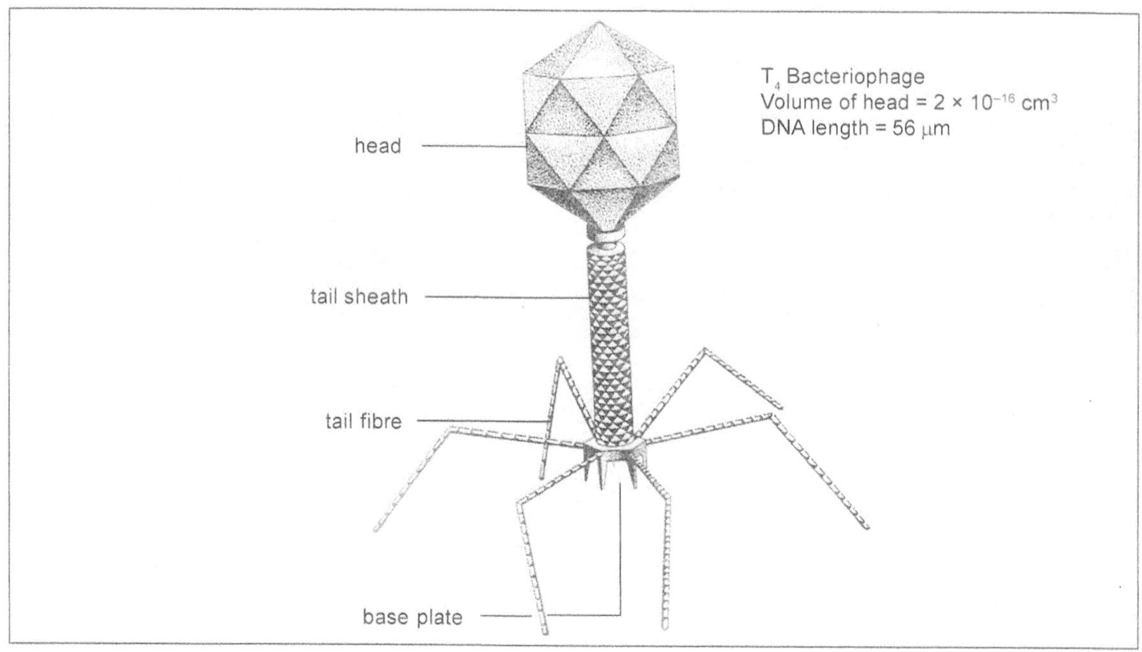

Fig. 1.10 Structure of a T_4 Bacteriophage

The dsDNA molecule is circular and tightly packed inside the head. All the T–even phages contain 5 – hydroxymethylcytosine instead of cytosine.

1.6.3 Multiplication (Life cycle)

The multiplication involves several stages such as adsorption, penetration, synthesis, replication, assembly and release. The phage particles interact with the *E. coli* cell wall and tail fibers recognize the specific recognition sites present on the cell surface. Specific receptor may be a lipopolysaccharide (T_3, T_4 and T_7), lipoprotein (T_2 and T_6) or both (T_5). The process of interaction of phage to recognition of specific receptor site is known as *landing*. Landing is followed by adsorption which is known as *pining*. It is an irreversible process while prior to pining are reversible. Penetration is affected by mechanical or enzymatic digestion of the cell wall at the receptor site. The DNA is injected into the cell while the phage head and tail remain outside the cell. The empty protein coats are known as '*ghosts*' (Fig. 1.11).

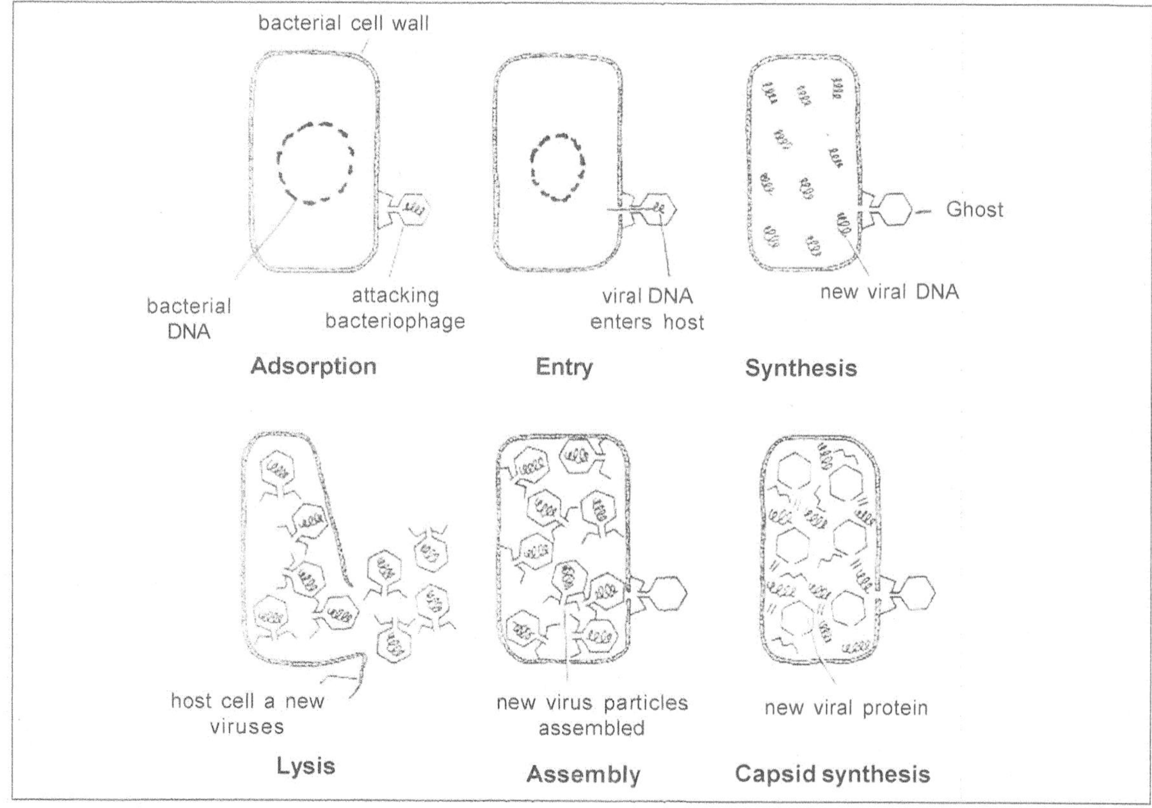

Fig. 1.11 Virulent bacteriophage life cycle

Synthesis of macromolecules leading to phage multiplication essentially involves the degradation of bacterial chromosome, protein synthesis and DNA replication (Fig. 1.11). The bacterial chromosome gets unfolded soon after the phage infection. The nucleoid is disrupted and DNA helix gets attached with plasma membrane at different points. Protein synthesis by bacterial host is stopped after 3 minutes of infection and partial break down of bacterial chromosome occurs. The RNA polymerase and ribosomes modify to initiate protein synthesis of phage but not of bacteria. The phage DNA contains early and late genes. The delayed early genes synthesize the phage enzymes to produce 5-hydroxy methyl cytosine which inturn replace cytosine of the bacterial DNA. At this stage bacterial restriction enzymes could not degrade the phage genome. The late phage gene products include lysozyme and structural components of new phages (head, tail and tall fibres).

1.6.4 Replication

No phage could be recovered during the first 10 minutes of phage infection from the infected bacterium and this period is known as '*eclipse period*'. Replication involves five important events such as

(a) Degradation of host DNA

(b) Synthesis of 5-hydroxy methylcytosine (5 HMC)

(c) Prevention of incorporation of cytosine into the T-phage DNA

(d) Glucosylation of T_4 DNA and

(e) Enzymology of DNA replication.

Bacterial DNA is degraded by phage coded exonucleases. Replication of T_4 DNA begins after six minutes and reaches the maximum after 12 minutes. After the synthesis of structural proteins and phage DNA, assembly of these components begins to give rise to the mature phages. Phage particles accumulate inside the bacterial cell after the maturation or assembly. The progenies are released by the lysis of the host cell wall. Lysozyme is attached to the tip of the tail of phages. The time span between infection and lysis is known as *latent period*.

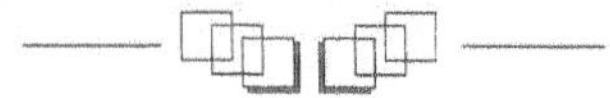

Essay Type Questions

1. Give a detailed account of structure of bacteria.
2. Describe the transfer of genetic material in bacteria.
3. Give an account of transduction in bacteria.
4. Describe the distinguishing features of viruses and add a note on classification.
5. Describe multiplication of viruses.

Short Answer Type Questions

1. Structure of bacteria
2. Bionary fission
3. Transformation
4. Classification in fungi
5. Structure of bacteriophage

2 Introduction to Plant Biology

2.1 Introduction

Diversity of living organisms or biodiversity is the inseparable facet of interaction and evolution of organisms. Species forms the basic unit of biological diversity. Haeckel grouped living organisms into plants, animals and protista. All primitive organisms were included in protista. Stamir and Van Niel (1962) established the concept of prokaryote and recognised bacteria as prokaryotic organisms. Distinction of prokaryotic and eukaryotic organisms was based on the cell anatomy which includes nuclear organisation and presence or absence of certain cell organelles. In 1969, Whittaker proposed a five kingdom system including plants, fungi, animals, protista and monera (Fig. 2.1). based on energy-yielding systems and cell anatomy.

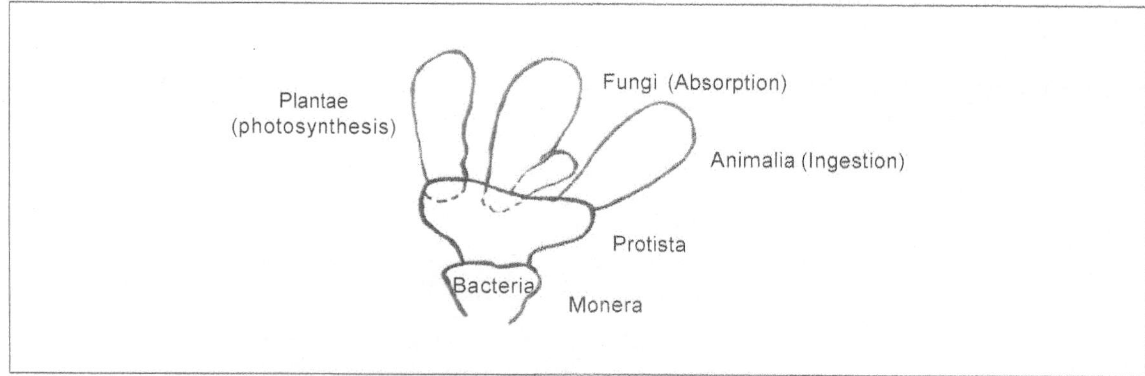

Fig. 2.1 Five kingdgom system of Whittaker (1969)

Recently, Woese (1990) introduced a new concept of domains over kingdoms such as Archaea, Bacteria and Eukarya. In this classification, bacteria, cyanobacteria and actinomycetes are grouped under bacteria while the methanogens, extremophiles (extreme thermophiles, halophiles etc.,) are included in the domain Arachaea. Eukarya includes all the eukaryotic organisms such as fungi, protozoa, plants and animals.

Kingdom plantae is further differentiated into cryptogams (non flowering plants) and phanerogams (flowering plants) based on the nature of flowering. Cryptogamic plants include algae, bryophytes and pteridophytes and phanerogams include gymnosperms and angiosperms (Table 2.1).

Table 2.1 *Outline Classification of Plant Kingdom*

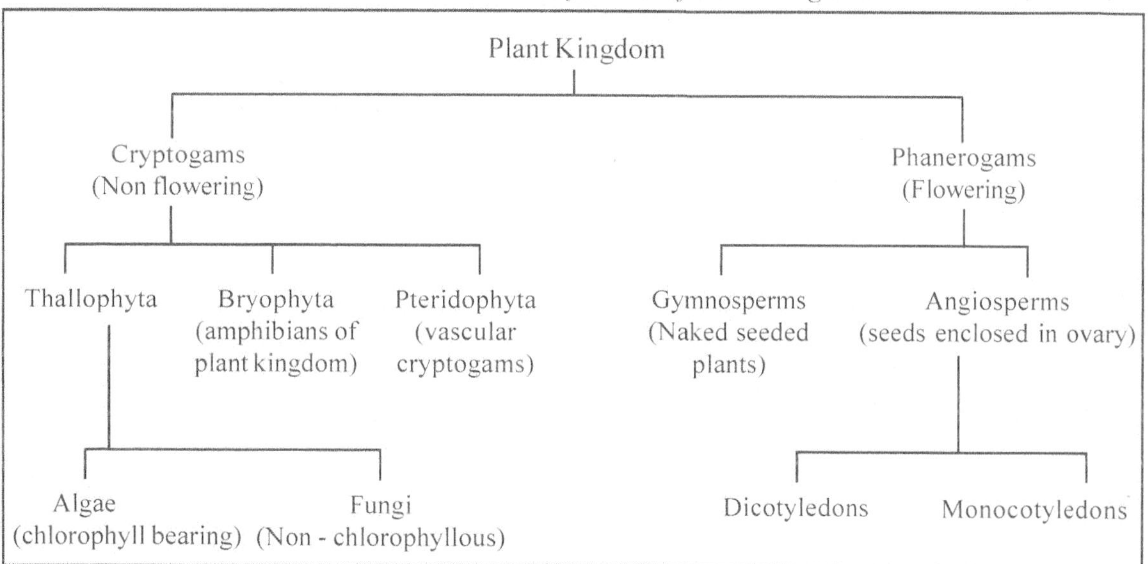

Kingdom 'plantae' include organisms of great diversity which synthesize their food from CO_2 and utilizing sunlight as a source of energy in chlorophyll pigments, hence considered as autotrophs. Plants exhibit great variation in their form, structure and reproduction. Unicellular organisms (*e.g. Chlamydomonas*) belonging to the group of lower plants called algae are simple forms compared to the elaborate, structurally complex, evolved higher plants. Besides, the form and structure, plants also exhibit a wide range of nuclear organisation from primitive prokaryotes to advanced eukaryotes. Blue green algae belonging to cyanophyta are prokaryotic while rest of the plants are eukaryotic in nature. Bryophytes stand at an evolutionary level higher than algae with multicellular sex organs with an outer layer of sterile cells and permanent retention of the zygote within the archegonium. Pteriodophytes have evolved further and acquired heterospory which led to seed habit. The seed plants (spermatophytes) include two major groups on the basis of the protection provided to the ovule before and after fertilization. These are the gymnosperms and the angiospersms. In gymnosperms, the ovules are freely exposed before and after fertilization as they are not enclosed in any ovary wall. In angiosperms, the ovules are compltely enclosed within the ovary. Angiosperms are considered as advanced due to the protection afforded to the ovule and the seed. Gynmosperms are generally considered as intermediates between the pteridophytes and the angiosperms.

Gymnosperms are most ancient seed plants that originated during the late Palaeozoic Era and reached the climax during the Mesozoic Era.

The evolutionary high status of angiosperms and their success over other groups of plant kingdom is attributed to : (i) diversified habit and vegetative forms, (ii) conducting system with xylem, tracheids containing wood vessels and the phloem possessing companion cells, (iii) adaption to diversified habitats, (iv) adaption of flower to insect pollination, (v) bisexual flower ensuring self pollination in case of failure of cross pollination, (vi) development of ovules within the ovary ensuring protection of developing ovules and seeds (vii) efficient dispersal of seeds by many means (insects, birds, animals, wind, water) and specialized mechanisms and (viii) vegetative propagation for rapid multiplication.

Alternation of Generation

Zygote, the fusion product of two haploid gametes, is the starting point of sporophyte development. A fixed chromosome number is maintained for every plant species which forms the characteristic feature of that particular species. Maintenance of a fixed chromosome number is achieved through an ingenious event called meiosis or reduction division during sexual reproduction. The stage of life cycle at which meiosis occurs varies in different groups of the plant kingdom. In lower plants such as algae (*Chlamydomonas*, *Chlorella*, *Spirogyra*), meiosis occurs before the germination of zygote into a new individual. Similarly in fungi also meiosis takes place at the time of germination of the zygote. In these forms the plant body is haploid and referred as gametophyte (as it bears gametes). The diploid spore-bearing phase (sporophyte) is represented by the zygote. This phase is very brief in lower plants.

A distinct sporophyte occurs only in the bryophytes (liverworts and mosses) as a result of simple mitotic divisions leading to a multicellular, well differentiated and specialized sporophyte. Sporophyte is completely dependent on gametophyte in bryophytes. Meiosis takes place in the specialized cells (spore mother cells) of the sporophyte resulting in the formation of haploid spores. The spores germinate and form an independent gametophyte which is the dominant phase in bryophytes. In bryophytes, two distinct generations, the gametophyte and the sporophyte, alternate with each other in a definite sequence.

Pteridophytes including ferns also exhibit the phenomenon of alternation of generations. In these plants, sporophyte attains independence from the gametophyte despite of its initial development on the gametophyte. The sporophyte shows an elaborate differentiation into true roots, shoots, leaves and vascular tissues. In pteriodophytes and seed-bearing plants (gymnosperms and angiosperms), the gametes are produced while still enclosed in the sporophytic tissues. The dominant gametophytic phase in lower plants has gradually reduced as the plants evolved while sporophyte which was completely dependent on gametophyte in bryophytes has evolved and reached a climax in the angiosperms.

Sexuality in plants also evolved from undifferentiated gametes (isogamy) in lower plants (*Chlamydomonas* and *Spirogyra*) to anisogamous and oogamous mode of reproduction where male and female sex organs are differentiated. The male sex organs of bryophytes and pteridophytes are referred as antheridia and the female sex organs are called archegonia. The plants with distinct sex organs (bryophytes, pteridophytes and some gymnosperms) produce male gametes called sperms and female gametes are called the eggs. Fertilization is effected by the male gametes or entire male gametophytes (pollen grains) through a phenomenon called pollination. In seed plants, on reaching the pistil, the pollen grain germinates and transport the male gametes to the egg.

Gametophytic tissue provide nutrition for the development of new sporophyte in bryophytes, pteridophytes and gymnosperms. In flowering plants, the endosperm, a special tissue produced due to triple fusion provides nutrition to the growing sporophyte. The embryo remains enclosed in the surrounding tissues of the sporophyte which constitute the seed. Seeds carry the embryonic sporophyte to long distances by means of insects, animals, wind and water to establish in new localities.

2.2 Algae

Algae are simple thallophytic plants comprising of a large, heterogenous groups which lack distinct organization like roots, stems and leaves as in the case of higher plants. It is difficult to define this highly diverse group of organisms. However, Fritsch (1935) defined algae in more appropriate terms as "*Unless purely artificial limits are drawn, the designation algae must include all holophytic organisms (as well as their numerous colourless derivatives) that fail to reach the level of differentiation characteristic of archegoniate plants*".

Algae are chlorophyllous and exhibit a wide range of thallus organisation. Algal bodies vary from simple single cells to complex structures present in giant kelps. Unicellular, colonial, filamentous, foliose and tubular type of plant bodies together with highly complex structures resembling root, stem and leaves are of common occurrence in nature. Algae may be motile (flagellate) or non motile (non-flagellate) and filamentous (branched or unbranched), net like or dendroid (Fig. 2.2).

Absence of multicellular wall around gametangia or sporangia differentiates algae from other cryptogams. Sporangia and sex organs in algae are unicellular or when multicellular, all the cells are fertile unlike other higher plants. However, algae have many common features with other plants. They have the same metabolic pathways and biological molecules like nucleic acids, chlorophylls, carbohydrates, proteins and lipids as in higher plants and other living organisms.

Algae are ubiquitous in distribution and inhabit all habitable environments. They are more common and abundant in stagnant waters. They habitat damp soils, rocks, tree trunks and old walls. Algae also occur in extreme environments like thermal springs. Algae are classified based on reproductive structures, cell wall structure, type of pigment and storage products. Algae are classified into 11 classes and salient features of different classes are presented in Table 2.2. Algae belonging to chlorophyta (green algae) are a highly diverse group and share features of higher plants. There are unicellular and multicellular forms. Sexual reproduction involves isomorphic alternation of generations or with either the sporophyte or gametophyte which is much reduced.

Red algae (Rhodophyta) are mainly multicellular and occur in fresh as well as sea water. Most of the seaweeds inhabit deep offshore waters in dim light. These are used for food (*Porphyra*) or as a source of agar (*Gelidium*). Reproduction involves a gametophyte and two different sporophyte generations. Phaeophyta dominate many intertidal regions. These are largest, multicellular seaweeds. Reproduction involves a gametophyte and sporophyte, with both generations looking similar. While in some gametophyte is reduced to a gamete.

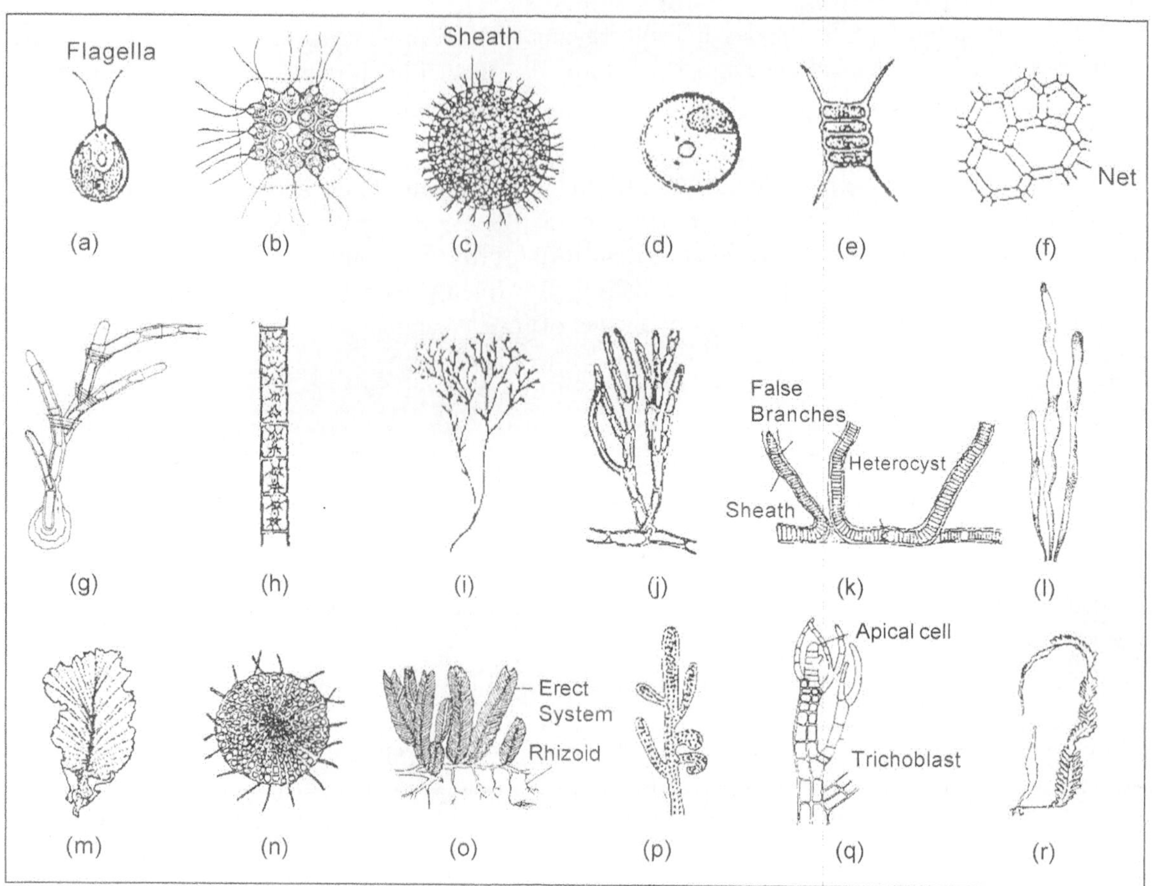

Fig. 2.2 Thallus types in algae.

(a) Unicellular biflagellate *Chlamydomonas*
(b) Coenobial *Gonium*
(c) Motile colonial *Volvox*
(d) Nonmotile unicellular *Chlorococcum*
(e) Non-motile coenobial *Scenedesmus*
(f) Nonmotile net-like *Hydrodictyon*
(g) Dendroid *Ecballocystopsis*
(h) Unbranched filamentous *Zygnema*
(i, j) Branched filamentous *Cladophora*
(k) False-branched *Scytonema*
(l) Tubular parenchymatous *Enteromorpha intestinalis.*
(m) Foliose parenchymatous *Ulva*
(n) Discoid *Coleochaete scutata*
(o) Siphonaceous *Caulerpa*
(p) Siphonaceous simple branched *Vaucheria.*
(q) Polysiphonaceous *Polysiphonia*
(r) Complex thallus of *Laminaria agardhii*

Reproduction

Algae reproduce in three different ways viz., 1. Vegetatively, 2.Asexually and 3. Sexually.

Table 2.2 *Distribution of pigments, cell wall composition, food reserve and presence or absence of flagella in 11 algal classes*

Class	Chlorophyll	Bili protein(s)	Carotenoids	Main food reserve(s)	Cell wall composition	Flagellum (number and insertion)
PROCARYOTA		PC, PE	Myxoxanthophyll, zeaxanthin	Glycogen, cyanophycin	Diaminopimelic acid, glucosamine muramic acid	Absent
Prochlorophyceae	b		α-carotene, zeaxanthin, cryptoxanthin	?	Diaminopimelic acid, glucosamine, muramic acid	Absent
EUCARYOTA Rhodophyceae		PE, PC	α-carotene, lutein, zeaxanthin, cryptoxanthin	Floridean starch	Cellulose, xylans, galactans	Absent
Cryptophyceae	c	PC	α-carotene, Σ-carotene, alloxanthin	Starch	Absent	2, unequal subapical
Dinophyceae	c		Peridinin, neoxanthin diaxtoxanthin, diadinoxanthin	Starch, rarely rarely oil	Cellulose or wall absent	2,1 trailing, 1 girdling
Chrysophyceae	c		Fucoxanthin, α-carotene tene, zeaxanthin, violaxanthin, neoxanthin, diatoxanthin, diadinoxanthin	Chrysolaminarin, oil	Cellulose or wall absent	1 or 2, unequal or equal, apical
Tribophyceae	c		Vaucheriaxanthin, cryptoxanthin, neoxanthin, diatoxanthin diadinoxanthin, heteroxanthin	Leucosin, oil		2, unequal 1 smooth, 1 with stiff hairs
Bacillariophyceae	c		Fucoxanthin, diatoxanthin, diadinoxanthin	Chrysolaminarin oil	Silica	1, with stiff hairs
Phaeophyceae	c		Fucoxanthin, violaxanthin	Laminarin, mannitol	Cellulose, alginic acid, fucoidin	2, unequal lateral
Euglenophyceae	b		Diadinoxanthin, heteroxanthin, diatoxanthin, zeaxanthin	Paramylon	Wall absent	1-3, apical, subapical
Chlorophyceae	b		Lutein, α-carotene, zeaxanthin	Starch	Cellulose	2 or more, apical, smooth
Charophyceae	b		Lutein, α-carotene, zeaxanthin, neoxanthin	Starch	Cellulose	2 or more apical, smooth

Chlorophyll - α and β-carotene are present in all classes. PC, phycocyanin; PE, phycoerythrin

Asexual and sexual means of reproduction is most common in algae. Vegetative reproduction is less common and may take place either by fragmentation (in filamentous forms) or by cell division (in unicellular forms).

Akinetes and spores are the means of asexual reproduction. Akinetes are resistant to unfavourable conditions and germinate into new individuals at the onset of favourable conditions. Spores germinate to give rise to new individuals without fusion unlike the gametes of sexual reproduction. Various kinds of spores are observed in algae, *viz.*, zoospores, apalanospores, autospores, hormospores, heterocyst, endospores, exospores, and other spores (Fig. 2.3).

Zoospores

These are flagellated spores produced inside zoosporangium and aplanosporangium or from vegetative cells. Zoospores are motile and swim in water for some time before they rest and shed their flagella and germinate into new individuals.

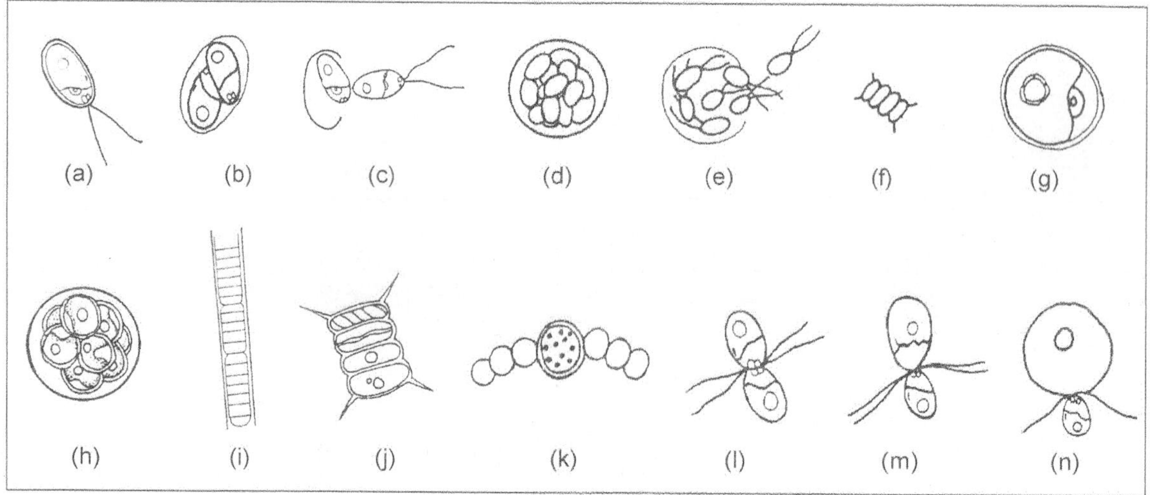

Fig. 2.3 Methods of reproduction and formation of different spores in algae
(a-c) Binary fission or bipartition, (d-e) Zoospore formation, (f) Aplanospore formation, (g-h) Autospore formation, (i) Fragmentation or formation of hormogonium, (j) Autocolony formation, (k) Akinete formation, (l) Isogamy, (m) Anisogamy and (n) Oogamy.

Aplanospores

These are non-motile and lack flagella. They can withstand unfavourable conditions.

Autospores

These spores look like aplanospores but differ in their ontogeny inability for mobility.

Hormospores

Fragments of some filamentous forms develop thick sheath to overcome unfavourable conditions and referred as hormospores.

Heterocyst

Heterocysts are large, transparent cells with thick layers found in some filamentous blue-green algae. They may be terminal and intercalary which are the points of fragmentation of the filaments.

Endospores

Endospores are non-flagelette thin walled spores formed internally within the vegetative cells of some unicellular algae.

Exospores

These are thin walled non-flagellate spores produced externally from the protoplasts of vegetative cells.

Sexual reproduction

It involves three main steps, the plasmogamy (union of cells), karyogamy (union of nuclei) and meiosis. Gametes fuse to form a zygote. Gametes are formed by gametogenesis. Sexual reproduction is of different types.

Isogamy

Gametangia are the cells that produce the gametes. Morphologically similar gametangia are known as Isogamatangia. Isogamy involves the fusion isogametes (Fig. 2.3 (*l*)) which are morphologically and physiologically similar. Isogametes may be motile or non-motile Gametes are not externally distinguishable either as male or female. They may be referred as plus (+) or minus (–) and may be produced by vegetative cell of the plant body.

Anisogamy

Gametes are motile or nonmotile and are morphologically dissimilar. Larger gamate is considered as female and the smaller one as the male. Female gamate is generally non-motile. These are known as anisogametes or heterogametes. Fusion between these gametes is referred as anisogamy. Sometimes physiological anisogamy is observed between morphologically identical gametes (Fig. 2.3(m)).

Oogamy

Oogamy is distinguished by the fusion of a large non-motile egg or ovum with a smaller motile sperm. Eggs are formed within an oogonium and the sperm within the antheridium. Female gamete or egg confines to oogonium, either permanently or until maturity. Antherozoid released from the mature antheridia swim around in water and reach the oogonium to affect the fusion of gametes (Fig. 2.3(n)).

2.2.1 Chlamydomonas

Chlamydomonas is a unicellular, motile, fresh water alga found commonly in pond water rich in ammonium salts. *C. ehrenbergi* occur in saline water while *C. nivalis* inhabit arctic and alpine regions. Some species are epiphytes on other algae such as *Volvox* and *Ulothrix*.

The thallus of *Chlamydomonas* is either oval, spherical, ellipsoidal or pyriform - measuring 20-30μ. Two equal whip like flagella are present at the anterior end (Fig. 2.4). Each one arising from the basal granule or blepharoplast. Contractile vacuoles are present at the base of the blepharoplasts which are useful in osmoregulation. Rhizoplast connects the centrosome of the nucleus with one of the basal granules. The cell wall is made up of cellulose covered by a layer of pectose. The cell is occupied by a cup shaped chloroplast. Pyrenoids are present inside the chloroplast. There is an eye spot or stigma at the anterior side of the chloroplast. It is sensitive to light and helps in migration of the organism from high light intensity to low light intensity.

Reproduction

Chlamydomonas reproduces by asexual and sexual methods. Asexual reproduction takes place by the formation of zoospores, aplanospores and hypnospores.

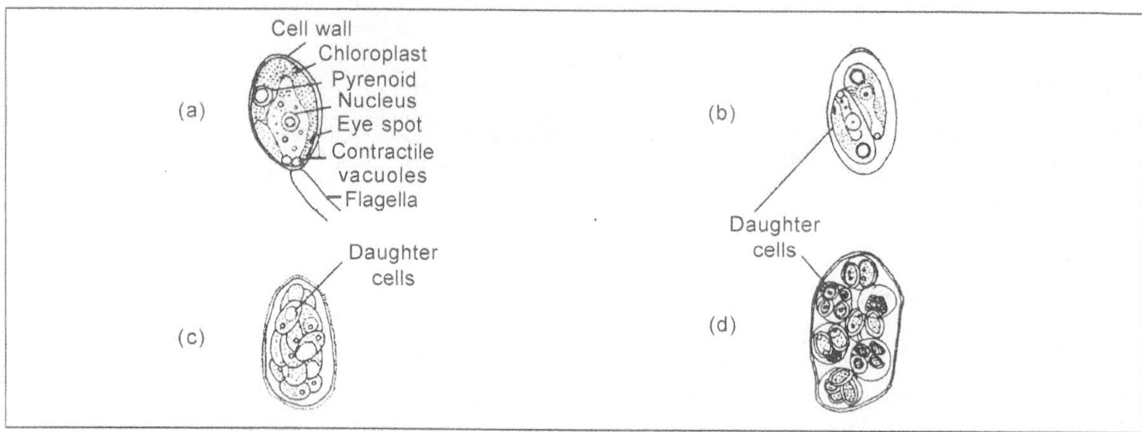

Fig. 2.4 *Chlamydomonas sp.* (a) A single cell showing different parts, (b-c) stages of cell division, (d) Palmelloid stage showing groups of cells within a common wall

Asexual Reproduction

Zoospores

The parent cell loses flagella and becomes non-motile under favourable conditions. The protoplast divides to form 4, 8 or 16 daughter protoplasts. Each daughter protoplast develops cell wall and flagella before liberating from the parent cell. They swim and grow into new *Chlamydomonas* (Fig. 2.3 (e)).

Palmella Stage

Under unfavourable and dry conditions, zoospores are not released but remain in the parent cell wall. No flagella are developed and these spores are non-motile and referred as *aplanospores*. The parent cell gelatinises to form mucilage sheath. Mucilaginous sheath protects these spores from desiccation. These daughter cells divide and produce large number of cells embedded in the sheath. This stage is referred as palmella stage (Fig. 2.4 (d)). Under favourable conditions, mucilage wall dissolves and the daughter cells develop flagella to form new individuals.

Hypnospores

Development of thick wall around the daughter protoplasts forms the *hypnospores*. A red pigment, haemotochrome accumulate in these spores. e.g. *C. nivalis*.

Sexual Reproduction

Chlamydomonas shows a diversity of reproduction methods such as isogamy, anisogamy and oogamy.

Isogamy

Isogamy involves the fusion of similar gametes. The protoplast of the cell undergoes longitudinal cell division producing 8-32 biflagellate gametes. These are similar to the zoospores but smaller in size. Gametes liberate and gather as a group which is referred as clumping. Pairing between similar gametes takes place from different individuals (Fig. 2.5(a-f)).

Anisogamy

Anisogamy is the fusion between dissimilar gametes. Gametes may be morphologically

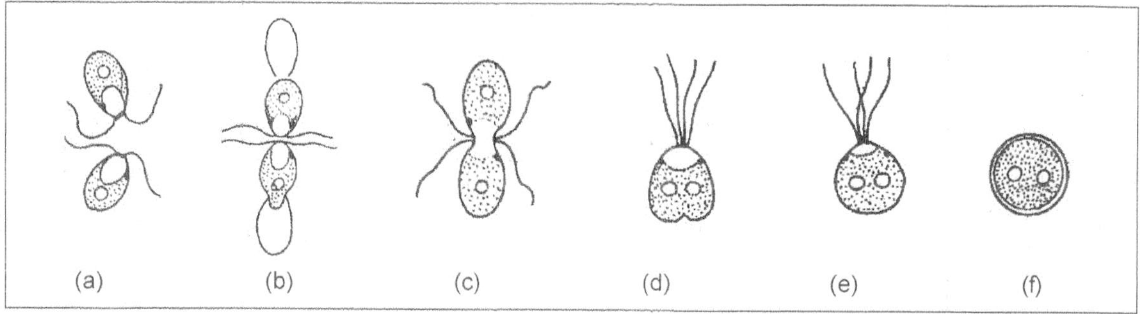

Fig. 2.5(a-f) Sexual reproducti on in *Chlamydomonas sp.* (a) two motile cells, (b-d) Gametic Union (e) motile zygote (f) Resting zygote

dissimilar or they may similar in size but show physiological anisogamy. e.g. *C. monoica.* The anisogametes come together and fuse to form a zygote. The zygote may secrete a thick wall and perrennates (Fig. 2.5(g)).

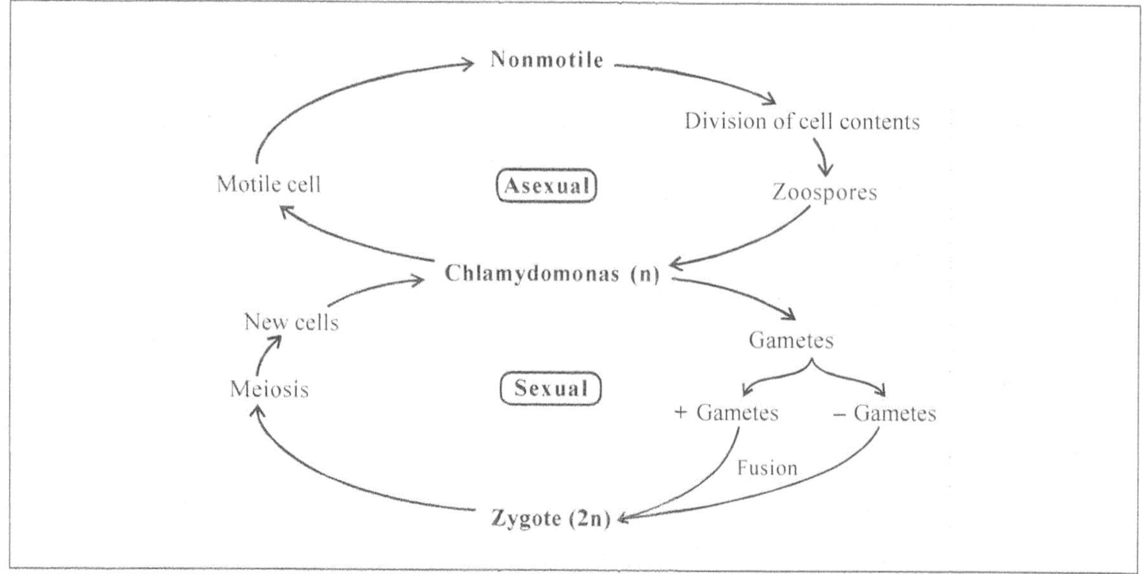

Fig. 2.5(g) Life cycle of *Chlamydomonas*

Oogamy

In some species of *Chlamydomonas* (*C. coccifera*), the female cell loses its flagella and directly function as non-motile female gamete. The male cell produces 16-32 biflagellate gametes, which swim to reach the female cell and fuse (Fig.2.4). Zygote, the fusion product develops thick, smooth or spinous wall and sinks to the bottom of the pond. Zygote is rich in reserve food materials and undergoes a period of rest. At the onset of favourable conditions, it germinates by undergoing meiosis resulting in the formation of four zoospores. Zoospores on release develop into adult individuals.

Life Cycle

The life cycle is haploid or haplontic. The cell belongs to haploid phase while zygote represents diploid phase in the life cycle. There is an alternation of haploid and diploid generations in the life cycle.

2.3 Fungi

Fungi are typically filamentous. Each individual filament is known as a *hypha (*pl. *hyphae)*. Mostly the hyphae are surrounded by a wall. Chitin is the major component of fungal cell wall. Fungi exhibit apical growth and branch periodically behind the tips resulting in a network of hyphae called mycelium.

Fungi are eukaryotic and have membrane bound nuclei and other cell organelles. Ribosomes are of 80s type in contrast to the prokaryotic bacteria. Fungi are achlorophyllous heterotrophs and they can utilize preformed organic materials as both the energy source and as carbon-skeletons for cellular synthesis. Fungi are characteristically absorptive in nutrition.

Reproduction in Fungi is by both sexual and asexual means with spores being the end product. Therefore, fungi may be defined in simple terms as "*Eukaryotic, characteristically mycelial or yeast like heterotrophs with absorptive nutrition*".

Classification

Fungi have traditionally been considered as a sub-kingdom under the plant kingdom, but recently they are elevated to a separate kingdom, equivalent to plants and animals (Whittaker 1969). Fungi are distinguished as a separate kingdom because of their characteristic mycelial nature, absorptive nutrition and dispersal through spores. The true walled fungi (Eumycota) can be differentiated from the wall-less fungi (Myxomycota). Eumycota is further divided into five main groups, such as mastigomycotina, zygomycotina, ascomycotina, basidiomycotina and deuteromycotina. Zygomycotina include mostly saprophytes like *Mucor* and *Rhizopus*, only few members are plant pathogens.

Deuteromycotina includes the imperfect states of many fungi that are known to reproduce asexually. In this group sexual stages or perfect states are unknown. Most of the microfungi and plant pathogens causing rusts and smuts are included in Basidiomycotina. Mushrooms and toadstools also belong to this group.

Ascomycotina comprises of many important fungi such as *Neurospora*, Yeasts (*Saccharomyces*), *Aspergillus* (*Eurotium*) and *Penicillium* (*Talaromyces*). The group ascomycotina is characterized by the development of sexual spores (ascospores) in a cell called Ascus. The asexual spores (conidia) are not formed in a sporangium but develop directly from hyphae. A majority of yeasts and antibiotic producing fungi are included in this group. Life history of a typical yeast, *Saccharomyces* is described here.

2.3.1 Saccharomyces

Saccharomycetaceae is the family of great industrial importance. Different species of the genus *Saccharomyces* are used for fermentation processes. Yeasts are ubiquitous, saprophytic organisms occurring in air, soil, and on all sugary substrates. They are used in brewing, baking and acid industries. Of the known 40 species of *Saccharomyces*, *S. cerevisiae* is the most common bakers yeast. Certain special strains of this species are used in making beer, wine and fruit juices.

Vegetative Structure

Thallus is unicellular, cells are elliptical and measure 6-8 × 5-6 μm. Fine structure of yeast cell reveals the presence of three layers of cell wall, outer layer, middle layer and inner layer (Fig. 2.6). Outer layer consists mainly of mannan-protein and some chitin and middle layer is largely made up of glucan while innermost layer contains protein glucan. Cell organelles of yeast cell are like that of a typical eukaryotic cell : endoplasmic reticulum, ribosomes, mitochondria, golgi apparatus and nucleus. Lipid granules are present. Large vacuole is prominent in the middle of the mature yeast cell.

Asexual Reproduction

Yeasts reproduce vegetatively by budding. A small protruberance or bud starts growing mainly at the poles of the cell (Fig. 2.7). The bud enlarges and subsequently the nucleus divides by constriction without the breakdown of nuclear membrane. A portion of the constricted nucleus enters the bud along with other cell organelles. The new cell wall material blocks the cytoplasmic connection between the parent cell and the bud. Finally the bud separates from the parent cell leaving a scar at the point of attachment. Sometimes many bud scars may be present on a single cell. In *Schizosaccharomyces*, asexual reproduction occurs by fission.

Sexual Reproduction

Many strains are heterothallic and the ascospores belong to two mating types. Formation of ascospores can be induced by growing the yeast on a nutrient-rich presporulation medium. Presporulation medium contains sugar, nitrogen source and vitamin B complex. The vegetative diploid cells develop directly into asci (sing. *ascus*) within 12-24 hours on sporulation medium. Sporulation medium contains a low concentration of sugar with Na or K at 0.1–1.0% w/v. Meiosis or reduction division precedes spore formation resulting in four haploid spores. Ascospores are spherical in shape thick-walled and during meiosis nuclear membrane remains intact. Life cycle of bakers yeast is outlined in Fig. 2.8. Ascospores may give rise to haploid buds on a nutrient medium and these can be maintained indefinitely in culture. From the haploid ascospores diploid state may be reestablished in the following ways.

1. *Fusion of ascospores :* This takes place outside the ascus. The wall separating the spores may breakdown leading to plasmogamy and nuclear fusion. Otherwise a short conjugation tube may develop bringing the cytoplasm of the two spores into contact. Karyogamy follows and zygote develops diploid buds (Fig. 2.7).
2. Diploids may develop by simple fusion between haploid cells
3. By fusion between the haploid cells and ascospores leading to the formation of diploid yeasts.

The life cycle in yeasts is considered as diploid or haplobiontic type of life cycle.

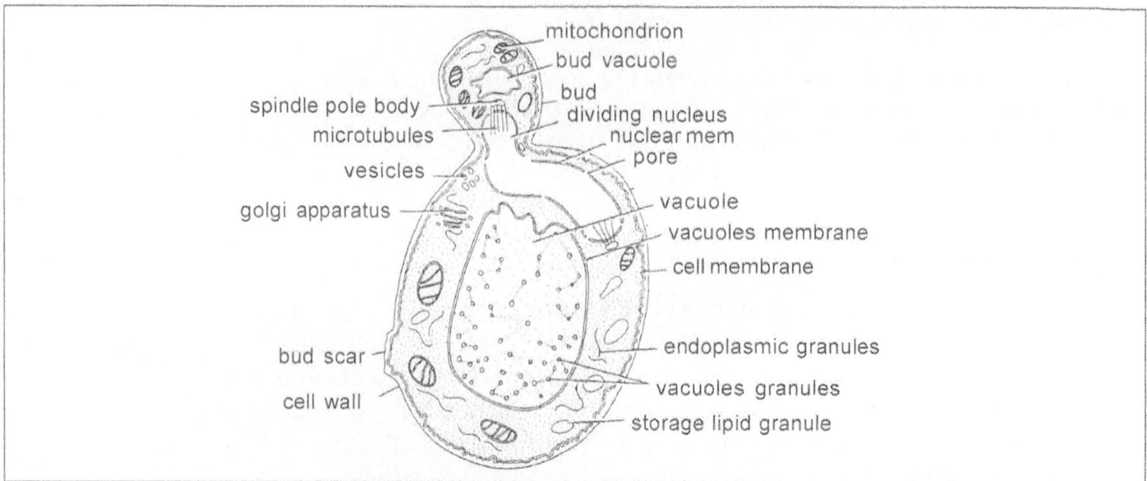

Fig. 2.6 Diagrammatic representation of a section of a budding yeast (*Saccharomyces cerevisiae*)

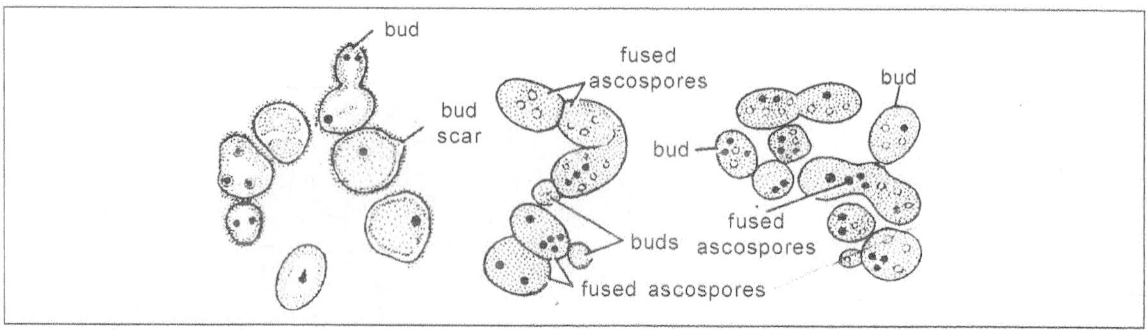

Fig. 2.7 Stages of reproduction in *Saccharomyces cerevisiae*. Asexual reproduction, budding, diploid vegetative cells are budding to form diploid cells

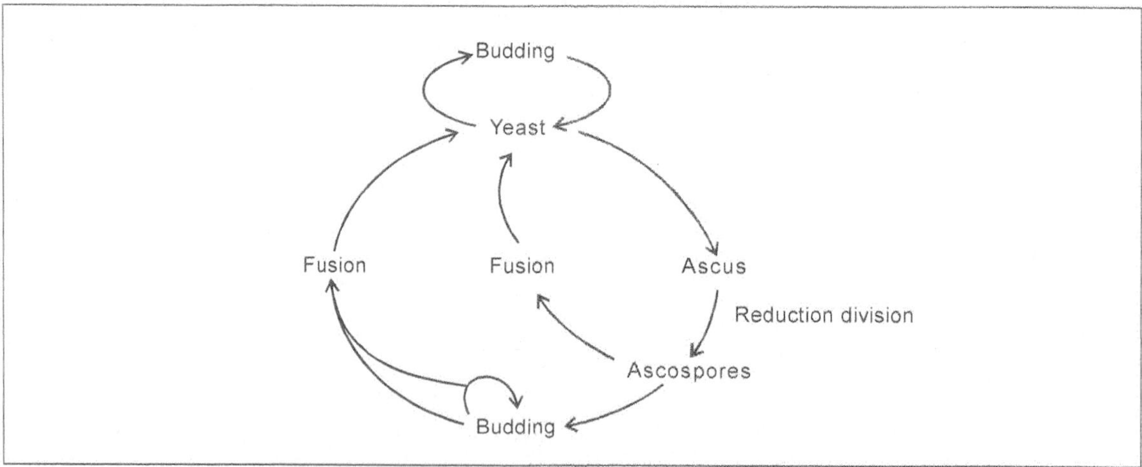

Fig. 2.8 Outlines of life cycle of *Saccharomyces* (yeast)

2.4 Lichens

The lichens are unique organisms of dual nature comprising of two dissimilar life-forms, a fungus and an alga. They live in a symbiotic association. The fungal partner is known as a *mycobiont* and the algal partner is called a *phycobiont*. Mycobiont is a dominant partner of the association as it makes the bulk of the thallus. It produces sexual reproductive structures. It is considered that the alga synthesizes the food and the fungus absorbs it.

Mycobiont mainly belongs to Ascomycotina while a few are from Basidiomycotina. Phycobionts are from chlorophyceae (*Trentepohlia, Coccomyxa*), xanthophyceae (*Heterococcus*) and cyanophyceae (*Nostoc, Scytonema*).

Lichens are normally classified on the basis of the nature of the mycobiont and the kind of fruit body.

1. *Ascolichens :* Mycobiont is an ascomycete fungus. Majority of them are Discomycetes and others being Pyrenomycetes or Loculoascomycetes.
2. *Basidiolichens :* The fungal component is a basidiomycete.

Lichens grow in a variety of habitats such as roofs, leaves, tree bark, soil and on barren surface of rocks. Lichens grow luxuriantly on trunks of forest trees in India. Depending on the habitat, lichens are distinguished as:

1. *Saxicoles* which grow on rocks and mostly in temperate areas.
2. *Corticoles* grow on bark, mostly in tropical and sub-tropical areas of the world.
3. *Terricoles* are soil inhabiting lichens from hot areas with scanty rain and dry summer.

Lichens are mostly perennial and slow-growing. These are the pioneer colonizers on barren rocks. Acids produced by lichens disintegrate the rocks and help in the formation of soil paving way for the colonization of first land plants, the mosses. Lichens are extremely sensitive to atmospheric pollutants, such as SO_2. Thus, lichens can serve as reliable biological indicators of pollution. Lichen are also able to accumulate radio active fallout. Beside ecological significance, lichens are used as food and fodder as they contain polysccharides, enzymes and vitamins. *Lecanora, Cetraria,* and *Umbilicaria* are in use by man in Israel, Egypt and Japan. *Parmelia* species are used as curry powder in India. Lichens are used in confectioneries for making delicious chocolates and pastries in France.

Lichens are also used in several kinds of industries such as brewing, tanning, dyeing, cosmetics and perfumes etc. Lichens have been in use for their medicinal value in curing jaundice, fever, diarrhoea and skin diseases. *Parmelia sexatilis* is used for epilepsy while *Peltigera comoma* is used against hydrophobia. Usnic acid present in *Usnea* and *Cladonia* is a broad spectrum antibiotic useful in the treatment of infections.

Thallus

Plant body of a lichen is a thallus. Some species may have yellow, red, orange or brown pigments. Based on their growth, form and nature of attachment to the substratum three main categories of thalli are distinguished.

1. *Crustose* : These are encrusting lichens. Thallus is thin, flat, without lobes and closely attached to stones, rocks and bark (Fig. 2.9). *Eg.* : *Rhizocarpon, Lecanora, Graphis.*
2. *Foliose* : There are leafy lichens with flat, lobed horizontally-spreading dorsiventral structure (Fig. 2.9). It is attached to the substratum by rhizoid–like structures called rhizines. *Eg.* : *Xanthoria, Parmelia, Hypogymnia.*
3. Fruiticose : These are shrubby lichens. Thallus is cysindrical , flat or ribbon– like, branched structure (Fig. 2.9). *Eg.* : *Cladonia, Usnea.*

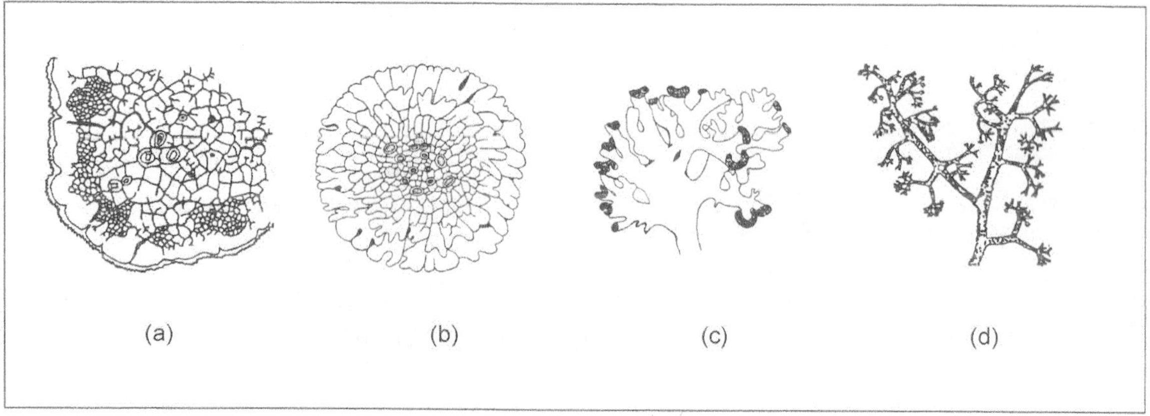

Fig. 2.9 Three major growth forms of lichens
(a) Crustose : part of a thallus of *Ochrolechia tartarea* dried out on a rock; (b,c) Foliose (b) *Xanthoria* sp. with apothecia, growing closely applied to a rock; (c) part of a *thallus of Hypogymnia sp.* with reflexed tips of branches bearing powdery soredia underneath); (d) Fruticose : part of an erect thallus of *Cladonia rangiformis*.

Internal Structure

Lichen thallus is of two types, Homoiomerous and Heteromerous. In Homoiomerous lichens, algal cells and fungal hyphae are uniformly dispersed throughout the thallus (Fig. 2.10 (a)). Phycobiont is gelatinous and belongs to cyanophyceae. *Eg.* : *Leptogium, Collema.*

In *Heteromerous* lichens algal component is restricted to a distinct layer, usually the lower or upper side of the thallus. Thallus is differentiated into distinct layers as cortex, algal zone and medulla (Fig. 2.10(b)).

Cortex is the protective covering both on upper side (upper cortex) and lower side (lower cortex) which is compactly interwoven by fungal hyphae. The cortex is covered with a thin homogenous gelatinous layer (Fig. 2.10 (b)).

Algal zone occurs just below the upper cortex, inter woven loosely with fungal hyphae. The algal cells of green or blue-green algae. *Chlorella, Pleurococcus, Cystococcus* (chlorophyceae); *Nostoc* and *Rivularia* (Cyanophyceae) constitute the photosynthetic layer. Medulla is the central zone of thallus which is less compact and has loosely interwoven fungal hyphae with large space between them. The hyphae are thick walled in different directions.

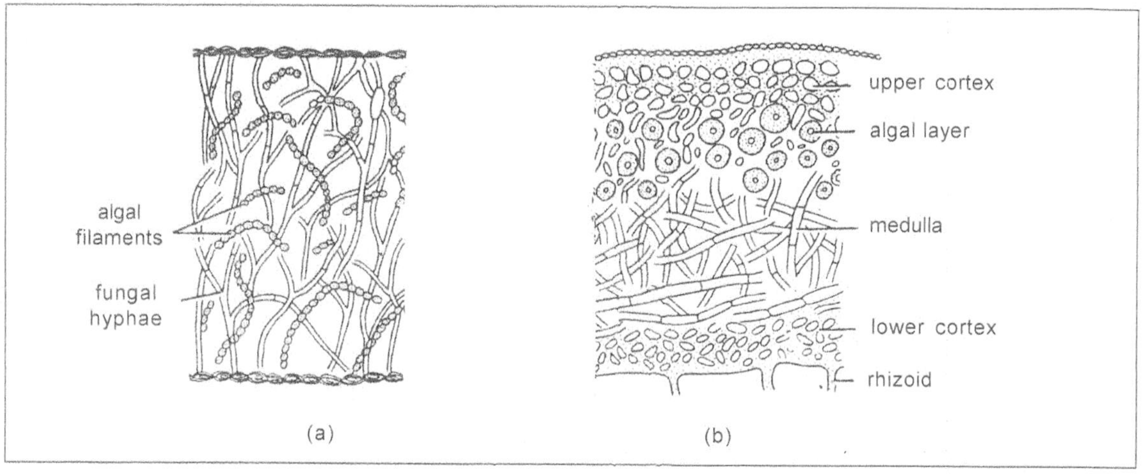

Fig. 2.10 Internal structure of lichen thallus. (a) Homoiomerous thallus (b) Heteromerous thallus

Reproduction

Lichens multiply by its individual algal or fungal components and also by other asexual means such as Fragmentation, Soredia, Isidia and Spermatia.

Fragmentation occurs by breakdown of adult thallus into segments that develop into new thalli. Ageing and accidental injury are the other two ways by which fragmentation takes place.

Soredia : These are small bud like out growths over the upper surface or edges of the thallus. Each soredium contains a few algal cells enclosed by a little weft of fungal hyphae (Fig. 2.11(a)). If soredia develop in a more organised manner occurring in localized pustules, they are known as Soralia. *Eg.* : *Parmelia* and *Physcia.*

Isidia : These are small, conical, coral-like structures, usually constricted at the base and can be easily broken off under favourable conditions. It encloses the same algal and fungal components as those in the thallus (Fig. 2.11(b)). They are in various shapes as rod-like in *Parmelia*, coralloid in *Umblicaria*, scale-like in *Collema* and tiny coral-like in *Peltigera.*

Spermatia : Some lichens produce spermatia or pycniospores inside the pycnidium (Fig. 2.12). It normally behaves as male gametes but in some cases germinate to form fungal hyphae that may come in contact with appropriate algal partner to form a new lichen thallus.

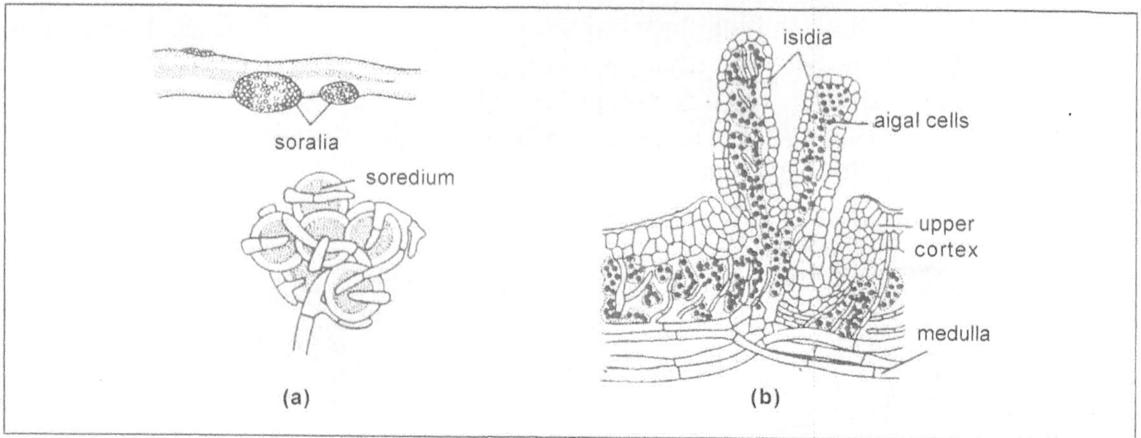

Fig. 2.11 Specialised-structures associated with lichen thalli. (a) soralia and soredium, (b) isidia.

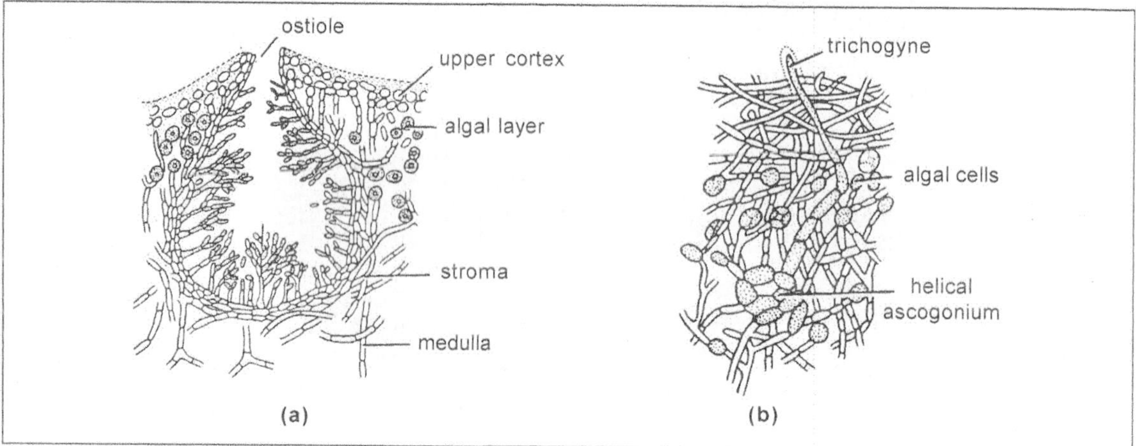

Fig. 2.12 Reproductive structures of lichens. (a) pycnidium and pycniospores (Spermatia). (b) carpogonium

Sexual Reproduction

Only mycobiont of the lichen thallus reproduces sexually and form fruit bodies on the thallus. Sexual reproduction corresponds with the type of mycobiont in the lichen thallus. Ascolichens follow the general pattern of reproduction like ascomycetes while basidiolichens follow basidiomycetes. These are either monoecious or dioecious. Male sex organs are pycnidia (spermagonia) and the female sex organs are carpogonia.

Pycnidia

It is a flask shaped structure often behave as male sex organ in some lichens (Fig. 2.12(a)). It is immersed in the upper cortex and medulla of the lichen thallus. Pycnidium is ostiolate and produce aseptate or septate spermatophores from the inner layer. Each spermatophore produces small, unicellular spermatia at its apex. Spermatia are released into a slimy mass that oozes out of the ostiole.

Carpogonia

Carpogonium is helically coiled at the base with an upper multicellular trichogyne (Fig. 2.12(b)). The coiled part is the ascogonium proper. Ascogonium is embedded in the thallus near cortex. It is multicellular structure and each cell is either uni or multinuclear trichogyne tip is sticky.

Spermatia on dissemination arrives at the tip of the trichogyne and the nuclei migrate into the ascogonium. After fertilization, fertilized cell of the ascogonium produce ascogenous hyphae which form asci and ascospores as in higher ascomycetes. Along with the asci, a large number of sterile hyphae called paraphyses develop in between the asci. The ascocarp may be a perithecium as in *Dermatocarpon* and *Verrucaria* or an apothecium as that of *Parmelia.*

Apothecia are disc-like flattended structures and are of two types.

1. *Lecideine type :* The apothecium is covered only by the fungal hyphae and there is no algal component in any part of the fruit body. E.g. *Lecidea, Cladonia.*
2. *Lecanorine type :* In this type, apothecia are more developed and possess algal component along with fungal hyphae in forming the thalline margin or amphithecium. A mature apothecia of a lecanorine type lichen have three distinct parts *viz.*, the thecium (= hymenium), epithecium and hypothecium (Fig. 2.13). Thecium is the fertile zone with a layer of asci and paraphyses. The paraphyses project beyond the level of asci to form a layer called the epithecium. The hypothecium is a loosely arranged hyphae below the thecium.

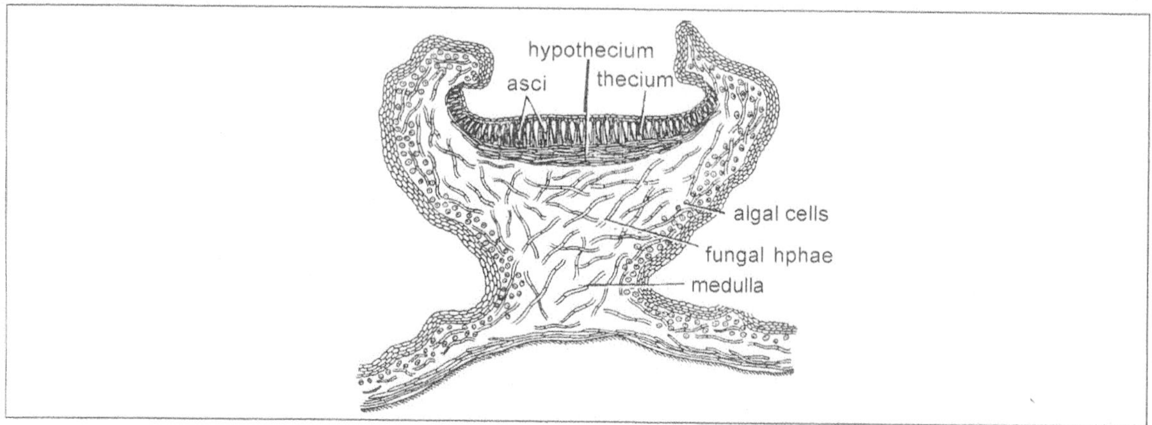

Fig. 2.13 Vertical section of Apothecium of a lichen

2.5 Bryophyta

Bryophyta include the simplest and the most primitive land plants. They are small inconspicuous green plants occurring in all habitats, particularly in wet places and deep shade. They are predominant in some sub-polar regions and bogs where they form peat. Bryophytes comprise of three groups commonly known as liverworts, hornworts and mosses. Accordingly, the sub-kingdom Bryophyta is divided into Hepaticopsida, Anthocerotopsida and Bryopsida or Musci. Hepaticopsida include all the liverworts while anthocerotopsida include a small group of hornworts. The mosses belong to Bryopsida and show relatively advanced features over other two groups.

Bryophytes are the first land plants mostly growing on moist soils, hence referred as amphibians of plant world. They are incompletely adapted to land conditions because almost all of them still require water for performing the act of fertilization.

The gametophyte is highly developed, independent, conspicuous long lived structure. It is an undifferentiated flattened thallus (liverworts and hornworts or a definite rootless leafy shoot (Mosses). However, the leaves and stem are not comparable to that of the vascular plants since in this case such organization represent the gametophyte generation. A true root is absent in bryophytes and the function of the root is taken up by a unicellular or multicellular structures called the rhizoids.

The gametophyte of liverworts is flat, thallose, prostrate, dichotomously branched structure with a dorsiventral symmetry (Fig. 2.14). The thallus has a conspicuous midrib and along the midrib on the dorsal surface, cup like structures are present. These are known as gemma cups and they contain special reproductive bodies called the gemmae. From the ventral side arise the rhizoids and scales. These serve as absorbing and anchoring organs.

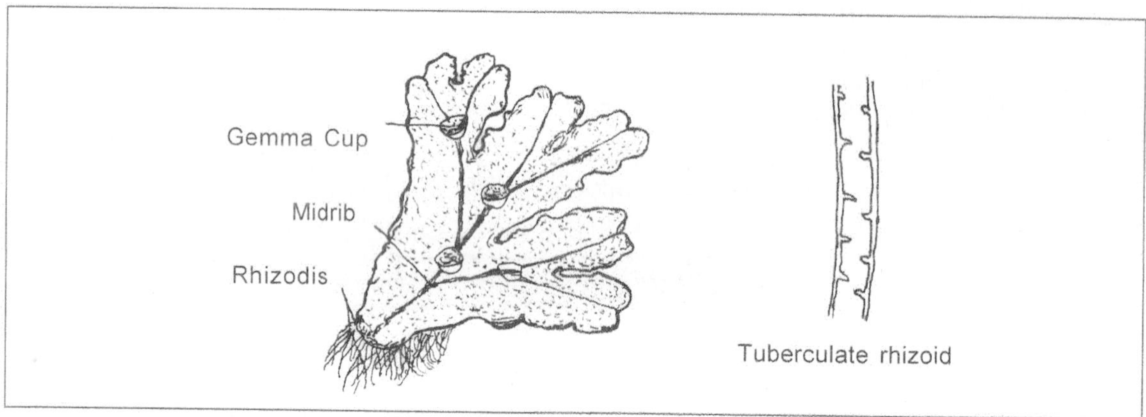

Fig. 2.14 Gametophyte thallus of a liverwort (*Marchantia*)

Reproduction

Sexual reproduction is oogamous. All bryophytes have antheridia which produce motile antherozoids and archegonia contain eggs. These structures are fairly uniform in all the three groups of bryophytes. Sex organs are antheridium (male) and archegonium (female). The sex organs are multicellular and covered by a sterile wall layer (Jacketed). The male gamete or antherozoid is motile and then female gamete is a non-motile egg or oosphere. The male sex organ is the antheridium and antherozoids are produced from antherozoid mother cells.

The female sex organ, the archegonium is characteristic as that of pteridophytes and gymnosperms. The archegonium is a multicellular, flask-shaped structure with a well defined venter and a slender neck. The venter wall enlarges with developing embryo to form a protective envelope called the calyptra.

Special upright branches develop from the growing apex. These branches bear the sex organs. These are of two kinds, *antheridiophores* and *archegoniophores* and are borne on different thalli (Fig. 2.15) hence referred as dioecious. Antheridia and archegonia are borne on these erect, stalked branches. The structure of the antheridia and archegonia is fairly uniform across the three groups of bryophytes. The antheridia are near spherical to ovoid sacs on short stalks. The jacket of the body consists of a single layer of thin walled cells (Fig. 2.19). Antheridium encloses sperm mother cells which produce number of antherozoids. The mature antherozoid possess two flagella and can swim a very short distance.

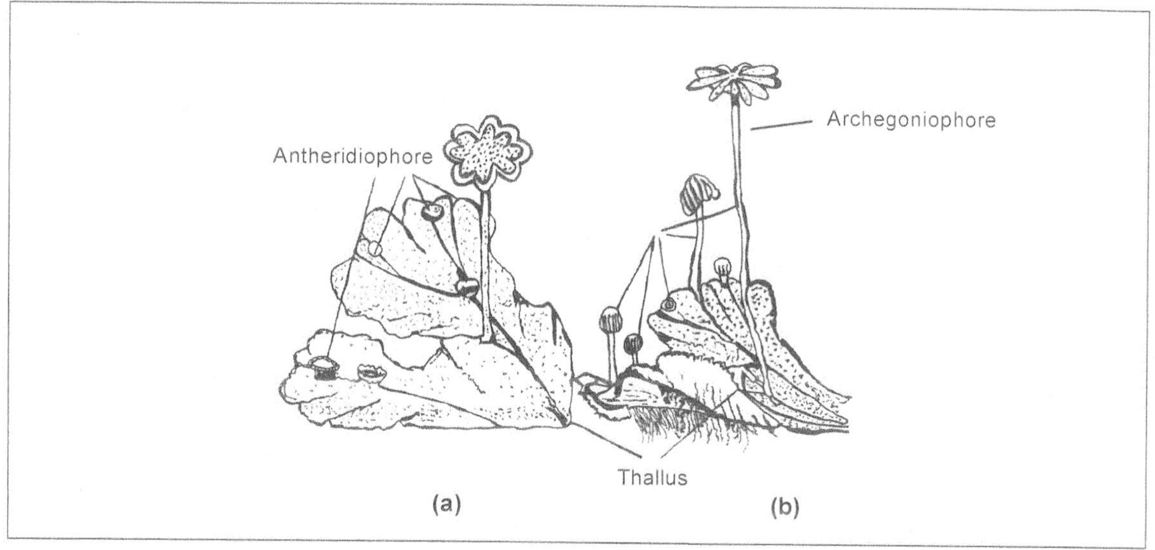

Fig. 2.15 *Marchantia polymorpha.* (a) portion of thallus of *Marchantia*, bearing antheridiophores, (b) portion of thallus bearing archegoniophores.

The archegoniophore is similar to the antheridiophore in having a stalk and disc. The disc of the archegoniophore becomes a lobed structure resembling a rosette with eight lobes.

The archegonia are cylindrical with an inflated base called venter and a long neck (Fig. 2.19). A jacket of sterile cells encloses the egg in the inflated basal part. Neck contains approximately 10 neck canal cells which degenerate at maturity. Archegonia are borne on short stalks.

Antherozoids upon release from the antheridia swim in a film of water available in the dorsal furrow and reach the archegonia. Antherozoids are chemotactically attracted to the open neck of the archegonium which may swim down the neck canal and fertilize the egg resulting in a diploid zygote.

The Sporophyte

The zygote is the first cell for the onset of sporophyte or asexual generation. The zygote is retained in the venter of the archegonium and develops into a sporophyte. It is always typically diploid. There are three parts in a liverwort sporophyte a foot embedded in the gametophyte a colourless seta and a capsule at the tip (Fig. 2.16). The sporophyte draws nutrients from the gametophyte

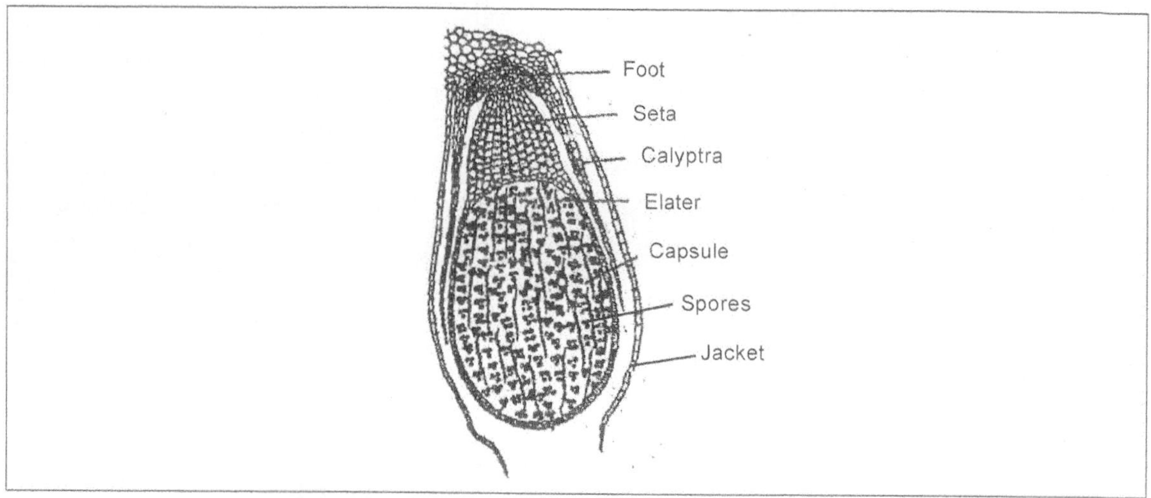

Fig. 2.16 *Marchantia polymorpha*. Longitudinal section of mature sporogonium.

through the foot. When the spores are mature the capsule wall splits into four valves. Spore dispersal is assisted by the sterile hairs known as elaters. In some liverworts like *Riccia*, the sporophyte is just a sac of spores embedded in the gametophyte and the spores are liberated on its decay.

The Sporophyte of Hornworts

In Anthocerotopsida, the sporophyte is a long cylindrical capsule. It has a large bulbous foot and a smooth slender capsule (Fig. 2.17). There is no seta between erect cylindrical 2-4 cm long capsule. In between foot and capsule there is an intermediate meristematic zone which helps in continuing the growth of the capsule several weeks unlike the liverworts. The capsule is green and photosynthetic with stomata on the outside attaining a degree of independence from the gametophyte. There is a column of sterile cells with attached elaters which are called as pseudoelaters in the middle. The capsule wall, the central column and the elaters all twist to aid spore dispersal.

The Sporophyte of Mosses

Typical moss sporophyte lasts for several weeks (Fig. 2.18) sporophyte consists of a foot embedded in the gametophyte a tough elongated stalk and a capsule. The capsule is photosynthetic and has stomata and central column of sterile tissue. There is a lid at the tip of the capsule. It is thrown off once the capsule is mature. One or two layers of peristomial teeth, is present inside the lid. They facilitate the opening of the capsule for spore dispersal in dry weather and close it in wet conditions.

The life cycle of bryophytes show a typical and sharply defined heteromorphic alternation of generations. The two distinct phases, the gametophyte and sporophyte, follow one another in regular succession. The gametophyte phase represents the haploid phase with one complete set of chromosomes and is a sexual generation, as it produces gametes (antherozoids and eggs). The sporophyte or diploid phase intervenes between syngamy and meiosis. A typical life cycle is diagrammatically represented in Fig. 2.19.

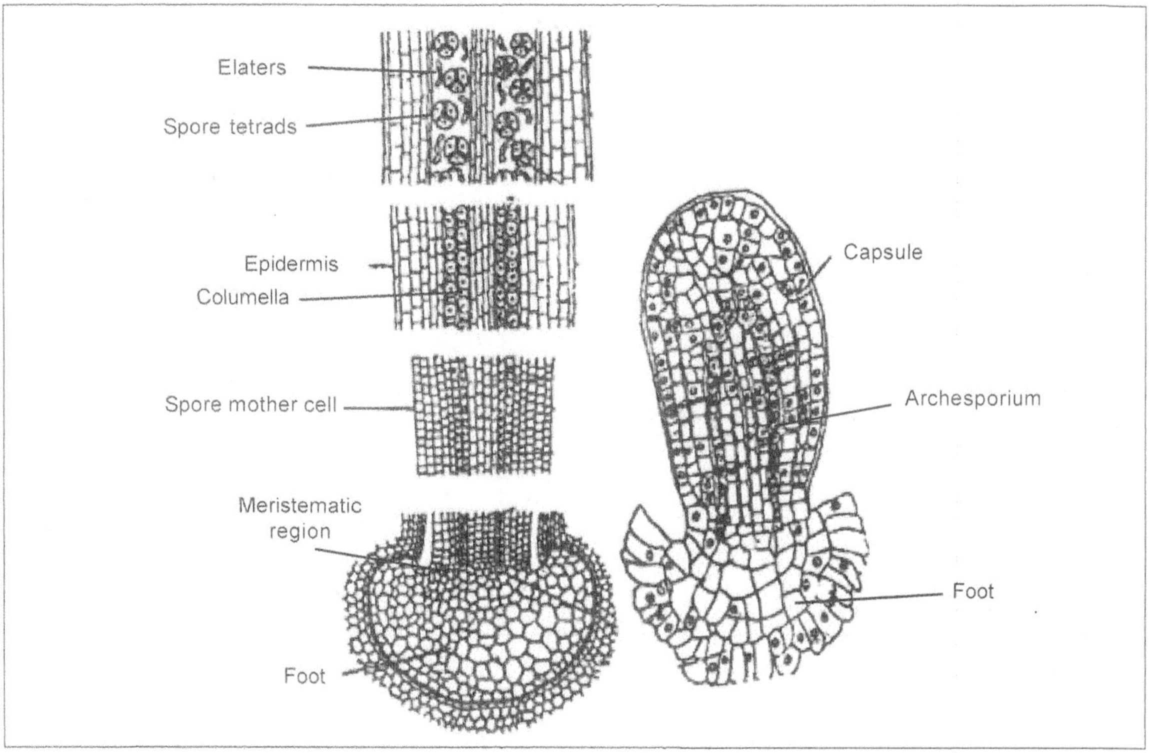

Elaters
Spore tetrads
Epidermis
Columella
Spore mother cell
Meristematic region
Foot
Capsule
Archesporium
Foot

Fig. 2.17 *Anthoceros* Longitudinal section through different portions of a sporogonium

Sporogonium
Operculum
Peristome
Annulus
Rim
Archesporium
Sporesac
Columella
Lacuna
Epidermis
Stometa
Apophysis
Seta
(a)
(b)

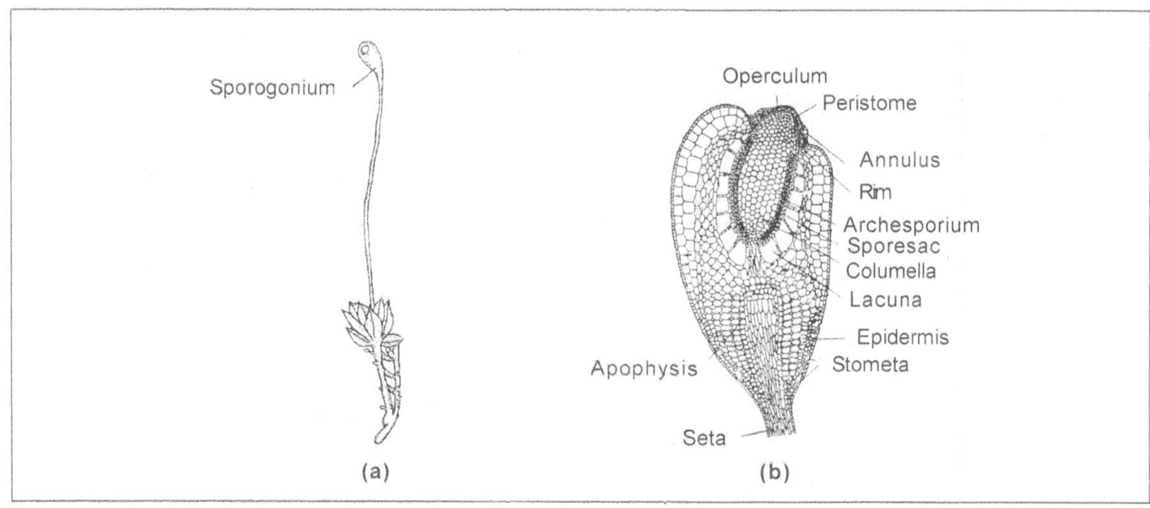

Fig. 2.18 *Funaria hygrometrica* (Moss) (a) a plant, (b) L.S. of Capsule

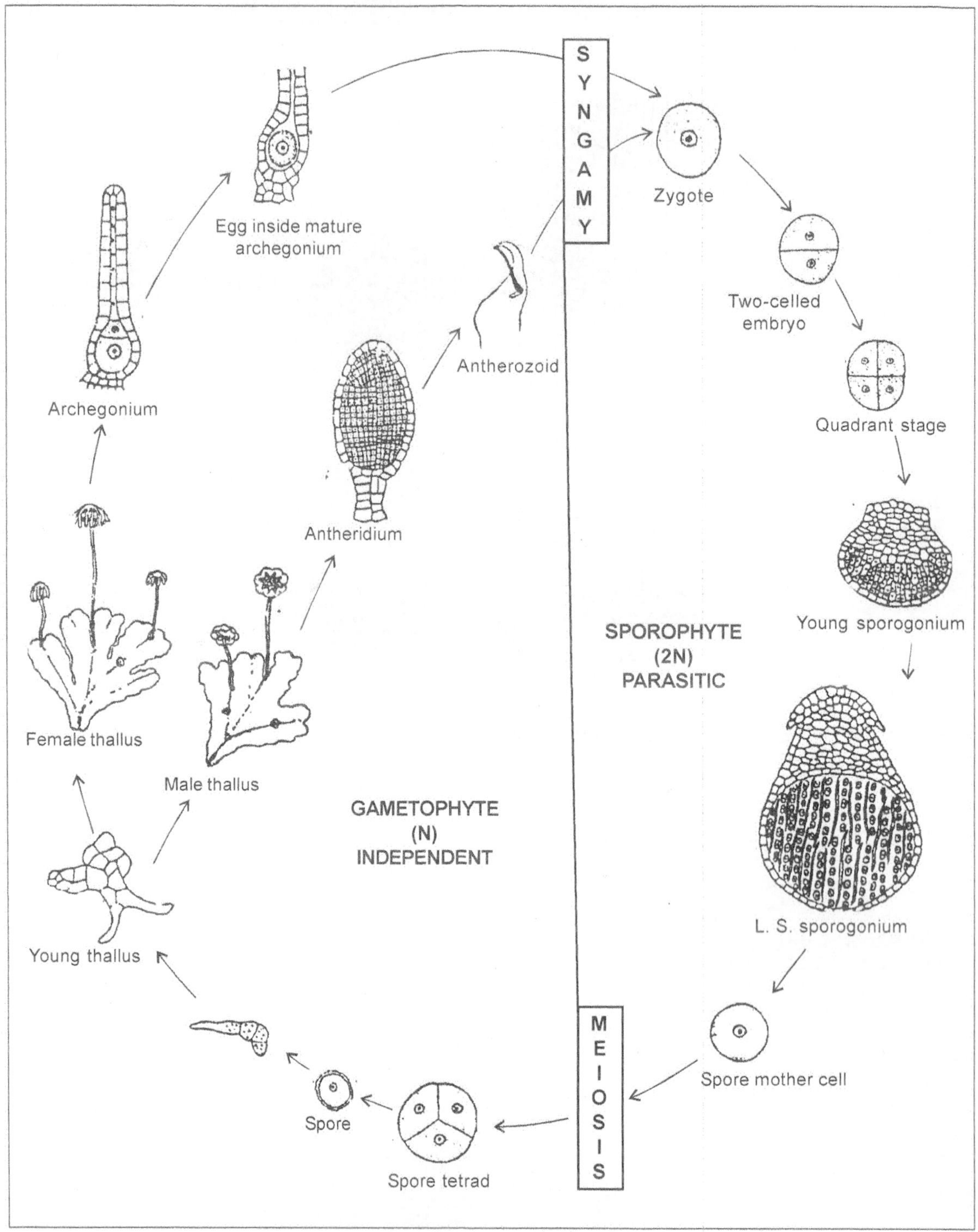

Fig. 2.19 Diagrammatic representation of the life-cycle of a liver wort (*Marchantia*)

2.6 Pteridophyta

Pteridophytes are the early vascular plants first appeared in the Silurian period. The fossils recorded are little over 400 million years old. The oldest fossil is that of *Cooksonia* (*C. hemispherica*) which had dichotomously branched photosynthetic stem with no leaves, but with only rhizoids anchored to soil. The sporophytes of the pteridophyta have developed a tissue for conducting system, consisting of xylem and phloem.

The pteridophytes have certain similarities with bryophytes and as well as with spermatophytes, thus, occupying a position between the bryophytes and the spermatophytes.

Pteridophytes resemble bryophytes in having the egg cell protected by a large number of sterile investing cells. The archegonium is the female sexual reproductive structure. The male reproductive structure is the antheridium (Fig. 2.20). It consists of a single layer of sterile jacket cells enclosing many androcytes or antherozoid mother cells, which produce motile antherozoids. Pteridophytes require water for the fertilization and exhibit distinct alternation of generation in a regular succession which is similar to that of bryophytes.

The opening of antheridia and subsequent fertilization is still dependent on the presence of water as in bryophytes. Alternation of generations is seen with two distinct phases which occur in regular succession. The gametophytic phase or sexual generation comprises of one complete set of chromosomes in the cells. The haploid phase (n) bears sexual reproductive organs which produce the gametes. This stage occurs between meiosis (reduction division) and syngamy (fertilisation). The diploid phase (2n) is commonly referred as the sporophyte or non-sexual generation

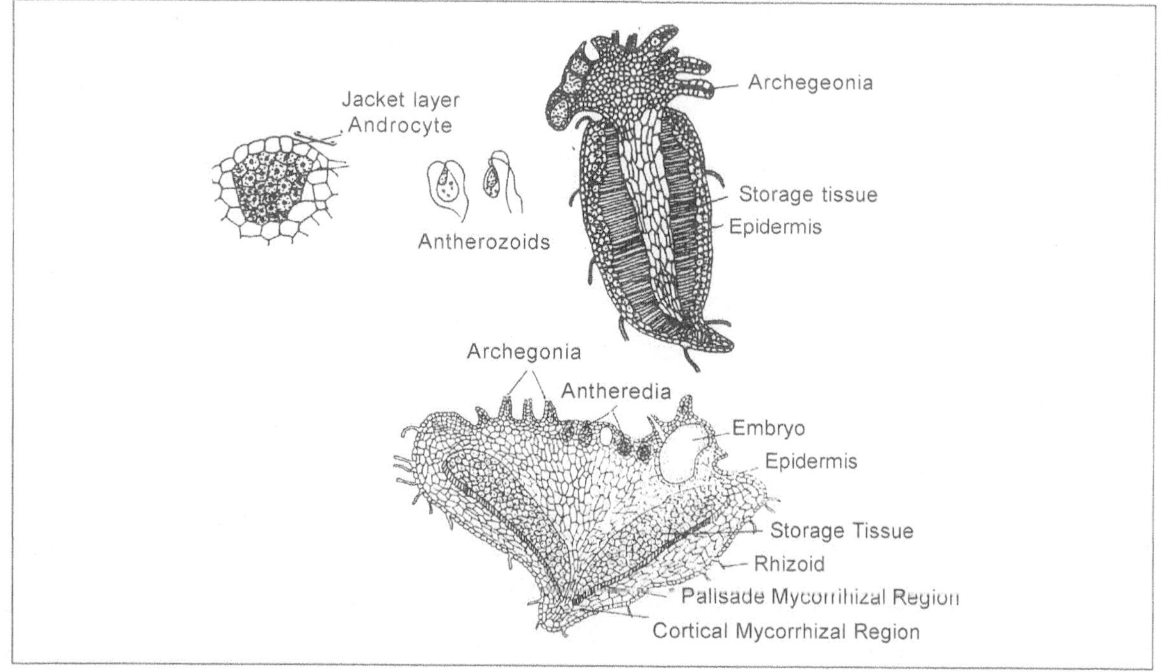

Fig. 2.20 Gametophytes of *L. clavatum, L. cernum, L. complanantum*

as it produces non-sexual spores. Sygamy and meiosis are intervened the sporophyte generation. Spore mother cells produced by the last division of the sporogenous tissue divide meiotically and form tetrads of spores. Formation of spores is similar in both bryophytes and pteridophytes.

However, Pteridophytes show distinct advance over the bryophytes in other features. The sporophyte is independent and dominant, typically photosynthetic phase of the life cycle. Sporophyte is organised into the stem, leaves and roots. The true roots have appeared for the first time in plant kingdom making it nutritionally independent of the gametophyte. Stem and root show apical growth and tissue differentiation. The root and the shoot are provided with conducting system while the function of photosynthesis is restricted to leaves. Leaves are provided with an efficient epidermis with stomata and chlorophyll bearing tissue. Because of these features the sporophyte resemble much more closer to the sporophyte of the spermatophytes. In contrast, the gametophyte is inconspicuous and similar to that of simpler thallose liverworts.

Initially, the vascular plants have been divided into pteridophyta and spermatophyta. Pteridophytes differ from other spermatophytes by lacking seeds. Later on, taking into consideration the origin and development of vascular system, a new term Tracheophyta was given to all vascular plants on par with bryophyta and thallophyta. The Tracheophyta have been divided into four groups : Psilopsida, Lycopsida, Sphenopsida and Pteropsida. Sporne (1969) classified pteridophyta into 5 groups *viz.*, Psilophytopsida, Psilotopsida, Lycopsida, Sphenopsida, and Pteropsida (Table 2.3).

Table 2.3 *Classification of Pteridophytes as proposed by Sporne*

	Pteridophytes			
A.	**Psilophytopsida** * Psilophytales *			
B.	**Psilotopsida** Psilotales			
C.	**Lycospsida**			
D.	**Sphenopsida** 1. Hyeniales *	2. Sphenophyllales*	3. Calamitales*	4. Equisetales
E.	**Pteropsida** (a) Primofilices * 1. Cladoxylaces* (b) Eusporangiatae 1. Marattiales (c) Osmundidae Osmundales (d) Leptosporangiatae 1. Filicales	 2. Coenopteridales* 2. Ophioglossales 2. Marsileales	 3. Salviniales	

* fossil groups (Sporne, K.R, 1966).

The members of Lycopsida (Lycopods) include about 1000 species of herbaceous plants commonly known an clubmosses and quilworts occurring throughout the world. It includes both living and extinct plants. Fossil types such as *Lepidodendron* and *Sigllaria* which are large , arborescent plants while *Selaginellites* and *Lycopodites* are simpler herbaceous forms. Club mosses are terrestrial or epiphytic with branched stems, microphyllous leaves and roots. The members of quilworts or Isoetales are aquatic plants. Lycopods are represented by the present day flora such as *Lycopodium*, *Phylloglossum Selaginella*, and *Isoetes*.

There are about 20 species of horsetails belonging to Eiquisetopsida which are distributed throughout the world. These are homosporous and spores have elaters. Ferns are another large group of pteridophytes distributed all over the world.

2.6.1 Lycopodium L.

Lycopodium is one of the two living genera of the family Lycopodiaceae represented by about 180 species and distributed worldwide. It grows in arctic, temperate and tropical lands in both the hemispheres. *Lycopodium* is predominant in substropical and tropical forests as hanging epiphytes or form part of the forest undergrowth. Majority of species are terrestrial with prostrate stems creeping on or beneath the soil surface and bear erect leafy branches. In India, about 35 species of *Lycopodium* grows in cool, moist shady places are reported. The genus *Lycopodium* is divided into two sub genera viz., Urostachya and Rhopalostachya (Fig. 2.21(a)).

The shoot system of *Lycopodium* is dichotomously branched. The stem has a central vascular system (Fig. 2.21(a)) with no pith and covered with microphylls. Microphylls are small leaves which are characterised by a single unbranched median vein which does not reach the apex.

Internal structure of the stem is differentiated into epidermis, cortex and stele (Fig. 2.21(a)) Epidermis is the outermost layer and is one celled in thickness. Stomata are present on epidermis. Cortex varies in thickness and structure from species to species. Cortex may be parenchymatous or divided into three concentric zones. The outer and inner cortex is sclerenchymatous while the middle portion is thin walled and parenchymatous. A single layered endodermis separates the stele and the cortex. Casparian thickenings are seen on the radial walls of the endodermis in the early stages. The endodermis becomes lignified at maturity. Pericycle is 3-6 layered and lies beneath the endodermis. Cylindrical vascular tissue occupies the centre of the stem. The stele in *Lycopodium* is a protostele. Protostele is characterized by the absence of the pith. Variation in the arrangement of xylem and phloem is seen in different species of *Lycopodium* and sometimes in different parts of the same individual.

The xylem is exarch with protoxylem towards the periphery. The protostele of the *Lycopodium* exhibits variation in different species *viz*.

1. *Actinostele :* The central mass of xylem radiate and the radiating arms give a star like appearance to the stele. The protoxylem mass is present at the tops of the arms and the metaxylem elements are towards the centre (Fig. 2.22(a)). The phloem is present in between the rays.
2. *Plectostele :* The xylem is broken into a number of horizontal plates in *L. volubile*. The xylem is present in the form of plates and alternate with bands of phloem. This type of stele is called a plectostele (Fig. 2.22(b)).

3. *Mixed Protostele* : The xylem is scattered in groups and are embedded in the phloem. This type of stele is called a *mixed protostele* Fig. 2.22(c). The xylem consists of tracheids and the phloem consists of sieve cells and phloem parenchyma cells.

Leaf traces originate from the protoxylem points and pass through pericycle and cortex and enter the leaf base.

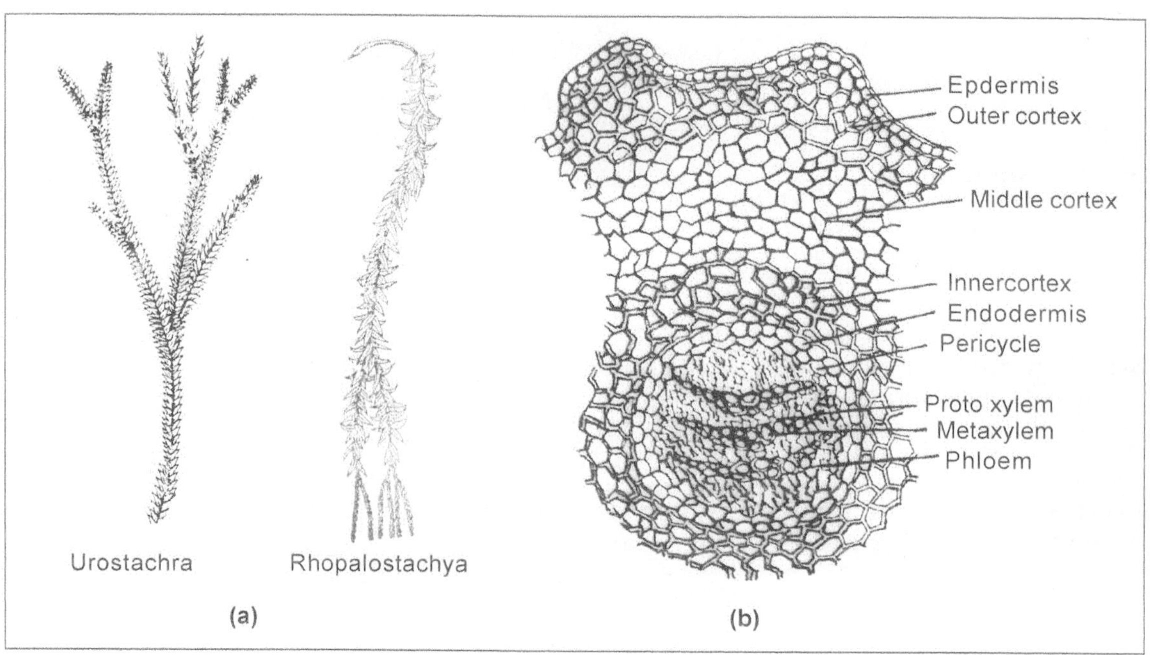

Fig. 2.21 (a) Lycopodium habit (Urostachya and Rhopalostachya), (b) Transverse section stem

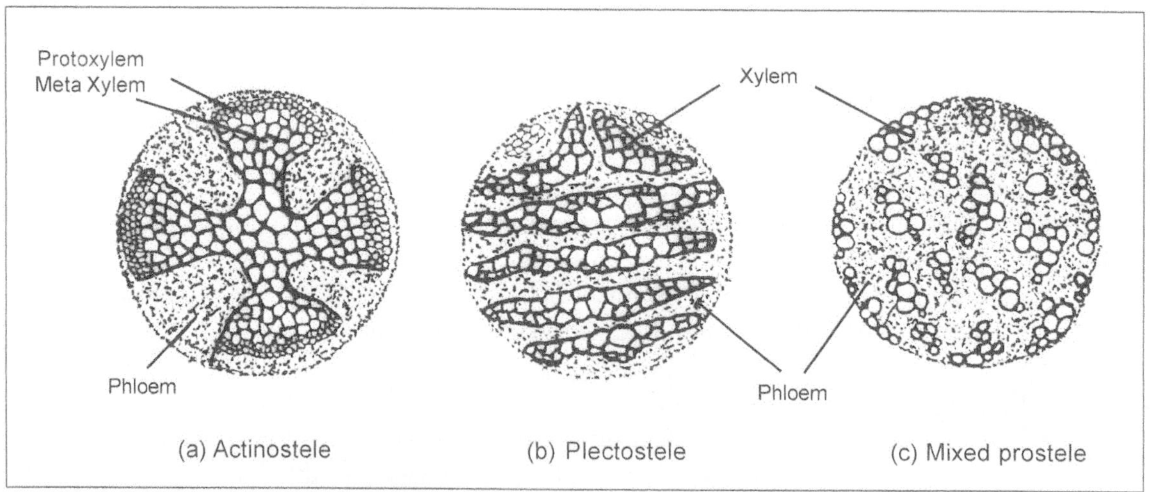

Fig. 2.22 Types of steles in *Lycopodium*

Reproduction

Lycopodium reproduces both vegetatively and sexually. Vegetative reproduction is by means of gemmae or bulbils. Bulbils are small bud like structures present on the stems. Bulbils are present only in one of the subgenus of *Lycopodium*, the *Urostachya*. *Lycopodium* is homosporous but *Selaginella* and *Isoetes* are heterosporous. Homosporous forms produce only one kind of spores in sporangia. Heterosporous forms bear two kinds of spores. Sporangia are produced in strobili in the leaf axils. The leaves that bear sporangia are fertile and called as sporophylls. Sporophylls are organised into a compact structure referred as strobili or cones at the tips of the branches or on the main stem which are physiogically and morphologically different in majority of species. Each strobilus has a central axis which are covered by the spirally arranged sporophylls Fig. 2.23(a). Sporangia are borne singly on the upper side of the sporophylls near the base in the leaf axils.

The sporophylls differ from vegetative leaves by being smaller in size, paler in colour and in having dentate margins Fig. 2.23(b). A mature sporangium is kidney shaped, short stalked and yellow or orange in colour. Sporangial wall is multilayered and the inner most layer is called the tapetum. Spores are produced inside the sporangium. Sporangia dehisce transversely along with the breaking line the stomium and release the spores. The spore is the first cell of gametophyte generation. The spore germinates within few days. It gives rise to the prothallus or gametophyte. The spore begins to divide even before the exospore is ruptured but the development of the gametophyte beyond 5 cell stage is solely dependent on the entry of mycorrhizal fungus into the basal cell. Further development of gametophyte is stalled if the mycorrhiza fail to establish in the prothallus. The fungus supplies certain nutrients to the growing gametophyte. Gametophyte exhibits great diversity in form and structure. Based on the habit and nutrition, three main types of prothalli are recognized.

1. The first type of prothallus is inconspicuous (2-3 mm in height) and grown on the surface of the ground e.g. (*L. cernuum*). It has a short cyllindrical mycorhizatlbasal portion below the soil and the upper green leaf like lobes. Sex organs are formed between the leaf like lobes.

2. The second type of prothallus is a subterranean saprophyte found in northern creeping species e.g. (*L. clavatum*). The prothalli are colorless or brownish tuberous structures (1-2cm long or wide) resembling a top or carrot or disc internally the central region with elongate cells serves for storage and cortical region is occupied by *endophytic*. The sex organs are borne on the flattened upper surface while long rhizoids arise from the lower region.

3. The third type of prothallus is saprophytic and colourless e.g. (*L. phlegmaria*) which grow on the trunks just below the surface of humus. The prothallus consists of a tuberous central body (2 mm in a diameter) that produce several slender, cylindrical colourless branches (1-6mm length) away from the apex of main body. The entire prothallus is associated with a mycorrhizal fungus. The organs are borne on upper surface interspersed with paraphyses.

Antheridia and archegonia develop on the prothalli large numbers. Mature antheridia exhibit variation in size, shape and number of antherozoids. Antheridia are sunken or slightly projected (Fig. 2.23). Antheridium contains androcytes which produce antherozoids. Antherozoids may be oval and bearing two long flagella.

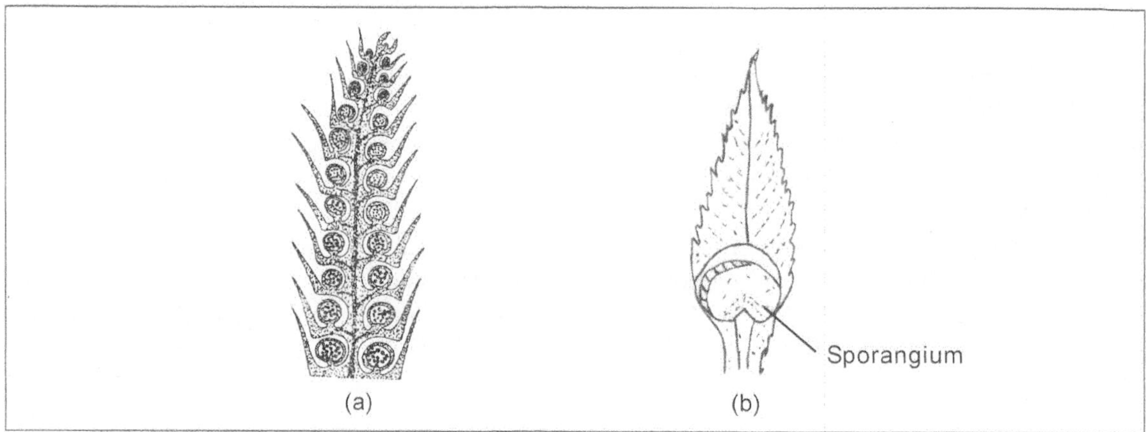

Fig. 2.23 Gametophytic prothallus. (a) Strobilus, (b) Sporophyll

The archegonial neck protrudes out from the gametophytic prothallus (Fig. 2.23 (a), (b)). Subterranean gametophytes have archegonia with a slender and long neck (6-13 neck canal cells) while the surface living type have short neck with 3-4 cells height. The venter contains a venter canal cell and an egg. Venter canal cell and neck canal cells disorganise to enable the smooth passage of antherozoids after the separation of neck cells at the tip.

Fertilization occurs in the presence of water. Antherozoids are attracted chemotactically to the archegonium by oozing out viscous fluid from the archegonia. Zygote is formed as a result of fusion between the egg and the antherozoid. The zygote undergo divisions producing an embryo with a suspensor and two tiers (four cells each) of cells. The lower tier of cells forms the foot and the upper tier forms the remaining portion of the embryo. These four cells of the upper tier emerges out of the prothallial tissue as a free growing structure and enlarge into an extraprothallial, undifferentiated tuberous body called *protocorm*.

Protocorm is spherical in form consisting of parenchymatous tissue occupied by a symbiotic mycorrhizal fungus. It has rhizoids on ventrl side erect, green leaf-like outgrowths known as protophylls on the upper surface. It differs from the typical young sporophyte in lacking roots and vascular tissue. A young sporophyte plant body develops from the distal end of the protocorm sooner or later. The protocorm is green and the protophylls have the stomata. The whole protocorm separates from the prothallus as soon as the first protophylls are formed. Protocorm is an independent intermediate phase between the normal embryo and the definite leafy shoot.

The morphological nature of the protocorm has been viewed differently by various botanists. Treub (1905) regards it as the remains of a primitive undifferentiated structure originally possessed by the pteriolophytes which has disappeared and replaced by a definite leafy shoot in most existing pteridophytes of the present day. Protocorm is a tuberous structure helping the young and delicate sporophyte to over come an unfavourable season. According to Eames, (1947) it represents a physiological specialisation rather than of morphological significance.

Life cycle

Schematic representation of the life cycle of *Lycopodium* is represented in Fig. 2.23(c).

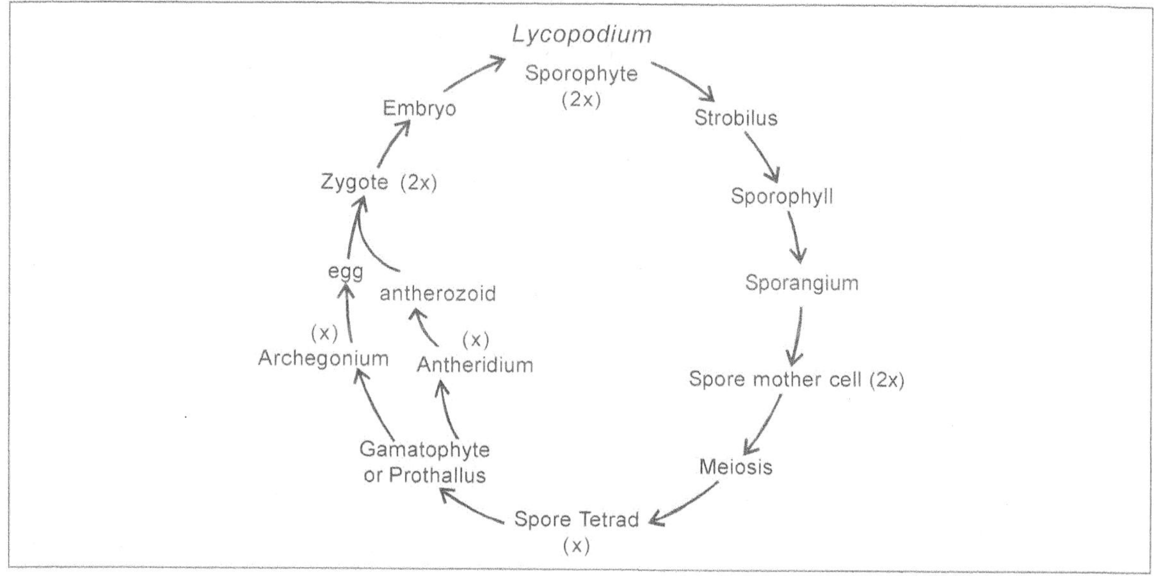

Fig. 2.23 (c) Life cycle of *Lycopodium*

2.7 Gymnosperms

Theophrastus in his book 'Enquiry into plants' (300 B.C.) used the term 'Gymnosperm' for describing the plants with unprotected seeds. The gymnosperm ovules are freely exposed before and after fertilization. The seed plants or spermatophyta include two main groups *viz*., the Gymnosperms and the Angiosperms. Gymnosperms are the most ancient seed plants that originated during the late Paleozoic Era and flourished in the Mesozoic Era. The leaves of first true seed plants resemble ferns. The Pteridosperms or seed ferns are the earliest seed plants appeared from Devonion to Permian period. The Cordaitales, an extinct group of plants, resemble Conifers of today. It is one of the most abundant fossil seed plant groups in the Carboniferous period. The Bennettitales with unbranched trunk and large pinnate fern-like leaves resembling the living cycads.

Gymnosperms are distributed both in Tropical and Temperate regions of the world. In India, they are distributed in hilly areas. The living Gymnosperms are mostly trees and shrubs showing xerophytic characters. Members of Gnetales are woody climbers while herbs and annuals are completely absent.

The plant body is a sporophyte and differentiated into root, stem and leaves. Root system include tap roots and lateral roots which may have blue-green algae in symbiotic association (corolloid roots of *Cycas*). Ectomycorrhizal association is seen in the roots of Conifers (*Pinus*). Stem is aerial, erect branched or unbranched structure covered with persistent leaf bases. Leaves may be small (microphyllous) or large (megaphyllous), evergreen, simple or compound. Young leaves show circinate vernation in cycads.

The modern gymnospersms are classified under five orders : The Cycadales, the Ginkgoales, the Coniferales, the Taxales and the Gnetales. The Cycadales and Ginkgoales include living members having long fossil history. Ginkgoales is represented by a single species, *Ginkgo biloba* which is considered as a living fossil. Conifers include the most familiar and economically important plants like *Pinus*, *Cedrus*, *Abies*, *Thuja*, *Cupressus* etc. The Gnetales are represented by *Gnetum*, *Ephedra* and *Welwitschia*. The taxales are represented by *Taxus*.

Gymnospersms have two types of secondary wood : (i) *Manoxylic* or (ii) *Pycnoxylic*. The Manoxylic wood is soft, porous, more parenchymatous and found in cycads. Medullary rays are wide and not of commercial use. Pycnoxylic wood is compact with narrow medullary rays and found characteristically in conifers. Pycnoxylic wood is of great commercial use. The xylen lacks wood vessels (except in Gnetates) and phloem is devoid of companion cells. Tracheids of secondary wood have bordered pits, sometimes with scalariform thickenings. The medullary rays may be uniserate or multiseriate and homogenous (*Ginkgo*) or heterogenous (*Pinus*). The protoxylem possess spiral thickenings in the seedling stages while it consists of scalariform thickening in the later stages. The phloem consists of sieve tubes on lateral walls. A ring of discrete vascular bundles with narrow or medullary rays. The bundles are described as conjoint, collateral, endarch and open.

The pollen gains or microspores produce male gametophytes while the single megaspore enclosed in the megasporangium produces the female gametophyte. The female gametophyte bears two or more archegonia or female sex organs. Sporophylls are arranged spirally along an axis forming strobili or cones. The cones bearing microsporophylls and microsporangia are referred as male cones. The cones bearing megasporophylls with ovules or megasporangia are called macrosporangiate or female cones. The two types of cones may present on the same plant (*Pinus*) or on different trees (*Ginkgo*, *Cycas*). The microsporangium produces numerous microspores (Pollen grains) while the megasporangium contain only one megaspore. These are the pioneer structures of haploid male and female gametophytes.

2.7.1 *Cycas*

The cycads are palm-like plants that are nearest living relatives to the Pteridosperms. Fossils of this group are more widespread and abundant between the Triassic and Cretaceous periods. Living Cycads belong to about 11 genera with 180 species distributed in Australia, South Africa, Central America and East Asia including India. The family Cycadaceae include 10 living genera and one fossil genus. *Paleocycas* is the fossil genus described from upper Triassic rocks of Sweden. The cycads resemble the ferns and the Pteridosperms on one hand and Bennettitales, Pentoxylales, Cordaitales and Ginkgoales on the other hand.

Morphology

The Cycads externally look like palms. The stem is thick, trunk-like normally unbranched, or occasionally with adventitious branches, cylindrical slow-growing and typically about 2m tall structure. The leaves have circinate vernation. The leaflets have a distinct mid-rib, with no side veins. The leaves are dimorphic with green, pinnate vegetative leaves and zones of brown cataphylls or scale leaves (Fig. 2.24 (a),(b)). Scale leaves are simple, hairy and sharply pointed. These plants have a deep tap root and surface roots often associated with blue green algae which help the plant in nutrition. The starchy central pith is eaten as sago in some places and some species are grown as ornamentals.

Internal structure of Cycas stems show a roughly circular outline due to the persistent leaf bases. There is massive peripheral cortex and central pith. Mucilage canals are present across the cortex (Fig. 2.24 (c)) the endodermis and pericycle are not distinct. The primary vascular bundles

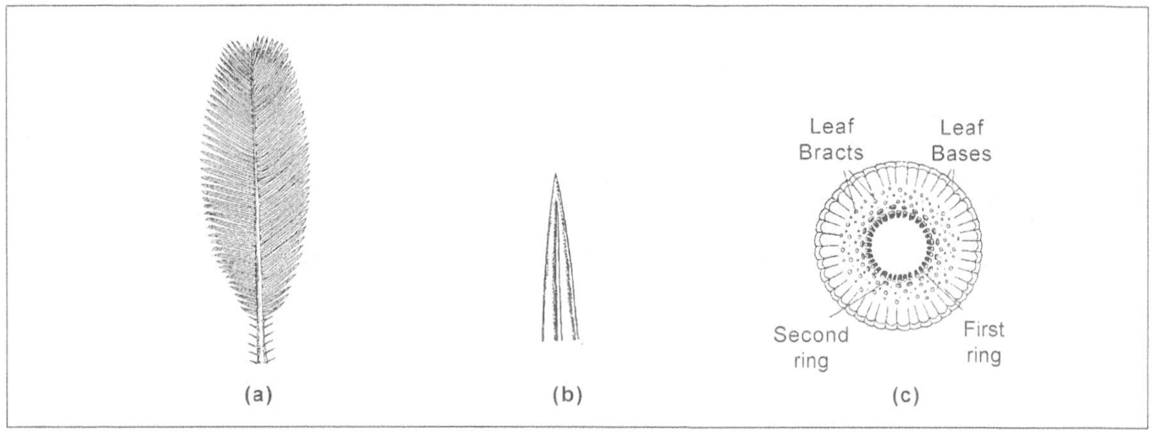

Fig. 2.24 *Cycas revoluta.* (a) Leaf, (b) Leaflet, (c) and Transverse section of stem (outlines)

consists of an interrupted ring of conjoint, collateral, open vascular bundles. Some leaf traces originate singly from vascular cylinder quite close to the point of attachment of the petiole while others may originate from distant points and girdle the stem. The girdling leaf traces are characteristic of the family cycadaceae. The leaf trace bundles are diploxylic exhibiting both centripetal and centrifugal xylem (Fig. 2.25).

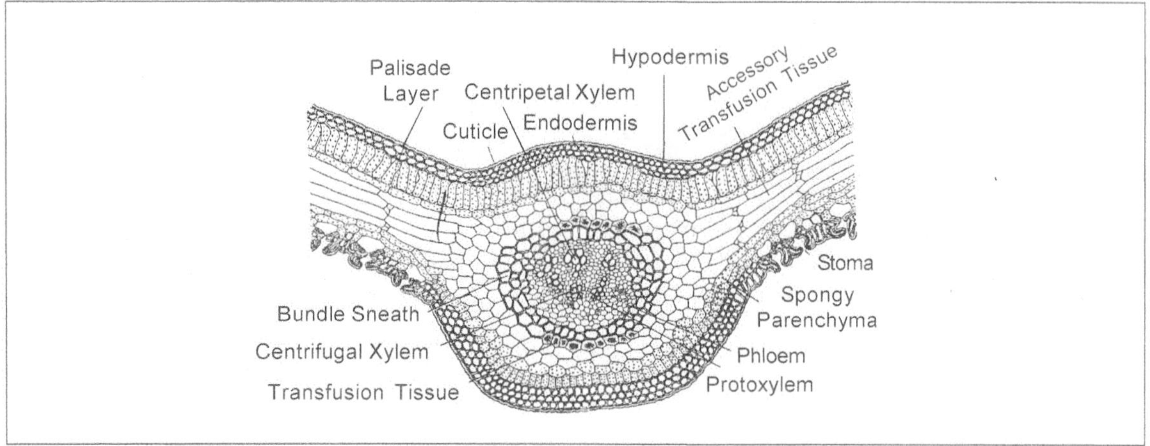

Fig. 2.25 T.S. of *Cycas circinalis* leaf showing detailed internal structure

The secondary xylem is very small and manoxylic. There is a single cambial ring that functions all through the plant life. But *Cycas*, wood is polyxylic i.e., it consists of successive rings of secondary wood formed by a succession of cambial rings developed outside the original ring. Many rings are present in secondary xylem extending up to pith.

The epidermis of the leaf is highly cutinised and contains sunken stomata. The vascular bundle of the midrib and leaf traces is diploxylic.

The roots of cycads are very thick at the base they are polyarch and have up to eight xylem bundles alternating with eight phloem bundles. Number of bundles decrease towards the apex until a diarch condition is reached. The primary xylem is exarch coralloid roots are present and contain an endophytic alga like *Anabaena* (Fig. 2.26).

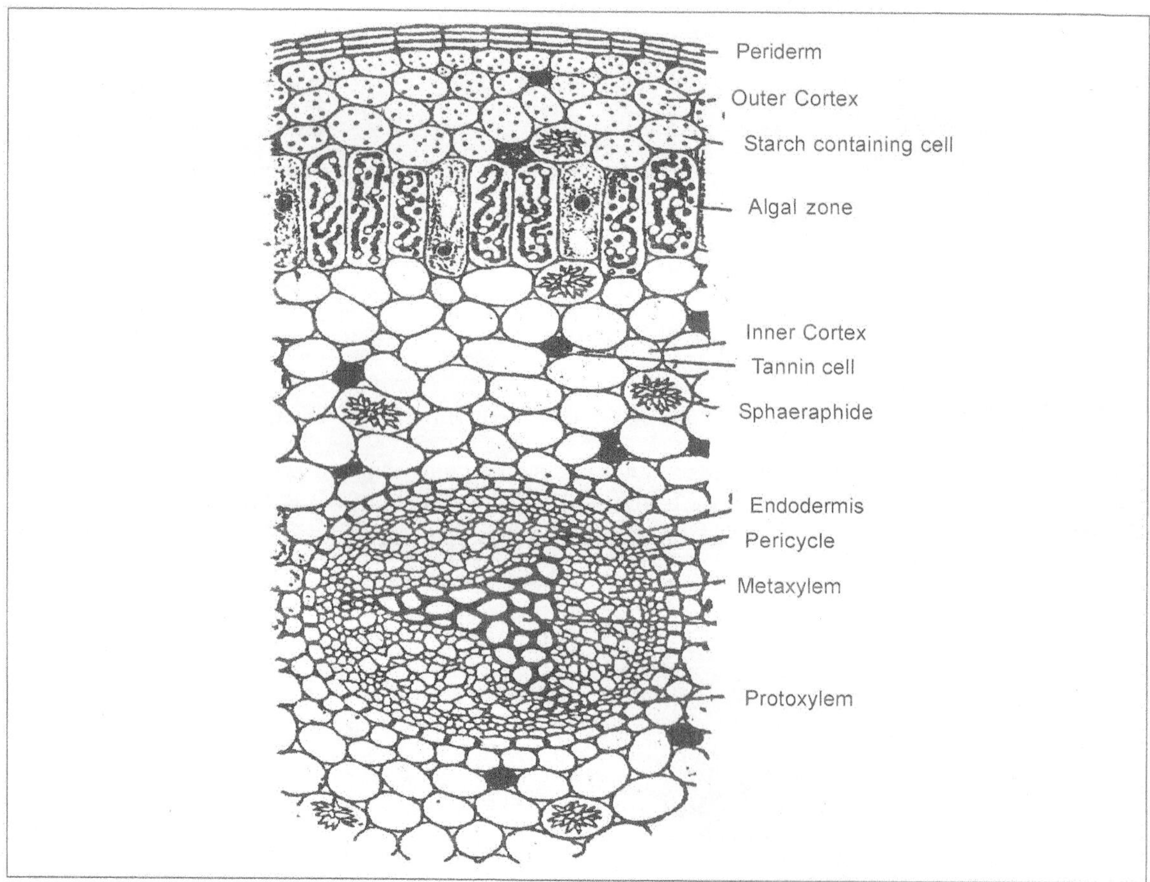

Fig. 2.26 Transverse section coralloid root of *C. revoluta*

Reproduction

Cycas is a dioecious plant. Megasporophylls are spirally arranged in female plants which alternate with the cataphylls and the foliage leaves. They appear in a sequence i.e., vegetative leaves - cataphylls - vegetative leaves and so on. Megasporophylls have a pinnate upper sterile part and slender proximal part bearing pairs of ovules. The pinnate nature of the sterile part reduces and almost lost in *C. rumphii* and *C. circinalis*.

The *Cycas* megasporophyll is considered as the most primitive due to its leaf like nature (Fig. 2.27 (a)). Megasporophylls are formed once in a year.

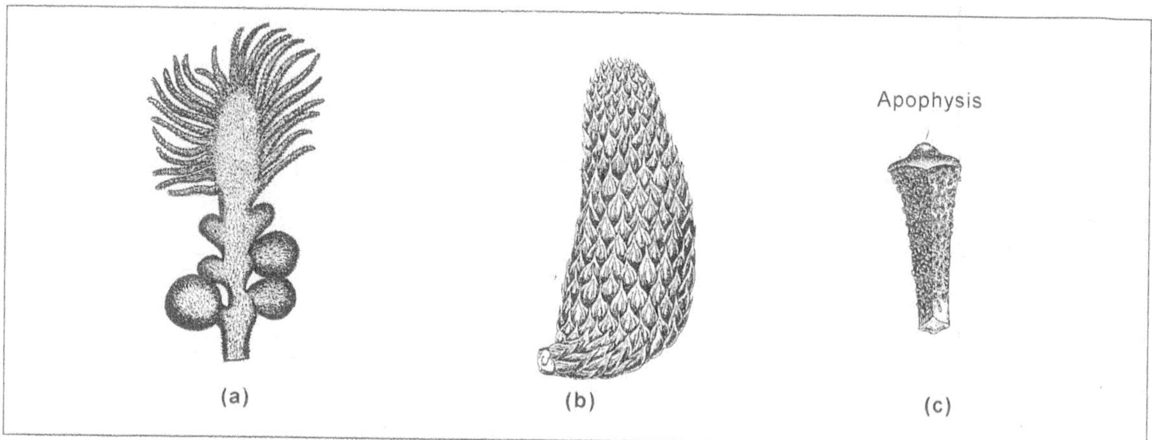

Fig. 2.27 *Cycas circinalis.* (a) Megasporophyll (*C. revoluta*), (b) Malecone and (c) Microsporophyll

Megasporangium (ovule)

The ovule of *C. circinalis* is the largest ovule among the living gymnosperms. The ovule is orthotropous and stalked. Ovule surface may be smooth or covered with brown hairs. The ripe seeds are fleshy and bright orange is colour (Fig. 2.28).

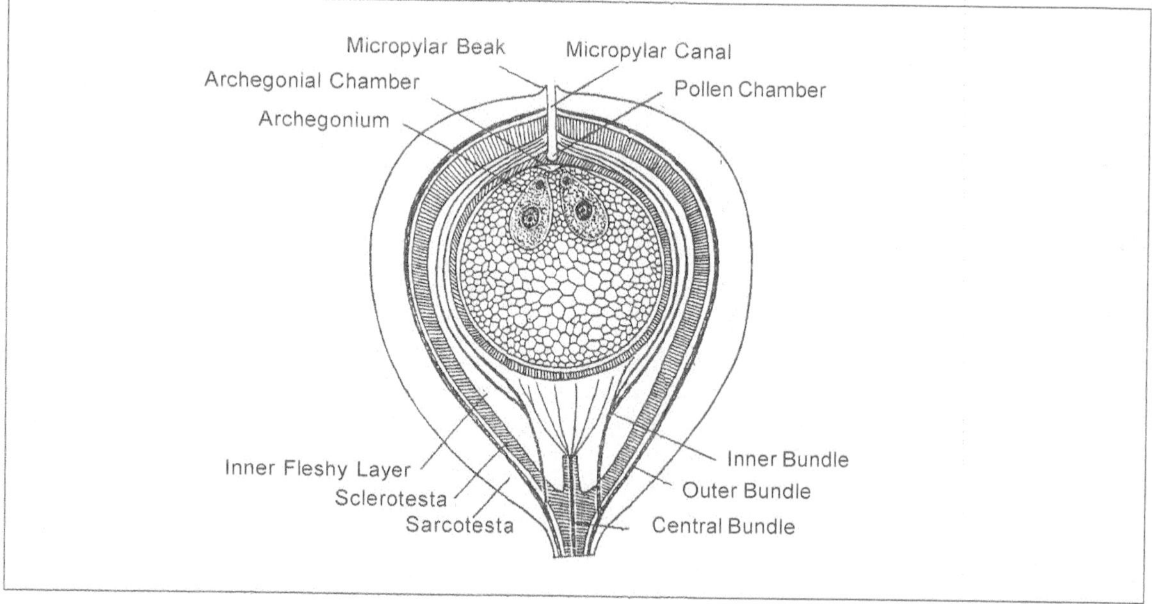

Fig. 2.28 *Cycas revoluta* Vertical section of ovule showing internal structure before fertilization and its vascular supply

Male cone

Male cone develops terminally which is surrounded by a crown of young leaves. The cone is solitary, compact, conical or oval in form and or woody in texture. Microsporophylls or stamens are arranged spirally along the axis (Fig. 2.27 (b)).

Microsporophyll

It is a minute, flattened and wedge shaped outgrowth. The flat distal part of the microsporophyll is fertile and the basal part is sterile. The upper sterile triangular portion beyond the fertile part is known as *apophysis* (Fig 2.27 (c)). Microsporangia are borne on the lower surface of the microsporophyll. Microsporangia are arranged in groups of 3-5 sori. The mature microsporangia are oval, sac-like structures with short stout stalk. Exothecium or a thick walled epidermis covers each sporangium. Exothecium helps in the dehiscence of sporangia.

Female Gametophyte

Megaspore is the beginning of the female gametophyte. Megaspore wall is two layered, a thick papillate outer wall i.e., the exospore and fibrillin inner wall, the endospore. Gametophyte development takes place while it is within the ovule. In *Cycas* endosperm is formed before fertilization unlike the angiosperms. After the formation of endosperm a few cells at the micropylar end enlarge and function as archegonial initials. Archegonial chamber is formed by the disorganisation of nucellar tissue above the initials. The number of archegonia vary from species to species. Archegonium has a two celled neck and venter. No neck canal cells are present in *Cycas*. Venter contains a ventral canal nucleus and egg. The venter is surrounded by a nutritive archegonial jacket cells. Endosperm cells form the jacket layer. *Cycas* egg is the largest amongst all living plants.

Male Gametophyte

The boat shaped pollen grain is with a longitudinal slit represents the first cell of the male gametophyte. The Microspore is two walled, the outer thick wall (exine) and the inner thin wall (intine). Pollination occurs through insects (beetles and bees) i.e., entomophilous. The cycads and *Ginkgo* are unique among seed plants in having multiflagellate male gametes which are largest among the plant kingdom. As the motile sperm touches the neck cells, it is sucked in violently to take it near the egg resulting in fertilization.

After fertilization, the embryo matures in three phases i.e., proembryogeny, early embryogeny and late embryogeny. The mature embryo contains two cotyledons as in most of the gymnosperms. The cotyledons are closely appressed and not obvious. The embryo takes about one year for its full development.

The *Cycas* show a range of chromosome numbers (n = 8, 9, 21, 12, 13). The basic chromosome number in *Cycas* is 11.

2.8 Angiosperms : Growth and Development

The origin of Angiosperms is still continued to be a topic of "*abominable mistery*" as described by Charles Darwin (1857). About 2.5 lakhs angiosperms recorded till date are divided into nearly 400 families comprising the modern angiospermic flora. Angiosperm plants have acclamatised various ecological habitats covering diversified environments. This group of plants are highly diversified, extremely plastic and extraordinarily adaptable.

The angiosperms possess exclusively the following characteristic features :

1. Double fertilization and triple fusion leading to triploid endosperm
2. Conduplicate carpels (which are totally absent in gymnosperms)
3. Germination of pollen grains on stigmas (in gymnosperms they germinate on nucellus).
4. Pollen wall stratification constitute columellate and tectate condition

In addition to the above characters which are extremely confined to the angiosperms, several other characters are commonly associated with majority of angiosperms. Thy are

1. Presence of highly reduced male gametophyte
2. Presence of bitegmic ovules
3. Presence of accessory whorls such as sepals and petals
4. Presence of vessels in xylem
5. Presence of libriform fibers (absent in gymnosperms)
6. In leaf two types of reticulate venation patterns are observed; they are single branched unicostate venation and branched multicostate venation. Of these, dicots exhibit multicostate venation.

All the above characters need not be present in a single plant but, a "*typical angiosperm*" plant may possess the bulk of advanced characters as well as primitive characters in a ratio of 9:1. Regarding the origin of angiosperms, two different schools of thought are in vogue. Baroghoorn and Scott (1958) opined that the angiosperms originated in the early Cretaceous period which was supported by Hughes, Doyle and others, Angiosperms occupied diversified habitats and dominated during the upper most part of Cretaceous (55-75 million years ago) period. The protagonists of the second school proposed an early Triassic/Mesozoic period (135-220 million years ago) as the time of origin of angiosperms. Place of origin is again controversial as it may be either up lands or mountain tops.

Angiospermic plant, such as a neem tree or a rose bush or a rice plant, is a sporophyte, differentiated into root, stem and leaves with a well developed vascular system. In these, reproduction occurs by vegetative means or by sexual reproduction. Vegetative reproduction takes place through runners, rhizomes, tubers, bulbs and corms or by human intervention through grafting and layering etc. Sexual reproduction involves fusion of haploid male gametes (male nucleus) with haploid female egg cell produced in a flower. The diploid zygote develops into the embryo of the seed.

The Flower

The flower is a modified compact dwarf shoot bearing four whorls of floral parts on a short axis called *receptacle* (Fig. 2.29 (d)). Of these four sets, stamens (microsporophylls) and carpels (megasporophylls) are called as *essential whorls* and involve in the process of reproduction. *Accessory whorls* (calyx and corolla) surround the stamens and carpels, and are protective in nature. In monocots (e.g., onion, lilies, and palms) all the perianth leaves are similar. In dicot flowers (e.g. *Datura*, Pea), outer layer of the perianth is differentiated into sepals which form the calyx and the inner whorl of petals that form the corolla. Sepals are usually green and protective, petals are brightly coloured and attract pollinators.

The stamen (Microsporophyll)

Stamen arises from the receptacle and constitutes part of the *androecium*. It consists of a slender thread like structure called filament, bearing a cylindrical or ovoid pollen bearing part. The filament continues above into the connective which is traversed by a vascular strand and connects the anther lobes of the anther (Fig. 2.29 (a)). Each anther lobe contains two pollen-sacs or microsporangia (number vary among different angiosperms). Each pollen sac contains a number of microspore mother cells surrounded by a layer of nutritive cells called the *tapetum*. Each microspore mother cell after two successive nuclear divisions produce four pollen grains (microspores). Pollen grains are liberated after dehiscence of the anthers, called *anthesis*.

Carpel (Megasporophyll)

Carpel typically originates from the tip of the receptacle (thalamus) and constitutes a part of the gynoecium or pistil. Carpel comprises of three parts (Fig. 2.29 (c)). 1. The swollen basal part, the ovary which constitutes one or more ovules; 2. Elongated stalk called the *style*; 3. The tip stigma, modified apex of the style adapted for the reception of the pollen grains. Pollen grains are deposited on the stigma while the ovules are enclosed in the ovary.

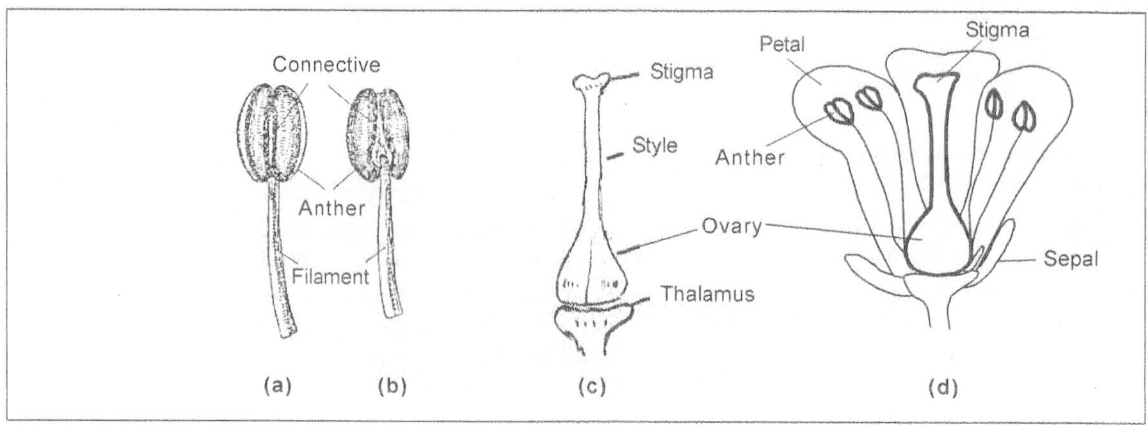

Fig. 2.29 Parts of a stamen. (a) Face view, (b) back view, (c) carpel and (d) L. S. of flower

Ovule (Megasporangium)

Each ovary may have one or several ovules. Ovules are borne on a distinct stalk called the *funicle*, from the placenta at the base of the ovary or on the inner surface of the ovary. Funicle attaches to the body of the ovule at a point called *hilum*. They have a central embryo sac which includes the egg surrounded by a nucellus of parenchymatous cells within two layered protective integuments in monocots and single layered integument in higher dicotyledonous families. A third integument, *aril* is found in plants like *Asphodelus* and *Trianthema.* In the family Euphorbiaceae, *Ricinus* and several other plants show an integumentary proliferation known as *Caruncle*. In parasitic plants like *Santalum* (sandal-wood) and *Loranthus* integuments are absent.

The integuments arise from *chalaza,* the basal part of the nucellus. They grow upwards enclosing the nucellus. There is a small opening in the integuments called the *micropyle* through which the pollen tubes penetrate. The embryo sac, the most important part of the ovule, is embedded in nucellus at the micropylar end. The embryo sac bears three protoplasts constituting the *egg-apparatus* towards the micropylar end. Of these, one is the egg and the other two are synergids or help cells. There are three *antipodal cells* lie at the chalazal end and secondary nucleus in the centre of the embryo sac (Fig. 2.30).

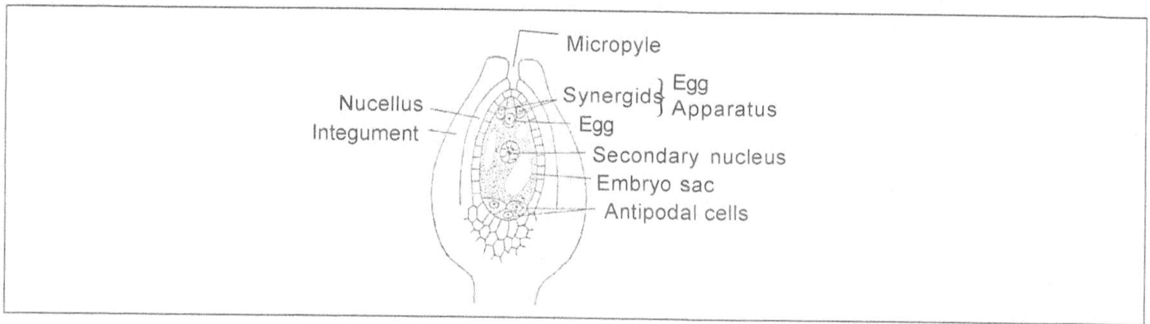

Fig. 2.30 A typical mature ovule

Mature ovules are of four types (Fig. 2.31) such as Orthotropous, Anatropous, Amphitropous and Campylotropous.

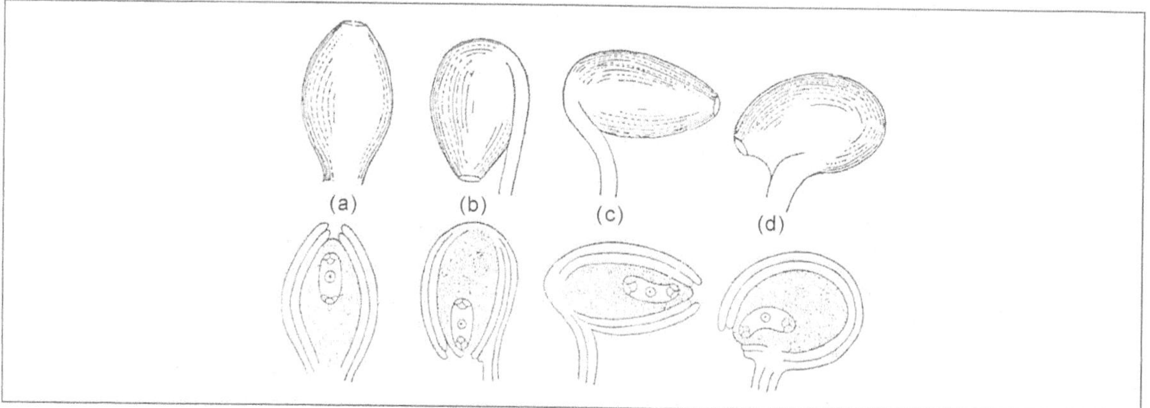

Fig. 2.31 Forms of ovules. (a) orthotropous, (b) anatropous, (c) amphitropous, (d) campylotropous Upper row shows entire ovules, the lower in section

1. *Orthotropous* (or straight).

In this type, the body of the ovule is straight as the funicle, chalaza and micropyle lie in one and the same vertical line (straight). This is the most primitive and rare type of ovule found in families Polygonaceae, and Piperaceae *e.g. Polygonum,* and *Piper.*

2. *Anatropous (or inverted)*

The ovules usually turns through 180^0 with the micropyle facing the stalk. The inverted ovule gets fused with the funicle and the fused portion is called the *raphe.* The micropyle and hilum lie near each other while chalaza lies toward the other end. This is the most common and universal type found in both monocots and dicots.

3. *Amphitropous (or transverse)*

The ovule is bent over or twisted only half way so as to lie at right angles to the funicle. The hilum, chalaza and micropyle all lie apart from one another. It is a rare type found in Ranuculaceae and some members of Brassicaceae.

4. *Campylotropous (or curved)*

The body of the ovule is bent upon itself resulting in a horse-shoe shaped ovule with the micropyle coming near the funicle. The hilum, micropyle and chalaza all lie near each other. It is not a common type and occurs in some members of Fabaceae, Brassicaceae and Poaceae.

Development of the Ovule (Megasporangium)

The ovule primordium (a small protuberance) on the surface of the placenta develops into a mass of tissue known as *nucellus or megasporangium proper*. From its basal or chalazal portion, the inner integument initially, second integument later on grows upwards and cover the nucellus completely except for a narrow pore at the apex where the micropyle appear. The primary archesporium cell differentiates from the nucellus at the apex just below the epidermis and produce an outer primary parietal or wall cell and an inner megaspore mother cell. Megaspores are produced by a reduction division followed by mitosis resulting in four potential megaspores in a linear tetrad. Development of megaspores from the megaspore cell varies from genus to genus. Tetrahedral tetrads and isobilateral tetrads of megaspores are uncommon while T - shaped or ⊥ shaped arrangement of megaspores is frequent in angiosperms. The genus *Musa* exhibits all the four different kinds of tetrads *viz.*, linear, T-shaped, ⊥ - shaped and isobilateral in the ovules of the same species. As a rule, the chalazal megaspore of the linear tetrad becomes functional while others degenerate. The functional megaspore or the embryo sac cell enlarges by absorbing abortive megaspores and surrounding nucellus tissue.

The Female Gametophyte

The haploid functional megaspore is the starting point of the female gametophyte. In gymnosperms *(Pinus)* and pteridophytes (*Selaginella)*, female gametophyte development is initiated by free nuclear divisions, but unlike gymnosperms and in pteridophytes the nuclear division in angiosperms *does not proceed beyond 8-nucleate stage*. The megaspore divides into two nuclei and move to opposite poles of the embryosac due to the formation of a large vacuole between them. Two succeeding divisions of each daughter nucleus result in the formation of four nuclei at each end. One nucleus from each polar group moves to the centre and are called the *polar nuclei*. The polar nuclei generally unite at once to form a diploid *secondary nucleus* or *fusion nucleus*. The three nuclei at the micropylar end of the embryo form the *egg apparatus*. Of the three nuclei, one is the egg, ovum or oosphere and the others are synergids or help cells (Fig. 2.30).

The remaining three nuclei at the chalazal end of the embryo are called the *antipodal cells* which are small and naked. This embryo sac is eight nucleate, 7-celled and considered as a normal type. It was first described in *Polygonum*, thus it be designated as the *Polygonum type*. The antipodals are supposed to be nutritive in function until the endosperm is formed and represent vegetative cells of a greatly reduced female gametophyte.

Male Gametophyte

Each anther lobe generally contains two pollen sacs or microsporangia (Fig. 2.32) which produce a number of microspore mother cells surrounded by conspicuous layer of nutritive cells called the *tapetum*. Each microspore mother cell after two successive nuclear divisions produce four *pollen-*

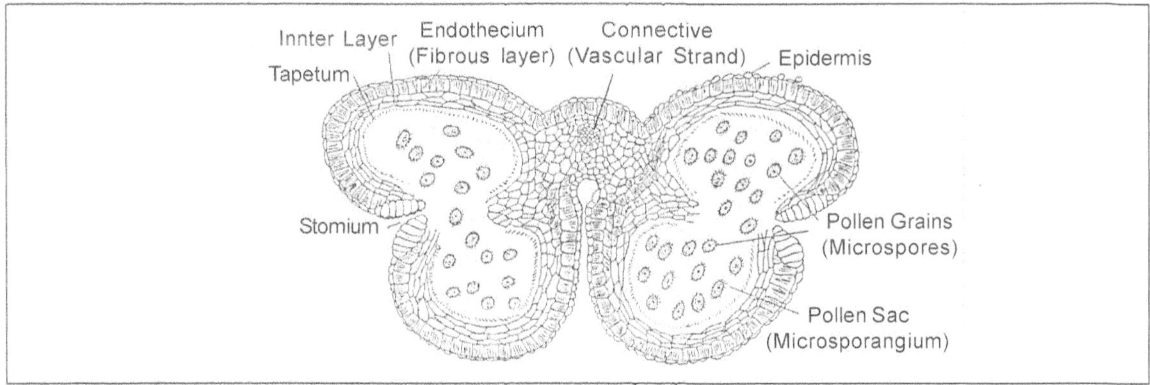

Fig. 2.32 A cross section of anther

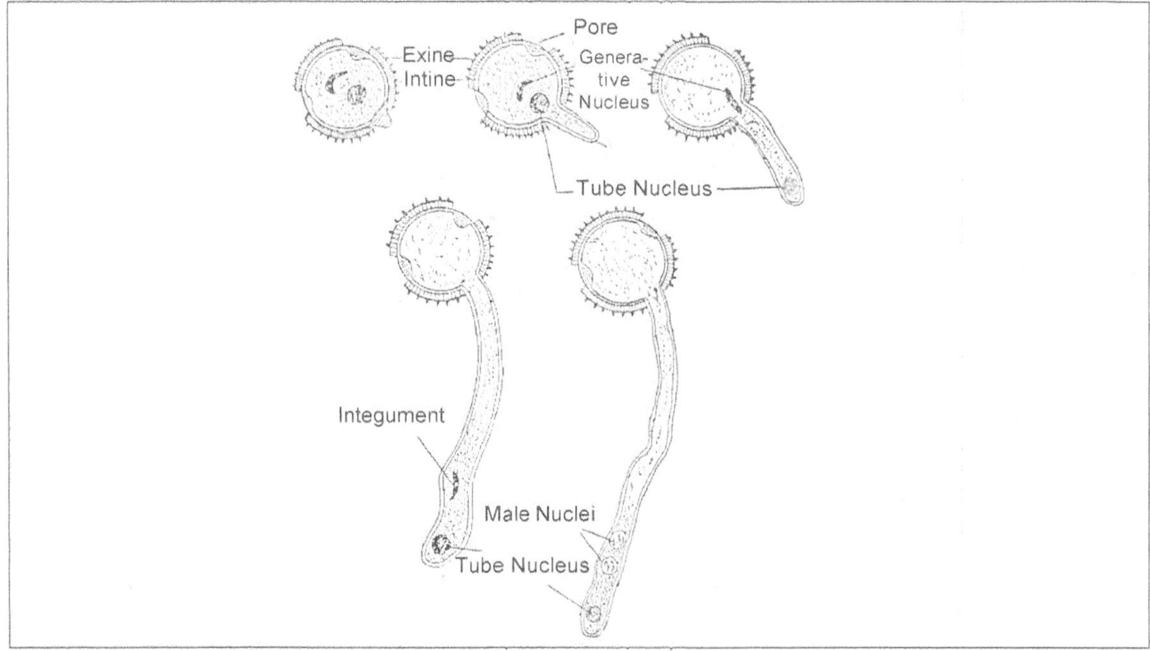

Fig. 2.33 Germination of pollen grain

grains (microspores). The haploid microspore (pollen grain) is the beginning of male gametophyte. The pollen nucleus begins to divide almost immediately in tropics while it often undergoes a period of rest in temperate (colder) countries ranging from few days to several weeks. Pollen grain germinates even before its liberation from the pollen sac (Fig. 2.33). The nucleus divides mitotically into a small generative nucleus and a large *vegetative* or *tube nucleus*. A cell wall between the two nuclei forms a small, elliptical or spindle shaped *generative cell* and a large *vegetative* or *tube cell*. Soon after the cell wall disappear and the two nuclei lie freely in the microspore. In angiosperms, the male gametophyte is highly reduced *prothalial cell* is not produced. The formation of male gametes is delayed for a long time as the generative cell usually divides in the pollen tube. The generative cell function as a *primary spermatogenous cell* and divide to form two equal non-motile male cells (gametes).

The pollen grain is usually liberated at two-celled or 3-celled condition (partially developed male gametophyte) from the pollen sac. Pollen grains may be carried to the stigma of the same flower (self pollination) or stigma of the another flower (cross pollination). This phenomenon of transportation of pollen grain is known as *pollination*. Pollination is effected by wind or by insects and other agencies. Pollen grain absorbs water and other exudates secreted by the stigma and style comprising sugars and organic acids to swell up and rupture through a germ pore into a slender pollen-tube (Fig. 2.33). Pollen tube gradually elongates and passes through the style to reach the ovule. The generative nucleus divides to form two male gametes as the pollen-tube is developing. Normally pollen-tube enters the ovule through the micropyle and is referred as *porogamy* (Fig. 2.34). In some angiosperms, the pollen-tube enter by boring through the tissue near the chalaza region and is called *chalazogamy*. In cucurbits the pollen-tube enters the ovule through the integument. This is known as *mesogamy*.

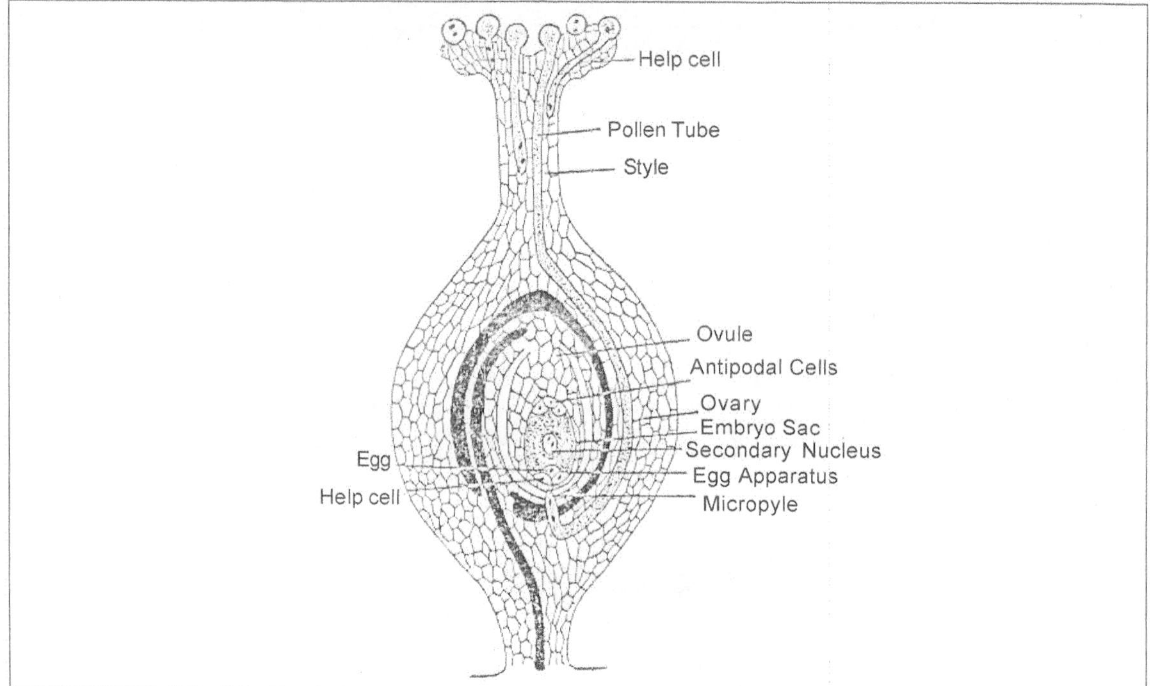

Fig. 2.34 Porogamic fertilization.
Note : several pollen-grains germinating on the stigma and two male nuclei at the tip of a pollen tube entering the ovule (Schematic)

Fertilization

Pollen tube up on entry into the ovule passes into embryo sac and comes in contact with the egg - apparatus. The synergids play an active role in the entry of the tube. Pollen tube discharges its contents into the embryo sac. Of the two male gametes, one penetrates the egg and fuses with female nucleus and is referred as syngamy or fertilization. The fusion product is termed as zygote. The fertilized egg surrounded by a cell wall develops into embryo of the seed. The second male gamete moves down to the centre of the embryo sac and fuses with secondary nucleus (fusion nucleus) to form the *primary endosperm nucleus* which give rise to the endosperm. Thus, in the same embryo sac, the two male gametes bring about fertilization twice, giving the fertilized egg (zygote) and the primary endosperm nucleus or triple fusion nucleus. This unique phenomenon is called *double fertilization* and *triple fusion* was first described by Nawaschin in 1898. The zygote develops into the embryo and the primary endosperm nucleus form the endosperm tissue. The integuments form the seed coats and the fruit tissue is formed by the ovary wall.

Endosperm

The development of endosperm generally begins even before the division of the zygote. The endosperm development is mainly of two types : 1. Cellular and 2. Nuclear. In cellular type of endosperm, each division of primary endosperm nucleus is followed by cell wall formation e.g., *Datura, Petunia.* In nuclear endosperm, the divisions of primary endosperm nucleus and a few subsequent divisions are free nuclear divisions without any wall formation, e.g. coconut, *Acalypha.* In coconut, few to several thousand free nuclei are suspended in its sap and becomes partly cellular later resulting in both cellular and free nuclear endosperm called *liquid synchatrium*.

Endosperm serves as storage tissue with large quantities of reserve food in its cells. The endosperm may be starchy as in wheat, rice and other cereals or oily as in castor and coconut. The endosperm supplies nourishment to the developing embryo.

Development of Embryo in Dicots

The oospore start dividing simultaneously with the endosperm to give rise to the embryo and each division is accompanied by wall formation. In dicots, the oospore elongates and divides by a transverse wall into two cells, a suspensor cell (near the micropyle) and an embryo cell (towards the cavity of the embryo sac). Suspensor cell further divides transversely to produce a 8-10 celled suspensor which pushes the embryo cell down into endosperm. The upper most cell of the suspensor is much larger than others and is attached to the micropyle end of the embryo sac to serve as haustoria. The lower most cell of the suspensor is known as *hypophysis* and gives rise by further division to the radicle or root apex.

The embryo cell enlarges and becomes spherical in outline. It divides longitudinally into two cells. The embryo becomes eight-celled by two further divisions, one longitudinal and the other transverse. This 8-celled embryo is referred as *proembryo*. The four cells next to the suspensor constitute the *hypobasal* or posterior octants while the other four constitute the *epibasal* or anterior octants. The epibasal octants give rise to the two cotyledons and the plumule (stem-apex) while the hypobasal octants give rise to the hypocotyl except its tip.

Each octant divides periclinially (parallel to the surface) into an outer and inner cell. The outer cell subsequently divide (anticlinially) to form the *dermatogen* from which epidermis is formed. Periclinal divisions of the inner cells produce the central tissue called the *pleurome* that gives rise to the stele. The cells between the dermatogen and pleurome constitute the *periblem* that forms the cortex. The dermatogen is incomplete at the root end of the embryo as the apex of the root is formed from the lower most cell (hypophysis) of the suspensor. As growth continues, the free end of the embryo becomes heart shaped and each lobe acts as the primordium of a *cotyledon.* The *plumule* is differentiated in the groove between the cotyledons. Thus in dicots, the plumule or the stem-tip is terminal and the cotyledons are lateral.

Development of Embryo in Monocots

As in dicots, in monocots also the oospore gives rise to a proembryo but there is a lot of variation in the embryogenesis of different monocots. In *Sagittaria,* a generalised condition is observed where the oospore divides to form a three celled filament of proembryo consisting of a large basal cell, a middle cell and a terminal cell. The basal cell near the micropyle enlarges (does not divide) greatly and becomes a conspicuous part of the suspensor. The middle cell divides repeatedly to produce a few suspensor cells, radicle (root-tip), hypocotyl and plumule (stem-tip). Single cotyledon is produced from the terminal cell as a result of a series of divisions in different planes. *In monocots, the cotyledon is a terminal structure while the plumule is lateral in position rather than terminal as in the dicotyledons.* A number of variations from the *Sagittaria* type are observed in other members of the monocots.

Ovule increases markedly in size and loses moisture as the embryo and endosperm are developing. The integuments dry up and form the *testa* or seed coat while the inner when present, forms the thin coat *tegmen.* The ovule ripens into a seed that encloses the resting embryo with ample stored food which is protected by the seed coats. The micropyle of the ovule persists as a small pore on the seed while *hilum* is seen as a small scar on one side of the seed.

Development of the Fruit

The ovule after transforming into a seed, stimulate the wall of the ovary and other parts of the flower. The petals and stamens quickly wither and fall off. The style and stigma also dry up and disappear but the ovary enlarges leading to the development of fruit. Therefore, the fruit is regarded as a mature ovary enclosing one or more seeds. The ripened ovary wall transforms into fruit wall or *pericarp* which may be soft and fleshy or dry and hard. When the fruit develops only from the ovary of the flower, it is referred as true fruit. If other parts of the flower such as receptacle or perianth also contribute in fruit formation, it is known as the false fruit or *pseudocarp* (e.g. apple, pear, mulberry, fig and pineapple).

Alternation of Generations

An angiospermous plant represents the sporophyte generation with diploid chromosome number (Fig. 2.35). The gametophytephase is highly reduced and lacks chlorophyll. Both the gametophytes are dependent upon the surrounding tissue of the sporophyte. The male gametophyteis suppressed and is represented by a germinating pollen grain. Male gametes are produced as two male nuclei in the pollen tube. The female gametophyte produces the egg. Usual sex organs, the antheridia and archegonia, are lacking in the gamatophytes of angiosperms.

The female gametophyte of an angiosperm is eight nucleate with only three of its nuclei (the egg and two polar nuclei) actually taking part in reproduction. The remaining 5 nuclei (2 synergids and 3 antipodals) are considered as the remnants of archegonium and of a prothallus which are no

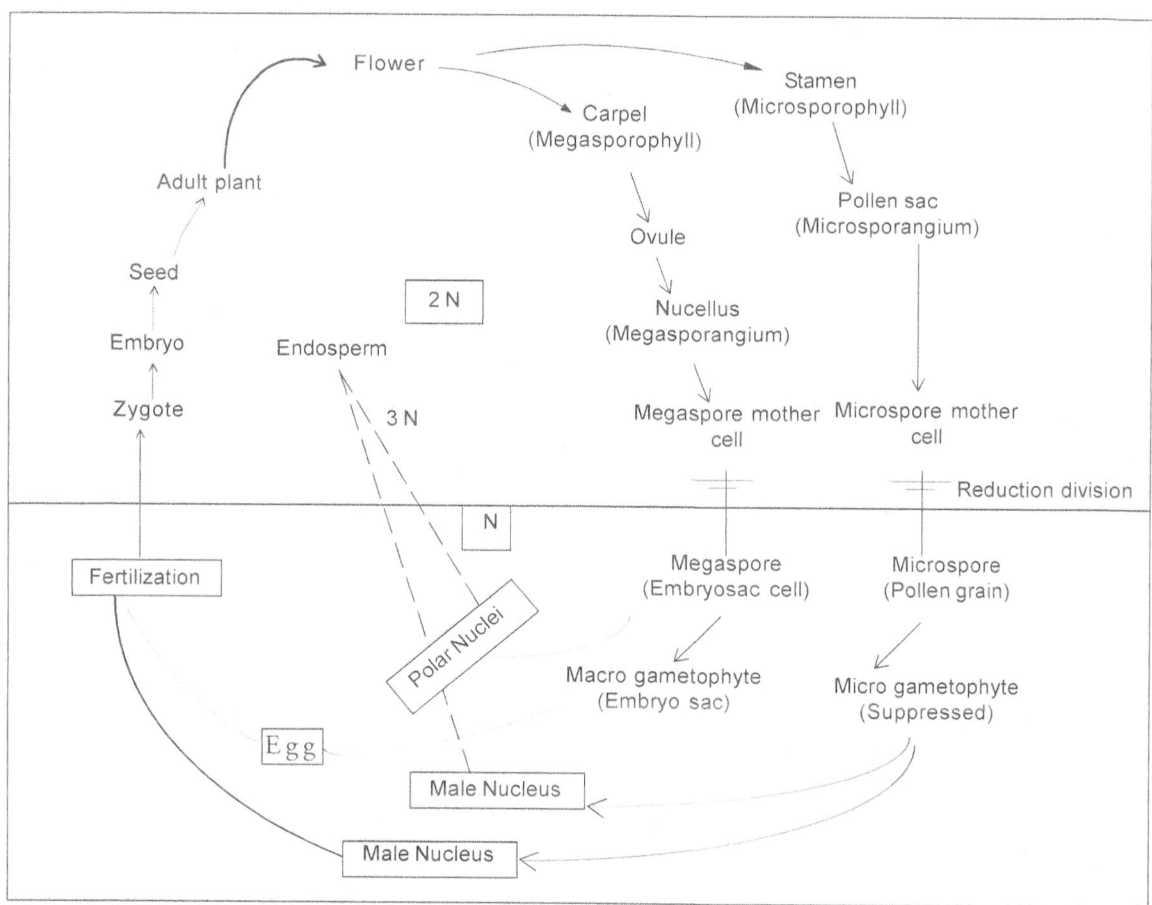

Fig. 2.35 Life cycle of an angiosperm (Schematic)

longer functional in angiosperms. The fertilized egg or oospore is diploid and represents the sporophyte generation, which develops into a seed (Fig. 2.35).

Alternation of generations is not well pronounced in angiosperms, with an independent, dominant sporophyte and an inconspicuous, highly reduced and dependant gametophyte phase. An evolutionary trend in the alternation of generations has been observed in the life cycles of various plant groups across the plant kingdom. In lower plants such as algae and bryophytes, the gametophytic generation is dominant, more complex and autotrophic while the sporophytic generation is less complex and heterotrophic as it is partially or totally dependent on the gametophyte. This situation is reversed in the higher plant groups such as angiosperms. The sporophyte in angiosperms is more prominent, complex autotroph differentiated into root, stem and leaves provided with vascular supply while the gametophyteis smaller and simple structure. The expansion of the sporophyte and reduction of the gametophyte has culminated in the angiosperms (Fig. 2.36).

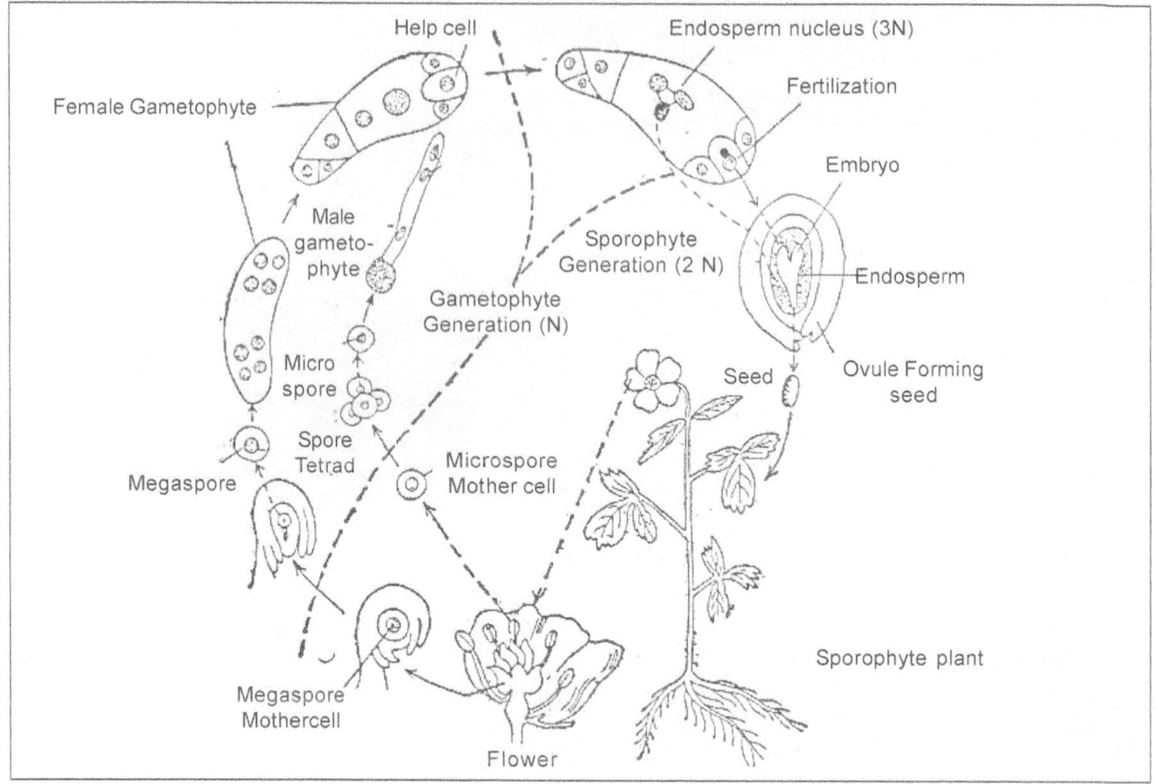

Fig. 2.36 Diagrammatic life-cycle of an angiosperm

2.8.1 Arabidopsis thaliana

Arabidopsis thaliana is a dicotyledonous angiosperm belongs to the mustard family, Brassicaceae. It is a tiny weed with small white flowers which grows low to the ground in meadows and cultivated in laboratories throughout the world (Fig. 2.37 (a)). *A. thaliana* possess several features that make it an ideal experimental model like *Drosophila* in animal kingdom. It resembles other higher plants in growth, development, flowering and seed production. In addition, it has a shorter generation time than most other angiosperms with only six weeks for seeds to germinate and develop into mature plants that produce more seeds. It reproduces mainly by self-fertilization. A wild type *Arabidopsis* plant can produce a very large number of seeds from 10,000 to 40,000 with high rate of germination. This abundant and rapid reproduction enables the geneticists to screen large populations of seedlings for specific phenotypes. *A. thaliana* grows well under laboratory conditions which makes it a model plant to understand the breeding, genetics, physiology, biochemistry, growth and development of a plant at the molecular level.

Genome Structure and Organisation

A model organism for genetic studies should possess relatively a small genome size which represents the genomes of other organisms. *A. thaliana* consists of 100 Mb genome, is one of the smallest genomes known in the plant kingdom. It is smaller than the genomes of most other angiosperms. Comparatively the genome of the angiosperm maize *(Zea mays)* is 45 times larger than *Arabidopsis*. *Arabidopsis* genome is 60% the size of the *Drosophila melanogaster* genome. The nuclear DNA of *Arabidopsis* is carried by five pairs of small chromosomes. It contains of much less repetitive DNA than the genomes of other angiosperms. About 20% of the genome consists of noncoding repetitive DNA while the remaining 80% genome is a single copy DNA. An estimated 15,000 to 20,000 genes are present in the five chromosomes. The genes are similar in structure to the genes of other plants and are more tightly packed containing only small introns.

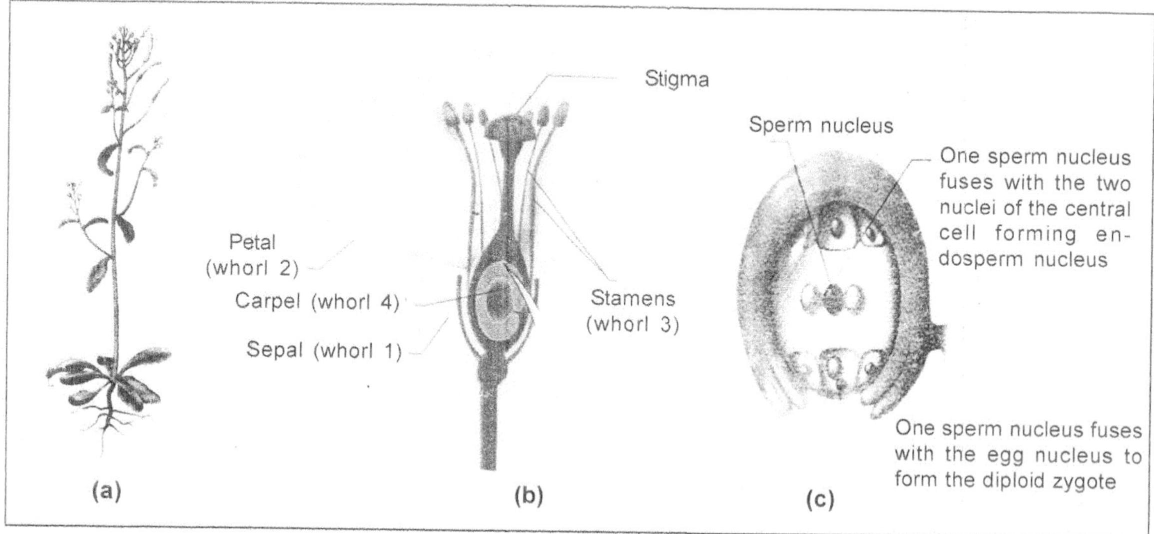

Fig. 2.37 Arabidopsis *thaliana*. (a) whole plant, (b) flower, (c) double fertilization

Anatomically, like most higher plants, *A. thaliana* root and stem contain several cell types, organised into the outer epidermis, the middle cortex and the inner vascular cylinder comprising water conducting xylem vessels and the food conducting phloem elements. Leaf epidermis possess trichomes and stomata. The mesophyll layer performs the photosynthesis and vascular tissues help in conducting water and nutrients between the leaf and stem.

Growth take place by the continuous division of cells of apical meristem of stem and root. The root meristem consists of a central zone of meristem cells and a peripheral zone of differentiated cells. The cylindrical root is produced through cell divisions, cell expansion and cell differentiation. In shoot meristem, cell division in the central zone also maintain the meristem while peripheral zone gives rise to mature organs such as leaves and lateral shoots.

Reproduction

The shoot apical meristem metamorphose into an inflorescence meristem which in turn produces a series of lateral floral meristems. A floral primordium is produced from each floral meristem that gives rise to a flower. The flowers are arranged in a spiral around the inflorescence stalk (Fig. 2.37 (a)).

The flower is a modified shoot with four concentric whorls (Fig. 2.37 (b)). The first whorl consists of four green leaf like sepals and the second whorl is occupied by four white petals. The third whorl contains six stamens bearing the male gametes in the form of pollen. The fourth whorl is a cylinder of two fused carpels containing female gametes in the form of ovules. The fused carpels are part of a cylinder known as pistil that consists of pollen receiving stigma at the top with short neck or style and a ovary. After fertilization the pistil develops into the seed bearing fruit.

Life Cycle

Arabidopsis is bisexual and capable of self fertilization. However, cross pollination by artificial means is easy to accomplish. Fertilization being the initial step in a life cycle includes embryonic development. Seed germination, vegetative growth and reproductive development followed by senescence are the other steps in the life cycle of the plant.

Double Fertilization

Double fertilization is a process unique to higher plants and *Arabidopsis* as well. Each pollen grain possess two coupled sperm cells while embryo sac is eight nucleate. In each ovule, among six mononucleate cells three cells each are distributed at opposite polar ends while the central cell contain two nuclei. The germinating pollen migrates through the style or transmitting tract to an ovule in the ovary. In the ovary one sperm nucleus (1N) from the pollen fuses with the egg nucleus from the ovule to form a diploid zygote (2N) while the second sperm nucleus (1N) fuses with the central nuclei (2 + 1 N) resulting in a triploid (3N) endosperm nucleus (Fig. 2.37 (c)). After fertilization, the zygote mitotically divide through mitosis to form endosperm tissue. Endosperm tissue nourishes the developing embryo as the other remaining 1N cells degenerate. The outer layer of the embryo sac hardens to form a seed coat.

Embryo Development

The embryo develops within the protective seed coat following a sequence of well defined cell divisions leading to a series of stages (Fig. 2.38). Embryo development in *Arabidopsis* occurs in five different stages *viz.*, the octant stage, the globular stage, the triangular stage, the heart shaped stage and the torpedo. The first division of the zygote results into two cells, one small and one larger. The small cell constitutes embryo proper and the larger cell divide to produce the suspensor. The embryo undergoes few more divisions resulting in the octant stage embryo. Additional divisions parallel to the surface of the embryo result in discrete outer layer that will give rise to epidermis of mature embryo and as innermost group of cells which acquire an elongated shape. This is the first evidence of differentiation towards the vasculature found in the mature embryo. The embryo attains the globular stage by now. The embryo rapidly loses the globular shape and acquires a more triangular stage of embryo. The embryo becomes self sufficient for growth at triangular stage. The embryo moves through the heart stage to the torpedo stage as two *protuberances* expand and differentiated into two discrete, well defined cotyledons.

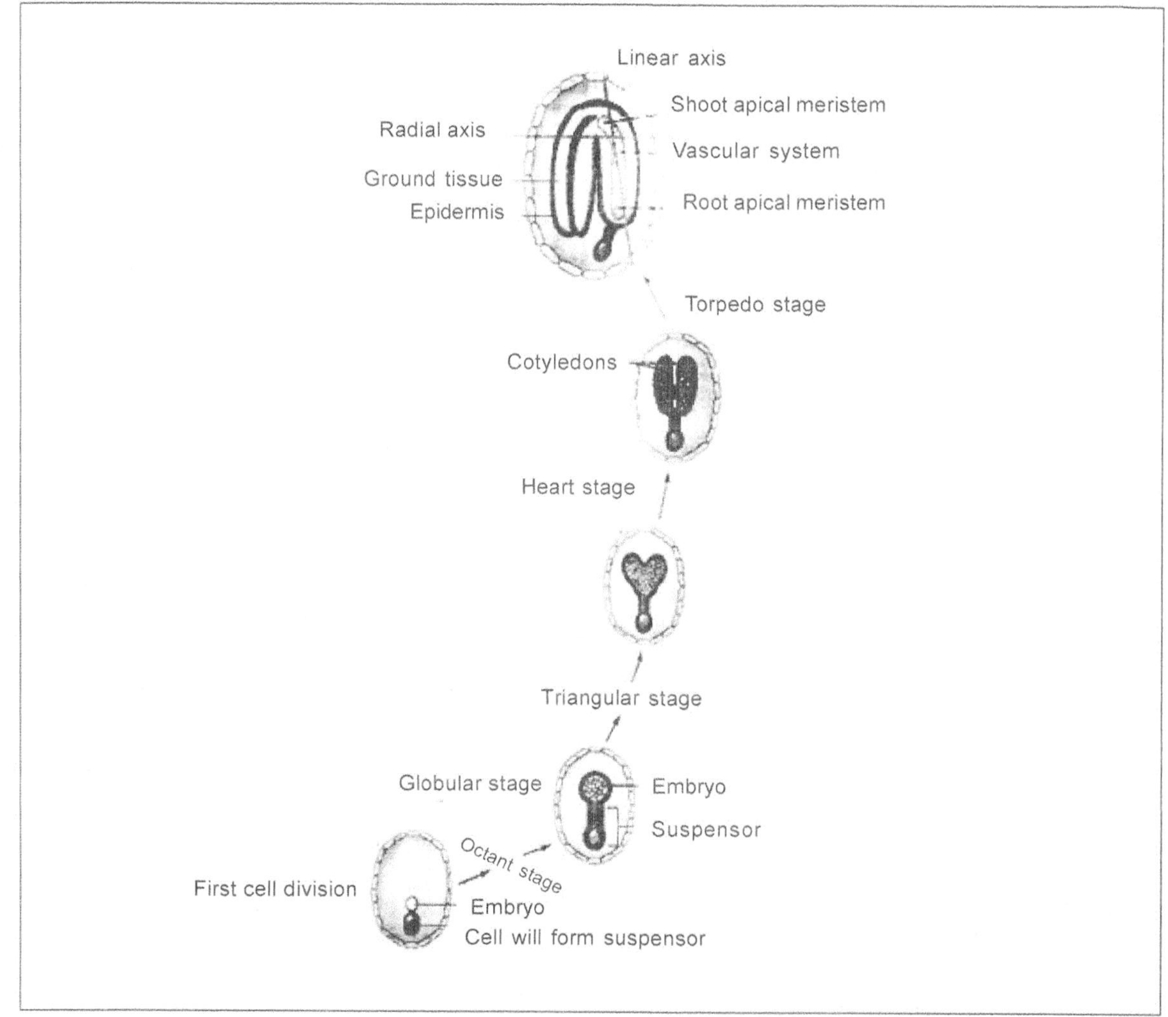

Fig. 2.38 Stages of embyonic development in *A. thaliana*

At the end of the torpedo stage, all the major tissues required for future vegetative development of the plant along two main axes are differentiated. The linear axis makes up the shoot/root continum and the radial axis moving from the center out to the edge of that continuum. The shoot/root axis constitutes two cotyledons that store nutrients in the mature seed and will differentiate into the first rudimentary leaves after seed germination. The axis also includes a hypocotyl, an embryonic stem between the two cotylendon and the embryonic root and the root and shoot meristems. The radial axis produces the central vascular system, a middle layer of ground tissues and the outer epidermis.

As the embryo reaches maturity, growth and development ceases and desiccation ensures, ultimately entering into a dormant state to withstand the adverse conditions.

2.9 Growth

The plant body of vascular plants consists of branched or unbranched axis with lateral appendages. The axis is differentiated into an aerial stem bearing lateral appendages called *leaves* and a subterranean part-root. The aerial stem with leaves is known as the *shoot system*. The subterranean root with lateral roots is known as *root system*.

The stem is organised during the development of the embryo. The fully developed embryo consists of the hypocotyl-root axis bearing one or more cotyledons at the upper end which serves as the primordium of the shoot. At the lower end of the axis, the primordium of the root covered with root cap. The embryonic shoot consists of an axis with unexpanded internodes and one more leaf primordia. The embryo shoot or the first bud is commonly known as a plumule and its stem as epicotyl. The beginning of shoot organisation is found in the hypocotyl-cotyledon system where hypocotyl represents the first stem unit of the plant and the cotyledons are the first leaves.

During seed germination and growth of the embryo into adult plant, the root meristem forms the first root and the shoot meristem differentiates into new nodes and internodes with leaves at the nodes.

Meristems

All cells undergo division in the embryo during the early stages of development. At later stages cell division and multiplication become restricted to special parts of the plant in which the tissue remain embryonic as the cells retain the ability to divide. These embryonic tissues in adult plant are called *meristems*. Meristem cells continue to divide indefinitely to add new cells continually to the plant body contributing to its growth. Thus meristem may undergo temporary resting phase as seen in perennial plants that are dormant in certain seasons and also in dormant axillary buds even during the active phase of the plant.

The process of growth and morpho-physiological specialisation of cells produced by meristems is referred as *differentiation*. Differentiation may lead to the gradual loss of embryonic nature of the meristem which acquires the mature state and such tissues are referred as *mature or permanent* tissues.

Meristems are classified on the basis of various criteria such as their position in the plant body, their origin and the tissues they produce, their structure, their stage of development and their function.

According to the position of the meristems in the plant body they are divided into 3 categories : 1. *Apical meristems*, at the apices of main and lateral shoots and roots; 2. *Intercalary meristems*, found between mature tissue such as in the bases of inter nodes of grasses; 3. *Lateral meristems*, situated parallel to the circumference of the organ in which they are present - e.g. vascular cambium and phellogen.

Based on the origin of meristems, two types are distinguished 1. *Primary meristems* and 2. *Secondary meristems*. Primary meristems develop directly from the embryonic cells maintaining a direct continuation of the embryo. Secondary meristems develop from the mature tissues which have already undergone differentiation (e.g. phellogen, cambium and callus tissue in tissue culture).

The apical meristem is divided into two main regions, a *promeristem* and a meristematic zone that has undergone certain degree of differentiation. *Promeristem* comprises of the apical initials and neighbouring cells. The partly differentiated meristematic zone consists of three basic meristems *viz*., the *protoderm*, which gives rise to epidermal system, the *procambium* that produces primary vascular tissues and the *ground meristem* from which ground tissues of the plant develops. Ground tissues include parenchyma and sclerenchyma of the cortex and pith and the collenchyma of the cortex.

Hanstein (1868) proposed the *histogen theory* which distinguishes three zones (Fig. 2.39) in the shoot apex of angiosperms *viz*., an outermost zone (*dermatogen*), a central zone (*plerome*) and a hollow cylindrical zone of several layers of cells between the dermatogen and plerome. Hanstein stated that the dermatogen, plerome and periblem develop from independent groups of initials that act as direct histogens. This theory states that the type of tissues the meristems produce is determined from the beginning itself i.e. the epidermis develops from dermatogen, the cortex and internal tissues of the leaf from the periblem and central cylinder from the plerome. Histogen theory was disproved later in view of the following : 1. Distinction between periblem and plerome is not possible in most spermatophytes; 2. No predetermination of the mature tissues can be traced in various initials.

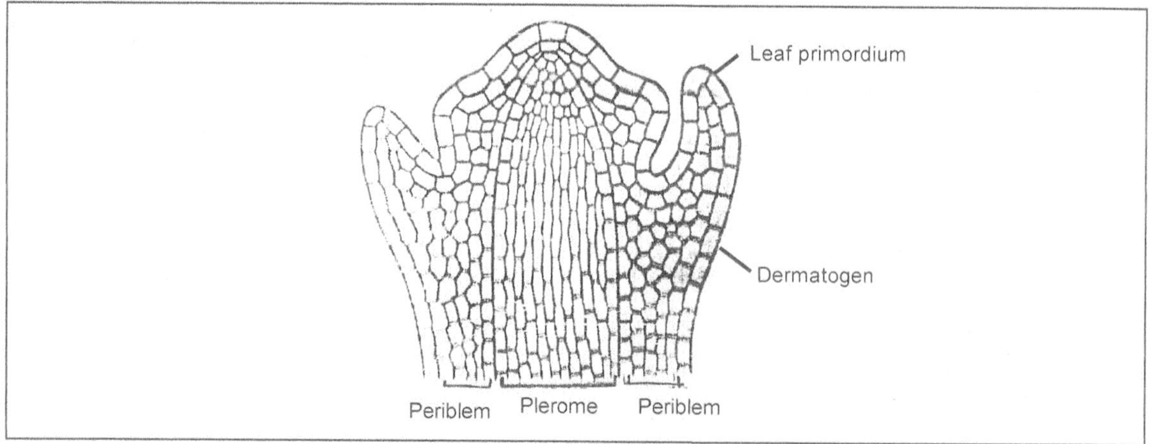

Fig. 2.39 Shoot apex of angiosperms showing three distinct zones (Hanstein, 1868)

Schmidt (1924) postulated a new theory called *Tunica-Corpus theory* which divides the apex into two regions, the *tunica* and the *corpus*. The two regions recognised by this theory are distinguished by the planes of cell divisions in them. The tunica represents the outermost layer or layers of cells which surround the inner cell mass called the *corpus* (Fig. 2.40). In tunica, the place of cell division is mainly anticlinal while in the corpus, the planes of cell division is in all directions. The tunica enlarges in surface area and the corpus in volume. The *tunica-corpus theory* is generally accepted in literature and it draws no constant relationship between particular initials of the promeristem and the inner tissues of the shoot. Tunica consists of a single or few layers of cells surrounding the inner meristem. These number of layers is not always constant but vary even within a family, genus, species and at different growth stages within a single plant.

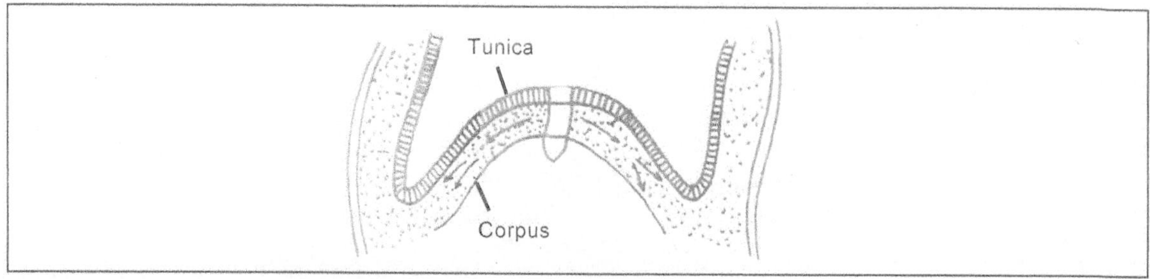

Fig. 2.40 Concept of apical meristem (Schmidt, 1924)

Plant Growth Regulation

Charles Darwrin (1890) observed in canary grass *(Phalaris canariensis)* some tropic movements and confirmed that these are due the presence of some specific substances. He stated that "*Some influence is transported from the stem tip to the lower parts where it causes growth*". Subsequently several workers discovered the existence of small molecules that play a vital role in growth and development.

The movements and growth in plants are controlled by chemical compounds which are called *phytohormones* or growth regulators. Growth is a complex phenomenon. This involves many important processes like cell division, cell elongation, cell differentiation and morphogenesis. All these processes involve regulation of specific chemical substances which may promote or inhibit the growth. Hence, these are called *growth regulators.* To distinguish from animal hormones often these are also referred as *plant hormones* or *phytohormones.*

Phytohormones may be defined as "*an organic substance produced naturally in plants which controls the growth and other physiological functions at a site away from its place of synthesis*".

Types of Growth Regulators

Plant growth regulators are broadly classified into two categories : 1. Growth promoting hormones. *e.g* : Auxins, Gibberellins, Cytokinins, 2. Growth inhibiting hormones. *e.g* : Abscissic acid, Ethylene

Growth Promoting Hormones

1. *Auxins*

Hormones that promote longitudinal growth in plants are called *auxins*. Auxins are naturally available in plants as well as they may be synthesized artificially. Auxins that occur in plants are called *natural auxins,* they are : (i) Indole-3-acetic acid (ii) Indole-3-acetaldehyde and (iii) Indole-3-pyruvic acid. These may be synthesized artificially and have properties like natural *auxins.* Some of them are : (i) Naphthalene acetic acid (NAA) (ii) Indole butyric acid (IBA) and (iii) Indole propionic acid. Auxins bring about wide variety of physiological effects in different parts of the plant body. Some of the important effects are :

(a) ***Growth of various plant parts :*** Auxins cause growth in leaves, roots, stems by inducing cell enlargement and cell elongation. Auxin is synthesized in the apices of root and stem.

(b) ***Root formation :*** Auxins like IAA, IBA, NAA induce the development of adventitious roots on stem cuttings, which are very helpful in vegetative propagation of many ornamental plants and fruit trees.

(c) ***Parthenocarpy :*** Hormones like IAA, IBA and NAA induce parthenocarpy in many plants.

(d) ***Cell elongation :*** Induction of cell elongation is the most characteristic effect of auxins. IAA is thought to act on DNA and promote the synthesis of mRNA which directs the synthesis of cell wall softening enzymes like glucanases, pectinases. Due to softening of the cell, wall pressure decreases and more water enters into the cell. Cell wall becomes stretched and finally results in cell elongation.

(e) ***Tissue and organ culture :*** IAA along with cytokinins promote cell enlargement and cell division leading to callus formation. IAA induces root initiation in callus. High concentration of auxin inhibits bud initiation in callus cultures.

2. *Gibberellins*

Gibberellins are another class of growth regulators that are mainly concerned with stem elongation and flowering. Gibberellins were discovered in Japan in connection with *Bakanae disease* or foolish seedling disease of rice caused by *Gibberella fujikori*. More than 40 types of gibberellic acids were discovered and they are usually written as GA_1, GA_2 and GA_3.

Some of the physiological effects include :

(a) ***Stem elongation :*** Gibberellins bring about internodal elongation in many plants. Gibberellins accelerate cell division and cell elongation.

(b) ***Leaf expansion :*** The leaves become broad. This results in the increase of the total photosynthetic area and biomass.

(c) ***Seed germination :*** In cereal grains like maize, barley, wheat etc., the endosperm is surrounded by a luring, protein-rich layer called *aleurone layer*. During germination of these grains the embryo produces gibberellins.

3. *Cytokinins*

Cytokinins (cytos = cell; kinesis = division) are hormones that promote cell division in plants. The presence of cell division inducing substances in plants was demonstrated by G. Haberlandt in coconut milk. Some naturally occurring cytokinins are zeatin, ribosyl zeatin, dihydrozeatin etc.

2.10 *In Vitro* growth : Tissue Culture

The term *Tissue Culture* includes *in vitro* culture of plant cells, tissues and organs while '*Cell culture*' is used for *in vitro* culture of single or relatively small groups of plant cells (suspension cultures). In tissue culture, cultivation of plant cells results in an unorganised mass called '*Callus*'. When plant parts like root tips, shoot tips etc., are used as explants and cultured *in vitro* to obtain their development as organised structures, it is referred as '*Organ culture*'.

During the last two decades, plant cell, tissue and organ cultures have developed rapidly and became a major biotechnological tool in agriculture, horticulture, forestry and industry. The problems which were not feasible to tackle hitherto through conventional methods have now been made possible by tissue culture. The techniques which are generally practised are : plant cell, tissue and organ cultures with a specific objective of plant regeneration, somatic embryogenesis and production of synthetic seeds. Secondary metabolites of therapeutic interest, genetic transformation and transgenic plants with desirable agronomic traits are produced through tissue culture.

Great deal of research is being carried out in many countries on plant cell culture, differentiation, regeneration and transformation in tropical grain legumes, woody legumes and cereals besides several other economically important plant species. Tissue culture enabled to improve growth under stress conditions, resistance to pests and diseases, to improve nutritional quality, including increased ability for nitrogen fixation etc.

Totipotency

In vitro techniques demonstrated the ability of individual plant cell to develop into a total plant. Thus each living cell of a multicellular organism, is capable of independent development when provided with suitable conditions (White, 1963) and it is known as *totipotency*. The term totipotency was coined by Morgan (1901). **Totipotency** is the ability of a plant cell to perform the functions of development in toto which is the characteristic feature of a zygote (i.e., the ability to develop into a complete plant).

Haberlandt (1902) for the first time envisaged the concept of cell culture in which he attempted to cultivate the isolated plant cells *in vitro* on an artificial medium (Knop's solution containing peptone, asparagine and sucrose). Callus was developed from bud, root and shoot fragments of about 1.5 mm in size without using any nutrient medium. R. J. Gautheret in France and P. R. White in USA contributed to the development of techniques for cultivation of plant cells under defined conditions. Skoog and his group provided most of the modern media in vogue.

The term '*tissue culture*' can be applied to any multicellular culture growing on a solid medium or on liquid medium. Maheswari and Guha (1964) produced haploid plants for the first time from the pollen grains of *Datura* by culturing anthers.

In plant cells plasma membrane is bound by a rigid cell wall (unlike animal cells). Cocking (1960) produced naked cells called '*protoplasts*' by enzymatic degradation of cell walls. In the last few decades the isolation, culture and fusion of protoplasts has developed into a new technology in producing transgenic plants. Cultured protoplasts are used extensively in somatic cell fusions, for taking up foreign DNA, cell organelles, bacteria and virus particles.

The benefits from protoplast culture currently availed are :

(i) Intraspecific, interspecific and intergeneric protoplast fusion,

(ii) Transfer of mitochondria and plastids into protoplast,

(iii) Uptake of certain beneficial genes of blue-green algae, bacteria and viruses by protoplasts and

(iv) Transfer of genetic information into isolated protoplasts.

Maintenance of aseptic conditions are essential in tissue culture (Table 2.4) at every stage.

Table 2.4 *Tissue Culture Facilities (Modified from Freshney, 1987)*

Minimum requirements (essential)	**Desirable features (beneficial)**	**Useful additions**
1. Sterlie area : clean and quiet, no thorough traffic, separate from animal house and micro-biological labs	1. Filtered air (air conditioning)	1. Piped CO_2 and compressed air
2. Preparation area	2. Hot room with temperature recorder	2. Store room for bulk plastic
3. Storage areas :	3. Microscope room	3. Containment room for biohazard work
(a) liquids-ambient, 4°-20 °C	4. Dark-room	4. Liquid N_2 storage tank (500 *l*)
(b) glassware (shelving)	5. Service bench adjacent to culture area.	
(c) plastics (shelving)	6. Separate preparation room	
(d) small items (drawers)	7. Separate sterlizing room	
(e) specialized equipment (slow turnover), cupboard (s)	8. Cylinder store	
(f) chemicals ambient, 4°-20°C (share with liquids but keep chemicals in sealed container over desiccant)		

Laminar Flow Cabinets or Hoods

This is a clean bench which facilitates and maintains aseptic or sterile conditions so that no separate arrangement for sterile room is required. If the culture has to be done in a room, that should be provided with filtered air supplied from the ceiling, so that the whole room is regarded as a sterile working area. Incubation is done in incubators or thermostatically controlled hot rooms where racks are designed for the placement of cultures (Tubes, Petri plates, Flasks, etc).

Nutrient Media

Vital activity of a cell is the absorption of nutrients through cell membrane and rapid multiplication of cells resulting in drastic increase in number. White (1934) observed the unlimited growth of isolated root tissues when they were provided with nutrient media containing inorganic salts, sucrose, vitamins, growth hormones and few amino acids.

Inorganic minerals – include macronutrients (*e.g.*, nitrogen, phosphorus, potassium, calcium, magnesium and sulphur in the form of salts) and micronutrients (boron, molybdenum, copper, zinc, manganese, iron and chloride). A stock solution of iron is prepared in a chelated form and is added to medium when required. *Growth Hormones* - Cyokinins promote cell division and regulate growth and development similar to kinetin. Auxin stimulates shoot elongation. Gibberellins are not very important. *Vitamins* - Vitamins are required in trace amounts as they collapse the enzyme system. Vitamin B_1 (Thiamine) is used for all plant tissue culture - niacin (nicotinic acid), vitamin B_2 (riboflavin) ascorbic acid, vitamin H (biotin) and vitamin B_{12} (cyanocobalamin) are commonly used vitamins. *Amino acids* – Inspite of the nitrogen source which are present in the inorganic salts various amino acids and amides are used in plant tissue culture media. Mostly L-aspartic acid, L-asparagine, L-glutamic acid, L-arginine are used. Some important nutrient media are Murshige and Skoog medium and Eriksson medium (Table 2.5).

Table 2.5 *Composition of Nutrient Media required for in vitro Culture*

Constituent chemical	M S (mg/1) medium*	ER medium* (mg/1)	B5 (mg/1) medium*
1. *Macronutrients*			
NH_4NO_3	1650	1200	–
KNO_3	1900	1900	2500
$CaCl_2 2H_2O$	440	440	150
$MgSO_4 6H_2O$	370	370	250
KH_2PO_4	170	340	–
$(NH_4)_2SO_4$	–	–	134
NaH_2PO_4	–	–	150
2. *Iron*			
Na_2 EDTA	37.3	37.3	37.3
$FeSO_4 7H_2O$	27.8	27.8	27.8
3. *Micronutrients*			
$MnSO_4 4H_2O$	22.3	2.3	–
$MnSO_4H_2O$	–	–	10.0
$ZnSO_4 4H_2O$	8.6	–	2.0
Zn versanate	–	1.5	2.0
$Na_2Mo_4 2H_2O$	0.25	0.025	0.25
$CuSO_4 5H_2O$	0.025	0.0025	0.025
$CoCl_2 6H_2O$	0.025	0.0025	0.025
KI	0.83	–	0.75
H_3BO_3	6.2	0.63	3.0
4. *Vitamin*			
Glycine	2.0	2.0	–
Nicotinic acid	0.5	0.5	1.0
Pyridoxine - HCl	0.5	0.5	1.0
Thiamine - HCl	0.1	0.5	10.0
5. *Cytokinin*			
Kinetin	0.04 – 10.0	0.02	0.1
Myo-inositol	100.0	–	100.0
IAA	1.0 – 30.0	–	–
NAA	–	1.0	–
2.4 D	–	–	0.1 – 1.0
Sucrose (g)	30.0	40.0	20.0
p^H	5.7	5.8	5.5

* *M. S. Murshige and Skoog (1962), E. R. Eriksson (1965). B5. Gamborg et al., (1968)*

Culture of Plant Material Explant

Seed bearing plants exhibit varying diversity (trees, herbs, grasses) and show morphological differences in root, stem and leaves which in turn vary in cell tissues, and their totipotency. Of all the tissues parenchyma is capable of division and growth. The development of the tissue is characterized by three types of cell growth : cell division, cell elongation and cell differentiation. Healthy and young part of the explant is used in tissue culture parenchyma from stems, rhizomes, tubers and roots easily and quickly respond to culture conditions *in vitro*.

Callus Formation and its Culture

Development of callus takes place due to stimulation by endogenous growth hormones, the auxins and cytokinins. A 2-5 mm sterile segment is excised from stem, tuber or root and is transferred into a nutrient media and is incubated at 25 - 28^{0}C in an alternate light and dark regimen of 12h. The nutrient medium is supplemented with auxins which induce cell division. After some time the upper surface of explant is covered by callus. A callus is an amorphous mass of loosely arranged thin walled parenchyma cells. The unique feature of callus is that it has biological potential to develop normal root, shoots and embryoids, which in the end forms a plant.

Formation of callus is determined by various factors such as the source of explant, nutritional composition of medium and environmental factors. Callus formation includes three developmental stages : 1. Induction 2. Cell division and 3. Cell differentiation.

Induction

It depends on the metabolic rate of cells while duration depends on physiological status and nutritional and environmental factors. Induction increases the metabolic rate, cells accumulate high content and finally divide to form many cells. After callus grows on nutrient medium it is essential to subculture it within 28 days on fresh medium otherwise there is nutrient depletion in original medium which results in accumulation of toxic metabolites.

Organogenesis

Root, shoot and leaves (not embryo) are the organs that are induced in plant tissue culture. Embryo is not considered to be a plant organ as this is an independent structure and it does not have vascular supply. Organogenesis (i.e., development of organs) which starts with stimulation caused by chemicals in the medium. The composition of the medium invariability influence the type of organ it induces. Shoot induction takes place at higher concentrations of cytokinins while roots are induced at higher levels of auxins in the medium (Table 2.6).

Table 2.6 *In vitro control of organogenesis by Auxins and Cytokinins*

Auxin (mg/l)	Cytokinin (mg/l)	Organogenesis
0.0	0.2	No growth
0.03	1.0	Shoots
3.0	0.02	Roots
3.0	0.2	Callus

Meristem Culture

Culturing of axillary or apical shoot meristems is known as '*meristem culture*'. It involves the development of shoot meristem and regeneration of adventitious roots from the developed shoots. Shoot apical meristem lies at the shoot tip beyond the leaf primordium (youngest leaf). It differs from shoot apex in having shoot apex and a few leaf primordia. This selected piece of tissue of organ used for culturing is called as *explant*. Explants of various size are used for rapid clonal propagation. The size of the explant may be large (5-10 mm) for vegetative propagation. For eliminating virus infection the apical meristem should be excised along with a minimum of surrounding tissue. Explants taken from actively growing plants at the beginning of the season are more promising.

Shoot tip culture is extensively applied in horticulture, agriculture and forestry. As the size of the propagules are minute in size, this *in vitro* propagation technique is known as *Micropropagation*.

Fig. 2.41 Initiation of callus and plantlet regeneration (a) Callus of early maturing mutant (d) Regenerating root and shoot (c) Regenerating plantlets (d) Regenerating seedling in pot cultures

Somatic Embryogenesis

Embryo production is characteristic feature of the flowering plants. Embryoids have also been artificially induced in cultured plant tissues besides zygotes. This was first done in carrot. Somatic embryogenesis can be initiated in two ways.

1. By inducing embryogenic cells within the callus which has been already formed.
2. Directly from pre-embryonic determined cell (without callus) which are ready to be differentiated into embryoids.

In somatic embryogenesis two nutritional media of different composition are required for obtaining embryoids. The first medium contains auxins which helps to initiate embryogenic cells. Second medium lacks auxins or may be in reduced level for development of the embryonic cells into embryoids and plantlets. In both the media 'N' is required in reduced amount. The embryogenic cells pass through 3 different stages e.g., globular, heart shaped and torpedo shaped, to form embryoids (Fig. 2.42). The embryoids are separated and isolated mechanically by using glass beads. When plant reach torpedo stage they are transferred to filter paper bridge which is a sterile and pluged culture tube containing 10 ml MS liquid medium with kinetin (0.2 mg/l) and sucrose (2%A\V) on which Whatman No.1 filter paper is placed. Some of the plants cultured by somatic embryogenesis are : *Atropa belladonna, Brassica oleracea, Citrus chinesis* etc.

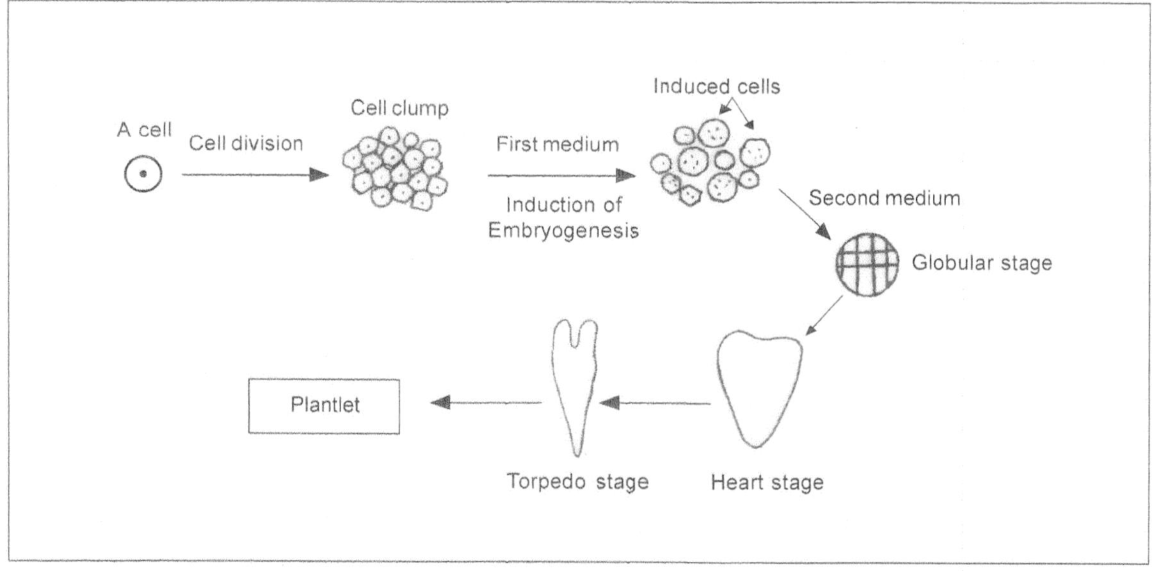

Fig. 2.42 Events of Somtic embryogenesis

Somaclonal Variation

Somaclonal variation is an alternative tool for plant breeder for generating new varieties with disease resistance, improvement in quality and yield. Plants like cereals, legumes, oil seeds, tuber crops, fruit crops are developed by this method.

Protoplast Culture

Isolation of Protoplast

Protoplasts (without a cell wall) are the biologically active and most significant material of cells. Isolated protoplast is known as '*naked plant cell*'. Techniques of Isolation and culture of protoplast are given in Fig. 2.43.

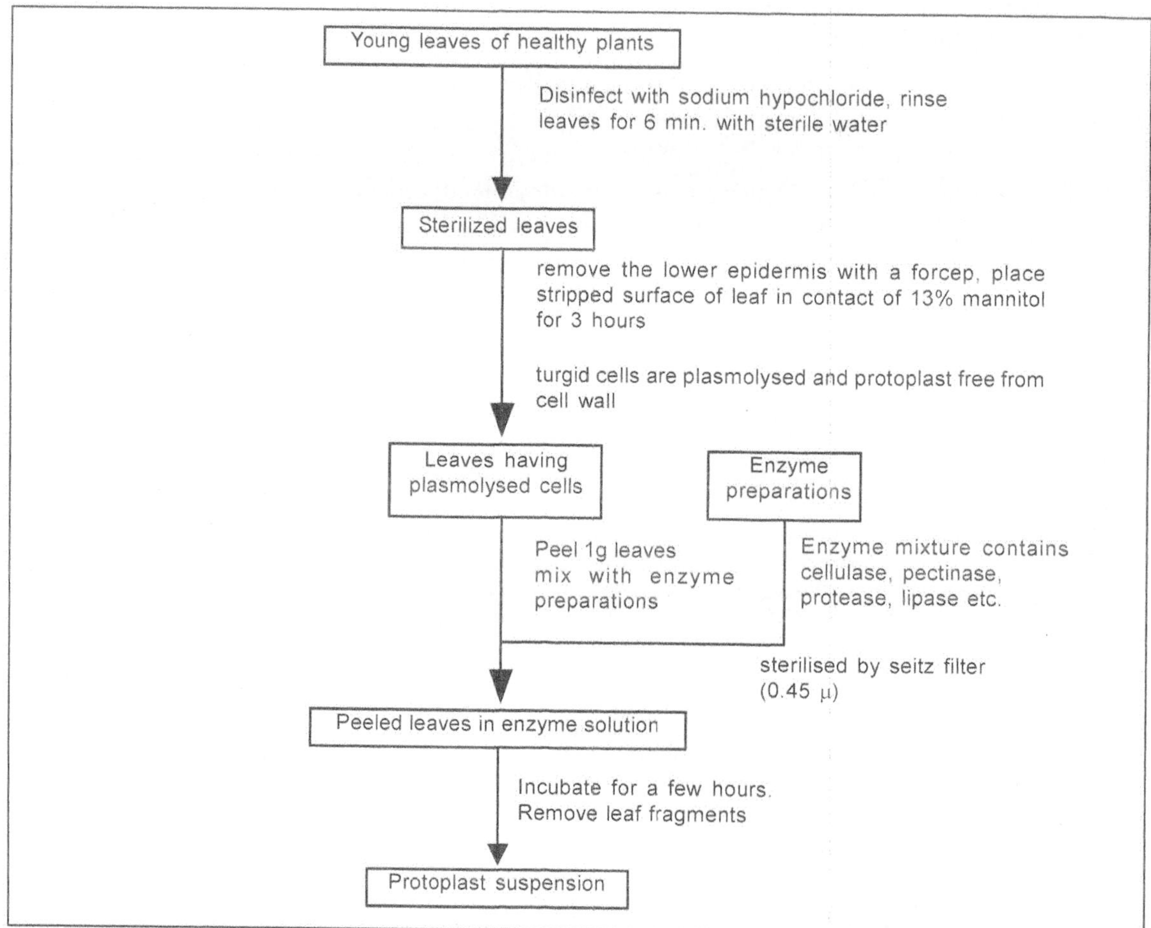

Fig. 2.43 *In Vitro* Culture Techniques of Plant Cell, Methods of Isolation of protoplast

Protoplast Culture and Regeneration

From the protoplast solution of known density (about 10^5 protoplasts/ml) about 1 ml suspension is poured on sterile and cool nutrient medium in Petri dishes. The plates are incubated at 25 ^{0}C in a dim white light.

The protoplasts regenerate a cell wall, undergo cell division and form callus. Embryogenesis begins when it is placed on nutrient medium. The embryo develops into seedlings and finally into mature plants (Fig. 2.44).

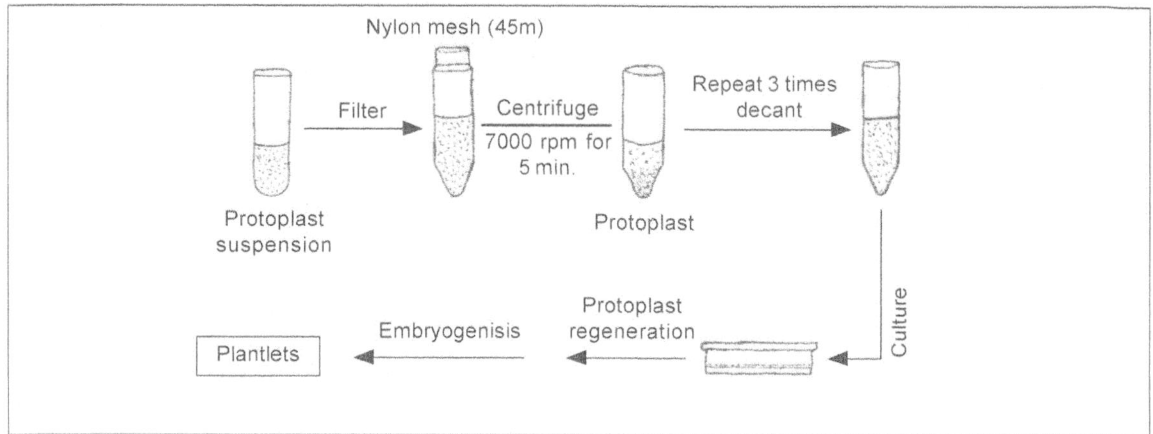

Fig. 2.44 Purification, culture and regeneration of protoplasts

Protoplast Fusion and Somatic Hybridisation

Protoplast isolation and regeneration is a new tool of genetic manipulation of plants. The fusion of protoplasts of genetically different lines or species is also possible. Somatic hybridization of crop plants represents a new challenge to plant breeding and crop improvement. Pest and disease resistance and transfer of C_3 photosystems into C_4 crop plants.

Protoplast Fusion Involves :

1. Production of fertile amphidiploid somatic hybrids of sexually incompatible species which results in a new variety of homo as well as heterokaryotic colonies.
2. Production of heterozygous lines within one plant species which normally will be propagated only vegetatively *e.g* : potato.
3. The transfer of only a part of genetic information from one species to another by using chromosome elimination technique.

Hybrids and Cybrids

Often the term hybrid refers to the fusion of genetic information present in the two nuclei of hybridizing plants. When two cells (protoplasts) are fused, the resultant hybrid cell not only contains the genetic information present in the nucleus but also has the genetic information of the cytoplasm of both the parents. Hybrid cells are those which retain the cytoplasmic information of one of the parental cell only but has the fusion of genetic information contained in the nuclei of both the parents. On the other hand cybrids (organellar genome) contain cytoplasmic information of both the parents but has the nuclear genetic information (genome) of one of the parents. The cybrids are produced by special techniques after protoplast fusion in following ways.

(i) Fusion of protoplasts wherein one of the parental cell is devoid of nucleus (enucleate) or having inactivated nucleus.

(ii) Elimination of one of the parental nuclei.

(iii) The process of chromosome elimination during cell division.

Fusion of protoplasts and production of hybrid plants is depicted in (Fig. 2.45).

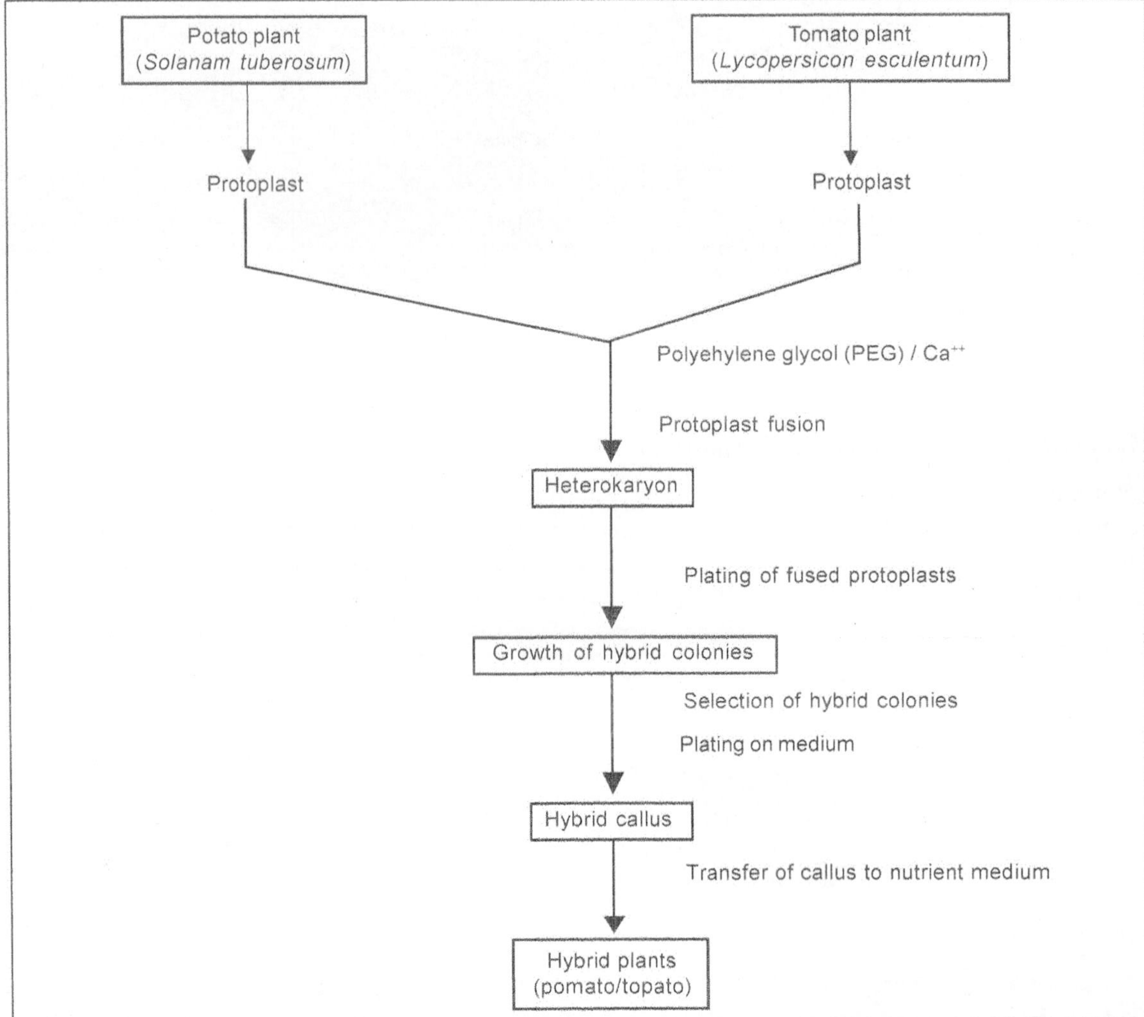

Fig. 2.45 Fusion of protoplasts of potato and tomato, and production of hybrid plant (pomato)

Methods of Somatic Hybridization

(i) Isolation of protoplasts from suitable plants.

(ii) Protoplast fusion is facilitated by the presence of fusigenic chemicals i.e., chemicals promoting protoplast fusion, such as polyethylene glycol (PEG)(20%W/V), sodium nitrate ($NaNO_3$), maintenance of high p^H (10.5) and temperature of 37°C viable heterokaryons are produced (as a result of fusion of protoplasts. PEG induces fusion of plant protoplasts resulting viable Heterokaryons.

(iii) Wall regeneration by heterokaryotic cells.

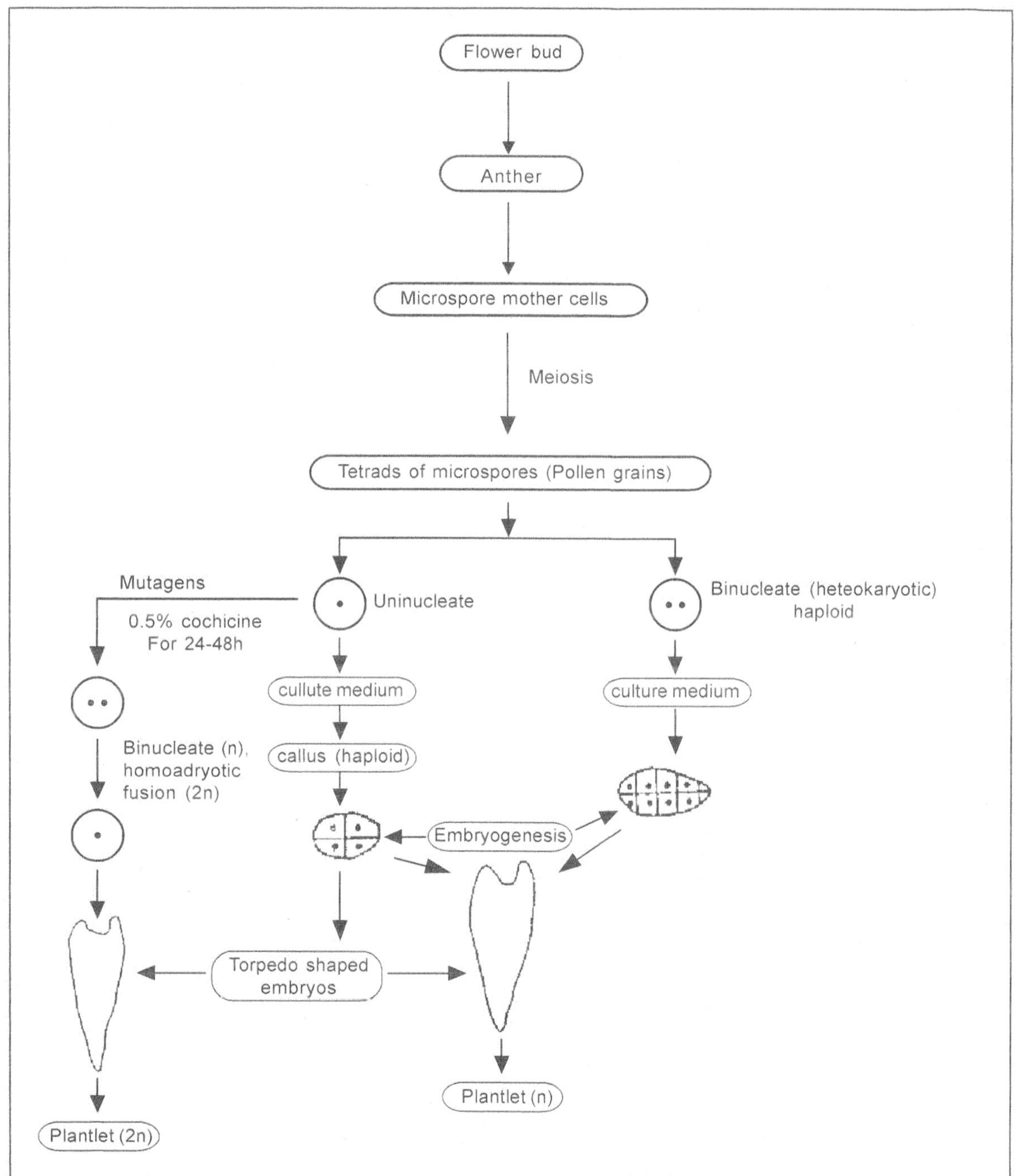

Fig. 2.46 Methods of anther culture and *in vitro* production of haploids and diploids

(iv) Fusion of nuclei of heterokaryons to produce hybrid cells.

(v) Planting and production of colonies of hybrid colonies.

(vi) Selection of hybrids, subculture and induction of organogenesis in the hybrid colonies.

(vii) Transfer of mature plants from the regenerated callus.

Anther and Pollen Culture

Development of haploid plants from pollen grains by placing anthers or pollen grains on a selective medium is known as anther or pollen culture.

Anther is a male reproductive organ and chromosomes are diploid in number. During microsporogenesis tetrads or microspores are formed from a single spore mother cell. They are called Pollen grain after release from tetrads. Pollen grains possess a single set of chromosomes. The main aim of anther and Pollen culture is to get haploid plants by induction of embryogenesis. Haploid plants have single complete set of chromosomes which are useful in the improvement of crop plants. The chromosome set of these haploids can be doubled by mutagenic chemicals (colchicine) to obtain fertile homozygous diploid inbreds.

At present more than 247 plant species belonging to 38 genera pertaining to 34 families of dicots and monocots have been regenerated using anther culture technique. They include economically important crops such as rice, wheat, maize, and trees like coconut and rubber etc.

Culturing Techniques

Methods of anther culture is shown in the Fig. 2.46. Anthers are surface sterilised and washed with double distilled sterile water. They are excised from the flower buds and their development stages are studied under microscope.

2.11 Biology of Pests

Insect Pests and their Significance

An insect whose population increases to the extent that it starts causing annoyance, inconvenience or injury to man, animals, plants and material possessions is called a "*Pest*".

In their attempts to secure food, insects inflict considerable damage to almost every part of plants and the effects of feeding may be direct, causing outright injuries or indirect through helping in the transmission of diseases-bacterial, fungal or viral. Biting insects by feeding on the growing points of plants cause retardation of growth and often cause defoliation. They may notch the edges of leaves or make holes by feeding or roll up the leaves and feed from inside or only feed on a layer of surface tissue. Sometimes they may live concealed under loose bark of plants or cut the tender stems of plants at the time of germination. By feeding on the flower buds and flowers of plants, a reduction in seed production is often caused. In addition, seed production is also affected. They may also nibble and cut off entire ear heads. Insects with sucking habits cause a general chlorosis of leaves or a silver whitening of the leaf surface, or yellow specking or brownish necrotic lesions. Crinkling and curling of leaves is a common effect. Infestation in numbers on the shoot and fruits often cause a premature shedding of developing fruits or piercing of the rind of fruits and sucking their juice cause premature fruits fall. Injuries are also caused by internal feeders, feeding from within the plant tissues, some called the *borers*, boring the internal tissues and others called the *miners*, mining the leaf tissue. Many plant tissues react with salivary toxins during feeding, by the formation of galls or abnormal outgrowths and due to this toxaemia the growth of plants may be impaired and the setting of fruits, seed and grain adversely affected. Another aspect of injury is through feeding by subterranean insects living in the soil, feeding on the roots of plants by chewing, or boring, sucking, or by formation of galls. In general, the attacked parts result in stunting, discolouration, withering and death of the plants. The serious damages caused by insects to stored products are well known.

The indirect effects of feeding are evident in the loss of quality of produce through reduction in the nutritive value or marketability. Through mechanical or biological transmission of disease agents such as bacteria, fungi and viruses, insects are indirectly responsible for the severe damage and loss caused. Injuries are also caused by oviposition.

When pest responsible for 5 per cent of the loss of yield and are designated minor and major pests when the loss ranges from 5 to 10 per cent and more respectively. According to the periodicity of their occurrence insect pests are said to be regular, occasional, seasonal persistent or sporadic. Severe infestations often result in an epidemic. The major causes for the outbreak of insect pests is through destruction of forests or by bringing them under cultivation, destruction of their natural enemies predators and parasites, intensive cultivation of crops, introduction of new crops and new strains, accidental introduction of foreign pests, etc., to mention a few.

Types of Phytophagus Insects

Three categories of pests of insects are recognized : *monophagous*, *oligophagus* and *polyphagus*. *Monophagus* insects strictly feed on a single species of plant e.g., mulberry silkworm. *Oligophagus* insects characteristically feed on a group of botanically related plants, potato moth on solanaceae family. *Polyphagus* insects are those that accept many plants from a diverse range of plant families, even preference, still exist eg., locusts, termites, noisy caterpillars etc.

Pests of Important Crops

A. *Pests of Sugarcane*

Sugarcane is one of the most important cash crops in India being grown in over 5 million acres of land of which largest chunk belong to Uttar Pradesh (55%), followed by Punjab (8.9%), Bihar (8.3%), Maharashtra (5.3%), Tamil Nadu (3.3%), Andhra Pradesh (3.1%), and rest in other areas. Over 170 insects have so far been recorded from the sugarcane plants in India of which about 32 can be regarded as real pests that attack every part of the plant and cause enormous loss to sugarcane industry. Some of the more important sugarcane pests have been shown in Table 2.16 and few representative pests life history and control methods are described in detail.

Table 2.7 *Pests of Sugarcane*

Common Name	Scientific Name	Order	Family
Stem borer			
1. Top shoot borer	*Scirophophaga nivella*	Lepidoptera	Pyralidae
2. Shoot Stem borer	*Chilo infuscatellus*	Lepidoptera	Crambidae
3. Stalk borer	*Chilo auricillus*	Lepidoptera	Crambidae
4. Internode borer	*Chilo tumidisco*	Lepidoptera	Crambidae
5. Stem borer	*Acigona stenilla*	Lepidoptera	Crambidae
6. Ragi pink borer	*Sesamia inferens*	Lepidoptera	Noctuidae
7. Green borer	*Raphimetopus ablertellus*	Lepidoptera	Pyralidae
Sap Suckers			
8. Leaf hopper	*Pyrilla perpusilla*	Hemiptera	Fulgoridae
9. White fly	*Aleurolobus barodensis*	Hemiptera	Aleyrodidae
10. Spotted fly	*Neomaskella bergii*	Hemiptera	Aleyrodidae
11. Meally bug	*Saccharicoccus sacchari*	Hemiptera	Coccidae
12. Scale insect	*Aulacarpis tegalensis*	Hemiptera	Diaspididae
13. Sugarcane aphid	*Aphis sacchari*	Hemiptera	Aphididae
Leaf feeders			
14. Sugar cane beetle	*Anomola biharensis*	Coleoptera	Scarabaeidae
15. Sugar cane beetle	*Alissonotum simile*	Coleoptera	Scarabaeidae
16. Rice grasshopper	*Hieroglyphus nigrorepletus*	Orthoptera	Acridiae
Root feeders			
17. Root borer	*Emmalocera depressella*	Lepidoptera	Pyralididae
18. White grub	*Holotrichia serrata*	Coleoptera	Meblonthidae
19. White grub	*Holotrichia consanguinea*	Coleoptera	Meblonthidae
20. Termite (White ant)	*Odontotermes obesus*	Isoptera	Termitidae

1. *Pyrilla perpusilla (Sugarcane Leaf hopper)*

The sugarcane leaf hopper is commonly distributed throughout India. Though Pyrilla is a major pest of sugarcane, it also attacks maize, wheat, barley, bajra, oats, sorghum, sudan grass etc., and sometimes on bhendi, karela and cauliflower. The adult insect is pale straw in colour and its body length is about 8 to 10 mm. The head is prolonged anteriorly into snout like structure and has prominent red eyes. A pair of whitish brown anal processes, covered with white mealy wax, which help in up and downward active movement of the insect (Fig. 2.47). It breeds throughout the year. The female lays eggs in clusters (300 to 500 eggs) on the lower surface of the leaves during summers (April) and inside the leaf-sheaths during winter (October and November). The eggs are oval, shining and pale - white or greenish in colour. The eggs are hatched into nymphs after 7 to 22 days of laying in summers but in winters the hatching period increases. The nymphs are dirty white in colour and are provided with a pair of anal trufts of 1mm. The nymphs starts suck the cell sap of the leaves. After five moults, the

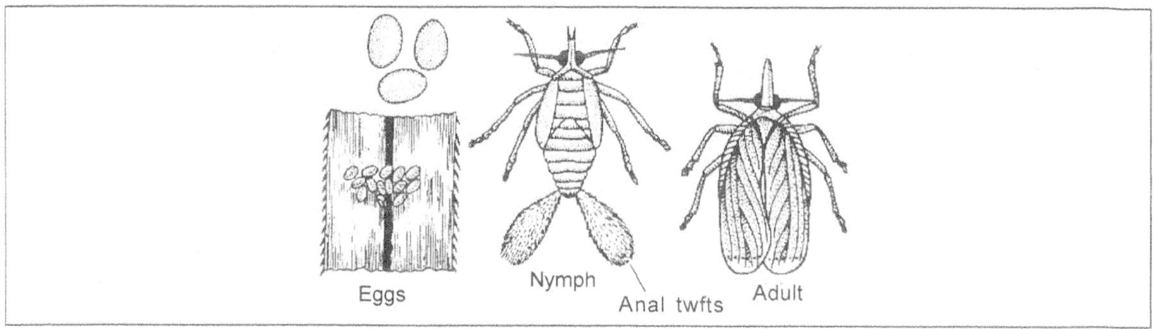

Fig. 2.47 Life History of Pyrilla perpusilla

nymphs are changed into imagoes and this takes about 6 to 8 weeks in summer and about 4 months in winter. The life span of the adult male is 5 to 7 weeks and that of the female about 5 to 8 weeks. In monsoon season the whole life cycle is completed in 40 to 60 days only.

Both nymphs and adults suck the cell sap of succulent leaves of sugarcane by their rostrum. As a result, the leaves become pale yellow and dry up. They secrete a sweet sticky transparent liquid which attracts and helps in growth of the harmful fungi resulting in reduction of rate of photosynthesis. It results in the reduction of sugar content to extent of 35%.

The pest is controlled through destroying the egg masses by burning, burying or spraying phenyl water, spraying of 0.25% Endosulfan or 0.025% Fenitrothion infields; the dusting of infested crop by 5 to 10% BHC dust and the spraying of 0.12% to 0.25% wettable powder of agrocide at the rate of 30 to 60 gallon per acre.

2. *Emmalocera depressella (Sugarcane Root Borer)*

This is a serious pest of sugarcane in eastern part of India. It is a major pest of sugarcane but in the absence of this crop the root borer feeds on sorghum, maize, sarkanda etc. The moth is dirty, pale brownish in colour and is not easily detected even in the infested crops. It is about 20 mm in length with wing-expanse. The white hind wings are shorter and wider in comparison to the fore-wings. The female moth on an average lays 230 to 285 eggs on the leaves, stems or on the ground. These eggs are scale-like, creamy-white and laid in the month of May. After 5 to 7 days the egg hatches into a young caterpillar of about 2mm in length. The caterpillar is pale yellow in colour and provided with 5 pairs of prolegs. Just after emergence it bores into the stem below the soil surface. With the help of their cutting and chewing type of mouth parts the caterpillars cut right across the stem resulting into the formation of dead-hearts. The larval stage is of about 35 to 45 days during which they attain maximum length of about 25 to 30mm. The borers, when left underground, attack the other sugarcane shoots and plants. The full-grown larva moves above the soil surface in the stem, makes an emergence hole in the form of a silken-tube above the soil surface and pupates inside the sugarcane. After 9 to 14 days of pupation period the moth emerges from the holes. The life cycle is completed in about 50 days. This pest causes heavy damage to the young sugarcane plants particularly from April to June. They feed voraciously on the stem below the soil surface resulting into the formation of dead-hearts. The young ones die and the older ones dry and fall down (Fig. 2.48). The attack of this pest causes 10 per cent reduction in cane production. The damage caused by this pest can be controlled by

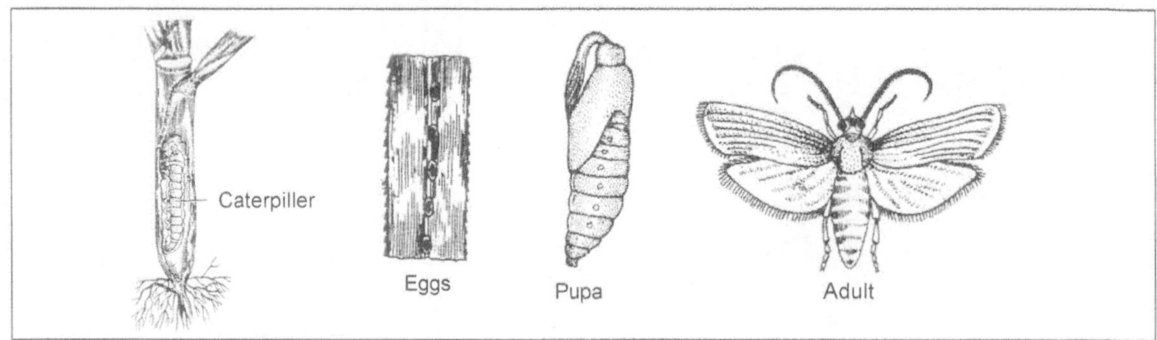

Fig. 2.48 Life History of Emmalocera depressella

removal as well as destruction of attacked shoot; raising resistant varieties of sugarcane, the moths should be light-trapped and destroyed, and killing the caterpillars; 10% BHC dust at the rate of 8kg per acre should be applied in the rows of cane and Aldrin at the rate of 10 litre per acre, mixed in the soil, can kill the eggs, larvae and pupae.

3. *Chilo infuscatellus (Sugarcane Shoot Borer)*

This pest is found throughout India and also causes damage to sugarcane crop in Afghanistan, Burma, Taiwan, Pakistan and Philippines. Sugarcane is the main host of this insect but it is also found feeding on bajra, maize, sarkanda and some other grasses. The moth has straw-coloured fore wings and hind wings are white in colour with apical light buff areas. The moth is about 30 to 40 mm in length with wing expanse. The female moth lays 10 to 35 eggs in clusters on the lower surface of sugarcane leaves. The eggs are scale - like and creamy-white in colour. The total number of eggs laid by a single moth is 300 to 400. After 4 to 5 days of laying, the egg are hatched into larvae called Caterpillars. The capterpillars, with the help of their cutting and chewing mouth parts, bore inside the shoot at the base of cane plant. They feed voraciously and attain full-grown stage in 3 to 4 weeks time. After full growth the larvae start the formation of a chamber for pupation and pupate there. After 5 to 7 days of pupation period the adult moths emerge from the pupae which come outside through an exit previously prepared by the caterpillars. The life span of moth is about 2 to 4 days. The whole life cycle is completed in 35 to 45 days (Fig. 2.49).

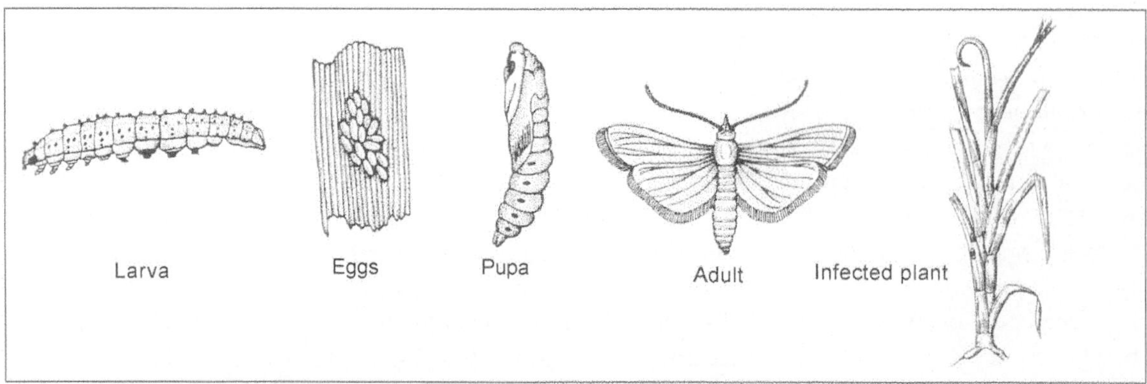

Fig. 2.49 Life History of Chilo infuscatellus

The caterpillar stage of this pest causes damage to sugarcane resulting in the formation of dead-heart. This pest causes 70% loss of young shoots of sugarcane. The shoots attacked by this pest should be removed from the field and destroyed; Moths should be trapped and destroyed 10% BHC dust, mixed with the soil, should be sprayed twice during April to June which can control the pest and the pre-sowing treatment with lindane or chlordane causes heavy mortality of the pest are some of the control methods.

4. *Tryporyza nivella (Sugarcane Top Borer)*

This is distributed throughout India, sugar cane is main host of this pest but it is also found feeding on sarkanda, kahi and other grasses. The moths are silvery white in colour and males are comparatively smaller than the females. Females are about 25 to 40mm across the wings when spread and provided with a tuft of yellow, orange or brownish silken hairs over the tip of the anal segment. (Fig. 2.50). The adult female moth lays about 500 elongated and oval eggs in clusters of 30 to 60 eggs on the lower surface of leaves of sugarcane. After 5 to 7 days of laying, eggs are hatched into caterpillars which are about 2 mm in length and are black headed. Just after emergence the young caterpillar starts boring into the mid-rid of a leaf and enters into the young terminal shoot through the rolled base of the top leaves. They feed voraciously upto 10 to 15 cm and destroy growing buds which results into dead-hearts. They attain full grown stage through five stages after 30 to 40 days of larval period. The full grown larva is about 30 mm in length, sluggish and creamy- white in colour. Full-fed caterpillar forms a characteristic chamber with an emergence hole just above the node. Now, larvae pupate in this chamber and after 7 to 10 days of pupation period they emerge out as moths. The life span of moth is of 4 to 5 days.

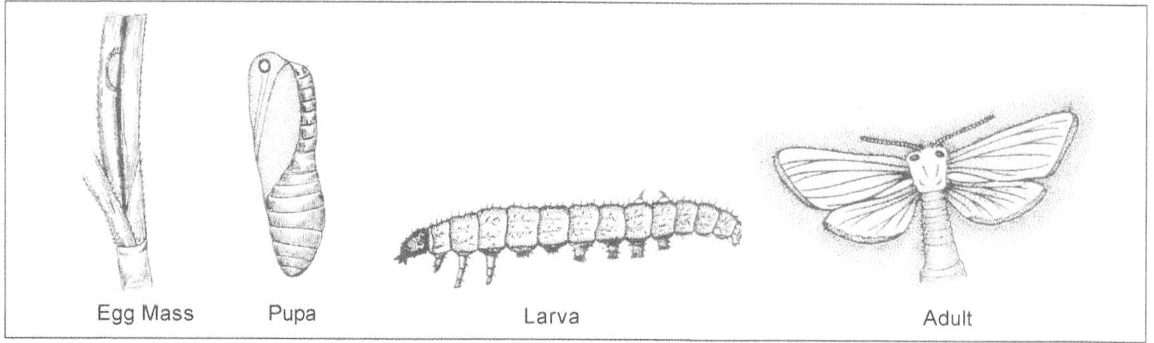

Fig. 2.50 Life History of *Tryporyza nivella*

The caterpillar feeds upon the growing buds in the months of March and April and further bores up to 4 to 5 nodes of the top-shoot resulting in drying of the leaves forming dead-hearts. The attack of this pest causes reduction in the cane crop and the quality of sugar is also affected. The pest is controlled by handpicking of eggs from green leaves and their destruction; attacked stems of canes containing dead-hearts should be removed and destroyed along with the caterpillars; moths should be trapped by light and destroyed; after harvesting in March, the strobbles and leaves should be burnt; the spraying of 0.02% endrin at the time of egg hatching is quite effective to kill the eggs and young larvae and dusting of 5% BHC is also effective.

5. *Aleurolobus barodensis (Sugarcane White Fly)*

The pest is distributed throughout the Indian subcontinent specially Uttar Pradesh, Madhya Pradesh, Bihar, West Bengal and Orissa. Its main host plant is sugarcane but it can also feed on sarkanda, wheat, barley and other grass plants. Adults are small fragile and pale-yellow insects provided with prominent black eyes. It is about 3mm long with wings expanded. The female is longer with heavy body in comparison to the males (Fig. 2.51). The fertilized females just after copulation lay 60 to 70 creamy white conical eggs in the months of November to December. After 8 to 10 days of laying. eggs are hatched into 0.36 mm long young nymphs of pale yellow colour. The nymphs with the help of their piercing mouth parts suck the cell sap of the leaves. They attain full growth through 4 instar stages which takes 25 to 30 days. The nymphs now enter into the pupal stage which is an exception in Aleurodidae because there is usually an incomplete metamorphosis in the Hemiptera. The pupal stage lasts for 8 to 10 days and then the adults are emerged. The life span of adults is about 2 days only.

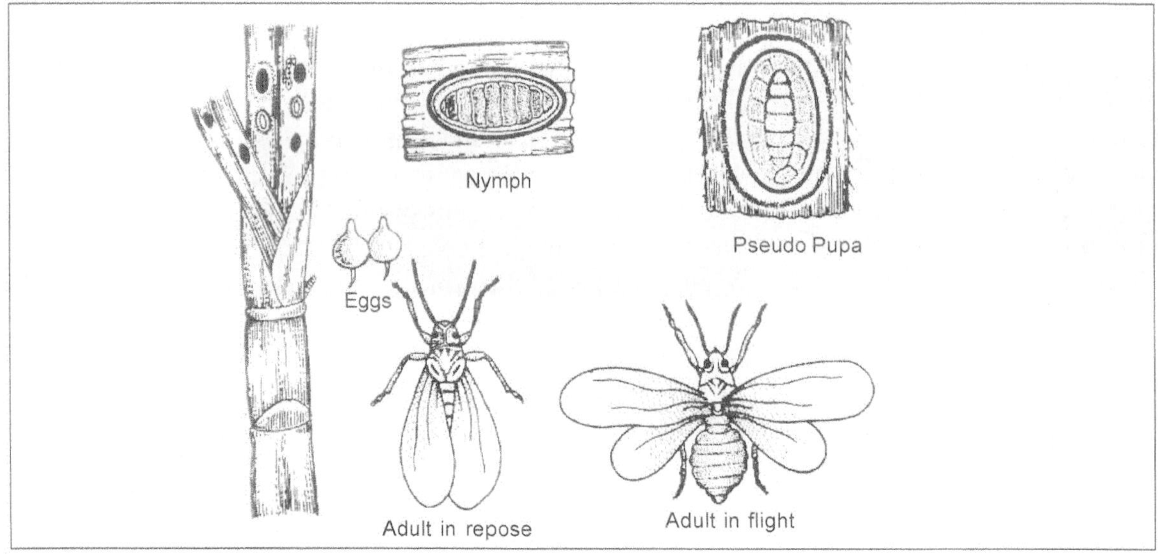

Fig. 2.57 Life History of *Aleurolobus barodensis*

The nymphal stage causes damage to canes by sucking the cell sap of the leaves with the help of piercing mouth parts. The maximum attack is recorded in the months from July to November. The sucrose content is decreased up to great extent. The pest control measures include balance of nitrogen content in field, attacked leaves should be collected and destroyed, resistant varieties of sugarcane should be planted; water-logging should be avoided; the spraying of 0.2% BHC, 1% malathion or 0.12% endrin is quite effective and three c.c. of folidol in one gallon of water should be sprayed to kill 90% population of the pest.

B. *Pests of Rice*

Rice is one of the important staple foods of India. There are many insects which feed on paddy causing damage. The important pests of rice are given in Table 2.17 and biology of major pests is described.

Table 2.8 *Pests of Rice*

Common Name	Scientific Name	Order	Family
Stem borer			
1. Yellow stem borer	*Sciropophaga incertulus*	Lepidoptera	Pyralididae
2. Paleheaded striped borer	*Chilo suppressalis*	Lepidoptera	Pyralididae
3. Dark headed striped borer	*Chilo polychrysa*	Lepidoptera	Pyralididae
4. Pink borer	*Sesamia inferner*	Lepidoptera	Noctuidae
5. Rice stem borer	*Tryporyza incertulus*	Lepidoptera	Pyralididae
Leaf feeders			
6. Rice gall midge	*Pachydiplosis oryzae*	Diptera	Cecidomyiade
7. Paddy gall fly	*Orseolia oryzae*	Diptera	Cecidomyiade
8. Rice case worm	*Nymphula depunctalis*	Lepidoptera	Pyralididae
9. Rice army worm	*Mythimna separata*	Lepidoptera	Noctuidae
10. Rice hipra	*Hispa (Dicladispa) armigera*	Coleoptera	Chyrosomelidae
11. Rice leaf hopper	*Cnapalocrocis medinalis*	Lepidoptera	Pyrallididae
12. Rice grass hopper	*Hieroglyphus banian*	Orthoptera	Acrididae
13. Rice butterfly	*Melanitis ismene*	Lepidoptera	Nymphalidae
14. Swarming caterpiller	Spotoptera mauritia	Lepidoptera	Noctuidae
Sap Suckers			
15. Rice bug	*Leptocorisa varicornis*	Hemiptera	Coreidae
16. Green leaf hopper	*Nephotettix virescens*	Hemiptera	Cicaedellidae
17. White backed plant-hopper	*Sogetella furcifera*	Hemiptera	Delpharidae
18. Rice mealy bug	*Heterococris rehi*	Hemiptera	Coccidae
19. Rice thrips root feeder	*Thrips oryzae*	Thysanoptera	Thripidae
20. Paddy root-weevil	*Echinonemus oryzae*	Coleoptera	Cuscilionidae

1. *Leptocorisa varicornis* (Rice Bug)

This pest is widely distributed in India, and tropical as well as subtropical parts of oriental region. It mainly feeds on paddy but has also been found feeding on millets, maize, bajra, jowar, ragi, sugarcane and some grasses. It is green or light brown in colour and about 20 mm in length. It has long legs, 4 jointed antennae and 3 jointed tarsi. They are very active in the morning and evening hours. The abdomen of the females is slightly swollen at the apex (Fig. 2.52). It breeds throughout the year and attacks paddy crops in the months of August to October. The adult female lays about 24 to 30 eggs in rows on the leaves and each row may have 6 to 20 eggs. The eggs are black, oval, flattenned having concave margins. These eggs are fixed on the leaf by a gummy substance secreted by them. It is about 1mm in length. After 6 to 7 days of laying, the eggs hatch into young nymphs which are pale green, long legged and devoid of wings. Within 15 to 18 days, young nymphs grow to become adult and during this process it pass through six stages. The life span of the adult bug is of 33 to 35 days. The nymphs and adults are found clustering on paddy crop, although they are not of gregarious nature. Five generations have been recorded in paddy growing season. The whole life cycle is completed in 4 to 5 weeks. This pest multiplies from July to November, hibernates from December to February and feeds on grasses from March to June.

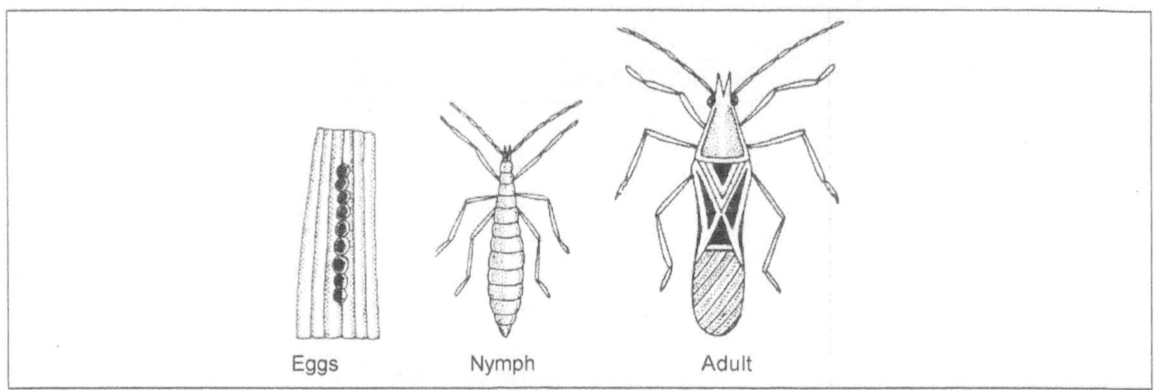

Fig. 2.52 Life History of *Leptocorisa varicornis*

This is a major pest of rice and attacks the paddy field at the milky stage. The nymphs and adults both suck the juice from developing grains in months of August to October as a result of which the ears become white and do not form the mature grains. The pest is controlled by following measures :

Insects should be picked up by hand net and killed by strong-crude oil emulsion, spraying crude oil emulsion in the fields, pests should be attracted by light trap and destroyed, 0.25% BHC dust should be sprayed before flowering and the dusting of paddy fields by 2.5% aldrin, 5% Chlordane and 5-10% BHC has been found to be useful for the control of the pest.

2. *Tryporyza incertulas* (Rice stem borer)

This pest is widely distributed throough-out the oriental region. It is a major pest of paddy in Andhra Pradesh, Orissa, Tamil Nadu and West Bengal. This yellow moth borer is a specific pest of rice and feeds only on paddy crop. The moth is 15-20 mm in length and has a wing expanse of 24-45 mm. The female moth is bigger than the male and has distinct black spot on each of the bright yellowish forewing. The male moth has pale whitish yellow coloured front wing without any spot (Fig. 2.53). The moths become active, fly and mate after dusk. The female moth lays about 200- 300 eggs during its life span of 5-7 days. The eggs are laid on the underside of tender paddy leaves in 2-5 clusters of 60 to 100 eggs each and are covered with a tuft of buff coloured hairs of the female. The freshly laid eggs are oval, flattened and dirty white in colour which become darker and turn brown at the time of hatching. The eggs after 6-8 days of incubation period, hatch into the tiny black-headed caterpillars. The newly hatched caterpillar suspends with the fine threads and disperses within a radius of 10cms with the help of air. The caterpillars, upon reaching the leaf sheath, enter through the space between leaf blade and inner sheath and soon bore into the stem. On reaching near the root they bore into the inner stem and start feeding inside causing 'Dead Heart'. During the heading stage of paddy the caterpillar bores directly into the peduncle and damages the inner walls of stalk causing '*white ear head*'. The larval period lasts for 20-30 days passing through 6 stages of different instars. The well grown caterpillar is smooth, yellowish white in colour and about 2.5 cm in length. Now the full grown caterpillar prepares an emergence-hole just above the water level in the field, pupates inside the attacked stem and emerges out as a moth after 9-12 days through the same hole. Whole life cycle takes 38-55 days.

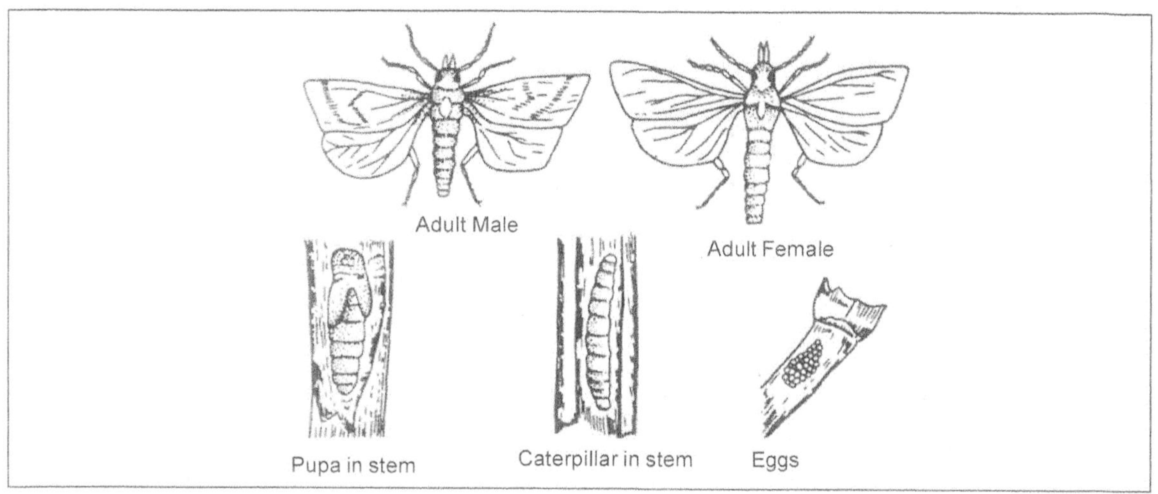

Fig. 2.53 Life History of *Tryporyza incertulas*

The caterpillar stage of this pest causes damage to paddy crops. They bore into the stem near the root causing 'dead heart' of central shoots which easily comes off when pulled. The plants attacked in early stages are killed altogether, whereas, in the plants infested during flowering stage, the ear heads dry up which are devoid of grains and are called as 'white ears'. An average loss of about 100-500 kg. of paddy per hectare has been recorded from Andhra Pradesh. The damage caused by this pest is severe in southern states of India but in Norther states the damage is not serious. The control measures of this pest are the spraying of 0.08% methyl parathion, 0.05% phosphamidon, 0.6% diazion granules after 20- 30 days of intervals can control the pest, Twelve kg. of 6% lindane granules should be broadcast in the paddy field containing waterup to 15 cm deep. Three replicates are necessary at the interval of 30 days after harvesting, paddy fields should be ploughed in order to destroy the larvae in the stubbles, stubbles should be burnt to destroy the hibernating larvae, and resistant varieties of paddy should be used for sowing.

3. *Hieroglyphus sp.* (Kharif grasshopper)

Kharif grasshoppers are widely distributed in the paddy areas of India, they are major pests of paddy crops. But due to polyphagous nature they feed also on maize, millets, sugarcane, jowar, bazra, sunhemp, arhar and grasses, the adult is greenish dry grass coloured yellow hopper 4-5 cm in length. There are three black lines running laterally on either side of the thorax in *H. nigrorepletus* but only 2-3 black markings in *H. banian.* Both the species are found in full winged and small winged forms (Fig.2.54). Female lays egg in masses (Egg pod), each containing 30-40 eggs, by inserting her abdomen in the soil 5-8 cm deep along the sides of the field bunds in the months of September to November. The hatching of eggs takes place in the month of June or early July of the following year, a few days after the first shower of the monsoon. The newly emerged nymphs are generally green. They start feeding actively on grasses and attain full grown nymphal stage by the middle of September within 8-10 weeks by casting off their skin 5 or 6 times at every interval of 10-15 days. The young

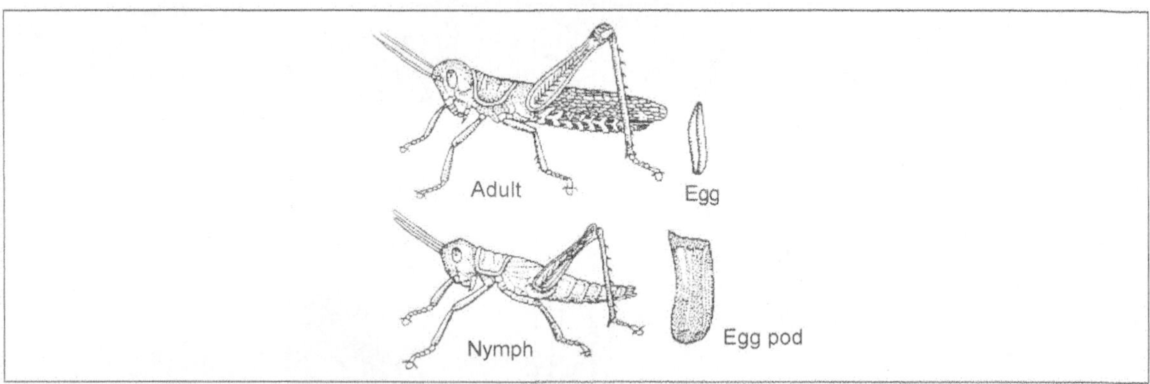

Fig. 2.54 Life History of *Hieroglyphus nigrorepletus*

hoppers are like the adults except for the absence of wings. Only a few days after attaining adulthood they start mating in September or early October. The male dies after copulation while female starts egg laying and dies after oviposition.

Both, adult and nymph damage paddy crop by feeding on leaves and shoot. At the time of severe infestation the plants are completely defoliated which affect the growth of plant thereby resulting into low yield. In epidemic form, they completely damage the crops moving from field to field over large areas. Millets, sugarcane, grasses and other graminaceous plants are also damaged. The control measures of this pest include :

Hoppers and adults should be collected and destroyed, 0.02% dieldrin, 0.04% heptachlor and methyl parathion are used for effective control, the affected field should be ploughed to a depth of 10-15 cm after harvesting the crop destroying eggs, 30% aldrin at the rate of 25 ounces per hectare has been found to be much effective.

4. *Pachydiplosis oryzae* (Rice gall midge)

The rice gall midge is a sporadic major pest, found in India, Maharashtra, Mysore, Orissa, Uttar pradesh and West Bengal. It feeds only on paddy crop. It is a mosquito sized small fly. The body of the female is bright red coloured and provided with stout abdomen, whereas, males are more slender and dark coloured. The adults are found to be very active during night and hide themselves during day time (Fig. 2.55). The adult female lays reddish elongated eggs singly or in clusters of 2-6 eggs on

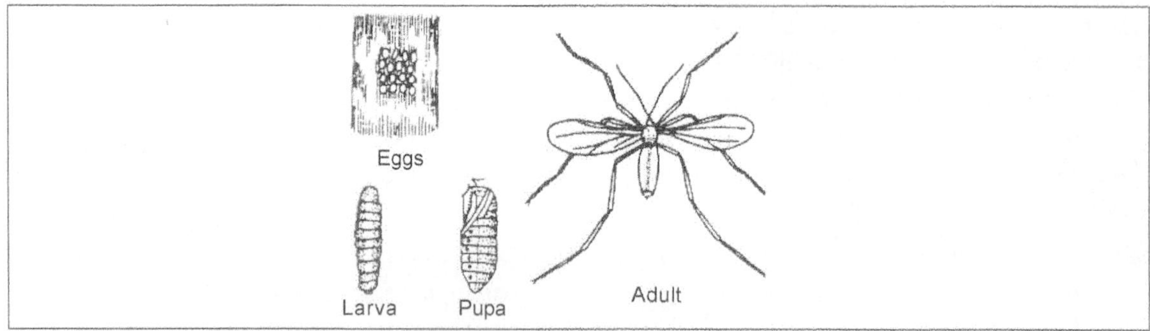

Fig. 2.55 Life History of *Pachydiplosis oryzae*

the undersurface of the paddy leaves. The newly laid eggs are reddish, elongated and broader at one end, measuring 0.5 - 0.6 mm in length. After few days, eggs become chocolate coloured and after 2-4 days of incubation period they hatch into coloured larvae of about 1 mm length. The tiny larva crawls on the leaf sheath till it reaches the base of the shoot. Now larva bores into the stem near the growing point (bud) and after feeding for about 10 days it pupates. The pupa wriggles its way up with abdominal spines and cuts a hole on the tip of gall from which the adult emerges after 4- 7 days. The life span of adult is of 3-4 days. The life cycle is completed within 15-23 days.

The larval stage causes damage to the crop. The extent of damage caused by this pest varies from region to region. The maggot reaches the apical point of shoot for feeding which causes a number of physiological changes thereby hampering the growth of plants. The central shoot of tiller becomes hollow which is followed by the gall formation on the basal portion. The gall formed in young plants are not visible due to leaf sheath while in advanced stage of plants it is visible very clearly like projection. Now entire central shoot is turned into a hollow tube of silvery green colour (Silver shoot) and the tip remains green. Heavy attacks are recorded in water logged areas. The control measures include : Early planting of crop can minimize the damage, adults are attracted towards light, so light trap should be managed in the field, granular application of Diazinon, Sevidol and Thimet 10 G 1.25 kg per hectare is very effective and the spraying of phosphamidon 0.05%, carbaryl 0.2%, diazinon 0.05%, dimethoate 0.02% and lindane 0.5% reduces pest population very effectively.

5. *Mythimna separata (= Pseudoletia separata)* (Armyworms or cutworms)

It is distributed throughout the Oriental region. Cutworm is a polyphagous insect found and feed on paddy, barley, maize, sugarcane, wheat and other millets. The full grown moth is pale brown in colour with a wing expanse of 3.5 cm in length. Front wings are provided with white spots in its centre (Fig. 2.56). They are found to be very active during night and remain hidden during the day time. Female moth, after mating, starts egg laying. The eggs are laid in clusters on the lower leaves and are covered with greyish hairs. After 4-5 days of incubation period eggs are hatched into young caterpillars of pale green colour. The caterpillars, up to half grown stage, are very specific because of its looping habit of crawling. They are found feeding in large number in field crops and grasses also.

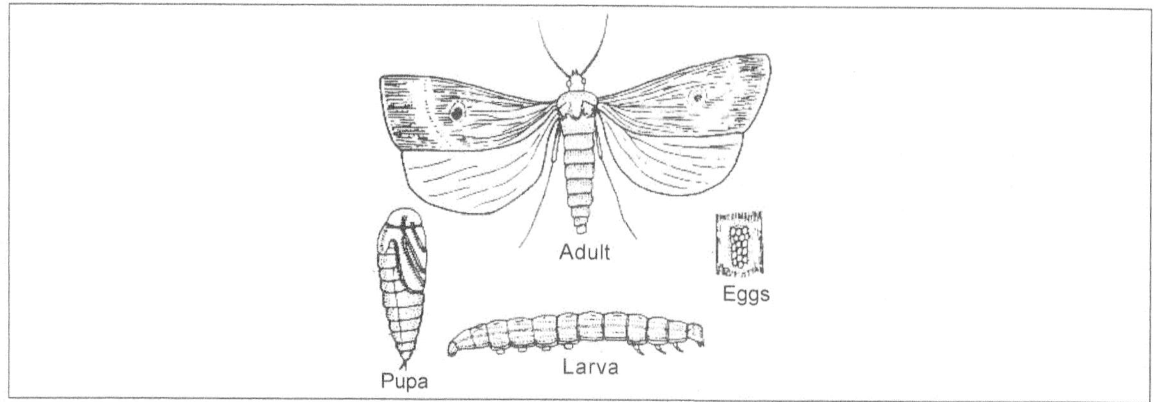

Fig. 2.56 Life History of *Mythimna separata*

The crawlers take rest at the base of clums during day time and become very active for feeding with sunset. They climb on the plants, cut rachides and maturing ear-heads resulting in severe damage of the field crops. When they feel shortage of food in the damaged area they move together in large numbers like armies and further attack the paddy crop in the nearby fields.

The caterpillars after maintaining the full grown stage measures about 3.5 cm in length. The larval period lasts for 21-28 days passing through five successive larval instars. The full grown fifth instar caterpillars leave the plant and construct earthern chamber after entering into the soil and get pupated therein. After 10-12 days of pupa period, moth emerges out. Five generations have been recorded in single year. Its larval stage is the destructive phase of the life cycle. A number of crops are destroyed by army worms. The appearance of pest has been noticed after heavy rains and flood. The caterpillars attack the crop in the night and cut off half ripe ears. In day time they keep themselves hidden in the folds of leaves and cracks in the soil. The infestation period is from September to the second week of October. The control measures include :

Caterpillars should be hand-picked and killed. Grasses should be removed from the field. Planting should be adjusted for the harvesting of crop by the middle of October. Dusting with 2% toxaphene or 5% aldrin dust @30-35 kg. per acre is effective, spraying with 0.04-0.05% of diazine or malathion or parathion or phosphomidan @800 litres per acre is very effective to minimize the pest population.

6. *Spodoptera mauritia* (Swarming caterpillar)

The swarming caterpillar are found throughout Tropical and Sub-tropical parts of the world. They feed on paddy, jowar, maize, wheat and sugarcane. The adult moth is medium-sized black spotted and greyish brown in colour. The forewings are provided with an irregularly waved light line. Female moth after mating, starts egg laying, in clusters of 100-400 eggs; on the lower surface of leaves. After 6-8 days of incubation period eggs are hatched into caterpillars and start feeding at once on the young leaves of the host plant. The young caterpillar, after 4-5 moults, attains full grown stage and measures about 3cm in length. The caterpillar stage lasts for 3 weeks and full fed caterpillars crawl down to the ground and pupate in the soil. After 10-12 days of pupation period adult moth emerges out from the pupa (Fig. 2.57).

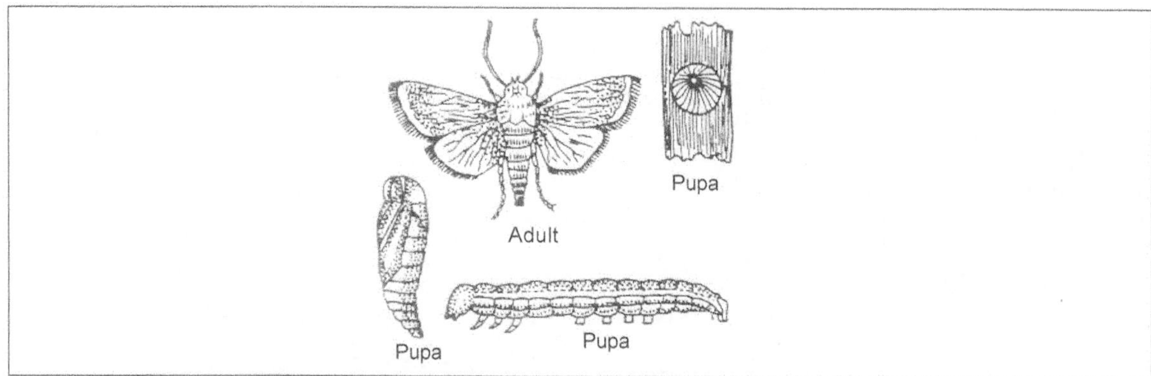

Fig. 2.57 Life History of *Spodoptera mauritia*

The larval stage causes destruction of the crops. The caterpillars feed on the foliage of young seedlings in the nursery and transplanted plants. The caterpillars are nocturnal and feed only at night. The caterpillars appear in swarms, therefore, entire crop of the field is damaged severly. The control measures of pests are 30 to 35 kg. per hectare.

C. *Pests of Cotton* (Gossypium spp.)

Cotton is one of important commercial crops of India. The following are the major pests on cotton (Table 2.9).

Table 2.9 *Pests of Cotton*

Common Name	Scientific Name	Order	Family
1. Pink Boll worm	*Pectinophora gossypiella*	Lepidoptera	Gelechiidae
2. Spotted Boll worm	*Earias vitella*	Lepidoptera	Lymbidae
3. Cotton jossid	*Amracia devastans*	Hemiptera	Cicadellidae
4. Red cotton bug	*Dysdercus koenigii*	Hemiptera	Pyrhocoridae
5. Red cotton bug	*Dysdercus cingulatus*	Hemiptera	Pyrhocoridae
6. Dusky cotton bug	*Oxycarenus laetus*	Hemiptera	Lygaeidae
7. Cotton white fly	*Bemisia tabaci*	Hemiptera	Aleyrodidae
8. Cotton leaf roller	*Sylepta derogate*	Lepidoptera	Pyraustidae
9. Cotton grey wevil	*Myllocerus maculosus*	Coleoptera	Corculionidae
10. Bud moth	*Phycite infusella*	Lepidoptera	Pyralididae
11. Cotton aphid	*Aphis gossypii*	Hemiptera	Aphididae
12. Cotton stem weevil	*Pempheris affinis*	Coleoptera	Curculionidae
13. Cotton semi looper	*Tarache notabilis*	Lepidoptera	Noctuidae

The biology and control measures of three important pests of cotton are described below :

1. *Pectinophora gossypiella* (pink boll worm)

It is most destructive pests of cotton, cosmopolitan in distrubution. Cotton is the main food plant of this pest. It is a small grey brown moth measuring 3-9 mm and possess black spots on forewings. The antenna are filliform. The moth hides during day and active afterwords. The female lays elongated flattened eggs (300) in clusters of 2-10 on the boll, leaves, flowers and buds. After 4-23 days of incubation, eggs are hatched into caterpillars with darkhead and greyish prothoracic head. The caterpillar bores into bud, boll, shoot and flowers. It grows to the length of 12 mm with pink body and brown head. After 10-30 days, they pupate in a silken cocoon in soil or under fallen leaves. After about 7 days of pupal period adult moth emerges out and starts mating the next night, oviposits by 4th night after which female dies. The whole life cycle is completed in 3-8 weeks (Fig. 2.58). The pest passes winter as larva in the cocoon and larva pupate in March, adult emerge in first week of April. The caterpillar is the infective stage and during months of December and January 7-100% bolls are damaged. The attacked seed cotton gives poor cotton. The measures of prevention and control are; the seed of cotton should be treated with fumigants before sowing, cotton plants infested should be destroyed, and the pest population can be reduced by spraying carboryl 0.1%, and endosulfan 0.05% at 15 days interval during August and September.

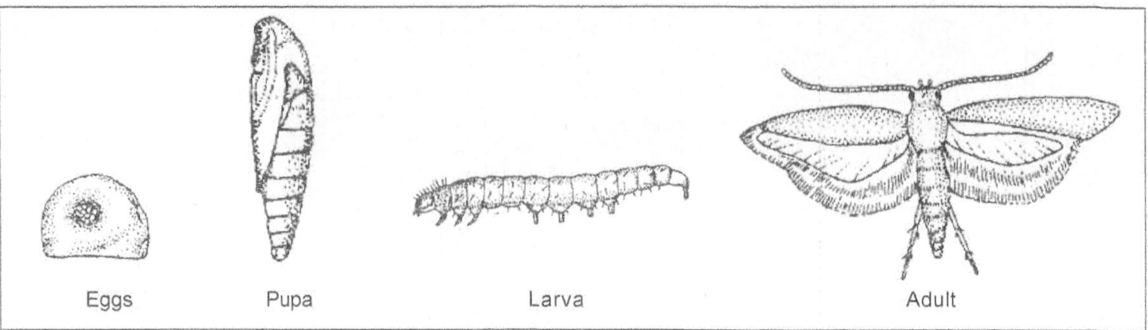

Fig. 2.58 Life History of *Pectinophora gossypiella*

2. *Dysdercus koenigii (D. cingulatus)* (Red cotton bug)

Cotton is the main host plant of this pest and it is commonly found in India (is a serious pest of cotton in Andhra pradesh, Uttar pradesh, Bihar, Tamil Nadu and Maharashtra), Australia, Phillippines and Srilanka. It is a blood red pest. Female are longer (15 mm) than the males (12 mm). They are characterized with white bands across abdomen, black colour of the forewings, antennae and scutelium. Moth pest is active throughout year and adult stage passes in cluster of 70-30 eggs each under moist surface. After about 7 days in warm and moist weather, eggs are hatched into active 1 mm long red coloured nymphs. The nymphs live gregariously on the cotton bolls. The nymphs attain adult stage passing through 5 moults within 50-90 days (Fig. 2.59). The bug sucks cell sap of green bolls of cotton and leaves, resulting in lint of poor quality. The infested seeds have low percentage of germination. The control measures of pest include bugs are handpicked and killed in kerosenised water, spraying of malathion or phosphosmidon, Fenithrothion etc.

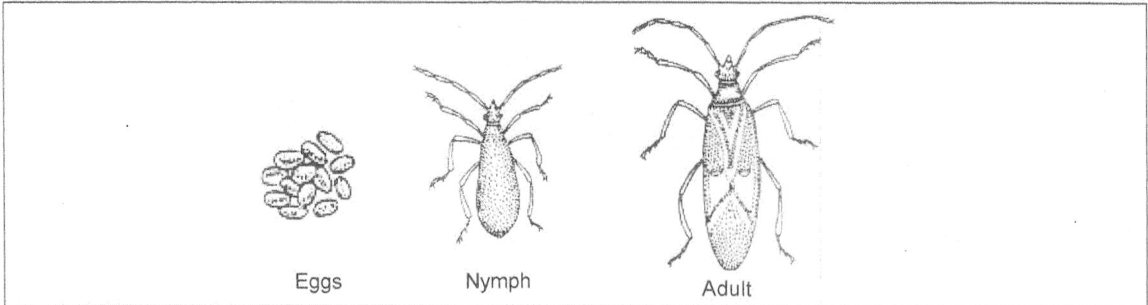

Fig. 2.59 Life History of *Dysdercus koenigii*

3. *Earias vitella* (Spotted boll warm)

The two species of spotted boll worms (E. vitella, and E. insulana) are distributed in India, Pakistan and North Africa. *E. vitella* is a small moth of about 12 mm in length. The fore wings are of greenish and hind wings are of white colour. The adult moth appear in April, lays 200-400 eggs singly on bolls, shoots, flowers and buds. The eggs are spherical and after on incubation period of 3-4 days in summer and 7 days in winter, eggs are hatched into brownish white caterpillars provided with a dark head and prothoracic sheild (13 mm in length). They start boring into the tender shoots and later on into flowers and bolls. The caterpillar passes through 6 stages of instars and becomes full grown in 10-15 days, (20 mm in length). The full red caterpillar moves out of boll and pupates in tough grey silken cocoon. The pupat periods last 15-21 days (in water 80 days) and moth emerges out (Fig. 2.60). The whole life cylce is completed in about one month in summer. The pest carried from year to year by roots of cotton. The infective stage is caterpillar. The caterpillars attack the flower, bud and boll. The infested bolls open prematurely and produce poor lint. The control measures of pest are : the attacked shoots and bolls should be collected and destroyed, resistent varieties should be grown, and Carboryl (0.1%) or Fenitrothion can be sprayed.

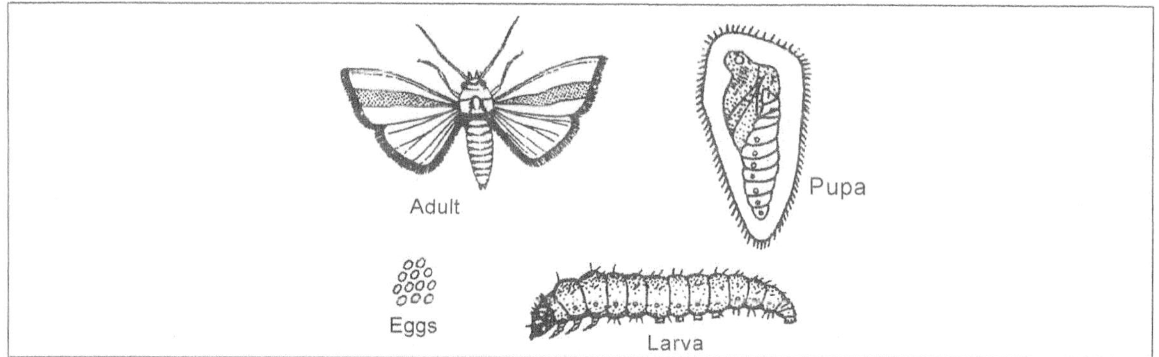

Fig. 2.60 Life History of *Earias vitella*

D. *Pests of Groundnut*

Groundnut is one of the major oil seed of India. The following table gives the list of important pest on groundnut (Table 2.10).

Table 2.10 *Pests of Groundnut*

Common Name	Scientific Name	Order	Family
1. Groundnut aphid	*Aphis craccivora*	Hemiptera	Alphidae
2. White grub	*Holotricha consanguinea*	Coleoptera	Melolonthidae
3. Groundnut leaf miner	*Stomopterx nertaria*	Lepidoptera	Geleciidae
4. Grondnut-stem borer	*Sphenoptera pesotetti*	Coleoptera	Buprestidae
5. Red hairy caterpiller	*Amsaeta moorei*	Lepidoptera	Arcitidae
6. Groundnut earwig	*Euborellia stali*	Dermaptera	Forficulidae
7. White ant	*Odonotermes obesus*	Termitidae	Isoptera

The biology of major pests of groundnut and their control measures is described below :

1. *Aphis craccivora (Groundnut aphid)*

It occurs throughout India, more prevalent in Gujarat, Maharashtra, M.P, A.P and Tamil Nadu. Besides groundnut it also attack peas, pulses and beens etc. It has minute soft body with greenish black colour. They are viviparous and produce young ones which undergo four moults to become adults. Adult start giving birth within 24h and have a life span of 10-12 days. They breed throughout the year. Both nymphs and adults suck plant sap and prevent pod formation (Fig. 2.61(a)). It also results in stunted growth of plants, falling of this insect include use of organophosphorous insecticides like 0.04% monocrotophos or 0.03% metastox or 5% malathion is also effective.

2. *Stomopteryx nertaria* (Groundnut leaf miner)

It is distrubted all over India, particularly in A.P, Karnataka and Tamin Nadu. Besides groundnut, it also attacks soyabeen, pigean pea or arhar. The adult moths are small, 1 cm long, dark brown insects with white spots on margins of forewings (Fig. 2.61(b)). The eggs, in hundreds are laid on leaves and shoots. Newly hatched larvae mine into the leaves. They entangle several leaves together and feed inside leaf. Reddish brown pupa is formed inside the mines. The insects are controlled by spraying DDT.

3. *Holotrichia consanguinea* (white grub)

It is found all over India. It is large sized, copper colored and nocturnal beetle. The grubs are white coloured. The eggs are laid in loose sandy soil on the onset of monsoon. Full grown larva (grub) migrates underground and pupates (Fig. 2.61(c)). Adults remain at a depth of 10-20 cm and come

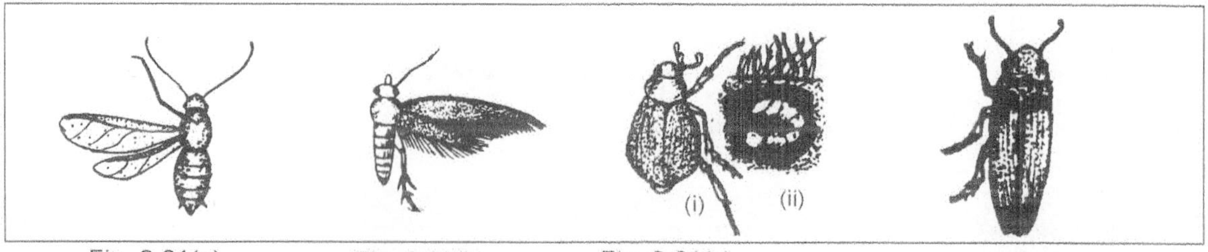

Fig. 2.61(a) Aphis craccivora

Fig. 2.61(b) *Stomopteryx nertaria*

Fig. 2.61(c) *Holotrichia consanguinea* (i) adult beetle (ii) underground grub

Fig. 2.61(d) *Spenoptera perotetti*

out during the night for feeding. The damage is done by both adults and grubs. The adult beetles feed on the foliage and are capable of defoliating plants and even trees of banayan and neem. The grubs feed on roots and rootlets killing the plants. The control measures include - collection and destruction of beetles, field sanitation and use of soil insecticides.

4. *Spenoptera perotetti* (Groundnut stem borer)

It is distrubuted throughout India. Besides groundnut, it also attacks sesamum and pulses. The adult is metallic brown in color and small in size (upto 1 cm). The full grown grub is whitish in colour (Fig. 2.61(d)). The eggs are laid on the main stem of the plants. The slender pale white grubs, (flat-headed grubs) bore into the branches, eat their way into the main stem and ultimately reach the top root. Pupation occurs inside the burrows from which the adult beetles emerge by cutting their way out. Grubs damage the plant by feeding on contents of stem, branches and roots of the plant. The control measures include - the beetles should be collected and destroyed, infected plant should also be destrobyed or chemical treatment include spray of 0.25% of BHC or DDT at larval stage.

2.12 Photosynthesis

Introduction

Photosynthesis is the most important metabolic reaction of the plants, by which they release molecular oxygen in to the atmosphere and produce chemical energy required for their activity. Historically it was Priestly (1770) who for the first time performed the experiments with leaves of mint in a closed chamber with living mouse, burning candle as well, and demonstrated the release of the oxygen that served the oxygen need of the burning candle and mouse. He concluded that the process that is occurring in the green plants is exactly reverse of the respiration where the oxygen is consumed. Engelmann (1880) discovered that chloroplasts are responsible for the evolution of oxygen. By the microscopic studies in bacteria and other microbes including algae, it was shown that the microorganisms present. It was concluded that the chloroplasts are the oxygen producing sites hence were attracting the bacteria. Subsequently Ingenhouz noted that the green parts of plants produce oxygen only in the presence of light. During early nineteenth century quantitative studies of oxygen and carbon dioxide were performed. Together with Robert Mayer's recognition that sunlight supports the formation of the organic matter has led to the general equation :

$$CO_2 + H_2O \rightarrow O_2 + \text{Organic matter}$$

Plant cells have specialized cell organelles known as the chloroplasts which trap the solar energy and convert it in to the chemical energy (ATP, NADH). This photochemical reaction occurs in two stages 1. Production of ATP, NADH and 2. Reduction of carbon dioxide. These two results in the production of glucose, which in turn undergoes a catabolic reaction to produce ATP which will serve the plant's energy needs.

Photosynthesis occurs not only in Eukaryotes but also in Prokaryotes like blue-green algae, green sulphur bacteria and purple bacteria Chlorobium (aerobic) and Chromatium (anaerobic) respectively, also occurs in the phytoplankton like diatoms, dinoflagellates and microscopic algae.

Photosynthesis is essentially a photochemical reaction mediated by solar energy, converting atmospheric carbondioxide and water into carbohydrates and release molecular oxygen into atmosphere. The reaction can be represented as :

$$6CO_2 + 6H_2O \xrightarrow{\text{light}} C_6H_{12}O_6 + 6O_2$$

Site of Photosynthesis

The Chloroplast is the structural and function unit of the photosynthetic machinery, which accommodates the apparatus required for the photochemical reaction to occur.

Chloroplast

It is a membrane bound organelle specific to the plant cells, which is autonomous with its own genome (self replicating). They have the dimensions of 1-10 mμ in diameter. The shape varies depending upon the location and can be globular or discoid, ribbon like (*Spirogyra*). It has a fragile outer membrane, a continuous folded structure called lamellae enclosing a compartment called stroma. Lamellae form the flattened membranous structures called thylakoids; they occur in the form of the stacks and called as grana, the paired membranes between the grana are called as grana integrenal lamellae. The thylokoid membrane and the integrenal lamellae are equipped with the photosynthetic pigments, and the enzymatic machinery required for the photosynthetic process. The photosynthetic bacteria do not have the chloroplasts, so the light receptor systems are located in cell membrane of the special structures known as chromophores derived form the cell membranes.

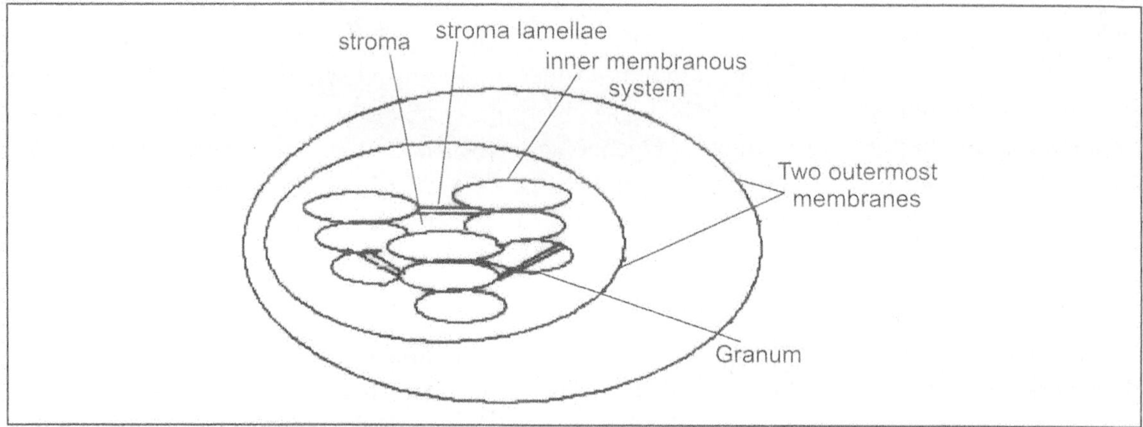

Fig. 2.62 Structure of chloroplast

Pigments in Photosynthetic cells

Photosynthetic cells contain primary photosynthetic pigments known as chlorophylls which are green in color, and also contain some accessory pigments which are blue, red, purple or yellow. Chlorophyll absorbs the light primarily and passes to the photoreaction centers that the present in the chloroplast. The accessory pigments phycobilins (blue, red) and carotenoids (purple, yellow) also help in trapping the solar energy and passing it to photo reaction centers indirectly.

Chlorophyll

Chlorophyll, a green colored pigment that is present in the photosynthetic cells, generally occurs in two forms; in higher plants Chlorophyll *a* and chlorophyll *b*. Chlorophyll *a* is a porphyrin ring with a Mg^{+2} ion forming a coordination complex, which is stable planar. It also has a terpenoid attached to the 7th carbon of the IVth pyrrole ring, with an alcohol (Phytol) esterified to propionic acid.

Chlorophyll *b* is predominant in the blue - green algae, diatoms and chlorophyll *c* is dominant in dinoflagellates. Photosynthetic prokaryotes also contain chlorophyll but they do not evolve oxygen. The chlorophyll in the photosynthetic bacteria is known as *Bacteriochlorophyll* which differs from the chlorophyll *a* with ring I and ring II which are acetylated and reduced respectively unlike chlorophyll *a*. Due to the conjugated and reduced respectively unlike chlorophyll, plants absorb visible light efficiently and support the photosynthetic process of autotrophs.

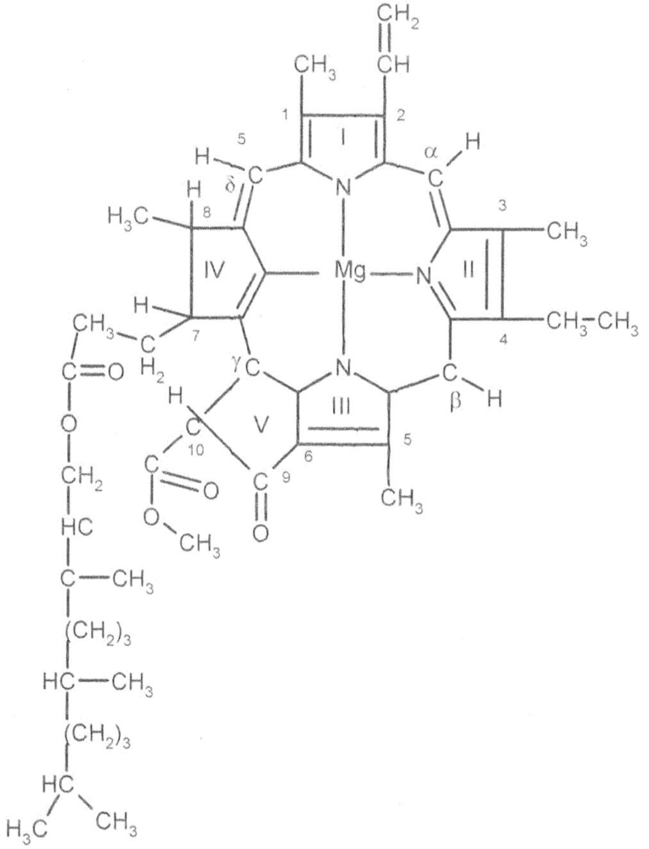

Fig. 2.63 Chlorophyll *a*

Phycobillins

The phycobilins are the major light absorbing pigments in the red algae, which contain small portions of the chlorophyll, but have *phyco-erythrobillin*, a red phycobillin. As these are the accessory pigments, have absorption at different wavelength than that of chlorophyll they serve as supplementary light absorbing systems. The light thus absorbed should ultimately be transferred to the chlorophyll to

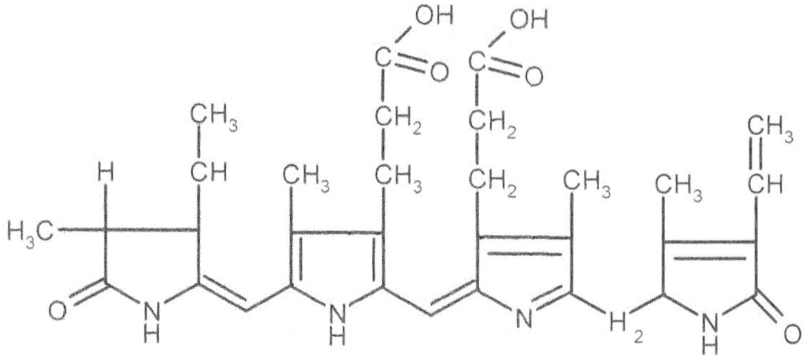

Fig. 2.64 Phycoerythrin

support the photosynthetic process. They are linear tetrapyrrols like that of the chlorophylls, but they lack the Mg^{+2} ions and are conjugated with the specific proteins, common examples are phycoerythrin (red) and Phycocyanin (blue).

Carotenoids

Carotenoids are multiple isoprene molecules having an unsaturated cyclohexane ring at the end. Two major classes of carotenoids are carotenes, xanthophylls which are deoxygenated and oxygenated isoprenoid hydrocarbons respectively.

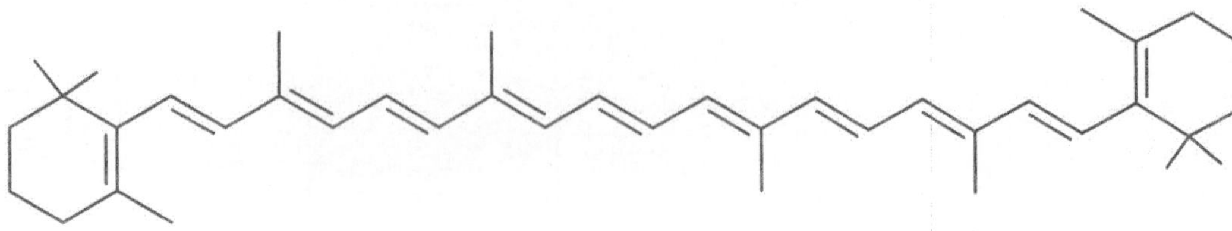

Fig. 2.65 β-Carotene

Mechanism of the Photosynthesis

The photochemical reaction occurs in the two stages :

1. Production of ATP and NADH (Light Reaction).
2. Reduction of carbon dioxide utilizing the ATP and NADH (Dark reaction)

Production of ATP and NADH

The hydrogen atoms from the water molecules are removed leaving the molecular oxygen and is utilized in the reduction of the $NADP^+$, simultaneously the ADP is phosphorylated to ATP.

$$H_2O + NADP^+ + Pi + ADP \xrightarrow{\text{light}} O_2 + NADPH + H^+ + ATP$$

These light reactions occur in the thylakoids, because of the two photoreactions centre present in them called as **Photosystem-I** and **Photosystem-II** (PS-I and PS-II). These PS-I and PS-II absorb the visible light of range 700nm and 680nm respectively and utilize this energy in the excitation of the electrons. The excited electrons are transferred from one to the other electron acceptors, cyclically or non-cyclically, and lead to the formation of the NADH and ATP.

Hill Reaction

This process of spliting of water molecule (*Photolyses*) resulting in release of oxygen and simultaneous reduction of the eletron acceptors is known is Hill reaction, named after its discover, Robin Hill. Hill and Scarisbrick showed that the isolated chloroplasts could release O_2 in the presence of light, provided a suitable electron acceptor (EA) like Ferric salt (FE^{3+}) is available.

$$H_2O + EA \xrightarrow{\text{light}} \tfrac{1}{2} O_2 + \text{Reduced EA}$$

Red drop

The sharp drop in the efficiency of photosynthesis when the photosynthetic cells are illuminated with the far red light above 680 nm in known as Red Drop.

Emerson's Enhancement Effect

The enhancement in the efficiency of the photosynthesis with supplementation of the light at 710nm with 670 nm is known as Emerson's Enhancement Effect. The drop in the efficiency of photosynthesis at 680nm and enhancement at 710nm when supplemented with 670nm indicated that there are two Photosystems that are responsible for the above mentioned effects. L.N.M Duysens designated the system which is near to 710nm as PS-I and the one near to 670nm as PS-II.

Activation of the Photosystems

Each photosystem of the photoreaction centers are activated by different range of the visible light, an the electron transfer occurs from the PS-II to PS-I. The transfer of electrons from PS-II and PS-I is known as a *non-cyclic* electron transport or Z-scheme, where the electrons ejected by the activation of PS-II with the 380nm visible light passes through a chain of the electron carriers to the plastoquinone molecules, results in the conversion of plastoquinone to plastoquinol, which reduces the cytochrome b_6 and cytochrome f with the simultaneous transfer of protons into the thylakoid lumen. Then the cytochrome transfers of protons into the thylakoid lumen. Then the cytochrome transfers the electrons the plastocyanin, which activates the PS-I oxidized with the 700nm of light. The electron thus released from the PS-I encompasses the chain (A0, A1, A, B, Fd) of electron carriers and reduces $NADP^+$ to NADPH. The protons thus released from the splitting of water molecule by the oxygen evolving complex for the release of an electron which is transferred to PS-II and the protons released by the reduction of the cytochrome b6 f powers the production of chemical energy ATP by the ATP synthase complex.

In Cyclic electron transport, the electrons return from the chain of electron carriers in PS-I to the cytochrome b_6 f complex.

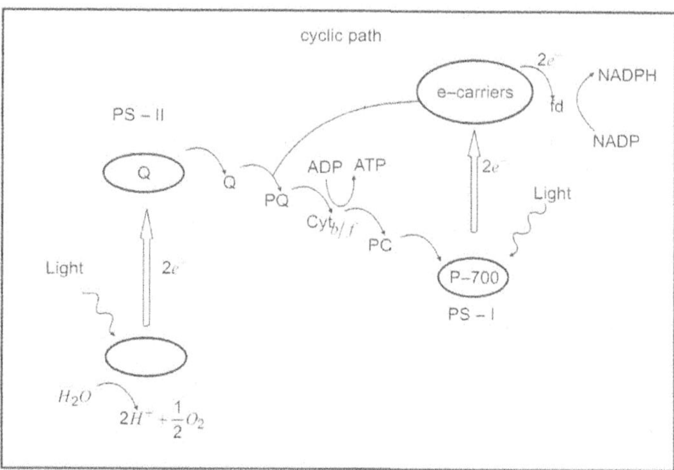

Fig. 2.66 Non-cyclic and cyclic electron transport; Q = quinol Pool; PQ = plastoquinone; PC = plastocyanin; fd = ferridoxine

Reduction of Carbondioxide / C_3 Cycle

The NADH and ATp thus produced in the light induced reactions are utilized in the reduction of carbon dioxide. The reduction of Carbon dioxide does not require the light hence termed as the Dark Reaction, these reaction can be divided in to two two phases

1. Energy input phase, where the ATP and NADH are utilized in the production of GAP (glyceraldehyde 3-phosphate), often is termed as productive phase.
2. Reforming phase, where the starting compound the Ru5P (Ribulose-5 phosphate) is reformed with the shuffling of the carbon atoms of GAP in the precise array of reactions.

The first reaction of the first phase of this cycle is phosphorylation of the Ru5P by *phosphofructokinase* to form Ribulose-1, 5 biphosphate (RubP). The RubP undergoes carboxylation (Carbon fixation), resulting in the formation of 3PG (3-phosphoglecerate) which is further phosphorylated to form 1, 3-BPG (1,3-bisphophoglycerate) which reduces to form GAP.

The second phase is initiated by the isomerization of the GAP to DHAP (Dihydroxyacetone phosphate) by *troise phosphate isomerase*, thus formed DHAP is tracked to two different paths where it undergoes an aldol condensation followed by the phosphate hydrolytic reactions which are catalyzed by fructose bisphophatase and sedoheptulase respectively, the remaining reaction of the Calvin Cycle are catalyzed by the enzymes of the pentose phosphate pathway like *transketolase*, *phosphopento epimerase*, *ribose phosphate isomerase* and *ribulose* bisphosphate *carboxylase* etc. To yield respective products depending upon the substrate and produce to the final compound Ribulose5-phosphate which is recycled in to the Calvin pathway i.e.

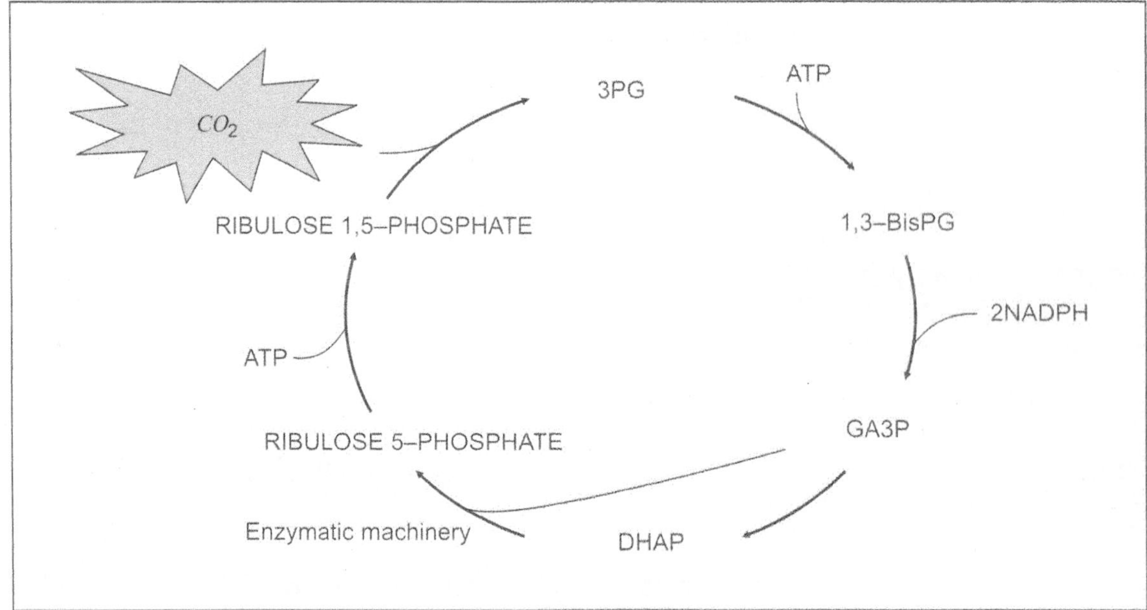

Fig. 2.67 Reduction of the Carbon dioxide;
PG = phosphoglycerate, Bis-PG = Bis phosphoglycerate, GA3P = Glyceraldehyde 3 phosphate, DHAp = Dihydroxy acetone phosphate.

C4 Cycle

The C4 cycle occurs in certain plants like corn, sugar cane, weeds etc., in two types of cells known as bundle sheath cells and the mesophyll cells. The mesophyll cells lack RubP carboxylase, and this take up the carbon dioxide by condensing it as HCO_3-with phosphoenol pyruvate (PEP) to yield oxaloacetate. Further the oxaloacetate is reduced by NADPH to Malic acid which is transported to the bundle sheath cells where the Malic acid is decarboxylated to the bundle sheath cells where the Malic acid is decarboxylated by $NADP^+$ to form CO_2, pyruvate and NADPH. The carbon dioxide thus produced is transported to the Calvin Cycle, again the pyruvate returns to the mesophyll cells to continue the cycle where it is phosphorylated to phosphoenol pyruvate. It costs around 2 ATP's per each molecule of CO_2 to be concentrated in the bundle sheath cell.

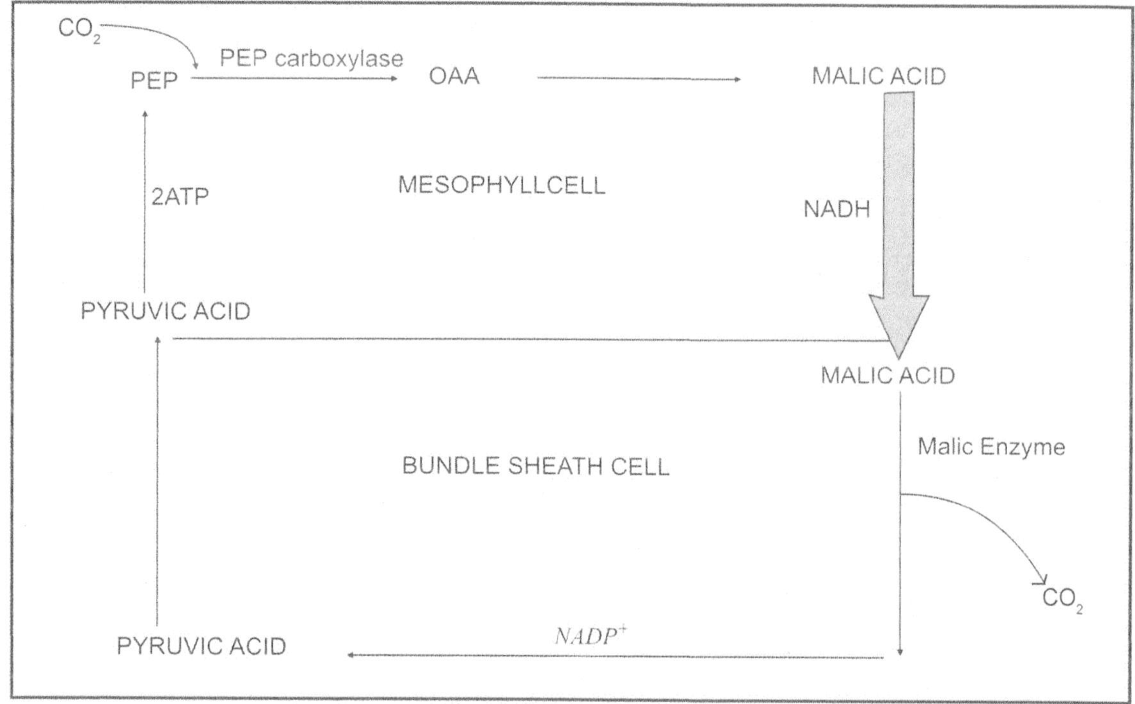

Fig. 2.68 C4 Cycle

Crassulacean Acid Metabolism (CAM)

This is a special type of pathway that occurs in the crassulaceae family of plants and the succulent xerophytes, this is a similar type of the path way in the use of CO_2 by the C_4 plants. There is a separation in the time rather the compartmentalization. PEP which is necessary to store a day's supply of CO_2 is obtained from the glycolytic pathway, during the day malate is broken down to CO_2 which enters the C_3 cycle and pyruvate is used to resynthesize starch. Thus these plants can carry out photosynthesis with the minimal loss of water.

Bacterial photosynthesis

Some Prokaryotes are capable of producing their own energy by photosynthetic process due to the presence of the bacteriochlorophyll in them, which generally absorbs the light at a near infrared region because the bacteria inhabit the murky stagnant water where the visible light can not pass-through. These bacteriochlorophylls are associated with the prosthetic groups like ubiquinone, Menaquinone etc., and contain Fe^{+2} between ubiquinone and Menaquinone (Bacteriopheophytin). And this bacteriochlorophyll have different subunits of the proteins H, L and M which could collectively bind to the two molecules of chlorophyll and two molecules of the Bacteriopheophytin. The photoreaction centre of the bacteria have a special pair (named because they absorb light in the 870nm range) PS-870, which gets excited when a 870 nm range of light is absorbed and thus the electron released from the photo-oxidation of the PS-870 is transferred to the Bacteriopheophytin and transferred to the Quinone compounds Qa, Qb (menaquinone or ubiquinone) gets reduced and reaches the Q pool from where it is targeted to the cytochrome bc1 complex and then to cytochrome C_2 which carries the electron back to the PS-870. Every two electron transferred from the Qb to cytochrome four protons enter the periplasmic space, thus there is a formation the transmembrane proton gradient, the dissipation of this proton gradient powers the production of the ATP by the photophosphorylation. (The mechanism of the photosynthesis differs from the species to the species of bacteria, the above given mechanism is given taking the consideration of *Rhodopseudomonas viridis*).

Essay Type Questions

1. Describe methods of reproduction in algae.
2. Describe different types of spores on algae.
3. Write in detail about the reproduction in *Chlamydomonas*.
4. Describe yeasts and reproduction in *Saccharomyces*.
5. What are asexual means of reproduction in lichens and describe their structures?
6. Describe the life cycle of a typical bryophyte.
7. Describe the structure of sporophyte in mosses
8. Discuss the types of stele present in *Lycopodium*.
9. Describe reproduction in *Cycas*.
10. Describe various types of Ovules in angiosperms.
11. Write about alternation of generations in an angiosperm.
12. Describe the development of embryo in monocots and dicots.
13. Describe the life cycle of *Arabidopsis thaliana*.
14. Write in detail about the theories on apical meristems.
15. What is a growth regulator and describe different plant growth hormones?
16. Write an essay on tissue culture and its role in agriculture.
17. What is a protoplast and discuss various aspects of porotoplast culture?
18. Define hybrids and cybrids and discuss their methods of production.
19. Explain diagramatically *in vitro* production of haploids.
20. What is organogenesis and explain the meristem culture?
21. What is pest? Explain the biology of one each from pests of Rice, Sugarcane, Cotton and Groundnut.
22. Write on essay on the importance of controlling insect pests of plants to increase the productivity.
23. Briefly discuss the mechanism of photosynthesis.

Short Answer Type Questions

1. Thallus types in algae.
2. Isogamy and anisogamy.
3. Life cycle of yeast.
4. Apothecium of a lichens.
5. Type of thallus in lichens.
6. Gametophyte of a liverwort.
7. Differentiate strobilus and cone.
8. Megasporophyll of Cycas.
9. Male cone of Cycas.
10. List out are the unique features of an angiosperm?
11. Porogamic fertilization.
12. Endosperm .
13. Campylotropous ovule.
14. Double fertilization and triple fusion.
15. Genome structure and organisation of *Arabidopsis*.
16. Meristems
17. Somatic embryogenesis.
18. Somaclonal variation.
19. Hybrids and Cybrids.
20. Anther culture.
21. Hill's reaction
22. Photosynthetic pigments
23. Structure of chlorophyll
24. Photosynstems : I & II

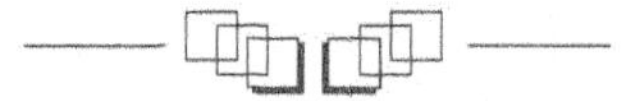

3 Introduction to Animal Biology

3.1 Introduction to Animal Kingdom

An enormous variety of plant and animal species inhabit the earth with mind boggling diveristy although many kinds of plants and animal have become extinct during past geological periods. Animals differ from one another in size, form, structure, manner of life and many other ways. According to conservative estimate, more than one million extant animal species are known presently, and several new species are being constantly discovered from time to time, further adding up to this massive number and bewildering diversity. In terms of population, some are present in enormously abundant numbers, others are in moderate numbers yet others are of rare occurrence. To study and analyse the relationship between different kinds of species, Animal kingdom is divided into various groups (class, phyla, genera, species etc.), some of them are large and some others are small. The art of classification is known as *taxonomy*. The classification serves two purposes, the first is to identify basic taxonomic units, or species and give their description as completely as possible. The second is to device an apt way of pigeon - holing them into different categories.

Species

The Units of Classification

The concept of species is quite complex. Broadly speaking, a species comprises of an assemblage of individuals that share a high degree of resemblance morphologically (externally) and functionally, coupled with a significant dissimilarity from other assemblages of the same general kind. The notion that it species is the basic unit of classification has been workable right from the beginning of the development of biology, because of the existence of sharp discontinuities in variations of organisms, some time coupled with the reproductive isolation, especially in the case of plants and animals that reproduce sexually. In these cases, species can be defined in terms of genetic and evolutionary relationship (phylogeny). As long as a sexually reproducing population is free to interbreed randomly, its entire gene pool undergoes incessant reasortment, and new mutations if and when arise, are dispersed throughout the population. Such an interbreeding population might evolve adoptively in relation to and for as a response to changes in the eco-geographical environment with reasonable uniformity. The emergence of new species is sequel to the process of *divergent evolution.* This is possible only when a segment of population becomes reproductively isolated in an environment that is altogether different from that occupied by the rest of the (parental progenitor) population. Reproductive isolation at first may be geographic in the form of some kind of physical barrier like mountain ranges or a large water bodies dividing the continuity of the initially contiguous population of organisms. The genetic mutations supply the necessary variation, which is compounded by every cycle of reproduction and the recombination. Within each of the sub-populations a common gene pool is maintained by interbreeding randomly, but by the joint action of chance mutation and selection the two sub-populations are free to evolve on different lines largely depending upon their adaptation (adaptive evolution) to a given environment. This continues to diverge as long as they are geographically separated. Eventually the cumulative differences become so accentuated that these manifest as physiological changes, and the physiological isolation gets superimposed on geographical isolation. Such members of sub-population gets reproductively isolated to such an extent that they fail to reproduce even if these are brought together. Even if these population co-mingle once again they fail to interbreed because their gene pools are separated permanently.

The Species Characterization

Ideally, species should be characterized by a set of descriptors, which should enable us to describe in toto of their *phenotype*-the sum total of external appearence or even better-of their *genotypes*-the genetic makeup of species group. However, the taxonomic practices fall short of these ideals and in most of the biological groups, even the external morphology is fragmentally described and genetic characterization is not done or incomplete.

As a general rule, form, structure (including anatomical) and functions (physiological and/or biochemical) can be most easily be determined by direct measurements and observations. Most of the present day classification is based on the above criterion. However, the refined molecular biological approaches are being increasingly used in understanding taxonomical status and phylogeny.

Naming of the Species

According to a convention known as the binomial system of nomenclature, every biological species consists of a latinized name that comprises of two words. The first word indicate the taxonomic group of immediate higher order, genus (plural genera) to which the species belongs to, and the second word identifies it as a particular species of that genus. The first letter of generic name is capitalized and the first letter of second word is written in small letter and the whole name is italicized : *Homo sapiens*. In situations where there is no scope for confusion the generic name is often abbreviated to its initial letter : *H. sapiens*.

In the taxonomic treatment of animals or plants (biological groups) the individual species are grouped in a series of categories of successively higher order : genus, family, order, class and division (or phylum). This type of classification is referred to as a *hierarchial* one, since each category in the ascending series unites a progressively larger number of taxonomic units in terms of a progressively diminishing number of shared characters. However, the genus has a position of special importance, because according to the rules of nomenclature a species cannot be designated unless it is assigned to a genus. The allocation of a species to a higher taxonomic category beyond the genus doesn't carry any essential nomenclatural information. It indicates the position of an organism relative to other organisms in the system of arrangement adopted.

The Problems of Classification in Taxonomy

In dealing with a wide variety of a large number of species it is necessary to have a system of classification for orderly arrangement of organisms for the purpose of storage of data and retrieval. The earlier systems of biological classification were mainly artificial in design since the system of classification was based on arbitrarily chosen criteria. However, as knowledge about the form and anatomical structure, of plant and animals increased, it became obvious that these organisms conform to a number of major patterns or types each of which shares many common features, including those that are not evident upon superficial examination. Examples of such types are mammalian, avian, reptilian and amphibian types among the vertebrate animals. Linnaeus was the first naturalist to develop a system of classification that endeavoured to provide grouping of organisms in terms of typological resemblances and dissimilarities in the middle of eighteenth century. The Linnean way of classification, the taxonomic position of an organism provides a large body of information about its properties. For instance to say that an animal belongs to vertebrate class Mammalia, immediately gives us an insight that it possesses all those characteristics which distinguish mammals collectively from other vertebrates. Since the Linnaean classification depicts biological nature of the objects that is classified, it is called as *Natural system of classification* as opposed to the earliest system of artificial classification.

The Phylogeny Approach in Classification

When the fact of (biological) organic evolution as propounded by Charles Darwin was recognized, another paradigm shift was immediately become evident in the concept of natural classification. The typical grouping of organisms by the biologists of eighteenth century merely depicts resemblance. However, for post Darwinian biologists, they, revealed relationships. In 19th century, the natural system of classification changed to a system that grouped organisms in terms of their evolutionary affinities and relationships. The taxonomic hierarchy became in certain sense the reflection of a family tree, and taxonomy assured a new goal; the restructuring of hierarchies to reflect evolutionary relationships. Such a taxonomic system is known as a *Phylogenetic system*.

Numerical Taxonomy

An alternative approach to group or classify organisms based on the qualification of the similarities and dissimilarities of traits among organisms have come to be known as *numerical taxonomy*. This approach is an empirical one. Michel Adanson was the first to suggest this system of classification and hence it some times bears his name as *Adanonian (or numerical) taxonomy*. The basic principle underlying is that, if each phenotypic trait is given explicit weightage, it should be possible to express numerically the taxonomic distance between organism in terms of the number of characters shared by them, relative to the total number of characters employed. The significance of the quantitative relationships, arrived at among a group of organism is largely influenced by the number of characters evaluated, these should be as numerous as varied as possible to obtain representative sampling of phenotypes.

Because of the magnitude of computations involved in numerical taxonomy, until recently the numerical taxonomy appeared impractical. However, with the advent of computers and availability of sophisticated software for rapid computation, it is now possible to compare a large number of characters and organism and compute the degree of similarity. For any pair of organism, the calculation of similarity can be made in slightly two different ways :

The similarity coefficient (S_j) is based on positive matches and do not take into account characters negative for both organisms whereas the matching coefficient (S_s) takes into account both positive and negative matches in the calculation (Table 3.1).

Table 3.1 *The determination of similarity coefficient and matching coefficient for two organisms characterized with respect to many different traits.*

Number of characters positive in both strain	: a
Number of characters in organism 1 and negative in organism 2	: b
Number of characters negative in organism 1 and positive in organism 2	: c
Number of characters negative in both	: d
Similarity coefficient - $(S_j) = \frac{a}{a+b+c}$	
Matching coefficient - $(S_s) = \frac{a+d}{a+b+c+d}$	

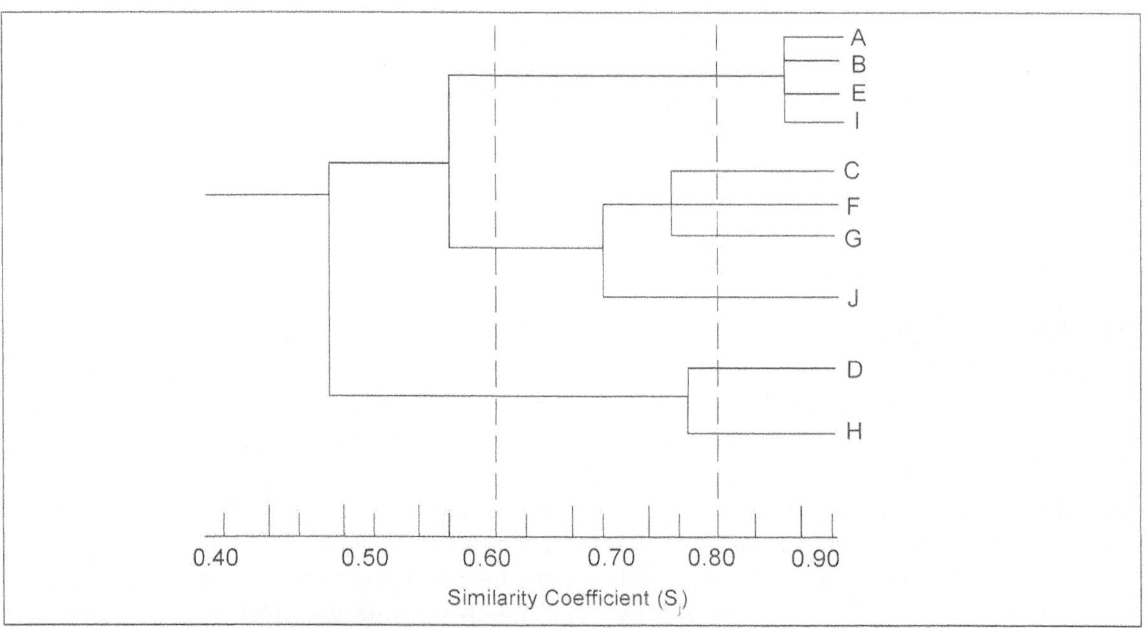

Fig. 3.1 A dendogram depicting similarity relationship among 10 organisms. The two dotted lines indicate possible similarity levels at which successive ranks in the numerical taxonomy may be given.

The data can then be transposed into dendogram (Fig. 3.1) as a basis for determining taxonomic arrangement in terms of numerical relationship. The dotted vertical lines show similarity level that might be considered appropriate for recognizing two different taxonomic rank e.g., genus and species etc.

Numerical taxonomy doesn't show evolutionary connotation of phylogenetic taxonomy. However, it provides an objective as well as stable basis for the construction of taxonomic groups. It cannot be employed, until a large mass of data and number of characteristics are considered. Its use encourages the thorough examination of phenotypes. It provides a chance for refinement and for constant revision based on the generation of new data. Further details are discussed elsewhere in this book.

Molecular Taxonomy

Molecular taxonomy is the new approach for the classification and phylogenetic study of (living) organisms. The growth and developments in molecular biology has opened up (a number of) new vistas to the classification and to study the phylogeny of organisms. The perpetual sparing match between '*Clumpers*' and '*Splitters*' can be resolved with the molecular approach. In recent time the unraveling of sequences of genomes (Genomics) has fuelled the molecular biological analysis of origin and evolution and interelationship of biota in toto. Some of the approaches are : the study of base composition of DNA, nucleic acid hybridization, analysis of reassociation kinetics, nucleic acid sequencing (genomics), molecular markers and DNA and RNA finger printing, homology of proteins etc. A great deal of data on gene (genome sequences, protein sequences) and is being available in public domain for the use of scientists. Coupled with computation biology and bioinformatics, the molecular data may unravel new insights into the taxonomy.

3.2 Functional Morphology and Classification

3.2.1 Introduction

The predominantly occurring species on the planet earth are complex eukaryotic life forms belonging to the animal kingdom. According to an estimate, anywhere between 2-12 × 10^6 species are extant today. Animals are essentially *heterotrophs* and reproduce mainly sexually (hermophrodite or bisexual). Each and every kind of animal species has evolved its own form, structure and physiological functions, as per their requirements for perpetuating themselves in a wide ranging environment by the process called *Adaptive radiation*. Surface of the earth is uneven, covered by saline and freshwater, with wide variation of soils, rocky land masses, besides widely varying climatic factors, precipitation of rain etc., Tropical regions receive more heat from the sun, humidity etc., than the temperate areas especially polar regions. Amount of humidity, snowfall and rainfall varies significantly, which influence the plant and animal life, outside of boom and bust of population. The dynamic interaction of physical and biological factors greatly influence the *organisms* including man. The sum total of these interaction constitute the *web of life* or *balance of nature*. The developments of techniques in physiology, embryology, biochemistry and molecular biology added new dimension to comparison of homologies as the basis of classification instead of relying on only on *Linnean* ways of classification based in mere resemblances and dissimilarities of external morphology. Various degrees of resemblances and differences are present in any heterogenous assemblage of organisms. Based on some total of inherent characteristics including structural features, size, segmentation, appendages, skeleton, mode of development, larval characteristics, physiological attributes and biochemical features, the animal kingdom is further classified in the Taxonomy.

Broadly, the Animal kingdom (Animalia) has been grouped into two major categories : Non-chordata (Invertebrata) or Chordate (Vertebrata) based on the absence or presence of vertebral column. As a rule, all these animal species which do not have vertebral column are grouped under phylum non-chordata and those of the species which have true vertebral column are classified under phylum Chordata (vertebrata). Though this kind of grouping is artificial, but serves in systematic studies. In fact the higher non-chordate taxons share many structural features with chordates *viz.*, body plan, symmetry, triploblastic-coelomic conditions, segmentation and level of organization etc., Despite these facts, the two groups also depict basic differences. A comparison of diverse features of vertebrates and invertebrates is presented in Table 3.5. The specific differences between lower and higher invertebrates is shown in Table 3.4.

Categories of Animals–Animal Taxonomy

Attempt to classify animals and plants in an orderly way goes back to the time of Greek Philosopher Aristotle. Nonetheless, advancement towards the universal classification has not progressed until Carols Linnaeus paved the way in the eighteenth century. The objective of classification is to depict true *phylogenetic relationship* of organisms with that of ancestral ones, such a system is always based on homologies. An useful classification should be based on cautious observation and the use of features to reflect genuine ancestral relationship, organization and naming of all animals, also follows the Linnaeus system and binomial nomenclature. The first name indicating the *genus* to which the organism belongs to i.e., a genus is a congregation of similar species. The second term indicates the *species* of the organism. As earlier defined it is a basic unit of classification where all the closely related individuals with ability to interbreed and generate fertile progeny are put together. As more and more number of animals were discovered and new knowledge of animal science has advanced more ways of identifying differences led to erection of broader groups in animal taxonomy. To reiterate once again the basic design of grouping animals is: *families* are created to organize the similar *genera* into one group; similar families are put into the *order*, the orders which bear likeness go into the same *class*; and the classes with a resemblances goes into a *phylum*, and ultimately all those phylum, which have similarity goes to make the *kingdom*. The major groups of phyla in *Animal Taxonomy* along with the approximately known number of species in each phyla is given in Table 3.2 and Table 3.3 lists the principal traits of different Animal phyla. The classification in an over simplification of the work of stalwarts of Animal taxonomy like Hymen, Barnes and Storer with modification. An account of characteristic features of each taxon is given below :

Table 3.2 *A broad outline of animal classification, showing major groups and phyla*

Major Groups of Animals	Phyla	Number of Species
Subkingdom I Protozoa	1. Protozoa	50,000
Branch 1. Mesozoa	2. Mesozoa*	50
Branch A. Radiata	3. Porifera	5,000
Grade A. Radiata	4. Coelenterata	11,000
	5. Ctenophora*	90
(i) Subdivision Acoelomata	6. Platyhelminthes	15,000
	7. Rhynchocoela* (Nemertinea)	750
(ii) Subdivision Pseudocoelomata	8. Acanthocephala*	500
	9. Entoprocta*	60
	10. Rotifera*	1,500
	11. Gastrotricha*	175
	12. Kinorhyncha*	100
	13. Nematoda*	12,000
	14. Nematomorpha*	100
(iii) Subdivision Lophophorate Coelomata	15. Phoronida*	15
	16. Ectoprocta* (Bryozoa)	4,000
	17. Brachiopoda*	260
(iv) Subdivision Schizocoelous Coelomata	18. Priapulida*	08
	19. Sipunculida*	275
	20. Mollusca*	80,000
	21. Echiurida*	60
	22. Annelida*	8,700
	23. Tardigrada*	180
	24. Onychophora*	73
	25. Arthropoda*	900,000
	26. Pentastomida*	70
Subdivision Enterocoelus Coelomata	27. Chaetognatha*	50
	28. Echinodermata*	6,500
	29. Pogonophora*	80
	30. Hemichordata	80
	31. Chordata	49, 000

Hierarchy labels: Kingdom Animalia; Subkingdom II. Metozoa; Branch 3. Eumetazoa; Grade B. Bilateria; Dovision 1. Protostomia (subdivisions i–iv); Division 2. Deuterostomia (Subdivision Enterocoelus Coelomata).

* minor phyla

Table 3.3 *Distinctive diagnostic characteristics of principal animal phyla*

Cells	Germ layers	Organization	Symmetry	Segmentation	Digestive tract	Excretory organs	Coelom	Circulatory system	Repiratory system	PHYLA	Distinctive features
1		Proto-plasmic								PROTOZOA	Microscopic, single or colonies of like cells.
	2. Diploblastic	Cellular	Radial biradial	a	Incomplete without anus	a	a	a	a	PORIFERA	Bodywall perforated by pores and canals.
		Tissue grade		a		a	a	a	a	COELENTERATA	Nematocysts present. Digestive tract sac like.
		Organ grade		a		a	fc	a	a	CTENOPHORA	Biradial comb plates for locomotion.
				a	compl.	ps	fc	a	a	PLATYHELMINTHES	Flat, soft. Digestive tract much branched.
				a	Complete with anus. Protostomic	a	fc	+	a	ACANTHOCEPHALA	Worm-shaped. Proboscis hook-bearing.
				a		ps	a	a	a	NEMERTINEA	Slender, ciliated. Proboscis soft, eversible.
				a		ps	fc	a	a	NEMATOMORPHA	Thread-like. Mouth may be absent.
				a		ps	+	a	a	ROTIFERA	Microscopic. Cilia on oral disc.
				a		ps	fc	a	a	GASTROTRICHA	Microscopic. Surface spiny. Cilia ventral.
				a		lo	a	a	a	NEMATODA	Cylindrical. Cuticle tough. Cilia absent.
				a		lo	+	+	a	BRYOZOA	Moss-like colonies. Lophophore present.
				a		h sc	+	+	+	MOLLUSCA	Soft. External limy shell of 1, 2 or 8 parts.
				+		sc	+	+	+a	ANNELIDA	Metamerically segmented. Setae present.
				+		ps	+	+	a	TARDIGRADA	4 pair of unjointed legs bearing claws.
				a		ps	+	+	a	ONYCHOPHORA	Skin soft. Paired nephridia and tracheae present.
				+		h sc	+	+	+	ARTHROPODA	Chitinous exoskeleton. Jointed apendages.
				a	Complete Deuterostomic	en	+	a	a	CHAETOGNATHA	Arrow-shaped. Transparent Lateral fins.
			Radial	a		en r	+	+	+	ECHINODERMATA	5-parted radial symmetry. Tube feet. Spiny endoskleton
			Bilateral	+		en	+	+	+	CHORDATA	Notochord. Dorsal tubular nerve cord. Gill slits. Limbs or fins for locomotion.

(a) Absent, (+) Present. **(en)** Enterocoelom. **(fc)** Flame cells. **(h)** Haemocoel. **(lo)** Lophophorate coelomata. **(ps)** Pseudocoelom. **(r)** Reduced. **(sc)** Schizocoelom. **(v)** Various or none.

All animals are grouped in one or other group or class or phyla, of animal kingdom like *Protozoa, Porifera, Coelenterata, Ctenophora, Platyhelminthes, Nemertina, Annelida, Arthropoda, Mollusca, Ehinodermata, lower chordates and vertebrates*.

The concept of major and minor phyla depends upon two factors, 1. Number of species and individuals and 2. Their participation in ecological communities. The minor phyla include *Rotifera, Ectopracta, Endopracta, Bryozoa* etc.

Lower and Higher Invertebrata

Lower invertebrates are generally smaller in size and simple in body organization. These include *Protozoa, Porifera, Coelenterata, Platyhelminthes* and *Nematoda*. On the other hand higher invertebrates are generally larger in size and complex in body organization. These include Annelida, Arthropoda, Mollusca, and Echinodermata. The differences in lower and higher invertebrata are shown in Table 3.4.

Table 3.4 *Comparision of lower and higher invertebrates.*

Lower Invertebrates	Higher Invertebrates
1. Generally smaller in size	Generally larger in size
2. Body organization simple	Organization complex
3. Radial, biradial or no symmetry	Bilateral symmetry
4. Germ layers wanting or 2 and 3 germ layers	Germ layers 3
5. No Coelom or a pseudocoelom	True coelom
6. Generally no separate coelom	Mouth and anus separate
7. No muscular gut	A true muscular gut
8. Blood vascular system not well developed	A well developed blood vascular system

Invertebrates and Vertebrates

The entire animal kingdom has been broadly grouped into two major categories - Invertebrata or Vertebrata. The invertebrates include those animals that do not possess a vertebral column while vertebrate is characterized by its presence. Though higher invertebrates share many structural peculiarities with the vertebrates viz. axiate body plan, bilateral symmetry, triploblastic and coelomate condition, metameric segmentation and organ system grade of organization. The two groups have several fundamental differences. A comparisons of different-attributes of invertebrates and vertebrates is given in Table 3.5.

3.2.2 Phylum Protozoa

The protozoa are a diverse group of organisms. There are over 30,000 known species and they are found in all environments where usually water is present. Each protozoan functions as an independent unit and is able to perform effectively all the activities necessary for life. However, protozoa are sometimes separated from Animal Kingdom and are placed in protoctista (Protista) with some algae. As, protozoans possess numerous animal features they are included in animal kingdom.

Table 3.5 *Differences between Invertebrates and Vertebrates.*

Features	Invertebrates (Non-Chordates)	Vertebrates (Chordates)
1. Symmetry	1. Radial, biradial or lacking.	1. Bilateral.
2. Metamerism	2. True or pseudometamerism or lacking.	2. True metamerism.
3. Post-anal tail	3. Lacking.	3. Usually present projecting beyond anus.
4. Grade of organisation	4. Protoplasmic to organ-system.	4. Organ-system.
5. Germ layers	5. 2(diploblastic) 3(Triploblastic) or lacking	5. 3 (triploblastic)
6. Coelom	6. Acoelomate, pseudocoelomate or truly coelomate.	6. Truly coelomate.
7. Limbs derivation	7. From same segment.	7. From several segments.
8. Notochord	8. Notochord or backbone lacking.	8. Present at some stage or replaced by a back bone made of ring like vertebrae.
9. Gut position	9. Dorsal to nerve cord.	9. Ventral to nerve cord.
10. Pharyngeal gill-slits	10. Absent	10. Present at some stage of life.
11. Anus	11. Opens on the last segment or absent.	11. Differentiated and opens before the last segment.
12. Bloodvascular	12. Open, closed or absent.	12. Closed and much developed.
13. Heart	13. Dorsal, lateral or absent.	13. Ventrally placed.
14. Dorsal blood vessel	14. Blood flows anteriorly.	14. Blood flows posteriorly.
15. Hepatic portal system	15. Absent.	15. Present
16. Haemoglobin	16. In plasma or absent.	16. In red blood corpuscles.
17. Respiration	17. Through body surface, gills or tracheae.	17. Through gills or lungs.
18. Nervous system	18. Solid.	18. Hollow.
19. Brain	19. Above pharynx or absent.	19. Dorsal to pharynx in head.
20. Nerve cord	20. Double, ventral, usually bearing ganglia.	20. Single, dorsal, without ganglia.
21. Segmental nerve roots	21. Dorsal and ventral roots not separate.	21. Dorsal and ventral roots separate
22. Reproduction	22. Asexual reproduction predominant.	22. Sexual reproduction predominant.
23. Regeneration power	23. Usually good.	23. Usually poor.
24. Body temperature.	24. Cold - blodded.	24. Cold or warm-blooded.

The Characteristic features of Protozoa are :

1. Small usually one celled, some in colonies of few to many similar individuals, symmetry none, bilateral, radial or spherical.
2. Cell form usually constant, oval, elongate spherical, or otherwise varied in some species and changing with environment or age in many.
3. Nucleus distinct single or multiple, other structural parts are such as organelles, no organs or tissues.

4. Locomotion by flagella, pseudopodia, cilia or movements of cell itself.
5. Some species with protective housing or tests; many species produce resistant cysts or spores to survive unfavorable conditions or for dispersal.
6. Mode of life free living, commensal, mutualistic or parasitic.
7. Nutrition various modes : *Holozoic* - subsisting on other organisms, *saprophytic*, living on dissolved substances in their surroundings, *saprozoic* - subsisting on dead animal mater and *holophytic or autotrophic* - Producing food by photosynthesis.
8. Asexual reproduction by binary fission, multiple fission or buddding, some with sexual reproduction by fusion of gametes or by conjugation.

The protozoa are divided according to the structures they posses for locomotion into five classes.

1. *Mastigophora or Flagellates* with one or more whip like flagella
2. *Sarcodina or Rhizopoda*, with pseudopodia
3. *Sporozoa* with no locomotor organelles
4. *Ciliata or ciliates* with cilia throughout life and
5. *Suctoria* with cilia in young and tentacles in adult stages.

Different Classes of Protozoa

Class : Mastigophora	Class Sarcodina	Class : Sporozoa	Class : Ciliata	Class : Suctoria
Some possess chromotophore, other do not. Semi rigid cell, covering is a pellicle.	With psudopodia for locomotion and food capture. No chromatophore	No locomotor organs or contractile vacuoles. Reproduction by multiple asexual fission or sexual phases usually producing spores.	Cilia usually present-throughout macro and micnucleus	Cilia in only in adult attached by stalk with tentacles - for feeding.
Adult movement by one or more flagella	Variable shape One nucleus	all internal parasites usually with intracellular stages.		No cytostome.
One nucleus	Asexual reproduction by binary fission.			mostly sessile, in salt or fresh water
Asexual reproduction by longitudinal binary fission. multiple fission in cyst. eg. *Euglena, Trypanosoma*	May sporulate eg. *Amoeba, Arcella Polystomella*	e.g. *Monocystis, Plasmodium, Nosema*	e.g. *Opalina, Paramecium, Nosema*	e.g. *Podophyra Trichophyra.*

Some of the salient features of the classes along with examples are given in following table and representative animals are shown in Fig. 3.2.

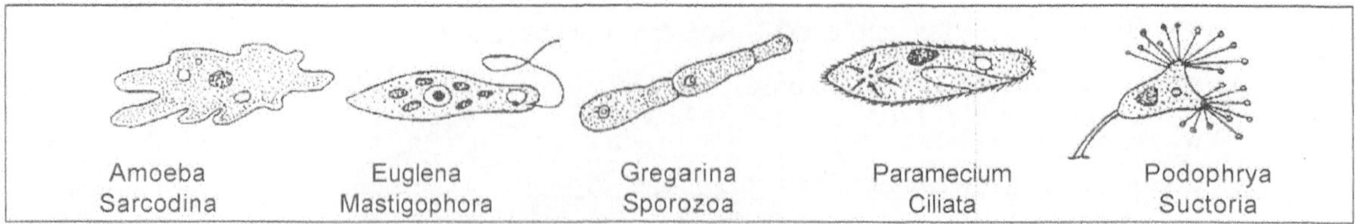

Fig. 3.2 Common representatives of the Phylum Protozoa

The extreme age of phylum is evident by finding the hard remains of radiolaria and foraminifera in precambrian rock. Many of flagellates of mastigophora have cells embedded in a common gelatinous matrix and show physiological coordination between the individuals. Some are connected to one another by protoplasmic threads and in volvox there is differentiation into vegetative and reproductive cells. These conditions parallel the *formation of tissues* and segregation of somatic and germ cells in Metazoa. Certain protozoans ordinarily free living, are occasionally found living within the bodies of other animals, showing derivation of parasitic forms. Some chlorophyll bearing flagellates resemble the green algae in structure and physiology and suggest common origin for plants and animals. The mastigophora are probably the most primitive, the ciliate most specialized and sporozoa are probably simplified as a result of their strictly parasitic manner of life.

3.2.3 Phylum : Porifera

Nearly 10×10^3 species are known, majority of them are marine and about 150 species of fresh water. Size variation ranges from few mm to 1m. Structural grades vary from simple vase like asconoid through folded syconoid to the leuconoid condition with ramifying channels to vastly enhance the surface area for feeding. Organism of thin phyla have asymmetrical body architecture with two cell layers separated by mesenchyme (Gelatinous) with amoeboid cells and skeletal spicules and fibers. Incurrent pores (Ostia) and large excurrent pores (Oscula) connect the interior cavity (Spongocoel).

Feeding is through spongocoel which is lined with choanocytes which push water with flagella through ostia, the incurrent canals. Food sticking to the collars of the choanocytes and is passed on to the cell body or is engulfed by the amoeboid cell lining the canals. Food is partly digested in choanocytes and undigested portion is passed on to the amoeboid cell which engulf and digest and the soluble products are diffused through the sponge.

Most of the organisms in the group are sessile, though few free swimming are also present. The skeleton in sponges is essentially made up of silica or calcium carbonate with or without protein fibres, spongin. Respiration is simple gaseous exchange through the cell membranes by the diffusion process. Contractile vacuoles help in osmoregulation of each cell and the excretion is again simply by diffusion.

Reproduction is asexual and or sexual. When asexual it is mainly vegetative, the fragment may get loosened and float along with water current and gets established to multiply and grow. Alternatively, gemmules, amoeboid cells (archaeocytes) laden with rich nutrients, surrounded by epithelial cells may present as 'seeds' for reproducing sponges. Sexual reproduction, hermaphrodite or sexes separated on different individual organism. Mesenchymal cell develop into gametes. They spermatozoa are motile and traverse through excurrent and reach another individual and gets trapped by choanocytes which then becomes amoeboid cells and are carried to ova in mesenchymous where the zygote is formed after fertilization, which subsequently differentiate and becomes motile, multicellular, flagellate larvae and subsequently metamorphose into an adult by eversion to bring flagellate surface inside.

Many of the sponges are symbiotic with algae. Sponges are known from the time of early Cambrian Era. This phyla constitute a discrete group relationship with metazoan, though a relationship exists with choanoflagellate and Sarcomastigophora.

Characteristic Features and Classification of Phylum Porifera (Pore bearing) - Sponges :

1. Some cellular differentiation, but no tissue organization. Two layers of cells - outer pinacoderm and inner chaonoderm
2. Adults sessile and all marine.
3. Body frequently lacks symmetry
4. Single body cavity
5. Numerous pores in body wall.
6. Usually a skeleton of calcareous or silicious spicules or horny fibres.
7. No differentiated nervous system.
8. Asexual reproduction by budding.
9. All are hermaphrodite and protrondrous.
10. Embryonic development include blastula and larval stages.
11. Great regenerative power.
12. Dead-end phylum it has not given rise to any other group of organisms.

Classes of Porifera

Class calcarea	Class Hexactinellida	Class Demospongiae
Calcareous spicules	Siliceous six-rayed spicules.	Siliceous spicules not six rayed; or spongin fibres or without skeleton elements.
eg. Leucosolena, Sycon, Scypha	eg. Euplectella, Hyalonema, Regadrella	eg. Euspongia, Microciona Haliclona, Poterion

Representatives of Porifera are shown in Fig. 3.3.

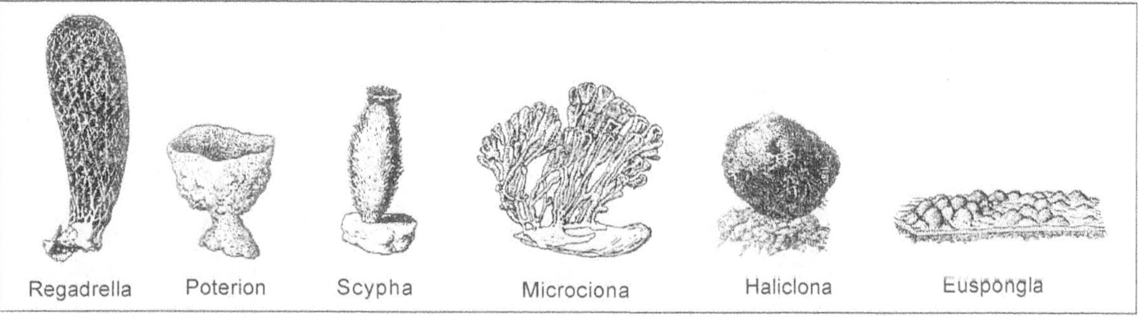

Fig. 3.3 Phylum porifera–certain examples

3.2.4 Phylum Cnidaria (Coelenterata)

The members of this group are mostly marine forms (sea anemones, jelly fish, corals) and few fresh water forms (Hydra) also occur. Basically two forms of organisms-columnar *polyps* with the mouth facing upward, and medusa with umbrella like shape pelagic (free swimming), mouth being present downwards surrounded by tentacles. The size of the organism vary widely, ranging from microscopic to 2m in the case of polyp's and medusae (sea blubber) types may reach 3.5m diameter with 3 mm tentacles. Coral reefs are made up of $CaCO_3$ in shallow sea surfaces. The earliest organism of this group Ediacara is known to exists 7×10^2 million years ago, and almost all the cnidarians were extant from the Cambrian period, although the phylum has reached the dead blind alley of evolution. Body is made of bilayer of cells (diploblastic) separated by mesoglea. The outer layer ectoderm or, epidermis and an inner endodermis (gastrodermis) line the gut, between these two lies gelatinous mesoglea containing loose cells, (Fig. 3.4). The motile cells present in mesoglea and radial and circular muscle help in movement of medusa. Polyps are generally sessile, and can show movements like burrowing, somersaulting or pedal creeping.

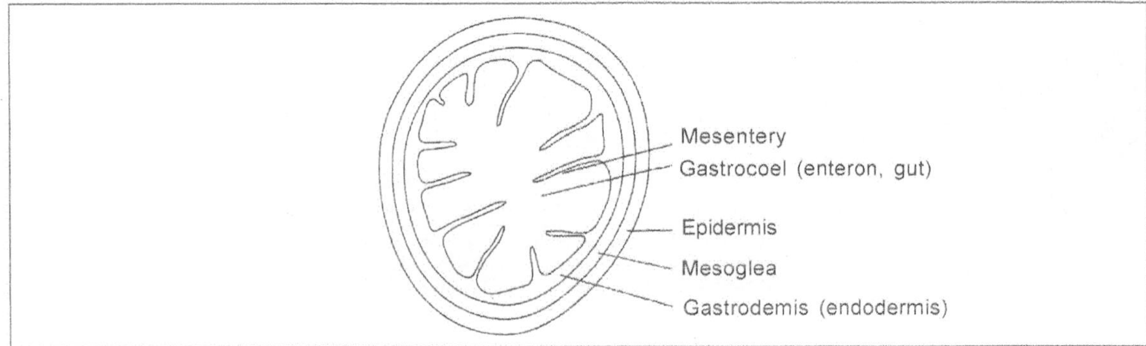

Fig. 3.4 Cross-section through a typical Cnidarian

Cnidarian body is characteristically radially symmetrical. These animals possess body cavity, *gastrocoel* or *coelenteron* lined by endodermal cells, which also serves as gut. Cnidarians are carnivores; the nematocysts on tentacles help in feeding, food enter the gut and after digestion it is absorbed by surface area of gut which is increased by mesenteric folds. As there is no anus, waste is eliminated through mouth. The cnidarians are known for symbiotic relation with intracellular, photosynthesizing dianoflagellate. Unmyelinated, naked nerve cells with highly branched fibres form a nerve net which regulates the integration function. Reproduction is asexual by budding and sexual reproduction through gamete formation; Fertilization is external and external development usually through planula larva.

Coelentrate or Cnidarians are known for tissue level of organization, for *polymorphism, alteration of generation* and *coral formation.*

Characteristic Features and Classification of Cnidaria

Salient features : 1. Diploblastic animals - outer ectoderm, inner endoderm, with medial mesoglea, 2. radially symmetrical 3. tissue level of organization achieved 4. single cavity - Coelenteron single opening for ingestion and egestion. 5. polymorphic zooids - polyp forms solitary and free swimming 6. nervous system is a collection of cells forming net or plexus. 7. Asexual reproduction by budding or strobilation 8. sexual reproduction with planula larva.

Coelentera is divided into three classes, Hydrozoa, Scyphozoa and Anthozoa and representatives are shown in Fig. 3.5.

Class Hydrozoa	Class Scyphozoa	Class Anthozoa
Polyp dominant	Large medusa dominant	Polyp present
Medusa simple	Polyp present	No medusa
Gonads ectodermal	Mesentaries present in young polyp	Large mesentaries present
Polyps solitary or colonial	No gullet Gonads endodermal	Gullet lined by ectoderm Gonads endodermal
No mesentaries and gullet e.g. Obelia, Hydra	e.g. Aurelia, Vellela	Polyp solitary or colonial e.g. Actinia, Madrepora

Phylum - Ctenophora

Ctenophora is a minor phyla representing about 90 species. These are marine, diploblastic with mesoglea like Coelenterata. The ctenophoran are characteristically have eight paddle - like combplates (ctena) and stinging colloblasts equivalent to that of nematocysts. Ctenophorans are commonly called as Sea Combs or Comb jellies. Example is Cestum (Fig. 3.5).

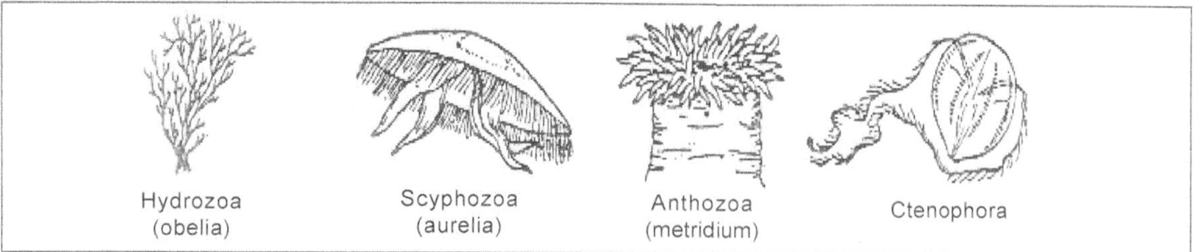

Fig. 3.5 Phyla Coelenterata and Ctenophora–*Examples*

3.2.5 Phylum : Platyhelminthes

Flatworms and their relatives, numbering about 15,000 species, found as marine, freshwater, or terrestrial free living or parasitic forms belong to this phylum. The size of animals is between 1 mm to 35 m long. All animals are acoelomate (no enclosed body cavity) bilaterally symmetrical with triploblastic condition, the outer ectoderm surrounds mesoderm which inturn surrounds the endoderm (Fig. 3.6). The alimentary canal without anus extensively ramified to increase large surface area for absorption of food and distribution of nutrients.

The free living platyhelminths exhibit locomotion by means of cilia and with the help of muscles. The integument with glycoproteins and turgor of tissues serve as hydrostatic skeleton particularly in parasitic forms. As there is no specific respiratory and blood vascular system, the body surface facilitates the transport of oxygen, carbondioxide, ammonia etc. The osmoregulation and excretion is performed by ciliated flame cells as well as protonephridia eliminating waste material through excretory pores. The nervous system is simple with rudimentary cephalization and tactile chemoreceptors are present laterally on head.

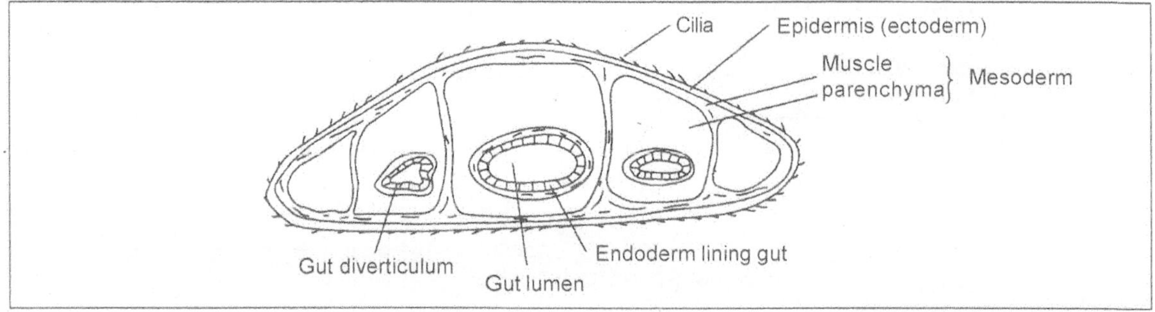

Fig. 3.6 Cross-section through a typical platyhelminth

Flatworms reproduce asexually by dividing themselves into two parts and also have excellent power of regeneration. Most worms are also hermaphrodite, and exhibit sexual reproduction with self fertilization. Yolky eggs in free living forms hatch into miniature adults while parasitic forms exhibit complex life histories with successive larval stages and in more than one host species. Some flatworms also exhibit parthenogenesis.

The Characteristic Features and Classification of Platyhelminthes are :

1. Triploblastic
2. Bilaterally symmetrical, dorsoventrally flattened with mouth but no anus
3. Unsegmented,
4. Acoelomate
5. Anteriorly placed central nervous system with simple network and ganglia
6. Excretory system with branching tubes ending in flame cells
7. Complex hermaphrodite reproductive system, usually larval forms present.

The platyhelminths are grouped into three classes namely Turbellaria, Trematoda and Cestoda and the typical representative are shown in Fig. 3.7.

Class Turbellaria	Class Trematoda	Class Cestoda
Free living, aquatic	Endoparasitic	Endoparasitic
Delicate, soft leaf like body	leaf like body	Elongated body divided into proglottids
Ciliated outer covering, cuticle absent	Suckers present	Suckers and hooks on proscolex
Enteron present	Enteron present	Thick cuticle, no cilia in adult
Sense organs present	Sense organs only in free living stage	No enteron
Simple life history. *e.g.* ***Planaria***	Complex life history. *e.g.* ***Fasciola***	Complex life history. *e.g.* ***Taenia***

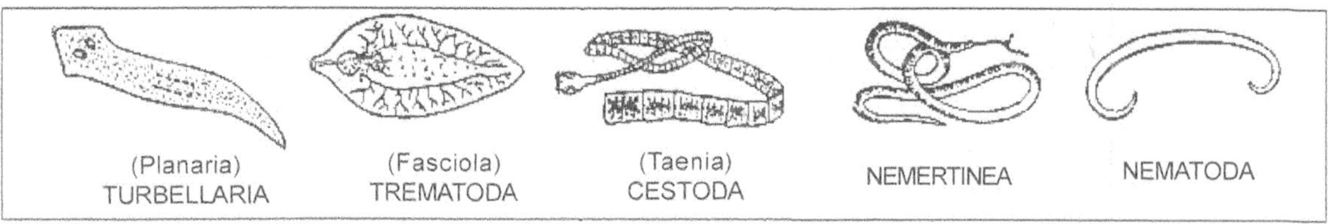

Fig. 3.7 Phyla Platyhelminthes (three classes), Nemertina, and class Nematoda

Phylum Nemertina or Rhynchocoela

This is one of the minor phyla, representing about 900 species, varying in size from 0.5 mm to 30 m, mainly marine burrowing worms like ribbon worms and proboscis worms. Few are fresh water nemertians like Geonemertes sp. Nemertians have bilaterally symmetrical, triploblastic, acoelomate, and unsegmented body. These are characterized by a long, anterior, eversible proboscis, known as *Rhynchocoel* (Fig. 3.8). These have gut mouth and anus. This is carnivorous in feeding. Locomotion (burrowing, crawling and swimming) is by cilia as well as peristalsis like waves of body musculature. Nemertians possess a closed blood vascular system, blood with pigments like haemoglobin. Respiration is by integumentary diffusion. Osmoregulation is effected by ciliated flame cell lining tubules. Nervous system with considerable cephalization with lobed cerebral ganglia and longitudinal lateral nerve cords. Asexual reproduction is by fragmentation. Sexual reproduction is also seen with external fertilization and development either direct or through diplodium larva.

Examples : Emplectonema, Tubulanus sp. *Lineus* sp.

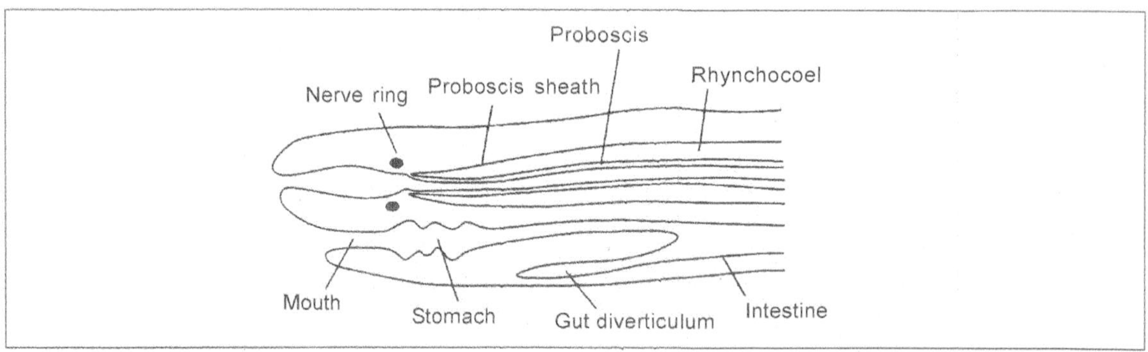

Fig. 3.8 Longitudinal section through a nemertine, anterior end, e.g., Lineus sp.

3.2.6 Phylum Nematoda

About 80,000 nematodes are described, which occur as marine, freshwater, terrestrial but mainly as parasitic worms. The body size varies from 100 mm to more than 100 cm, with long cylindrical body tapering or rounded at the ends. Cilia absent but adhesive glands or hooks are present. Free living nematodes have important role in cycling of minerals, food components and aeration of soil. Parasitic forms exist in animals, human as well as in plants.

Nematodes are bilaterally symmetrical and triploblastic animals with tough elastic cuticle. The muscular system is well developed with longitudinal muscle, which along with epidermis and endodermis surround the body cavity which is *pseudocoel* (Fig. 3.9). The Nematodes have long straight gut with pharynx and specialized mouth parts such as elaborate, predatory teeth and associated with hooks for attachment. These animals exhibit anguilliform movement by contraction of longitudinal muscles since circular muscle are absent. The muscle act against hydrostatic skeleton of body because of cuticle, pseudocoel and structure of body.

There are no respiratory organs as well as blood vascular system and gas exchange occurs through integument. The excretory system consists of lateral, longitudinal canals with a pore on the anterior end. The nervous system comprises a circumpharyngeal commissure and longitudinal nerve cords passing along the body. The sign of neuromuscular coordination was observed.

The sexes are separate, with male usually being smaller than female. The gonads (ovaries or testes) have ducts which exit through gonopore in female and via cloaca in male. Fertilization is internal. They are ovoviviparous. The hardy and desiccation resistant millions of eggs are produced and development is direct or with larva depending on favorable conditions. The phylum is not divided into classes. The examples are : Human parasitic form, *Ascaris*, Phytoparasite - *Rotylenchus*.

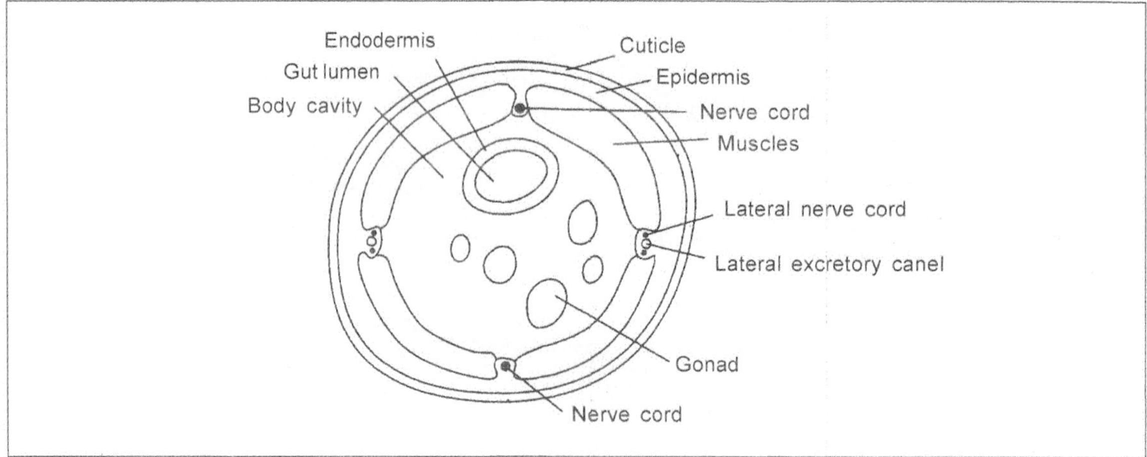

Fig. 3.9 Cross-section through a nematode, e.g. Ascaris sp.

Phylum : Rotifera

Rotifera represents about 1500 species, mostly free living (few marine) forms in damp mosses or soil with size ranging from 0.1 mm to 1.0 mm, is a minor phyla. Body is elongated with about 1000 cells, divisible into trunk and posterior foot. The anterior end bears corona of cilia arranged in two discs which produce circular wheel like waves (wheel animalcules) helping in movement. However, some rotifers are sessile or planktonic. Well developed pharynx with mastax (teeth), large stomach with short intestine leading into anus is present. Reproduction can be sexual with internal fertilization. Parthenogenesis also exists. An example is *Philodina* sp. (Fig. 3.10).

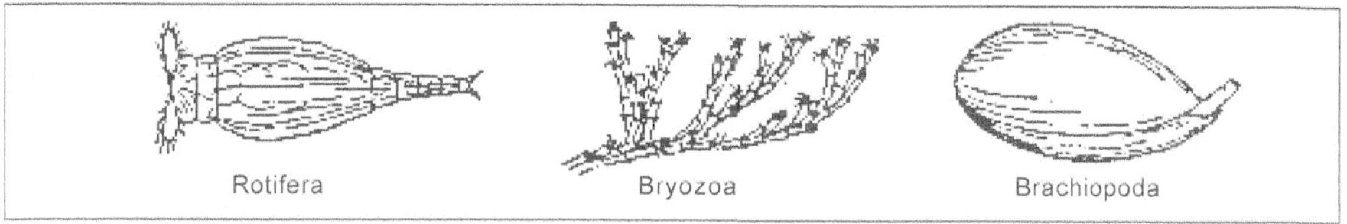

Fig. 3.10 Miscellaneous phyla and classes

3.2.7 Phylum Annelida

Annelids are marine, freshwater or terrestrial segmented worms or leeches mostly predators or scavengers representing about 8300 species. Their body size ranges between 500 mm to 3m. All animals are protostomes. Annelids are bilaterally symmetrical triploblastic animals. An ectoderm present on the outside, endoderm lines the gut lumen and between two is mesoderm and it holds muscles, reproductive organs, blood vessels and excretory organs. The body cavity is true coelom formed by the splitting of mesoderm (Fig. 3.11). The body has repeated segments, *metameric segmentation*, separated with segmental septa corresponds to annular rings seen on body.

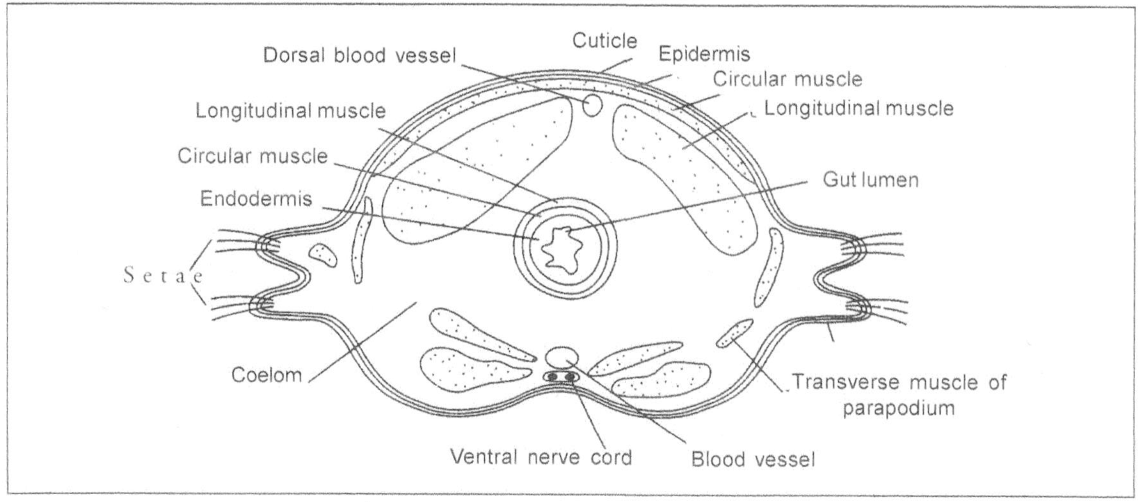

Fig. 3.11 Cross-section through a polychaeta, Nereis sp.

Annelids exhibit diversified feeding habits, *viz.*, predators, scavengers and filter feeders. The gut is a straight muscular tube with mouth and anus capable of peristaltic constrictions. The antagonism of circular and longitudinal muscle acting against hydrostatic skeleton facilitates movements with the assistance of structures like parapodia polychaeta, chaeta (oligochaeta) and suckers (Hirudinea). The circulatory system is closed type carrying blood which may contain respiratory pigments (haemoglobin, chlorocruorin). Dorsal vessel pumps blood anteriorly and ventral vessel pumps posteriorly with the help of valves. Segmental vessels return blood to the dorsal vessel via bodywall or gut. Gas exchange taken place through the moist body wall.

Excretion and osmoregulation is through ciliated segmental nephridia, two for each segment. The nervous system has cerebral ganglia which are linked to a double ventral nerve cord by circumpharyngeal commissure. Cephalization is clearly seen. Eyes with retinal structure and lens and statocysts are also seen. Polychaetes have separate sexes with gametes produced in several segments, development through well developed trocophore larva in marine polychaetes, oligochaetes and hirudineans are hermaphrodites with direct development.

Characteristic Features and Classification :

1. Triploblastic, coelomate, bilaterally symmetrical, metamerically segmented animals.
2. Perivisceral coelom.
3. Preoral prostomium.
4. Central nervous system of paired cerebral ganglia connected to ventral nerve cord by commissure.
5. Segmental, ectodermal origin, ciliated nephridia.
6. Closed circulatory system.
7. Definite cuticle secreted by ectoderm.

The Phylum is divided into three Classes, polychaeta, oligochaeta and hirudinea and their representative animals are shown in Fig. 3.12.

Class polychaeta	Class Oligochaeta	Class Hirudinea
Marine	Inhabit freshwater or	Ectoparasite with suckers.
Distinct head	No distinct head	No distinct head
Setae on parapodia	Few chaete, No parapodia	Fixed number of segments
Dioecious	Hermaphrodite	No chaete or parapodia
Gonads, throughout body	Gonads in few segments	Hermpahrodite
Fertilization is external	Cross fertilization	
Free swimming troco-phore larva	Clitellum with egg laid in cocoon	Cross fertilization Eggs laid in cocoon
eg. *Nereis* (clam worm)	Development direct. eg. *Pheretima* (earth worm)	development direct eg. *hirudo* (leech)

There are three related minor phyla to Annelids. Those are Echiura, Sipuncula and Pogonophora.

Phylum Echiura

The Echiurians (spoonworms) found in rock crevices, or burrows in mud and sand of shallow seas, and with body size of 15 to 50 cm long. There are about 130 species. They are bilaterally symmetrical, triploblastic coelomates with cuticle covering body, however segmentation is absent. Spoon worms have spoon shaped extensible proboscis. The sexual dimorphism is shown and fertilization is external.

Example : Echiurus.

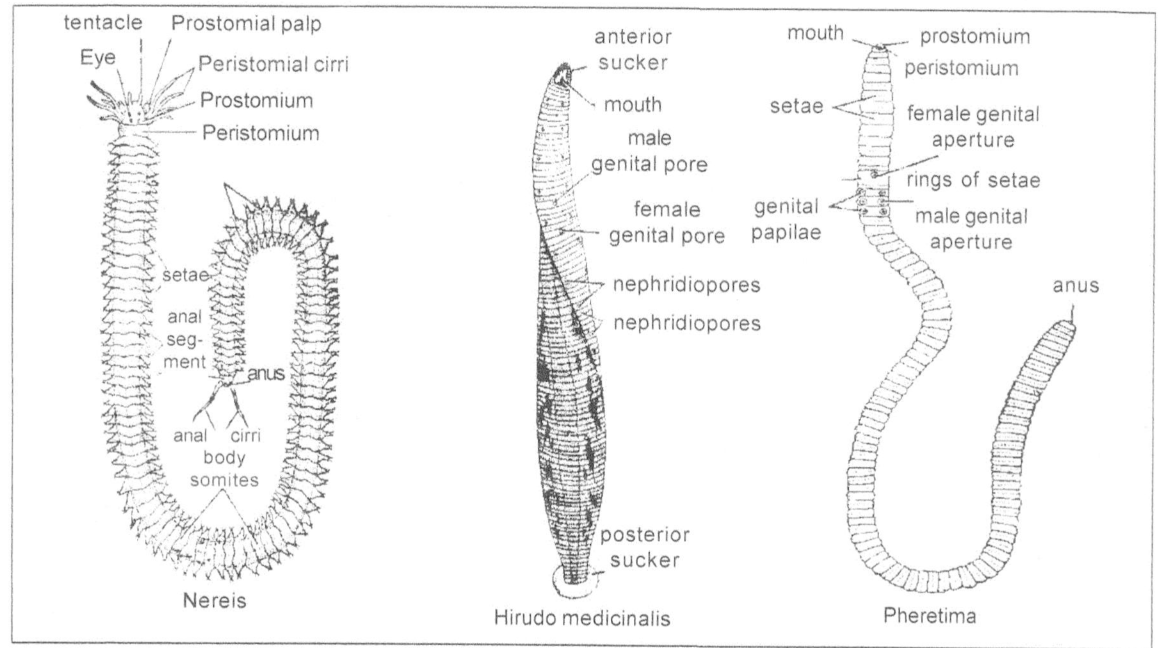

Fig. 3.12 Representatives of Annelida

Phylum Sipuncula

These (Peanut worms) are unsegmented, bilaterally symmetrical, triploblastic coelomate, burrowing marine worms, about - 300 species, and body size ranges from 2 mm to 75 cm. The sexual dimorphism is shown; fertilization is external, the gametes are released to outside through excretory metanephridia.

Example : Themite lageniformes.

Phylum Pogonophora

The tube dwelling pogonophores (Beard worms), known species are 120, lives on edges of continental shelves of marine waters and are bilaterally symmetrical triploblastic coelomates. The body length varies from 10 cm to 2m. They have characteristic anterior tuft or beard of tentacle. These are unusual in lacking a gut and food is taken through epithelium. Sexes are separate. Though it is classified under protostomes, some of the characters of this group have affinities with deteurostome - hemichordates.

Example : Siboglinum sp. (Fig. 3.13).

3.2.8 Phylum Onychopora

Phylum onychopora represents small number of about 70 existing species, which have not changed characters from long time (living fossil), are small caterpillar like animals living in moist habitats of tropical and temperate regions, These are bilaterally symmetrical, triploblastic coelomates with protostome pattern of development. The body structure is similar to that of Annelids and with reduced coelom. They are active predators possessing complete gut with mouth and anus. Locomotion is through numerous fleshy limbs. The exoskeleton has a cuticle containing chitin. Growth like arthropods is effected by successive moulting. Respiration is by trachea and open, internal transport vascular system is present. The excretion is by nephridia.

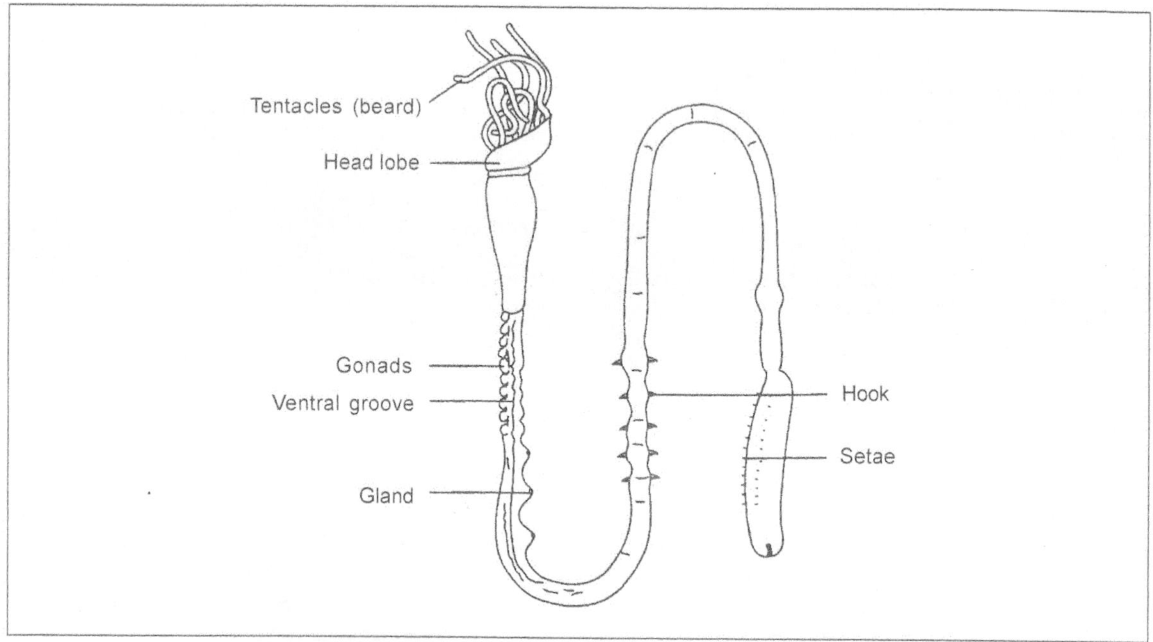

Fig. 3.13 A pogonophora (beard worm), e.g. Siboglinum sp.

The structure of body wall, nephridia, thin flexible cuticle and nonjointed appendages are annelidan characters. However, onychoporans are more like arthropods in characters like reduced coelom, moulting of cuticle, paired appendage for feeding, tracheae for gas exchange, open circulatory system with dorsal tubular heart etc. Therefore, it is grouped as independent phylum and treated as evolutionary *missing link* between the Annelids and Arthropods.

Example : Peripatus sp. (Fig. 3.14).

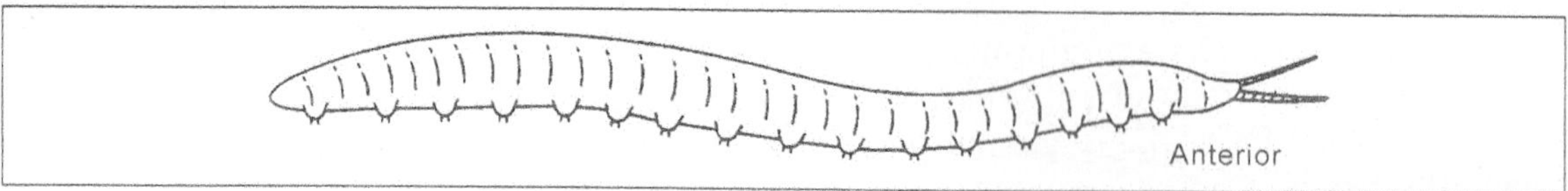

Fig. 3.14 An onychophoran, e.g. Peripatus sp.

3.2.9 Phylum Arthropoda

Arthropoda represents jointed-legged animals with chitinous exoskeleton, represents largest number (about 10 millions) of animals, with body size between 0.1 mm to 60 cm long and with diversified habitats like freshwater, marine and terrestrial. The animals include crustaceans, millipedes, centipedes, insects and arachnids. Arthropods are bilaterally symmetrical, triploblastic, coelomate, segmented animals with body usually divisible into distinct-*head, thorax* and *abdomen*. Each segment usually possess a pair of *jointed appendages* for feeding, locomotion, sensory or reproductive functions. The coelom is reduced and represents the cavities of reproductive organs and pericardium. The main body cavity is *haemocoel*.

The jointed limbs are used for locomotion on land and in water and insects have wings for powered/gliding flight activity of neuromuscular system. Arthropods are enclosed in an exoskeleton of chitin with calcium and wax. These have complete gut with a mouth and anus, foregut and hindgut are lined with chitin. Arthropods have well developed mouth parts, predacious carnivores, herbivores, fluid feeders and parasitic forms found. These animals have discrete respiratory organs like *gills, trachea, booklungs* etc. The contractile dorsal tubular heart with neuorgenic control along with haemocoel and blood pigments constitute vascular system. The excretion and osmoregulation is performed through specific structures like *green gland, malphigian tubules*. The nervous system is well developed in pattern of annelid, including with double ventral nerve cord and sensory structures like simple and compound eye, statocysts, etc. The sexes are separate, fertilization is internal and eggs are centrolecithal, cleavage is spiral but superficial. Life histories are characterized by many stages, involving ecdysis or moulting process and metamorphic changes.

Salient Features and Classification upto Classes are given below :

1. Triploblastic, coelomate, segmented, and bilaterally symmetrical.
2. Secreted exoskeleton of chitin.
3. Each segment with a pair of jointed appendages.
4. Cilia absent.
5. Coelom in reduced form of perivisceral cavity-haemocoel.
6. No nephridia but malphigian tubules and, green glands are present.
7. Dorsal heart with open vascular system.

Class Crustacea	Class Chilopoda	Class Insecta	Class Arachnida
Mainly aquatic	Mainly terrestrial	Mainly terrestrial	Terrestrial
well defined cephalothorax	Clearly defined head	Head, thorax, abdomen	Prosoma and opisthosoma
Head, with 2 pairs of antennae	Head with one pair of antennae	Head with 1 pair of antennae	No antennae
Three pairs of mouth parts, gnathites	One pair of gnathites	Pair of compound eyes or simple eyes	Simple eyes
Abdomen with 11 segments	Numerous similar segments	3 pairs of gnathites	Segments of 4-7 of legs.
Genital aperture thoracic	Median genital opening	3 thoracic segment with a pair of apendages and 2 pairs of wings	Abdomen with not seen externally
Development is direct or nauplius or other larval form	No larval form	Abdomen of 11 segments	No larval form
Gills - respiration. e.g., Astacus (crayfish)	Trachae - respiration	Complicated metamorphosis, development direct or indirect (larval form)	Tracheae or book lungs respiration e.g. Scorpio
Daphnia (water flea)	e.g., Julus (millipeda) Lithobius (centipede)	Trachae -respiration e.g. Periplanata (cockroach) Apis	(scorpion) Araneus (spider)

8. Central nervous system of paired cerebral ganglia connected by commissures to ventral nerve cord.
9. Development with ecdysis or larval forms.

The phylum is divided into four major classes, Crustacea, Chilopoda, Insecta and Arachnida and representative animals are shown in Fig. 3.15.

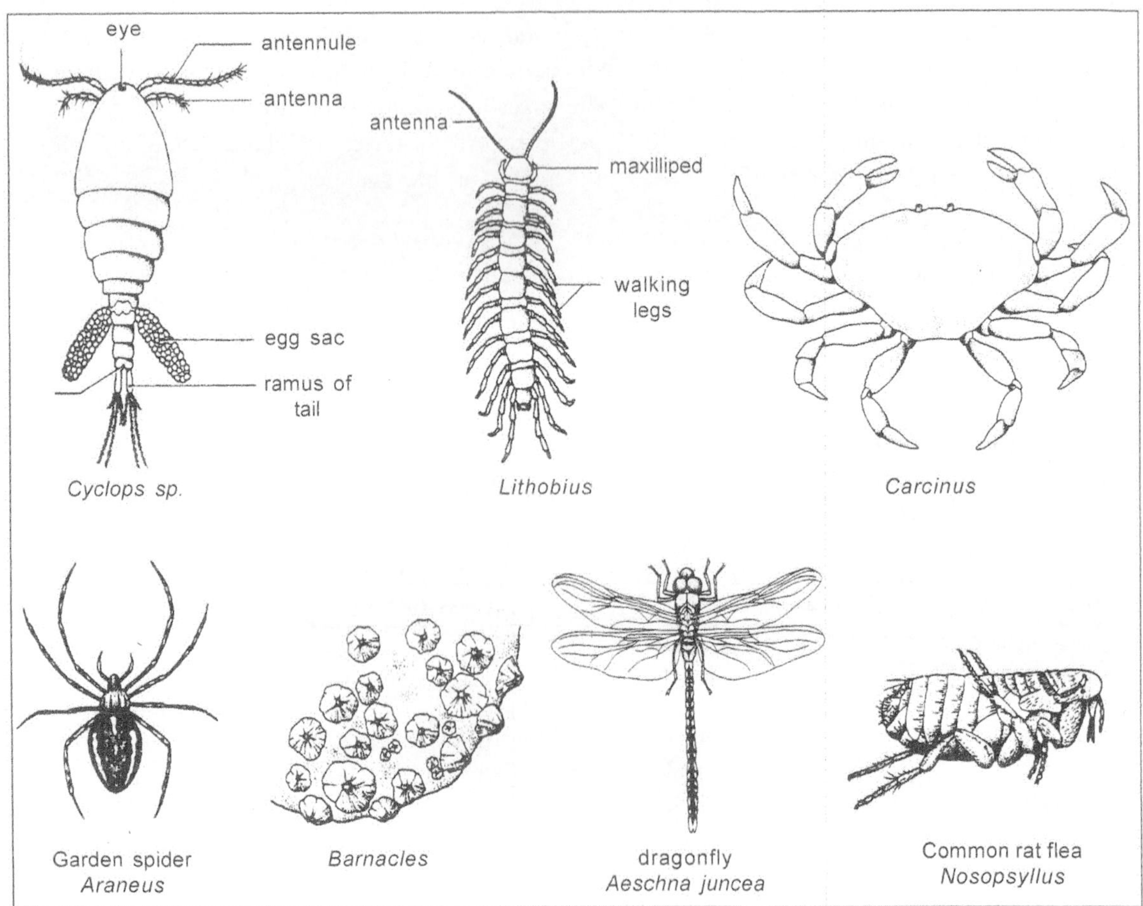

Fig. 3.15 Representatives of arthropods

There are two minor phyla *Pentastomida* and *Tardigrada*. *Pentastomida* (tongue worms)- these are about 90 species of endoparasitic form of carnivorous vertebrate especially reptiles. An example is Linguatula.

Tardigrada (water-bears) there are about 500 species, of size up to 1mm in length, live in intertidal zones, on edges of fresh water habitats and surface water films, lichens and bryophytes.

Example : Echiniscus.

3.2.10 Phylum : Mollusca

This phylum includes diversified forms about 1,10,000 species and also well preserved extinct forms, with variable body size 1mm to 1.3m, (The largest known invertebrate, the giant squid, *Architeuths* sp. with tentacles of 20 m length). Most of animals have internal or external shell. Shell is secreted by mantle which covers the body. All mollusca are aquatic, with some having damp terrestrial habits. The body of mollusca is unsegmented, triploblastic, coelomate comprising head-foot and a visceral mass, bilaterally symmetrical during development only, the foot is large muscular organ, used variably for locomotion, burrowing, attachment, or extended into arms and tentacle has seen in cephalopods. The visceral mass contain all internal organs. The haemocoel, open circulating system is predominant with reduced coelom.

The gut is complete, most are sedentary filter feeders (bivalves) or large molluscans like squids are active macrophageous predators. Molluscans characteristically possess chitinous rasping structure, *radula* in buccal cavity. Foot or modified foot is used for movement. Respiration is through body surface, (respiratory tufts), gills-ctenidia or vascularised cavities with blood and pigments (haemocyanin). Open circulatory system and dorsal heart constitutes the circulation. Excretion is by metanephridia or simple kidneys.

Extensive cephalization is observed and sensory organs include statocysts, head tentacles, ospharadia, paired eyes etc (except in bivalves). The brain along with paired pedal and visceral cords is present in nervous system. The most sophisticated of all invertebrate nervous system is found in cephalopods. The sexes are separate, and fertilization is external in aquatic mollusca. Eggs hatches to form a trochophore which metamorphose into a veliger larva, however, cephalopods and land snail directly develop into miniature adults.

Characteristic, features and classification of Mollusca

1. Unsegmented triploblastic coelomates,
2. Body is divided into a head, ventral muscular foot and dorsal visceral hump
3. Skin soft and mantle forms shell
4. Respiration through gills or body surface
5. Open haemocoelic system with heart
6. Nervous system with cerebral, pleural ganglia, pedal cords and visceral connective
7. Oviparous with trocophore larva

There are five classes of Mollusca-Gastropoda, Pelcypoda, Cephalopoda, Monoplacophora and Polyplacophora; and the representatives are shown in Fig. 3.16.

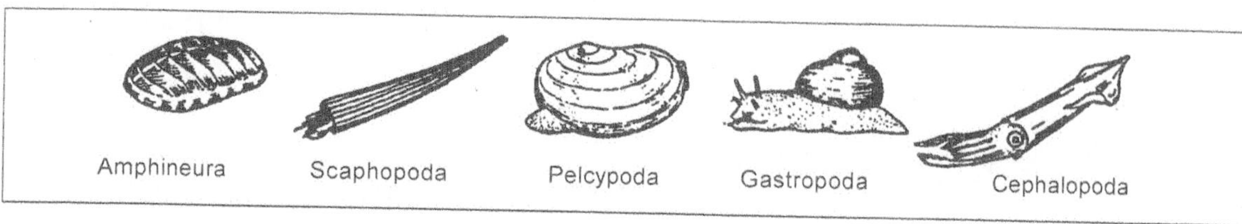

Fig. 3.16 Phylum Mollusca. The five classes

Class Gastropoda	Class Pelcypoda	Class Cephalopoda
Terrestrial, Aquatic	Aquatic	Aquatic
Asymmetrical	Bilaterally symmetrical	Bilateral symmetrical
Torsion and detorsion	Body laterally compressed with two valves, hence Bivalvia	Shell is internal reduced or absent Head, tentacles and eyes highly developed.
Anus is anterior	Head reduced, tentacles absent	Gills
Coiled one piece shell	Plate like gills	Radula and horney beak
Head with eyes and tentacles		
Radula	Filter feeder	
e.g. Limax (slug) Helix (land snail) Patella (limpet)	External fertilization e.g. Mytillus (mussel) Ostrea (Oyster)	Internal fertilization e.g. Sepia (cuttle fish) Loligo Octopus

3.2.11 Phylum Echinodermata

Echinoderms are marine forms, about 6000 species are extant and many fossil species are known. These are bilaterally symmetrical in early development on which pentaradiate symmetry about oral-aboral axis is imposed in adult forms. They are also unsegmented and triploblastic coelomates and coelom modifies into specific water vascular system and main perivascular system. The gut is complete with mouth and anus, and they have predatory, scavenging or suspension feeding. Locomotion is affected by tube feet in some forms (star fishes) and other forms, locomotion is with spines associated with muscular movements (brittle stars). The surface of echinoderm is covered with epidermis embedded with endoskeleton of calcareous plates and with pedicellarie.

Respiration is through body surface and in some individuals it is affected by papillae, tiny gills, tube feet or specific respiratory trees. As there is no true blood, coelomic fluid serve respiratory, excretory (no specific excretory organs) and circulatory function. Amoebocytes present may have equivalent function of lymphocytes. Nervous system is very simple with circum oral nerve ring, radial nerves and nerve net and cephalization is lacking. Fission and impressive regeneration process are seen. The sexes are separate with external fertilization and development is deuterostome type. The ciliated, echinoderm larva-dipleurula and other larvae-with bilaterally symmetry metamorphose to pentaramous adult. Development pattern and larvae shows the echinoderm's link with chordates.

Characteristic Features and Classification

1. Pentaradiate, triploblastic coelomates
2. Calcareous exoskeleton

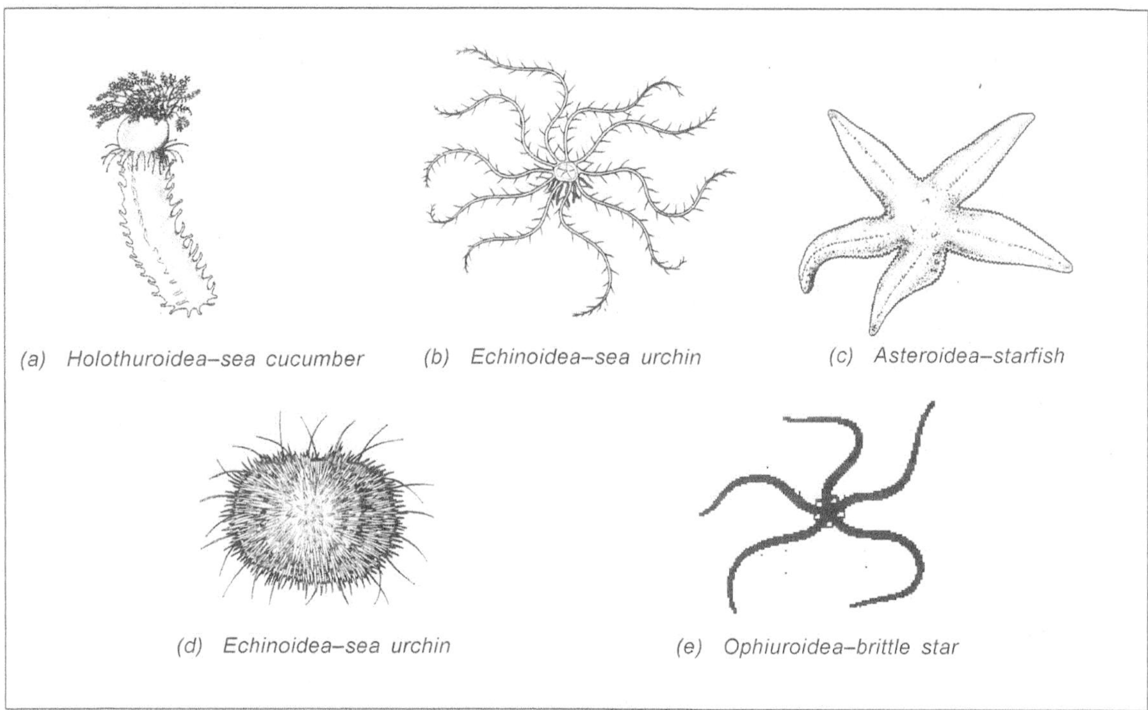

(a) Holothuroidea–sea cucumber (b) Echinoidea–sea urchin (c) Asteroidea–starfish (d) Echinoidea–sea urchin (e) Ophiuroidea–brittle star

Fig. 3.17 A variety of echinoderms

3. Tube feet, pedicellariae, water vascular system
4. No excretory organs
5. Simple nervous system
6. Basic ciliated bilaterally symmetrical pelagic larva.

The phylum echinodermata is divided into five classes : Asteroidea, Echinoidea, Crinoidea, Holothuroidea and Ophiuroidea and the representative animals are shown in Fig. 3.17.

Class Asteroidea	Echinoidea	Holothuroidea	Ophiuroidea	Crinoidea
Free living Star shaped, flattened	Free living Globular	Free living Cucumber shaped	Free living Star shaped	attached by a aboral stalk star shaped
Arms - not distinct from disc	No arms	No arms	Very long arms	arms are present
Few calcareous plates in body wall, movable spines	Numerous plate in body wall	No external spines	demarcated from disc	
e.g. *Asterias* (star fish)	e.g. *Echinocardium* (Sea urchin)	e.g. *Holothuria* (Sea cucumber)	Spines and calcareous disc plates present e.g. *Ophiothrix* (brittle star)	no spines e.g. *Antedan* (feather star)

Minor Phyla

1. ***Chaetognatha*** – Represent about 100 species of arrow worms, marine forms, with characteristic grasping spines on head (chaetognath). *Example - Sagitta*
2. ***Bryozoa or ectopracta*** – Colonial marine forms with test, and have lophophore (U - shaped ridge with 1 or 2 rows of hollow cilliated tentacles) commonly known as moss animals. e.g. - *Plumatella*
3. ***Brachiopoda : (lampshells)*** - These animals have two-valved shells and lophopore lies within shell. These animal lie buried in sand or mud and protrudes above and supported by worm like stalk. Excellent fossil record is available. e.g. *Lingula*
4. ***Phoronida :*** It represents about 15 species, marine worms living in chitnous tubes in mud or sand or attached to rocks. Cilia and lophopore help in feeding. e.g. is *Phoronis*.

3.2.12 Phylum Hemichordata

All are hermaphrodites, and marine, about 90 species, live in open ocean or muddy sediments. Earlier it was included under phylum chordata (protochordata) because they posses dorsal nerve

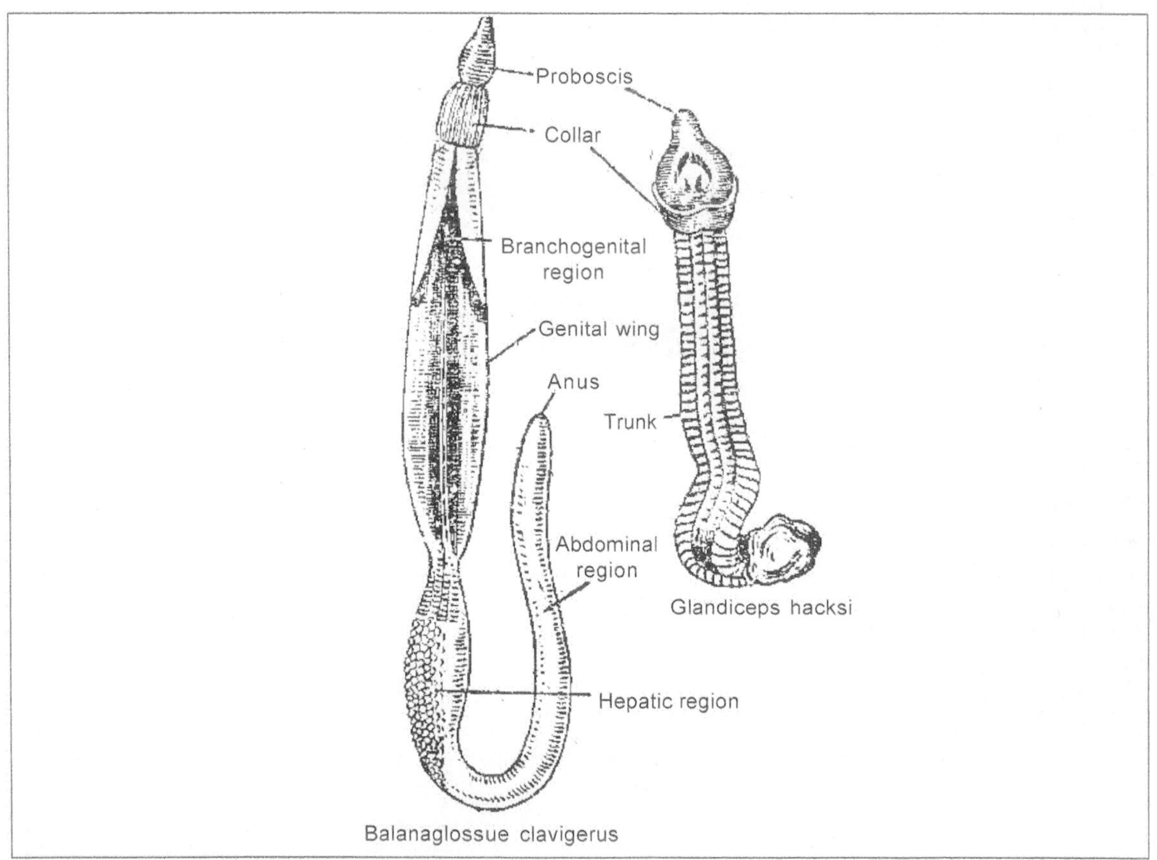

Fig. 3.18 Two species of scorn or tongue worms

cord often hollow developed from epidermis. The longitudinal stiffening stomatochord, homologous to notochord of chordata is present hence it is separated from chordata and treated as independent phylum. The Hemichordates are triploblastic, bilaterally symmetrical coelomates with unsegmented body divided into three regions. The body is divided into prosoma, mesosoma and metasoma. These animals have pharyngeal gill slits. Hemichordata are either filter feeders using lopophorate (pterobranch) or detritus feeder (enteropneuts). They are sessile (pterobranchs) or burrowers (enteropneusts). Respiration is through gas exchange across vascularized membrane of gill slits or body surface. Circulatory system consists contractile ventral and dorsal vessels linked to open sinuses and pulsatile heart which pumps blood through open circulatory system. Elimination of wastes is linked through general body surface. Nervous system consists of mid dorsal and mid ventral nerve cord by neural rings and sub-epidermal nerve nets. Sexes are separate, fertilized eggs shown duterostome type of development and also include tornaria larva. Pterobranch also show asexual reproduction by budding.

Hemichordates are divided into two classes, pterobranchiata and Enteropneusta.

Pterobranchiata

1. Colonial, tubiculous marine forms of size upto 8 mm,
2. Have lophopore like ciliated tentacles filter feeder *Eg.* Rhabdopleura, cephalodiscus.

Enteropneusta

1. Marine solitary acorn worms
2. Proboscis, coller, vermiform body with numerous gill slits.
   Eg. Balanoglossus sp. (Fig. 3.18).

3.2.13 Phylum Chordata

The chordata include about 45000 species of diverse nature and live in fresh, brackish, marine waters and terrestrial habitats. All animals are broadly grouped into proto chordates, (or invertebrate chordates), urochordates and cephalochordates, chordates which are more prevalent, diverse and successful animals.

Chordates are bilaterally symmetrical, triploblastic, segmented coelomates with dueterostomic mode of development and they characteristically show at some stage of development, (i) Dorsal tubular nerve cod, (ii) Notochord (iii) Pharyngeal gill slits (iv) Segmented muscle and (v) Usually a tail.

Chordates are characterized by an *endoskeleton*, which is made up with *cartilage* (elasmobranches) and replaced by harder *bone* in further evolved animals like teleost-fish, amphibians, reptiles, birds and mammals. In lower chordates, *notochord*, consisting of gelatinous matrix with connective tissue, is an anti-telescopic cartilaginous rod dorsal to gut and ventral to nerve cord. In vertebrates, it gradually replaced by segmental vertebrae surrounding nerve cord as vertebral column. Most of chordates actively swim using muscles with fulcrum of notochord or vertebral column. The gradual evolution of paired, median and caudal fin, like in fishes, control the swimming. Further, in land vertebrates fins are replaced by jointed limbs which act in movement of body. Chordates have evolved flight with suitable modification of skeleton and muscles like in birds. In addition, there are secondary modifications like sessile nature in urochordates, loss of limbs in snakes, modification of limbs to flipper as seen in aquatic mammals etc.

Protochordates are filter feeders and jawless vertebrates are semiparasites on jawed fishes and most of other chordates are herbivorous or carnivorous macrophagous feeders. Presence of teeth is universal in vertebral jaws. The gut is complete, with associated elaborate development of digestive glands. Endostyle present in protochordates is homolog of thyroid glands of vertebrates. Respiration is by pharnygeal gill slits in protochordates, replaced by gills in amphibian larvae and fishes. Terrestrial vertebrates use lungs for respiration which develop from gut similar to that of swim bladder of fishes. Most chordates have a high pressure, closed circulation with powerful pumping ventral heart and elaborate arterial and venous system, with single circulation in fishes and double circulation in other vertebrates.

The excretory and osmoregulatory function in protochordates is either by diffusion (Urochordates) or with solenocytes (cephalochordates), while vertebrates have efficient excretory organ, kidneys. The characteristic hollow nerve cord shows (except degeneration in urochordates) progressive cephalization forming central brain and spinal cord with sensory, motor and integrative functions and peripheral sensory, motor and autonomic nervous system. In addition, the ductless endocrine system, consisting of pitutory, thyroid, parathyroid etc., developed and function together with nervous system for well coodrinated, complex living of higher animals. The vertebrates also developed immune system for protection. Reproduction is normally sexual (except parthenogenesis and budding seen in some lower chordates) with amniotic and anamniotic egg formation and excellent parental care.

The phylum chordata is divided into three subphyla urochordata, cephalochordata and craniata (vertebrata).

Sub-Phylum – Urochordata (tunicates)

Tunicates, about 1300 species, have larva with chordate features (notochord, dorsal nerve cord, segmentally muscled tail) which loses these features in *retrogressive metamorphosis* and develops into adult. Urochordata have the following salient characters, no trace of notochord and nerve cord (reduced to round ganglion-neural gland); adult sessile filter feeder, nothing like a chordate, pharynx with branchial basket of gill slit for filter feeding and respiration, presence of endostyle, body enclosed in a test, rudiments of immune systems, blood cells with vanadocytes. Reproduction by asexual budding and sexual with a tadpole like larva.

There are three classes, the Ascidacea, Larvacea and Thaliacea.

Ascidacea – Sessile sea squirts – *Herdmania* (Fig. 3.19).
Larvacea – Adult with some larval features – *Oikopleura*
Thaliacea – Motile with jet movement – *Salpa*.

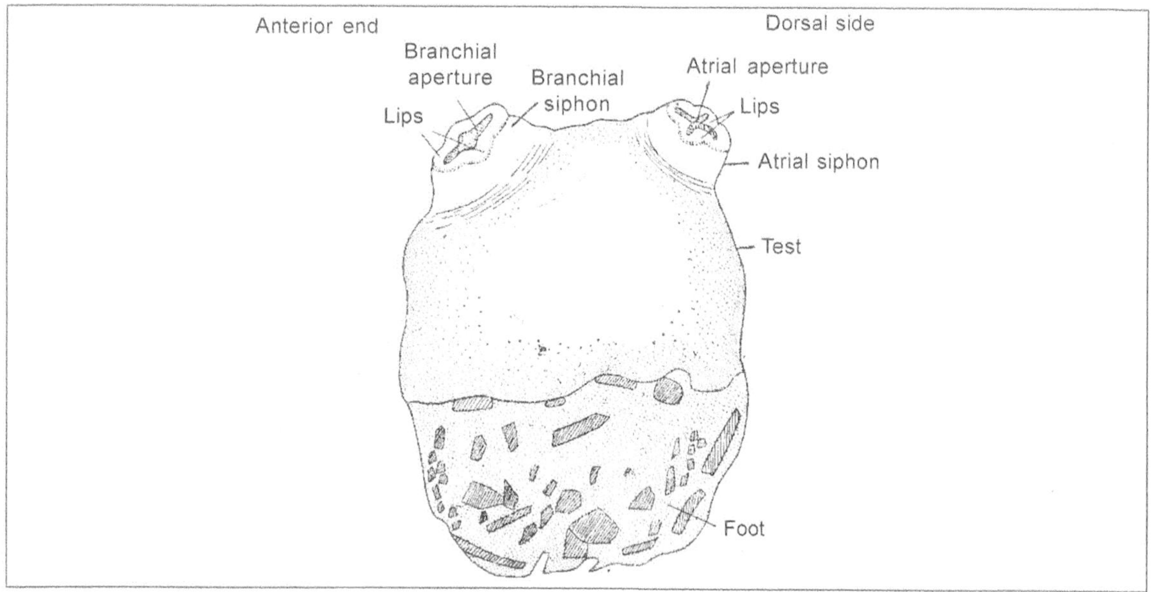

Fig. 3.19 Herdmania

Sub-Phylum – Cephalochordata

It represents two genera. These are fish like animals showing basic chordate characters, like persistent notochord, pharynx with ciliated gill slits, segmental myotomes. The special features of this group is asymmetrical body architecture. No head, with little cephalization filter feeding, excretion by solenocytes, contractile blood vessels without heart and no blood cells are some of salient feature of this group. Example : Amphioxus or Branchiostoma (Fig. 3.20).

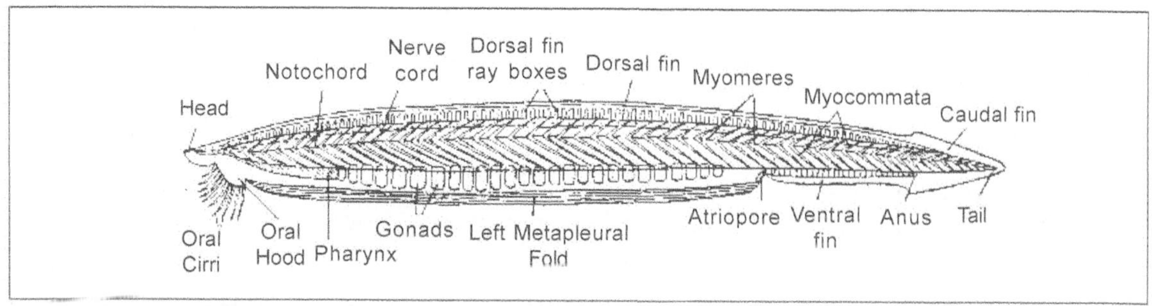

Fig. 3.20 Branchiostoma

Sub-Phylum – Craniata (Vertebrata)

Craniata possess characteristic chordate features of notochord, hollow dorsal nerve cord, pharyngeal gill slits and a post anal tail. All craniates are characterised by well developed brain in cranium (Box like structure with skull bones) and elaborate presence of sensory organs in head. Notochord is replaced by vertebral column. Besides these animals have internal skeleton, two pairs of limbs, muscular ventral heart, kidneys for nitrogenous excretion and osmoregulation.

The subphylum is divided into seven classes : Agnatha (cyclostomata), chondricthyes, osteicthyes, amphibia, reptilia, aves and Mammalia.

Class Cyslostomata or Agnatha

The absence of jaw is distinctive feature. These are earliest known vertebrates with many extinct ostracoderms and present living cyclostomes comprising lampreys and hag fishes (Fig. 3.21). They are semiparasitic with streamlined body, slimy skin, persistent notochord, seven pairs of gills, lack of paired fins, cartilaginous skeleton, simple eyes etc.

e.g., Lamprey - Lampretta; Petromyzon, Myxine.

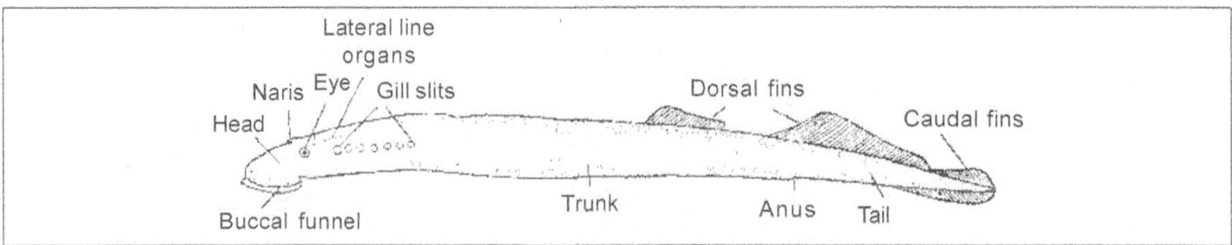

Fig. 3.21 Petromyzon – the sea lamprey

Class Chondricthyes

These are mostly marine forms, cartilaginous fishes, first seen in devonian period have following characters; skin is with placoid scales; cartilaginous skeleton; paired fleshy pectoral and pelvic fins; gills with separate opening and visceral clefts, lateral line system developed, only inner ear, internal fertilization, eggs produced and no larval stage.

e.g., Dogfish - Squalus, ray fish - Raja, chimera-Chimeara (Fig. 3.22).

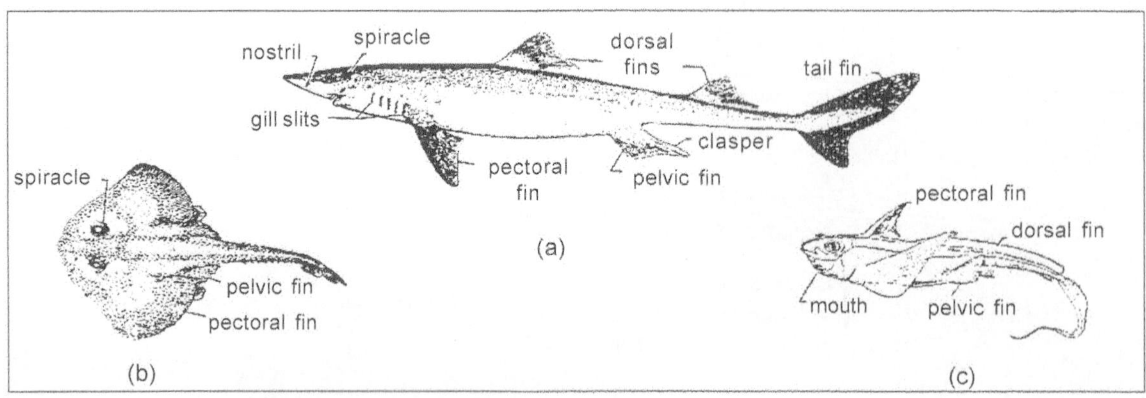

Fig. 3.22 Class Chondrichthyes. (a) Spiny dogfish or shark (Squalus), (b) Ray (Raja) (c) Chimaera (Chimaera)

Class Osteicthyes

The bony fishes, more than 30,000 species (this number is more than all vertebrate put together), are aquatic, having following features. Skin with cycloid scales, bony skeleton (except embryonic cartilage skeleton), paired pectoral and pelvic fins supported by rays, visceral clefts as openings but covered by operculum, presence of swim bladder, lateral line system well developed, only inner ear present, external fertilization, eggs are produced with larval stage. Osteicthyes represents ray-fined fishes, teleosts, crossoptergions (fleshy or lobe fined fishes), dipnoi (lung fishes). The examples are Hippocampus (sea horse), Latimaria (coelocanth), Neoceratodus (lung fish) (Fig. 3.23).

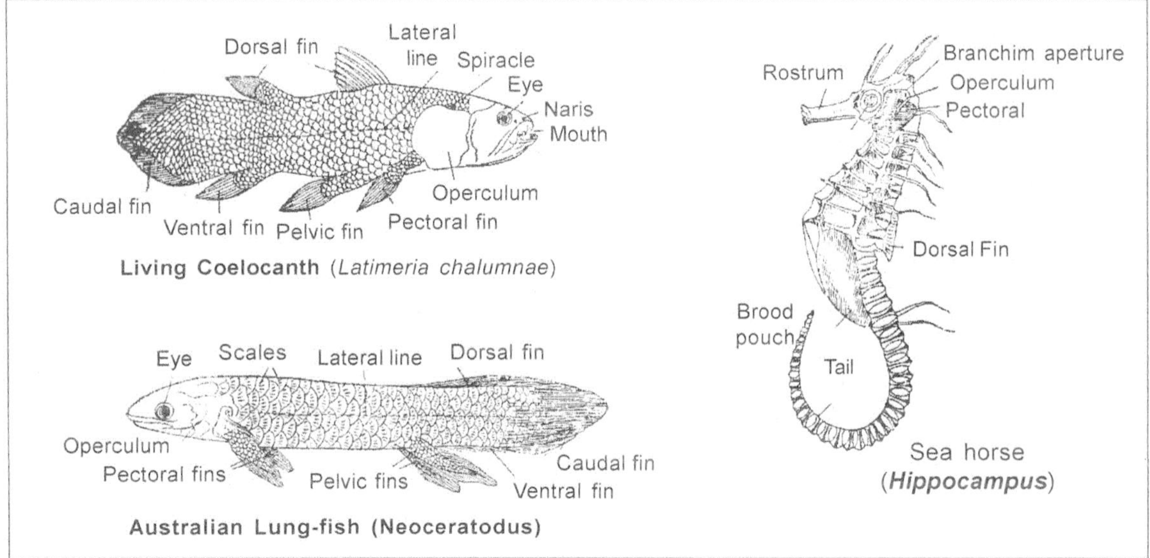

Fig. 3.23 The bony fishes

Class Amphibia

Amphibians evolved as labrinthodontia from crossoptergian fishes during early devonian period and successfully established as Lissamphibia the modern forms of frogs, toads, salamander etc. Amphibians are characterised by soft skin, paired pentadactyl limb, lungs in adult while visceral clefts and lateral system only in tadpoles, inner and middle ear and no external ear, external fertilization, eggs are telolecithal, development with larval stages, the amphibians in addition to their specialization in amphibious life and four chambered heart, evolved with strong pelvic and pectoral girdle, with vertebral column toward tetrapod locomotion. Some of the forms (urodela) exhibit *neoteny* (retention of larval characters in adult) and *paedogenesis* (sexual maturity in the larval stage).

The living amphibians (Fig. 3.24) are classified into three orders (i) *Anura* - The toads and frogs in which adult lack tail e.g., leopard frog, Rana, toad-Xenopus. (ii) *Urodela* - newts and salamandar with long tails and neoteny, e.g., tiger salamandar, Ambystoma. (iii) *Apoda* - limbless burrowing amphibians, e.g., *Icthyopis*.

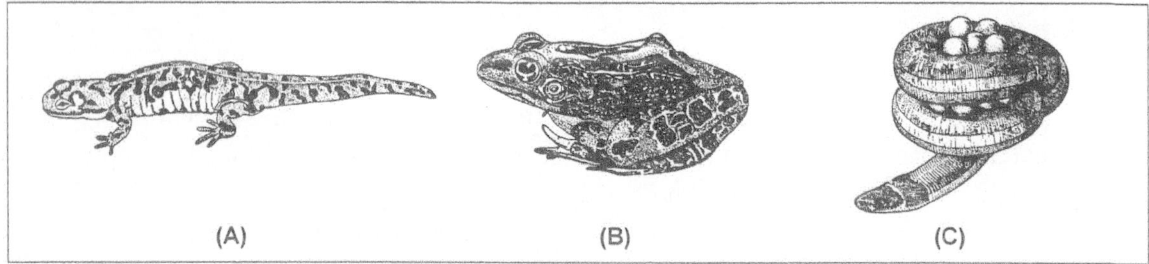

Fig. 3.24 Representative amphibians A. Tiger salamander (Ambystoma tigrinum). B. Leopard frog (Rana pipiens). C. tropical caecilian or limbless amphibian (Ichthyopis glutinosus).

Class - Reptilia

The labrinthodont amphibians gave rise to ancestral stem reptiles (cotylosauria) during late devonian period and reaching peak in mesozoic. This group evolved into various taxa, the chelonia (tortoise and turtles), lepidosauria (lizards, snakes), synapsida (mammal like reptiles), archosauria (crocodiles and flying pterosaurs, dinosaurs) etc. The characteristic features of reptile are dry skin with horny scales and body plates, paired pentadactyl limbs, no gills at any stage only lungs, no lateral line, inner and middle ear and no external ear, internal fertilization, yolk eggs in calcareous shell and no larval stage.

Reptiles, so also birds and mammals, called as *amniota*, which have an amniotic egg, have three extraembryonic membranes. Reptiles are divided into four orders : (i) *Rhyncocephalia* - known for its primitive reptilian characters and living fossil - example - Sphenodon (ii) *Squamata* : includes animals with scales - lizards and snakes boa, viper. (iii) order - *chelonia* - body enclosed in shell of bony plates, horny beek in place of teeth - eg. painted turtle - Chrysemus, green sea turtle - Chelonia. (iv) *Crocodilia* : skin is covered with body plates completely divided into four chambers, example - Alligator (Fig. 3.25).

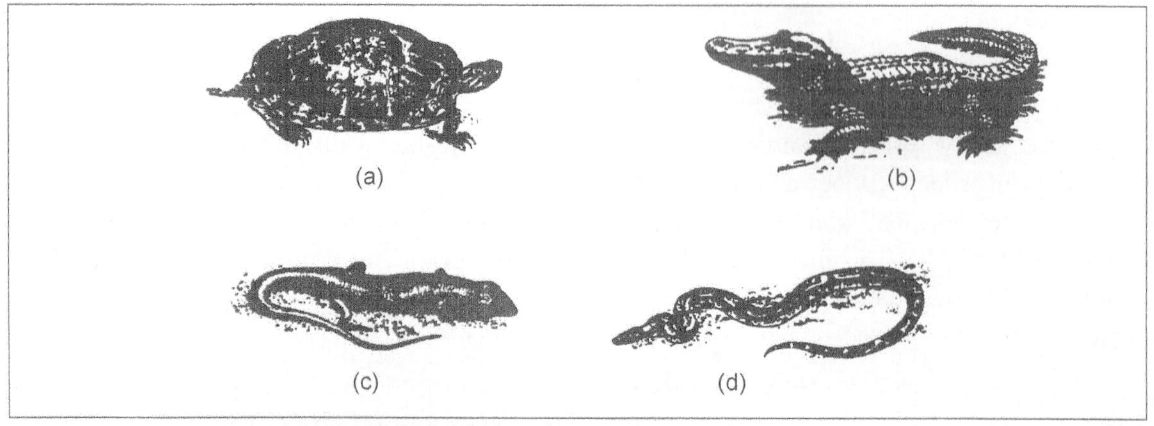

Fig. 3.25 Types of living reptiles (class reptilia). (a) Painted turtle (Chrysemys). (b) Alligator (Alligator) (c) a lizard (Eumeces), and (d) Boa, a snake.

Class Aves

The birds are evolved from saurischian archosaurs - lizard, hipped dinosaurs of jurassic period and fossil birds (archaeopteryx) show remarkable similarities to the same. Reptilian scales have evolved into feathers in birds. The fore limbs modified to form wings help in flight and the bones are pneumatic. The lungs are large, efficient, associated with air sacs, heart is four chambered and along with circulatory system, 90% of oxygen is utilized. Skull is large with big brain and large eye orbits. There are no teeth in birds and beak is developed to suit its feeding. All birds are oviparous, with parental care. The distinctive features of Aves are, skin bear feathers, legs with scales, paired pentadactyl limbs, front pair form wings, pneumatic bone, flight muscles, four chambered heart, lungs with air sacs, yolky eggs in calcareous shell, homiothermic animals.

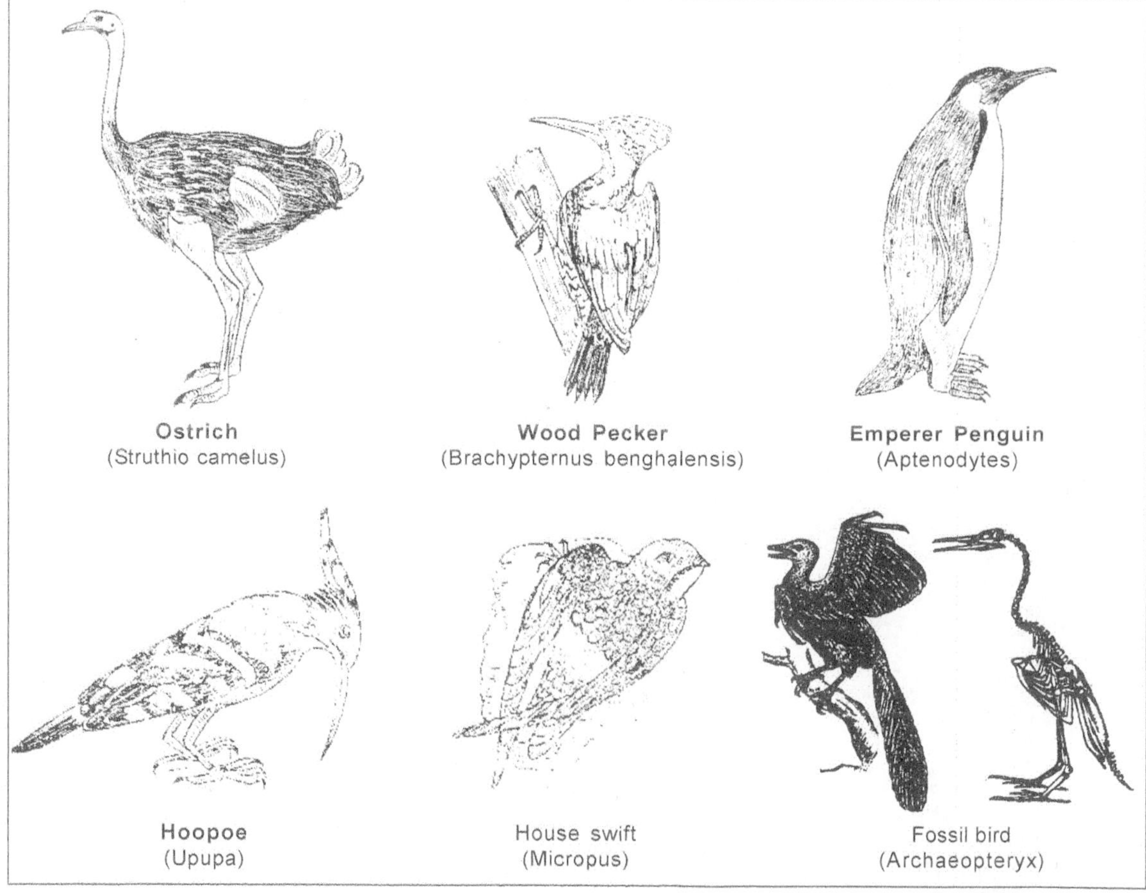

Fig. 3.26 Representative of Aves

The class is divided into two important superorders. (i) *Paleognathae ratitae* - Usually flightless birds, palate is specific e.g., Ostrich, Strutio, Kiwi, Apteryx. (ii) *Neognathae*-There are 20 orders, this includes all birds except ratitae.

e.g., House swift - (*Micropus*), penguin - (*Aptenodytes*), wood pecker, (*Brachypternus*).

Class Mammalia

These are highly diversified and evolved animals including primates and man. Mammals are originated from synopsid reptiles - cyonodonts and icthieosaurs. Mammals have two types of glands sebaceous and sudoriferous gland, presence of gland for suckling, single dentory bone in lower jaw, three auditory ossicles, external ear present, teeth in upper and lower jaw diversified and adapted to diet. The lungs have alveoli, respiration assisted by muscular diaphragm separating thoracic with abdominal cavities. Like birds, heart is four chambered with pulmonery and systemic circulation, internal fertilization, viviparous, and parental care.

Fig. 3.27 Representatives of Mammals

The class is divided into two subclasses :

1. *Monotremata or Prototheria* – lays eggs and milk is produced by gland - 3 living species from Australia, e.g., Duck billed platypus *Ornithorhynchus*, Spiny ant eaters *Tachyglossus*
2. *Theria include all other mammals* – grouped into subclass (i) *metatheria* (marsupialia) - young are born immature and kept attached to nipple in body pouch, marsupium, and marsupials have epipubic bone. e.g, American opassum-Didelphus, Kangaro - *Macropus* and (ii) *Eutheria* (placentalia)-the young grow within uterus, using chorio-allontoic placenta. This subclass is divided into a number of living orders which are given with an example (Fig. 3.27).

   1. *Insectivora* – common shrew
   2. *Chiroptera* – bats
   3. *Primates* – monkey, apes
   4. *Carnivora* – lion, walrus
   5. *Cetacea* – dolphins, whales
   6. *Proboscidea* – elephants, dugongs
   7. *Artiodactyla* – camel, deer
   8. *Perissodactyla* – horses, zebra
   9. *Rodentia* – rats
   10. *Logomorpha* – rabbits

3.3 Adaptive Biology

The evolution and development of multicellular organisation, increasing size of the organism brought about many physiological and anatomical changes which were quite essential for animals to survive and reproduce successfully in order to continue the lineage as well as organic evolution. Thus, the adaptation may be defined in simple terms as "Morphological and physiological modification in an organism to adjust itself in a particular or demanding environment". Since the environment is ever changing, the organisms' ability to acquire changes for various causes, and accordingly change or adapt in order to survive the rigours of new requirements or otherwise ready for extinction. Therefore, the succession of above mentioned changes is paralleled by development of adaptive features, including morphological, anatomical and physiological as seen in plenty in animals. Some of the important biological changes are summarised below.

1. Large animals with many cells require much food than the unicellular protozoans. Animals have become entirely heterotrophic and in most cases holozoic. Development of an *alimentary canal and associated digestive glands* and structures has enabled the animals to ingest large food particles, digest them and absorb soluble products. A variety of *feeding habits* have been developed incorporating herbivorous, carnivorous, omnivorous, saprozoic, and parasitic modes of life.
2. *Integumentary, muscle and skeletal* (Exo and Endoskeletons) systems have developed to maintain general body shape, protect and scafold inner structure, provide efficient propulsive forces for the movement and provide internal movement of material within body.

3. The efficient system for *locomotion* was evolved to enable the organism to meet demands of food, as well as protection, breeding or other activities.
4. The *body cavities* (acoelomate, pseudocoelomate, eucoelomate) and *symmetry* (asymmetry, radial, bilateral) have evolved as adaptive feature to provide a compact, organized body facilitating efficient functions. For instance, cnidarians and echinoderms due to sessile or slow moving nature, radial symmetry and pentaramus is developed to countenance environmental changes from all directions. Similarly, the acoelom in platyhelminths, pseudocoelom in nematodes and eucolemate in chordates elaborate the adaptive significance of coelom in body organization.
5. The impermeability of outer covering of animals, require some specailized areas for exchange of materials and also increase in size result in spatial problems of separation of central tissues from body wall and environment. Thus, the animals have responded by development of *internal transport system*. This consists of fluid tissue, like blood and their pumping system in the form of vessels, aorta, veins, contractile heart, or respiratory tracts, gaseous exchange structures like gills, trachea, and renal ducts, genital ducts etc. The transport system provides a means by which oxygen, carbondioxide, soluble food, waste materials, cells, biomolecules, like hormones, vitamins, mineral etc., are transported throughout the body. They may serve their destined function or expelled from body.
6. The *nervous system* was evolved from a simple net like structure in cnidarians to highly complex nervous system found in mammals. The stimuli from environment is received by sensory organs (effectors), and effectors produce appropriate response. The major sense organs and nerve centres were concentrated at the anterior end, in head, to encounter any environmental changes ahead. This process resulted in *cephalization*. In order to have effective coordination, nervous system has evolved into distinct structure and functions as brain, spinal cord, cranial, motor, sensory and autonomic nerves. The development of *endocrine system* in the form of various glands and hormones complement the nervous system as coordinating system to maintain the animals steady state and behavior.
7. The development of a multicellular animal from a single celled zygote is often long complex process and has to face many adverse influences. Thus, there are many adaptive changes in animals like ability of prolific asexual reproduction, (sponges, polychaetes, echinoderm), high fecundity, (insects, helminths, fishes) evolution of internal fertilization and complete development upto young in parent (viviparous), parental care, protection and facilitation of embryonic development with extraembryonic membrane (reptiles, birds), metamorphosis (insects), larval stages, (crustaceans, echnidoerms, molluscs), parthenogenesis (insect) etc.
8. The plasticity or adaptivity in the organization of animals is best seen in adaptive radiation or divergence. The animals of unrelated groups which occupy the same habitat exhibit features in common (*Convergence*) while animals of same group or closely related group exhibit great differences in their morphology when they are found in different habitats. (*Divergence*). The limb structure of mammals provide a classical example of adaptive radiation or evolutionary divergence. The basic pentadactyl limb of mammals have adopted or evolved into five different habitats with corresponding five specific modifications in their limb structures, as seen in Fig. 3.28.

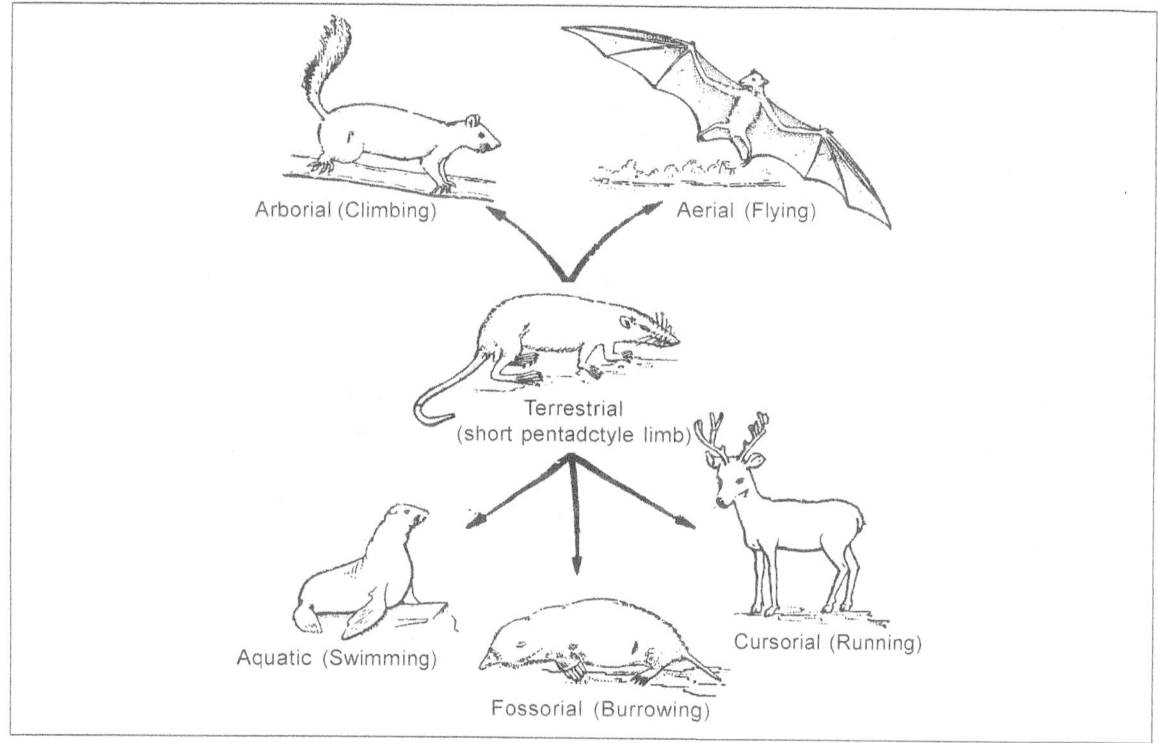

Fig. 3.28 Adaptive divergence seen in limbs of Mammals

1. *arborial* modification in tree dwelling (squirrels and primates).
2. *aerial modification* in animal adapted for flight (bat)
3. *cursorial adaptation* in mammals adapted for fast running (horse and deer).
4. *fossorial or burrowing* mammals (spiny ant eater, moles)
5. mammals adapted for *swimming* in aquatic bodies - (whales, sea lions).

Another example for adaptive divergence seen is adaption of mouth parts of insects to different feeding habits like biting type, chewing type, siphoning type, sucking type and lapping types.

The another phenomenon of adaptive significance is *convergence* where analogus similarities in divergent groups of organisms are seen. The best example is development of wings in insects, flying reptiles, birds and flying mammals. Yet another example is development of fusiform or stream lined body in cartilaginous fish, bony fish, icthyosaurus reptile, aquatic mammals.

9. The animals have developed *adaptations for protection* from its natural enemies. These include : (i) Protective coluration, animal gets concealed its body from environment with suitable adaptation, for example, polar bear, foxes are white in colour and not visible in white background, or chameleon colour changes with its background. (ii) Mimicry : the animal resembles other animal, plant or object thereby getting the advantage. An example is Kallima, the dead leaf butterfly looks like a leaf, or carausis, stick insect looks like stick.

10. Animals of different groups or same group found living together, as in animal associations, exhibit variable degree of adaptive features. These are seen in *commensalism* (e.g., between hermit crab and sea anemone), *symbiosis* (e.g., termites and flagellate trichonympha) and *parasitism* (e.g., Ascaris in men, liverfluke in sheep), etc.

3.4 Phylogeny of Invertebrates and Vertebrates

Though the purpose of classification is for convenience but another important aim is to show relationships. Animals are classified into various groups. They have been shown that such arrangements bring together animals otherwise greatly different. The modern natural system of classification uses all data, structure, physiology, embryology, distribution and other attributes, and each group is distinguished by many characteristics. However, these phyla are related, and these relationship known as phylogeny and genealogical relationship. The genealogical relationship of all phyla or major group are shown in Fig. 3.29(a).

Phylogenetic Origin of Animal Kingdom

It is probable that multicellular animal originated from protoctista but it is not certain which group. Heackel's opinion is that certain protozoans divided repeatedly into daughter cells and failed to separate. Some anatomical and physiological differences arose between the collection of cells leading to specialization. This produced a multicellular organism like *porifera*, *cnidaria* etc. Some poriferan cells are almost identical to a family of flagellates called *Choanoflagellate* and likely flagellates evolved from them (colonial theory). The other hypothesis Hadzi opines that the nucleus of protozoans divided repeatedly give multinucleate protozoans. This is seen in ciliated opalinids and in cnidospora. Subsequent internal division produced a multicellular condition (Syncitial theory). For example, the platyhelminth might have been evolved as a result of internal division of multi-nucleate protozoa.

Phylogeny of Invertebrates

Based on evolutionary history of invertebrate animals, it is assumed that the diverse phyla of invertebrates, have evolved as the result of gradual changes spread over in an immense period of geological time. In the absence of the fossil record of Archaeozoic era, it is rather impossible even to guess about the animal ancestry. It is believed that the phylum protozoa owes its origin to primitive algae from which first arose flagellate protozoa or mastigophora. In course of time they gave rise to other protozoans as well as the sponges (Porifera). The coelenterates are believed to have evolved from a primitive multinucleate ciliate protozoan (Proterospongia). The syncytial structure, later on became cellular and achieved bilateral form in this process and it resembled like a planula larva. From such a planula-like structure probably arose the coelenterates, the ctenophores and the acoelom flatworms. From primitive acoelom flatworms arose other animal groups like platyhelminths, aschelminths and nemerteans. Since a trochophore larva appears in the development of annelids and molluscs, it is believed that a trochophore-like animal was the ancestor of both these groups. The trochophore like ancestor itself seems to have evolved from the planula-like metazoan ancestor. Further, the Onychophora, which are connecting links between Annelida and Arthropoda, have led us to presume that the arthopods have evolved from annelids (Fig. 3.29(a)) and (Fig. 3.29(b)). The dipleurula stage is of common occurrence in evolution of echinoderms. It is established fact that dipleurula-like creature must have been the ancestor of all the echinoderms. Like trochophore, the dipleurula-like ancestor is also considered to be derived from the planula-like ancestor. The hemichordates and the chordates also have their ancestral links with Dipleurula.

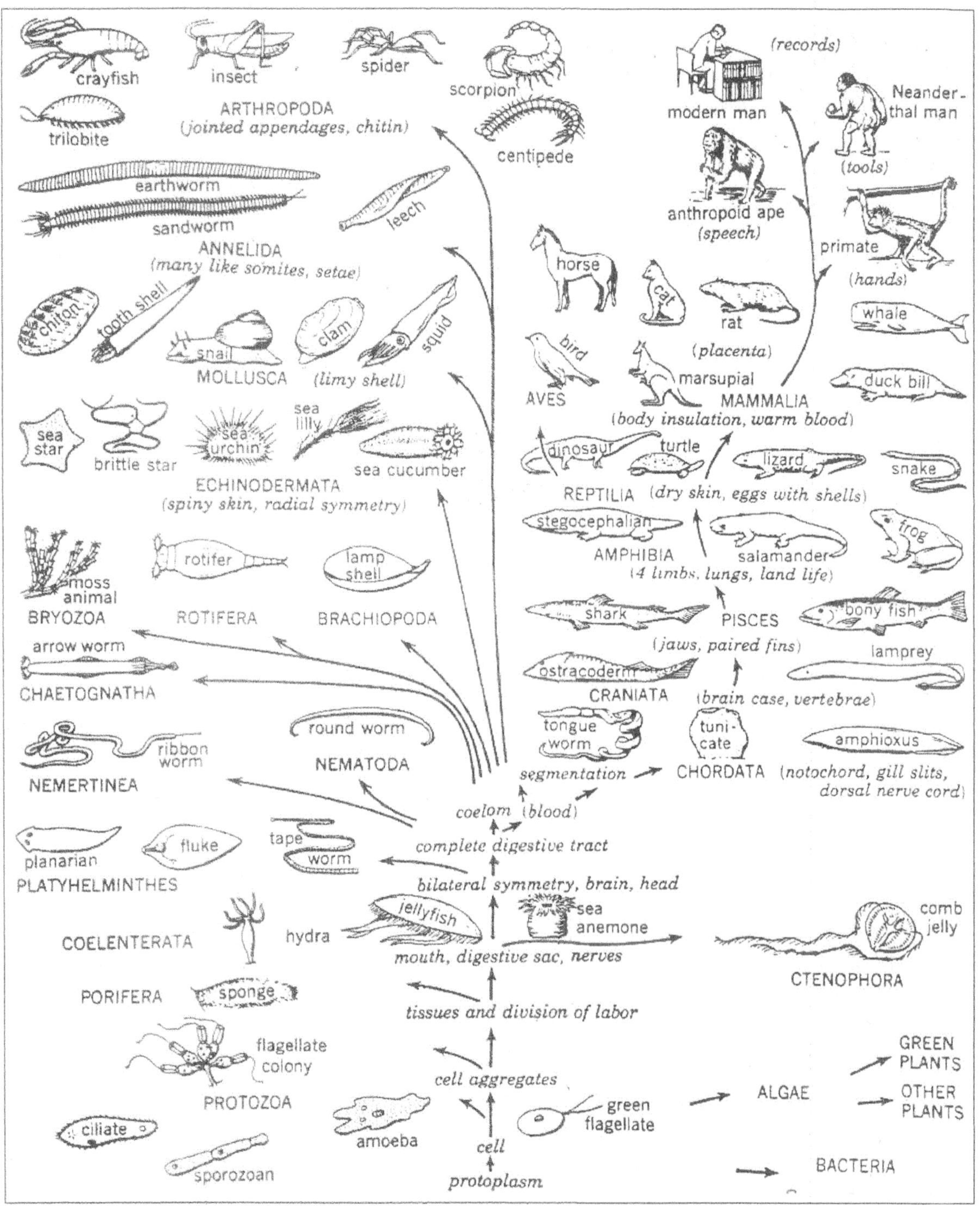

Fig. 3.29(a) A "Genealogical tree" of animal kingdom to indicate the probable relationships and relative position of the major groups

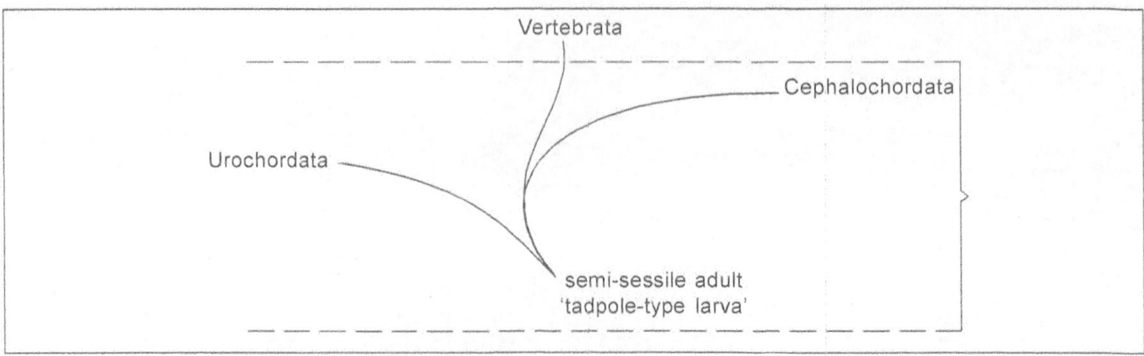

Fig. 3.29(c) Evolution of chordates

Non-vertebrate chordates

A brief review of the major features of the three non-vertebrate, classes Hemichordata, urochordata and cephalochordata, shows a gradual trend towards possession of all chordate features throughout-entire life cycle. It establishes link between the protostome (non-vertebrate) phyla, the echinoderm and vertebrates and suggest a course for chordate evolution.

Chordate Phylogeny

It has been proposed that the Urochordate ***ascidian tadpole*** (Fig. 3.29 (d), (e)) evolved into a pelagic fish like chordate by ***neoteny***, a process whereby the organisms becomes sexually mature and reproduces whilst retaining the body form of the larva stage (Fig. 3.29 (d), (e)).

It is viewed that ascidian larva shared a common ancestor with the echinoderm larva. The ciliated circum oral band of echinoderm larva, in moving to a dorsal position and rolling inwards with its associated nervous tissue gives rise to the ***dorsal hollow*** nerve cord. The tail muscle and ***notochord*** evolved and increased the locomotory power and internal support of the larva, increasing the organisms size and activity. The basic chordate features are shown in Fig. 3.29(d).

During the late devonian and lower carboniferous periods, land generally rose whilst the sea level was lowered. Consequently, there was a redistribution of the aquatic medium. It was this increasingly lack of water and drying up of large areas forced the *Crossoptergian* (lobe fined fish-Rhipidistians) onto land to seek new aquatic habitats. As a result, they began to spend more time on land. In order to exploit the terrestrial environment vertebrates had to overcome the following major problems :

1. ***Breathing gaseous oxygen :*** crossoptergians possessed well developed lungs, however they still possessed gills which were their main respiratory organs.
2. ***Desication :*** there is evidence that early amphibia retained the fish ancestor's scales and they are never to have ventured far from water. It was during the permian that they evolved resistent-body coverings.
3. ***Increased efficiency of gravity :*** The apparent increases in body weight in air is meant, new stresses on the vertebral column. It changed from a compression strut to being a girder. Limbs and girdles were developed.

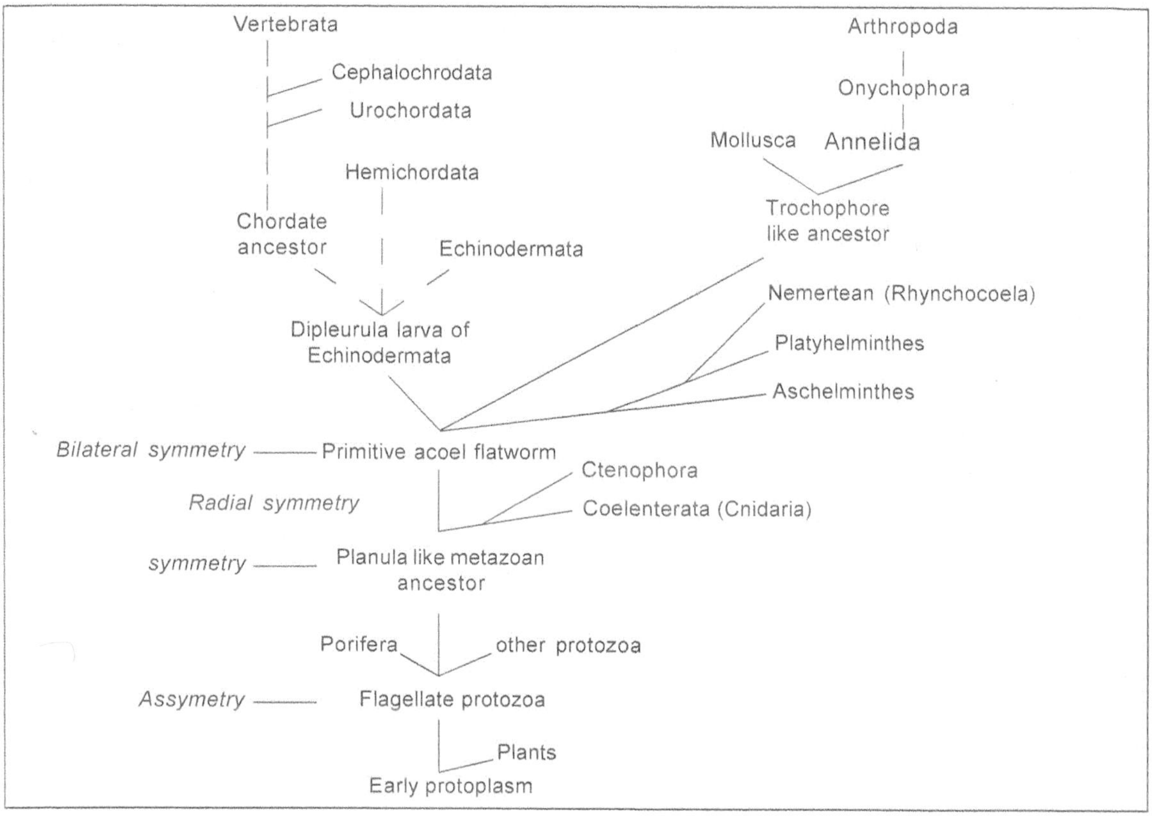

Vertebrata
Cephalochrodata
Urochordata
Hemichordata
Chordate ancestor
Echinodermata
Dipleurula larva of Echinodermata
Arthropoda
Onychophora
Mollusca
Annelida
Trochophore like ancestor
Nemertean (Rhynchocoela)
Platyhelminthes
Aschelminthes
Bilateral symmetry
Primitive acoel flatworm
Ctenophora
Radial symmetry
Coelenterata (Cnidaria)
symmetry
Planula like metazoan ancestor
Porifera
other protozoa
Assymetry
Flagellate protozoa
Plants
Early protoplasm

Fig. 3.29(b) Hypothetical phylogenetic tree of major animal groups

dorsal nerve tube
notochord
dorsal vessel
segmental myotomes
post-anal tail
pharyngeal slit
pharynx
ventral vessel
heart
ductus cuvieri
anus

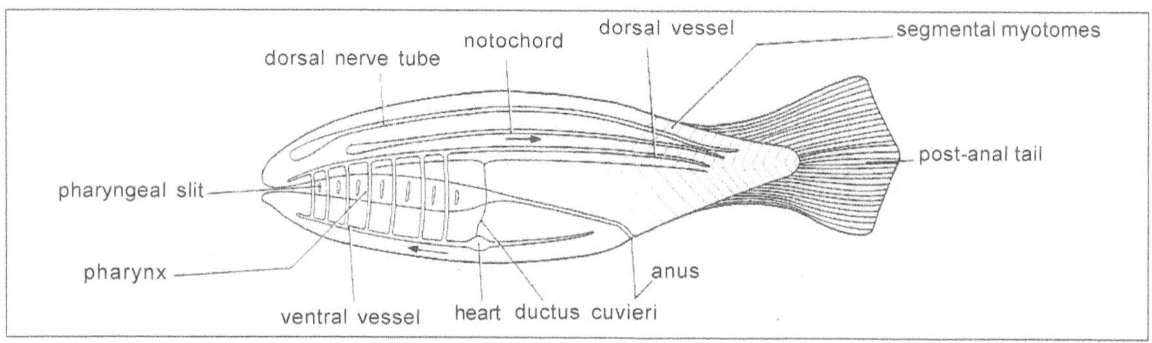

Fig. 3.29(d) Diagram showing basic chordate features

4. ***Change in locomotion :*** Paired appendages become the main locomotory structures with the tail used for balance. In fish, locomotion is affected by the body and tail with the paired fins being used for balance.

5. ***Reproduction :*** Tetropods, (Reptiles, Aves and Mammals) developed methods for protecting eggs from desiccation or return to water to produce. Amphibia have not solved this problem and adapt the latter alternatives.

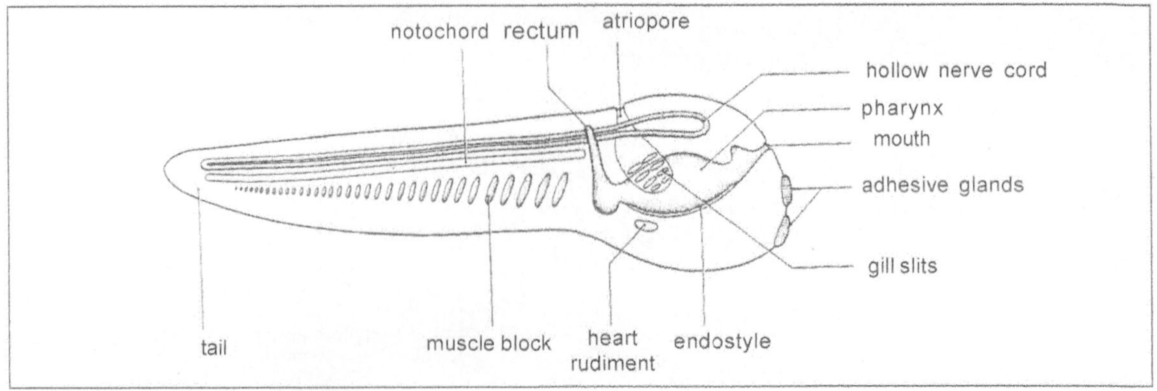

Fig. 3.29(e) Ascidian tadpole

An overall pattern of evolution can be postulated, based on theories of ***Walter Garstrang*** and ***A.S. Romer***. The chordate ancestor was a sessile arm feeder which gave rise to echinoderm and another group gave rise to *Hemichordates* (acorn-worms). The hemichordates with number and complexity of gill slits gave rise to *Urochordates*. The Urochodate larva developed notochord, nerve cord and tail muscles. Through a process (*evident in group of Urochordata, larvacea, okiopleura*) of neoteny and paedogenesis these could have given rise to an ancestral filter feeding vertebrate. The *cephalochordates* appear to be on evolutionary side shoot (Fig. 3.29(b), (c)).

The earliest ancestors of vertebrates, as there is no fossil record, before Cambrian period, were soft bodied, without hard skeletal elements. The first vertebrates are several groups of *fishes* in silurian and ordovician time. The earliest sharks and bony fishes appear in deposits and later both groups invaded the seas. The amphibians probably derived from crossopterygian fishes ***Riphidistia***. The amphibian appear in Devonian period and were differentiated into three orders by carboniferous period. Reptiles appeared from ***Labrinthodont*** amphibians during permian time and expanded widely into a great variety of dominant types in Mesozoic era. The first bird *Archeopteryx* appeared in upper Jurassic. The mammals arose from stem reptiles (*seymouria-link*) beginning in the Triassic and differentiated widely in Tertiary period. The evolution and phylogenetic relationship of various groups of vertebrates is depicted in (Fig. 3.29(f)).

In addition, animal phyla are related evolutionarily through symmetry, coelom, germinal layers etc. Considering radial or bilateral symmetry, the former being considered as more primitive (asymmetry $\rightarrow$ radial $\rightarrow$ bilateral). There is almost certainty in evolutionary progression from the diploblastic to the triploblastic condition and from an acoelomate body plan to one with body cavity, the true colom being regarded as most advanced. Reduction of coelom at expense of haemocoel is also observed. The *Protostome* group of phyla would seem to form natural or evolutionarily related assemblage as would the *Deuterostome* group.

Recent studies using molecular techniques such as nucleotide sequencing of ribosomal RNA, mitochondrial DNA, DNA hybridizations, aminoacid sequencing of proteins, together with serology confirm the more classical findings of evolution and phylogenetic relationship of animal groups based on fossil record, comparative embryology and anatomy.

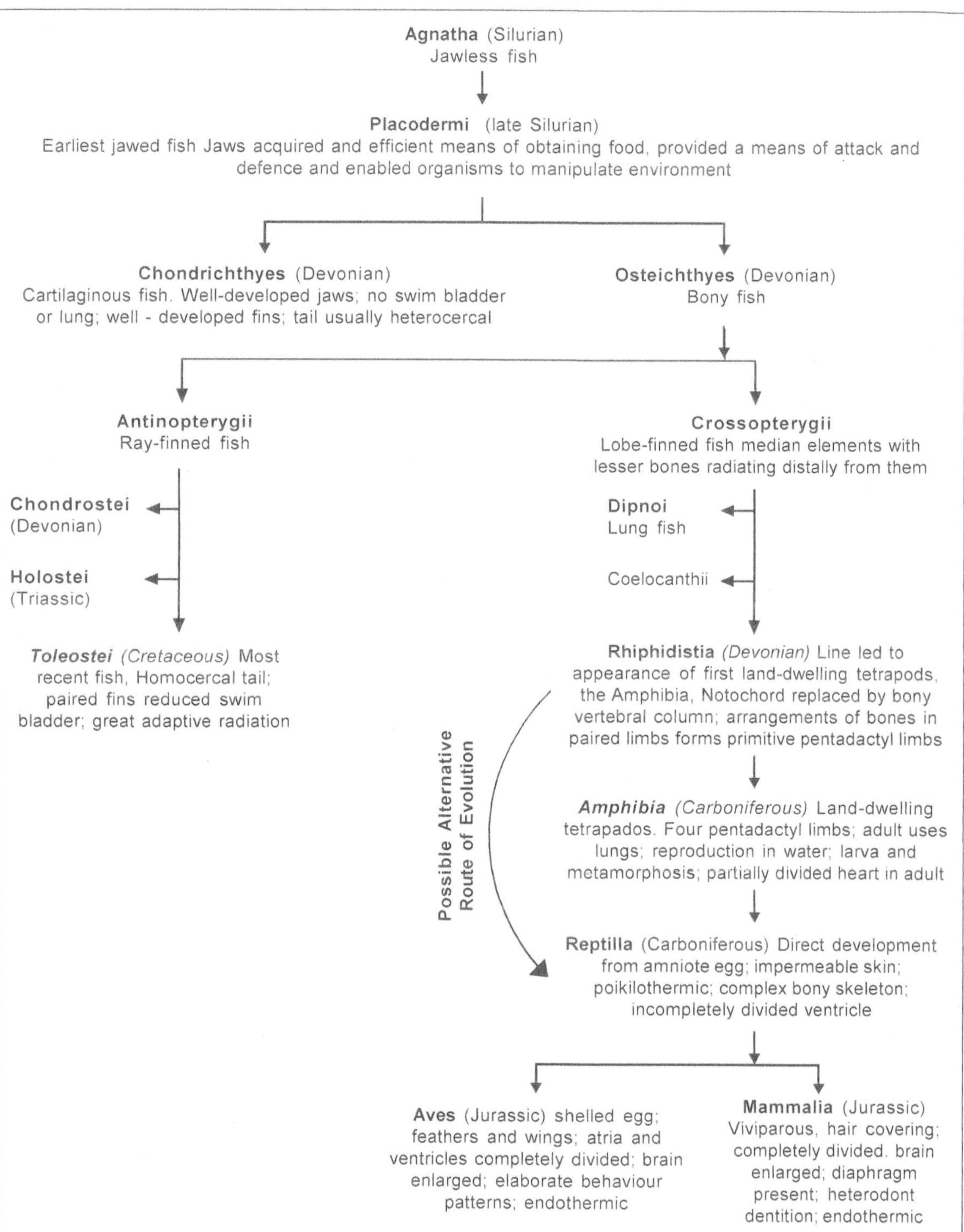

Fig. 3.29(f) Shows the evolution and pylogenatic relationship of various groups of vertebrates

3.5 Concept of Species, Taxonomy, Speciation and Models of Species Composition

3.5.1 Concept of Species

The foremost task of a taxonomist is to know the different kinds of animals occurring in nature. These kinds are actually the species. A species represents the lowest taxonomic group. To define a species, it has been one of the major problems of a taxonomist. Various definitions or concepts have been put forward independently by various workers like Cuvier, Darwin, Thompson, Dobzhansky, Huxley, Zimmerman, Borgemeir, Wilmoth Grant, Paterson, Lambert etc. The concept of species broadly can be grouped into typological or essentialist, species concept, nominalistic species concept and biological species concept. Further, based on reviews of works on "***Species***" problem, it can be inferred into various categories of concepts of species as shown below in Table 3.6.

Table 3.6 *Alternative ways of defining species*

Biological concept based on	***Definition***
Breeding behaviour	A group of organisms capable of interbreeding and producing fertile offspring
Ecological behaviour	A group of organisms sharing the same ecological niche, no two species can share the same ecological niche
Genetic composition	A group of organisms showing close similarity in karyotype of their genome
Evolutionary lineage & trends	A group of organisms sharing a unique collection of structural and functional characteristics

Among all the concepts on species and the most widely accepted and adopted one is "*Biological Species Concept*" which was introduced by *Mayr*. It is defined as "*A group of interbreeding natural population that is reproductively isolated from other such groups*". The biological species has three separate functions :

1. It forms *reproductive community*, the individuals of an animal species recognize each other as potential mates and seek each other for the reproduction, this ensures intraspecific reproduction.
2. The species is also an *ecological unit*. Regardless of the individuals composing it, they interact as a unit with other species, with which such individuals share the environment.
3. The species is also a *genetic unit* consisting of a large intercommunicating gene pool whereas the individual is merely a temporary vessel holding a small portion by the contents of gene pool for a short period. These three properties show that species are biological population and have all the properties ascribable to individuals.

Although this concept has wide following, it poses problems when applied in certain situations with reference to :

(i) *Apomictic or asexual groups :* Where the criteria of interbreeding is not fulfilled, which is foremost-characteristics of biological species concept. *Parthenogenesis* is regarded as a degeneration process of sexual mode of reproduction.

(ii) *Sibling and Cryptic species* these are feebly or not at all separated morphologically.

(iii) *Gradual speciation* - species passes through intermediate stages like biotypes, races, subspecies, ecotype or semispecies.

(iv) *Hybrid complexes* - the biological species concept also fails to give satisfactory explanation when applied to hybrid complex like hybrid swarms (syngameon or "semispecies"). Moreover, no other species definition can provide a foolproof system for correct assignment of isolated populations or cases of evolutionary intermediary.

In spite of these shortcomings the biological species concept has become the working definition of species among most populations and for evolutionary biologists.

In recent years, cladistic analysis has given the species problem a particular new vitality. Christofferson (1995) defined theoretical species concept as "*single lineage of ancestor-descendent-sexual population, genetically integrated by historically contingent events of inter breeding*". The biochemical definition of a species is given as "*groups of individuals with more or less similar combinations of sequences of purine and pyrimidine bases in their DNA and with a system of regulator elements leading to biosynthesis of aminoacid sequences*"

The modern field of biology, genomics also give us much more valuable information. Molecular data on nucleic acid (RNA & DNA) and protein sequence would reveal many interesting and hidden characters of a species in future.

Other Forms of Species

Organisms belonging to a given species rarely exist naturally as a single large population. It is usual for a species to exist as small population with observable phenotypic variation due to variety of factors like isolation, selection etc. In order to describe and characterize range of variations in species, different terms are used which are described below.

1. ***Sibling species :*** Pairs of groups of similar or closely related species which are reproductively isolated but morphologically identical or nearly so, are referred as sibling species.
2. ***Sympatric and Allopatric species :*** Those occupying the same geographical area are referred as Sympatric, while those normally inhabiting completely different geographical areas are Allopatric Species. Related concepts are, 'Syntopic' and 'Allotropic' all these four terms are further described as :

(i) *Sympatric :* To be used when two or more related species have the same or overlapping geographical distributions, regardless of whether or not they occupy the same macrohabitat.

(ii) *Allopatric :* To be used with reference to two or more related species which have separate geographic distribution.

(iii) *Syntopic :* To be used with reference to two or more related species which occupy the same macrohabitat. The species occur together in the same locality, are observably in close proximity, and could possibly interbreed.

(iv) *Allotropic :* To be used with reference to two or more related species which occupy the same macrohabitat. These species are presumably not in close proximity, cannot interbreed and do not occur together in the same locality although they may have the same geographic distribution.

3. *Continental species :* Those living on the large land masses, as distinct from insular species.
4. *Insular species :* Those living in isolated islands which owe their fauna to dispersal methods other than overland migration.
5. *Cosmopolitian species :* Widely distributed species over the earth, in all majority zoogeographical regions.
6. *Tropicopolitan species or pantropical species :* An ambiguous term, used for species found throughout the tropics.
7. *Montane species :* Those which occur at high elevations on mountain ranges.
8. *Morpho-geographical species :* Species known from Linnaean times to the present and based on morphological and geographical data (basic species of taxonomy).
9. *Agamospecies :* Those species which consist of uniparental organisms, i.e., all those animal species which reproduce parthenogenetically (obligatory parthenogenesis). Such species pose major problems for the accepted single definition of true species.
10. *Panmictic species :* Species in which each sex is produced by a different individual (dioecious, having two homes) or species in which the two sexes are produced by the same individual (monoecious, having one home, hermophrodite) are panmictic if some of the progeny are the result of cross-fertilization between different individuals, e.g., lumbricid worms.
11. *Apomictic species :* Those in which there is no mixing of gametes between different individuals, mostly unisexual, i.e., producing only ova; others reproduce completely asexually by budding or fission and have no functional sexual stage in any part of the life history.
12. *Parapatric species :* Species whose ranges abut with at most a narrow area of overlap. Most of such parapatric ranges are due to competition, combined with critical ecological boundaries.
13. *Contemporaneous species :* Those which occur at the same time level, whatever it is. These species indicate the number of phylogenetic lines or lineages occurring at any particular time.
14. *Polytypic species :* Consisting of two or more subspecies
15. *Monotypic species :* Species with no subspecies
16. *Transient species :* Those existing contemporaneously, as a cross section of the lineage of evolutionary species.
17. *Palaespecies or successional species :* Temporally succession species in a single evolutionary line or lineage and intergrade smoothly with each other.
18. *Palaentological species :* A fossil species
19. *Paraspecies :* In paleontology these are parataxa at the species level. Fragments (isolated parts of animals) are named equally as species and genera although such fragment species or paraspecies may include objects belonging to several species.
20. *Demes :* Organisms belonging to a given species, rarely exist naturally as a single large population, exist as small interbreeding populations, called demes, each with its own

gene pool. These populations may occupy adjacent or widely dispersed geographical areas. Spatial separation of populations means that the species may encounter a variety of environmental conditions and degrees of selection pressure. Mutation and selection within the isolated populations may produce phenotypic variation within the species.

21. *Geographical races :* Populations which are distributed over a wide geographical range have occupied well-separated geographical habitats for a long period of time may show considerable phenotypic differences. These are usually based on adaptations to climatic factors. For example, the gypsy moth (*Hymantria dispar*) is distributed throughout the Japanese Islands and eastern Asia. Over this range a variety of climatic conditions are encountered, ranging from subarctic to subtropical. Ten geographical races have been recognized which differ from each other with regard to the timing of hatching of their eggs. The northern races hatch later than the southern races. The phenotypic variations shown by the ten races are thought to be the result of climatic factors producing changes in gene frequencies within their gene pools.

22. *Ecological races (ecotypes) :* Populations adapted to ecologically dissimilar habitats may occupy adjacent geographical areas.

23. *Clines :* The term was coined by Huxley to describe a character gradient. A species exhibiting a gradual change in phenotypic characteristics throughout its geographic range is referred to as cline. More than one cline may be exhibited by a species.

24. *Polytypic species :* Species exhibiting marked phenotypic variation within population according to their degree of geographic isolation are known as *Subspecies*. The species which are not divided into subspecies are *monotypic* species and those which contain two or more subspecies are called *polytypic* species. Presently the *subspecies* is the taxonomically lowest category and is defined as a geographically separate aggregate of local populations of the species.

25. *Superspecies :* It is defined as a monophyletic group of closely related and largely or entirely allopatric species. The population which could not be kept in species or subspecies are categorized as semispecies to mark intermediate nature.

3.5.2 Allopatric and Sympatric Speciation

Speciation is "*the process by which one or more species arise from previously existing species*". A single species may give rise to new species, known as *intraspecific speciation* or two different species may give rise to a new species, known as *interspecific hybridization.*

If intra specific speciation occurs whilst the populations are separated it is termed as *allopatric speciation.* Whereas the process occurs when the populations are occupying the same geographical area it is called *sympatric speciation.*

Process of Speciation

There are several factors involved in intraspecific speciation, but in all cases gene flow within populations must be interrupted. As a result of this each subpopulation becomes genetically isolated. Changes in gene and genotype frequencies within the populations as a result of natural selection on the range of phenotypes produced by mutation and recombination, lead to the formation of races and subspecies. If the genetic isolation persists for a long period of time and the subspecies subsequently come together to cohabit the same area they may or may not interbreeding. If the gene flow is successful there may still be exchange of gene pools by the same species. If the breeding fails, speciation has occurred and the subspecies may now be considered to be different species. This is the way, it is believed that evolutionary change have been taken place.

An initial factor in the process of speciation may be due to the laxity of selection pressure within the population. This eventually leads to increased intraspecific variability capable of conferring ability to some of the genotypes to increase. Their geographical range, if such members show adaptations to environmental conditions, found at the extremes of the range. There is no reduction in gene flow whilst showing the localized phenotypic variation (ecotypes), and still share the gene pool and continue to exist as a single species, such variants are called as cline.

The formation of barriers which lead to reproductive isolation between members of the population results in the process of speciation, different kinds of mechanism called as reproductive isolation is brought about by, reproductive isolating mechanism. These are ways producing and maintaining reproductive isolation within a population. This can be brought about by mechanisms acting at the prior to or after fertilization. Dobzhansky suggested different categories of isolating mechanisms as given in Table. 3.7.

Table 3.7 *Different Reproductive Isolating Mechanisms*

Prezygotic Mechanisms	Essentially creating barriers to the formation of hybrids
Seasonal Isolation	Occurs where two species have mating or flowering at different seasons of the year : e.g., *Pinus radiata* flowers in February whereas *Pinus attenuate* flowers in April in California, USA
Ecological Isolation	Occurs where two species occur in the same regions but have different habitat preferences;e.g., *Viola arvensis* grows on calcareous soils whereas *Viola tricolour* prefers acidic soils
Behavioural Isolation	Occurs where animals exhibit courtship patterns; mating occur only if the courtship display by one sex is accepted by the other sex; e.g., certain fish, bird and insect species
Mechanical Isolation	Occurs in animals where differences in anatomy of genitalia preclude copulation and in plants where related species are pollinated by different animals
Postzygotic mechanisms	Essentially the viability and survival of barriers affect hybrids
Hybrid inviability	Hybrids are formed but fail to develop to maturity, e.g., hybrids formed between northern southern races of the leopard frog (*Rana pipiens*) in North America
Hybrid sterility	Hybrids do not produce functional gametes; e.g., the μ (2n = 63) result from the cross between the horse (Equus equus. (2n = 60) and the ass (*Equus hemionus*. 2n = 66)
Hybrid breakdown	F_1 hybrids are fertile but the F_2 generation and backcrosses between F hybrids and parental stocks fail to develop or are infertile, e.g., hybrids formed between species of cotton (genus *Gossipium sp.*) and rice (*Oryza sativum* × *O.glaberrimm*)

Allopatric (Geographic) Speciation

Allopatric (allos, other, patria, native land) speciation is characterized by the spatial separation at some stage. Geographical barriers like mountain ranges, seas or rivers, or habitat preferences, may form a barrier to gene flow because of spatial separation leading to reproductive isolation. Adaptation to new environment or random genetic drift in small founder populations may lead to changes in gene and genotype frequencies. If this occurs continuously for a long time, tract of population may become genetically isolated even if brought together. In this way new species may arise. For instance the diversity of finch species on the islands of the Galapagos Archipelago are the resultant products of allopatric speciation.

It is opined that an original stock of finches reached the Galapagos Islands from the mainland of South America and in the absence of competition from endemic species got diversified into various species. The species are believed to have evolved in geographical isolation to the point and differentiated when dispersal brought them together on certain islands they were able to maintain their identity as separate species. There are two types of allopatric speciation :

(i) ***Dichopatric speciation :*** A reasonably large area is divided by a (geological, geographic or vegetational) event which splits the earlier continuous range into two isolated groups of populations (Fig. 3.30(a)). This is also known as speciation by splitting.

(ii) ***Peripatric speciation :*** A new population is founded outside the continuous species range by a fertilized female, a small founder group or / and remains isolated groups of populations (Fig. 3.30(b)).

Sympatric Speciation

In populations which have geographically isolated for a much shorter period of time, the genetic differences may accumulate allopatrically. Populations on bringing together, these differentiated hybrids may form where these overlap. For instance, both the carrion crow (*Corvus corone*) and the hooded crow (*Corvus corone cornix*) are found in the British Isles. The carrion crow is completely black is common in England and southern Scotland. The hooded crow is black with a grey black and belly and is found in the north of Scotland. Hybrids formed from the mating of carrion and hooded crows occupy a narrow region extending across central Scotland (Fig. 3.30(c)). These hybrids have reduced fertility and serve as an efficient reproductive barrier to gene flow between the populations of the carrion and hooded crows. In time, selection against crow-breeding may occur., leading to speciation. Since such speciation occurs finally in the same geographical area, this is called *Sympatric* (*sym*, together, patria, native land) *Speciation.*

Sympatric speciation does not always need geographical separation of populations for the genetic isolation to occur. It necessitates the development of some sort of reproductive isolating mechanism be structural, physiological, behavioural or genetic.

Sympatric speciation provides an explanatory mechanism of how closely related species, which probably arose from a common species within the same geographical area. In the Galapagos Archipelago the finch *Camarhyncus pauper* is found only on Charles Island, where it coexists with a related form C. *psittacula* which is widely distributed throughout the central islands (Fig. 3.30(d)). The finch species appear to choose their mates on the beak size. The range of beak size of C. *pauper* on Charles Island and C. *psittacula*, on Albemarle Island are approximately equal, but on Charles Island C. psittacula has a longer beak. The two species remain distinct and are able to coexist.

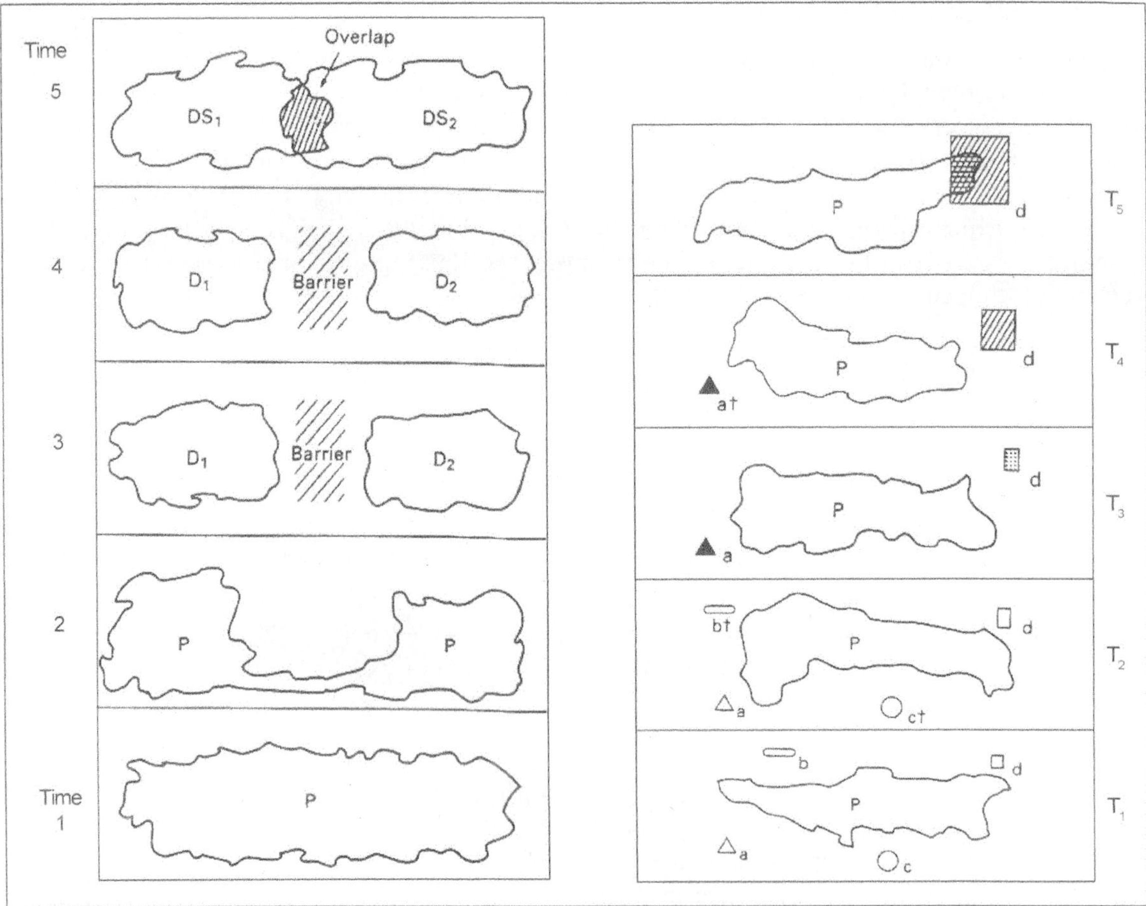

Fig. 3.30(a-b) Dichopatric speciation (speciation by splitting). The parental species P is separated at time 3 (T_3) by a barrier into two daughter populations D_1 and D_2. During their isolation (times T_3 and T_4) these separated populations evolve into two independent species, DS_1 and DS_2 which at time T_5 can overlap without interbreeding, (b) Peripatric speciation (speciation by budding). Parental species P with four peripheral isolates (a, b, c, d) at time (T_1). Isolates b and c become extinct at time 2. (T_2), isolate a at time 4 (T_4). Isolate d becomes a different species and has overlapped the range of the parental species P at time 5 (T_5)

A special form of sympatric speciation which occurs at the point where two populations at the extremes of a cline meet and inhabit the same area, thus closing the ring (Ring species) for example, gulls of the genus *Lorus* form a continuous population between latitudes 50 – 80°N, encircling the North Pole. A ring of ten recognizable races or subspecies exist which principally differ in size and in the colour of their legs, back and wings. Gene flow occurs freely between all races except at the British Isles. Here, at the extremes of the geographical range, the gulls behave as distinct species, that is the herring gull (*Larus argentatus*) and the lesser black-gull (*L. fuscus*). These have a different appearance, different tone of call, different migratory patterns and rarely interbreed. Selection against crossbreeding is said to occur sympatrically.

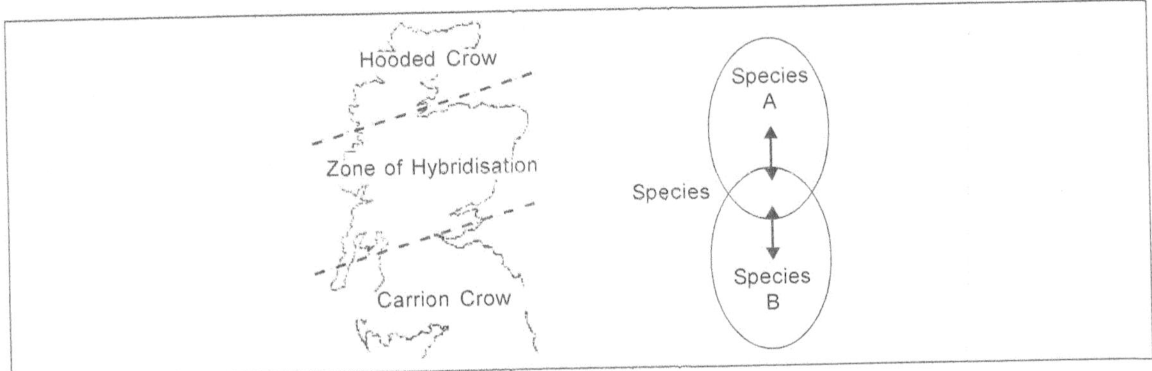

Fig. 3.30 (c) Hybrid barrier as a means of preventing gene flow between two populations. The maintenance of the two crow species is shown to due to the existence of a zone of hybridisaiton extending across scotland. The existence of hybrid barriers between adjacent populations is common and functions as follows. Where the geographical ranges of A and B overlap, mating produces a hybrid with lowered fertility. A will interbreed freely with AB and AB with B but the existence of AB prevents the interebreeding of A and B populations.

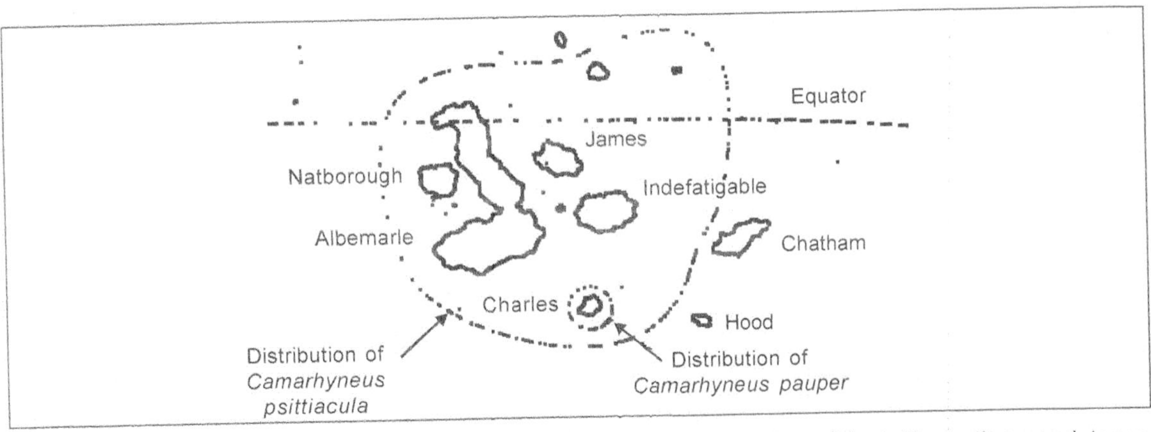

Fig. 3.30(d) The Galapagos Islands and the distribution of two species of finch illustrating coexistance following allopatric speciation

Sympatric speciation without geographical isolation in sexually reproducing species is unlikely. However, in asexually reproducing organisms, including vegetatively propagated angiosperm, a single mutant so different from its pattern population as to be genetically isolated could give rise to a new species sympatrically. An example is polyploidy in *Spartinta.*

3.5.3 Taxonomy

The aim of modern taxonomy is to describe, identify, and arrange organisms in convenient categories to understand their evolutionary histories and mechanisms. Earlier approaches were mainly based exclusively on observed characters without going into the question of specific differences. Presently the great attention is paid to the subgroupings of the species like subspecies, populations, clines etc.,. The old morphological species is now called a "Biological one" which also include ecological, genetical, biochemical and other characters. All new approaches have contributed a lot in explaining the true structure of the species and its evolutionary position. Most of the approaches used in taxonomy include morphological, embryological, behavioral, cytological, biochemical and numerical approaches to taxonomy. Among all approaches, numerical approach because of its procedure and involvement of number of characters is quite important for animal taxonomy.

Concept of Numerical Taxonomy

To overcome subjectivity, the taxonomists record similarities and differences on a large number of variables. Numerical taxonomy is quantitative evaluation of the affinity or similarity between taxonomic units and ordering of these units into taxa on the basis of their affinities. Taxonometrics and Taxometry in place of numerical taxonomy is also used in the literature. Numerical Taxonomy is defined as *"the methodology of assembling individuals into taxa on the basis of an estimate of unweighted overall similarity"*. This concept is based on principles of Adanson, on the use of maximum number of characters which are given equal weight. Principles of Numerical Taxonomy by Sokal and Sneath is classical work of this new school. The larger the number of characters better will be the result. The characters include any attribute of external (morphological), internal (anatomical) traits OTU (Operational Taxonomic Units) derived from biochemical, behavioral, cytological, ecological development etc. Distinct taxa can be constructed due to diverse character correlation in the various groups. The use of number of characters vary, it may be 60, (as per Sokal and Sneath), 135 to 146 (as per Moss) or 1000 characteristics (as per Steyskal).

Numerical Taxonomy is concerned with the classification in group-recognition sense, and is quite useful in solving the problems of biological classification. The best illustration of numerical Taxonomy is provided from studies of Moss on the mite genus, *Dermanyssus*, where 135 characters belonging to 15 OTUs from all areas of body are used (Table 3.8). Moss suggested the creation of some new subgenera and subgroups for the species of this genus. Recently, computerised method are used in devising dichotomaous keys for identification of various Taxa. For example, J.A. Peters of United States, National Museum, Washington had devised a computer program for identification of genera of Central and South American snakes.

Table 3.8 *Break-up of number of characters considered for taxonomy*

Body parts	Qualitative	Qualitative		Total	%
		Meristic	Continuous		
Gnathosoma	1	0	14	15	11.1
Idiosoma dorsal	8	3	34	45	33.3
Idiosoma ventral	5	4	31	40	29.6
(i) Sternal shield	0	4	14	18	13.33
(ii) Epigynial shield	1	0	7	8	5.92
(iii) Anal shield	1	0	7	8	5.92
(iv) General	2	0	1	3	2.22
(v) Peritreme	1	0	2	3	2.22
Legs	1	25	9	35	25.9
Total	15 5	32 4	88 31	135 40	99.9 29.6

Limitations

The orthodox taxonomists do not accept the claims of numerical taxonomists regarding their ability to establish biological relationships between organisms. Their approach is strictly typological and has no relation to historical reality and to causality of differentiation of organisms. The use of large number of characters probably does tend to reduce the effect of homoplasy on the result. The relatively few characters which provide clues to relationship are diluted by redundant characters and characters that are only 'noise' in computations. The another limitation is in giving equal weight to all characters. It does not allow for diversified patterns of evolution, special adaptations, convergence, parallelism, developmental and genetic homeostasis besides genetic developmental evolutionary, and phenomena that disturb the expected close relation between phenetic similarity and phylogeny. The use of complex mathematical and statistical methods by numerical taxonomists has gone against its due recognition because of great difficulties faced by biological taxonomists to understand them.

However, all systematics to some extent are numerical. Computer and numerical taxonomic programs are now standard resources in every museum and systematic laboratory.

Advantages

The advantages of numerical taxonomy are :

* Application method requires no previous knowledge of studies on taxon and its literature, only the ability to make observations and quantification is suffice.
* The classification is more natural, since it is based on more characters.
* It is not subjective or arbitrary like conventional classifications based on influence of phylogenetic or other theoretical consideration.

The future of Numerical Taxonomy

The popularity of this approach has greatly declined since the rise of cladistics. However, in combination with other methods or as a check against other approaches, this method is still much in use in the classification of objects whose groups are not the result of causation but simply are due to joint possession of similar characteristics. However, there is a resurgence in the use of numerical taxonomy in modern biology mainly with the availability of

1. The renewed wealth of molecular characters that can be supplemented to morphological traits for computing,
2. clustering is no longer the obligatory method, and
3. taxa should be relatively homogeneous.

An eventual synthesis of numerical taxonomy with other approaches to classification would be facilitated if numerical taxonomy adopts the following modifications :

1. The weighing of characters whenever the data allow for this,
2. testing of provisionally recognized taxa for monophyly and
3. the recognition that species have reality in nature and are not purely subjective, arbitrary inventions of taxonomists.

3.5.4 Multivariate Models of Species Composition

Major approaches of Species Analysis

Many of developments in Taxonomy in the last 200 years have led to a clearer separation of the procedure/methods of identification and those of classification. Earlier method of *logical division* in which organic world as a whole considered as genus, which as first step of identification was divided into two "*Species*". The *genealogical principle* of common descent, was later found in classification of species. Further, the tentative classification produced deeper understanding of the information content of character, possible homoplasies, other informations etc., – *hypothetic deduction* approach.

Three schools of macrotaxonomy have profound influence on method or approaches of species composition. These are (i) Phenetics (ii) *Cladistics* and (iii) *Evolutionary classification.*

Phenetics : *Degree of similarity, as displayed by a large number of characters, is the overriding criterion of this method, the groups delimited and ranked on the basis of principle are natural groups. This approach is not in conflict with approach of common descent.* The phenetic method, however, does not have any procedure for testing this assumption. The difficulties of phenetic approach are caused by homoplasy, mosaic evolution, absence of criteria for character selection etc. The method is laborious, dependence on availability of large assortment of morhpological characters, subjectivity, of methods, repeatability and choice of computational methods.

In combination with other methods, or a check against other approaches phenetic methods are still much in use. Because latest developments, with the inclusion of these modifications (i) weighing of characters whenever the data allow for this, (ii) testing of provisionally recognized taxa or group for monophyly and (iii) recognition that species have reality in nature and are not purely subjective, the modified phenetic method is found much use.

Cladistics : According to Henning, the taxa based exclusively on possession of synapomorphies i.e., shared derived characters should be recognized while ancestral (plesiomorphic) characters should be ignored. Furthermore, every taxon (can be traced back to semispecies) should be monophyletic, consisting of stem species and all its descendents, including all exgroups. The cladistic analysis consists of partitioning of characters into derived and ancestral ones. The recognition of taxa and their ranking entirely depend on inferred branching pattern of phylogenetic tree (*Cladogram*). Cladogram is branching pattern revealed by cladistic analysis and based on pattern of synapomorphies. Each branching point represents a speciational event which potentially given rise to a new holophyletic taxon. Sister groups are placed at the same distance from the ancestral species and sister taxon that has retained more primitive (Plesiomorphic) characters is placed on the left branch of the dichotomy.

The limitation of cladistic analysis are : (i) homoplasy (is a similarity in character that two taxa have acquired independentally) (ii) Convergence (a superficial similarity of two distant related taxa) (iii) parallelism (an independent acquisition of same characters in related taxa) (iv) reversal of previously required characters for consideration (vi) polytom (branching into more than two) (vii) placement of ancestral (stem) species and (viii) inability to deal with fossil groups.

Despite these factors with advent of suitable computer methods for cladogram construction, the cladistic methods tended to replace phenetic method.

Evolutionary Approach

It is based on two criteria : common descent and amount of evolutionary divergence. However, these two should be treated sequentially. Although, numerical and claddistical methods are used, wherever selection of most important evolutionary characters are considered as like mosaic evolution and homoplasy this method is useful.

All three schools present the results of their analysis in the form of tree like diagram, so called *dendograms*. The main differences of three approaches are : A cladist wants all the groups to be holophyletic, a pheneticist classification by overall similarity without regard for phylogeny, an evolutionary taxonomist uses taxa to be monophyletic and also reflect the information content of plesiomorphic characters in delimitation and ranking of taxa.

The synthesis of these approaches is the future hope to have better models for classification and analysis of species composition.

Species and Multivariate Analysis

In modern conception species is a relational term. It may also be called as non-dimensional because it lacks dimensions of space and time. Every species, taxon in nature consists numerous local populations which raises problem of how to treat these populations. Knowledge of variability of species is valuable not only for the taxonomists but for any one dealing with biology of species. Evolutionists in particular, but also ecologists and population biologists, are interested in nature and extent of variation within and between population of species.

Taxonomic research at the species level consists to a larger extent in comparison of population or more correctly samples from different population. It is based on such comparison that important taxonomic decision are made such as whether samples belong to same species or whether they belong to the same subspecies or two different ones. These differences have to be accounted in terms of *quantitative* rather than qualitative terms like more hairy or larger etc. Thus, it is essential to show in numerical differences. This led automatically for statistical approach. Therefore, statistical methods are used whenever they can contribute to taxonomic analysis. The best source for this is quantitative zoology by Simpson, Roe and Lewontin. A far more extensive treatment is provided by Sokal and Rohif, however no truely comprehensive treatment of multivariate methods is available.

Although, the measures of centrality (mean, median, mode) are also important in taxonomy, is usually of chief interest. The most commonly used statistics of variation are standard deviation (SD) and coefficient of variation (CV). The CV indicates the homogenity of sample. Zones of secondary intergradation of subspecies are often characterised by a greatly increased CV. However, measurement of regression analysis is better choice. When one is dealing simultaneously with more than two variable, methods of *Multivariable* analysis must be employed. Numerous methods of multivariate analysis, particularly with computer use are now available, which include *Principle component analysis* and *discriminant function analysis*.

Many methods have been proposed in the last 40 years but no unanimity is possible which method is the best for inferring the relationship or phylogeny. The available different numerical methods and computer programs may be used depending on task with little or parallel homoplasy, automorphy or apomorphies etc. The numerical methods were classified into phenetic methods can be converted to cladistic ones and vice versa. It is more informative to distinguish between *distance methods*, in which the similarity or distance of taxa from each other is determined and *character data methods* in which diagrams are constructed that consists of taxa that share derived characters. Some *molecular methods*, (like DNA - DNA hybridization) produce only distance data. One can also convert character data into distance data by calculating measures, of distance between taxa on the basis of the character data. *Character based method*, begin with a taxon-by-taxon character matrix of character states, include parsimony, compatibility and maximum likelyhood method. *Distance method* begin with a taxon-by-taxon matrix include method of pheneticists for classification and phylogeny inference like wagner distance method. These methods draw inference by construction of a tree - like diagram, *dendrogram*. The dendograms are two types,

(i) ***Cladogram*** - represent the branching pattern of phylogeny and

(ii) ***Phylogram*** - reflects not only branching pattern but amount of anagenetic divergence among the taxa.

The different methods, (hand analysis or with computer programs) are used for inferring cladograms and phylograms of analysis of small groups of taxa. These methods are broadly *Distance method, Character method, Compatibility method* and *parsimony method.*

Example : For easy comprehension of various methods, the same data based on a set of imaginary insects, A to H, shown in Fig. 3.31(a) and Fig. 3.31(b) is used to compare various methods. The specific characters, their mnemonic names which are used in cladogram, and their codes assigned a number from 0 to 3, are shown in Table 3.9 and the data matrix list the character and code for each of eight imaginary insects is given in Table 3.10.

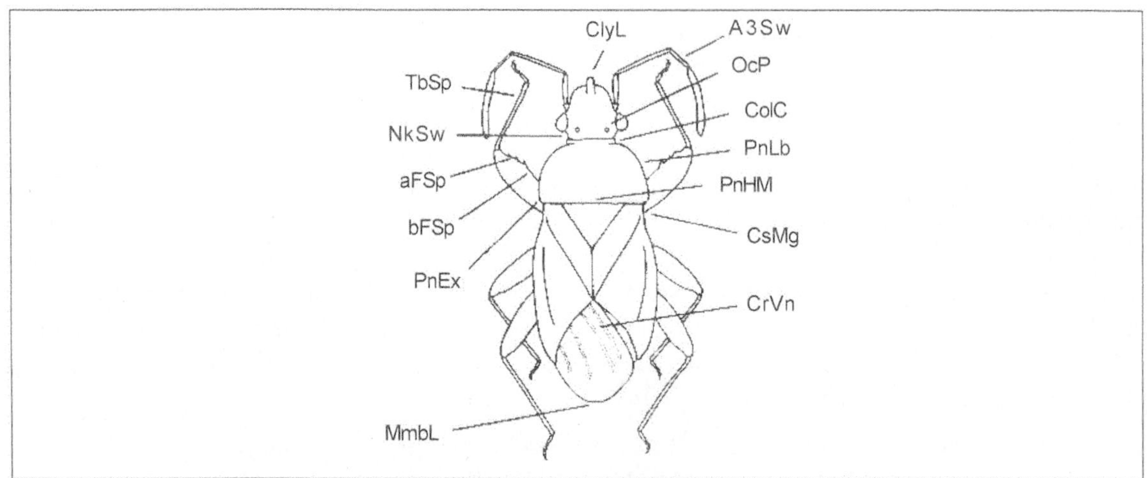

Fig. 3.31(a) Character locations for imaginary insects

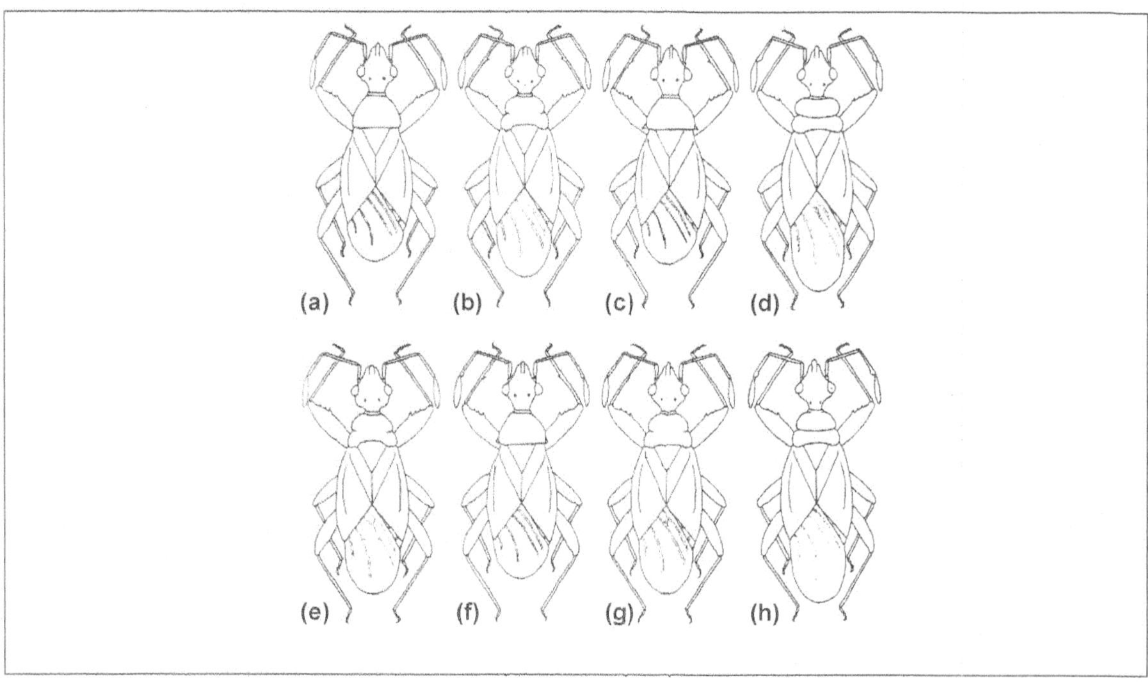

Fig. 3.31(b) Imaginary insects (a) through (h). Characters from these insects serve as data for illustration in this section.

Table 3.9 *Character Mnemonics, Codes and State Definitions*

Character		Code	State definition
No.	Mnemonic		
1	PnLb	0	Pronotum medially without a tranverse groove
		1	Pronotum medially with an incomplete transverse groove producing lobes
		2	Pronotum medially with a complete transverse groove producing lobes
		3	Pronotum with anterior and posterior lobes greatly swollen
2	CsMg	0	Costal wing margin curved in at base
		1	Costal wing margin curved out at base
3	NkSw	0	Neck curved in behind eyes
		1	Neck curved out behind eyes for about length of an eye
		2	Neck curved out behind eyes for more than length of eye
		3	Neck quadrately swollen behind eye
4	ClyL	0	Clypeus projecting beyond rest of head
		1	End of clypeus even with rest of head
5	ColC	0	Collar incomplete
		1	Collar complete

Contd.. Table 3.9

Character		Code	State definition
No.	Mnemonic		
6	PrEx	0	Pronotal expansion absent from posterior angle
		1	Pronotal expansion rounded apically
		2	Pronotal expansion acute apically
7	CrVn	0	Cross vein absent from wing membrane
		1	Cross vein present in wing membrane
8	MmbL	0	Wing membrane shorter than head and pronotum together
		1	Wing membrane longer than head and pronotum together
		2	Wing membrane longer than basal part (corium) of wing
9	PnHM	0	Pronotal hind margin straight
		1	Pronotal hind margin lightly curved
		2	Pronotal hind margin deeply curved
10	OcP	0	Ocelli on an imaginary line drawn between imaginary line on posterior eye margins
		1	Ocelli less than one eye length behind imaginary line on posterior eye margins
		2	Ocelli one eye length behind imaginary line on posterior eye margins
11	A3Sw	0	Antennal segment 3 not swollen
		1	Antennal segment 3 swollen apically
12	aFSp	0	Forefemur with two spines apical to large spine
		1	Forefemur with one spine apical to large spine
13	bFSp	0	Spine near base of forefemur absent
		1	Spine near base of forefemur present
14	TbSp	0	Foretibia without a spine
		1	Foretibia with a spine

Table 3.10 *Data Matrix*

		A	B	C	D	E	F	G	H
1	PnLb	0	1	0	3	1	0	1	2
2	CsMg	0	1	0	1	1	0	1	1
3	NkSw	1	2	0	1	3	0	2	1
4	ClyL	0	1	0	0	1	0	0	0
5	ColC	1	1	1	1	1	1	1	0
6	PnEx	0	0	2	0	0	1	0	0
7	CrVn	0	0	0	1	1	0	1	0
8	MmbL	0	1	0	2	1	0	1	2
9	PnHm	0	2	0	1	2	0	1	1
10	OcP	0	1	0	1	1	0	1	2
11	A3Sw	0	0	0	1	0	1	0	1
12	aFSp	1	0	0	0	0	0	0	0
13	bFSp	0	1	0	0	0	1	0	0
14	TbSp	0	1	1	0	0	1	0	0

A. *Distance methods*

Measures of distance between taxa can arise from the observation or by transforming character data into distance data. These data are arranged into a matrix of distances. The relationship among, taxa are determined from distance matrix by means of *Cluster Analysis*.

Distance measures can be calculated by comparing character states of several taxa to determine the overall similarity or difference between each possible pair of taxa. The taxa to be classified are thoroughly searched for differences, and the differences are coded and assembled into a data matrix. Then a coefficient of similarity (any of various distance or correlation measures) between each possible pair of taxa is calculated by comparing all character states of each taxon with those of every other taxon (or OTU). The coefficients are assembled into a *similarity matrix* (Table 3.11 and Fig. 3.32).

Both the columns and the rows in a similarity matrix are for taxa. The cells contain a measure of similarity or distance between each possible pair of taxa, exactly like the tables on many road maps that give the distances between cities. A matrix for *n* number of taxa requires calculation or acquisition of *(n2 -n)/2* similarity measures. A matrix for 10 taxa, then, requires 45 calculations, and a matrix for the 8 taxa of the data set requires 28 calculations.

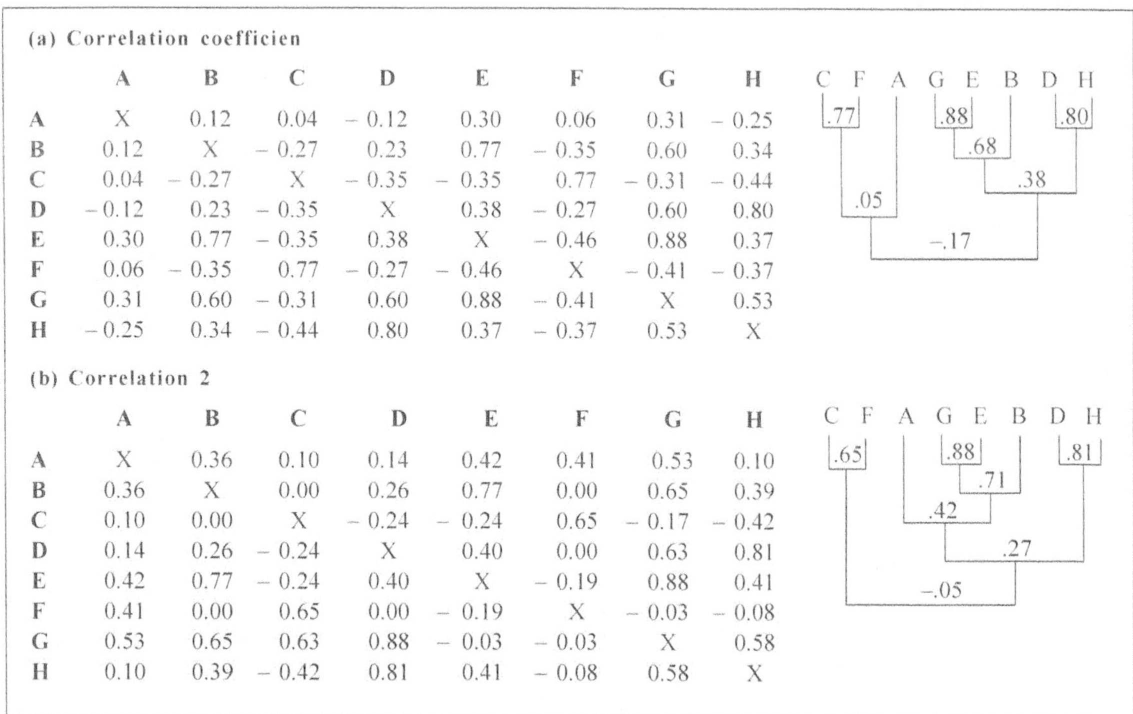

(a) Correlation coefficien

	A	B	C	D	E	F	G	H
A	X	0.12	0.04	−0.12	0.30	0.06	0.31	−0.25
B	0.12	X	−0.27	0.23	0.77	−0.35	0.60	0.34
C	0.04	−0.27	X	−0.35	−0.35	0.77	−0.31	−0.44
D	−0.12	0.23	−0.35	X	0.38	−0.27	0.60	0.80
E	0.30	0.77	−0.35	0.38	X	−0.46	0.88	0.37
F	0.06	−0.35	0.77	−0.27	−0.46	X	−0.41	−0.37
G	0.31	0.60	−0.31	0.60	0.88	−0.41	X	0.53
H	−0.25	0.34	−0.44	0.80	0.37	−0.37	0.53	X

(b) Correlation 2

	A	B	C	D	E	F	G	H
A	X	0.36	0.10	0.14	0.42	0.41	0.53	0.10
B	0.36	X	0.00	0.26	0.77	0.00	0.65	0.39
C	0.10	0.00	X	−0.24	−0.24	0.65	−0.17	−0.42
D	0.14	0.26	−0.24	X	0.40	0.00	0.63	0.81
E	0.42	0.77	−0.24	0.40	X	−0.19	0.88	0.41
F	0.41	0.00	0.65	0.00	−0.19	X	−0.03	−0.08
G	0.53	0.65	0.63	0.88	−0.03	−0.03	X	0.58
H	0.10	0.39	−0.42	0.81	0.41	−0.08	0.58	X

Fig. 3.32 The correlation similarity coefficient (a) Correlation coefficients and phenogram based on the data in Table. (b) Correlation coefficients and phenogram based on the same data but with the coding of character PnEx reversed (i.e., 0 = 2 and 2 = 0). Note the new placement of taxon A. This change in data has no effect on the similarity matrix or on a phenogram based on any distance coefficient.

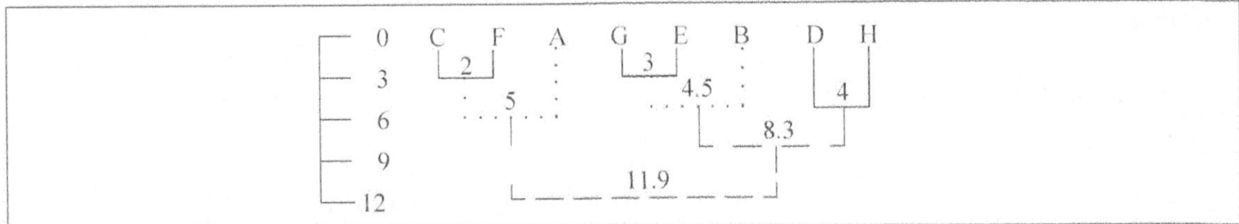

Fig. 3.33 Phenogram. Manhattan distance of the demonstration insects. Primary clusters are indicated by solid lines, secondary clusters by dotted lines, and teritary clusters by dashed lines. The scale on the left is a distance measure.

Table. 3.11 *Similarity Matrix : Manhattan Distance*
Taxa

	A	B	C	D	E	F	G	H
A	X	11	5	11	11	5	8	11
B	11	X	12	10	4	12	5	10
C	5	12	X	14	14	2	11	14
D	11	10	14	X	8	12	5	4
E	11	4	14	8	X	14	3	10
F	5	12	2	12	14	X	11	12
G	8	5	11	5	3	11	X	7
H	11	10	14	4	10	12	7	X

Manhattan distance is the simplest similarity coefficient to calculate and, most useful, it is calculated by finding the sum of the absolute differences between the character states of each character *for* each possible pair of taxa (Fig. 3.33). For example, in Table 3.11 the Manhattan distance between taxa A and B is 11 between C and F it is 2, and between D and H it is 4. These values are assembled into a taxon-by-taxon similarity matrix (Table 3.11).

Cluster Analysis

For hand cluster analysis, taxa in the similarity matrix must be fitted in one's best approximation of the final order of similarity. Rearranging Table 3.11, in which taxa are not so ordered, requires one to place taxa with the lowest similarity coefficients close together. Study of Table 3.12, which is Table 3.11, rearranged by similarity, shows C and F together (co-efficient 2), D and H together (coefficient 4), and G and E together (co-efficient 3) with B close to them (coefficients 4 and 5). This sorting is rough and will be refined as the analysis proceeds. Note that on a rear ranged matrix, the lowest distance coefficients fall adjacent to the diagonal row of X's. The cluster analysis method explained here is called the unweighted, pair-group method using arithmetic averages (UPGMA) and is a commonly used distance method. This cross-averaging method is also known as average linkage. The first step in hand cluster analysis is to find primary clusters, i.e., those which consist of only two taxa (Fig. 3.34).

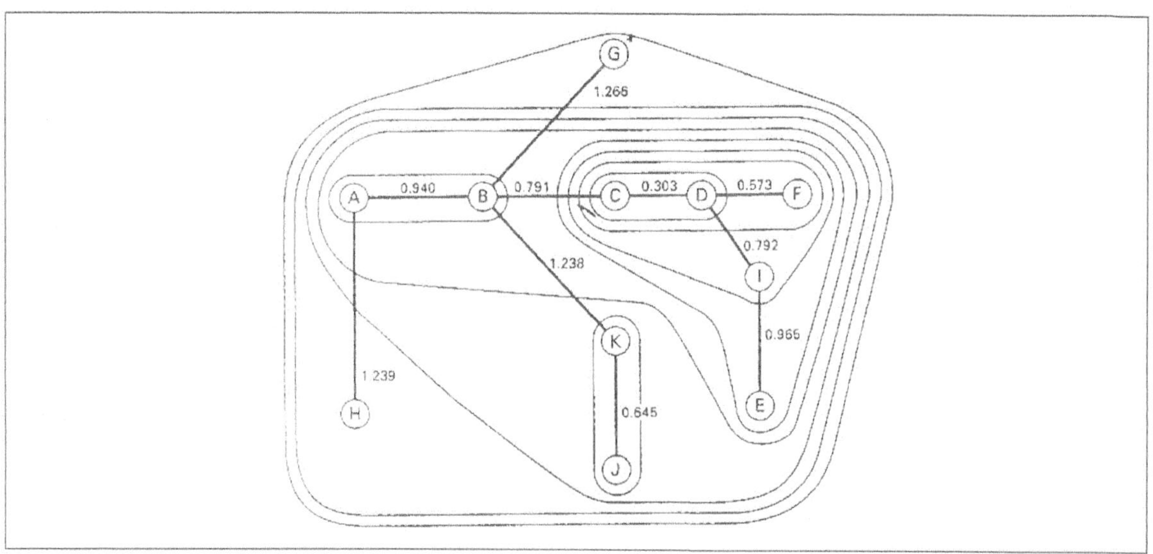

Fig. 3.34 Shortest spanning tree with UPGMA cluster analysis, expressed as contour lines, of the Hoplitis producta group of bees, showing faulty phenetic relationships determined by UPGMA cluster analysis

Table 3.12 *Similarity Matrix : Manhattan Distance Rearranged by Similarity*

Taxa

	C	F	A	G	E	B	D	H
C	X	2	5	11	14	12	14	14
F	2	X	5	11	14	12	12	12
A	5	5	X	8	11	11	11	11
G	11	11	8	X	3	5	5	7
E	14	14	11	3	X	4	8	10
B	12	12	11	5	4	X	10	10
D	14	12	11	5	8	10	X	4
H	14	12	11	7	10	10	4	X

Calculation of Similarity Coefficients from Character Data

Similarity coefficients are used in an attempt to measure the overall similarity between pairs of individual taxa or OTUs. They form several families, especially distance and correlation coefficients. However, only the distance coefficients have been widely employed; they measure dissimilarity but are complements of similarity coefficients. Association coefficients have been used primarily with binary (two-state) data.

Euclidean Distance, also known as *taxonomic distance*, this coefficient measures the distance between two taxa in multidimensional space and is widely employed in phenetic studies (Fig. 3.35).

The formula for Euclidean distance may be simplified so that for all the characters of each pair of taxa j and k, the expression becomes,

$$\sqrt{\sum_{i=1}^{n} (j-k)^2}$$

A related coefficient, Euclidean distance squared, expressed as

$$\sum_{i=1}^{n} (j-k)^2$$

has also been widely used. The differences between all these distance measures (including Manhattan distance) are due to the use or non-use of roots and squares.

Manhattan distance : The formula for this distance between each pair of taxa is expressed as follows :

$$\sum_{i=1}^{n} |j-k|$$

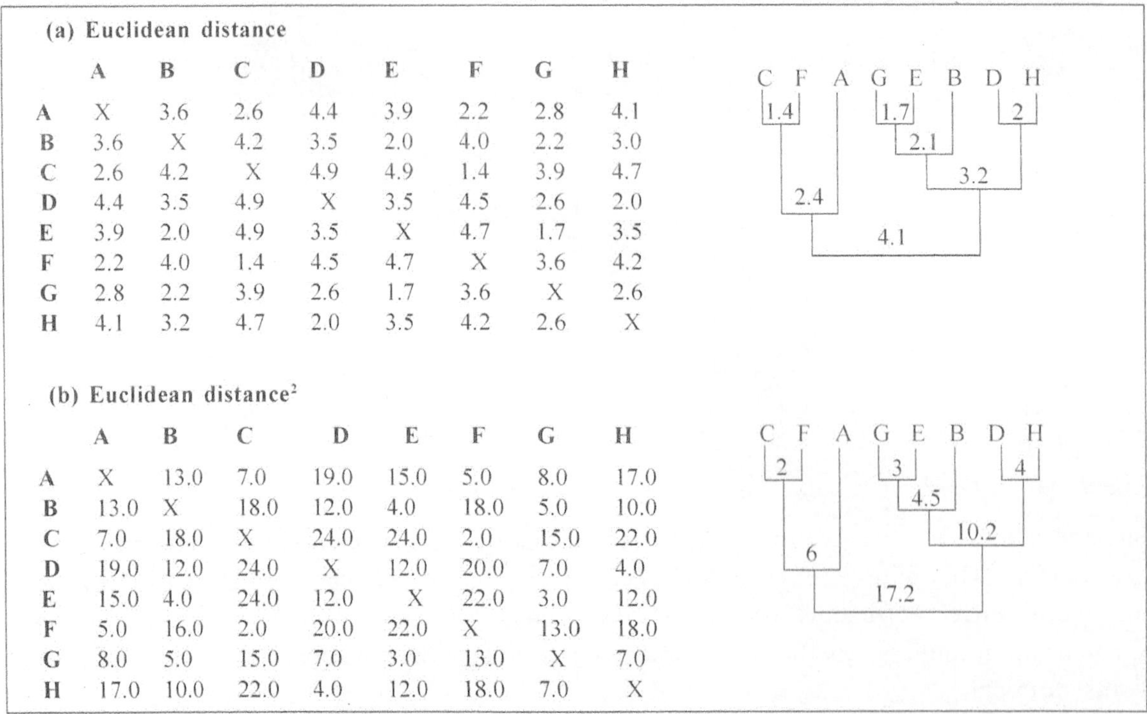

(a) Euclidean distance

	A	B	C	D	E	F	G	H
A	X	3.6	2.6	4.4	3.9	2.2	2.8	4.1
B	3.6	X	4.2	3.5	2.0	4.0	2.2	3.0
C	2.6	4.2	X	4.9	4.9	1.4	3.9	4.7
D	4.4	3.5	4.9	X	3.5	4.5	2.6	2.0
E	3.9	2.0	4.9	3.5	X	4.7	1.7	3.5
F	2.2	4.0	1.4	4.5	4.7	X	3.6	4.2
G	2.8	2.2	3.9	2.6	1.7	3.6	X	2.6
H	4.1	3.2	4.7	2.0	3.5	4.2	2.6	X

(b) Euclidean distance^2

	A	B	C	D	E	F	G	H
A	X	13.0	7.0	19.0	15.0	5.0	8.0	17.0
B	13.0	X	18.0	12.0	4.0	18.0	5.0	10.0
C	7.0	18.0	X	24.0	24.0	2.0	15.0	22.0
D	19.0	12.0	24.0	X	12.0	20.0	7.0	4.0
E	15.0	4.0	24.0	12.0	X	22.0	3.0	12.0
F	5.0	16.0	2.0	20.0	22.0	X	13.0	18.0
G	8.0	5.0	15.0	7.0	3.0	13.0	X	7.0
H	17.0	10.0	22.0	4.0	12.0	18.0	7.0	X

Fig. 3.35 Euclidean distance (a) and Euclidean distance squared (b) Phenograms derived from the demonstration data comparing Euclidean distance and Euclidean distance squared. Compare the Manhattan distance phenogram. Difference in the three phenograms are due to the use of versus, the use of squares and roots.

Other distance methods have been developed, particularly for molecular data. It is simply the sum of the absolute values of the character state differences; i.e., the differences are all positive numbers. This measure differs from the Euclidean distance in that it neither squares the character state differences nor takes the square root. In fact, both distances are cases of a general class of distance functions, Minkowski metrics. The Manhattan distance is not invariant when rotated, and so it is inappropriate for ordination. However, it is interpretable as the evolutionary distance between taxa, because it measures character state differences as steps. It is therefore the distance used in numerical cladistics.

Numerous other distance measures have been proposed, including multivariate distances such as Mahalanobis's generalized distance, and they continue to be developed for specialized problems. However, none have achieved as wide use as the Euclidean and Manhattan metrics, nor are they likely to.

Correlation Coefficient

The *correlation coefficient* was widely used as a measure of similarity. The expression for the correlation coefficient is

$$r = \frac{\Sigma\left(j-\bar{j}\right)\left(k-\bar{k}\right)}{\sqrt{\Sigma\left(j-\bar{j}\right)^2 \Sigma\left(k-\bar{k}\right)^2}}$$

Correlation values of 1 are supposed to indicate complete similarity, while values approaching 0 indicate greater difference. Negative correlations are also possible and indicate even less similarity. The process of clustering correlation coefficients by UPGMA is the same as that for distance values except that one works from higher to lower numbers.

The Wagner Distance Method

Farris developed a cladistic algorithm based on a taxon-by-taxon matrix of phenetic differences. A value is assigned to each comparison between pairs of taxa (OTUs). Such an approach is particularly suitable for data sets, such as those derived from immunological comparisons and DNA hybridization. It can be used for a molecular datum that is expressed as a degree of difference between taxa. Its advantage over other techniques for processing such data is that it does not make the assumption that the rate of evolutionary change is the same in all lineages.

The trees calculated by the Wagner distance method are undirected trees. Three techniques for the rooting of such trees are available :

1. Finding the nearest connection with an out-group, preferably a sister taxon,
2. Assuming a uniform rate of change within the holophyletic group, and
3. Using a minimization procedure. For example, using cladistically arranged data, roots trees by employing a hypothetical ancestor that is primitive in all character states. The use of different rooting methods may lead to rather different cladograms.

In the original Wagner distance method, synapomorphic as well as plesiomorphic character states were utilized in calculating the distance matrix. To make it a more strictly cladistic method, a coefficient of special similarity is calculated. All character states that are not synapomorphies of any included pair of taxa, thus converting the phylogenetic method back into a strictly cladistic one.

The disadvantage is that there is no guarantee that pair-wise methods reveal better topologies than other methods do. It must also be remembered that converting a data matrix into distances always entails a loss of information.

Other Distance Methods

The construction of phylogenetic trees from molecular data is often based on a matrix of distances. Some types of biochemical comparison, such as microcomplement fixation yield data directly in the form of distances. Non-distance data-allozyme frequencies and amino acid data, are usually converted to distances for purposes of analysis. The earliest and still most frequently used method was proposed by Fitch and Margoliash, but about a dozen other methods for constructing phylogenetic trees from molecular distance data have since been proposed. All these methods attempt to fit a tree to a matrix of pair-wise differences (or similarities) between taxa. The objective is to find a tree that comes as close as possible to predict the distances of the matrix of differences.

In the *Fitch-Margoliash method*, a provisional tree is formed by phenetic clustering but is then modified by a series of trial-and-error rearrangements. In each trial a "*length*" is assigned to each branch, and these lengths determine a matrix of tree-derived distances. The derived distance between a pair of taxa is the sum of the lengths of the branches that must be traversed in tracing a path on the tree from one member of the pair to the other. The "*best*" rearrangement is the one that gives the best fit of the tree-derived matrix to the data distance matrix.

Methods proposed in recent years to avoid the inefficient search for the maximum-parsimony tree among all possible branching patterns include, in addition to the Wagner distance method (Farris 1972), the following: Tateno, Nei, and Tajima (1982) (modified from Farris), Li (1981), Sattath and Tversky (1977), Fitch (1981), and Saitou and Nei (1987). None *of* these authors claim that his or her method necessarily.finds the correct tree, but all these methods are said to have special advantages. On the basis *of* computer simulations, Saitou and Nei claim that their "neighbor-joining method" and Sattath and Tversky's method are generally better than the other methods. Nevertheless, these uncertainties demonstrate that one should not automatically follow the findings of one particular numerical method but should also make use of all other available information for the construction of a classification. Distance methods are said to have two advantages over parsimony methods. They require much less computer time, and they usually give the branch length as well as the topology of the estimated tree.

B. *Character Methods*

In contrast to distance methods, character-based methods use data about the states found in taxa as a starting point.

It is helpful if the characters are given a mnemonic code (Table 3.9). For example, 'eye' would be suitable for eye colour and ABC for abdomen colour. Since all character states will be entered onto the final phylogenetic tree, the use of such mnemonic help us to understand the final phylogenetic hypothesis.

Given that the study group is monophyletic, a drawing of what one knows about the cladogenesis as one begins a study most resembles a fan. The fan is a polytom, and it is important to note that the members of a polytomy may be placed in any order. The first step in cladistic analysis is to find all characters that have a derived state in only one taxon. Table 3.10 shows that there are two such characters, both concerned with femoral spines: a FSp. These are autapomorphies; when placed on the fan diagram, they do not group any taxa.

Several principles have been employed to establish from a given set of data the cladogram that is most likely to be the right one. One is *compatibility*, heres tree is considered most probable which has the largest number of characters that are not in conflict with one another and with the tree. The second principle is *parsimony* (*e.g.*, Wagner trees). The most parsimonious tree is the one that, while accounting for all character states in the data matrix, has the fewest character states steps. Because incompatible characters force and require at least on extra character state step to arrive at the most derived state, the two principles - parsimony and compatibility-are related. Often but not always, the two methods dictate the same tree. Most frequently, especially on large studies, more than one equally parsimonious tree is found. Ranking characters by quality (weighting) helps avoid this problem. Neither the compatibility method nor the parsimony method is free of problems, and serious objections have been raised to both.

The drawing of individual trees of each characters, in cladistic pattern, the complete cladogram (Fig. 3.36) with the character states includes parallelisms, reversals and automorphies.

Overview of Character - Based Methods

Analogous to computer methods for distance data, computer methods have been developed for character data. Various scientists have proposed methods that primarily in terms of whether they

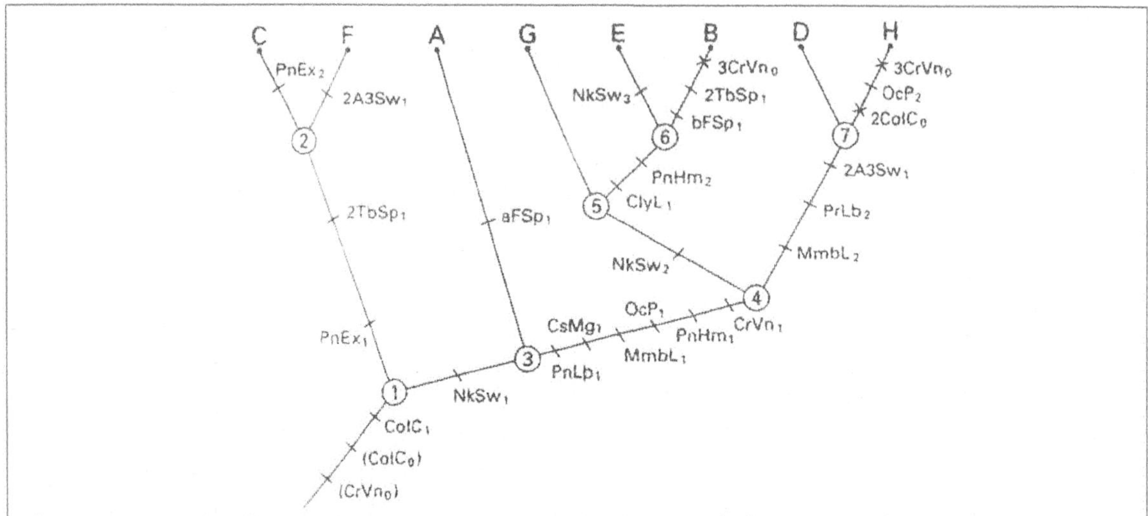

Fig. 3.36 The completed cladogram of the demonstration data with all characters entered, menmonics and the character codes in parenthesis on the stem of the tree.

emphasize compatibility of parsimony. As with distance methods, different methods sometimes produce different results; often they produce ambiguous results (more than one possible solution is found). Unlike the results of distance methods, however, the result of cladistic studies and the distribution of character states suggested by a given solution can be reviewed, and informed judgements can be made about their value.

Some of the older programs required the use of large mainframe computers, but most of the algorithms referred to here can be solved using the newer programs on a modern desktop machine. A large variety of software packages now exists. With new ones frequently becoming available.

The first character-based methods were developed by Camin and Sokal and Edwards and Cavalli-Sforza with additional path breaking work done by Farris and his colleagues. One of the major criticisms of pure cladistics was that it was interested only in cladogenesis, that is, in the location of the nodes in a cladogram, ignoring entirely all aspects of anagenesis. Wagner trees which consider the number of steps in a lineage, broke with the original cladistic tradition. Cladistic analysis thus became phylogenetic analysis. More important, the new approach permitted the development of algorithms that made it possible to incorporate quantitative characters. Since that time there has been a steady stream of improvements, modifications and additions to these methods.

Many currently employed methods are derived from Wagner's groundplan method but go considerably beyond the first numerical versions of it. They are parsimony methods. In that the tree judged to be optimal is the one which allows for the simplest (most parsimonious) explanation of the observed distribution of character states.

C. *Compatibility Methods*

A different approach is used in some of the compatibility methods developed by Estabrook and his followers. A compatibility method is one in which that tree is selected which has the largest number of characters perfectly compatible with it.

The best known method of compatibility analysis has been named *clique* by G. F. Estabrook, one of its early proponents. The system is based upon Le Quesne's observation that two two-state characters are cladistically incompatible only if their distribution among the taxa in the study group includes all four possible combinations of matches and mismatches. If anyone of the four is not found, the two characters are mutually compatible.

Since the method requires two-state characters, a multistate data matrix must be converted to a two-state data matrix (Table 3.13). The number of characters in the data matrix-l4-increases to 22 in the two-state matrix, in which each character in its derived expression is a true character state, not a character. Although increasing the number of required comparisons would seem to complicate the analysis, the reduction of characters to character states provides a balancing simplification.

Table 3.13(a) *Clique : Two - State Data*

Clique character no.	Character		Taxon								
	Character	Character state	Anc	C	F	A	G	E	B	D	H
1	NkSw	1	0	0	0	1	1	1	1	1	1
2		2	0	0	0	0	1	1	1	0	0
–		$3^{'}$	0	0	0	0	0	1	0	0	0
3	PnEx	1	0	1	1	0	0	0	0	0	0
–		$2^{'}$	0	1	0	0	0	0	0	0	0
4	ClyL	1	0	0	0	0	0	1	1	0	0
5	ColC	1	0	1	1	1	1	1	1	1	0
6	PnLb	1	0	0	0	0	1	1	1	1	1
7		2	0	0	0	0	0	0	0	1	1
–		3.	0	0	0	0	0	0	0	1	0
–	CsMg	1^{+}	0	0	0	0	1	1	1	1	1
8	CrVn	1	0	0	0	0	1	1	0	1	0
–	MmbL	1^{+}	0	0	0	0	1	1	1	1	1
–		2*	0	0	0	0	0	0	0	1	1
–	PnHM	1^{t}	0	0	0	0	1	1	1	1	1
–		$2^{\S}$	0	0	0	0	0	1	1	0	0
–	OcP	1^{+}	0	0	0	0	1	1	1	1	1
–		$2^{'}$	0	0	0	0	0	0	0	0	1
9	A3Sw	1	0	0	1	0	0	0	0	1	1
–	aFSp	$1^{'}$	0	0	0	1	0	0	0	0	0
–	bFSp	$1^{'}$	0	0	0	0	0	0	1	0	0
10	TbSp	1	0	1	1	0	0	0	1	0	0

Autapomorphy,

Duplicate of clique character number 6

Duplicate of clique character number 7

§Duplicate of clique character number 4.

First, since all autapomorphies (derived states found in single taxa) are compatible with all other character states, autapomorphic character states may be temporarily eliminated. Second, all character states with identical distributions are also compatible with one another. One example only of each such group of duplicate character states must be retained; the rest (six more two-state characters) may be eliminated. A new data matrix containing the 10 two-state characters remaining simplifies the generation of a compatibility matrix. Even with data in a primitive to derived sequence, the clique method produces unrooted trees unless an ancestor, which is coded 0 for every character, is included in the data matrix.

The clique matrix is concerned only with compatibilities, not with relative degree of incompatibility. The matrix will 3always be symmetrical on either side of the diagonal row of X's, and only half the matrix must be calculated.

Table 3.13(b) *Clique : Essential Data Matrix*

Clique character no.	Taxon								
	Anc	C	F	A	G	E	B	D	H
1	0	0	0	1	1	1	1	1	1
2	0	0	0	0	1	1	1	0	0
3	0	1	1	0	0	0	0	0	0
4	0	0	0	0	0	1	1	0	0
5	0	1	1	1	1	1	1	1	0
6	0	0	0	0	1	1	1	1	1
7	0	0	0	0	0	0	0	1	1
8	0	0	0	0	1	1	0	1	0
9	0	1	0	0	0	0	0	1	1
10	0	1	1	0	0	0	1	0	0

Table 3.13(c) *Clique : Character Ranking*

Rank	Character	Compatibilities	Incompabilities
1	3	8	1
2	2	7	2
3	4	7	2
4	7	7	2
5	1	6	3
6	6	6	3
7	5	5	4
8	8	5	4
9	10	4	5
10	9	3	6

Compatibilities are found by comparing each character with each other character (row 1 with rows 2, 3, etc.) to see whether the combinations 00, 0 1, 10, and 11 exist for any pair. Since the hypothetical ancestor, coded 0 for all characters, is included, the combination 00 will always appear. If anyone of the other three combinations is missing, the pair of characters is compatible. If all four combinations are found, the pair is incompatible.

Clique character no.	1	2	3	4	5	6	7	8	9	10
1	X									
2	:C.	X			C = Compatibility					
3	:C'	:C'	X		I = Incompatibility					
4	:C.	:C.	:C:	X						
5.	:|:.	'C:	'C:	'C:	X					
6.	:C.	'C:	:C'	'C:	:|:	X				

7.	:C.	:C'	:C'	:C'	:¦:	:C.	X			
8.	:C.	:¦:	:C'	:¦:	:C.	:C.	:¦:	X		
9.	:C.	:C'	:¦:	:C'	:¦:	:¦:	·C:	:¦:	X	
10.	:¦:	:¦:	'C:	:¦:	:C.	:¦:	:C'	:C'	:¦:	X
Compatibilities	6	7	8	7	5	6	7	5	3	4
Incompatibilities	3	2	1	2	4	3	2	4	6	5

Table 3.13(d) *Clique compatibility matrix*

It is convenient to picture the four corners of each cell in the matrix as representing the following chart of possibilities :

00

01

10

11

As a combination is found, a mark is placed in the appropriate corner of the cell. Any cell with four marks, then, indicates incompatible characters. The total number of compatibilities (C) and incompatibilities (I) for each character can be found by counting the row and column compatibilities for that character. The sum of compatibilities and incompatibilities will equal one less than the number of characters in the matrix Table 3.13(d).

Given the completed compatibility matrix, clique analysis locates groups of characters that are completely compatible with one another. These groups are called cliques. The clique with the largest number of compatible characters is the place where one begins constructing a dendrogram.

The ranked list of characters with the number of their compatibilities provides a means of finding character that cannot possibly be part of the largest clique in the study. When, the rank of a given character exceeds the number of compatibilities of that character, the first clique must have been exceeded. In the study group, the first six ranked characters have six or more compatibilities. The next two have five, but their rankings of 7 and 8 place them below the critical level. The first six characters are the only possible candidates for membership in the largest clique in the character set. To test these six candidates, one constructs a compatibility diagram (Fig. 3.37). The six characters are arranged in a circle, and a line is drawn between each pair of compatible characters.

This compatibility graph connects each character with each other character. Thus, the group is the largest clique. Note that the clique contains more than these six characters. The autapomorphic characters belong to all cliques; moreover, the six duplicate characters also belong to all this clique. The total number of characters in the largest clique is 18 of the original 22. Only four characters remain to be grouped.

The largest clique does not always contain all the characters that pass the rank-compatibility test. When it does not, some lines will be missing, in the compatibility diagram. If many lines are missing, it may be convenient to make an incompatibility graph that connects only the characters that are incompatible. Such a graph can be demonstrated with the four characters excluded from the first clique (Fig. 3.37). If the characters with the most connections indicating incompability is eliminated,

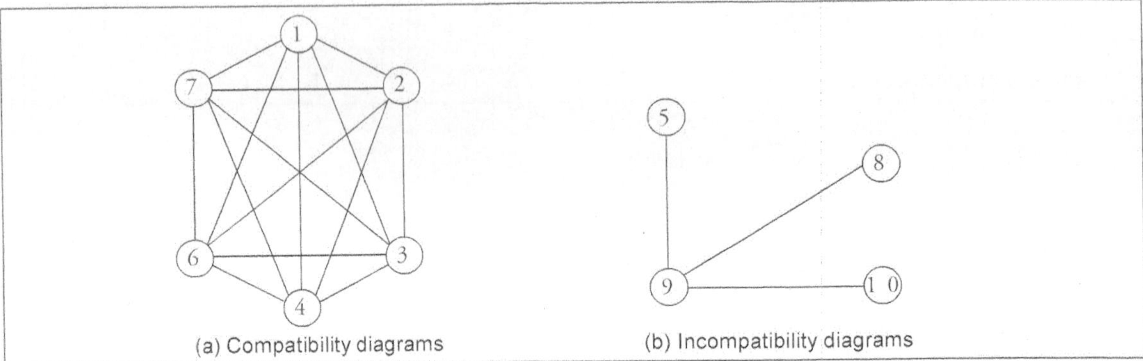

Fig. 3.37 Clique Diagrams

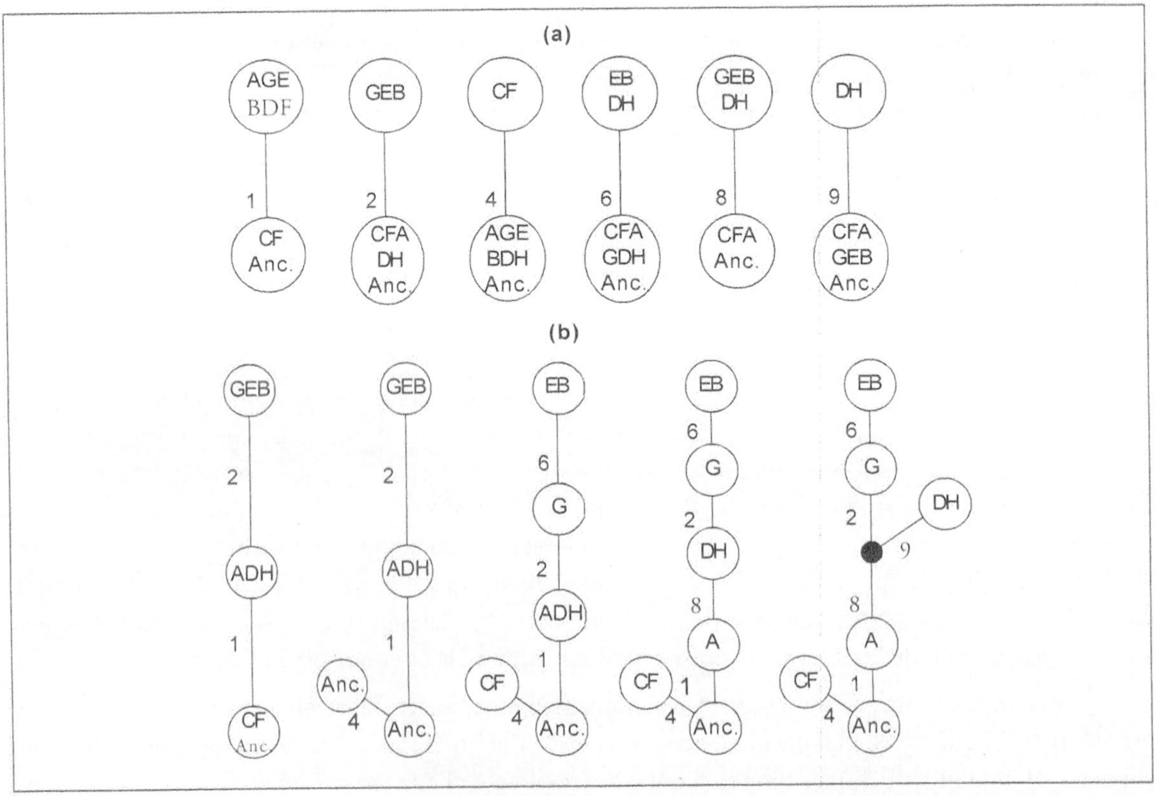

Fig. 3.38 Clique cladogram construction, (a) The six distinct two-character trees which form the first clique, (b) Construction of the cladogram by "*popping*" taxa.

a second clique is found (characters 5, 8, and 10). Character 9 forms the third clique. None of these four characters had duplicates, but the six autapomorphies belong to both cliques. Hence, the second clique consists of nine characters, and the third consists of seven. When one forms the tree, cliques are considered in order, from largest to smallest. Here the tree will be completely formed by the first clique; the other two are of no value in cladistic analysis. In larger studies, more than one clique is usually needed to form a cladogram.

The first six characters that form clique 1 may be diagrammed as in Fig. 3.38. Because each character defines a different holophyletic group and all the characters are compatible, they can be combined in any order, and each addition will develop the cladogram. The process of forming the tree is shown in Fig. 3.38.

Combining character trees is called popping. The tree as formed is essentially complete, but some relationships are not clear. Pairs CF, DH, and EB remain, and two taxa, A and G, rest within the tree. All these except taxon G will be popped when the autapomorphous characters are returned to the tree. Taxon G has no character states of its own and is unresolvable. Characters of the second and third cliques are incompatible with the cladogram; they must be placed in the most parsimonious manner, as parallelisms or reversals.

Note that the clique method clearly identifies the weakest characters and does so more precisely.

D. *Parsimony Methods*

Henning and his early followers constructed their cladograms by means of trial and error. Only synapomorphies were employed, whereas anagenesis was ignored.

A very early parsimony based method was that of Camin and Sokal. They constructed a compatibility matrix by placing the character states; of each character on a pattern or tree of each other character and then calculated the number of extra steps (if any) required for each character, character tree combination. Character trees were then ranked by quality, most compabilities and lowest number of extra steps and combined in order to produce a parsimonious tree. The original version have been corrected in modern computer programs. Kluge and Farris later developed a computer method, whose output is similar to that produced by the manual methods of Wagner. The algorithms are based on the construction of minimal or shortest spanning trees. Farris presented two algorithms, one to produce rootless trees and the other to produce trees with an ancestral root. The goal of these algorithms is the discovery of the most parsimonious trees, called Wagner trees. The two algorithms are not guaranteed to find the most parsimonious tree, and for most data they do not. The algorithms begin with a single taxon and sequently add more taxa to the growing tree until all taxa have been added. Many modifications and improvements of the methods of searching for parsimonious trees have been made in subsequent years. They include rearranging the branches of a tree if a more parsimonious tree can be produced and the discovery of multiple parsimonious trees. Also, the initial algorithms dealt with characters whose states were arranged into a linear array ("ordered" characters). Characters with unordered states, those obeying much more complex patterns of evolution, or those with the frequencies of states within the taxa specified can now be accommodated

To produce a rooted tree with Farris's original Wagner algorithm, one can add to the data matrix an ancestral taxon (anc) whose character states are all primitive and are coded 0 (Table 3.14(a)).

From this matrix, a Manhattan distance similarity matrix is calculated (Table 3.14(b)) which is identical to the Manhattan distance matrixed in the description of clustering methods except for the additional row and column for the ancestral taxon. The Wagner algorithm must with the ancestral taxon, and taxa are added to the tree in the order their distance from the ancestral taxon. Each new taxon is added at a position determined by its similarity to a taxon that is already in the tree, but it is placed on

Table 3.14(a) *Farris - Wagner Trees : Data Matrix*

Character	Taxa									Ancestral nodes						
	Anc	**C**	**F**	**A**	**G**	**E**	**B**	**D**	**H**	**1**	**2**	**3**	**4**	**5**	**6**	**7**
NkSw	0	0	0	1	2	3	2	1	1	0	0	1	1	2	2	1
PnEx	0	2	1	0	0	0	0	0	0	0	1	0	0	0	0	0
Clyl	0	0	0	0	0	1	1	0	0	0	0	0	0	0	1	0
CalC	0	1	1	1	1	1	1	1	0	1	1	1	1	1	1	1
Pnlb	0	0	0	0	1	1	1	3	2	0	0	0	1	1	1	2
CsMg	0	0	0	0	1	1	1	1	1	0	0	0	1	1	1	1
CrVn	0	0	0	0	1	1	0	1	0	0	0	0	1	1	1	1
Mmbl	0	0	0	0	1	1	1	2	2	0	0	0	1	1	1	2
PnHM	0	0	0	0	1	2	2	1	1	0	0	0	1	1	2	1
OcP	0	0	0	0	1	1	1	1	2	0	0	0	1	1	1	1
A3Sw	0	0	1	0	0	0	0	1	1	0	0	0	0	0	0	1
aFSp	0	0	0	1	0	0	0	0	0	0	0	0	0	0	0	0
bFSp	0	0	0	0	0	0	1	0	0	0	0	0	0	0	0	0
TbSp	0	1	1	0	0	0	1	0	0	0	1	0	0	0	0	0

a line drawn from the internode subtending the nearest relative, thus generating an ancestral node. When taxa in the example are ranked according to their distance from the ancestor, the list runs as follows : anc, A, C, F, G, H, B, D, E. The first segment of the tree consists of anc and the closest taxon, A (Fig. 3.39(a)). Taxa C and F are tied for second place, and either may be chosen. Here C is added to the tree by generating node 1 on the internode between anc and A. Taxon F differs from C by 2 and from it A by 5; it is added to the internode between C and node 2, generating 11: new node 2. The process continues until the cladograms is complete. The order in which taxa E, B, and D, which are equally distant from the ancestor, should be added to the tree presents a problem. One should choose the taxon that is closest , I to a taxon already on the tree. Here E differs from G by 3, while B and D both differ from their closest relative by 4. For this reason, E is entered first, and either B or D may be next. Since B and D will join different internodes, no conflict is found.

The more complete tree that includes character states expands the data matrix to provide descriptions of hypothesized ancestors. Simultaneously, it expands the similarity matrix to provide distance data between hypothesized ancestors (new nodes) and, taxa on other internodes. As shown on the data matrix, it differs from any by aFSPI, ColCl, NkSwl (Fig. 3.3.9(b)). As a second

Table 3.14(b) *Farris - Wagner Trees : Similarity Matrix*

	Taxa									Ancestral nodes						
	Anc	**C**	**F**	**A**	**G**	**E**	**B**	**D**	**H**	**1**	**2**	**3**	**4**	**5**	**6**	**7**
Anc	X	4	4	3	9	12	12	12	10	1	3	2	8	9	11	11
C	4	X	2	5	11	14	12	14	14	3	1	4	10	11	13	13
F	4	2	X	5	11	14	12	12	12	3	1	4	10	11	13	11
A	3	5	5	X	8	11	11	11	11	2	4	1	7	8	10	10
G	9	11	11	8	X	3	5	5	7	8	10	7	1	0	2	4
E	12	14	14	11	3	X	4	8	10	11	13	10	4	3	1	7
B	12	12	12	11	5	4	X	10	10	11	10	6	5	3	9	
D	11	13	11	10	6	9	11	X	4	11	13	10	4	5	7	1
H	10	14	12	11	7	10	10	4	X	11	13	10	6	7	9	3
1	1	3	3	2	8	11	11	11	11	X	2	1	7	8	10	10
2	3	1	1	4	10	13	11	13	13	2	X	3	9	10	12	12
3	2	4	4	1	7	10	10	10	10	1	3	X	6	7	9	9
4	8	10	10	7	1	4	6	4	6	7	9	6	X	1	3	3
5	9	11	11	8	0	3	5	5	7	8	10	7	1	X	2	4
6	11	13	13	10	2	1	3	7	9	10	12	9	3	2	X	6
7	11	13	11	10	4	7	9	1	3	10	12	9	3	4	6	X

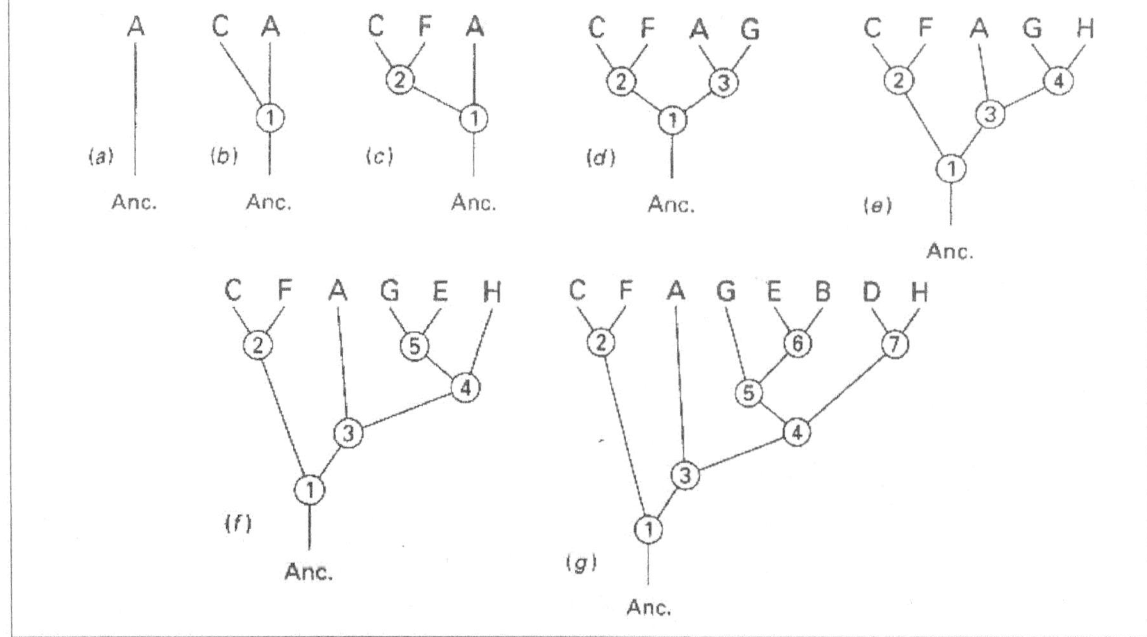

Fig. 3.39(a) Wagner tree construction using the simplified method of Whiffin and Bierner

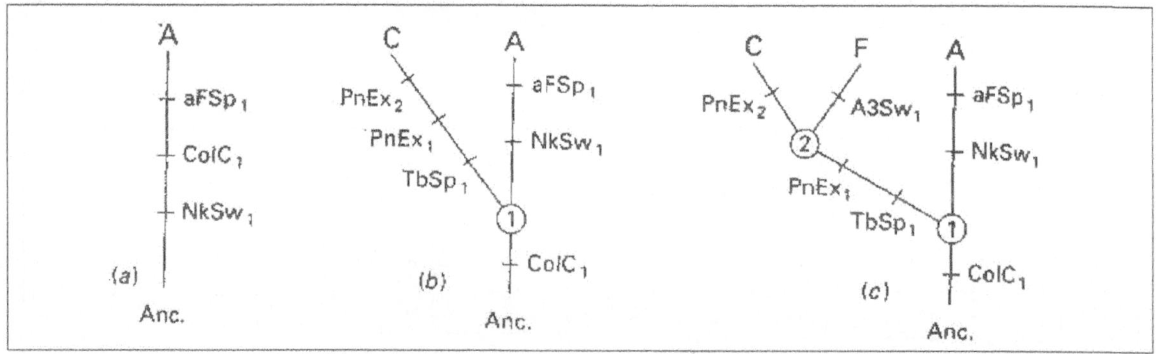

Fig. 3.39(b) Wagner tree construction using the method of Farris; only the first three steps are shown

taxon is added and node I is formed, these three character states must be sorted to see which belongs to A alone. Both A and C have ColC., but only A has aFSpl and NkSwl. The hypothetical ancestor at node I, then, differs from the original ancestor only by ColC I A new column is entered in the data matrix to reflect this difference. Treating node I as a new taxon, one adds a new row and column to the similarity matrix and calculates the distance of node I from each of the taxa. Finally, the other character states needed to describe taxon found in the data matrix and entered on the growing tree.

The final data matrix records the character states of each taxon and those inferred for each node, while the similarity matrix contains the distances between all combinations of taxa and nodes, accounting for parallel and reversed character states in the process.

These computational difficulties are a very minor problem compared with other difficulties inherent in the methodology of cladistic analysis. To justify the most parsimonious (shortest) tree, earlier cladists tended to minimize the frequency of homoplasies, particularly reversals.

The clique compatibility method and the Wagner tree parsimony method just described have been explained simply so that they can be attempted by hand by a patient person with a data set that is not too large: It must be emphasized that they are very primitive forms of these algorithms, useful for helping a student appreciate how a computer program approaches the difficult problem of machine cladistic analysis.

Current programs are far more sophisticated; they employ many variations on compatibility and parsimony techniques and incorporate many enhancements. For example, after the first run, a parsimony method may eliminate the weakest character states and be run again. The process is repeated until no further improvements can be found. Various branches can be moved about to achieve a better fit of character states and a more parsimonious tree. In clique analysis, an incompletely resolved tree can be broken into smaller trees, and any smaller cliques not used in the first run can be tried on the partial trees.

E. *Maximum – Likelihood Estimation*

Felsenstein has been the leading developer of statistical methods for inferring phylogenies. These methods have generally been applied only to molecular data. They presume that evolution follows a simple stochastic model; under this model, some patterns of distributions of character states among the taxa are more probable than others are. The tree for which the probability of observing the data is highest and is chosen as the maximum-likelihood estimate of the phylogeny. Trees are produced that may have less than the best obtainable parsimony scores but greater biological reality.

These methods have been criticized primarily on the grounds that the models of evolution upon which they are built, are incorrect oversimplifications of reality. Even if this is true, the statistical methods should not simply be abandoned but should instead be more vigorously explored and improved; they will play an important role in the future of numerical systematics.

A large array of computer programs are now available, some of them are NTSYS package for UGMA, PHYLIP – phylogeny inference package, PAUP – phylogeny analysis using parsimony, HENNING; Macclade – programm for manipulation and analysis of phylogenies and character evolution.

3.6 Ecosystem, Remote Sensing and Environmental Laws

Ecology is the study of the relationship of living organisms to each other and their surroundings. It adopts a holistic approach i.e., whole picture is built up which is more important than the parts. Among all approaches of study of ecology *viz*., ecosystem approach, community approach, population approach, habitat approach and evolutionary approach, the most fundamental one is *Ecosystem Approach*.

3.6.1A. Ecosystem Concept

The ecosystem was first defined by A.G.Tansley (1935) as the "*Living world and its habitat*". This also focuses on flow of energy and cycling of matter between living and non living components of ecosystem. Thus, the ecosystem is formed by interactions of individual organisms with each other and with their physical environment. Ecosystems are nearly self contained so that exchange of nutrients within the system is much greater than exchange with other systems. So, complete description of ecosystem must include the physical environment as well as biological components and the interactions between the two.

The elaborated definition of ecosystem was given by E. P. Odum as "*The basic functional unit of nature including both organisms and their environments, each interacting with the other and influencing each other's properties and both necessary for the maintenance and development of the system*". The ecosystem are diverse types (Fig. 3.40), mostly natural like lake, pond, forest, marine waters and few are man made or modified by man's activity like parks, aquaria, fish form, etc.

Components of Ecosystem

The ecosystem consists of biotic and abiotic components. The biological or *biotic components* of an ecosystem include both living organisms and products of these organisms. The microbes, all categories of plants and animals as well as their products are included. The biotic components

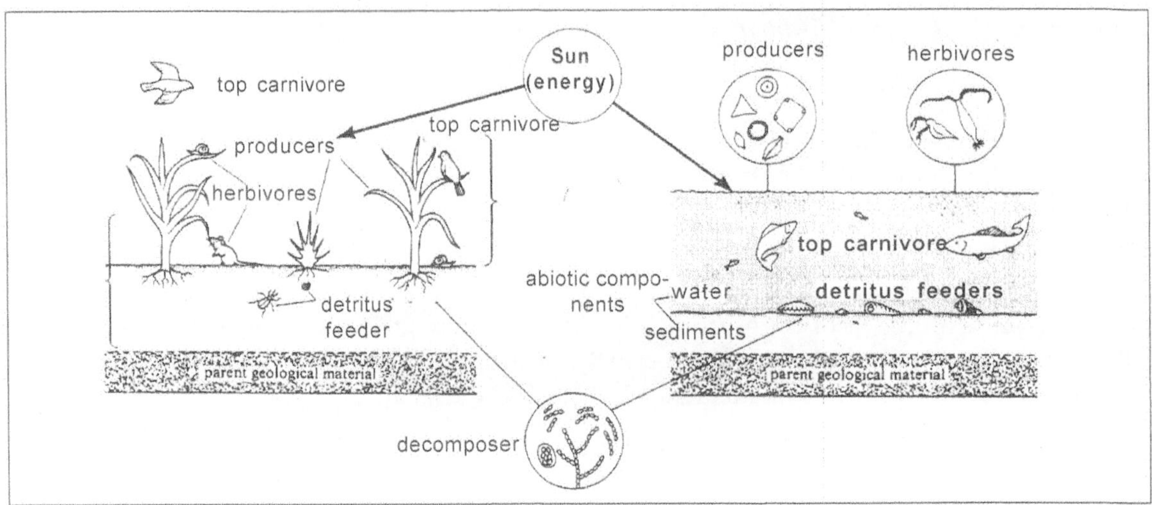

Fig. 3.40 A simple comparison of the gross structure of an aquatic (freshwater or marine) and a terrestrial ecosystem.

Necessary units for function are : abiotic components (basic inorganic compounds) producers (vegetation on land and, phytoplankton in water); animals - direct or grazing herbivores (grasshoppers, meadow mice etc. on land, zooplankton etc. in water); indirect or detritus-feeding consumers (soil invertebrates on land, bottom invertebrates in water); 'top' carnivores (hawks and large fish); decomposers (bacteria and fungi on decay).

are broadly categorised as *Producers* (plants and chemosynthetic bacteria) and *consumers* of different *classes viz., herbivores, (primary consumers)* animals that eat plants; *carnivores* (*secondary and tertiary consumers*), animals that eat-flesh of other animals, *omnivores*, animals that eat dead plant and animal matters. In addition there are *decomposer*, classes which include mostly *saphrophytic* organisms that help in nutrient and element recycling processes. The heterotrophic organisms, saprophytes feed only on dead organic matter, this saprophytic decomposition, is carried out by bacteria, fungi, protozoans etc.

The non-biological or *abiotic components* include climatic and edaphic factors, in particular climatic components like sunlight, temperature, air and water supply along with soil components including, organic and inorganic nutrients are essential contributing factors of ecosystem operation. The essence of ecosystems include how connections between the different organism and their abiotic environment work. Two important ways in which these connections found are *energy flow* and *nutrient cycling*.

Energy Flow and Nutrient Cycling

The two ecological processes of energy flow and mineral cycling involving interaction between physicochemical environment and biotic assemblage lie at the heart of ecosystem dynamics. They are fundamental to the structure and process that takes place within ecosystem, as shown in model, (Fig. 3.41) the movement of energy and matter within the biotic component and between it and the biotic nutrient pool takes place. The energy flows unidirectionally and also lost as heat from the system in several ways, whereas the minerals circulate within the ecosystem, (biogeochemical cycles) and there is net loss in several ways to sediments or other systems (Fig. 3.42).

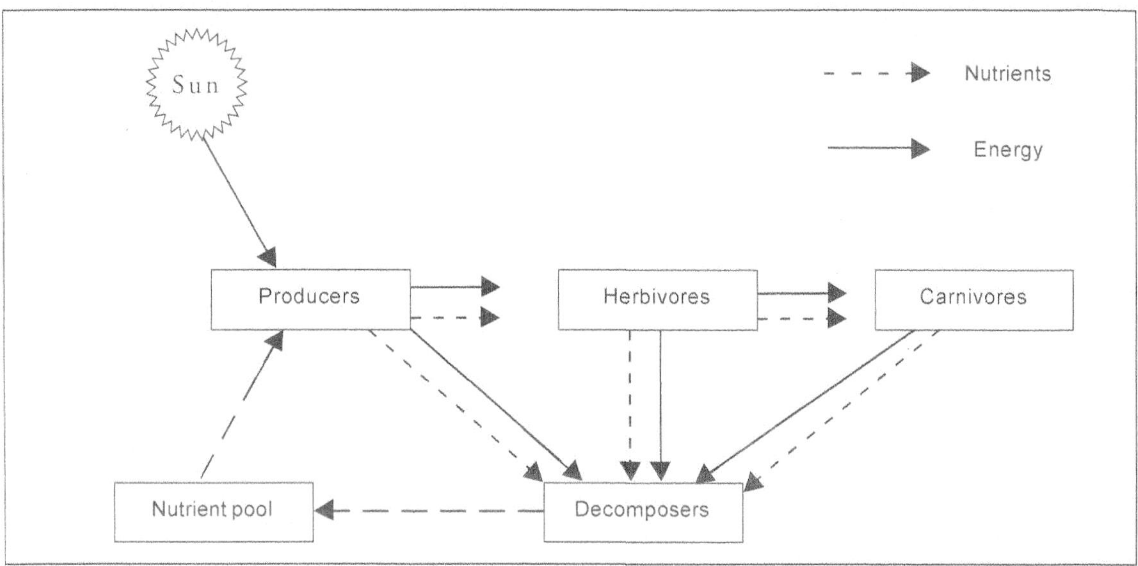

Fig. 3.41 Simplified model of energy and mineral movement in ecosystem. The energy flow is unidirectional and non-cyclic, while nutrient movement is cyclic.

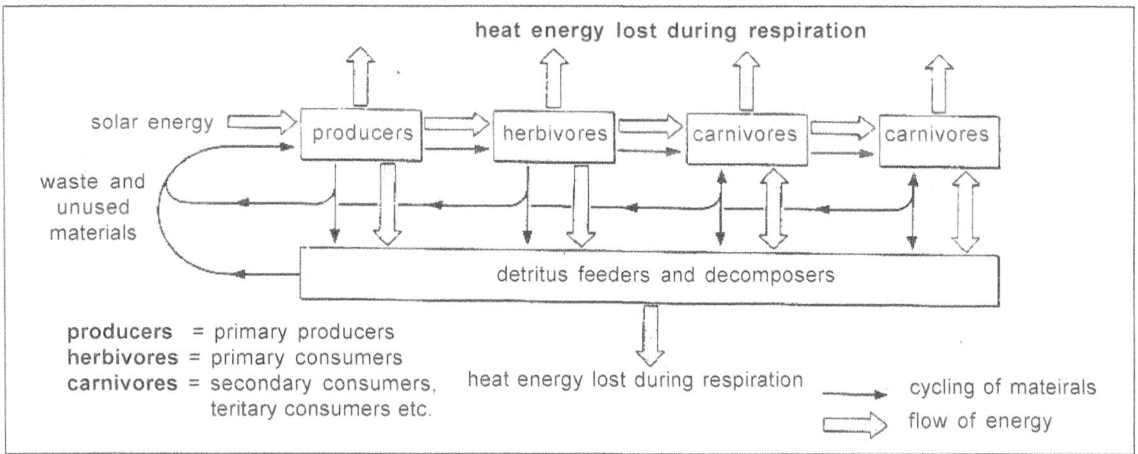

Fig. 3.42 Flow of enery and cycling of materials through a typical food chain. Two-way exchange is possible between carnivores and detritus feeders/decomposers. The latter feed on dead carnivores; carnivores may eat living detritus feeders/decomposers.

Productivity and Energy Flow

During the process of production and consumption, energy flows from one organisms to another. For example, solar energy is converted to chemical energy within leaves of plants. These can be eaten by some herbivore which inturn be eaten by a carnivore. For example, an ecosystem receives 1000 Kilocalories of light energy on a given day, most of this energy is not absorbed at all, some reflected back to space, some as heat and only a small amount is stored in plants.

Energy enters the biotic component of ecosystem through producers and the rate of which this energy is stored by them in the form of organic substances which can be used as food materials is known as *Primary productivity*. Some amount of photosynthetic materials is utilized for its respiration. The initial product is called as *Gross primary productivity* (expressed as kCal/m^2/yr). Only about half of gross primary productivity accumulates and rest is lost as heat as well as in respiration. The net gain of plant material is called as *Net Primary Productivity (NPP)*. The production by heterotrophs is called *Secondary productivity*. Thus, *Biomass* is the total dry mass of all organisms in an ecosystems and it provides a useful comparison between different ecosystems.

As shown in Fig. 3.43, there is decreasing quantity of energy available at each trophic level. As a consequence there will be a decreasing mass of organisms at each level. For example, for energy 10 Kcal plant-tissue available to herbivores, about 1 Kcal will be eaten and only about – 0.1 Kcal will be stored in the form of body weight. Carnivores that eat herbivores are likewise inefficient in inconverting food to biomass, so the energy available to the next carnivore is even less. Thus, it is obvious that the amount of usable energy decreases as it transferred from producers to higher trophic levels.

Food Chains and Trophic Levels

In the ecosystems, energy containing organic molecules produced by producers, autotrophic or

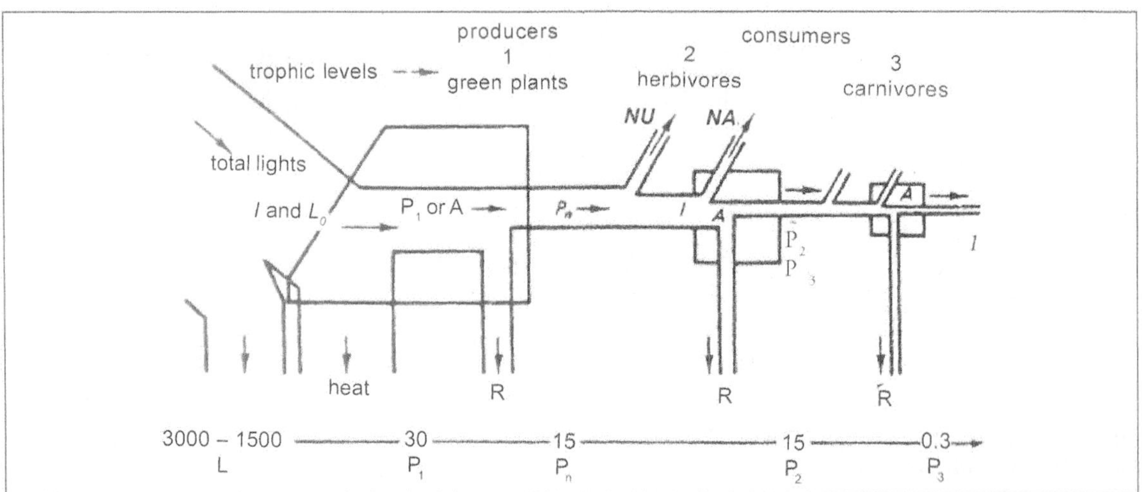

Fig. 3.43 A simplified diagram of energy flow in a food chain. The boxes represent the standing crop of organisms (1 : producers or autotrphs; 2 : primary consumers or herbivores, 3 : secondary consumers or carnivores) and the pipes represent the flow of energy through the biotic community. L - totle light : L_a = absorbed light; P_g = gross primary production; P_n = net primary production; p = secondary production at second (p_2) and third (P_3) trophic levels; I = energy intake; A = assimilated energy; NA = non-assimilated energy; NU = unused energy (Stored or exported); R = respiratory energy loss).
The chain of figures alone the lower margin of the diagram indicates the order of expected at each successive transfer starting with 3000 Kcal of incident light per m^2 per day.

chemotrophic organisms are source of food for heterotrophic organisms as seen in a plant being eaten by an animal. This animal may inturn be eaten by another animal and in this way energy is transferred through a series of organisms, each feeding on the preceding organism and provides raw material and energy for the next organism. Such a sequence is called as *Food chain.* Each stage of food chain is known as *Trophic level*, the first trophic level being occupied by autotrophic organism called as *Producers*, the organisms of second trophic level are usually called *Primary consumers*, that of third level as *Secondary consumers* and so on.

Some of the examples of food chain are :

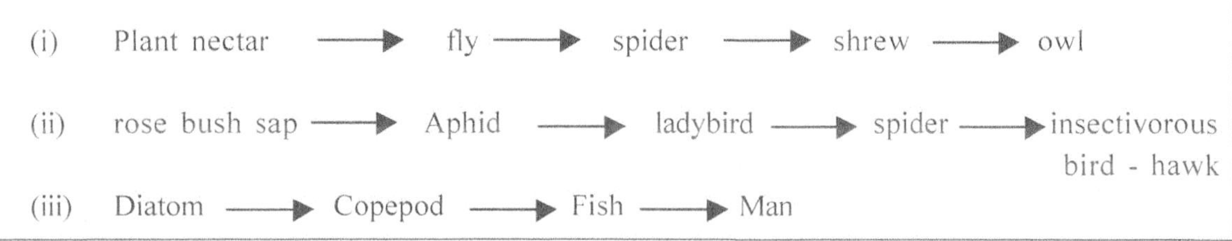

Two basic types of food chain exist, *grazing food chain* and *detritus food chain*. In *grazing food chain*, the first trophic level is occupied by a green plant, the second by grazing animal, herbivore, and subsequent level by carnivore as shown in the above examples. On the other hand, through activity of decomposers (micro organisms, fungi and bacteria) on plants and animals produce organic material. The fragment of decomposing material are called *Detritus* and many small animals feed on these are called detritivores. These detrivores inturn be fed upon by larger organisms, building another type of food chain. This is known as *detritus food chain*.

Detritus Food Chain

Dead animal → blow flies and their maggots → frog → grass snake

Grazing Food Chain

Plant → fly → Spider → Shrew → Owl

Food webs

The feeding relationship within an ecosystem are more complex and each animal may feed on more than one organism in the same food chain or may feed in different food chains. This is particularly true of carnivores and omnivores at the higher trophic levels. Thus, the food chain interconnect in a such away to maximize the energy transfer. This complicated pattern of interconnection of food chains is called as "*Food web*". Fig. 3.44 illustrates freshwater food web in which feeding relationships among organisms in an ecosystem and provide basis of energy flow and exchange of material in ecosystem.

Ecological Pyramids

Feeding relationships and energy transfer through the biotic component of ecosystems may be quantified and shown diagramatically as ecological pyramids (Fig. 3.45(a) & (b)). These give simple and fundamental basis for comparing different ecosystems or their other variations. Depending on the variable they represent like number, biomass, and energy, these are known as *pymarid of number, biomass or energy*.

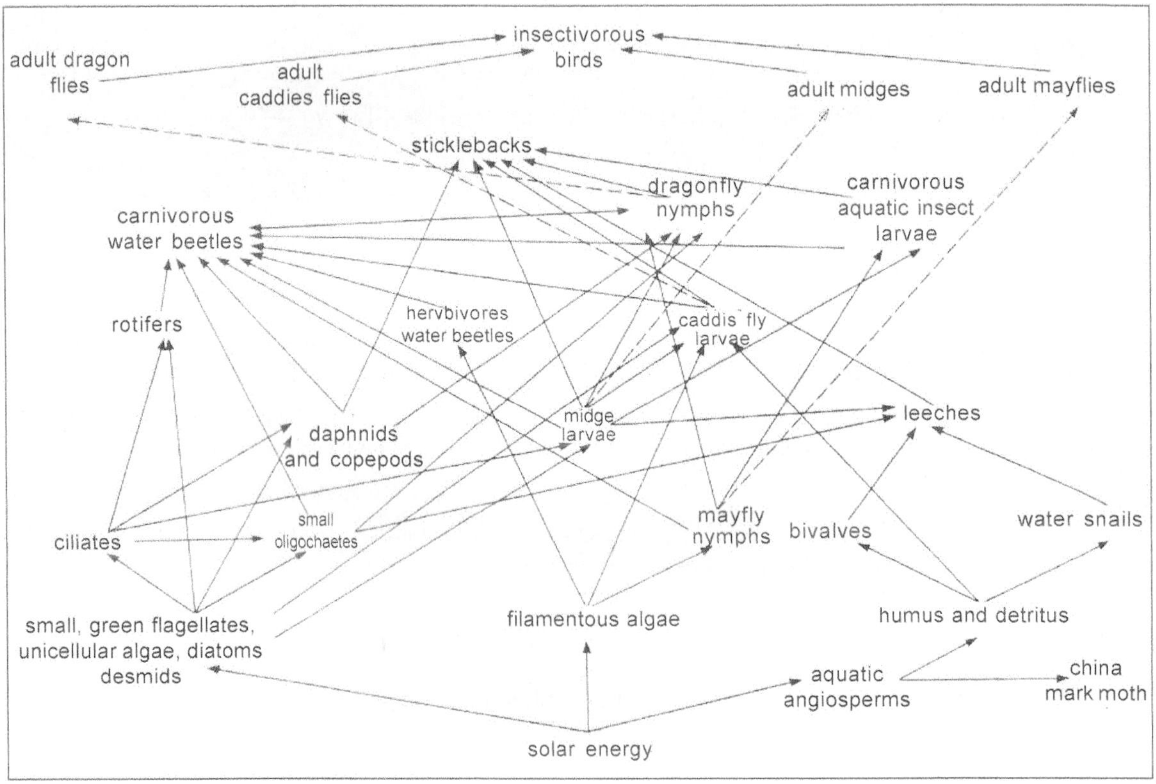

insectivorous birds
adult dragon flies
adult caddies flies
adult midges
adult mayflies
sticklebacks
dragonfly nymphs
carnivorous aquatic insect larvae
carnivorous water beetles
rotifers
hervbivores water beetles
caddis fly larvae
midge larvae
leeches
daphnids and copepods
ciliates
small oligochaetes
mayfly nymphs
bivalves
water snails
small, green flagellates, unicellular algae, diatoms desmids
filamentous algae
humus and detritus
aquatic angiosperms
china mark moth
solar energy

Fig. 3.44 Food web of a fresh water ecosystem

Teritary consumer
Secondary consumer
Primary consumer
Producers

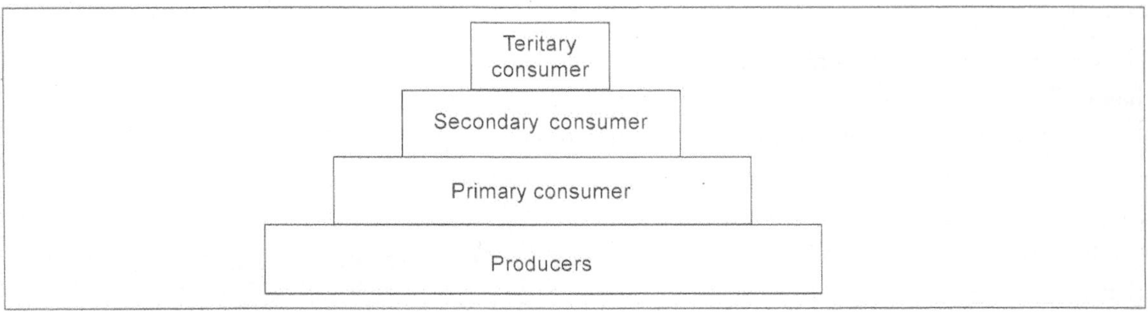

Fig. 3.45 (a) Pyramid of Number

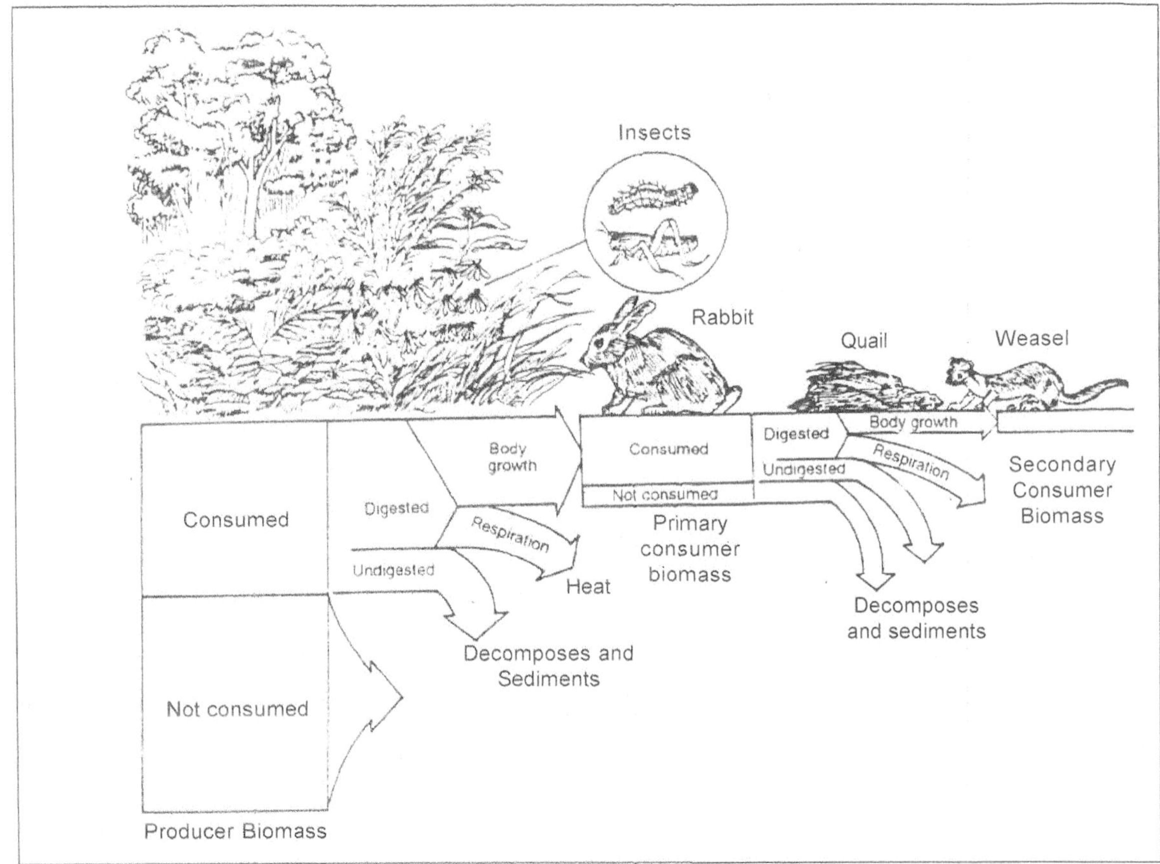

Fig. 3.45 (b) The energy pyramid

3.6.2B. Biogeochemical Cycles

Biogeochemical cycling is second major function of ecosystem, along with energy flow. Each cycle summarises the movement of chemical elements through the living component of ecosystem. The buildup in food chains of complex organic molecules incorporating the element and breakdown in decomposition to simpler organic and subsequently inorganic forms which can be used again to make the living reservoir pool. There are two types of biogeochemical cycles; *gaseous* cycle and *sedimentary* cycle. In *Gaseous cycle*, the element or compound can be converted to a gaseous form, diffuse through the atmosphere and they arrive over land or sea to be reused by biosphere in a much shorter time. The primary constituents of organisms, carbon, hydrogen, oxygen and nitrogen, move through gaseous cycle. In the *Sedimentary cycle*, the compound or element is released from rock by weathering, then follows the movement of water either in solution or as sediment. Eventually, by sedimentation and precipitation, these materials are converted to rock and when the rocks exposed to weathering the cycle is completed. These cycles include phosphorous, sulphur etc. The major elemental cycles are described briefly.

(i) *Carbon cycle*

The carbon cycle is very complex due to the fact that carbon can exist in a wide variety of different types of compounds, occupying in plants, animals, rocks, liquid, sediments and air. The most important carbonaceous materials are CO_2, CO, carbonates, and organic carbon materials lie as dissolved CO_2 in fresh water and saline water. They form carbonate or bicarbonate and thereby make deposition as sediment or crystal deposits like limestone and marble stone. However a major amount of CO_2 is fixed as organic carbon by the green plants as fossil fuel like coal and oil. On the other hand CO_2 is liberated back to atmosphere from burning of fuel, respiration of biota, volcanoes, and decomposition of organic waste. An overall scheme of carbon cycle is given in Fig. 3.46.

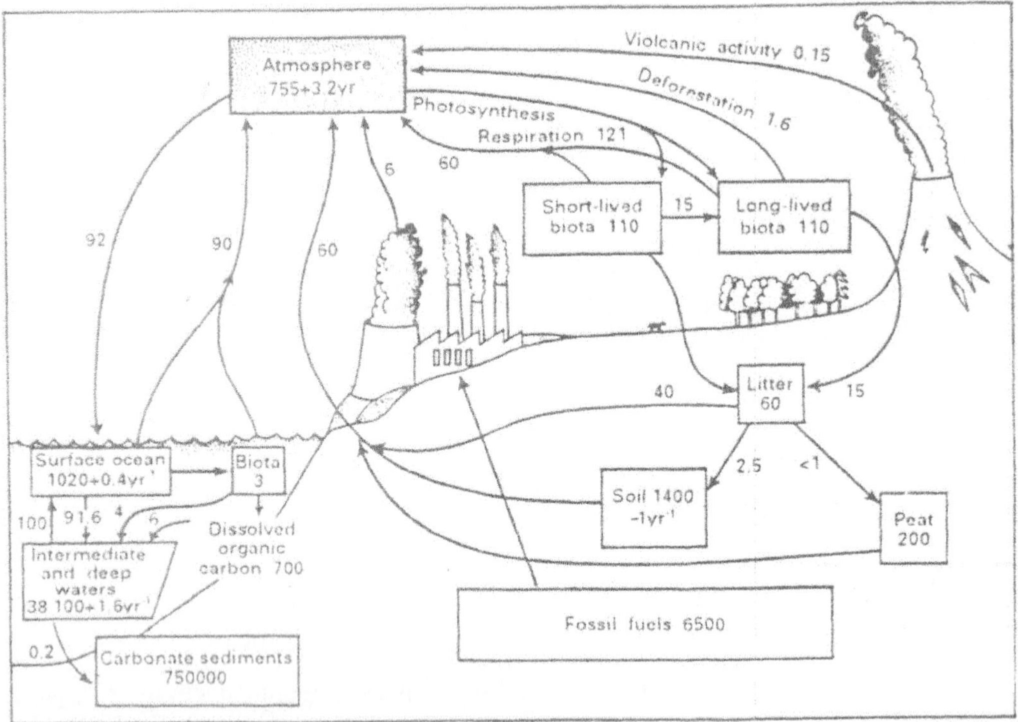

Fig. 3.46 Carbon cycle (number at arrows denote flux rate in Gt. y^{r-1}).

CO_2 cycle is very rapid as in plants which fix CO_2 within an hour by photosynthesis and also capable of releasing it at the same rate by respiration. But non-biological processes or complex biological and abiotic processes leading to carbonate rock formation, or fossil organic fuel deposit creation requires millions of years.

In modern times, human activities are making a significant impact on the global carbon cycle. When fossil fuels are burned, geological reserves of carbon are converted, into CO_2 and released into the atmosphere. Today, the release of CO_2 or discharge of CO_2 into the atmosphere is steadily increased, owing both to the burning of fossil fuels and to the destruction of forests.

As a consequence of global CO_2 rise over the years, green effects were noticed particularly during past couple of decades. As such rise of global warming problem takes place. An overview of global warming problem arising out of CO_2 cycle disturbances is shown in Fig. 3.47.

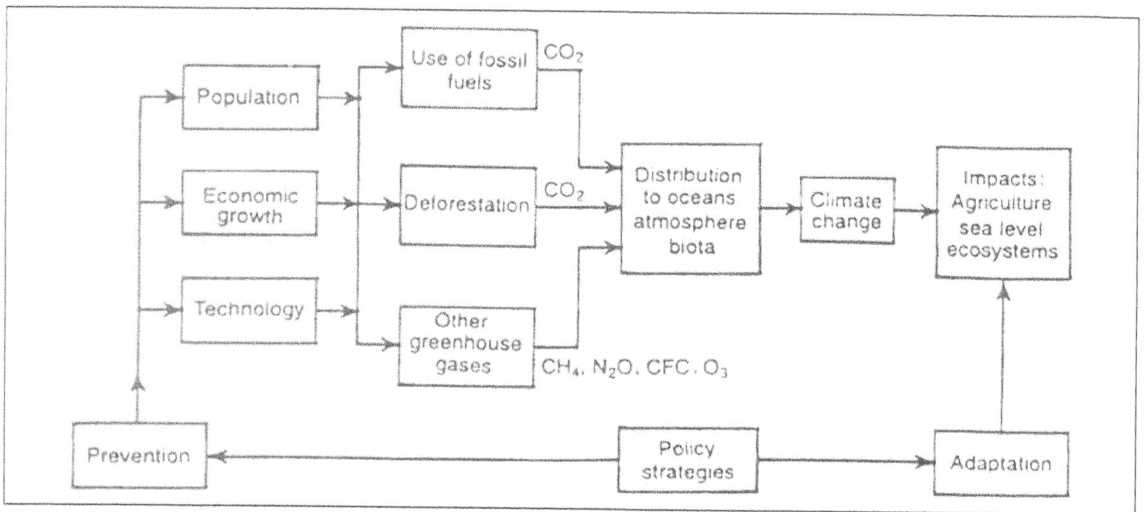

Fig. 3.47 An overview of the global warming problem

(ii) *Oxygen cycle*

This is one of the major gaseous components of atmosphere. The oxygen cycle involves movements and storage in different forms and its oxidative reactions with various elements. Oxygen enters this storage pool through release in photosynthesis both in the oceans and on the lands. Balancing the input to the atmospheric storage pool is loss through organic respiration and mineral oxidation. The oceans serve as a major storage pool for dissolved gaseous oxygen. Some oxygen is continuously placed in storage in mineral carbonate form in ocean floor sediments. The various process involved in oxygen cycle are shown in Fig. 3.48.

(iii) *Nitrogen cycle*

Nitrogen is one of the very important elements in biosphere. It remains in various forms in different environmental components, viz., as gas N_2, and NO as ion in soil and water, NO_3 NO_2, NH_4 or as organic material or insoluble nitrogenous materials. Among these atmosphere possesses 78% of nitrogen. Most plants absorb this vital nutrient when it is in the form of nitrites (NO_2^-), nitrates (NO_3^-) or ammonium ions (NH_4^+). Only some selected bacteria either in symbiotic or free living state fix atmospheric nitrogen to, NH_3 and then utilize the same for amino acid biosynthesis. The animals obtained the nitrogen from plants as food materials. The organic nitrogen thus available as extra from plants and animals or dead remains of organisms is decomposed, which may release N_2 gas, (NO) gas, NO_3 or NO_2 ions. Various processes thus involved in such cycling of nitrogen from environmental components to biological communities are shown in Fig. 3.49.

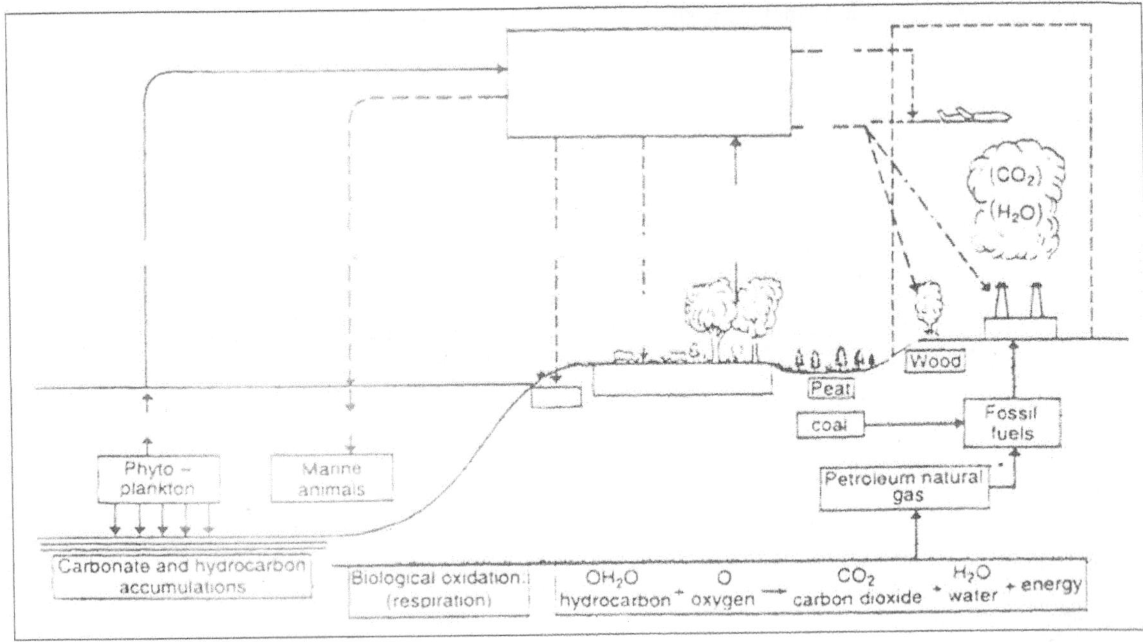

Fig. 3.48 The oxygen cycle

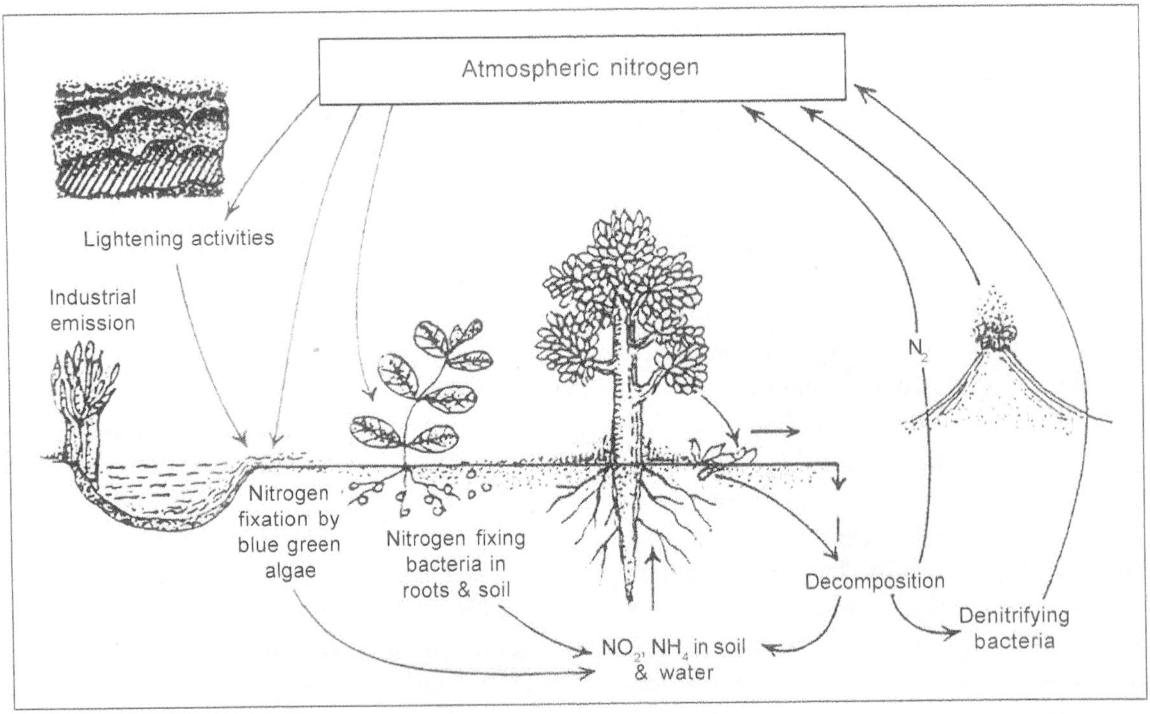

Fig. 3.49 The nitrogen cycle

(iv) *Phosphorous cycle*

The phosphorous does not occur naturally as gas, its cycle, unlike those of carbon and nitrogen, is a sedimentary cycle. Many rocks contain phosphorous, usually in the form of phosphates (PO_4) that are bound into the mineral structure. When rocks are weathered, minute amounts of these phosphates dissolve and become available to plants. Animals then absorb this element when they eat plants or other animals. Much of the phosphorous excreted by animals is also in the form of phosphate, which plants can reuse immediately. Thus, on land, phosphate cycles from plants to animals and back again. Land ecosystems preserve phosphorous efficiently, since both organic and inorganic soil particles absorb phosphates, providing a local reservoir of this element. In an undisturbed ecosystem, the intake and loss of phosphorous are small compared with the amounts of phosphorous that are internally recycled in the day to day exchange among plants and animals. Some phosphorous is inevitably lost by leaching and erosion of the soil into streams and rivers. When an ecosystem is disturbed, as by mining or farming, erosion can become so significant that large quantities of phosphorous and other nutrients are washed away. When phosphate reaches the ocean, it reacts with other minerals and sediment at ocean floor (Fig. 3.50).

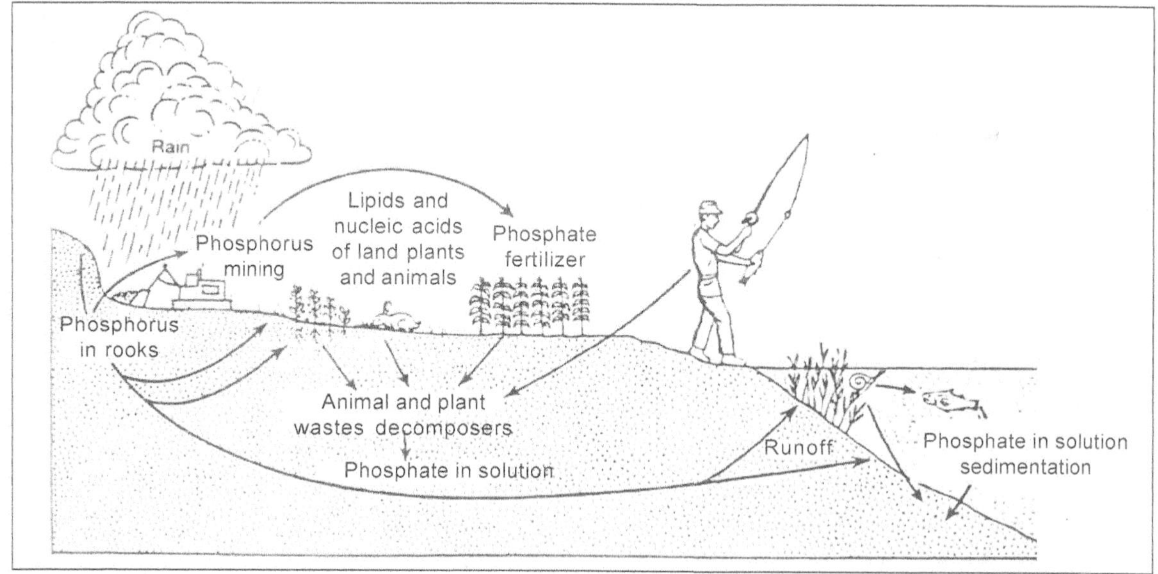

Fig. 3.50 Phosphorus cycle

(v) *Sulfur cycle*

Sulfur is one of the essential elements of life system and its major reservoir lies in the earth crust together with oceanic sediment. It acts as a component of several compounds like amino acids, protein, enzymes etc. Natural fuel like coal and oils also have some sulfurs (i.e., 1 to 3%). Burning these materials leads to formation of gas SO_2, the reactions of this gas with other compounds in the atmosphere are extremely complex, but one of the important products of its solution and further oxidation is sulfuric acid, which forms acid rain. An overall scheme of global sulphur cycle showing major reservoirs and the flux rates between these are presented in Fig. 3.43.

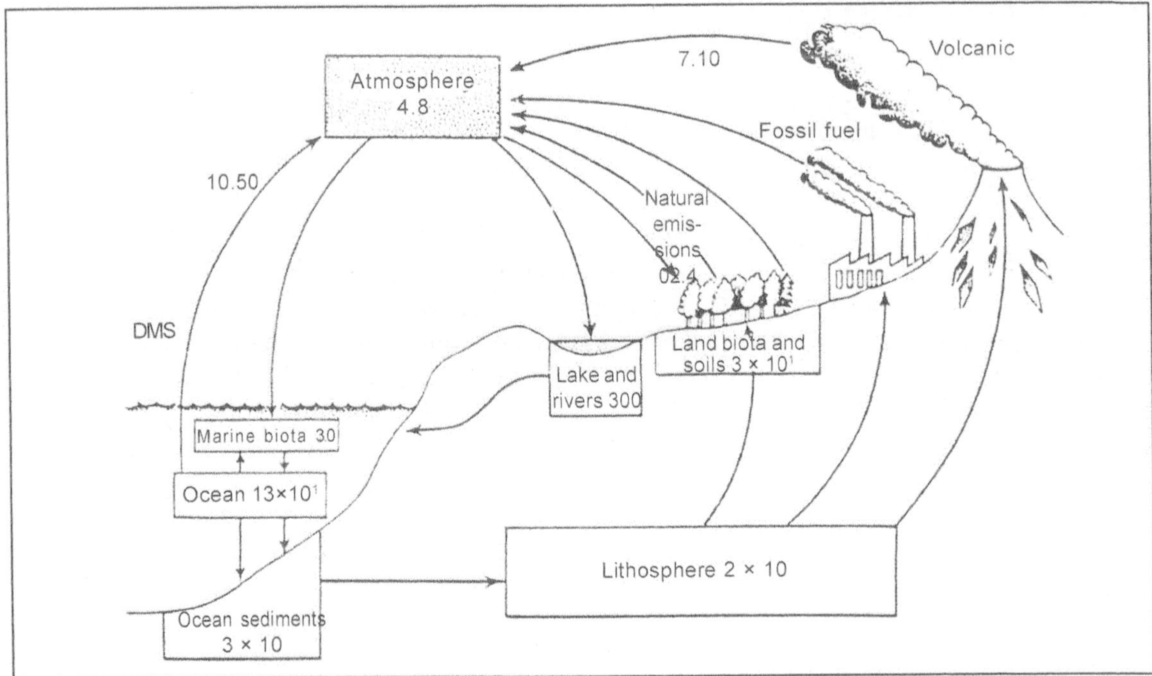

Fig. 3.43 Sulphur cycle

There is a strong analogy between sulfur and nitrogen in the way that microorganisms influence their geochemical cycling. When organic sulfur compounds are decomposed by bacteria, the initial excreted sulfur product is often hydrogen sulfide (H_2S). Many marine phytoplankton produce compounds that break down to produce dimethyl sulfide $(CH_3)_2S$. Dimethyl sulfide is a major biogenically produced sulfur compound.

(vi) *Mineral cycle*

Quite a good number of minerals are essential for biological life processes. They are either used as major or minor essential elements. They are mostly stored in rocks and soils (i.e., lithosphere). They are absorbed by the plants and used in metabolism, while animals obtained them from plants as food sources. They may be available in water bodies as soluble ions. In organic combination they may be stored in biomass. In particulate matters minerals also present in substantial amounts and thereby absorbed by plants. A schematic flow diagram of mineral cycle is given in Fig. 3.52.

3.6.3 Environmental Laws

Environmental law consists of all legal guidelines and procedures that are purported to protect the environment. It also enable judiciary to scrutinise the working of legislative and executive agencies. It covers the essence of environmentalism, from right of the biosphere and lithosphere to rights of citizens to protect their health, welfare and sustainability, guarantee of disclosure of information from all public agencies etc. The rules that govern environmental protection are two kinds :

(i) They are the ***Statuatory laws*** of constitution, national and regional legislatures and local governments, and

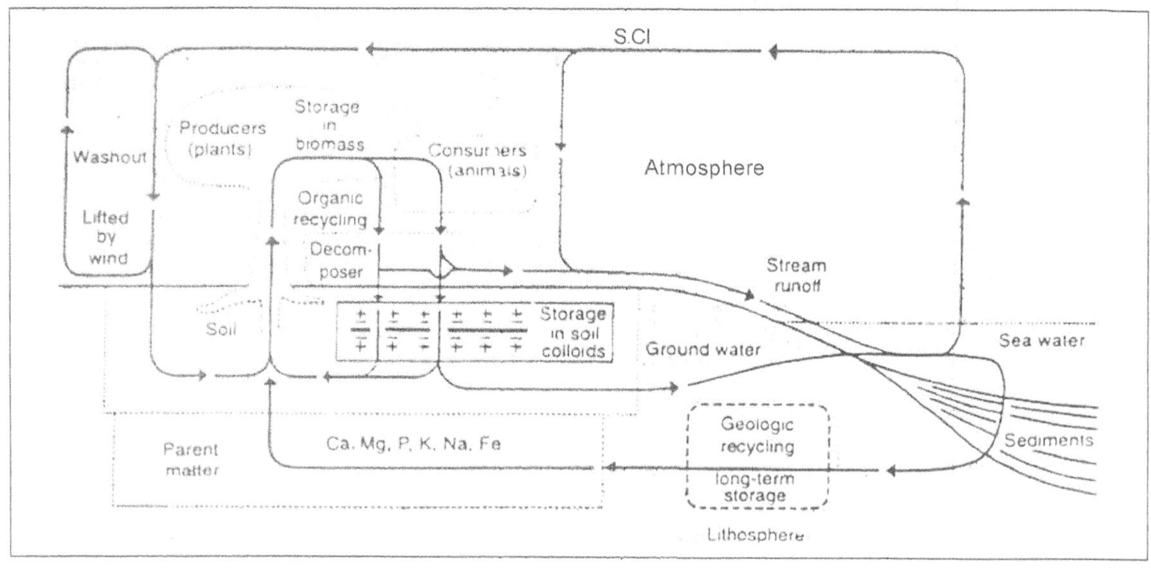

Fig. 3.52 Flow diagram of the mineral cycle

(ii) The common law, the body of judicial interpretation that create the precedents upon which future cases are judged, however, the common law relating environmental protection has number of drawbacks for its implementation.

Constitutional Measures

India is one of the very few countries in world which has provided for constitutional safeguard for the protection and preservation of environment. In the constitution of India specific provisions for the protection of environment have been provided. It is an obligatory duty of the state and every citizen to protect and improve the environment. The two articles in Indian constitution related to environmental protection are :

(i) Article 48 (A) (Directive principle of state policy) : The state shall endeavour to protect and improve the environment and safeguard the forests and wildlife of our country.

(ii) Article 51 A (Fundamental duties of citizen). It shall be the duty of citizen of India to protect and improve the natural environment including forests, lakes, rivers, and wildlife and to have compassion for living creatures.

In addition 73rd amendment Act of 1992 on Panchayat which adds a XI schedule to the constitution has 8 entries that are linked to environmental protection and conservation. The functions assigned to panchayats include soil conservation, water management, watershed development, social and farm forestry, drinking water, fuel and fodder, non-conventional energy sources and maintenance of community assets which are significant in view of environmental management. Similarly 74th amendment of constitution 1992, assigned to Urban local bodies, the function of "Protection of environment and promotion of Ecological effects" of Urban areas.

In addition article 246 of constitution deals with subject matter of laws made by parliament and legislatures of states.

Environmental Legislations : Status in India

In general there are three categories of legislation relating to environment, viz. protective legislation, planning legislation and preventive legislation. However, the planning legislation for environmental quality improvement, among them is highly neglected in India.

Bengal smoke nuisance Act 1905 is one of the oldest environmental protection Act in India. The factories Act 1948, Mines and Minerals (Regulation and development) Act 1957 are import environmental legislation of early post independence India. But most of environmental legislations were instituted after 1970s. A series of Acts thereafter were introduced in India, to facilitate the appropriate operation, rule and national policy on every aspect were drafted by high power committees. There are more than 20 statue have bearing on environmental matters in India. However, the major legal provisions made in the last 30 years are summarized below and shown in detail in Table 3.15.

- The Environment (Protection) Act 1986.
- The Water (Prevention and Control of Pollution) Act 1974, amended in 1988.
- The Wild Life (Protection) Act 1972, amended in 1983, 86 and 91.
- The Air (Prevention and Control and Pollution) Act, 1981, amended in 1987.
- The Hazardous Waste (Management and Handling) Rules 1989.
- The Public Liability Insurance Act, 1991.

Table 3.15 *Environmental Legislation in India*

Aspects	Acts	Rules
1. Water	1. The Water (Prevention & Control) Act, 19741. (amendment up to 1988).	1. The Water (Prevention and Control) Rules. 1975.
	2. The Water (Prevention & Control) Act.19772. (amendment upto 1991)	2. The Water (Prevention and Control) Cess Rules 1978.
2. Air	3. The Air (Prevention & Control) Act.1981 (amendment up to 1987)	3. The Air (Prevention & Control) Rules (1982 & 1983)
3. Environment	4. The Environment (Protection) Act. 1986.	4. Environment (Protection) rule 1992 and 1993. Environment statement.
		5. Environment (protection) rules, 1993 Environment standard.
		6. Environment (Protection) rules, 1994. Environmental Clearance.
4. Forest	5. The Forest (Conservation) Act, 1980 (amendment upto 1992).	7. Forest (conservation) rules, 1981
5. Wildlife	6. The Wildlife (Protection)Act, 1972 (amendment upto 1991)	8. Wildlife (Transection and Taxiderm) Rules 197.
		9. Wildlife (Stock Declaration) central Rules, 1973.
		10. Wildlife (Protection and Licensing) Rules 1983.

Contd... Table 3.15

Aspects	Acts	Rules
6. Minerals	7. Mines and Minerals (Regulation and Development) Act, 1957.	11. Mineral Conservation and Development Rules, 1988.
7. Monuments	8. Ancient Monuments and Archeological Sites and Remains Act, 1958.	–
8. Hazardous	–	12. The Hazardous waste (Management and Handling) Rules, 1989. 13. Manufacture, storage and import of Hazardous chemical Rules, 1989.
9. Microorganisms	–	14. Hazardous Micro-organisms and Genetically Modified Organisms (Manufacture, Use, Import, Export Storage). Rules.1999, (Previously Manufacture, use, Import, Export & Storage of Hazardous micro-organisms Genetically, Engineered Organisms or Cells Rules, 1989).
10. Public liability	9. The Public liability insurance Act,1991.	15. The Public Liability Insurance rules
11. Biomedical waste	–	16. Biomedical waste management and Handling Rules, 1998.
12. Coastal zone protection	–	17. Coastal zone Regulations, 1991.
13. Biodiversity	10. Biodersity Act, 2000	18. Biodiversity Rules, 1999.

3.6.4 Remote Sensing

Remote sensing is defined as the science of collecting and interpreting information about a target without being in physical contact with the object under study. The remote sensing include photography from aircraft and sensing from sensors of space shuttle.

Principle of Remote Sensing

The objects and constitutents of earth depending on physical and chemical features, reflect, reradiate or emit in various wavelengths. The measurement of reflected, radiated or emitted radiation forms the basis for characterising the objects. The selected position of electromagnetic spectrum which pass through earth's atmosphere with relatively little attenuation are used for remote sensing purpose. The selected region of the electromagnetic spectrum used in remote sensing include : 0.4 to 0.7μm, 0.7 to 30μm (IR band), 3 to 5μm and 8 to 14μm (TIR) and 0.1 to 30 cm (microwave).

Types of Remote Sensing

Remote Sensing technique, depending on the source of energy, are classified into two types : (a) *Passive remote sensing* and (b) *Active remote sensing*. In the former, the naturally radiated or reflected energy from objects is measured by sensors operating in different spectral bands on aircraft or spacecraft. In active remote sensing, system supplies its own source of energy to illuminate the objects and measures the reflected energy to the system.

Procedure in Remote Sensing

There are five distinct components which are involved in two major steps in remote sensing process. The first one is the *data acquisition* and second is *data processing and interpretation*. Data acquisition is made by *Sensors* from *Platforms*. This information is used for the formation of *product* which

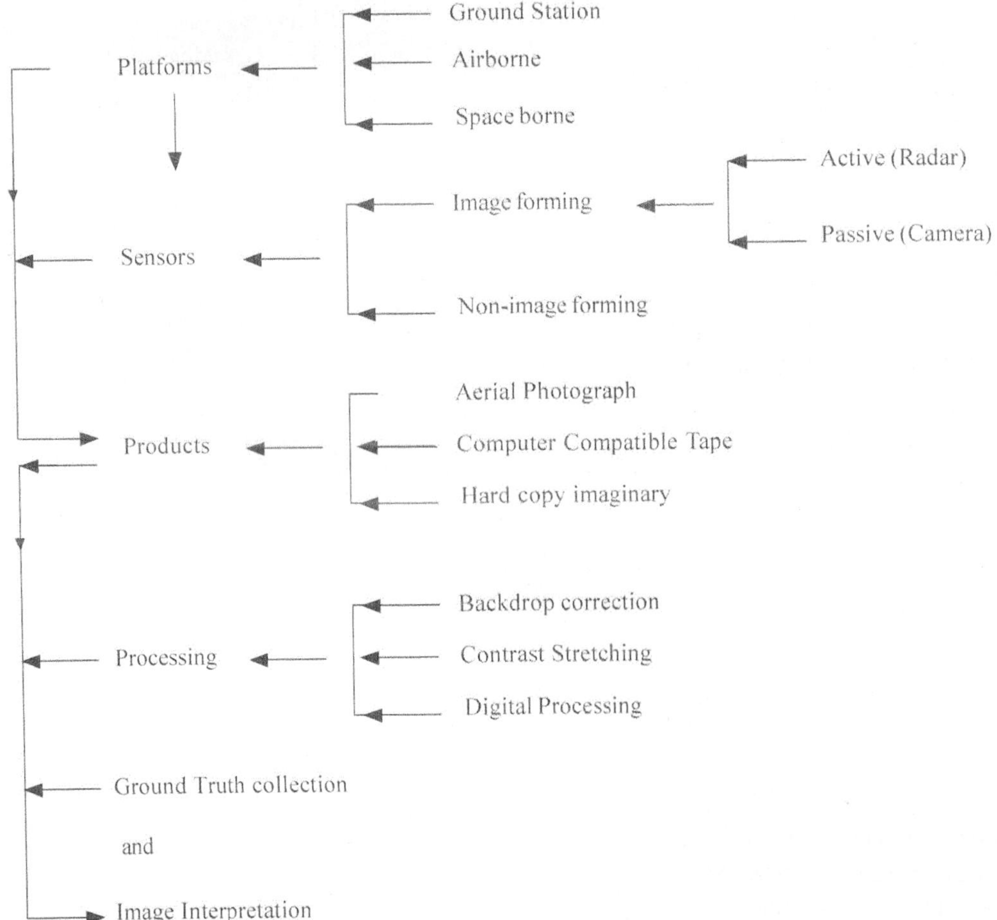

are finally used for *interpretation* purpose by comparing ground information, as shown below.

Steps Involved in Remote Sensing

The major tools used in remote sensing are satellites and radars.

(i) *Satellite Remote Sensing (Satellite and Sensors)*

Earth resource technology satellite (ERTS – 1), later named as LANDSAT – 1, was the first remote sensing satellite founded by NASA (KA) for surveying, mapping and monitoring of earth resources.

The other satellites used from different countries for remote sensing are SPOT (France), JERS (Japan), ERS (Europe) and IRS (India). Indian remote sensing satellite IRS-IA, IRS-IB and IRS-IC have been put into orbit in 1998, 1991 and 1995 respectively with sensors like LISS - I, II, III, WIFS and PAN. The remote sensing satellites are sun synchrnous and polar-orbiting type with a repetetive cycle of 16 to 26 days enabling repeated collection of data at the same place and time for continuous monitoring of earth's resources. The imaginary payloads of these satellites operate in different spectral bands, spatial resolutions and radiometric resolutions.

(ii) *Microwave Remote Sensing : Radar*

The Microwave portion on spectrum include wavelength within range of 1mm to 1m. Microwaves are capable of penetrating the atmosphere under virtually all conditions. The Radar (Radio detection and Ranging) uses radio waves to detect the presence of objects and determine their range. The process entails transmitting short pulses of a microwave energy in that direction of interest and recording the strength and origin of reflections received from objects within the systems field view. The spatial resolution of a radar system is determined among other things by size of its antenna. This was further improved with side-looking Radar (SLR) or side looking airborne radar (SLAR) system on airborne radar remote sensing technique.

Applications of Remote Sensing

The data obtained from Remote Sensing methods can be used for a variey of practical applications to biological, meterological, and earth resources. These include : application to

1. Atmospheric studies
2. Geospheric survey
3. Biospheric survey
4. Hydrospheric survey
5. Cryospheric survey
6. Environmental studies
7. Spatial Information Systems, land use patterns
8. Geobotanical, zoological exploration and
9. Oceanographic studies.
10. GIS applications

1. *Application to atmospheric studies*

Satellite pictorial data are particularly useful as they provide precision and detail for short-period weather forecasting. Thus, meteorologists have been making increasing use of weather satellite data as aids for the analysis of synoptic and small-scale weather systems. Many countries have now developed a good network of weather forecasting by the application of remotely sensed data termed as METEOSAT imagery system. The recent analysis of NIMBUS- 7 satellite picture for determination of ozone hole in the atmospheric region added a new dimension of application of remotely sensed data in forecasting atmospheric changes.

2. *Application in Geospheric Survey*

LANDSAT imagery, in particular, laid a profound application in analysis of geological structures, identification of rock types and their mapping and also in determining. the mineralized deposits on earth's surface. In addition to LANDSAT, TM data are also very useful for these kind of studies.

Geothermal mapping using scanner data is mainly carried out in geothermal energy exploration studies as part of investigations of alternative energy resources or for the monitoring of active volcanoes particularly in the case of eruption prediction studies. Recently SONAR, a major sensor, was used in sea floor mapping.

3. *Application to Biospheric Survey and Analysis*

To update and accurate assessments of the total average of different crops in production, anticipated yields, stages of growth and condition (health and vigour) are often incomplete or non-timely in relation to the information needed by agricultural managers. These managers are continually faced with decision on planting,fertilizing, watering, control of pests and diseases, harvesting, storage, evaluation of crop quality and planning for new cultivation areas. Remotely sensed information is used to predict marketing factors, to evaluate effects in case of crop failure to asses damage from natural disasters or to aid farmers in determining when to plough, water, spray or reap. The need for accurate and timely information is particularly acute in agricultural information systems because of the very rapid changes in the condition of agricultural crops and the influence of crop yield predictions in the world market.

In forestry, the LANDSAT MSS have proved effective in recognizing and locating the broadest classes of forest land and in separating deciduous, evergreen and mixed communities. More discrete classifications are possible with data from the Thermatric Mapper. Further possibilities include measurement of total average or forest and changes in this. Timber volume, age of forests and presence of disease or pest infections can also be determined. Heat sensitive channels on the LANDSAT Thermatric Mapper and the All VHRR are also used in forestry because of their ability to detect fires. Recently IRS-IA LISS-II scenes were very often used to forecast survey works in India.

4. *Application to Hydrospheric Studies*

Many of the hydrological features of interest for improved water resource management are easily detected and measured with remote sensing sytems. Efficient management of marine resources and effective management of activities within the coastal zone is dependent, to a large extent, upon the ability to identify, measure and analyse a number of processes and parameters that operate or react together in the highly dynamic marine environment including physical, chemical, geometrical and optical features of coastal and open zones of the oceans and the phenomenon of wind-driven coastal upwelling and the resulting high biological productivity.

5. *Application to Cryospheric Studies*

Floating ice in Arctic region interferes with, or prevent, a wide variety of marine activities, including ships in transit, offshore resource exploration and transportation and offshore commercial fishing. In addition, it can be a major cause of damage, resulting in loss of vessels and equipment, loss of life and major ecological disasters. Remote sensing techniques are particularly useful for gathering this ice information, both from aircraft and satellite platforms. The aircraft are specially equipped with transparent domes for visual observations. Side Looking Airborne Radar (SLAR) is used for gathering all weather information about capability and lesser profilometers for accurate measurement of surface roughness.

6. *Environmental applications*

Remote sensing provides information related in some way or other to the quality, protection and improvement of land and water resources. Remote sensing can also facilitate the timely response to naturally occurring phenomenon such as volcanic eruptions, earthquakes, hurricanes, tornadoes, and forecast fire. Examples of the use of thermal imagery for environmental monitoring include investigations of heat loss from buildings, septic tank seepage into water-supply and sewage outfall. Oil pollution through leakage in pipelines or accidental spillage is an all too familiar example of the harmful effects man can have on the environment. For oil pollution monitoring from satellites the spectral resolution is often a serious problem, although under favourable conditions it is possible to observe and monitor oil slicks using this data.

7. *Spatial Information Systems, Land Use and Land Cover Mapping*

In recent years satellite data have been incorporated into sophisticated information systems of a geographical nature all owing the synthesis of remotely sensed data with existing information. The growing complexity of society has increased the demand for timely and accurate information on the spatial distribution of land and environmental resources, social and economic indicators, land-ownership and value and their various interactions. Land and Geographic Information System (GIS) attempt to model, in time and space, these diverse relationships so that at any location, data on the physical, social and administrative environment can be accessed interrogated and combined to give valuable information to planners, administrators, resource scientists and researchers. Land use planning is concerned with achieving the optimum benefits for mankind in the development and management of the land for various purposes, such as food production, urbanization, manufacture, supply of raw materials, former production, transport and recreation. This planning aims to match land use with land capacity and specific uses with appropriate natural conditions so as to provide adequate food and material supply without significant damage to the environment.

8. *Application of Remote Sensing in detecting Geobotanical and Biogeochemical Anomalies*

Remote sensing has immense role in the field of mineral exploration in vegetated terrain. The use of remote sensing in this context is a recent addition to the sciences of geobotany and biochemistry. Geobotany is concerned with the affinity of particular plants, plant associations or plant morphological aberrations to soil containing higher than normal contents of havy metals, such as copper, lead, zinc, or tin. Biochemistry enables the analysis of plant tissue to asses its havy metal content. Geobotanical indicator species, plant association changes and biogeochemical anomalies can all be distinguished ftom the synoptic view of remote sensing platform. So during the last so many years, remote sensing, as an exploration tool, has become increasingly popular zoogeographical survey in relation to faunal data is also of significant value.

9. *Remote Sensing Application in Oceanography*

Satellites with their potential to cover vast areas of land and oceanic areas on a repetitive basis are ideal for coastal and oceanic studies. For a nation like India with a long coastline and extensive exclusive economic zone, satellite based remote sensing can potentially provide substantial economic benefits. With the use of satellite data it is now possible to determine the location of potential fishing grounds, river meandering pattern and upwelling zone of coastal shelves. Currently with the use of NOAA, IRS, LISS-I data, many features of coastal water and land mass were detected. Once the modern technique is perfected and relationships between various levels of production are established, thc remotely sensed data could be utilized for (i) studying migratory patterns of the pelagic fish; (ii) preparation of spatial distribution of chlorophyll charts and fishery forecasting; and (iii) ecosystem modelling.

10. ***GIS Application***

Related to remote sensing in recent times most commonly used system is geographic information system (GIS). GIS is defined as a means of storing, retrieving, analysing and displaying spatially related sets of resources data so as to provide management information to develop a better understanding of environmental relationship.

GIS may be manually done or computer aided. A manual GIS involves creating a base map and one or more layers of film to overlay over the base map. The combinations of layers indicate areas affected by different combinations of values within GIS.

The applications of GIS are :

1. Conducting resource inventory as an application of standard classification technique and integration of information for management
2. Resource monitoring as a more advanced application of classification with simple GIS integration to provide required information
3. Estimation of resource status using remote sensing
4. Resource modelling using remote sensing and GIS to predict potential outcomes to support the management of resources and
5. Estimation of crop production and natural resources of land and water with the aid of GIS data.

Essay Type Questions

1. Bring out the difference of invertebrates and vertebrates.
2. Discuss about the cause and effect of adaptations observed in Animal kingdom.
3. Write a critical account on phylogenetic relationships among different animal groups.
4. What is species? Explain the process of speciation?
5. Illustrate the process of speciation with suitable examples.
6. Explain the Multivariate models of composition of species.
7. What is an ecosystem? Explain functions aspects of ecosystem with illustrations.
8. Explain the process of energy dynamics in ecosystem.
9. Describe the remote sensing applications in resource management.
10. Write an essay on protection of our environment and laws governing them.

Short Answer Type Questions

1. Write brief notes on the following.
   (a) Concept of species
   (b) Allopatric speciation
   (c) Sympatric speciation
   (d) Multivariate models of species composition
   (e) Biogeochemical cycle
   (f) Ecosystem
   (g) Remote sensing

4 Introduction to Human Biology

4.1 Introduction

Human Biology deals with the study of Human body as a whole from the point of view of structure and function. In biological world, there is a vast range of organisms, from microscopic amoeba to the massive blue whale, from tiny moss to the gaint sequoi (red wood), including sub-microscopic bacteria, fungi and many bizzare and wonderful organisms along the way to Man. He alone occupies pre-eminent position, because of intellectual capacity, that the nature bred into him. He has mastered the control of not only his own life, but also the nature of living world to his advantage, since he is endowed with the capacity to obtain information, process, coordinate, learn about the environment and make decision to change and adapt as per needs.

Human body is a complex machine capable of thinking logically and has a conscience too. As other animals, living body is able to see, hear, keep warm, digest food and so on. A person walks gracefully, lifts his arms accurately, and perform so many functions including intellectual and often, one wonders what is happening to his nervous system, his muscles, his joints, his heart and to circulation system. His body essentially goes through a complex physico-chemical processes continuously in cells, tissues and organs of the body. The electrical events underlying the activity of the nervous system and the feed-back mechanism controlling the complex performance of human body as a whole essentially is a physico-chemical processes.

However, from the point of view of form and structure, the human body is highly organized composite system – Brain and the organs of sensory perceptions for the sight, hearing and sensing, thorax, lungs, heart, Abdomen, liver, stomach, intestine, kidneys, reproductive organs and limbs, hand and feet etc., each one of the organs in the human body performs vital function needed for the survival of human beings as a whole.

Organisation of the body

Brain along with the central nervous system (CNS) is at command position and co-ordinates all the body activities. Lungs are useful in the exchange of gasses, carbondioxide and oxygen and thus serve respiratory function. Heart pumps the blood and take care of circulation of blood. Stomach along with associated organs digests the ingested food and absorbed to let it pass into the blood. Liver along with pancreas regulates the amount of sugar in the blood and metabolized to produce urea. Besides, reproductive organs and endocrine system keeps the system healthy and performs a key role in reproduction (Fig. 4.1 and Table 4.1).

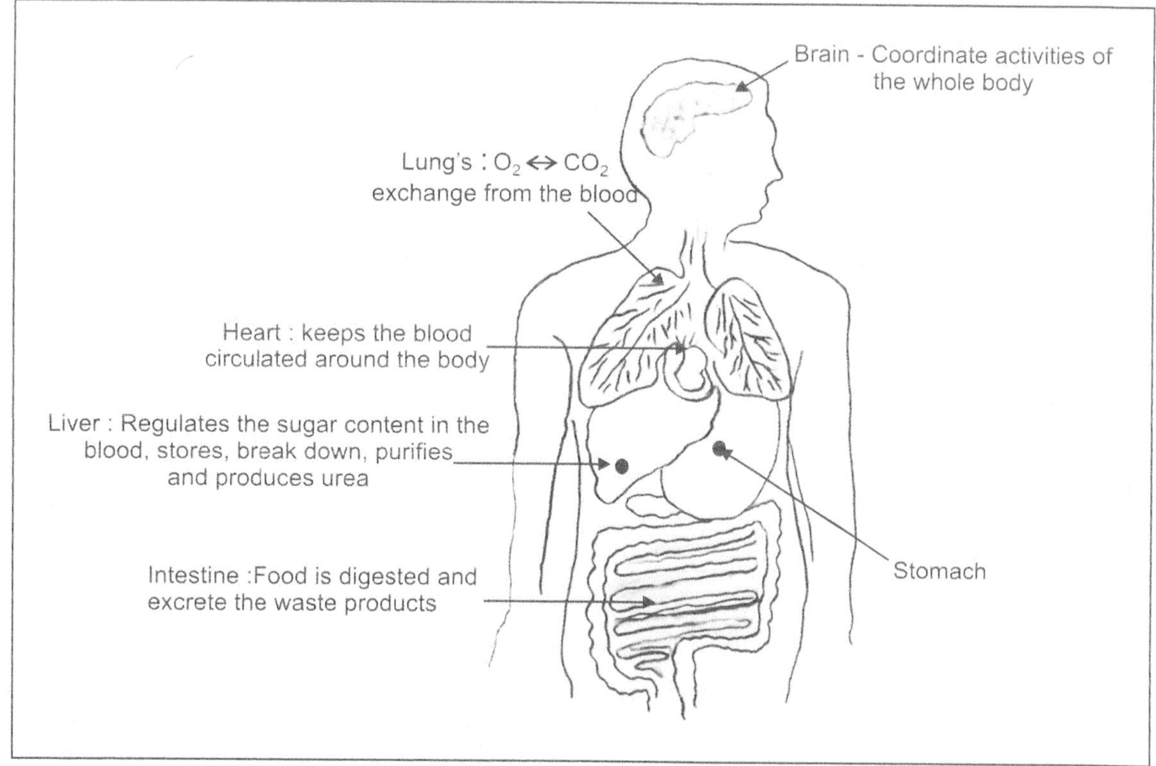

Fig. 4.1 Human Body

4.1.1 Cells, Tissues and their Organization

Just as all living organisms (excepting viruses) are composed of cells, human body is made up of trillions of cells. A typical human cell consists of outer cell membrane as a barrier or boundary called as plasma membrane which separates the interior of the cell or cytoplasm from the external environment. Inside the cell there is a nucleus and other organelles with their own membranes enclosing them.

Table 4.1 *Organization of Human Body parts and respective functions.*

System	Part	Organs Present
Nervous system	Head and Spine	Brain, eyes, ears, nose, spinal chord, nerves etc.
Respiratory system	Thorax	Lung
Circulatory system	Thorax	Heart
Digestive and excretary system	Abdomen	Liver, stomach, intestines, kidneys
Reproductive system	Abdomen	Reproductive organs for procreation
Skeletal system	Limbs	The entire body skeletal system

Cells come in all kinds of shapes and sizes, few of them, contain all the components of cells. In the body similarly differentiated cells combine together to form tissue. There are four basic types of tissues:

- Epithelial tissue - This forms the covering or lining of all body surfaces, both internal and external, including glands.
- Connective tissue - As the name indicates it connects and holds other tissues together including blood, lymph, bones and cartilage.
- Muscle tissue - Consists of adipose bodies, covers the skeleton.
- Nerve tissue - Consists of neurons.

These basic tissue types are in the human body are bound together in various combination to form structural and functional elements called organs. All the interacting organs are grouped into systems, like respiratory, nervous and circulatory systems etc.(Table 4.1 and 4.2).

Table 4.2 *Major system of the human body.*

System	Function
Digestive	To digest and absorb food and nutrients
Breathing	To take oxygen into the body and remove carbon dioxide
Excretory	To remove waste materials from the body
Circulation	To carry blood along with nutrients and gasses(O_2/CO_2) around body.
Nervous System	To carry nerve impulses and messages round the body.
Sensory	To receive information
Muscle	To bring about locomotion
Reproductive	To reproduce young

Some of the specialized cells into tissue are illustrated in Fig. 4.2.

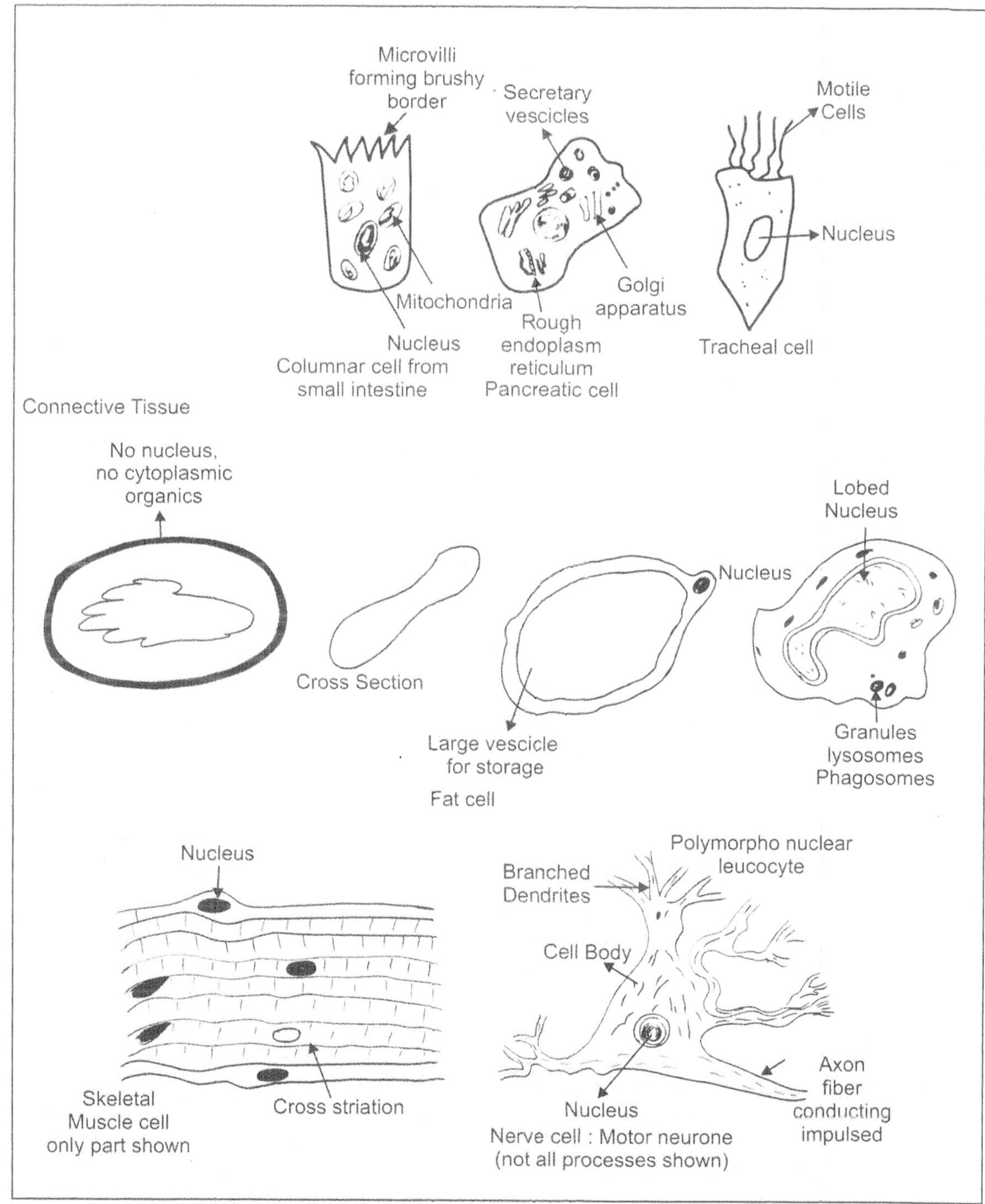

Fig. 4.2 Typical human cells and tissues.

4.1.2 Levels of Organization

In order to function co-ordinatedly human body is organized into cells. Cells with similar structure and function are grouped into tissues. Group of such tissues working together forms organs and systems. They form a network to perform certain specific bodily functions to make a human body.

When you view human body as a functional unit, at least eight major systems can be discussed as indicated. (Table 4.2). Further, all organsims made of cells, tissues, organs and systems constitute population, community, environment and ecosystems.

4.2 Body Fluids and Electrolytes

Human body including cells are filled with fluids. Infact about sixty percent of the average man's weight is water. This varies depending upon the degree of obesity, age and gender. The most significant factor for variation is an adipose tissue or fat content. Adipose tissue has much lower water content (10 - 30%) than other tissue except bones which are compact. Most tissues have very similar water content (more than 70%). The body fluids are of two types – intracellular and extracellular. The extracellular fluids is further divided into blood, plasma and tissue or interstitial fluids. In man, as with other mammals, there is a clear separation between the plasma and tissue fluids. While extracellular fluids come in contact with cells, the intracellular fluids are within the cells. Whereas, the blood plasma does not come into direct contact with the cells of the body apart from those which form the walls of the blood vessels besides the blood cells themselves (Fig. 4.3).

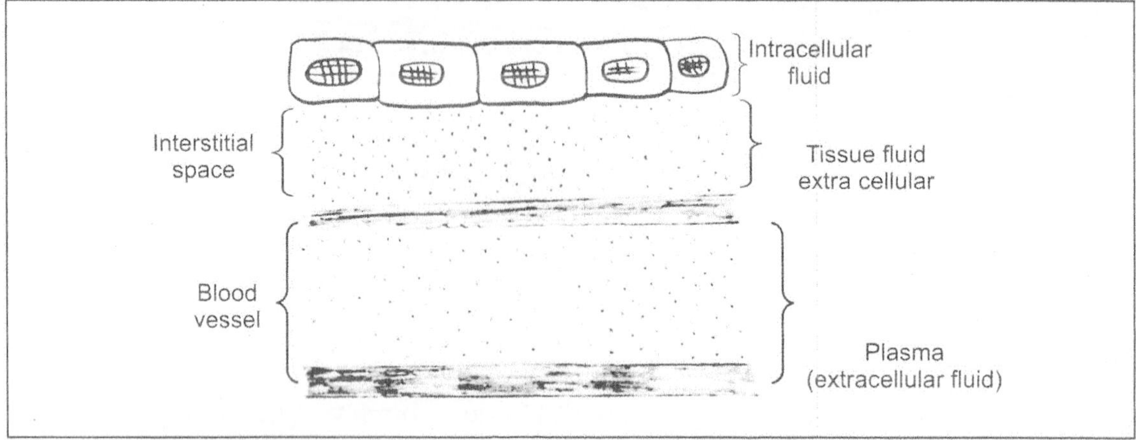

Fig. 4.3 Separation between plasma and tissue fluids in human body.

Two thirds of total body water is from intracellular fluids and remaining one third total body water is accounted from extracellular fluids. Thus the relative amount of fluids in the body fluid compartment expressed as the amount of fluids as a percentage of body weight, the total body water is 42 liters i.e., 40% of total body weight. In other words about 28 liters is composed of intracellular fluids, 10.5 liters of tissue fluids (15%) and 3.5 liters plasma (5%).

4.2.1 The Composition of Body Fluids

The composition of intracellular and extracellular fluid vary considerably. However, the tissue fluids and plasma are almost similar, except that the plasma has more protein. (Table 4.3).

Table 4.3 *Various components of body fluids.*

	Class of body fluids (m.kg.mo./l)		
Component Substances	**Intestitial fluids**	**Intracellular fluids**	**Plasma**
Sodium ions (Na^+)	145	10	140
Potassium ions (K^+)	4	160	4
Chloride ion (Cl^-)	115	3	100
bicarbonate ions (CO_2)	30	10	28
Proteins	10	55	16

4.2.2 The interrelationships of body fluids and their interactions with the outside environment

Since humans live in varying environment and go through diverse activity there will be a loss of water. Nonetheless the composition and volume of intracellular fluids must remain in balance. The regulation of both the volume and composition of body fluids needs the control of water intake and loss by a diverse means. It is important to note that, eventhough body fluids are compartmentalised as separate entities, the separation is not water-tight. There is an incesant movement and exchange of water and other components between them. The extracellular fluids and plasma act as protective barrier between the intracellular fluids compartment and the outside environment. The oral water intake and the loss of water through sweat and urine, excreta etc., has totally compensated and held in balance in order to retain the volume as well as composition of the extracellular fluids within the narrow range of limits. The constancy needed to protect the intracellular fluids and dependent on physiological functions.

The interactions of body fluids with the outside environment and inter relationship is shown in Fig. 4.4.

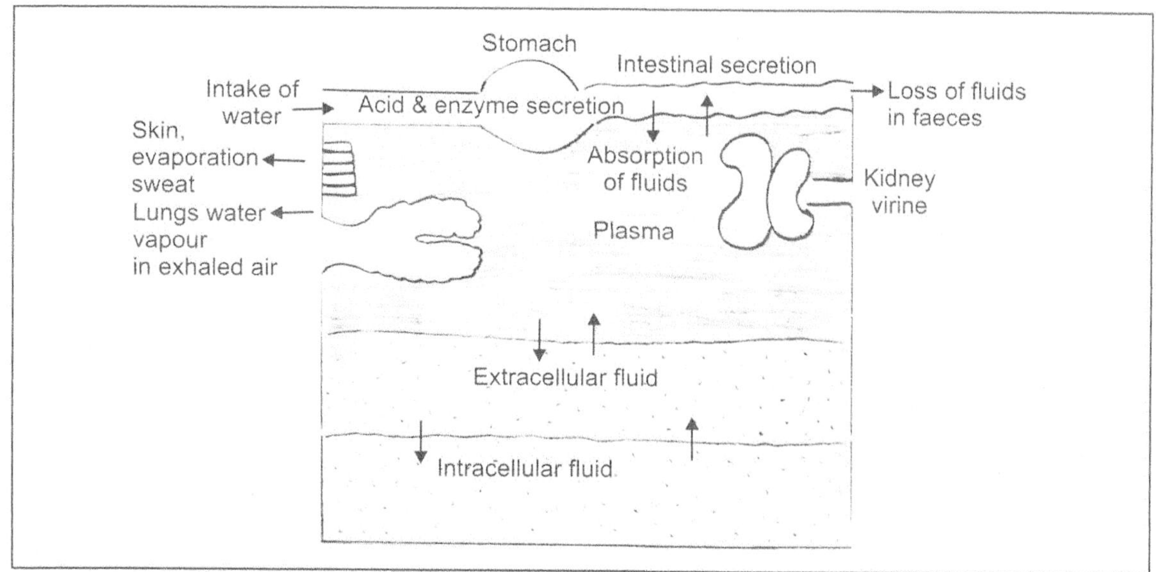

Fig. 4.4 Different body fluids compartmentalized, inter relationships and interactions with the outside environment.

4.3 Blood Physiology - Circulation System

Blood constitute the body's major transport system and does the following vital functions :

- Carries respiratory gases, nutrients and metabolic waste products.
- It distributes the body's heat all around to keep all the cells, tissues and organs healthy and warm.
- It helps the monitoring organs to detect variations in hormone levels, osmotic pressures, pH, temperature etc.
- It provides cellular transport.
- It contains antibodies (immunoglobulin) and other agents that protect against infections and tumours.

Blood circulates around the body in well defined blood vessels, arteries and viens, to form a closed circulatory system, separating from general tissue fluids.

4.3.1 Blood composition and structure

Blood is categorized as connective tissue. It is a thick suspension of cells in a watery electrolyte solution. When centrifuged in a glass tube, after taking care not to clot, it separates into liquid part called plasma and solid part comprising, Red blood cells, platelets and white blood cells (Fig. 4.5).

The percentage of red cells in the body by volume is known as the haematocrit. Normal range of values is 40-47%. When blood is allowed to clot it forms a jelly like substance leaving some fluid called as serum. Serum is nothing but plasma devoid of fibrinogen that cause blood to clot.

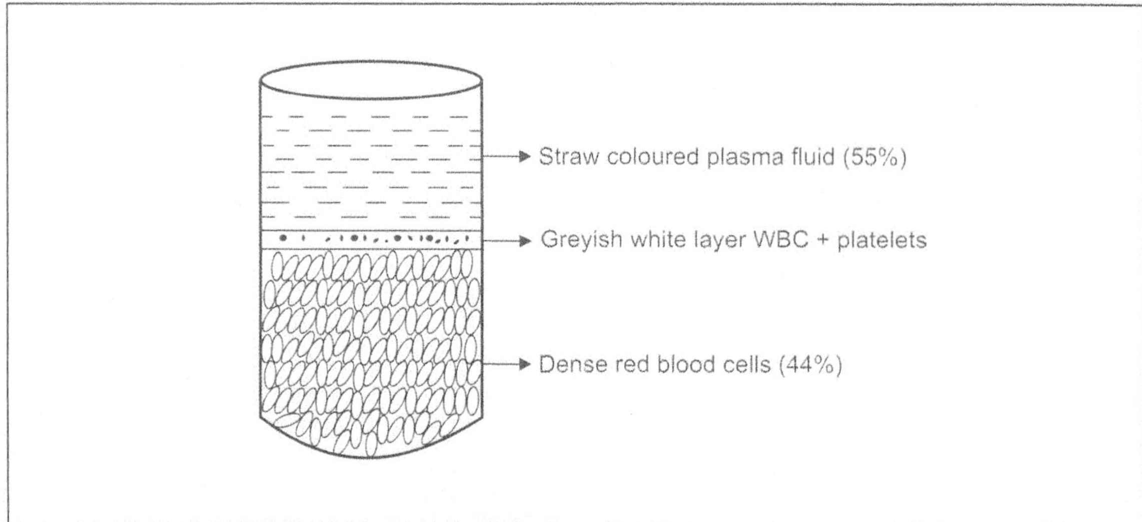

Fig. 4.5 Centrifugal separation of blood components.

4.3.2 Plasma

It is an yellowish clear liquid comprising of water (90%) and dissolved solutes (10%). Following a meal it may look opalescent or milky due to the presence of fatty substances. The normal constituents of plasma are:

- Water (9%)
- Protein (~ 8%)
- inorganic ions (~1%), mostly in the form of sodium chloride (NaCl) and sodium-bi-Carbonate.

Plasma Protein

The major kind of proteins present in the plasma are albumins, globulins and fibrinogen. The globulins include α and β globulins and immunoglobulin. All plasma proteins are formed in the liver with the sole exception of immunoglobulin which are synthesized by cells of immune system. (Table 4.4)

Table 4.4 Major Plasma Proteins.

Protein Class	Conc (g/l)	Mol. wt(D)
Albumin	35-45	~ 69,000
Globulin	15-32	15000-100,0000
Fibrinogen	2-4	400,000

Plasma Serves the following functions:

- Carries food i.e. glucose, amino acids, etc.
- Carries enzymes, proteins, hormones and antibodies.
- Carry urea formed in the liver to kidneys.
- Contains soluble fibrinogen responsible for blood clotting.

The plasma proteins, however serve a number of general functions:

- Plasma proteins are responsible for imparting osmotic pressure of blood. Because of this, blood proteins maintain the water balance between plasma and tissue fluids.
- Plasma proteins cause viscosity of blood. This affect the maintenance of normal blood pressure by the heart.
- Proteins provide about one sixth of total buffering capacity of the blood.
- Plasma proteins transport substances circulating in the blood. Many hormones, fatty acids, lipids, vitamins, minerals are carried in conjunction with protein.
- Some plasma proteins perform specialized functions. Fibrinogen takes part in blood clotting. Immunoglobulins (antibodies) act as body's defenders against bacteria, virus and alien organisms.

4.3.3 Electrolytes - Inorganic ions

The main electrolytes of plasma comprises of inorganic ions such as sodium, bicarbonates and chloride ions. Potassium, calcium and magnesium ions are also constituents, but in insignificant quantity. The precise concentration are controlled quantitatively mainly by the kidney.

4.3.4 Organic nutrients

Several organic substances like glucose, phospholipids, cholestral, amino acids and peptides are also present in the plasma. These serve as nutrients and transported along the blood circulation. These organic nutrients originating in gastrointestinal tract are shunted to storage areas like liver and other sites of the human body wherever they are needed.

Nitrogenous Waste products

The body metabolism generates several wastes as byproduct. Most of them are nitrogenous substances such as urea, ammonia and uric acid. The circulatory system transport these nitrogenous waste products to kidney for the effective excretion.

Regulatory Hormones

Human body produces several kinds of hormones which are very important regulatory chemicals. These are traverse across in the blood stream from their site of origin to their site of action.

Gases

Plasma contains in dissolved state several gasses like inert gas, nitrogen, besides life giving oxygen and waste gas, carbon dioxide, generated during the cellular respiration.

4.3.5 Structure and Function of Blood Cells

Blood cells can be categorized into three types :

(i) Red blood cells (RBC)

(ii) White Blood Cells (WBC)

(iii) Platelets

4.3.6 Red Blood cells or erythrocytes

Red Blood cells or erythrocytes are biconcave discs devoid of nuclei. The red colour is due to Heamoglobin and helps in carrying oxygen. In lungs oxygen enters RBC and joins with heamoglobin, while in tissues oxygen leaves red cells to pass on to respiring cells.

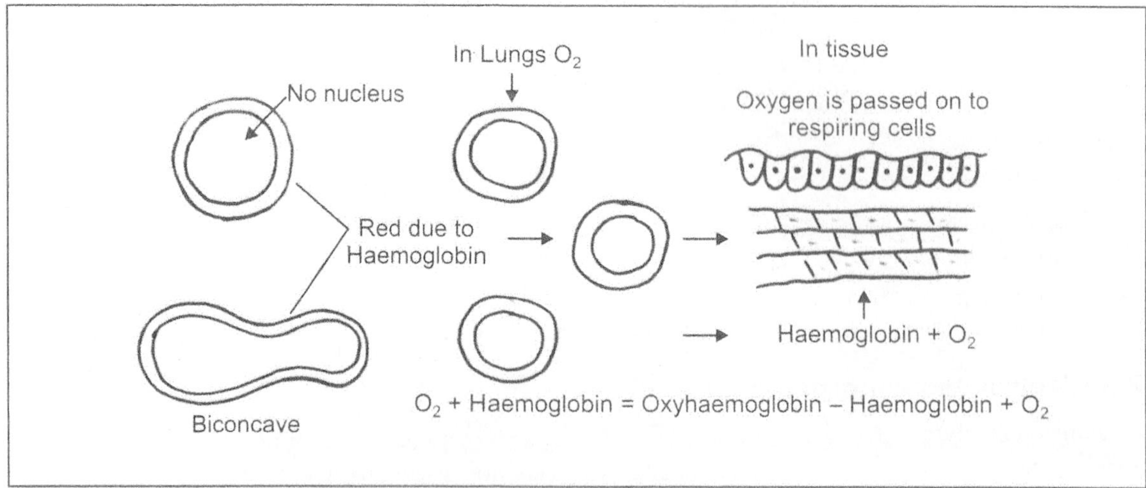

Fig. 4.6 Structure and function of RBC.

4.3.7 Platelets

Platelets are also called thrombocytes. These are small cells without nuclei, infact these are cell fragments containing fibrinogen that helps blood to clot whenever there is an injury or cut to skin. RBC get trapped in the thread like mesh caused by the fibrin and forms a clot. Blood clots help in the loss of blood when injury occurs and also prevents the entry of germs from such cut or wound. (Table 4.5).

4.3.8 White Blood Cell or Leucocytes (WBC)

White blood cell or Leucocytes (WBC) are colourless cells when compared to RBC which in general helps in fighting infections. These may be classified into granulocytic monocytes and lymphocytes. The granulocytes are distinguished by the presence of cytoplasmic granules. They are also called as polymorpho nuclear leucocytes (or polymorphs) as their nuclei cleaved into several lobes. Based on the presence of granules in the cytoplasm, the granulocytes are further classified as neutrophils, eosinophils and basophils. The leucocytes are essentially concerned with the defence of body against infection. These are not limited to blood vessels, but are found in massive quantities in the lymphatic system and connective tissues. The Lymphocytes produce antibodies which destroy bacteria and viruses. Phagocytes with lobed nucleus and ability to change shape engulf bacteria and digest them (Table 4.5).

Table 4.5 *Different types of blood cells in Humans.*

Blood cell type	**Appearance in a stained slide with blood film (diameter)**	**Number of cells per mm^3 blood**	**% of total number of WBC**
Red cell type (RBC)	7 μm	$4.8 - 5.4 \times 10^6$	-
Platelets	2-5 μm	$150\text{-}500 \times 10^3$	-
White Blood cells (WBC)			
Neutrophils	10-14 μm	2500 - 7500	60%
Eosinophils	10-14 μm	40-400	< 1%
Basophils	10-14 μm	15-100	<1%
Monocytes	15 - 25 μm	200 - 800	5%
Lymphocytes	7 - 14 μm	1500 - 3500	30%

4.3.9 Life Cycle and Differentiation of Pathways of Blood Cells

The formation of blood in the body is called haemopoiesis and the place where it is formed called haemopoietic centers. The differentiation pathway of blood cells is called haemopoetic system. In the fetus haemopoiesis takes place in the liver and spleen. But after birth and subsequent of growth, stages of growth occurs in the bone marrow, especially in the sternum ribs, vertebrae, cranium, and pelvis. The entire life history of blood cells is shown in the Table 4.6. Simplified pathway of blood cells differentiation omiting adjunct precursor cells is given in Fig. 4.7.

Table 4.6 Kinds of blood cells, their origin, life span and removal.

Type of blood cell	Origin	Life Span	Method of Removal
Red blood cells (RBC) Reticuloendothelial Platelets	Bone marrow Mega Karyocytes in bone marrow	120 days 8-14 days	Reticuloendothelial system, basic units, iron, amino acids are recycled
White blood cells(WBC) Neutrophils Eosiniphils Basophils	Bone marrow Bone Marrow Bone Marrow	Stored in bone marrow for 11days. When in blood only for 3-12 days while in transit. In tissues etc., may live for 10-12 days but cannot come back into blood stream.	Reticuloendothelial system or by leaving the body
Monocytes	Bone marrow precursors of macrophages which may also reproduce themselves.	Blood transit time is 32 h. May live several months in tissues as macrophage	Reticulo endothelial system, or by leaving the body
Lymphocytes	Bone marrow and other lymphoid tissues like thymus, spleen and lymph nodes	continuously recirculated between blood and lymphatic system. Long lived or average 4-5 years, some over 10 years.	Reticuloendothelial system.

4.3.10 Red Blood Cells and their Circulation in Body

One of the most important components of blood circulation system is red blood cell or erythrocyte itself. The vital function of these cells performance is to transport the respiratory gases, oxygen and carbon dioxide around the body. The erythrocyte may be regarded as sacs containing haemoglobin bounded by plasm membrane. The center of the erythrocyte is relatively less densely coloured when compared to periphery mainly due to biconcavity of the disc-shaped cells. Each RBC may contain about 280×10^6 haemoglobin molecules which comes to 33% of its weight. The normally cited sizes of red blood cells are 7.2 μm and 2.2 μm because these represents measurements made on dried blood film on the glass slides. The RBC may shrink in its diamensions by 8-16%. The biconcave shape of the RBC appears to be selected by nature for functional efficiency in permitting a much larger surface area for diffusion of gases and nutrients than would have been possible, if enclosed volume were to be sphere. Further more, it also provides for diffusion to happen across the shortest possible distance between the membrane and places in the cell interior (Fig. 4.8).

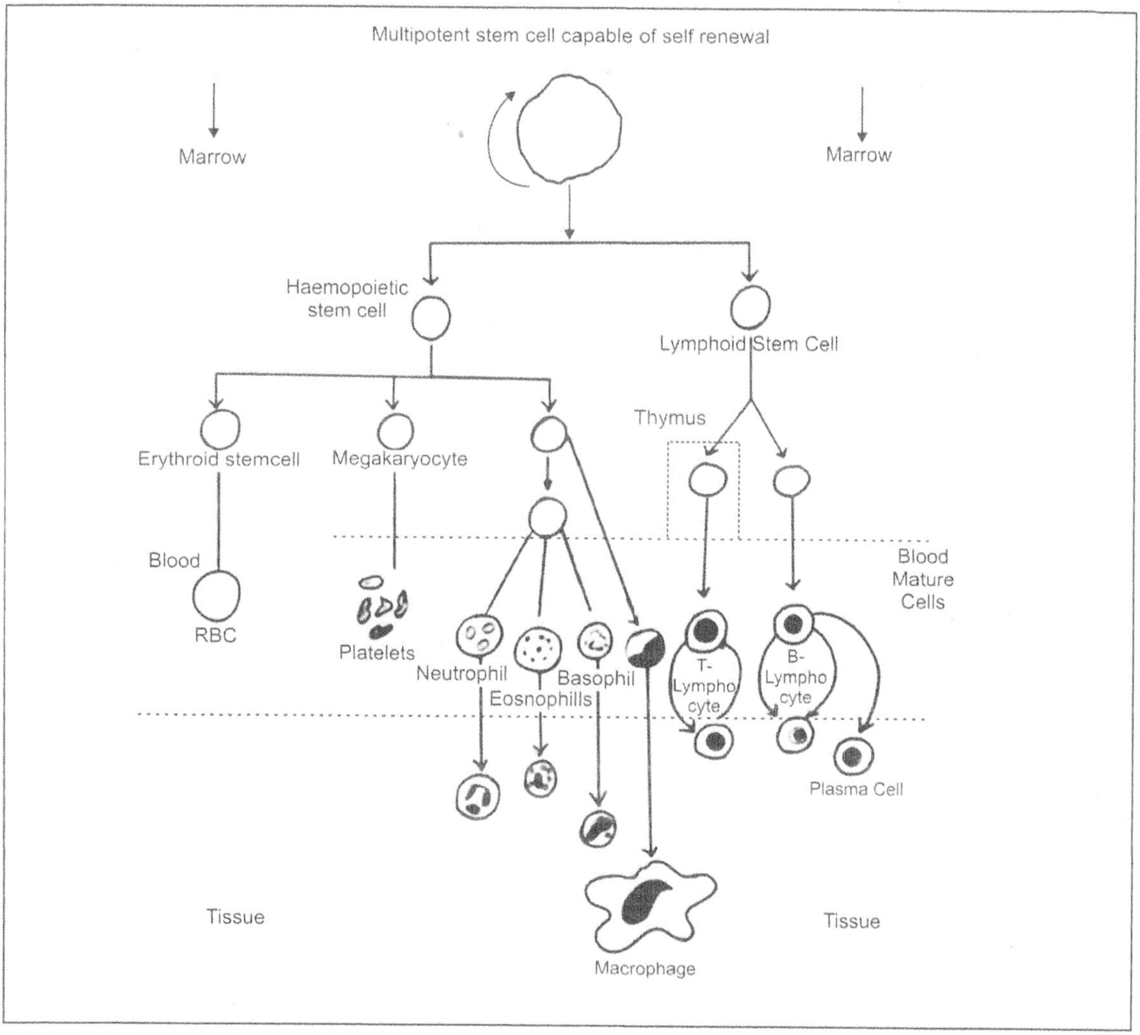

Multipotent stem cell capable of self renewal

Marrow

Marrow

Haemopoietic stem cell

Lymphoid Stem Cell

Thymus

Erythroid stemcell

Megakaryocyte

Blood

Blood Mature Cells

RBC

Platelets

Neutrophil

Basophil

Eosnophills

T-Lympho cyte

B-Lympho cyte

Plasma Cell

Tissue

Macrophage

Tissue

Fig. 4.7 Blood cells differentiation pathway constituting haemopoietic system; many details of intermediate steps in the differentiation pathway is not depicted

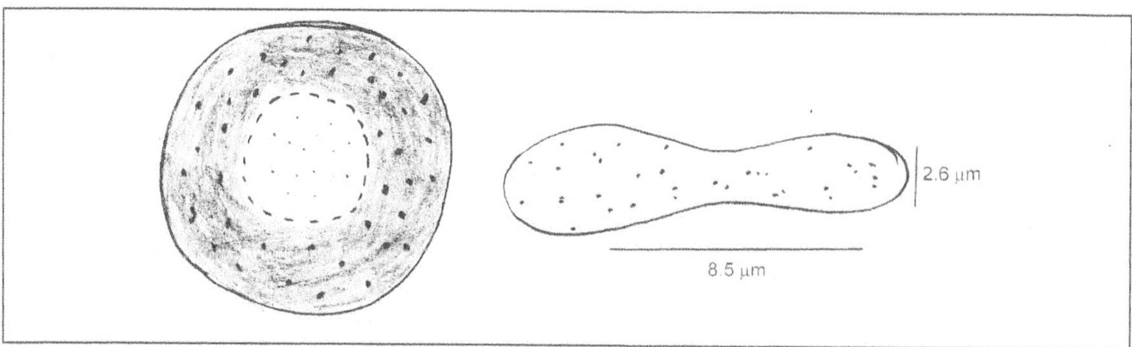

Fig. 4.8 Red blood cell, and its dimension.

4.3.11 Movement of RBC along even in Small Vessels is due to its Flexibility

Red blood cells have to circulate through all kinds of vessels, small and large. This is made possible because of a remarkable property of red blood cells to be deformed. They can be flexible and twist into different shapes, so that they are able to squeeze through the smallest blood vessels (capillaries) too (Fig. 4.9).

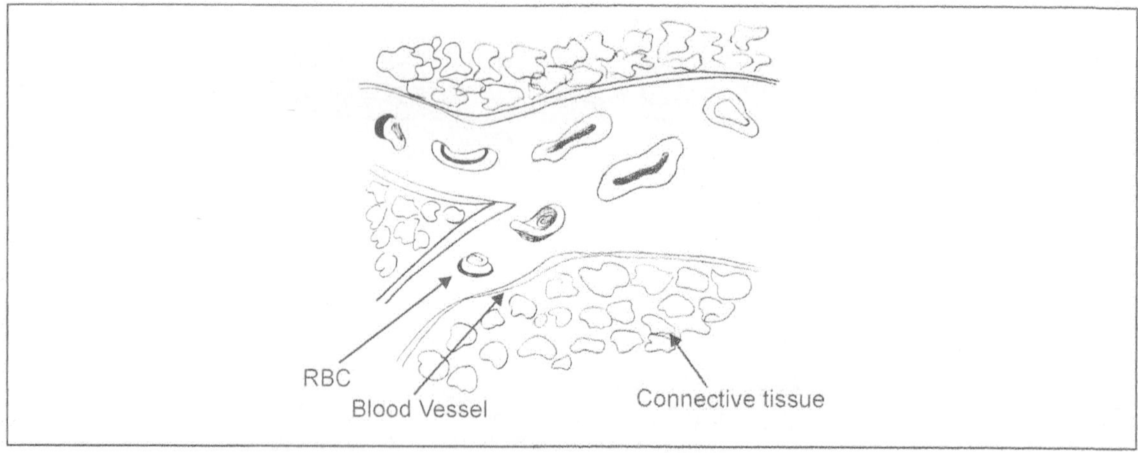

Fig. 4.9 RBC moving along a small versel - magnified for the clarity.

4.3.12 Erythropoiesis is the Formation of Mature RBC

In bone marrow, the RBC are formed from the normoblasts. The nucleus condenses and becomes smaller and extruded out to form a reticulocyte. Such an anucleated reticulocyte will have ribosomes that enables the synthesis and accumulation of haemoglobin. Once this is achieved cytoplasm losses all organelles. Consequently, the red blood cells become redder as more haemoglobin is synthesized.

4.3.13 RBC Production is well regulated

The production of RBC is well controlled. Old red blood cells have to be replaced continuously from the circulation system. About 2×10^{11} cells per day have to be produced in order to maintain a constancy of RBC circulating in the blood. A sort of regulatory network exists that senses the deficiency of RBC which promotes the production. A fine tuned regulatory system seperates the synthesis and breakdown of red blood cells to maintain a constant number of circulating RBC in the body. About 2×10^{11} cells have to be synthesized each day (Fig. 4.10 and 4.11). In certain disorders, there may be lowered RBC count than the normal circulating RBC. In some cases like sickle cell anaemia the blood cells themselves may be abnormal. There are several conditions that leads to lowered RBC count and attendant anaemia. Besides, there are several genetic defects like thalasemias that cause anaemia. Any such cases that lead to a lowered haemoglobin content, than normal is called anaemia.

The old RBC are broken down and released into the body, where reticuloendothelial system removes the dead cells and broken down components from the blood stream at the end of their life cycle (Fig. 4.12). This system consists of phagocytic cells, and macrophages distributed all over entire body.

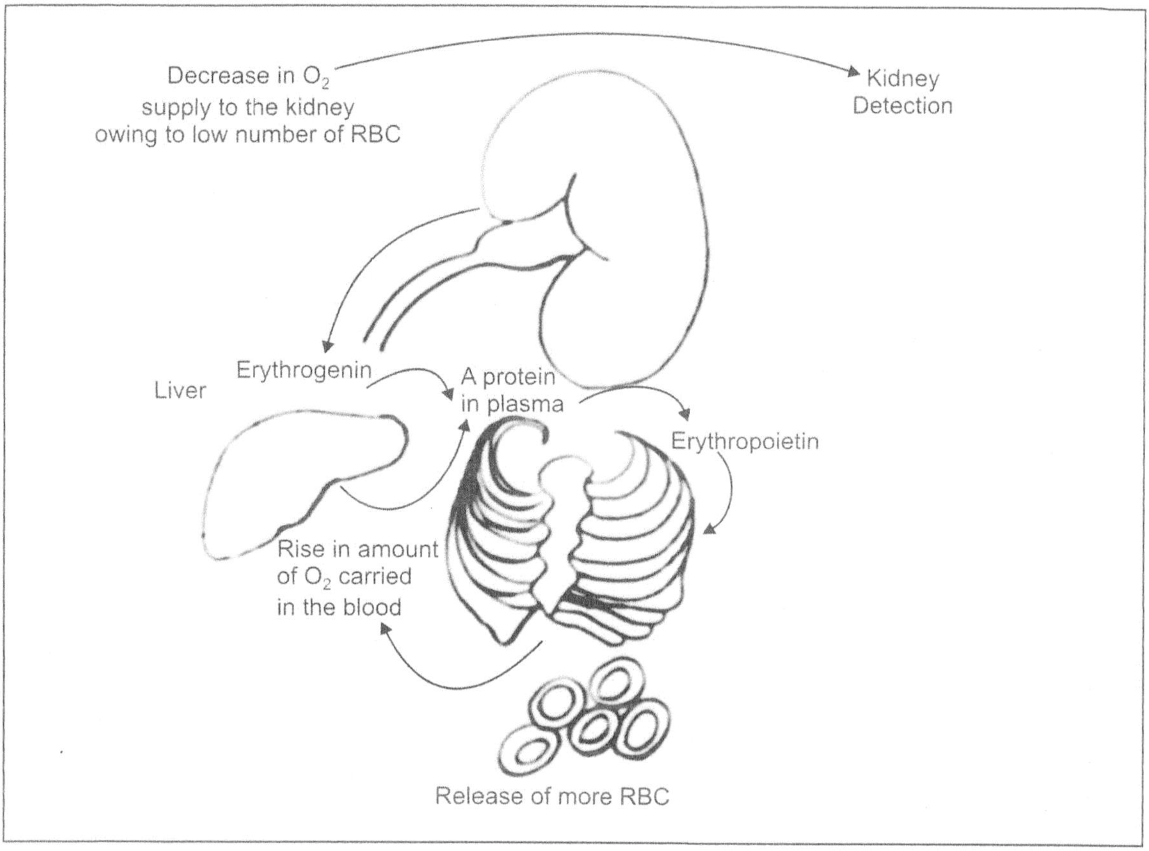

Decrease in O_2
supply to the kidney
owing to low number of RBC
Kidney
Detection
Liver
Erythrogenin
A protein
in plasma
Erythropoietin
Rise in amount
of O_2 carried
in the blood
Release of more RBC

Fig. 4.10 The control of RBC production.

Inito
Chondria
Normoblast
Reticulocyte
Mature RBC
Cells become red and
red as more haemoglobin
is made
Bone marrow
Blood Vessel

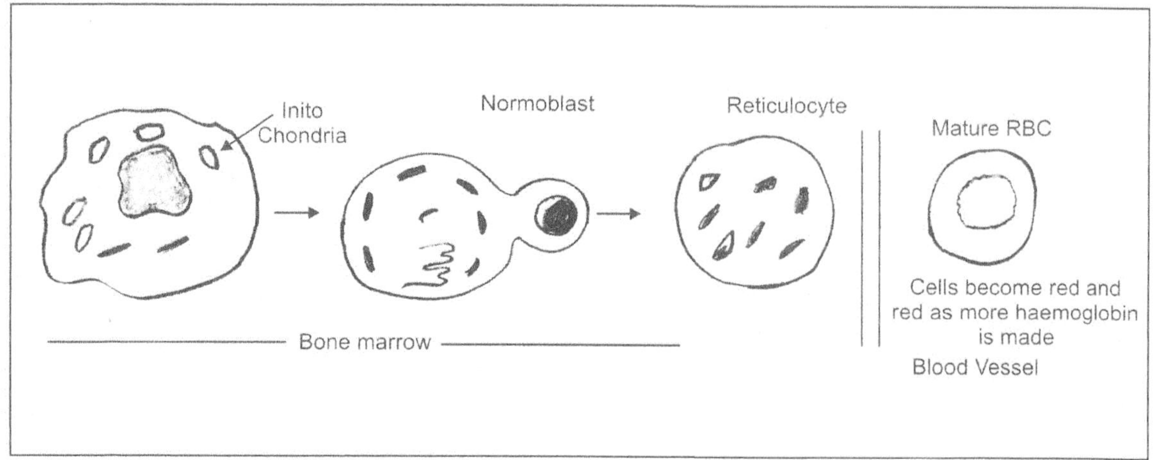

Fig. 4.11 Erythropoiesin in the bone marrow.

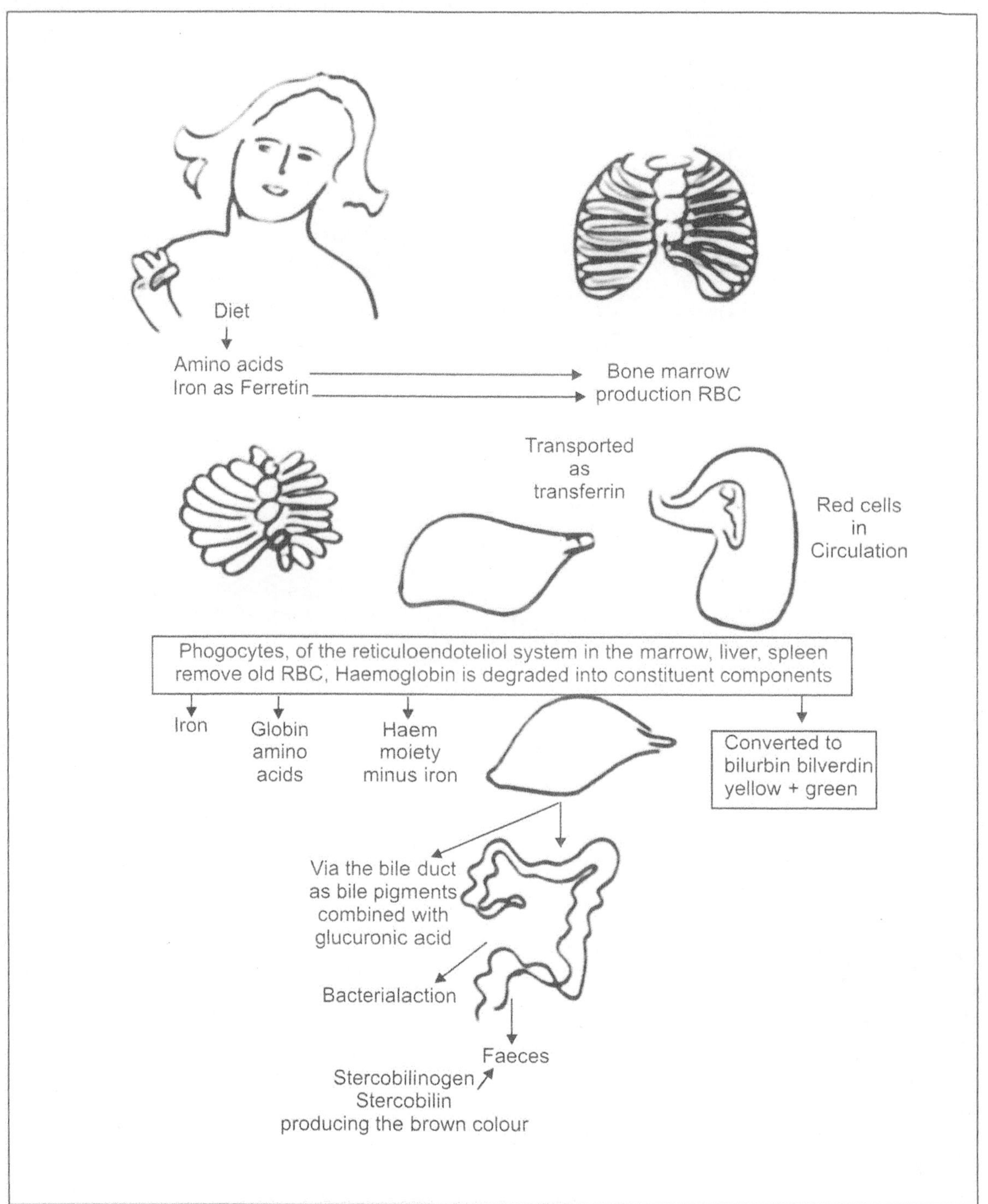

Fig. 4.12 Synthesis and break-down of RBC.

The haemoglobin is either broken down into its basic components haeme and globin and recycled or excreted as waste products through excretory system. If there is any impairment or failure in the process of the excretory pathway, the bile pigments, bilirubin and biliverdin accumulates in the blood and cause the skin to appear yellow. This condition is called as Jaundice and is generally indicative of diseases of gall bladder, liver, hepatitis etc. These pigments may also cause colouration of the skin, urine and eyes etc. (Fig. 4.13).

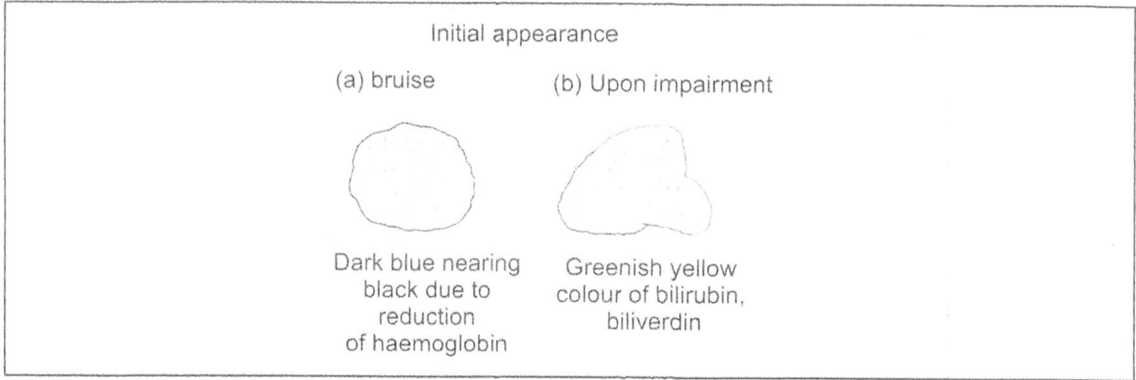

Fig. 4.13 The change in colour of skin due to (a) bruise and impairment like jaundice.

4.3.14 How does blood carry oxygen?

Oxygen combines loosely with the iron atom of haemoglobin present in RBC to result in oxyhaemoglobin. The process of reversibly combining of oxygen with the iron in haemoglobin is called as **oxygenation**. During this process iron remains in the ferrus state. This is a reversible phenomenon. As the blood goes through the lungs, haemoglobin combines with the oxygen, this oxygenated blood inundates the tissue through a network of capillaries; the oxygen is released into the cells for respiration. The deoxygenated haemoglobin becomes free and is called deoxyhaemoglobin which becomes dark blue almost nearer to black in colour. The bright red colour comes back again once the oxygenation takes place in the lungs during its circulation.

An oxygen is combined with haemoglobin, the haemoglobin molecule undergoes a change in conformation. Upon oxygenation it closes up, while deoxygenation it opens up.

Haemoglobin is a key element in the exchange of gases (O_2, CO_2 etc). Haemoglobin is an iron containing organic compound called porphyrin, a haem pigment, and a globin protein, all complexed together. There are four haem + globin subunits in each molecule of haemoglobin : 2 α type and 2 β type. However, these vary depending upon the stage of development. Two different gene families situated on 11th and 16th chromosomes control the production of haemoglobin, which are developmentally regulated. Any mutations in these genes result abnormal haemoglobin content, leading to hereditary diseases like thalassemia, sickle cell etc (Fig. 4.14).

Human adults have two major kinds of haemoglobin comprising of two α chains and two β chains; Fig. 4.14 in the embryonic stage the composition is different; fetus has two α and γ chains that change after birth.

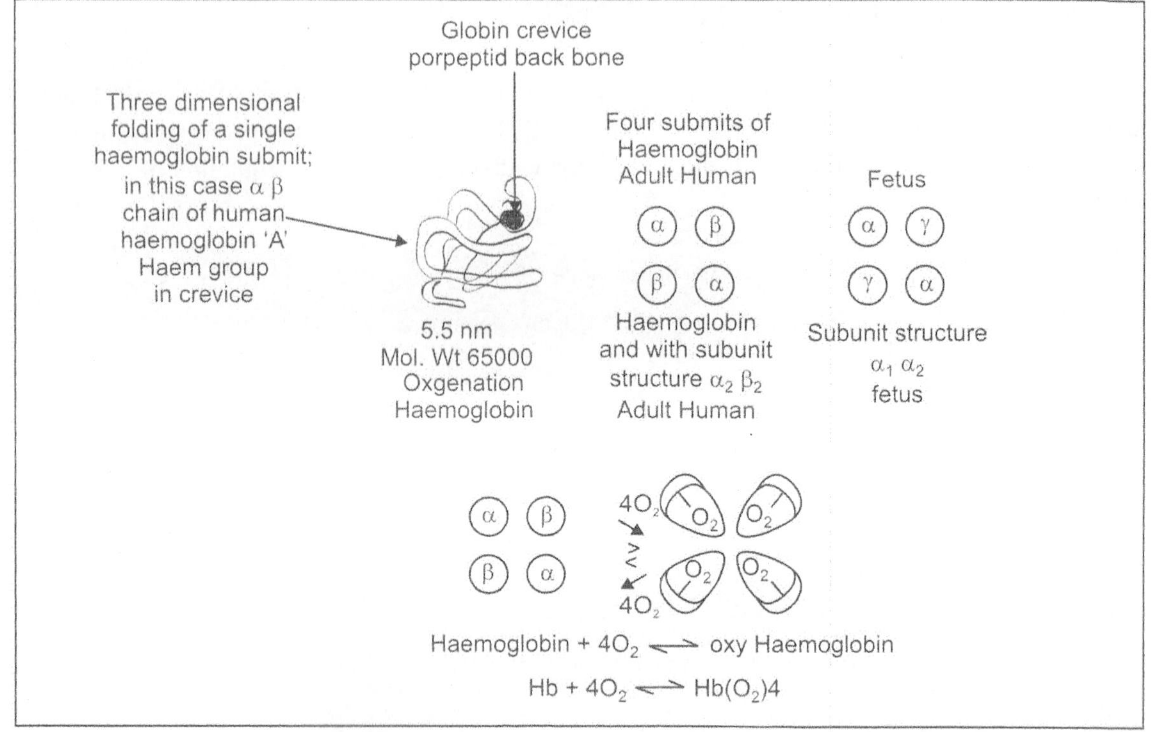

Fig. 4.14 Structure of Haemoglobin and the process of oxygenation

4.3.15 Transporting of O_2 and CO_2 around the circulatory system

During the processes of inhalation :

(i) the oxygen enters into the lungs and gets diffused into the blood and enters the RBC. Then

(ii) O_2 gets bound reversibly with the haemoglobin in the RBC to form oxyhaemoglobin.

(iii) The Red cells with oxygenated haemoglobin pumped through arteries by the heart around the body along with the blood circulation.

(iv) In the body, cells oxyhaemoglobin breaks down to release the O_2 for respiration producing CO_2

(v) RBC returns back with the load of CO_2 to the lungs along the veins to release CO_2 and collect more O_2 for the next round of circulation.

When O_2 uptake by the blood is impaired during its circulation through lungs, a condition called cyanosis results wherever arterial blood turns blue. Normally, in each one litre of blood there is about 145 g of haemoglobin (19.5%), about 1.3 ml of O_2 can combine with each gram of haemoglobin. That is about 200 ml of O_2 is carried by each litre of blood ($145 \times 1.34 = 200$ ml). If there were to be no haemoglobin in RBC, the blood could have carried only on average about 3 ml of dissolved oxygen in it. Nature has selected more effective way of transporting O_2, about 700 time more efficient by selecting evolution of haemoglobin in RBC (Figs. 4.15 and 4.16).

Haemoglobin also combines readily with other gases, besides O_2 and CO_2. The carbon monoxide also combines with haemoglobin to form a highly stable carboxyhaemoglobin which turns the blood bright cherry red and is toxic. This accounts for extremely lethal nature of carbonmonoxide inhalation.

Oxygenation and oxidation are two different reactions which should not be confused. Oxidation would occur if the iron changed from the reduced ferrus form into its oxidized ferric form. Oxidation can take place in certain diseases of blood. The oxidised form of haemoglobin methaemoglobin is incapable of transporting oxygen and hence such a state is harmful.

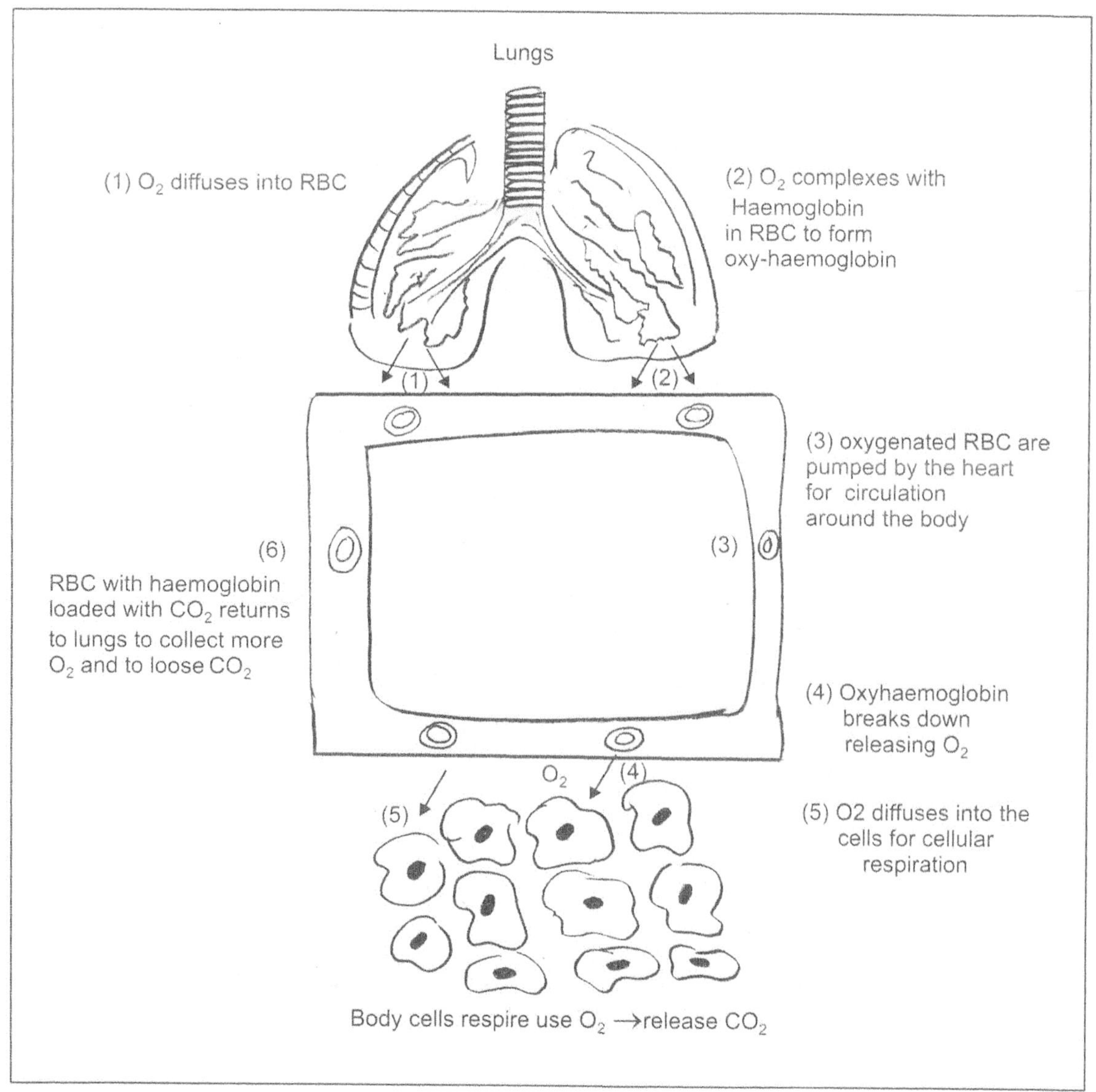

Fig. 4.15 Gaseous exchange process during respiration.

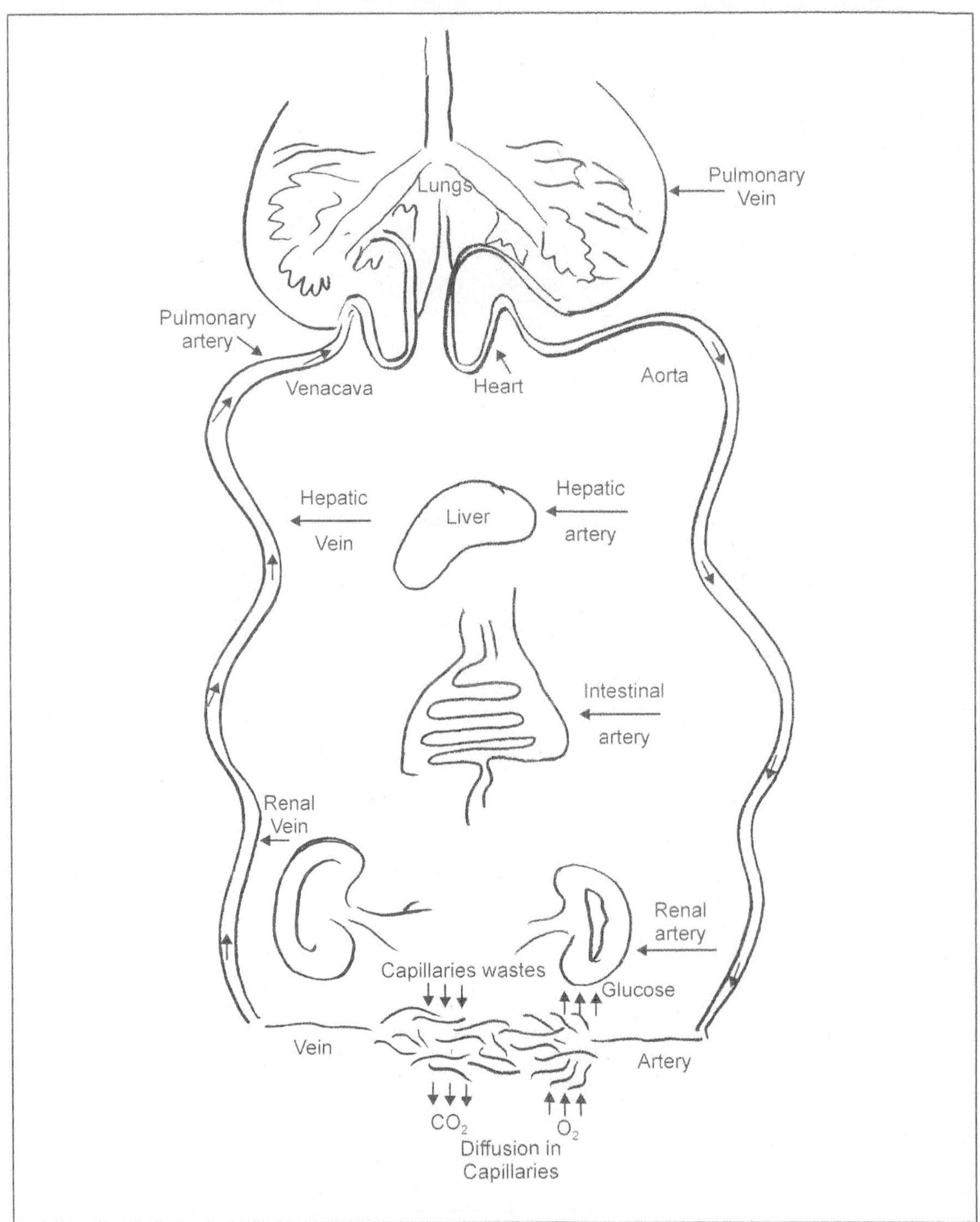

Fig. 4.16 Exchange of O_2 and CO_2 through blood circulation systems.

4.3.16 Factors Influencing the Oxygen Affinity

The Pressure of O_2 at which haemoglobin about is 50% saturation is known as P_{50}. This varies depending upon various factors:

- Changes in pH
- Changes in temperature
- Changes in concentration 2,3- diphosphoglycerates (2,3-DPG)

The downward change in pH changes the oxygen dissociation and affects haemoglobin to release more O_2. This mechanism ensures that haemoglobin to release more O_2 instantaneously in tissues where there is an increment in CO_2. This phenomenon is known as Bhor's effect. During the period of intense activity in muscles like exercises, this can be seen.

Increase in temperature reduces the O_2 affinity to haemoglobin and hence affects dissociation. Increased metabolic activity needs more oxygen and the effect of temperature upon O_2 affinity of haemoglobin serves the purpose to slow down the metabolism.

2, 3 - DPG is a very important metabolite produced during modified glycolysis in RBC. As the 2,3 - DPG combines with haemoglobin it changes the affinity to O_2. An increase in 2,3- DPG decreases the O_2 affinity of haemoglobin. In several situations this assumes importance in human life. For instance, the transfer of O_2 across the placental barrier or in the adaptation of living on high altitudes.

4.3.17 Transport of CO_2

The CO_2 results as a consequence of respiration and is carried in the blood to the lungs. Infact blood carries about three times more carbondioxide than oxygen. This is accompolish in the following ways.

- As bicarbonate ions, mostly combine directly with haemoglobin as carbamino - haemoglobin
- Dissolved state in plasma.

CO_2 combines with water to form carbonic acid (H_2CO_3) Fig. 4.17.

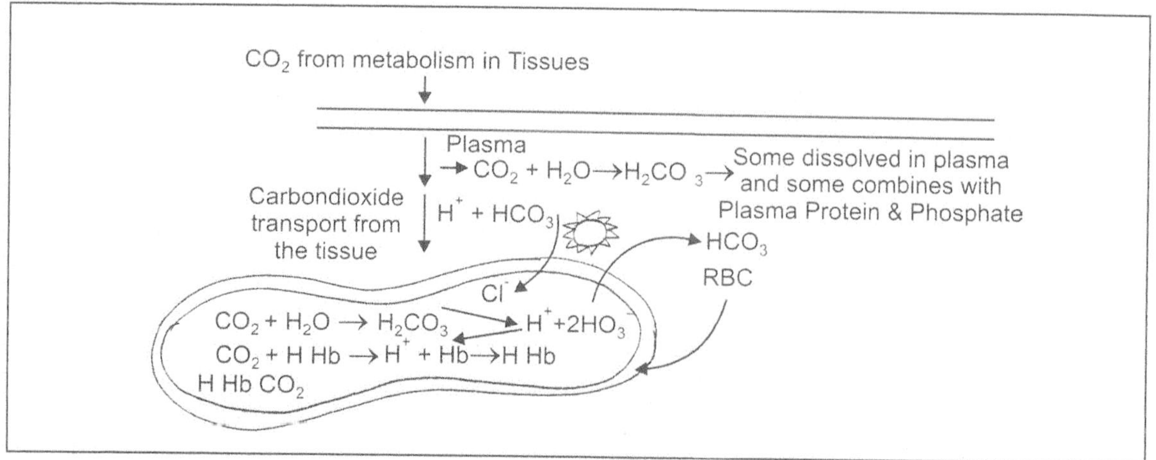

Fig. 4.17 Carbondioxide transport from the tissue.

This reaction increasingly takes place in RBC by an enzyme Carbonic anhydrase. The resultant Carbonic acid ionizes into hydrogen ions and bicarbonate ions [CO_2 - H.Hb → H $HbCO_2$; $CO_2 + H_2O \rightarrow H_2CO_3 \rightarrow H_2CO_3\ H^+ + HCO_3$].

These have to be removed immediately in order to give place for further bicarbonates being formed. Most of the H^+ ions are buffered by combining with deoxyhaemoglobin. The bicarbonate ions come out of RBC through diffusion process into the plasma by exchanging with chloride ions. This is called as chloride shift and takes place rapidly due to the presence of chloride-bicarbonate exchange carrier system in RBC membrane. This process is exactly reversed in lungs.

4.3.18 Platelets and Blood Clotting

Platelets are non-nucleated small oval shaped cells present in the blood. These are important in preventing loss of blood or haemorrhage when damage occurs to blood vessels. The natural stoppage of bleeding in the body also depends upon the blood vessels themselves. Various protein factors of plasma are released into the damaged tissues.

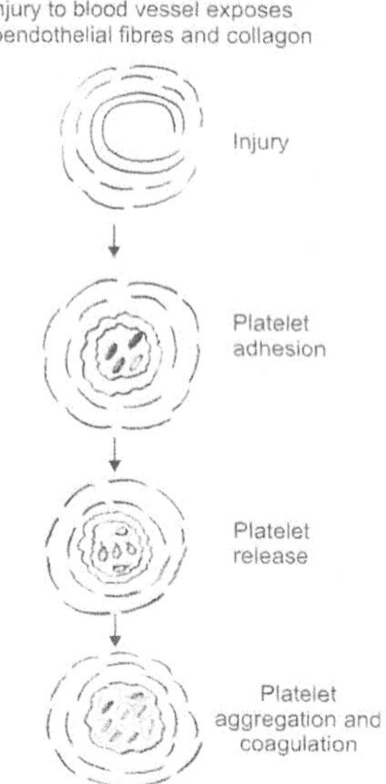

Blood clot takes place in two phases

Phase - I is temporary, causes bleeding to cease within minutes (2-6 minutes) and endures for an hour. It entails the constriction blood vessels cell walls and the blood leak is plugged by the platelets.

Phase - II is a permanent stage herald within 10 minutes after injury. It consists of formation of thrombus at the site of injury. A thrombus is a blood clot occuring within the circulation. This stage of stable blood clots is made by a mesh-work of sticky fibrin threads that hold blood cells. This process is also known as Plasma phase or secondary haemostasis.

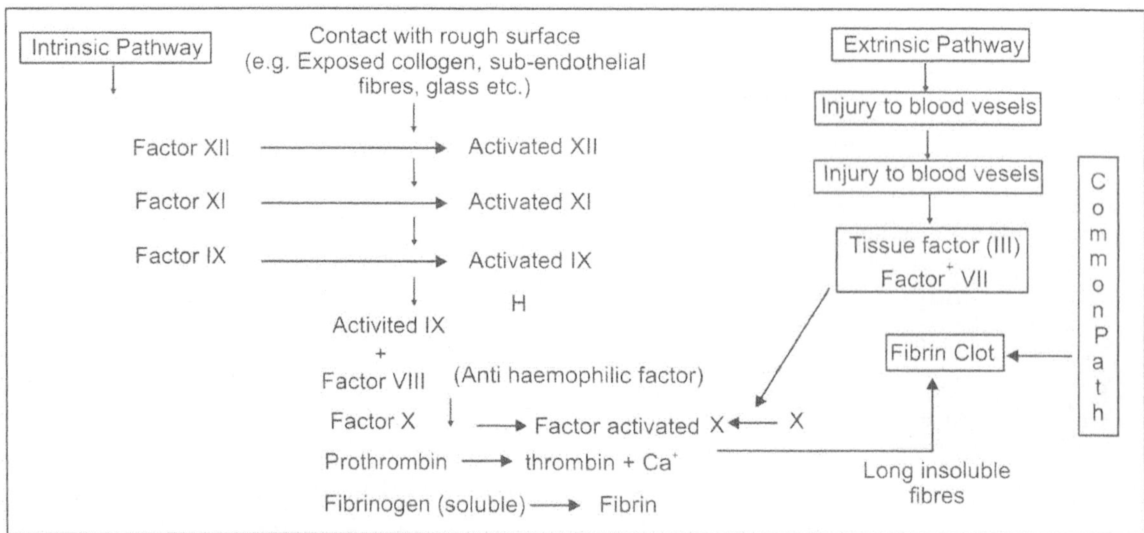

Fig. 4.18(a) Different pathways in the processes of blood clotting.

The cloting process *per se* is initiated by two different mechanisms.

- *Intrinsic pathway mechanism: A*s soon as blood comes in contact with a rough surface the clotting is triggered by the activation of components present in the blood.
- *Extrinsic pathway mechanism* consists of addition of substances which are not normal constituents of blood. Extracts of many tissues, particularly brain will induce clotting when added to plasma.

Both intrinsic and extrinsic pathways converge in the pathway to form blood clots.

The clotting pathways go through typically in a cascade fashion, each step produces an enzyme which will catalyse the next step. The advantage of cascade pathway is that a single initiation results in through numerous steps into amplification, which yields a large quantity of the final clotting factor (Fig. 4.18(a)). Probably, nature has opted for rapid responses to damage of blood cells, as the factors are needed for a proper clotting to take place. A single deficiency of a single protein in one of the pathways leads to severe bleeding disorder called as haemophilia. This gene controlled defect is irreccesive and haemophilics do not produce factor VIII (an antihaemophilic factor) in their blood.

Nutrition deficiency, especially vitamin K drugs, malabsorption etc., can also induce clotting disorders. The synthesis of prothrombin and other factors need vitamin K. Factor IV or calcium ions are also needed in most of the reactions including the critical step of conversion of prothrombin to thrombin.

Control of clotting must take place instantaneously. The process of coagulation of blood should be quick so that the blood limited to the wounded site only. Besides, the balance between clotting and thrombosis should be carefully controlled so that undesirable clotting does not takes place. This is accomplished in number of ways :

- Inhibition of blood clotting circulating in the system.
- Antithrombin III which inactivates thrombin by forming complex that is irreversible.
- Clearing of **activated** clotting factor by liver.
- Proteolytic degradation of fibrin – (fibrinolysin) removes the small fibrin clots by an enzyme, plasmin.

4.3.19 White Blood Cells (WBC)

There are five types of WBC or leucocytes in the human blood that can be distinguished through a thin film of stained blood under microscope. Neutrophils, Eosinophils, Basophils, Monocytes and lymphocytes. Each one of them have distinct shape and function.

A. Neutrophils

Numerically, neutrophils constitute the largest fraction of WBC. Typically, they have a 2-5 lobed nuclei, stain deep purple, when compared to cytoplasm that appear light pink with granules. The granules are of varying size but tiny. These are probably lysosomes (Fig. 4.19 (a)).

Neutrophils are phagocytes that are responsible for the first line of protection against invading bacteria into the body. Neutrophils are highly mobile and migrates quickly to the sites of infection. During infection the number of neutrophils increase considerably, mobilised from the bone marrow. Bone marrow stores about 50% of mature neutrophils within it until required. Neutrophils also act as scavengers removing the debris and dead cells.

B. Eosinophils

Eosinophils normally have two lobed nuclei and in their cytoplasm, there are many bright coloured reddish brown granules. These cells also are highly mobile. Their preferential place of location is epithelial surfaces. They are phagocytic but not as intensely as neutrophils. Hence do not play direct role against bacterial infection. However, they play a role in allergic reactions. They also play a role in the removal of parasitic worms. In such situations their number increases (Fig. 4.19(b)).

C. Basophils

The nucleus in basophils is akin to neutrophil but often obscured by the large violet or blue black coloured cytoplasmic granules. These contain histamines and are released whenever there is an allergy. Once histamines are released (the process called degranulation) to counter the allergens and their number increases significantly during allergies. The basophils have heparin, an anticoagulant and 5-hydroxytryptamine. Since these appear during healing process of the wound or injury, probably these may play a role in the prevention of coagulation of blood or lymph so that there is no obstruction in the tissue for circulation of blood. Like any other WBC, the bosophils are also extremely mobile. Basophils, however are not phagocytic and their precise role is not known. Their structure ressembles the mast cells found in the tissues of lungs, skin and gastro-intestinal tract, but they are not one and the same (Fig. 4.19(c)).

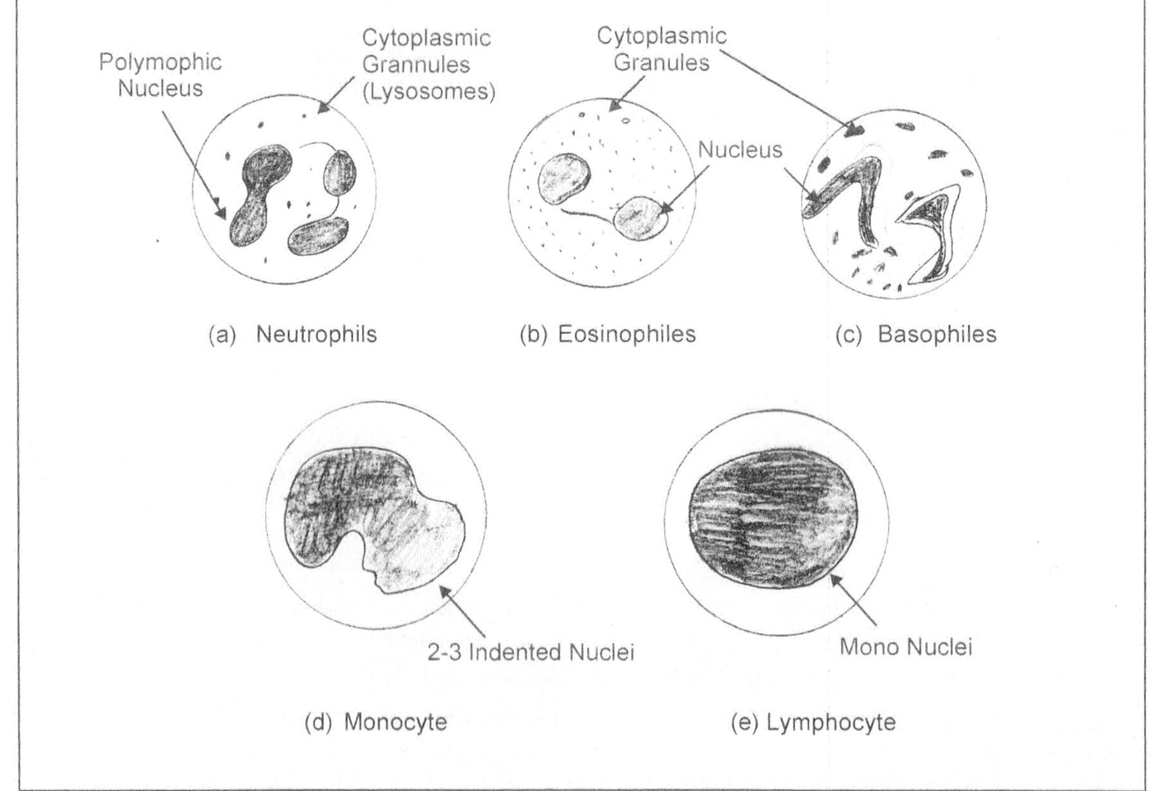

Fig. 4.19 Types of WBC.

D. Monocytes

Of all the WBC, monocytes are the largest cells. Their nuclei have slight indenture to appear two or three depression and stain blue-violet. Their cytoplasm is hazy and have few granules. The monocytes move in the blood and matures to become macrophage. These outnumber the monocytes by 400 times in an adult blood. When these becomes mature in different tissue the tissue macrophages are given different names.

- in brain tissue - microglial cell
- in liver tissue - Kupffer cell
- in lung alveolar tissue - Alveolar macrophage cells
- in lymph modes, spleen - macrophage.

Once in the tissue macrophages may go through division cycle to increase in their numbers. They may congregate at the **site** of inflammation or injured lesions to help mount immune responses. However, their main function remains to scavenge by phagocytosis to clear the bacteria, foreign bodies, damaged and dead cells, tumour cells besides other debris in the blood circulation system (Fig. 4.19(d)).

E. Lymphocytes

Lymphocytes are the second most numerous types of white blood cells and are larger than RBC. They have mononucleated but slightly indented large nucleus almost filling the most of the cell space. By stain deep purple, where as the cytoplasm stains light blue. Lymphocytes are the principle players in immune system (Fig. 4.19(e).

4.4 The Human Digestive System

The food that we eat has to provide not only energy but also repair and build the tissues, growth and development. Initially the form of food that we take is not in a position to pass through the wall of intestine. As a result it has to be broken down into simpler molecular components with the help of hydrolytic enzymes in the stomach and be digested. Hence the digestion is a process in which large insoluble substances are broken down into smaller molecules which can then be absorbed and circulated through blood to nourish the cells of the body.

The digestive system comprises of several organs : Mouth, Oesophagus, stomach, small intestine, liver, pancreatic secretion and large intestine . All of them helps in digestion and absorption of nutrients through gut wall and ensures that the product of digestion can be transported through blood circulation system to the cells for their use. Rest of the undigested food is ejected as faecal material (Fig. 4.19).

4.4.1 Mouth

First and foremost process in uptake of food is to break down the food into fragments. This is done by the mouth by chewing so that food material can be swallowed. Mouth secretes saliva as soon as food morsel is put into the mouth and chewing begins. The rate of secretion increases with larger bites and greater degree of chewing. About 1-1.5 litres of saliva is produced per day in the mouth. Several buccal glands and three pairs of salivary gland produces salivary secretion. Small buccal glands secrete relatively thick fluid called mucous that moistens and lubricates food and makes a bolus by binding making it amenable for swallowing through throat. Mixed glands of sublingual and submandibular regions along with **parotid** gives rise to serous, a watery fluid that contains amylase which breaks down the carbohydrates.

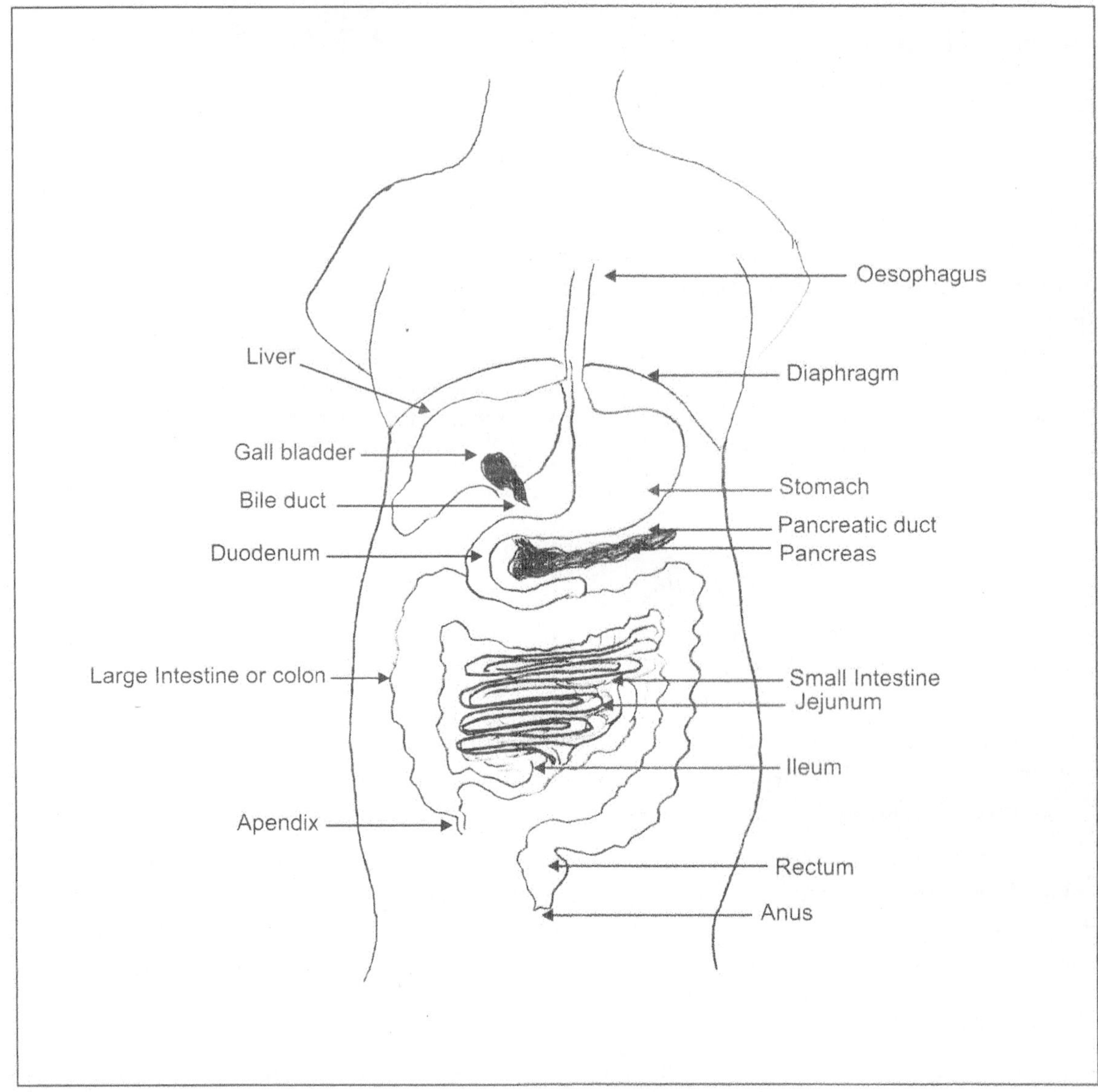

Fig. 4.20 Human digestive system

Saliva also does the job of cleaning the mouth and keeping it moist in the absence of food and so helps in speech. It also digests the soluble constituent of the food so that flavour is experienced.

4.4.2 Oesophagus

The Oesophagus connects the pharynx at the end of the mouth to the stomach. Upon swallowing, the food is propelled through the phayrynx and moves down the oesophagus with the help of gravity and circular contraction of the muscles (of oesophagus) generating peristalic waves ahead of moving food bolus. The speed with which these travel is about 5 cm per second and takes about 5-10 seconds to reach the stomach. The movement of food is made frictionless because of lubricating mucus. The presence of food in the oesophagus relaxes the *spincter* muscle, gastro-intestinal *spincter* and permits the food to go into the stomach (Fig. 4.20 and 6.20(a)).

4.4.3 Stomach

It is a temporary ware-house of food until the same is digested and passed on at a regulated rate into the duodenum. Nature has taken care of the problem of irregular intake of food in uneven quantities. The stomach stores the food several hours till digested and accomodates about 0.5 to 5.0 litres volume of food. The inner gastric mucosa of the surface of stomach has several openings called gastric pits dotted all along the inner surface and fundus of the stomach. Glands situated at the base of these pits open into them and are made up of three kinds of cells: Mucus neck secreting mucus; most abundant peptic releasing a digestive enzyme pepsinogen, parietal (oxynic) producing hydrochloric acid. The movement of stomach muscles, together with gastric secretions change the food into chyme. This is fairly uniform in consistency with respect to pH, osmolality, temperature etc. Gastric movements are generated by the pacemaker cells in the longitudinal layers of stomach muscle with having curvature producing waves of momentum called peristalic waves. These waves spread over the stomach with about 1cm per second at a regular frequency of 3 per minute. This goes on till the undigested food is excreted through the anus.

Gastric secretions consist of hydrochloric acid, pepsinogen, mucus and scores of intrinsic enzymes and factors like lipase that splits fat bodies and gelatinase that digests and liquifies gelantin. Each enzyme or factor has specific function that aids the digestion of food (Fig. 4.20 and 4.20(a)).

Hydrochloric acid has the following role :

- Breaks down connective tissues and muscles fibers.
- Provides a low pH medium for the action of pepsin.
- Bacterialcidal activity to kill any living cells or organisms that are ingested along with food.

Pepsin is an enzyme that breaks down protein molecules by hydrolysing the peptide bonds. The precursor of pepsin is inactive pepsinogen which in an acidic medium by an autocatalysis process is converted into an active enzyme, pepsin. The intrinsic factor is also a protein that binds to vitamin B_{12} to help its absorption by the system.

The whole process of secretion of acid and digestive enzymes is primarily regulated by several enzymes and hormones of the gastrointestinal tract: gastrin, pzcck secretion, gastric inhibitory peptide. These are not grouped into distinct glands. They are scattered throughout mucosal muscle cells that line the gut. Each one of them are different in terms of distribution as well as the kind of secretion of a specific polypeptide of hormone (Fig. 4.20(a), (b) and (c).

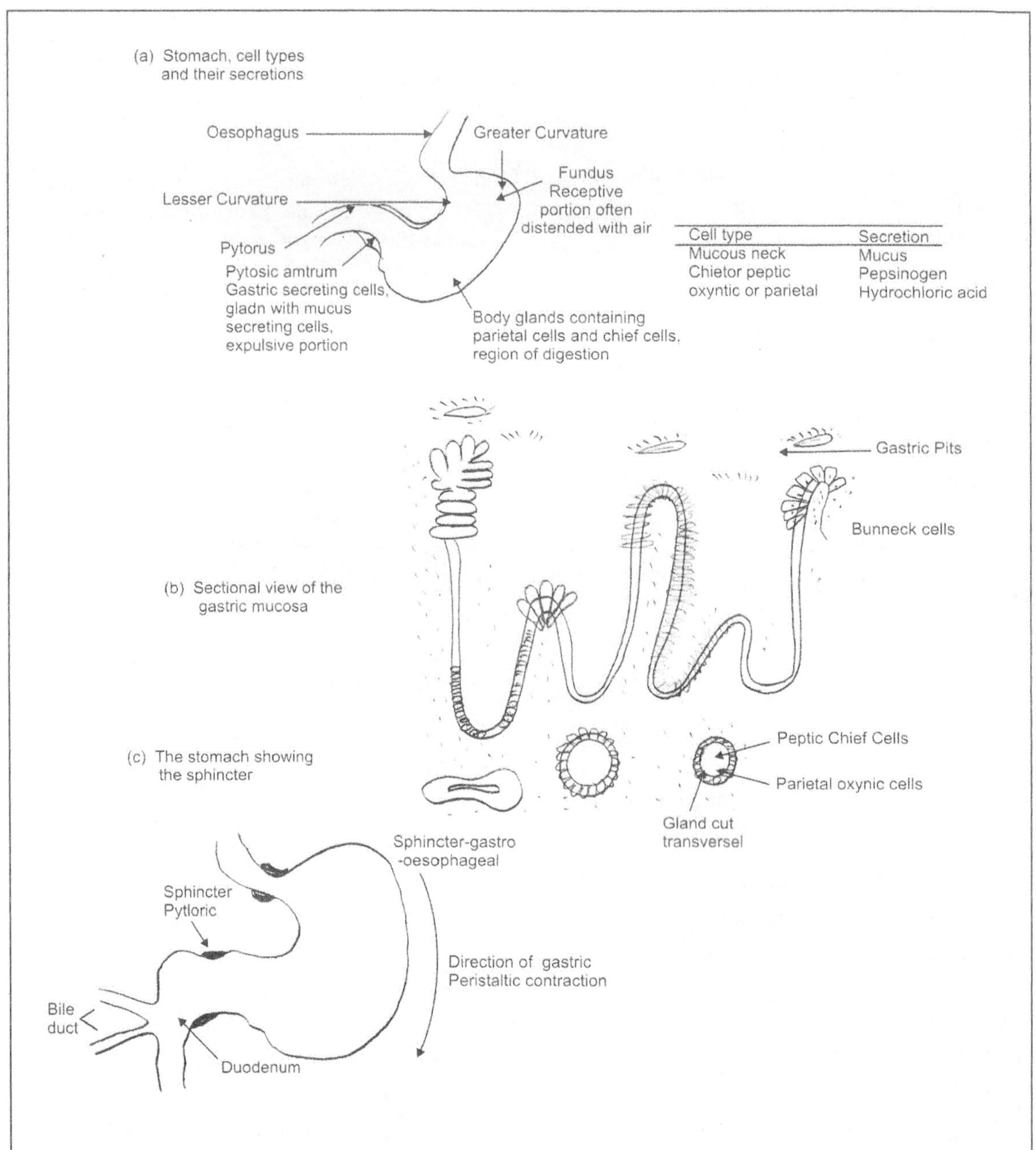

Fig. 4.20(a) (b) (c) Stomach and cell types; (a) their secretions; (b) gastric mucosa, sectional view; (c) stomach showing different sphincters.

The secretion of gut hormones in response to the contents of gut. Besides, autonomic nervous system also can induce their secretion. These exert action either in the classical endocrine style, or have a paracrine type of action. Their main target organs are pancreas and gut.

The rate at which the chyme passes out through the pyloric sphincter mainly depends on chemical and physical nature of the contents of the stomach. Solid food slows down the passage besides high or low osmolality, fats and acid in the duodenum. Chyme will be speeded up if the volume is large.

4.4.4 Small Intestine

Place where food is digested and absorb soluble food into blood stream.

The chyme enters first into the small intestine. The human small intestine is a tubular contraption of several meters in length. As soon as the chyme enters, its pH and osmotic pressure gets adjusted. The digestion process takes place in its lumen. However, the digestion is not entirely done there. Many chemical compounds with which the food is made of are broken down at the mucosal surface made of epithelial cells within their brushy confines or even intracellularly. As digestion is being completed the small fragmental molecules get absorbed into the blood stream through a fine network capillaries lining the cells.

The transfer of substances is bidirectional across the epithelial barrier formed by the intestinal walls. There is always a balance between a particular substance being transferred and substance secreted. The transfer across the intestinal wall occurs in one of the following way. (Fig. 4.20 (d) (e) (f) (g).

- Diffusion
- Passive transfer
- Active transport – an energy needing step
- A mix of all the above three

Different parts of the intestine may be involved for the absorption of food components. For instance, vitamin B_{12} is absorbed by a specific processes no where else but for the ileum.

Small intestine's movements

Small intestine is not a static place. The small intestine set a motility as a result the contents of it are pushed towards colon. Besides it also serves to mix the contents of intestine. The basic type of movement is segmentation, which primarily is a mixing process. The musculature around the intestine contracts in a series of circular motion (rings) each separately from the other by several centimeters. This takes place methodically with a gap of few seconds. The area that has contracted, relaxes and the region that was in relaxed state contracts next. These rings of alternative contraction and relaxation are distinct from peristalic movement and do not occur along the entire intestine and hence a distinct category of movement. However the lumenal contents movement is achieved by peristalisis. Peristalitic waves consists of ring like contractions each preceded by the relaxation. They move along the intestine at the rate of 1-2 cm/sec and extinguish themselves after traversing about 10 cm. Only when an acute enteritis occurs abnormal peristalitic movements extend to a long distance (Fig. 4.20 (h).

Small intestine's Secretions

Liver and Pancreas secrete bile juice and enzymes into the duodenum. The bile duct delivers these secretions. Bile consists of fluids that emulsifies fats and neutralizes acid. The enzyme secreted are,

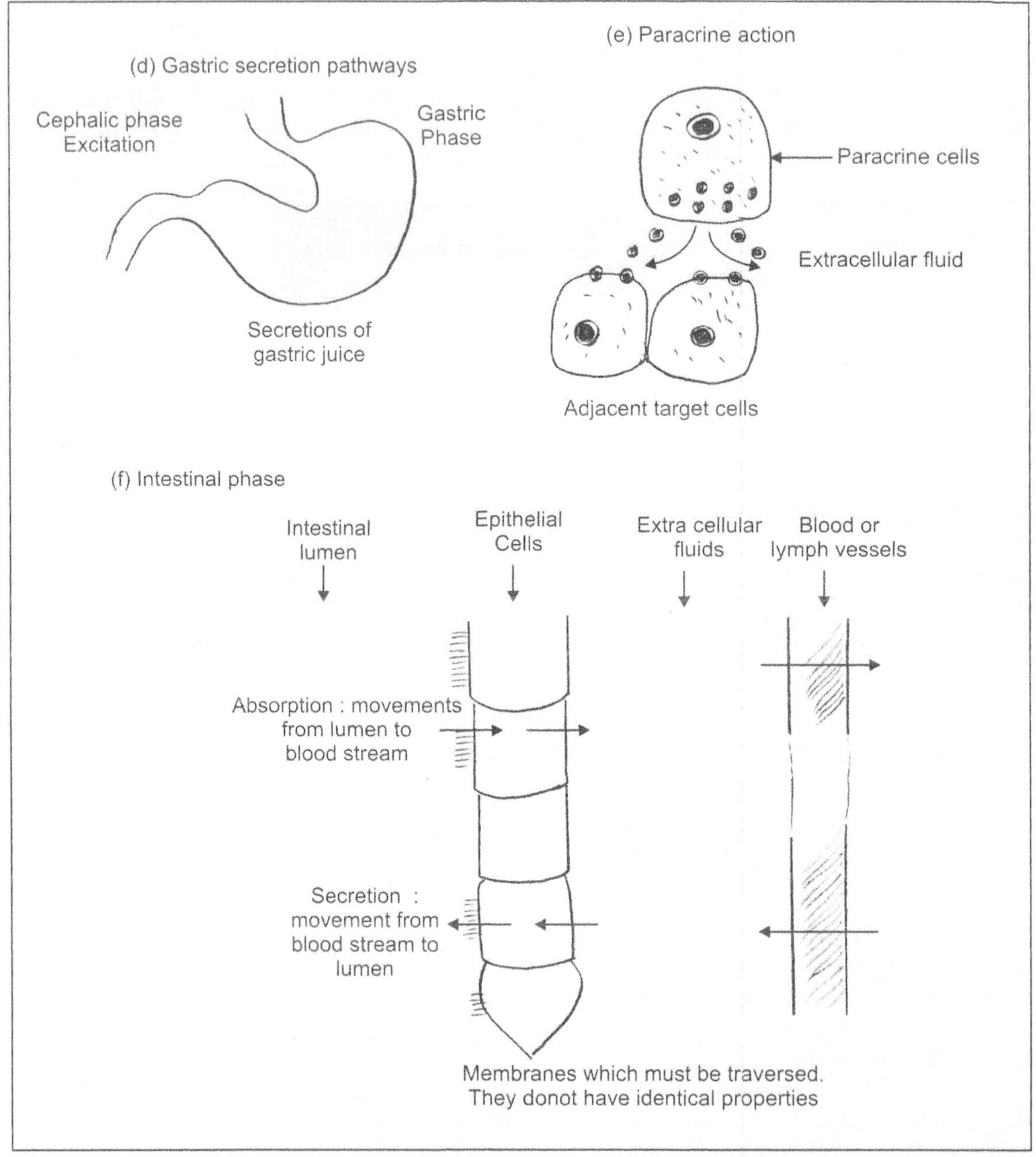

Fig. 4.20(d-f) (d) The process of digestion in the stomach with the help of gastric secretion; (e) paracrine action; (f) the digested food has to transverse across membrane of epithelial cells, blood or lymph vessels etc.

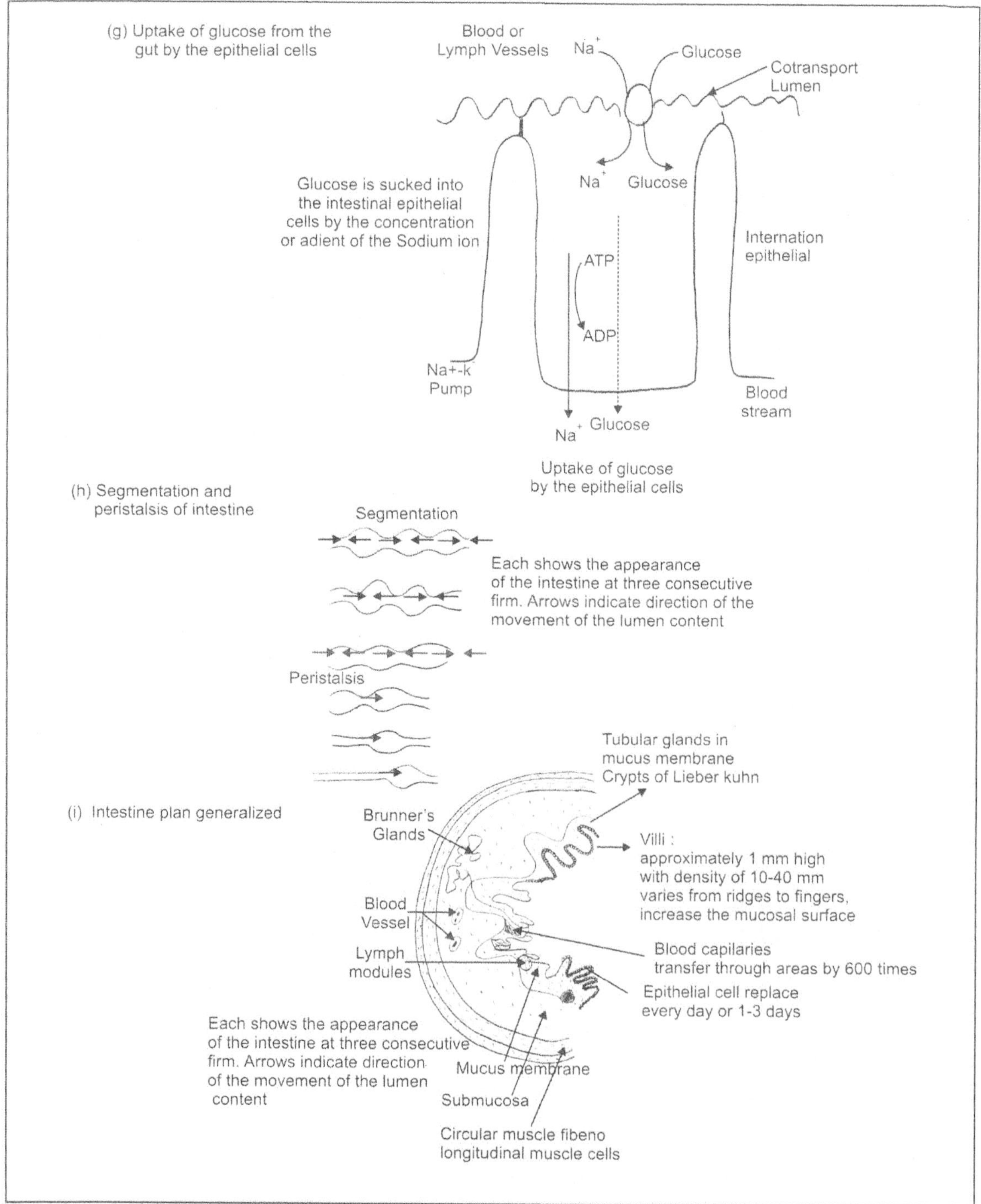

Fig. 4.20(g-h) (g) Glucose by the epithelial cells of intestine uptake from the gut, (h) segmentation and peristalsis of intestine, (i) generalized acount of the intestine.

protease that digests the protein; amylase that breaks down starch into glucose and lipase that degrades fats and lipids. Other secretions of intestine are mucus by the glands of intestine, Brunner's glands of the duodenum and Liberkuhn crypts throughout the small and large intestine (Table 4.6) (Fig. 4.20(g).

4.4.5 Secretions of Pancreas

The Pancreatic secretion contain two fractions:

- A large quantity of secretion at the rate of 200-800ml/day constitutes a high degree of biocarbonate in order to make it alkaline. Another identical type of fluid is quite distinct from bile acids is also secreted by the liver. However, the secretion of this fluid is under the common control with that of pancreatic fluids. These juices make the food alkaline prior to its entry into duodenum (Table 4.7).
- A small quantity of secretion containing enzymes is secreted by acinar cells of the pancreas.
- Some of the enzymes are converted from the inactive proenzymes into an active enzymes in the duodenum. eg : Pepsinogen is converted into pepsin.

Table 4.7 *Profile of Pancreatic enzymes.*

Nature	Enzymes of Pancreas	Substrates hydrolyzed
Protein degrading enzymes	Trypsin Chymotrypsin Carboxy peptidase	Protein
Starch degrading enzymes	α Amylase β amylase	Carbohydrate
Lipid degradation	Lipase phospholipase	Fats
Tissue degradation	elastase, collgenase	Elastic tissue and connective tissue Skin supportive protein
Nucleases	Ribonuclease Deoxyribonuclease	Nucleic acid

4.4.6 Secretions of Liver

Liver predominantly secretes bile juice. It is a neutral golden-yellow secretion with a variable mixture of water, organic and inorganic solutes including the substances/drug compounds metabolised by the liver. In an average man's daily output of bile is approximately 700-1200 ml per day. Bile juices contain the following :

- Bile acids
- Phospholipids - lecithin
- Cholesterol
- bile pigments - bilirubin

Bile acids are water soluble components of cholesterol. They have following major roles:

- Reduces the water-oil interface tension by emulsification process.

- Prevents the denaturation of lipase enzyme as it leaves the surface of emusified fats.
- Formation of micelles. These are aggregates resulting by combining with the byproducts of fat hydrolysis, monoglycerides. These micelles help in dissolving cholesterol, free fatty acids and fat soluble vitamins.

As the bile juice pass through the intestine, majority of it does not change (70%). They are reabsorbed without modification and recycled through the liver. Remaining 25% of the bile juice is acted upon by the gut bacteria to produce secondary products.

The major fraction of this also reabsorbed by the ileum and reconverted into acids by the liver. However, about 10% of the bile juice is lost every day (500 mg) through faeces. They are replaced by new bile acids synthesised by the liver once again in order to maintain the total amount of bile acids in the pool. This varies 2-4 g. Between meals bile is secreted in the gall bladder. The gall bladder concentrates the bile. During the process of digestion, the bile is delivered to the duodenum (Fig. 4.21).

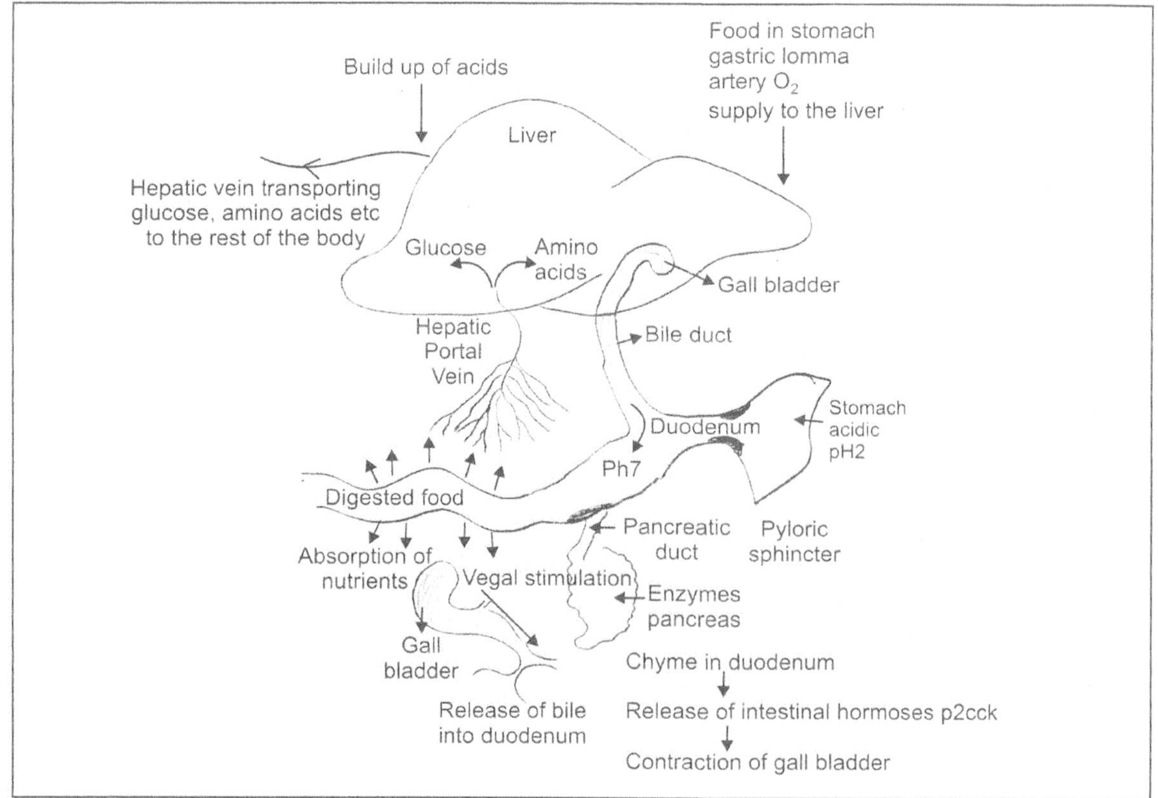

Fig. 4.21 Circulation of bile acids.

4.4.7 The role of water in digestion

Besides the addition of water to the food by drink, gastrointestinal secretion also add a large quantity of water to the digestive system. Though about 5-10 litres of water goes through the gut in this way, only 100 ml of water is excreted through stools. Since the intestinal mucosal lining is rich with pores that are highly permeable, water is quickly absorbed and thus balance of osmotic equilibrium is restored between the contents of duodenum and the blood (Fig. 4.22(a)).

4.4.8 Uptake of Sodium ions and vitamins

The walls of the small intestine absorbs sodium ions as well as vitamin molecules. The active uptake of sodium ions (active transport) conjointly absorb many nutrients like glucose, amino acids, some vitamins etc. In the case of fat soluble vitamins they are absorbed like a fat. Most of the water soluble vitamins are actively transported and absorbed. Vitamins like B_6 or pyridoxine is transported and absorbed simply by a diffusion process (Fig. 4.22(b)).

4.4.9 Colon (Large Intestine)

The large intestine is the place where an undigested food, water, salts etc. are received for ejection. The undigested food is passed out as stools, which contains little water. A large quantity of water and salts are reabsorbed. The colon on an average receives at irregular rate about 1.5 litres fluids each day. The ileum stimulates reflexly as the stomach is empty its content into the large intestine. Consequently the undigested food in semifluid state pass into the caecum at the beginning of the colon. Once in the colon, further absorption of water and salts takes place. The process of potassium ion absorption and the secretion of potassium ions is controlled by the hormone, aldosterone.

A large number of bacteria are present in the colon without which the development and survival of man is not possible. They play a vital role in man's sustenance as follows.

- They defend the body by stimulating immune process. Lymphoid tissue and inflammatory cells incidence is enhanced due to them.
- They help in recycling of certain endogenous proteins and enzymes by degrading them muco proteins, debris etc., are also degraded by them.
- They synthesize and contribute to the body certain essential vitamin like vitamine K, riboflavin and thiamine.

4.4.10 Movements of Large Intestine

The contents of colon are shuttled back and forth with a circular ring like movements of the muscles called, Laustral movement, which are very slow. These evenly spaced, ring like constrictions of the circular musculature of the large intestine divide the colon into what is called Laustral. The slow contraction is replaced by a slow relaxation and the neighbouring area tither to in a relaxed state may slowly contract. Thus, the contents of colon are shunted back and forth. This kind of movement is called laustral movement which is akin to segmentation movement of the small intestine, but differs in terms of duration. The segmental movements of the small intestine takes only few seconds while laustral movement lasts for several minutes. Occasionally multi laustral movement takes place at once. Besides, peristalsis also occurs at the rate of 1.2 cm/min. The net forward motion achieved is only 5 cm/hour.

The faeces, semi solid or solid undigested food matter is formed in about 36 hours after intake of food. The main composition faecal matter is :

- Water 60-80% of total weight.
- Nitrogenons substances (1-2 g/day). This includes millions of bacteria digestive enzymes, urea and de'squamated cells.
- Fatty material like fatty acids, neutral fats, phospholipids, sterols, and degraded acids of the bile.
- Fibres (20-60 g/day) like cellulose and other undigested food residues.

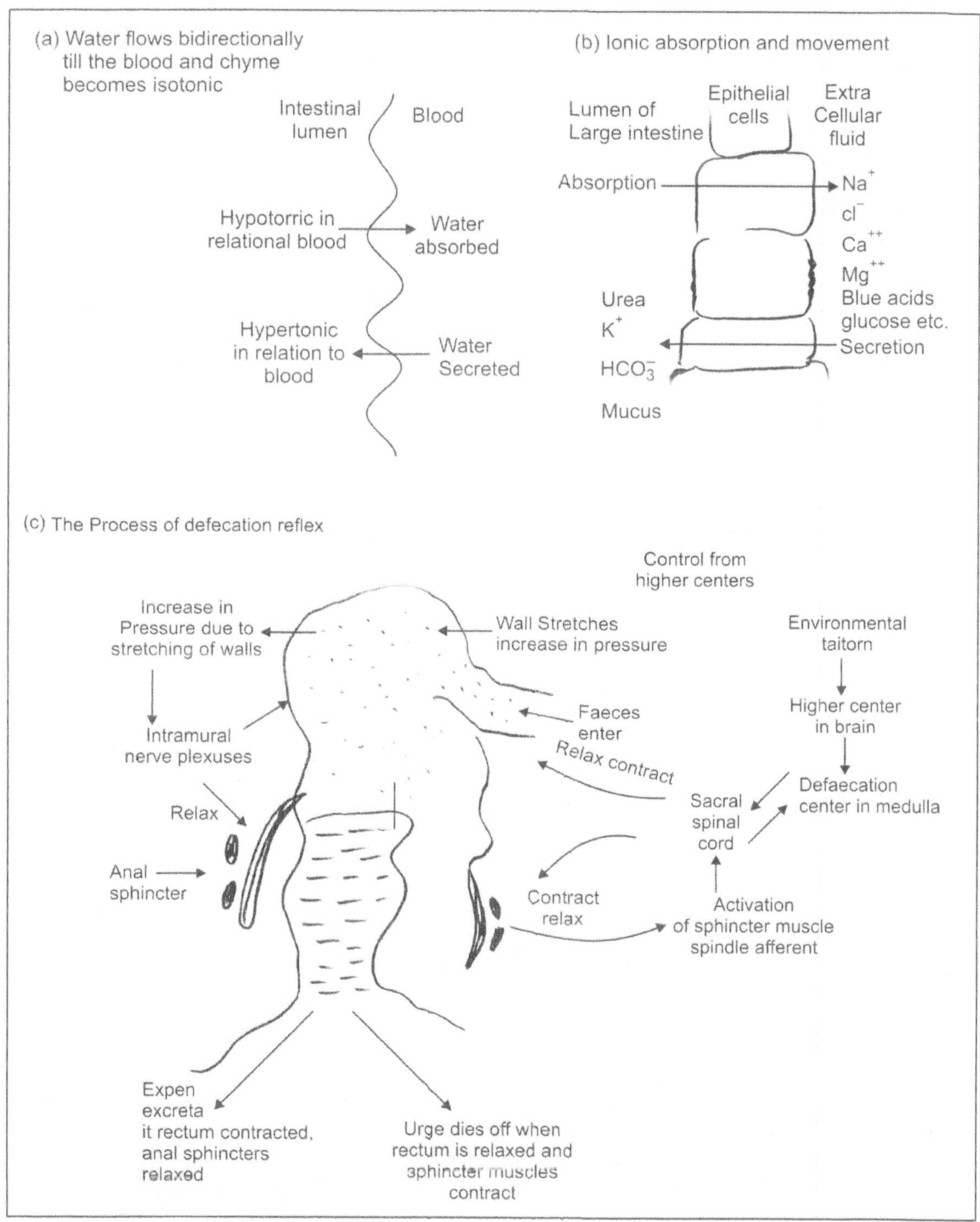

Fig. 4.22 Origins of defaecation reflexes and control of the procceses.

Motility of the colon increases after a meal and then faeces is pushed into the rectum. The distended rectum then arouses the urge to pass stools as due to defaecation reflex. This is generated locally, but is modified by the higher centers so that it takes place at appropriate time and place. The normal frequency of defaecation ranges from after every meal once every three days (Fig. 4.22(c).

4.5 Respiration System

4.5.1 Introduction

As in the case of complex multicellular organisms, humans too have with in them a specialized system to fulfill the needs of respiration or for exchange of gasses, called respiratory system. Lungs are part of the complex and provide mechanism that facilitate and ensure all the organs as well as tissues get adequate supply of fresh oxygen and the removal of carbondioxide from the body. The mechanism of the functioning of lungs is intimately connected to the heart as well as the central nervous system.

4.5.2 Structure and function of Lungs

Organs and tissues get oxygen from the external environment. The oxygen thus drawn has to be in intimate contact with the blood vessels over a vast surface area. The structure of lungs (Fig. 4.23) is designed towards this function. Basically, lung structure is made up of many branching tubular canalas leading to several minute sacks with thin membrane walls, each with a profuse supply of blood exposed to both sides. The blood flow is continuously supllied with and constantly moved through lung by the action of heart. The constant removal and supply of blood is achieved through the 'bellowing' action of the thorax. The nerve impulses originate from the central nervous system (Brain).

4.5.3 The volume of Respiration

When at rest, the respiration the volume is around 10 litre per minute. The volume of each breath, called as tidal volume about half litre (per breath). It varies slightly depending on the number of breaths one takes per minute depending on physiological status as well as age. The resting respiratory rate is 70 per minute at the time of birth, 25 per minute around 5 years of age, 20 per minute at teenage and 15 per minute from adult age of 25 years.

4.5.4 Defence of air passage

As the inhalation process goes on incessantly, foreign matter that might be harmful may also come along and this is prevented and is not allowed to easy entry into the lower respiratory tract. Even when pulmonary ventilation raises above 100 litres per minute during strenuous exercise, micro organisms do not enter tracha. Nose is provided with hairs, muscles and associated mucus that intercept any air borne foreign particles, cilliar activity of respiratory tract shifts mucus and particles, by a current induction out of lungs at the velocity of around 2 cm/minute. However, the cilliary activity of respiratory tract get affected by various factors. Nicotine, tar or sulphurdioxide etc. even if minute concentration as 2-3 ppm impair the process. When chemical or mechanical processes irritate the nerve endings, the reflex action comes in the form of cough or sneezing. Trachea its around bifurcation area is highly sensitive. Any particulate matter enters that is removed immediately through a cough that sends out a rapid explosion of expired air pushing out of the respirator tract.

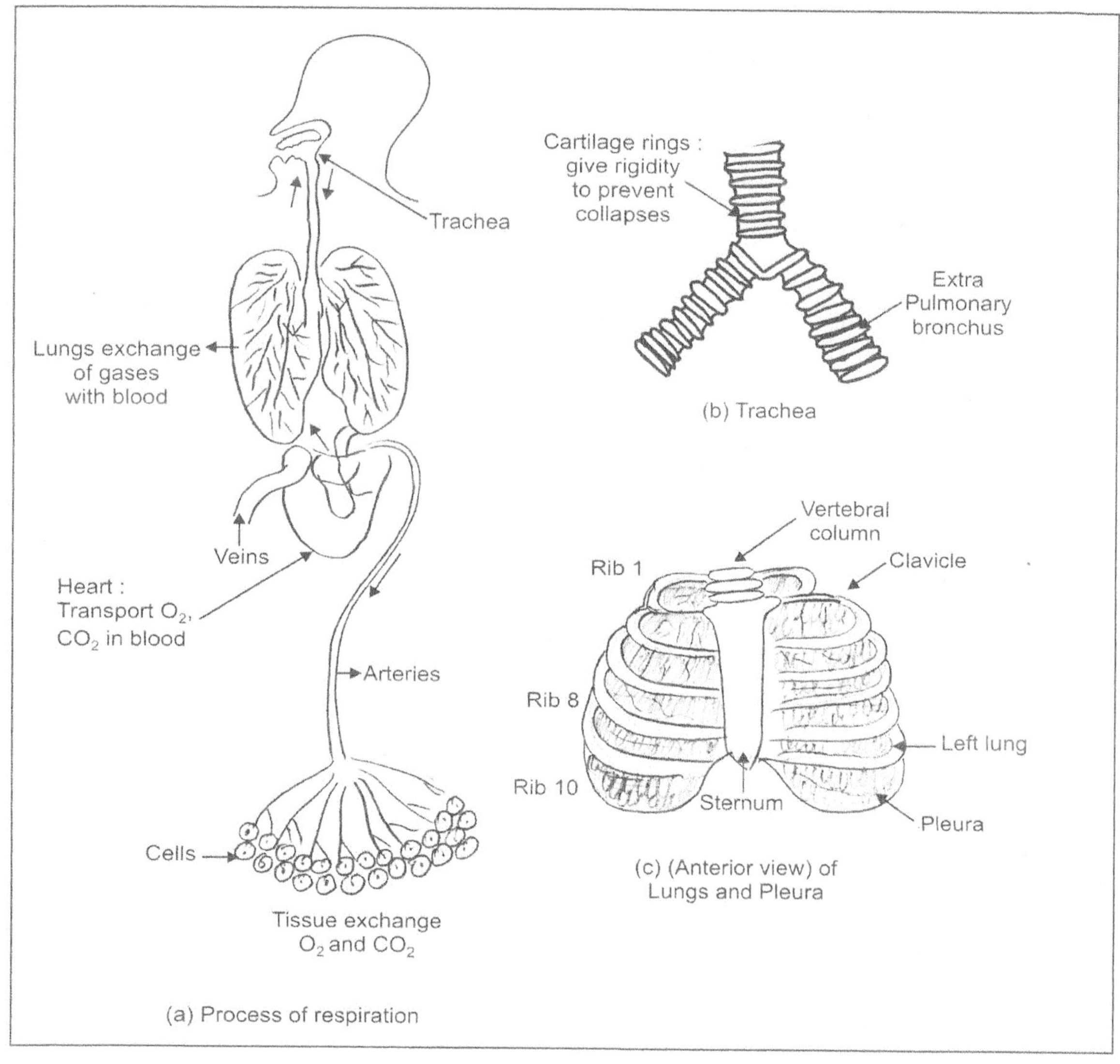

Fig. 4.23 Respiration process : (a) overview; associated structures: (b) trachea lungs and pleura.

4.5.5 Air Volume of the Lung

The terminology of respiration was standardized more than fifty year ago based on parameters of respiration.

- *Vital capacity (VC) of the lungs:* It is the volume of air that can be expelled during maximum expiration following minimum inspiration.
- *Expiratory reserve volume (ERV):* It is the volume of air expelled by a maximum expiration following normal expiration.
- *Inspiratory Reserve Volume (IRV)*: It has the maximum volume that can be inhaled after a normal inspiration.

- *Residual volum (RV)*: It is the volume of air that remains in the lung after a maximum expiration.
- *Functional Residual capacity (FRC)*: It is the volume of air remaining in the lung at the end of a normal expiration.
- *Tidal volume (V_t)*: It is the air shifted during one respiratory cycle.
- *Total lung capacity:* It is the sum total of IRV, V_t, ERV and RV.

Spirometer is the instrument that records the lung capacity. For a limited number of respiratory cycles, continuous recording can be made of lung volume for by means of a potentiometer placed on the pully spindle. Oscilloscope records the voltage changes which can be displayed and recorded as Akymograph on a drum (Fig. 4.24).

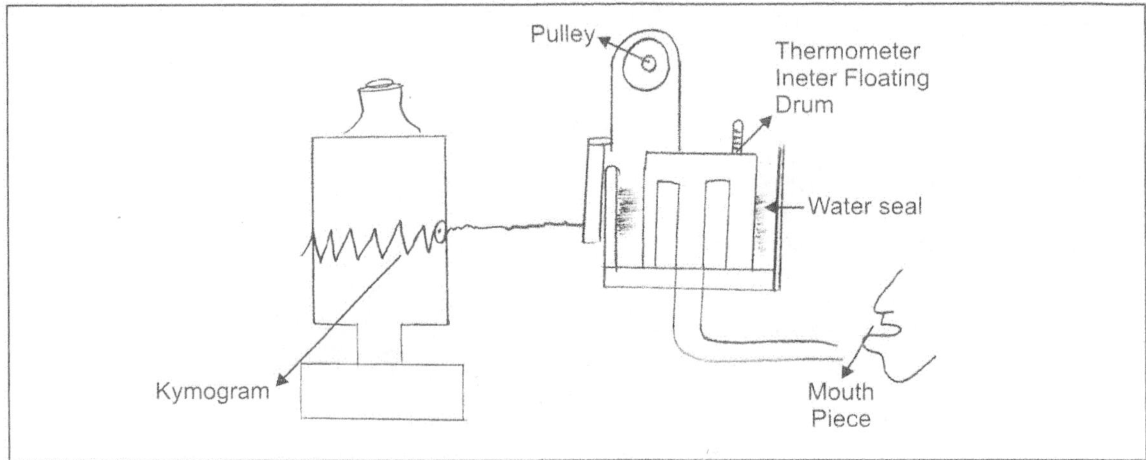

Fig. 4.24 Spirometer – a gadget that assesses the lung capapcity.

4.5.6 Exchange of Gases in the Alveoli

Gas exchange takes place between the alveolar gas and the tissue fluid (Fig. 4.25). All gas exchange surfaces are essentially moist because any surface that is permeable to water is also permeable to gas. Hence all surfaces that take part in interchange of gases are moist. The exchange of gases follows the fundamental gas laws. Dalton's law of partial pressures deals with gas mixtures. The law states that pressure exerted by any gas in the mixture is the pressure that the same gas would exert in the same volume, if no other gases were to be present. The total pressure in a gas mixture is the partial pressure of the separate gases in a mixture. In a dry weather, atmospheric air exerts total pressure of 760 mm Hg (100 kPa). The air comprises of mixture of gases – oxygen 21% and nitrogen 75% with traces of carbondioxide and rare gases. Hence the partical pressure exerted by oxygen (PO_2) is approximately 160 mm Hg ($21/100 \times 760 = 160$). In man the total surface area of all alveoli in lungs in nearly 80-100 sq. meters (which is roughly the area of tennis court).

The water vapour also follows the law of partical pressure, exerting its own pressure independent of gases, proportional to the quantity of its pressure. The amount of pressure varies with temperature of the air. Air at higher temperature contains more water vapour. The temperature in lungs is around 37 °C and the pressure exerted by fully saturated water vapour is 47 mm Hg (6.15 kPa). At room temperature, the same is 15.5 mm Hg when fully saturated. In case air is not fully saturated with water vapour, the pressure accordingly is lowered.

4.5.7 Mechanism of exchange of gases

In alveolar air, the pressure of gases is almost constant. However, slight variations may be present between expiration and inspiration. Air in alveoli is in equilibrium with systemic arterial blood. Several factors determine the constitution of alveolar air:

- Composition and pressure of expired air.
- Volume of tidal air, respiratory frequency, volume of the dead space.
- Volume of air remianing in the lungs at the end of expiration.
- The rate of absorption of oxygen by the blood.
- The rate of loss of CO_2 from the blood during expiration.

All the features of alveoli are intended for efficient exchange of gases during the respiration. The innumerous folded nature increases the surface area for the diffusion of gases. The thiners or cell thickness helps in speeding up the processes of diffusion within a short distance the gases have to travel; besides the large number of alveoli with which the lung is made of provides a huge area for the diffusion. The proximity of blood capillaries minimises the distance for diffusion (Fig. 4.25).

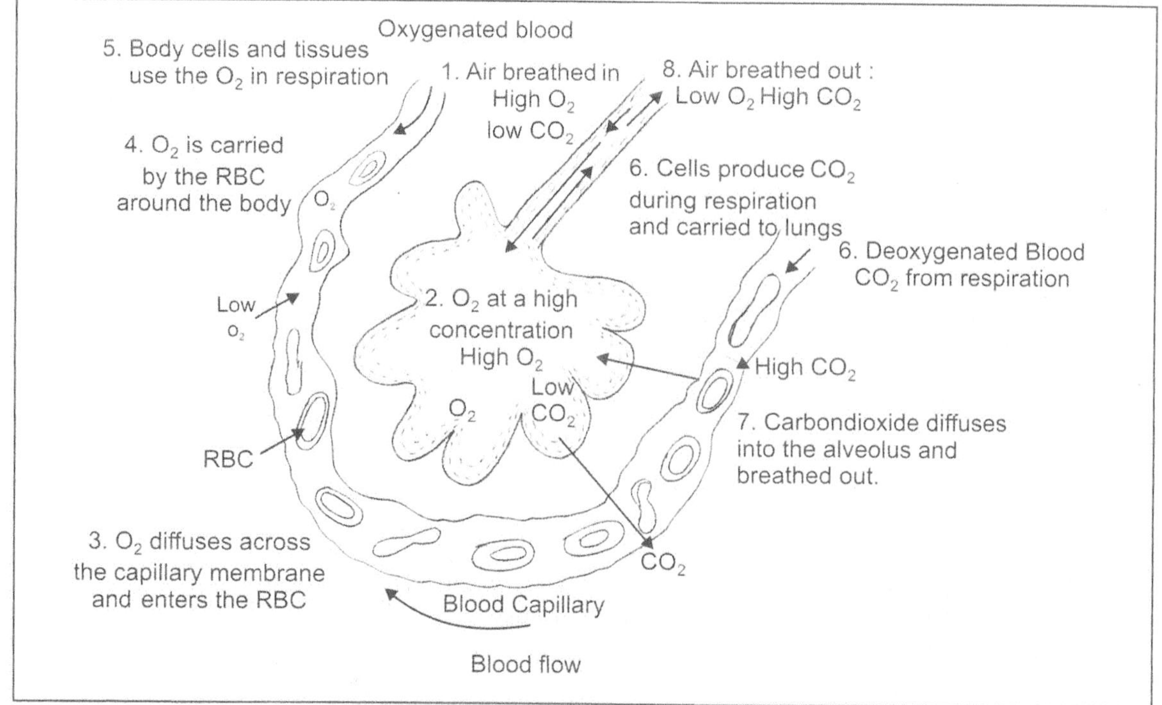

Fig 4.25 Exchange Process in one Alveolus.

4.5.8 Control of Respiration

Oxygen needs and the output of carbondioxide of the body varies depending upon the situation and must be controlled. The motor neurones in the cervical and thoracic spinal cord supplying the muscles involved in respiration send rhythmic signals for the initiation of respiration. The regulatory control of respiration is achieved by two neural mechanisms.

- A voluntary system in the cerebral cortex of the brain.
- An involuntary (autonomous) system in the pons and medulla of brain.

The voluntary system takes care of respiratory requirements during speech etc. while the automatic system regulates the overall gas exchange in relation to the metabolic status and requirement of the body. In the voluntary system of control, the nerve impulses are sent to the spinal chord through motor neurones involved in the respiration through cortico spinal tract, whereas, in the case of automatic system, reticular spinal pathways are pressed into the service.

4.5.9 Mechanism of Control Process

The whole process of breathing (respiration) is essentially an automatic process with a rhythm. It goes on even in deep sleep or moderately deep anaesthesia. The proof of CNS control comes from the observation that when all the nerve supply to the muscles involved in respiration are cut, the breathing stops. This imply that the respiratory muscles of the ribs and also the diaphagm do not have spontainity in their rythmic movement, something control them. They must be stimulated for the action by the CNS.

4.5.10 Respiratory Centre in the Brain

The centre of respiration is situated in the medulla, where a group of nerve cells control the rythmic respiration. It is modified by other regions such as pons. The stimulation experiments and subsequent microscopy and micro electrode recording demonstrate that there are certain nerve cells within the medulla, certain of it is responsible for inspiration and others for expiration. These form a diffuse cellular network. The insight was obtained by the affect of cuts made in the brain (Fig. 4.26).

There is no agreement about how rhythmic burst of action potential is generated for enabling the respiration. It seems that the respiratory neurons briefly activate the muscles involved in inspiration which is followed by relative lull period in which the neurones are quiet, relatively passive, expiration doesnot take place unless neurons for expiration are active. It appears that medulla by itself maintains inherently rhythmicity. However, this inherent rythmicity can be altered by changing neuronal and chemical inputs. The importance of spinal reflexes also play an important role in maintaining the large part of rhythmicity.

4.5.11 Breathing Changes in the Pattern of Breathing

A wide range of breathing operates under the respiratory system. The system controls both the intensity and rate of respiration. The depth of respiration is primarily controlled by the intensity of neuronic impulses to the pulmonary muscles. The high frequency of impulses depends on the activation of motor neurons. This kind of stimulation provide stronger respiratory contractions. The kind of control is akin to that takes place in ordinary skeletal muscles. The rate of respiration is controlled by the rate at which the bursts of action potential generated by the neurons in the medulla. This process

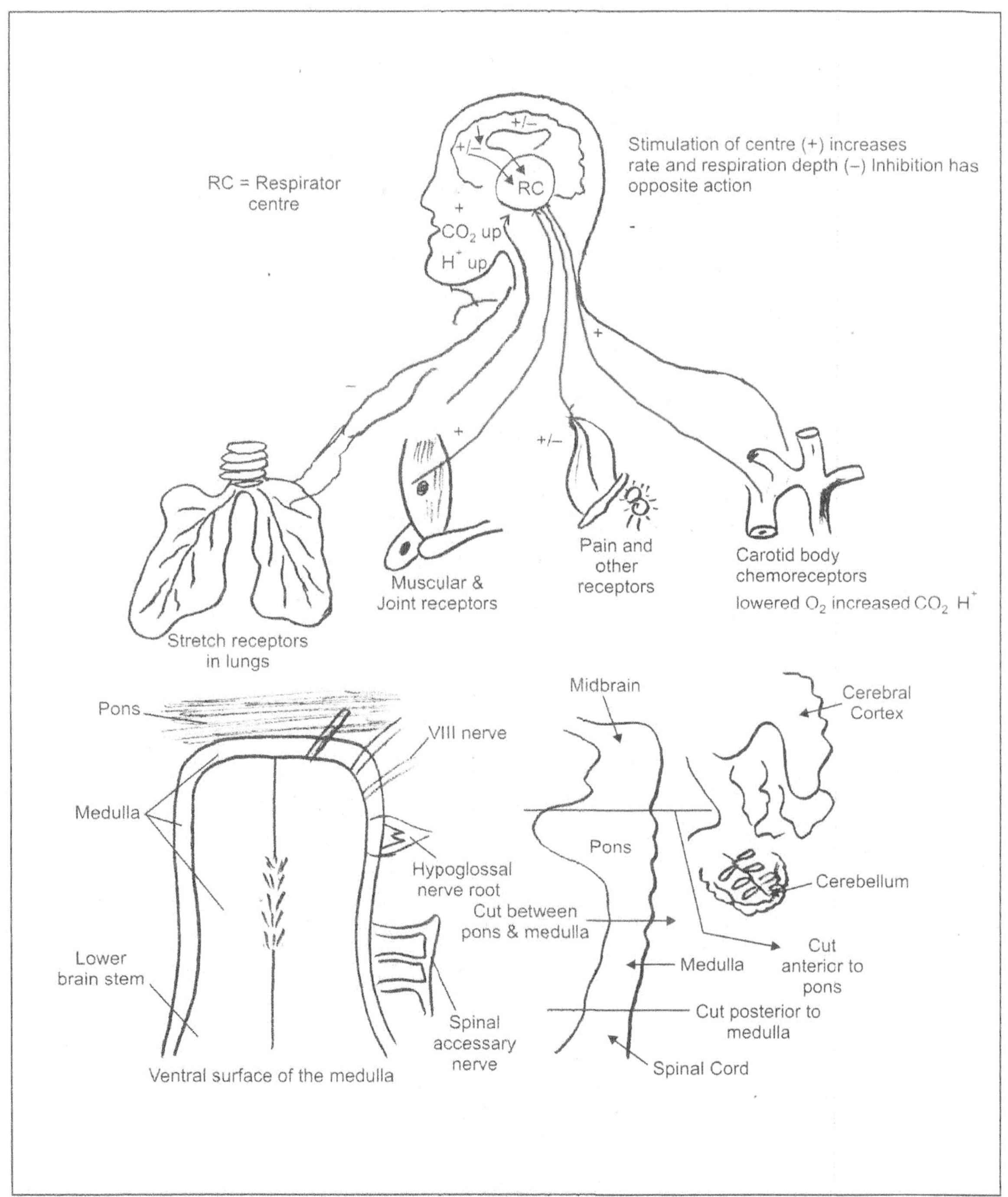

Fig. 4.26 Respiratory center in the brain – chemical and nervous factors.

is controlled by the chemical inputs. Deficiency of oxygen in the blood and excess of carbondioxide in the blood stimulates the respiration. When a body is in exercise mode, it needs more oxygen and take out carbondioxide, and hence necessitates more exchange of gases and the respiration enhances considerably.

This is accomplished by chemoreceptors. The receptors situated in the arch of the aorto and in the carotid body region sense the chemical content of the blood and well as cerebrospinal fluid in the brain, and regulate the respiration rate.

The effect of CO_2 is a bit more potent in regulating respiration rate than in O_2 deficiency. The oxygen deficiency acts on the peripheral chemo receptors. This is the precise reason as to why breathing donot begin again after hyperventilation till the CO_2 level reaches normal eventhough there is a considerable lack of oxygen during the same time. The want of oxygen effects the central chemoreceptors and depresses the same. The general effect of lack of O_2 in medulla is to depress it consequently the respirative cells.

4.6 Endocrine System

4.6.1 Introduction

The endocrine system is a unique system constituted by the comprising of several glands located in different parts of the body. They have a precise function in the growth and development of man. The endocrine glands secrete the chemical messengers called as hormones. Nervous system plays an important role in the production of the hormones by signaling the endocrine glands. The neuronal signaling is heirarchial. The signals from the neurons are passed on to hypothalamus which produces the releasing factors that act upon the anterior pituitary for the production of the trophic hormones

Table 4.8 *Hormones produced from anterior pituitary*

Trophic Hormone	Hormone from Anterior pituitary	Site of action
Growth hormone releasing hormone (Somatotrophin) (GHRH)	Growth hormone (GH)	Most of tissues and organs
Growth hormone releasing inhibitory hormone. (Somatostatin) (GHRIH)	Growth Hormone inhibitory (GHI) Thyroid stimulating inhibitory hormone (TSIH)	Thyroid gland Pancreatic islets
Thyroid releasing hormone. (TRH)	Thyroid stimulating hormone (TSH)	Thyroid gland
Corticotrophin releasing hormone. (CRH)	Adreno Cortico Trophic Hormone (ACTAH)	Adrenal Cortex
Prolactin releasing hormone (PRH)	Prolactin releasing hormone (PRH)	Breast
Prolactin inhibiting hormone (PIH)	Prolactin releasing inhibitory hormone. (PRIH)	Breast
Luteinising hormone releasing hormone (LHRH).	Follicle stimulating hormone (FSH)	Ovaries and Tests
Gonadotrophin releasing hormone (GnRH)	Luteinizing hormone (LH).	Ovaries and Tests

which in turn act on the specific endocrine glands which produces the hormone Table 4.8 and Fig. 4.27). Thus produced hormones are diffused in to the blood (fluid connective tissue) and are transported to their respective targets. The endocrine glands are also called ductless glands, which are associated with a complex network of the capillaries facilitating the diffusion of hormones directly into the blood.

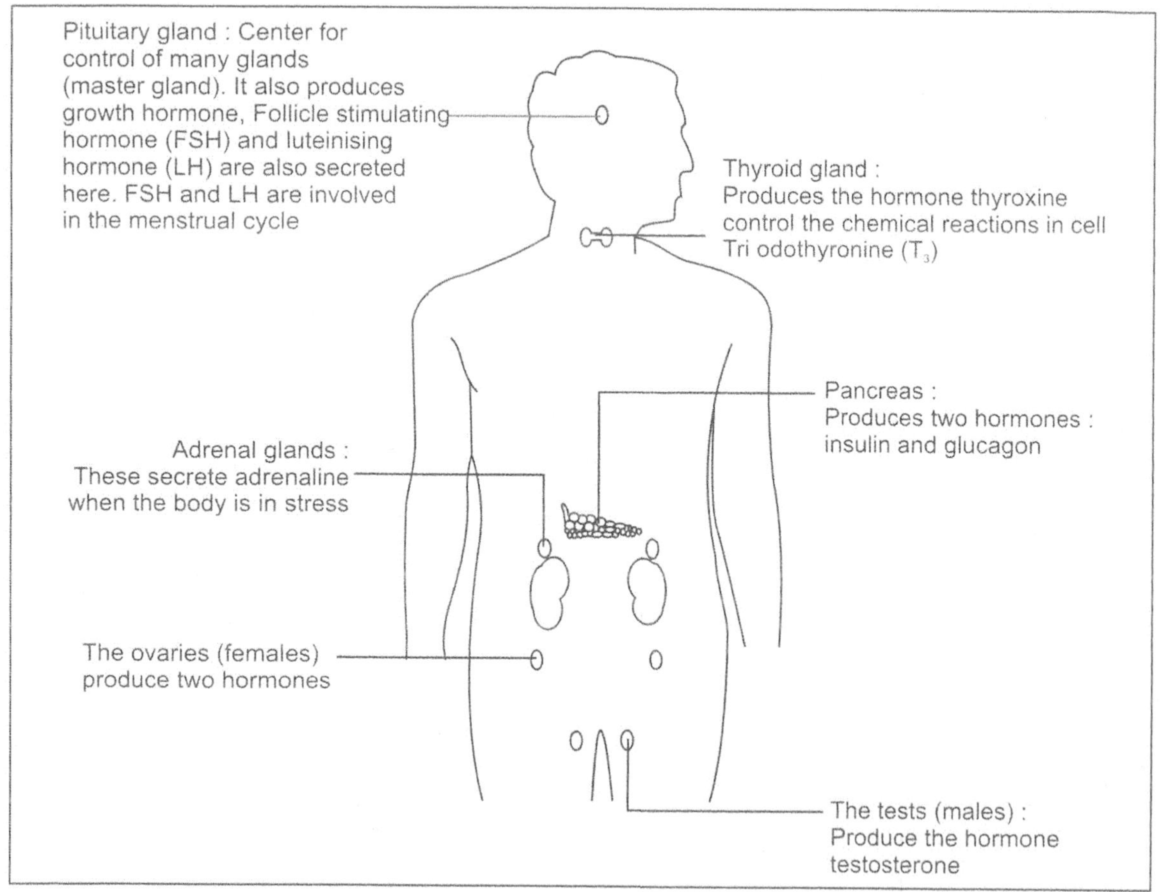

Fig. 4.27 Endocrine system : Different endocrine gland producing harmones.

4.6.2 The System

The endocrine system consists of the following glands:

- Pituitary gland (hypophysis).
- Thyroid gland.
- Parathyroid gland.
- Adrenal gland.
- Pancreatic islets (islets of Langerhans).

- Pineal body or Pineal gland.
- Thymus gland
- Ovaries in female.
- Testes in male.

4.6.3 Hypothalamus

Hypothalamus is the thickened floor of the third ventricle in the brain of the humans with nerve centers that controls the body temperature, besides playing a central role in the hormonal release from the endocrine glands. It stimulates the anterior pituitary to produce the releasing factors for the release of the trophic hormones from the anterior pituitary.

4.6.4 Pituitary Gland

Pituitary gland lies in the hypophyseal fossa of the sphenoid bone called **sella turcica** below the hypothalamus to which it is attached by a stalk called as **infundibulum.** It is about 1cm in diameter and weighs about 0.5 gm, the pituitary and hypothalamus acts constituently to regulate the activity of other endocrine glands. The pituitary gland has anterior lobe (*adenohypophysis*), posterior lobe (*neurohyphophysis*) called as anterior pituitary and posterior pituitary respectively. The anterior and the posterior lobes are separated by a thin strip of a tissue called as intermediate lobe (function unknown).

4.6.5 Growth Hormone (GH)

(GHRH) stimulates the release of (GH) and GH stimulates the growth and division in skeletal muscles, bones and in most of the body cells. It also regulates the metabolism in organs like liver, intestine etc., stimulates protein synthesis, promotes catabolism of fats, and also increases blood glucose levels.

4.6.6 Thyroid Stimulating Hormone (TSH)

TSH release is stimulated by TRH, in turn the growth and activity of thyroid gland, which produces thyroxine (T4) and triiodothyronine (T3). The regulation of this hormone is condition by the blood flow to the thyroid gland. In case the blood levels are too high to the thyroid gland, the TSH release is blocked and vice versa.

4.6.7 Adreno cortico trophic hormone (ACTH)

CRH stimulates the production of the ACTH, it increases the production of the steroids in adrenal cortex and thereby increases the production of the steroidal hormones like cortisol. Secretion of the hormone is controlled by the blood flow like that of the TSH.

4.6.8 Prolactin

Prolactin stimulates the breast for milk production (Lactation) after parturition (child birth). The release of prolactin is stimulated by the prolactin stimulating hormone.

4.6.9 Gonadotrophins

The gonadotrophins generally called as sex hormones which are secreted after puberty. GnRH or LHRH stimulates the production of the FSH and LH, in both the sexes, males and females. The LH and FSH are involved in the production of the *oestrogens* and *progesterone* during the menstrual cycles in females, and in males they stimulate the production of the testosterone.

4.6.10 Posterior Pituitary Hormones

The Oxytocin and Vasopressin are the hormones released from the posterior pituitary.

4.6.11 Oxytocin

It stimulates uterine muscles and lactating breast, during parturition which facilitates the uterine cervix expansion for the child birth. The suckling also generates the production of the Oxytocin.

4.6.12 Vasopressin

It is known as Antidiuritic hormone (ADH), which reduces the urine output. The ADH release is stimulated by the osmotic pressure on the osmoreceptors of the hypothalamus by the circulating blood.

4.6.13 Thyroid Gland

Thyroid gland located in the neck near trachea and larynx. It is an highly vascularised structure weighing around 25 to 30 gm with a fibrous capsule around it, resembling a butterfly. It consists of two lobes one on the either sides of the thyroid cartilage and upper cartilaginous rings of the trachea, these two lobes are joined by **isthmus.** The blood supply to the gland is by the superior and inferior thyroid arteries. It produces **Thyroxine (T4) and Tri-iodothyronine (T3).** Thyroid gland helps in iodine trapping from the different sources of iodinated foods. The thyroid gland selectively takes up the iodine. The release of T3 and T4 are regulated by the thyroid stimulating hormone (TSH) which is prior stimulated by the thyroid releasing hormone (TRH), the release of TRH is stimulated by the stress, malnutrition, low plasma glucose and sleep. Under the conditions of low iodine intake there is an enlargement of thyroid gland due to the continuous proliferation of the thyroid gland cells. T3, T4. regulate the metabolisms of the proteins, carbohydrates and fats, they also enhance the basal metabolic rate. These hormones effect the adrenaline and noradrenaline levels by exerting the action on their target cells. **Calcitonin** is secreted by the parafollicular cells of the thyroid glands, exerts its action on bone and the kidney to decrease the calcium levels by inhibiting the reabsorption of calcium from bones and also reabsorption by the renal tubules. It plays a central role in the child growth when they undergo a distinct change in size and shape.

4.6.14 Parathyroid gland

There exist a four small parathyroid glands, located in the posterior surface of the each lobe of the thyroid gland. They secrete parathyroid hormone which increase the calcium levels in blood by indirectly increasing the absorption of calcium from the intestine and by increasing the reabsorption from the by the renal tubules. The action of parathyroid hormone and the T3, T4 are exactly opposite and plays a major role in muscle contraction, blood clotting and nerve impulse production indirectly via calcium.

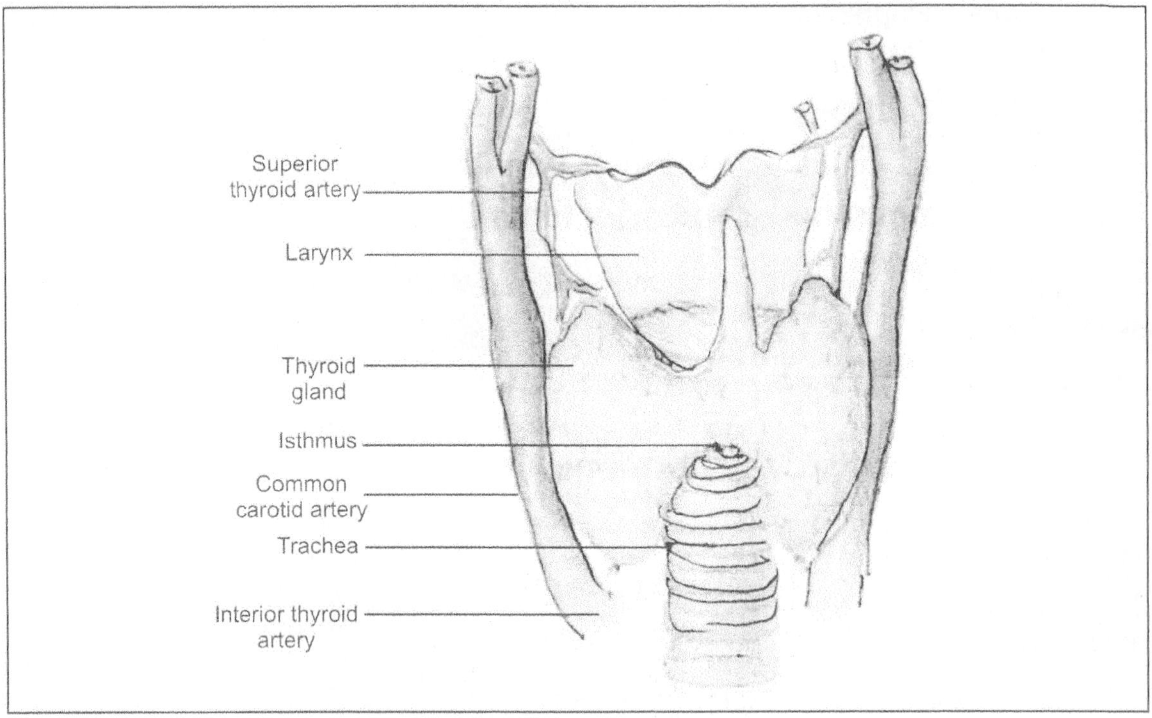

Superior thyroid artery
Larynx
Thyroid gland
Isthmus
Common carotid artery
Trachea
Interior thyroid artery

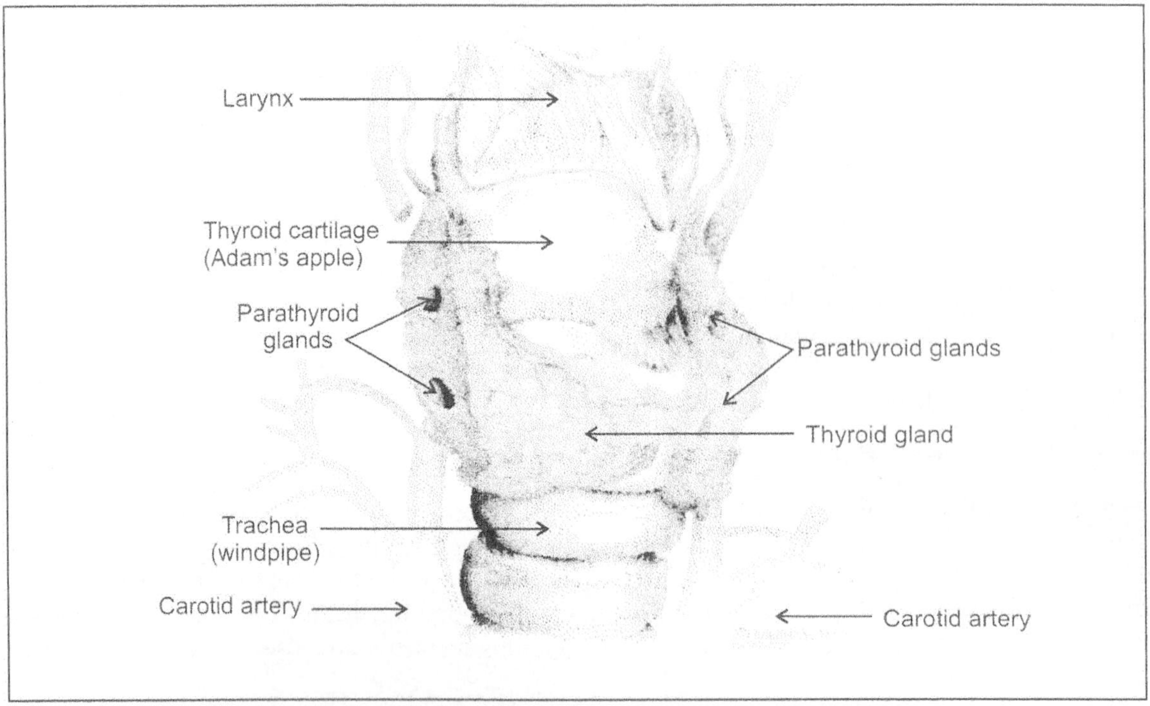

Larynx
Thyroid cartilage (Adam's apple)
Parathyroid glands
Parathyroid glands
Thyroid gland
Trachea (windpipe)
Carotid artery
Carotid artery

4.6.15 Adrenal glands

These are located on the upper side of the kidneys enclosed within the renal fascia and are about 4-5 cm long and 3-3.5 cm thick. The blood supply to the gland is by the abdominal aorta and renal arteries. It has two parts namely adrenal medulla and adrenal cortex. They are the inner and outer parts of the adrenal gland respectively.

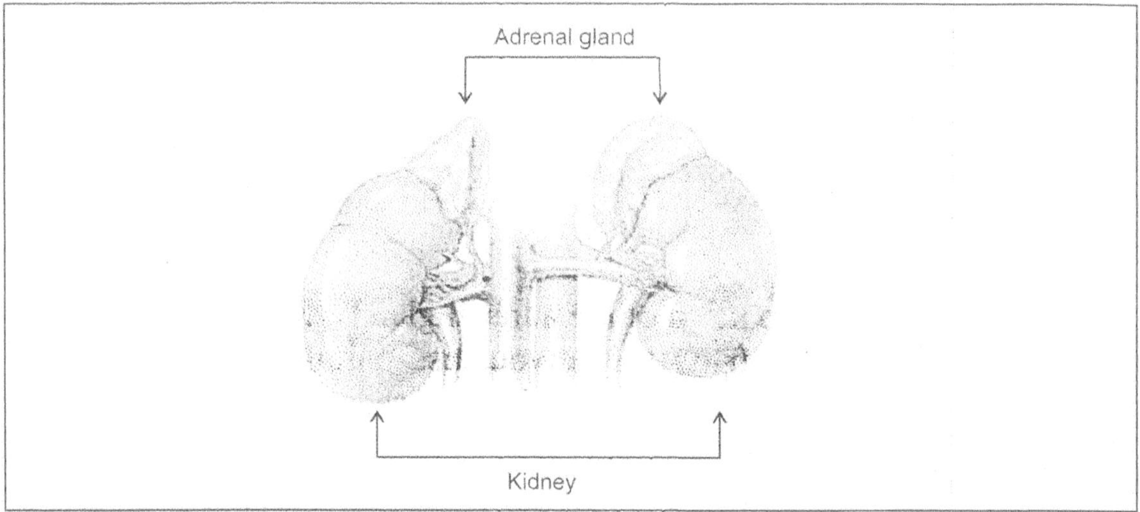

4.6.16 Adrenal cortex

Produces steroidal hormones called as adrenocorticoids, glucocorticoids, mineralocorticoids, sex hormones (androgens).

- **Glucocorticoids** are cortisol, corticosterone and cortisone, which plays a vital role in the regulation of the metabolism and the stress responses. The secretion of these glucocorticoids is stimulated by ACTH and stress. They have a wide range of effects like :
  - Gluconeogenesis (formation of glucose from non-carbohydrates) and hyperglycemia (raised blood glucose levels).
  - Lipolysis (break down of triglycerides).
  - Stimulation of the catabolism (break down) of proteins.
  - Stimulates the reabsorption of sodium and water by the renal tubules to an extent.

4.6.17 Mineralocorticoids

Aldosterone is the most important mineralocorticoid, which stimulates the reabsorption of the sodium ions from the renal tubules and potassium in the urine. It regulates the blood volume and blood pressure. The potassium levels in blood regulates the production of aldosterone. High level of the potassium, rises the aldosterone levels and vice versa. When the blood sodium levels fall down, the kidneys secretes rennin, an enzyme which converts the plasma protein to aldosterone, a pro-hormone, which in turn is converted by the liver and **angiotensins converting enzyme** to angiotensin1 and angiotensin2 respectively, which cause the vasoconstriction and increase the blood pressure.

4.6.18 Sex hormones

The sex hormones produced by the adrenal gland are androgens. This male sex hormones is produced in low levels and their role is unknown. In females the elevated levels of this hormone causes *masculanisation*.

4.6.19 Adrenal Medulla

The medullar is the inner portion of the adrenal gland which is completely surrounded by the cortex. It is developed from the nervous tissue in the embryo and is part of sympathetic division of the autonomic nervous system. It produces *adrenaline* and nonadrenaline. The adrenaline and noradrenaline are produced by stimulation of adrenal medulla during stimulation of the sympathetic nervous system. They are structurally similar and have a similar effects on the body, they are also called as the fight and flight hormones due to their effects under the stressed conditions and they proliferate the heart rate, increasing blood pressure, divertion of the blood supply to organs, increase in basal metabolic rate (BMR) and dilate the pupils.

Adrenal has a major effect on the heart and metabolic processes whereas the noradrenaline has influence on the blood vessels.

4.6.20 Pancreatic Islets

The endocrine part of a hexocrine gland (PANCREAS) have the *islets of Langerhans* which are called pancreatic islets and found in irregularly distributed clusters, produce pancreatic juice, secreted directly into the blood.

There are three types of cells in the pancreatic islets called as the *alpha cells, beta cells, delta cells*, which secrete the glucagons, insulin and somatostatin respectively.

4.6.21 Insulin

The function of insulin is to lower the blood glucose levels to maintain the homeostatic condition of the blood glucose. Insulin exerts its action by stimulating the uptake and use of glucose by the muscle and connective tissue cells, converts the glucose to glycogen in liver and skeletal muscle, accelerates the uptake of amino acids by cells and increases biosynthesis of proteins, promotes the fatty acid storage in adipose tissue (fat storing tissue), decrease glycogenolysis (break down of glycogen to glucose) and prevents gluconeogenisis. The secretion of insulin is stimulated by the increase blood glucose levels and amino acid levels and gastointesinal hormones like gastrin and cholecystokinin. And it is decreased by sympathetic stimulation, glucagons, adrenaline, cortisol and somatostatin.

4.6.22 Glucagon

The glucagons increase blood glucose levels by stimulating the conversion of glycogen to glucose in liver and skeletal muscle and by gluconeogenesis. The secretion of glucagon is stimulated by the low blood glucose levels, stress and decreased somatostatin levels and insulin.

4.6.23 Somatostatin

It is produced by the hypothalamus, which inhibits the secretion of both insulin and glucagons.

4.6.24 Pineal gland

It is small gland attached to the roof of the third ventricle and is connected to it by the small stalk and is of 10-12 mm long, which is reddish brown color and is also surrounded by a capsule.

It produces melatonin which is simulated by the amount of light entering the eye, stimulating the optic pathways, the level of the melatonin fluctuate between the day and night, the function of melatonin are not clearly understood but is known to co-ordinate the circadian and diurnal rhythms of many tissues, by influencing the hypothalamus and inhibition of growth and development of the sex organs before puberty, by preventing the synthesis of release of gonandotrophins.

4.6.25 Thymus gland

Thymosine is produced from the thymus gland and is required in the development of the T-lymphocytes for cell medicate immunity.

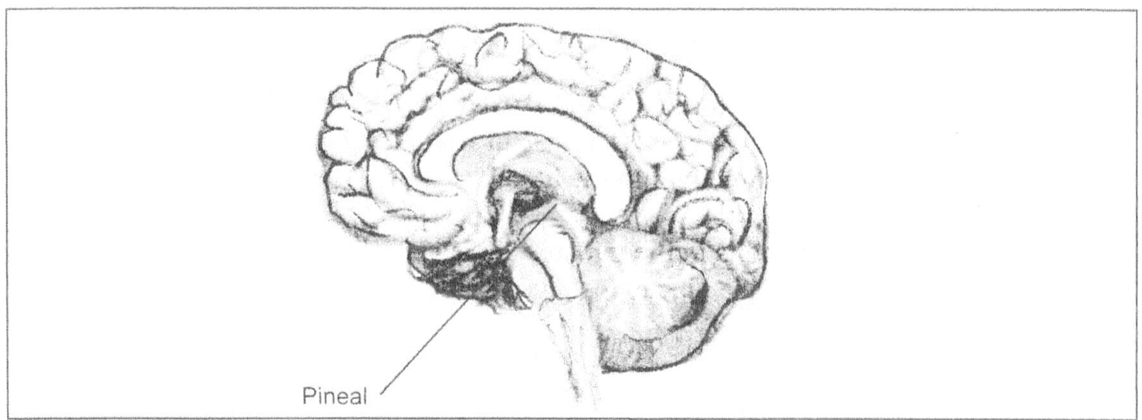

4.7 The Nervous System and its Physiology

4.7.1 Introduction

The preeminent factor in human life is the capacity of obtaining and coordinating information to learn about the environment and act effectively. The types of information are many and diverse. These include, auditory-communication and detection of sound, tactile-contact with objects, olfactory and gustatory-identification and choosing mate, food and environment, determination of mood, visual-information around us, position of body parts, status of internal organs etc. All these sources of information must be assessed, processed and converted into meaningful acts. In the course of evolution, human being have developed sensory organs capable of collecting all the information. The necessary neural processes to conduct the coded information to central neural process to decode and integrate the incoming information and distribute instructions to muscle, glands and other effector systems have been attained, besides a set of complex sensory receptors dealing with finer types of sensory modalities like sense of colour, odour, pressure and movement and body position and orientation surveillance of internal milieu etc. The specialized sense organs assumed even greater prominence and the central processing became larger and more complex to processes the available information more rapidly and accurately. Behavior was made incalculably complicated by interaction of several sensory systems. The *nervous system* is involved in all levels of human behavior, from the unconscious regulation of intracellular activity to coordinated of movement of whole organism, and to the elusive phenomena of consciousness, learning and memory.

4.7.2 Structural Elements – Neural Cells, Axons, Neuroglia

The basic structural element which carries the function of the nervous systems include ***neural cells*** supported by ***glial cells*** and ***sensory receptors***.

Structure and Functions of the Nervous System

The structures that make up the nervous system include the brain, cranial nerves and their branches, the spinal cord, spinal nerves and their branches, ganglia, enteric plexuses, and sensory receptors (Fig. 4.28).

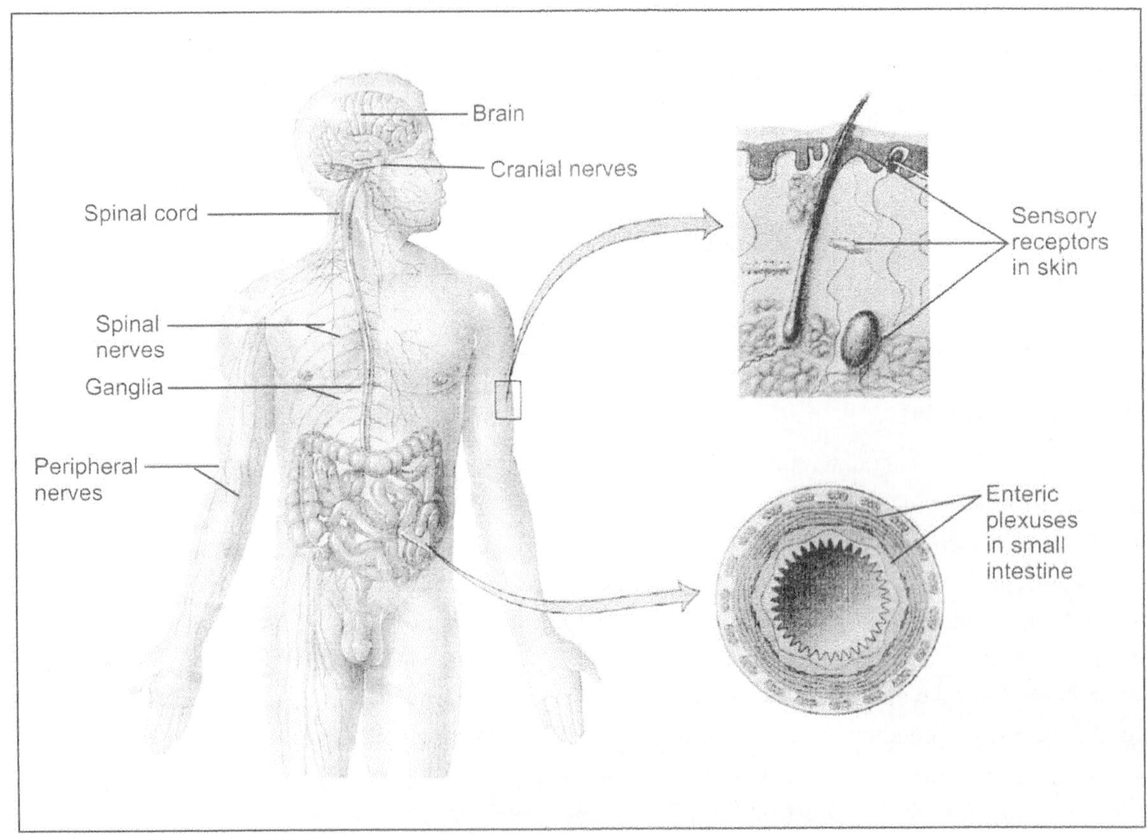

Fig. 4.28 Some of the components of nervous system.

The brain is housed within the skull and contains about 100 billion neurons. Twelve pairs (right and left) of cranial nerves, numbered I through XII, emerge from the base of the brain. **A nerve** is a bundle containing hundreds to thousands of axons plus associated connective tissue and blood vessels. Each nerve follows a defined path and serves a specific region of the body. For example, the right cranial nerve I carries signals for the sense of smell from the right side of the nose to the brain.

The spinal cord connects to the brain through the *foramen magnum* of the skull and is encircled by the bones of the vertebral column. It contains about 100 million neurons. Thirty-one pairs of spinal nerves emerge from the spinal cord, each serving a specific region on the right or left side of the body. **Ganglia** are small masses of nervous tissue, containing primary cell bodies of neurons that are located outside the brain and spinal cord. Ganglia are closely associated with cranial and spinal nerves. In the walls of organs of the gastrointestinal tract are extensive networks of neurons, called enteric plexuses that help regulate the digestive system. Sensory receptors are either the dendrites or sensory neurons. They are specialized cells that will monitor changes in the internal or external environment.

4.7.3 The nervous system has three basic functions

Sensory function

Sensory receptors detect internal stimuli, such as an increase in blood acidity, and external stimuli, such as touch. The neurons that carry sensory information into the brain and spinal cord are **sensory** or **afferent neurons.**

Integrative function

The nervous system integrates (processes) sensory information by analyzing and storing some of it and by making decisions regarding appropriate responses. The neurons that serve this function are inter neurons; these make up the vast majority of neurons in the body.

Motor function

The nervous system's motor function involves responding to integration of decisions. The neurons that serve this function are *motor* or *efferent neurons*, which carry information out of the brain and spinal cord. The cells and organs innervated by motor neurons are termed effectors; examples are muscle fibers and glandular cells.

4.7.4 Organization of the Nervous System

The nervous system is composed of two sub systems

1. The *central nervous system* (CNS) and
2. The *peripheral nervous system* (PNS)

The CNS consists of the brain and spinal cord, which integrate and correlate many different kinds of incoming sensory information. The CNS is also the source of thoughts, emotions, and memories. Most nerve impulses that stimulate muscles to contract and glands to secrete, originate in the CNS. The PNS includes all nervous tissue outside the CNS: cranial nerves and their branches, spinal nerves and their branches, ganglia, and sensory receptors.

The PNS may be subdivided further into a *somatic nervous system* (SNS), autonomic nervous system (ANS), and an enteric nervous system (ENS) (Fig. 4.29). The SNS consists of (1) sensory neurons that convey information from somatic and special sensory receptors primarily in the head, body wall, and limbs to the CNS, and (2) motor neurons from the CNS that conduct impulses to skeletal muscles only. Because these motor responses can be consciously controlled, the action of this part of the PNS is voluntary.

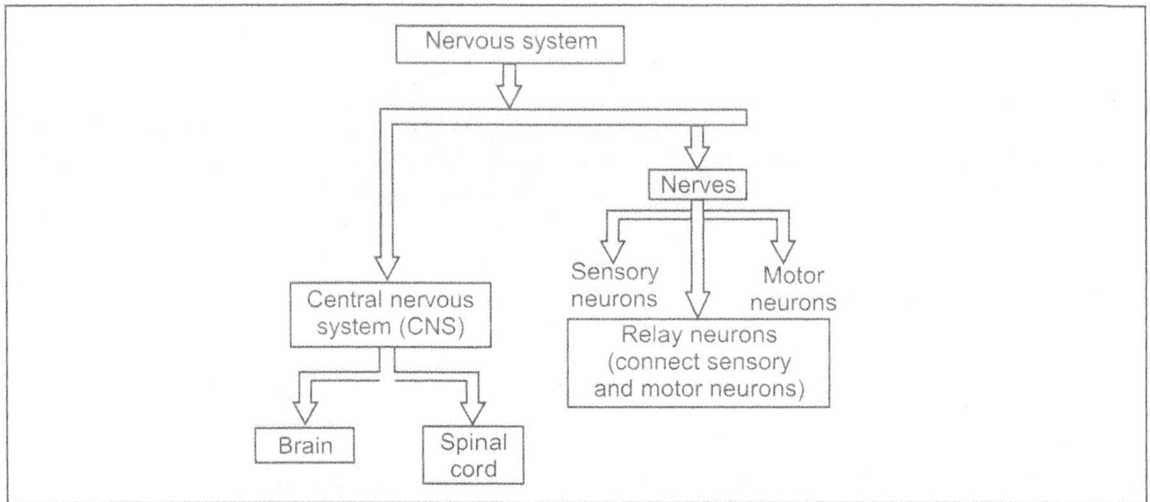

Fig. 4.29 General account of organisation of the nervous system.

The ANS consists of (1) sensory neurons that convey information from autonomic sensory receptors, located primarily in the viscera, to the CNS, and (2) motor neurons from the CNS that conduct nerve impulses to smooth muscle, cardiac muscle, glands, and adipose tissue. As its motor responses are not normally under conscious control, the action of the ANS is involuntary. The motor portion of the ANS consists of two branches, the sympathetic division and the parasympathetic division. With a few exceptions, effectors are innervated by both, and usually the two divisions have opposing actions. For example, sympathetic neurons increase heart rate, whereas parasympathetic neurons decrease it.

The ENS is the brain of the gut and its operation is involuntary. Once considered to be part of the ANS, the ENS consists of approximately 100 million neurons in enteric plexuses that extend the entire length of the gastrointestinal (GI) tract. Many of the neurons of the enteric plexuses function independently of the ANS and CNS to some extent, although they also communicate with the CNA via sympathetic and parasympathetic neurons. Sensory neurons of the ENS monitor chemical changes within the GI tract and the stretching of its walls. Enteric motor neurons govern contraction of GI smooth muscle, secretions of the GI tract, such as acid secretion by the stomach, and activity endocrine cells of GI tract (Fig. 4.30).

4.7.5 Nervous tissue consists of two principal kinds of cells

Neurons and *neuroglia*. Neurons are responsible for most special functions attributed to the nervous system: sensing, thinking, remembering, controlling muscle activity, and regulating glandular secretions. Neuroglia support, nourish, and protect the neurons and maintain homeostasis in the interstitial fluid that bathes neurons.

- ***Neurons***

  Like muscle cells, neurons have the property of electrical excitability, the capability to produce action potentials or impulses in response to stimuli. Once they arise, action potentials propagate from one point to the next along the plasma membrane due to the presence of specific types of ion channels.

- ***Parts of a Neuron***

Most neurons have three parts: (1) a cell body, (2) dendrites, and (3) an axon (Fig.4.31). The cell body contains a nucleus surrounded by cytoplasm that includes typical organelles such as lysosomes, mitochondria, and a Golgi complex. Many neurons also contain lipofuscin, a pigment that occurs as clumps of yellowish brown granules in the cytoplasm. Lipofuscin is probably a product of neuronal lysosomes that accumulates as the neuron ages, but it does not seem to harm the neuron. The cell body also contains prominent clusters of rough endoplasmic reticulum, termed **Nissal bodies**. Newly synthesized proteins produced by Nissl bodies are used to replace cellular components, as material for growth of neurons, and to regenerate damaged axons in the PNS. The cytoskeleton includes both neurofibrils, composed of bundles of intermediate filaments that provide the cell shape and support, and microtubules, which assist in involving materials to and from the cell body and axon.

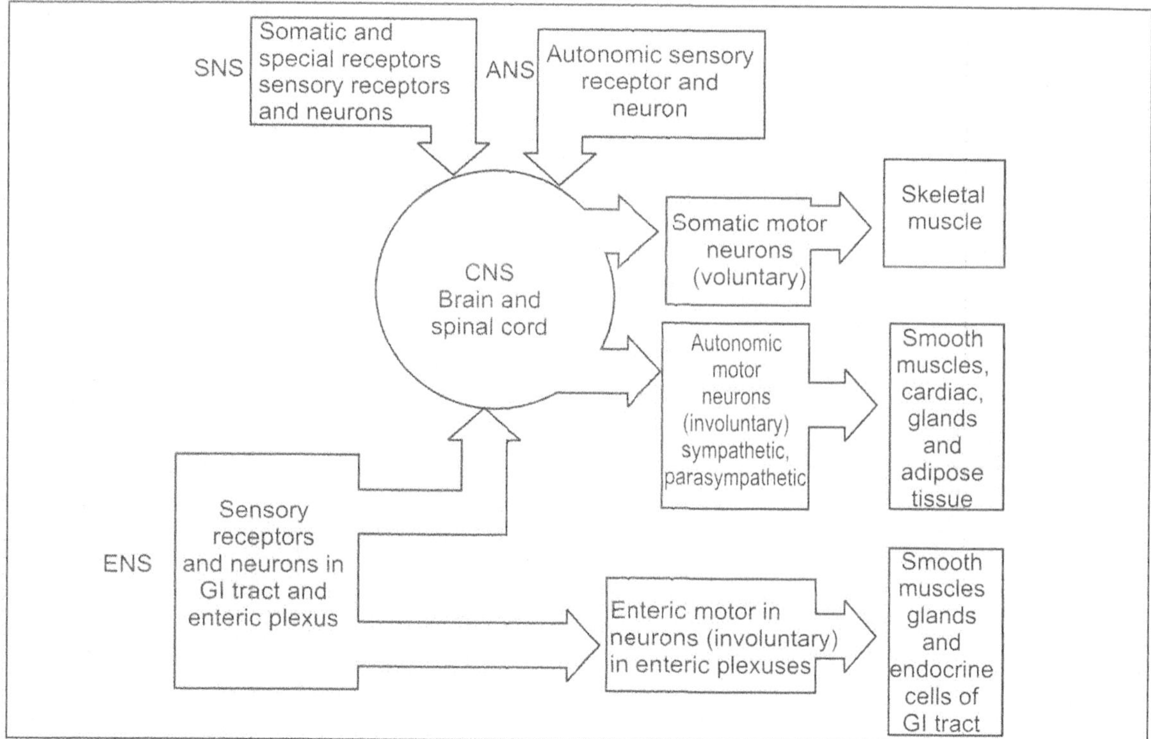

Fig. 4.30 Description of the network of nervous system.

Two kinds of processes (or extensions) emerge from the cell body of a neuron: multiple dendrites and a single axon. Nerve fiber is a general term for any neuronal process (dendrite or axon). **Dendrites** are the receiving or input portions of a neuron. They usually are short, tapering, and highly branched. In many neurons the dendrites form a tree-shaped array of processes extending off the cell body. Dendrites usually are not myelinated. Their cytoplasm contains Nissl bodies, mitochondria, and other organelles.

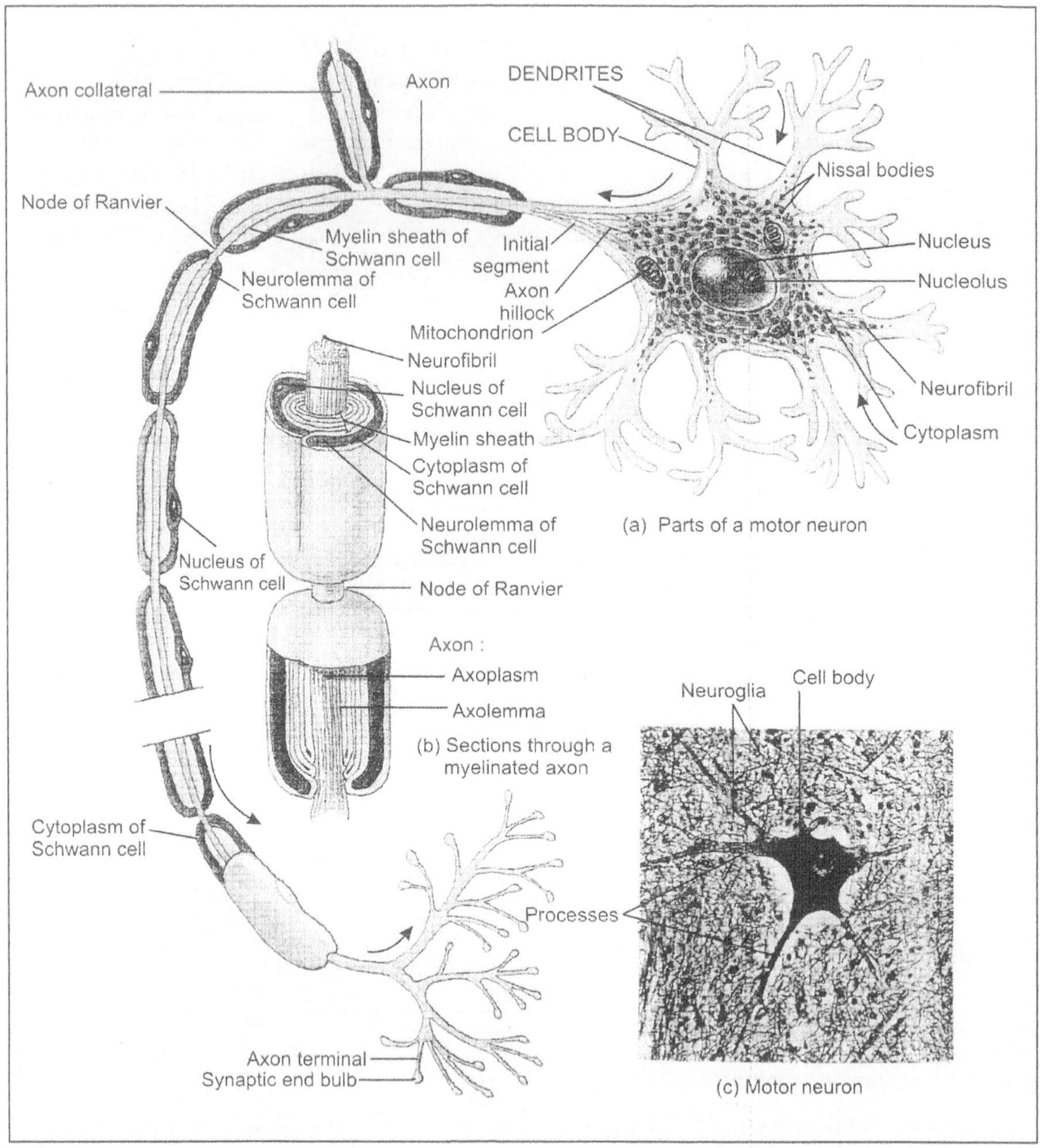

Fig. 4.31 Structure of typical neuron or nerve cell.

The second type of process, **the axon** propagates nerve impulses towards another neuron,or a muscle fiber of a gland cell. An axon is a long, thin, cylindrical projection that often joins the cell body at a cone-shaped elevation called the axon hillock. The first part of the axon is called the *initial segment*. In most neurons, impulses arise at the junction of the axon hillock and the initial segment, which is called the *trigger zone*, and then conduct along the axon. An axon contains mitochondria, microtubules, and neurofibrils. Because rough endoplasmic reticulum is not present, protein synthesis does not occur in the axon. The cytoplasm, called **axoplasm**, is surrounded by a plasma membrane known as the axolemma. Along the length of an axon, side branches called *axon collaterals* may branch off, typically at a right angle to the axon. The axon and its collaterals end by dividing into many fine processes called *axon terminals*.

The site of communication between two neurons or between a neuron and an effector cell is called a **synapse**. The tips of some axon terminals swell into bulb-shaped structures called *synaptic end bulbs*, whereas others exhibit a string of swollen bumps called *varicosities*. Both synaptic end bulbs and varicosities contain many minute membrane enclosed sacs called synaptic vesicles that store a chemical **neurotransmitter**. The neurotransmitter molecules released from synaptic vesicles then influence the activity of other neurons, muscle fibers, or gland cells.

The cell body is where a neuron synthesizes new cell products or recycles old ones. However, because some substances are needed in the axon or at the axon terminals, two types of transport systems carry materials from the cell body to the axon terminals and back. The slower system, which moves materials about 1-5 mm per day, is called slow axonal transport. It conveys axoplasm in one direction only – from the cell body toward the axon terminals. It supplies new axoplasm for developing or regenerating axons and replenishes axoplasm in growing and mature axons. The faster system, which is capable of moving materials a distance of 200-400 mm per day, is called fast axonal transport. It uses proteins that function as "motors" to move materials in both directions – away from and toward the cell body – along the surfaces of microtubules. Fast axonal transport moves various organelles and materials that form the membranes of the axolemma, synaptic end bulbs, and synaptic vesicles. Some materials returned to the cell body are degraded or recycled, and others influence its growth. They can cause damage. For example, the toxin produced by *Clostridium tetani* bacteria is carried by fast axonal transport to the CNS. There it disrupts the actions of motor neurons, causing prolonged painful muscle spasms – a condition called tetanus. The delay between the release of the toxin and the first appearance of symptoms is due, in part, to the time required for transport of the toxin to the cell body. For this reason, a laceration or puncture injury to the head or neck is a more serious matter than a similar injury in the leg. The closer the site of injury is to the brain, the shorter the transit time. Thus, treatment must begin quickly to prevent symptoms of tetanus.

4.7.6 Structural Diversity in Neurons

Neurons display a great diversity in size and shape. For example, their cell bodies range in diameter from 5 micrometers (μm) up to 135 μm. The pattern of dendritic branching is varied and distinctive for neurons in different parts of the nervous system. A few small neurons lack an axon, and many others have very short axons, but the longest neurons have axons that extend for a meter (3.2 ft) or more.

Both Structural and functional features are used to classify the various neurons in the body. Structurally, neurons are classified according to the number of processes extending from the cell body (Fig. 4.32). **Multipolar neurons** usually have several dendrites and one axon (see also Fig. 4.33). Most neurons in the brain and spinal cord are of this type. **Bipolar neurons** have one main dendrite and one axon; they are found in the retina of the eye, in the inner ear, and in the olfactory area of the brain. **Unipolar neurons** are sensory neurons that originate in the embryo as bipolar neurons. The trigger zone for nerve impulses in a unipolar neuron is at the junction of the dendrites and axon.

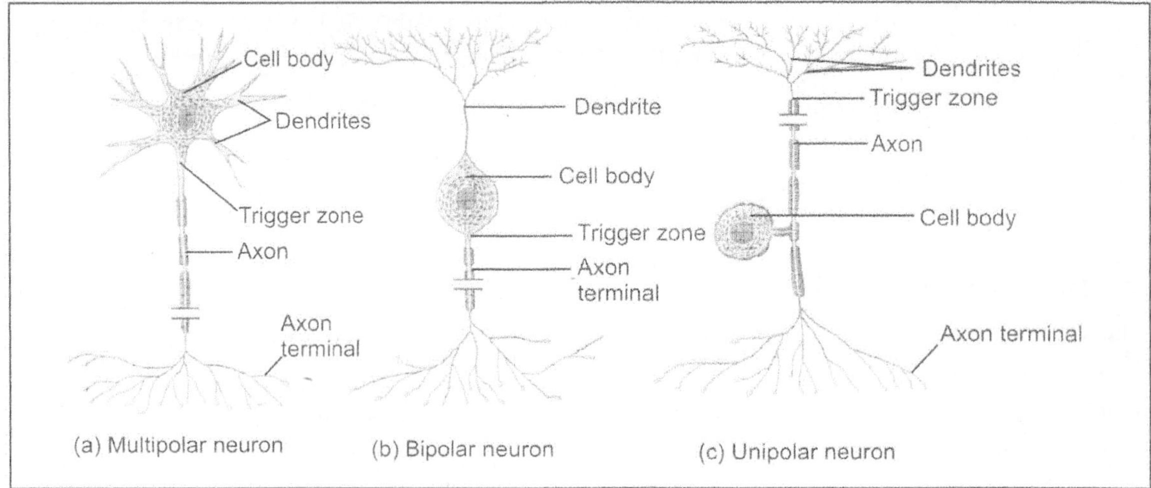

Fig. 4.32 Types of neurons

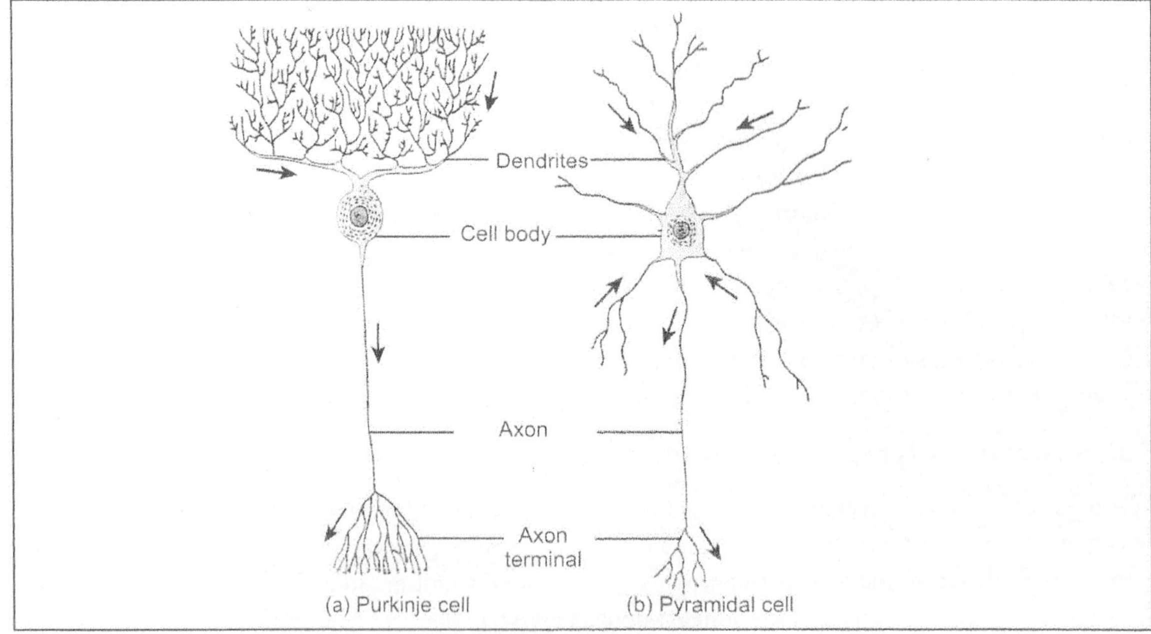

Fig. 4.33 Types of neurons.

During development, the axon and dendrite fuse into a single process that divides into two branches a short distance from the cell body. Both branches have the characteristic structure and function of an axon: They are long, cylindrical processes that may be myelinated, and they propagate action potentials. However, the axon branch that extends into the periphery has unmyelinated dendrites at its distal tip, whereas the axon branch that extends into the CNS ends in synaptic end bulbs. The dendrites monitor a sensory stimulus such as touch or stretching. The impulses then propagate toward the synaptic end bulbs.

Most of the neurons in the body – perhaps 90% – are **interneurons** of thousands of different types. Interneurons often are named for the histologist who first described them; examples are Purkinje cells in the cerebellum and Renshaw cells in the spinal cord. Others are named for some aspect of their shape or appearance. For example, pyramidal cells, found in the brain, have a cell body shaped like a pyramid.

4.7.7 Neuroglia

Neuroglia or glia constitute about half the volume of the CNS. Neuroglia are not merely passive bystanders but rather active participants in the operation of nervous tissue. Generally, neuroglia is smaller than neurons, and they are 5-50 times more numerous. In contrast to neuron, glia do not generate or propagate action potentials, and they can multiply and divide in the mature nervous system. In cases of injury or disease, neuroglia multiplies to fill in the spaces formerly occupied by neurons. Of the six types of neuroglia, four – astrocytes, oligodendrocytes, microglia, and ependyma cells – are found only in the CNS. The remaining two types – Schwann cells and satellite cells – are present in the PNS.

4.7.8 Myelination

The axons of most mammalian neurons are surrounded by a multilayered lipid and protein covering called the **myelin sheath**, it is produced by neuroglia. The sheath electrically insulates the axon of a neuron and increases the speed of nerve impulse conduction. Axons with such a covering are said to be myelinated, whereas those without it are unmyelinated (Fig. 4.34). Electron micrographs reveal that even unmyelinated axons are surrounded by a thin coat of neuroglial plasma membrane.

Two types of neuroglia produce myelin sheaths : Schwann cells (in the PNS) and oligodendrocytes (in the CNS). In the PNS, Schwann cells begin to form myelin sheaths around axons during fetal development. Each Schwann cell wraps about 1 millimeter (1 mm = 0.04 in.) of a single axon's length by spiraling many times around the axon. Eventually, multiple layers of glial plasma membrane surround the axon, with the Schwann cell's cytoplasm and nucleus forming the outermost layer. The inner portion, consisting of up to 100 layers of Schwann cell membrane, is the myelin sheath. The outer nucleated cytoplasmic layer of the Schwann cell, which encloses the myelin sheath, is the neurolemma (sheath of Schwann). A neurolemma is found only around axons in the PNS. When an axon is injured, the neurolemma aids regeneration by forming a regeneration tube that guides and stimulates regrowth of the axon. Gaps in the myelin sheath, called *nodes of Ranvier*, appear at intervals along the axon (see Fig. 4.34). Each Schwann cell wraps one axon segment between two nodes.

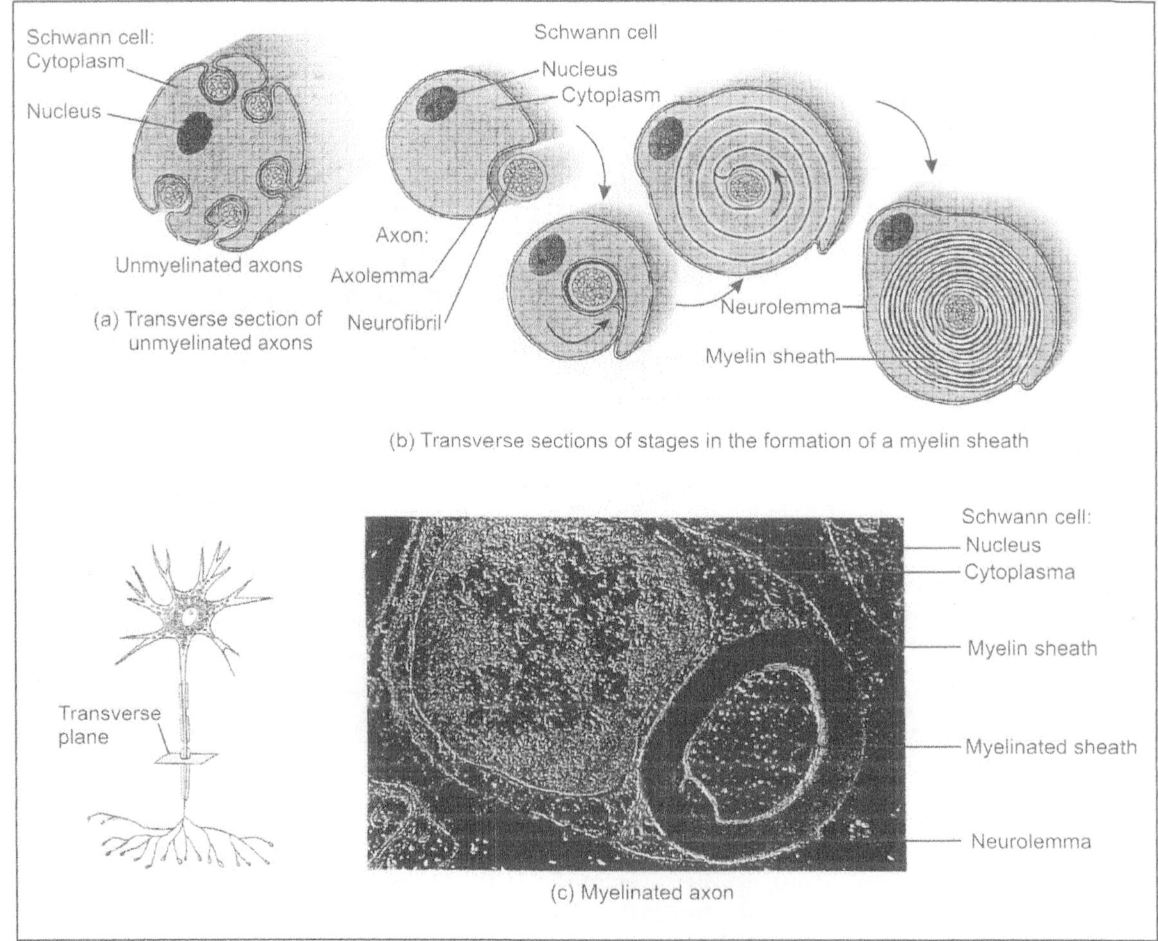

Fig. 4.34 Myelinated and unmyelinated axons and formation of myelin sheath.

In the CNS, an oligodendrocyte myelinates parts of many axons in somewhat the same manner that a Schwann cell myelinates part of a single PNS axon. It puts forth an average of 15 broad, flat processes that spiral around CNS axons, forming a myelin sheath. A neurolemma is not present, however, because the oligodendrocyte cell body and nucleus do not envelop the axon. Nodes of Ranvier are present, but they are fewer in number. Axons in the CNS display little regrowth after injury. This is thought to be due, in part, to the absence of a neurolemma, and in part to an inhibitory influence exerted by the oligodendrocytes on axon regrowth. The amount of myelin increases from birth to maturity, and its presence greatly increases the speed of nerve impulse conduction.

4.7.9 Gray and White Matter

In a freshly dissected section of the brain or spinal cord, some regions look white and glistening, whereas others appear gray (Fig. 4.35). The white matter is aggregations of myelinated processes from many neurons. The whitish color of myelin gives white matter its name. The gray matter of the nervous system contains neuronal cell bodies, dendrites, unmyelinated axons, axon terminals, and neuroglia. It looks grayish, rather than white, because there are little or no myelin in these areas. Blood vessels are present in both white and gray matter. In the spinal cord, the white matter surrounds an inner core of gray matter shaped like a butterfly or the letter H; in the brain, a thin shell of gray matter covers the surface of the largest portions of the brain, the cerebrum and cerebellum. (see Fig. 4.35). Many nuclei of gray matter also lie deep within the brain.

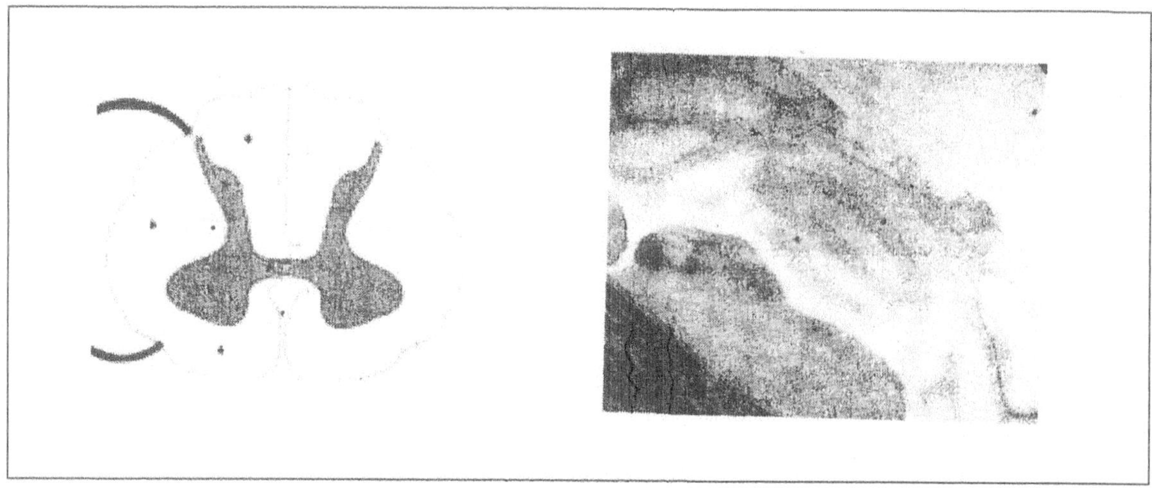

Fig. 4.35 Transverse section of spinal cord and human brain showing gray and white matter.

4.7.10 Nerve impulses : Basic properties of axons neurons

The nervous system carry nerve impulses very rapidly from one part of the body to another at speeds of 120 m s^{-1} in the form of electrical charge resulting from differences of potentials across the membrane of the nerve fibre. It is solely based on ion movement through a specialized protein pores and by an active transport mechanism.

4.7.11 Neuron at Rest

Axon membranes are partially permeable. The special conductance properties of neurons are due to the differences in permeability of the membrane of sodium and potassium ions, which sets neurons– from other cells. The axonic membrane is relatively impermeable to sodium ions but lets potassium ions to distinct permeate freely. Besides it also contains very active Na^+/K^+ pump, which uses ATP to move out sodium ions from the axon and lets potassium ions in. Consequence, is the reduction of Na^+ions concentration inside the axon, since they are pumped out and cannot diffuse back in the axon. Simultaneously the K^+ ions move inside, but can diffuse out again along the concentration gradient. Because of this there is a slight negative charge relatively inside the cell when compared to the outside. This is said to be polarised state (Fig. 4.36). The potential difference of from that occurs across the membrane is known as resting potential.

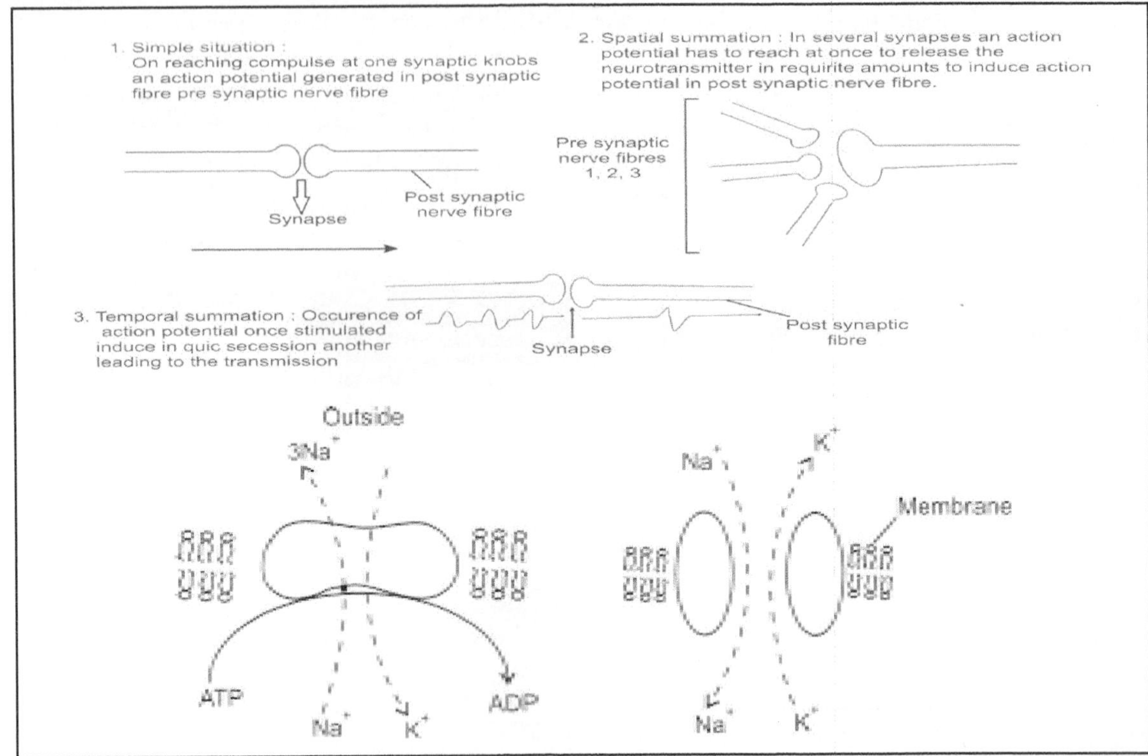

Fig. 4.36 The resting potential of the axon and the mechanism of Na^+/K^+ pump.

4.7.12 Neuron at Action

When an impulse travels the neurons get activated. The essential change is the altered permeability of the membrane to Na^+ions on the neuronic cell surface. This takes place due to stimulus of light, sound, touch, taste or smell. In the laboratory experiments, it could be well controlled electric impulse. When neurones spring into action, the membrane surface of neurone depict sudden and significant enhancement in its permeability to sodium ions by opening up sodium channel to gates. As a result Na^+ions rush all along and increase in concentration as well as electrochemical gradients consequently the potential differences across the membranes is briefly reversed making the exterior of the cell positive in relation to the interior. This is called depolarisation and lasts about 1 millsecond. This phenomenon is known as *action potential* and the potential difference across the cell membrance is about + 40 mV.

At the end of depolarisation the sodium gate closes again and the excess Na^+ions are quickly sent out by the sodium pump. Simultaneously the permeability of the membrane to K^+ ions is momentarily increased so that they diffuse out along an electrochemical gradient. It takes few millseconds before the resting potential is restored and once again the neurone is ready to carry out another impulse. This phase is called refractory period (Fig. 4.37) which ensures that the nerve impulse travel only in one-way direction since, the nerve fibre cannot conduct another impulse till the resting potential is restored, the nerve impulses continue to move in the same direction (Fig. 4.37).

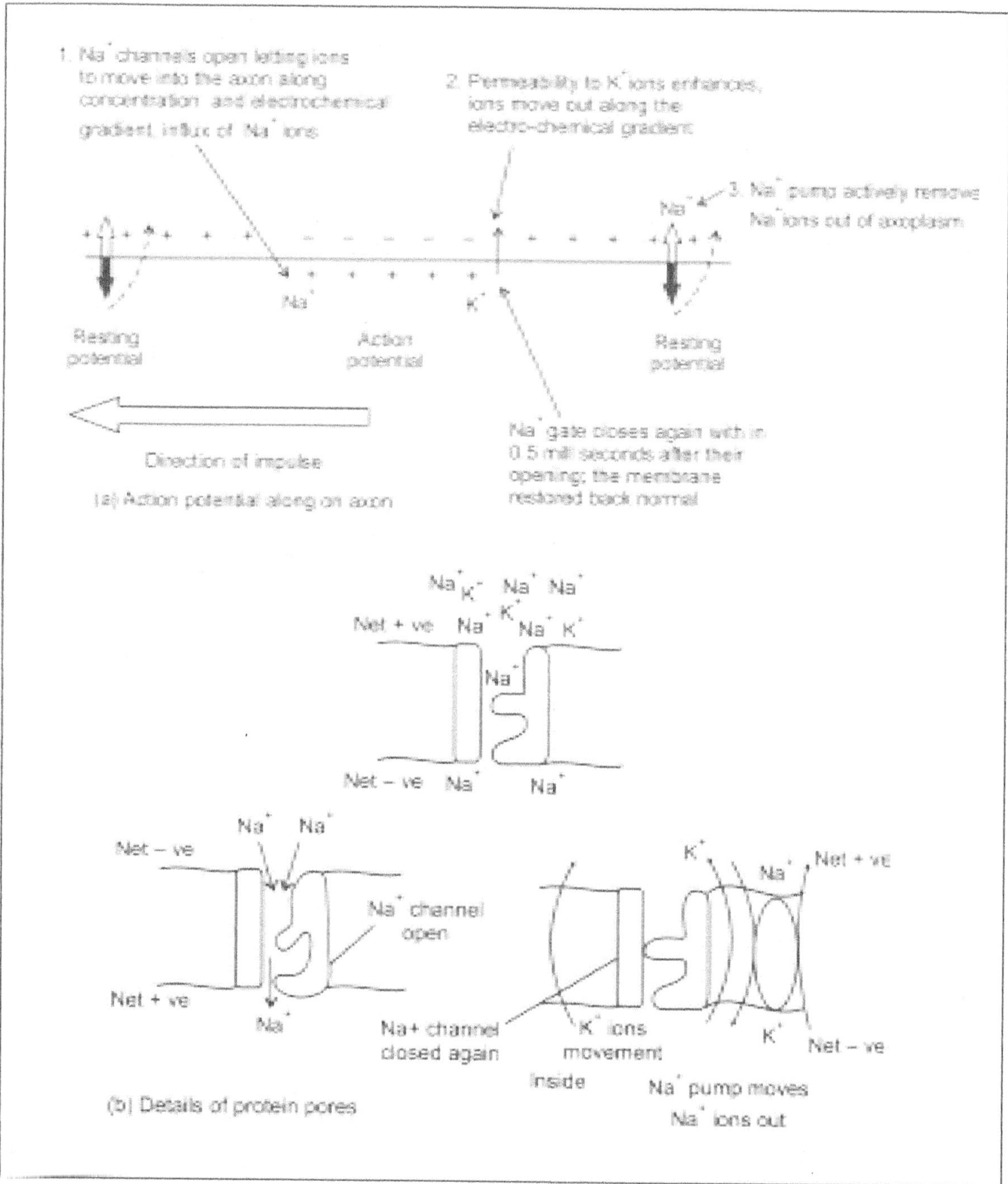

Fig. 4.37 The action potential.

Nerve impulses are investigated using a sensitive device called a cathode ray oscilloscope by placing a pair of electrodes on a nerve, one on the exterior another interior and passing a controlled stimulus. The electrode detect the impulse generated by the stimulant. When an action potential setup Na^+ channel or gate opens up permitting ions to move in. As a result the nerve fibre depolarised in the sense that exterior is more positive than the interior (around + 40 mV). The sodium gates are closed and ions sent out as usual. After a period the resting potential is restored. This period is called refractory priode, which could be absolute or relative. An action potential is on all-or-no response, it may or may not happen. Once the threshold is reached it gets activated. During the absolute refractory period the Na^+ gates are blocked and the resting period has not yet attained. At this stage neurone cannot be restimulated. At the time of relative refractory period, the stimulation is possible, but the threshold is enhanced. The refractory period limits the frequency at which impulses can travel along a neurone in one way only. This accelerates the conduction.

The action potential is propagated by local currents resulting from the movements of ions around the action potential. The opening of sodium channels is in front, but not behind due to refractory period. In the case of myelinated nerve fibres, the impulses jump from node to node freely in and out of the fibres. This type of conduction is called *saltatory conduction*.

4.7.13 Synapses link the nervous system and the body

The nerves are the fundamental units of nervous system. These are adapted for the rapid conductance of electrical impulses from one region to another. This is possible only when they inter communicate with each other. Receptors pass their information into the sensory nerves, which in turn relay the information to the CNS. The crux of the problem is to pass the information unimpeded around to central nervous system and the message has to pass on to the motor nerves which communicate to effector organs so that action takes place. This is accomplished by linking two nerves by the Synapses (Fig. 4.38)

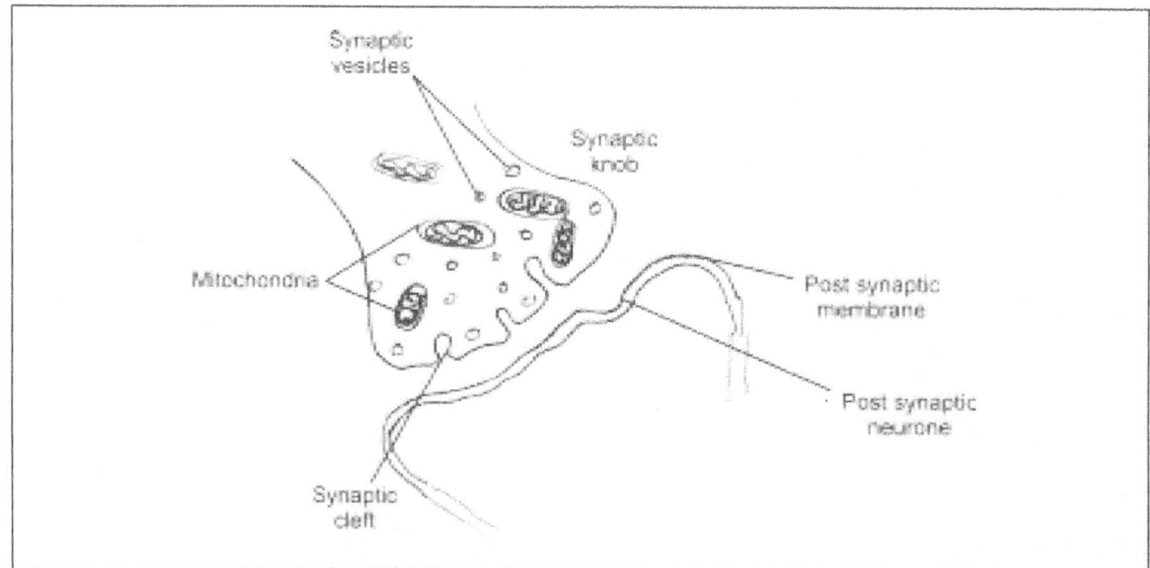

Fig. 4.38 The structure of a synapses

4.7.14 Synapses and their working

Each and every cell in the CNS is connected with synoptic knobs from other cells, some times the number goes up to several hundreds. The neurons never touch the target cell. Hence the synapse is a between two nerve cells across which the nerve message cross. (Fig. 4.38)

For a long time it was suspected that the transmission at the synapses was not electrical but chemical. The electron micrography enabled to visualize and measure the gap. The gap was simply too wide for action potential to jump across. It was concluded that the synaptic transmission has to be chemical and the subsequent experimental evidence confirmed the same.

The way synapse work

The impulse arrival at the synaptic knob increases the permeability of the presynaptic membrance to calcium ions. Along the concentration gradient that Ca^{+} ions move into the synaptic knob. This causes the vescicles containing synaptic transmitted substance to move to the presynaptic membrance.

Each synaptic vessels contain approximately about 3000 molecules of the transmitter substance. Some fuse with the membrane and may release the substance into the cleft of the synapse, which gets diffused across the gap and attaches to specific receptor protein on post synaptic membrance. Consequently, excitatory postsynaptic potential (EPSP) results due to opening up of ion gates and sudden influx of sodium ions, because of local repolarization. When sufficient action potential are present the positive charge is build up in the post synaptic cell and the action potential is set up which travels on the post synaptic neurone.

In some instances, the transmitter might have the opposite effect. The channels allow the inward influx of negative ions into the post synaptic membrane, thus making insides more negative than the normal resting potential, resulting in an *inhibitory post synaptic potential*. Such a situation makes it unlikely that an action potential will occur in post synaptic fibre.

Once the transmitter achieves its effect, it is chewed up by the enzymes. This is very important because unless the transmitter is replaced from the synaptic cleft, subsequent impulses cannot reach as all the receptors of post synaptic membrane are bound and occupied.

4.7.15 Nature of transmitter substances

Most commonly found transmitter substance at the majority of synapses is acetylcholine (ACh). It is synthesised in the synaptic knob using mitochondrial ATP. Nerves using ACh transmitter are called *Cholinergic nerves*. The enzyme cholinesterase hydrolysis rapidly once the acetylcholine does its job. It ensures that it no longer affects the post synaptic membrane. It also releases the components for recycling so that more acetylcholine is produced. Some parasympathetic nervous system produce noradrenaline in their synaptic vesicles and are called adrenergic nerve.

4.7.16 Neuromuscular junctions

Nerves have to communicate not only with each other but also with receptors and effectors motor nerves need to communicate with muscles. A special kind of synapse is present at the junction, where nerve and muscle fibre meet. Such ones are called *neuromuscular junction* (Fig. 4.39).

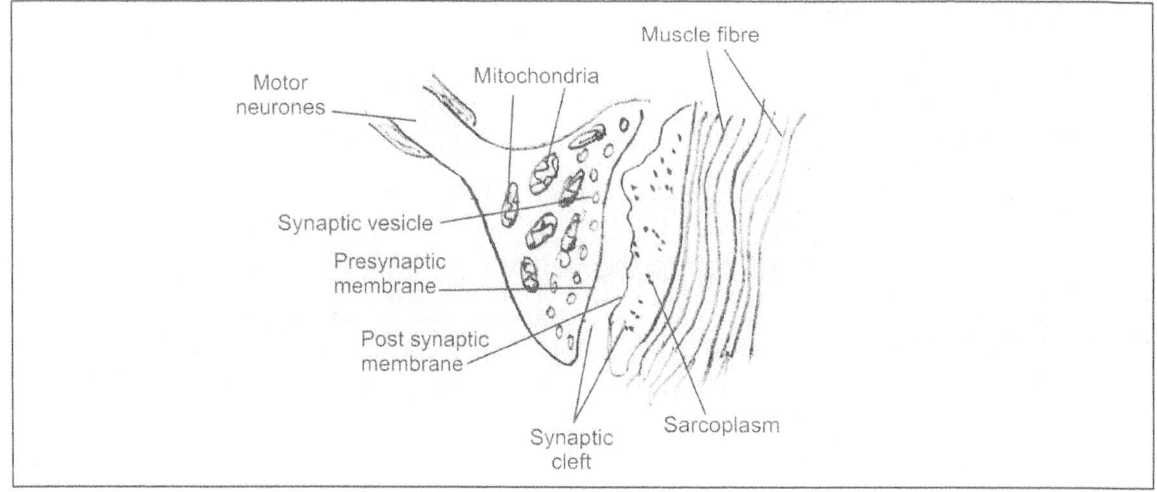

Fig. 4.39 The neuromuscular junction leading muscular contraction

The membrane of the muscle fibre - the sarcolemma is very folded and forms a structure known as *end plate*, to which end of the motor nerve is connected. Electron microscopic observation indicates the presence of innumerable mitochondria and synaptic vesicles containing acetylcholine. When an impulse reaches at the end of the motor neuron, acetylcholine is discharged into the synaptic cleft. Consequently on the post synaptic membrane an *end plate potential* is created which can be recorded.

In case sufficient end plate potential is generated the action potential is fired off in the muscle fibre, spreading through the T-tubules leading to a contraction of the muscles.

4.7.17 Coordinated Control of Neurons

The interaction of neurons is diverse and complex. Some times a single nerve fibre may carry an action potential to a synapse with another cell and the transmission across may set up another action potential. However, in many cases the position is not simple. Single synaptic knob may not release enough transmitter to induce action potential in the post synaptic fibre. A *spatial summation* may occur, several synaptic knobs release neurotransmitter on the same post synaptic membrane and their effects may add up to produce a post synaptic action potential (Fig. 4.40). When a single knob failed to release enough neurotransmitter substance to induce stimulation of the post synaptic nerve fibre, a second impulse it received from the same knob in quick succession an action potential may be caused. This phonemenon involving the same synaptic knob producing bursts of neurotransmitters in rapid succession is called *Temporal summation*. The first explosion makes easier the passage of the next impulse across the celft (Fig. 4.40). In other words, it lets in facilitation. On the other hand, when a nerve is repeatedly stimulated and looses its ability to respond, the process is called accommodation. This is due to depletion of neurotransmitter, and resynthesis might takes time.

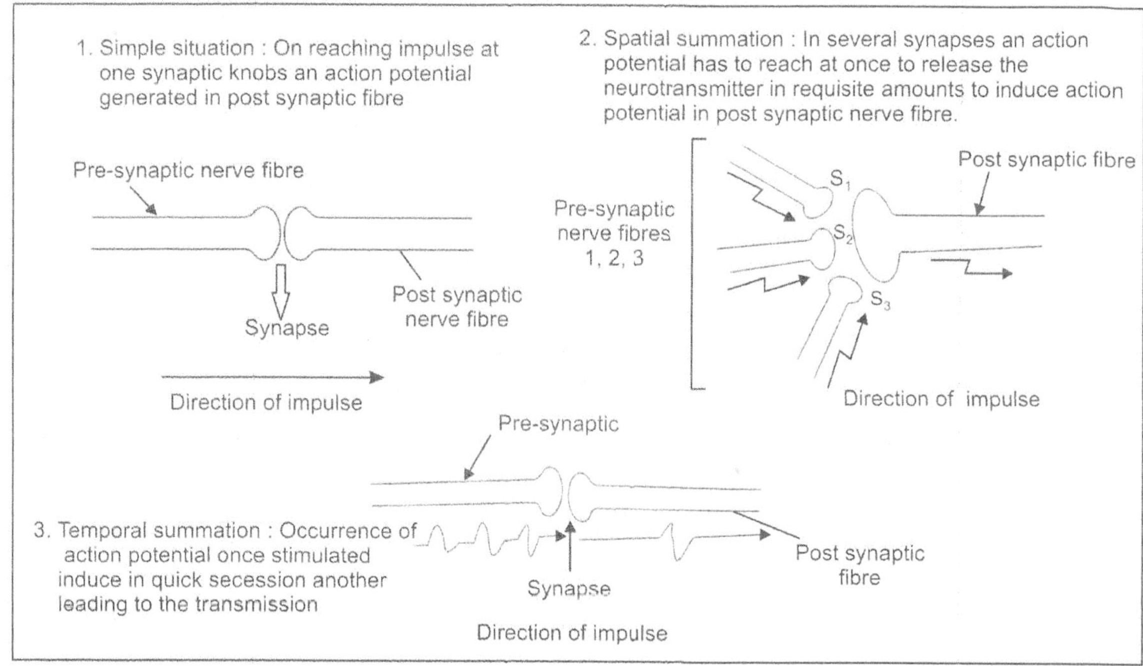

Fig. 4.40 Various kinds of neuro synaptic transmission

4.8 Sensory Receptors and Systems

4.8.1 Introduction

In order to survive and be aware of environmental changes as well as to perceive dangers within and outside, all biological organisms (pro-and-eukaryotes), have developed sensory systems and receptors. The pressure of natural selection is responsible for the evolution of a variety of sensory systems and receptors. Many kinds of specialized organs have evolved in diverse groups of living species to cater to the needs in the organisms in commensurate with the level of complexity in relation to its mode of living. Neurophysiology deals with the study of sensory systems and types of receptors. Two categories of sensory receptors were proposed: Exteroceptors, dealing with external and Interoceptors, concerned with internal stimuli. A classification of receptors into primary and secondary types has also been made. The categorization is mainly based on the kinds of sensory cells, which respond to stimuli. A Stimulus is one that produces a reaction in a living organism.

4.8.2 Features of Receptors

Structures which detect stimuli are called **receptors**. The classification into primary receptors or secondary receptors is essentially based on energy form of the stimulus i.e., chemical, mechanical or electromagnetic and the nature of response generated. Some of the generalization about the sensory receptors are:

- They tend to be concentrated on the body surface of bilaterally symmetrical animals, mostly towards the anterior end of the organisms.

- All including human receptors produce potentials in response to an appropriate stimulus.
- The receptors potentials differ significantly in their ionic mechanisms.
- Animals that are highly evolved for a restricted environment often show extreme development of one type of receptor at the cost of others. For instance, echo locating bats have large ears and hypertrophied auditory nervous system, but poor vision.
- The intensity of a receptor is more likely to be associated with importance of the organ to the animal, than with the animals' hierarchy phylogenetically.
- There are many cases of convergent evolution of receptor organs, where the nerve *impulses processing center remain quite different.*
- Very dissimilar sense organs apparently may provide essentially identical information.
- Highly evolved animals excel more in versatility of central nervous system development than in capabilities of receptors.

The special property of sensory cells and sense organs is that their ability to convert one form of energy to another form. The structure, which do this are called transducers. For instance the eyes can convert light energy (photons) into the electrical energy; in the case of ear vibrations of sound into nerve impulses. Thus sensory receptors can be considered as specialized structures that get stimulated by the environmental changes as well as by changes in the body. Sensory receptor can be defined as the terminal afferent endling that undergo depolarization in response to specific type of physical stimuli and the receptors are capable of transforming different types of sensory nerve impulses that travel through sensory nerve fibers towards the central nervous system (brain). Their functions are specific. One specific type of receptor perceives only one specific type of stimulus, such as touch, heat cold etc., percisely for this reason they are also called as the peripheral analysers. Whatever may be the type of perceptions, when a receptor responds to a stimulus, it sends a nerve impulse to the brain which makes the organism to be aware of the sensation.

4.8.3 Different Categories of Receptors

For the sake of convenience, different type of receptors present in the organism have been divided into the following classes:

- Photoreceptors
- Mechanoreceptors
- Chemoreceptor
- Proprioreceptors

Photoreceptors : These are those specialized cells that have capability to perceive visible light spectrum of 400-700nm. These receptors give sense of light. Rod and cone cells are best examples of photoreceptors present in vertebrates, including man. Wide variation exists in animal kingdom in the kinds of photoreceptors ranging from simple ocelli to highly complex eye of humans.

Mechanoreceptors : These respond to mechanical stimuli, such as touch, pressure, vibration and sound. A response is perceived by the group of cells or specialized organs. For example, hair cells in cochlea give sound sensation of animals. Similarly, hair cell in vertebrate utricle and semicircular canals gives angular position of the body. Somato-sensory receptors are present in dermis of the skin and these receptors perceive mechanical as well as other sensation as shown in the Table 4.7.

Table 4.7

Type of Stimulus	Specialized receptors present in the cells/tissues
Pressure	Pacinian corpuscle
Touch	Messeners corpuscle
Pain	Freenerve endings (nolireceptors)
Heat or temperature	Ruffni organelles
Cold	Kraures cells
Tactile (Tickling) sense	Tractile organs

Chemo receptors : These are the cells that responds to specific chemicals. These identify the chemical nature by the receptors present in the epithelium of the tongue and nasal cavity. For instance the taste buds have small hair like receptors cells to identify the taste of food and drink. The taste buds embedded in the tongue of humans (as with other mammals) posses four types of chemoreceptors, which respond and provide the sense of perception to the taste of food material and the drink and distinguish into four types of testes: sweet, slat, bitter and sour.

The human nasal epithelium, similarly, has a diverse variety of receptors that respond to chemicals in the environment and give us our sense of perception of smell or odour and flavours.

Proprioreceptors : These respond to the position and kinetic movements of the tendons. Some specialized receptors are present in the skeletal and muscle system performing this kind of function. For instance the Golgi tendon organs are present in tendons. Similarly, muscle spindle present in skeletal muscle, pacinian corpuscle in joints, nolireceptors (free nerve endings perceive the pressure, pain and disposition of the body. Some other free nerve endings seen in the visceral organs of the body perceive the pH, oxygen and carbon dioxide (CO_2) in the body fluids.

4.8.4 Sensory System

The coordinated activity of any organism mainly depends on a continuous monitoring of information from both inside the body as well as outside environment. This is vital for the survival of organism, since organism has to know whether a predator is lurking behind it or readying to pounce on. Similarly, the levels of vital gases like CO_2 have increased in the body, If undetected, either condition could be life threatening to the organisms. The sensory receptors play a vital role in giving informational inputs to organisms continuously about the internal as well as external environment. An assemblage of such specialized sensory receptors into organs constitute sensory systems.

4.8.5 Sensory Receptors and their Types

In its barest form, a sensory receptor is a neurone with a dendrite that is sensitive to one particular stimulus. As soon as the dendrite receives such a stimulus, chemical events takes place resulting in an action potential which will be transmitted through the neurone fibre to the central nervous system. This type of cell is called as primary receptor. A secondary receptor is a bit complex it consists of a modified epithelial cell which is sensitive to a specific type of stimulus. This subsequently synapses with a sensory nerve fibre, which in turn carries the impulse to the central nervous system (Fig. 4.41).

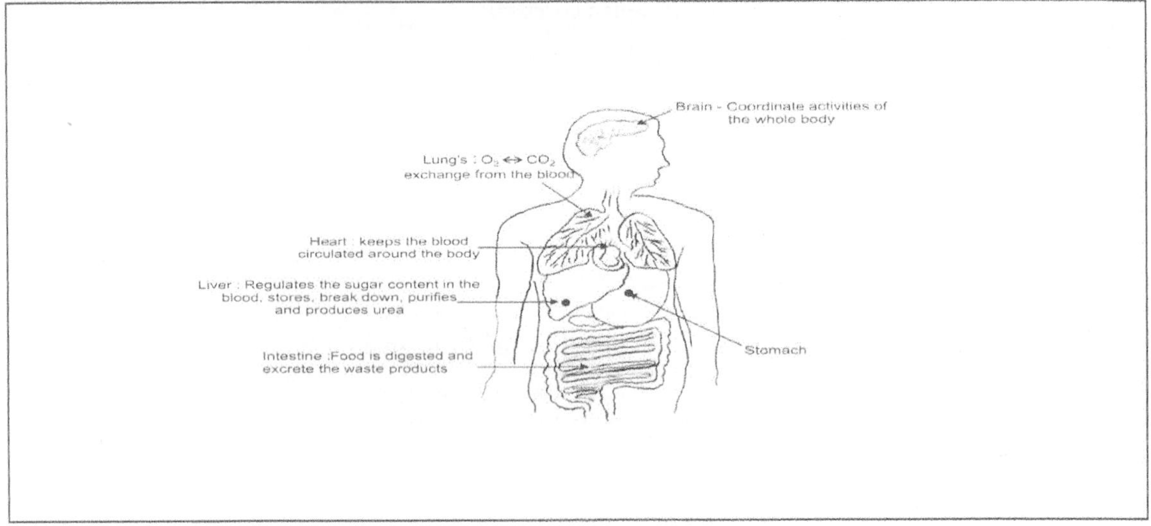

Fig.4.41 Two main types of sensory receptors.

While innumerable sensory receptors are found in isolated entities, in vertebrates including humans these sensor receptors are increasingly associated together in a system called as organs of sense.

4.8.6 The Ways of Receptors Working

Receptors have a resting potential, as in the case of rest of the nervous system, which is dependent on the sodium pump, to maintain electrically, negative situation in interior of membrane. When a receptor cell receives an appropriate stimulus, its membrane gets affected. As a result the mechanism of sodium pump is interfered with. Consequently the ions, more rapidly excit the cell membrane along the gradient intensity of electrochemical concentration. The influx of sodium ions is called as generator current and it result in the setting up of a generator potential. A small stimulus results in small generator potential while a large stimulus gives rise to a larger generator potential. The action potential of receptor neuron occurs only when the generator potential produced by the stimulus is large enough. Thus the action potential obey all or nothing rule.

The process is as follows and is common in one form or other to most receptors :

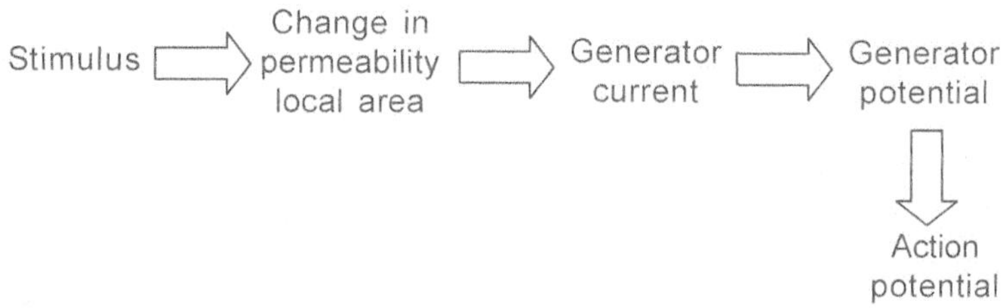

Stimulus ⇒ Change in permeability local area ⇒ Generator current ⇒ Generator potential ⇓ Action potential

4.9 Sensory System : Human Ear

Hearing is one of the sensory perception by which humans learn about the world around them. The advantages of detecting sound are many:

- Provides an awareness of impending danger; hunting and other activity of primitive humans relied on the sensory perception of hearing to find prey or food.
- Central to communication between persons
- The creative language, music, poetry, drama etc; mainly rests upon the aural communication

4.9.1 The Structure of the Human Ear

The ear is the organ of hearing in humans. It is designed to detect mechanical waves in the air through mechanoreceptors. Besides, ear also detects both gravity and movement using specially adapted mechanoreceptors. Its structure, external as well as internal, is closely related to its function.

The anatomy of the ear is presented in Fig. 4.42. For the felicity of description, the structure of human ear is divided in to three areas : 1. external ear, 2. middle ear and 3. inner ear.

External Ear

The external ear is made up of several constituent parts (Fig. 4.42)

- Helix
- Antihelix
- Tragus
- Antitragus
- Lobe

Pinna, a flap of skin-covered elastic cartilage helps to collect sound waves from the air and funnel them into in side of ear. Different parts of the external ear gets the nerve supply. For instance, upper lateral surface gets the nerve supply through auriculotemporal nerve from mandibular nerve; greater auricular nerve from C3 provides the nerve supply to lower lateral surface and medial surface. Lesser occipital nerve from C_2/C_3 supplies nerve connection to the skin posterior to ear.

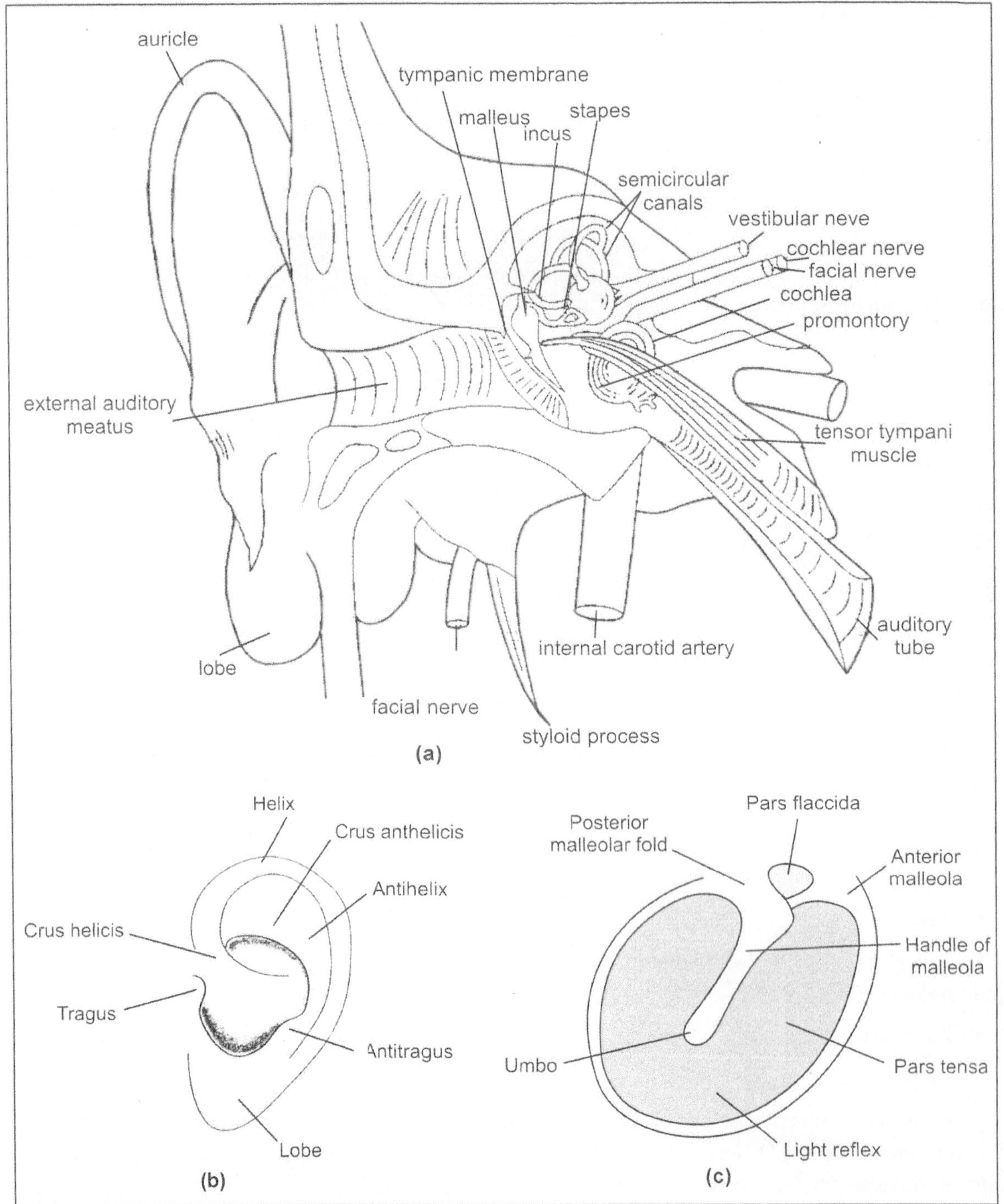

Fig. 4.42 The structure of the human ear : (a) ear anatomy (b) the external ear, (c) right side lympatic membrane.

4.9.2 Ear canal and tympanic membrane

The sound is perceived after pinna conducts sound waves funneled into ear canal which set up vibration on to ear drum. The ear canal is lined by stratified squamous epithelium and is consists of outer one third-cartilage, inner two thirds-temporal bone. The sound is usually resulted due to vibrations (mechanical waves) traveling through air; these sound waves even are capable of traveling through liquids. Sound do not make any sense until a mechanical wave in the air is picked up by the brain. The frequency of vibration of an object determines the frequency of the vibrations traveling through the air. The human ear is capable of recognizing and receiving frequencies between 40-20,000 Hz (cycles per seconds). Low frequency sound waves are perceived as deep sounds and high frequency sound waves as high pitched sound.

Sensory supply to the ear canal and external tympanic membrane is provided by the following: 1. auricultotemporal nerve to anterior, 2. facial nerve to superior lesser occipital nerve to posterior and 3. vagus nerve to inferior part of the ear canal. The relevant branch of the vagus known as Alderman's nerve that supplies the inferior part of the tympanic membrane as well as part of the canal. This is perhaps the reason cough is induced when ears are being suctioned.

The medial end of the canal is formed by the tympanic membrane and the anatomy is shown in Fig. 4.43. The medial surface of the tymphanic membrane is lateral limit of the middle ear. It is lined with pseudo-stratified columnar epithelium anteriorly and flat or cuboidal epithelium posteriorly. The pars tensa is a well ordered fibrous layer with peripheral thickening in the form of annulus where as the pars flaccida is a poorly organizing fibrous layer with no annulus.

Middle ear

The middle ear contains the following :

- Ossicles malleus, incus, stapes
- Stapedius – this acts to adjust the angle of the stapes and hence make it less sensitive to sound.
- Tensor tympani
- Corda tympani – this part of facial nerve communicates with the mandibular division of the trigeminal nerve. It pauses between the two layers of the tympanic membrane and over the handle of the malleus.
- Tympanic plexus of nerves – this part of the tympanic branch of the glasso pharyngeal nerve joins with parasympathetic fibres of the facial nerve and sympathetic fibres from the internal carotid artery. It ends up supplying secretomotor fibres to the parotid gland.

The ossicles are arranged in such a way as to increase the force but reduce the amplitude of the sound vibrations picked up from the tympanic membrane in a ratio of 1:2:1 anatomically which translates into a physiological ratio of 1:14 in practice.

The blood supply to the middle ear mainly is from anterior tympanic artery that branches off from maxillary artery; superior tympanic artery from middle meningeal artery; posterior tympanic artery from stylomastoid artery; inferior tympanic artery from ascending pharyngeal artery.

Glossopharyngeal nerve via Jacobsons nerve and facial nerve are the main sources of sensory nerve supply to the middle ear.

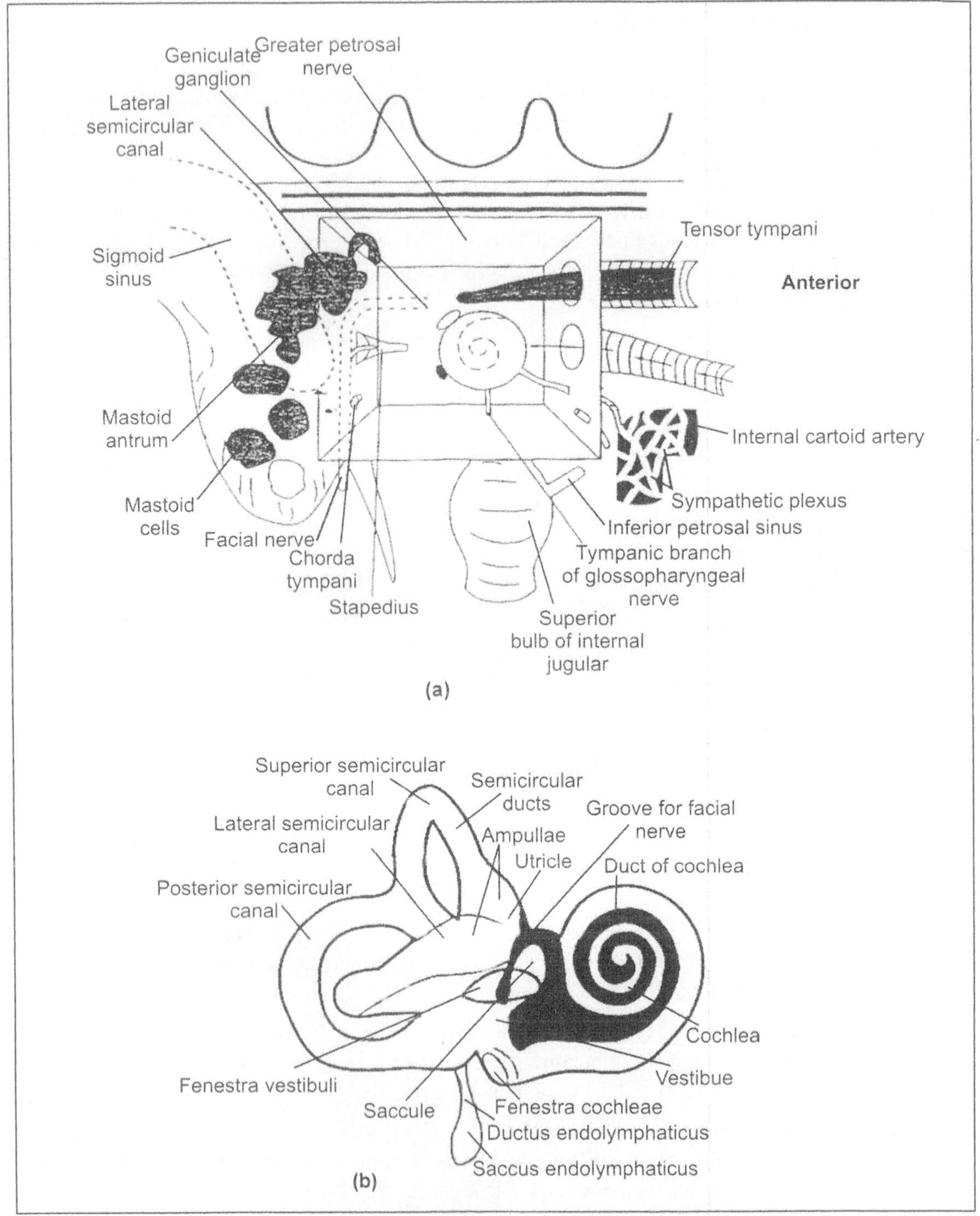

Fig. 4.43 The structure of (a) middle and (b) inner ear.

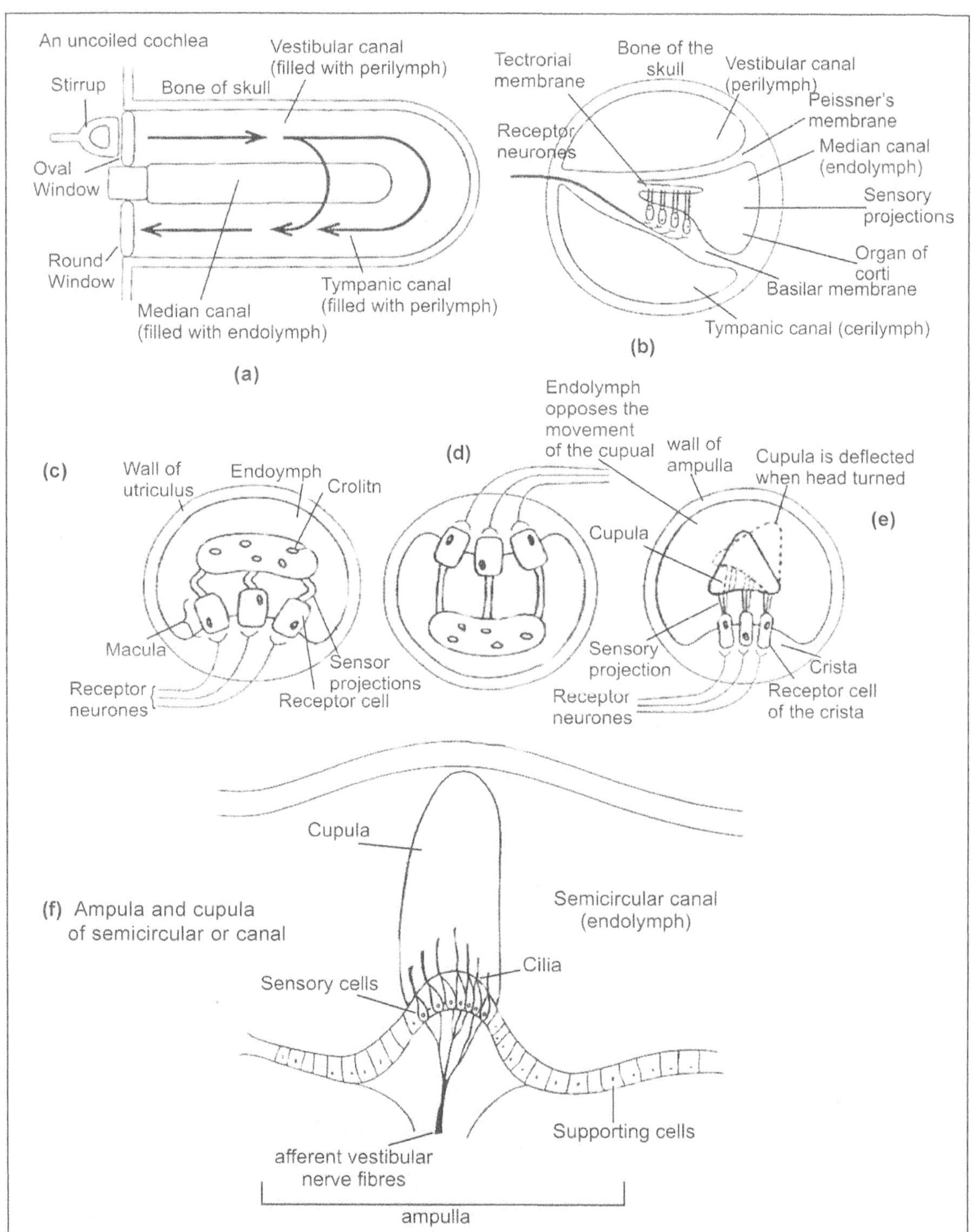

Fig. 4.44 Hearing appratus - (a) cochlea a delicate but complex system (b) T.S through a cochlea (c) cross-section of utriculus showing its owrking (d) & (e) simplify version of a section through a cupula (f) Ampula and cupula of a semicircular canal.

4.9.3 Inner ear

The inner ear (Fig. 4.43) comprises of a labyrinth of canals embedded in the temporal bone, which can be divided into two : 1. the vestibule and semicircular canal and 2. Chchlea.

The membranous labyrinth enclosed a hollow space that is a filled with endolymph to constitute the endolymphatic system. Close to the sigmoid sinus there is a blind sac, saccus endolymphaticus, into which the endolymph passes. The perilmpahtic system enclosed the membranous labyrinth and has a contact with the subarachnoid space via the cochlear aqueduct. Perilmphatic fluid is nothing but a filtrate of blood and cerbo-spiral fluid (CSF). Endolymphatic fluid on the other and is a filtrate of perilmphatic fluid but with a different sodium (Na^+) and potassium (K^+) concentration because of operation ionic exchange system.

The Cochlea is the main component of the organ of hearing contained in the inner ear (Fig.4.44). It consists of two and half turns around on a bony core called the modiolous. The endolymph transmits vibrations to the tectoral membrane. This results in a movement of the underlying hair cells which cause changes in polarization through alternations in K^+ ion permeability consequently there is a transmission information to the cochlear nerve.

The cross section of the cochlea reveals that the cochlear duct is sandwiched between the scala vestibule (Fig. 4.44) and scala tympani containing perilymph. They are separated by the upper vestibular membrane (Reissner's membrane) and lower basslar membrane.

The inner ear contains a labyrinth of three semicircular canals. The canals are positioned at right angles to each other; each includes a widening called as ampulla. Each ampulla has a gelatinuous mass known as the capulla, which surrounds a collections of hair cells embedded in the underlying membrane (Fig.4.44) endolyph caused by rotary motion leads to movement of capulla and of hair cells. The cells are depolarized and transmit impulses to the vestibular nerves. The saccule acts in a similar way to detect linear accelerations and decelerations. The utricle acts to detect gravity. The vestibular organs are responsible for about 15% of balance function. Their outputs are integrated with proprioceptic (15%) and visual (70%) outputs to be integrated in the brainstem to maintain balance.

4.10 Human Eye

Sensitivity to light has evolved among most of the basic living organisms. The specialized cells called photoreceptors are sensitive to electromagnetic stimuli like visible light. However, the ranges of light detected by animal species like insects transcend this, being able to detect UV light. Human eye developed far beyond a mere simple sensitivity to the absence or presence of light and possesses a mark sophisticated optical system, permitting clear and focused vision in a widely varying conditions. It is equipped with photoreceptors sensitive to electromagnetic reaction with wavelength between 400-700 nm.

4.10.1 Structure of the eye

The structure of eye is very intimately related to its function. The muscle attached to the tough outer coat of the eye ball permits it to swivel in the orbit to rotate and to point in many directions. The shape, size and contents of the eye ball lets the focusing of light on the photoreceptor cells constituting the retina. The iris and pupil regulate the quantum of light entering the eye. The lense along the muscles and ligaments helps for fine tuning the focus of objects. The sclera forms a robust outer coat and along with the eye lids and bony orbit helps to minimize the possibility of physical damage (Fig. 4.45).

4.10.2 The Working of the Eye

(*a*) ***The role of iris :*** When eye glances at something the eyeballs move in their sockets by muscles to position the pupil, at the centre of the iris is focused at the object of interest. Light from the object enters the eye though the pupil and the amount of light is controlled by size of the opening (Fig. 4.45).

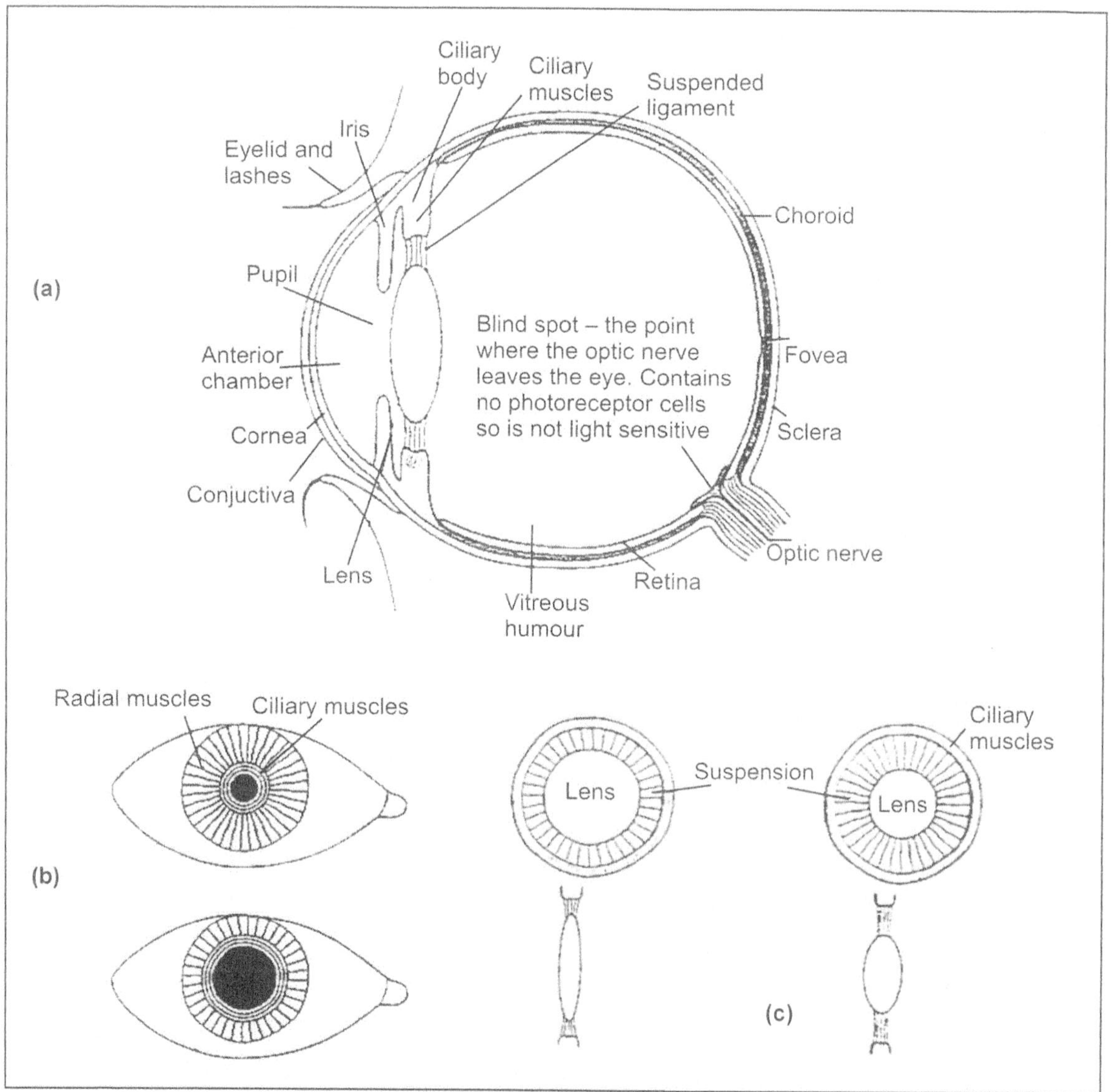

Fig. 4.45 The major parts of the eye (a) the major components of the eye (b) the varying sizers of the pupil due to contraction and relaxation of the muscles, (c) changing sizes of lens during the process of accommodation and fully accommodated vision.

Table 4.8 The main components of human eye and their function.

Component of the eye	Functions
Iris	It is a circular sheet of muscle dividing the eye into two chambers, anterior and posterior. The pigment present in it imparts colour to eye; genetically determined. The reflex contraction and relaxation of the muscles of the iris regulate the quantity of light entering into the eye.
Pupil	A circular opening in the iris through which light enters into the eye, the size of the aperture is controlled by the muscles of the iris.
Lens	A transparent, biconvex structure which is elastic to change shape. It separates the two regions of the eye ball. It helps in the focusing of light on to the retina.
Anterior	It is filled with aqueous humour, a clear solution consisting of salts that help to maintain the shape of the front part of the eye besides helping the process of focusing the light.
Cornea	Transparent area in front of the eye ball. It is also a part of the tough outer layer of the eye; plays an important role in focusing the light as it enter the eye.
Conjuctiva	It is a thin layer of transparent epithelial cell over the surface of the eye which protects the cornea.
Ciliar body ciliary muscles	Contains the ciliar muscles. These are bundle of smooth muscles which act to change the shape of the lens to enable it to focus light from objects both from near and distance away from the eye.
Suspensory ligament	Attaches the ciliary body to the lens.
Choroid	Pigment cells prevent any internal reflection of light within the eye and blood vessels supply the retina with nutrients and oxygen.
Fovea	This is an area of retina containing only cones, region of highest visual activity.
Sclera	This is a tough outer covering of the eye which maintains the shape of the eye ball and protects it forms physical damage.
Optic nerve	The bundle of nerve fibers carrying impulses from the retina to the brain.
Retina	It is layer of photo-sensitive cells consisting of rods and cones, the neurons from these photoreceptors to the optic nerve for processing information by the CNS.
Vitreous humour	It is clear, jelly like substances the eye ball its shape imparting.
Eye lids and lashes	These help to protect the eyes from the physical damage by closing in a reflex action when an object approaches; eyelids clean and moisten the surface of the eye by spreading tears over the surface by a reflex action called blinking.

4.10.3 The role of the Lens

The process of focusing of light on to the retina starts as soon as it enters the eye. To accomplish this, the light has to be refracted sufficiently to pass through the pupil and also has to be focused on the retina. Cornea performs the most of the job along with the fluid through which the light passess. However, the degree of refraction is same for light emanating from every object. Lens play on important role in providing fine accurate focusing of light coming from the object both near or far. The lense is not responsible for much of the refraction of the light.

The lens is a transparent elastic disc, whose shape can be changed by the action of ciliary muscles. These are arranged circularly around the ciliar body. The consequences of their contractions and relaxation are relayed to the lens by the suspensory ligaments. The lens by it self is elastic and in its natural state is relatively very short and fat.

A newborn baby can see clearly a certain distance, from his or her eyes, roughly a distance where the mothers face is positioned when the body is being breast-fed. However, with age, the ability to focus on objects at a very distant places develops. The lens shape is changed by the action of ciliary muscles. To permit light from objects at varying distances to be focused, the thickness of the lens can be varied between the two extremes (Fig. 4.45). As the ability to do this develops, the visual perception and understanding of the environemtn improves.

Light rays from an object diverge all around in all directions. When one looks at objects from distance the light rays are spreading less and appear almost parallel. When the object is viewed from a close range, a cone of diverging light enters the eyes (Fig. 4.45). The light entering the eye is bent by its passage via the conjunctiva, cornea aqueous humour and vitreous humour in exactly the same way irrespective of distance of the object. However, by changing the len's shape the degree of refraction of the light can be altered. Light from the distance object necessitates little bending to bring into focus on the retina. Hence the laens has to be thin. On the other hand, to bring light from near objects into focus on the retina, more bending of light is needed and so lens must become short and fat. This ability to focus light from at various distance is called as accommodation. The modus operandi is illustrated in the Fig. 4.45.

4.10.4 The Retina and its role

The retina is the screen onto which light from an object is focused. The retina perceives that light and communicate to the brain of its presence. To do this, retina contains nearly a 100×10^6 photoreceptor, along with the neurons with which they pair. Two principal kinds of photoreceptors are present in the retina: rods and cones (Fig. 4.46). These two are secondary type of receptors.

Rods are distributed across the retina except at fovea where these do not exist. The rods provide black and white vision only and are used mainly for visual perception in night, when the light intensity is very low. Accordingly to an estimation about, 1.2×10^8 rods are present in the retina. These are very sensitive to light even of low intensity. They process a single visual pigment called rhodopsin (visual purple). They are not tightly packed together. Several of them synapse with the same sensory neuron. They do not give a particular clear picture. However, they make them hyper-sensitive to light even of low intensity and movements in the visual field, since innumerable small generator potentials can trigger an action potential to the central nervous system.

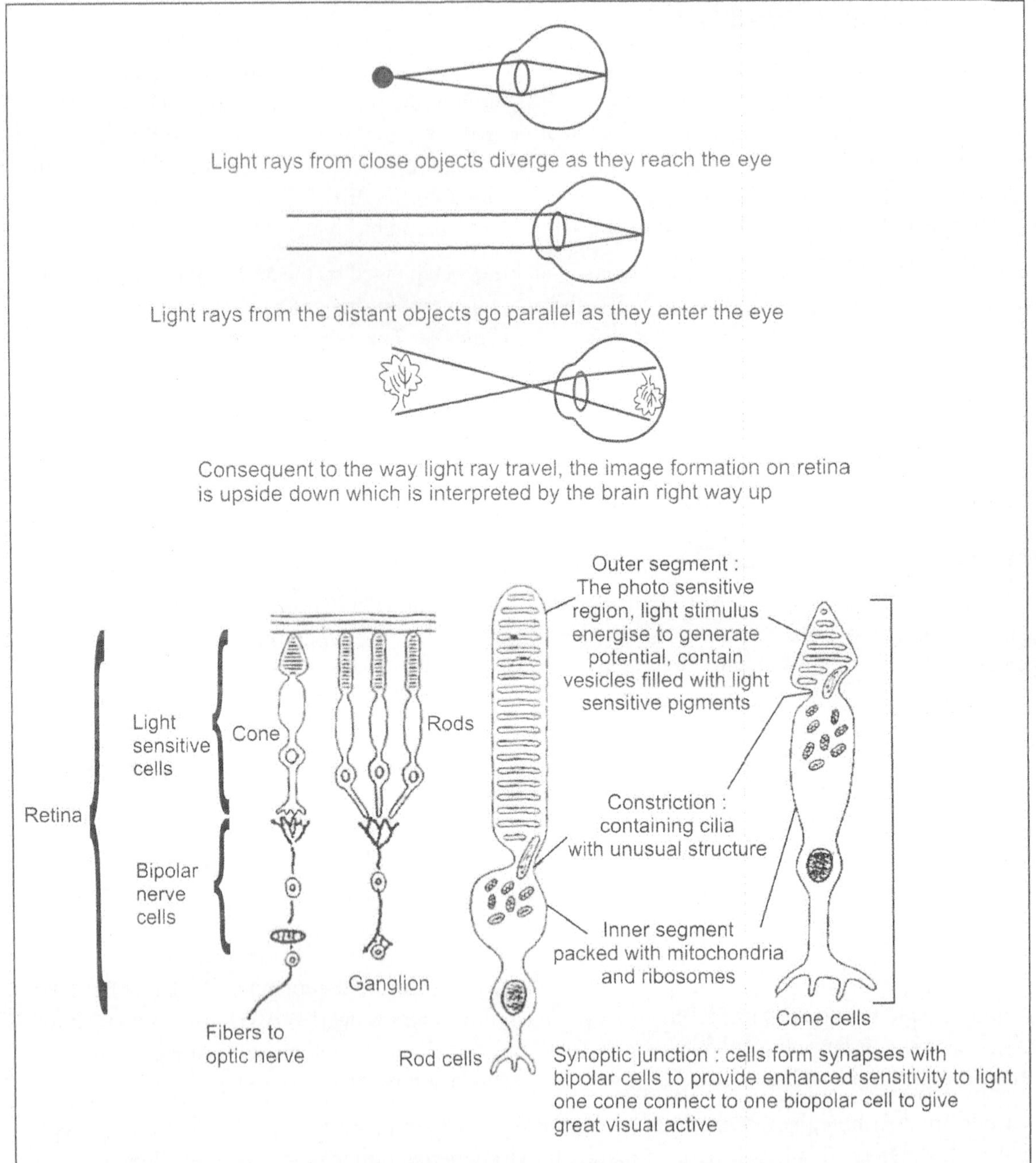

Fig. 4.46 The two types of photoreceptor cells with different arrangement of synapses in the retina to provide visual system combining the vast sensitivity to low level of light with increased visual activity and clarity of vision.

Cones are found tightly packed together in fovea. About 6×10^6 of them exist. They are mainly used for vision in bright day light, may contain one of the three visual pigments and hence provide a colour vision. As a consequence of close packing in the fovea and the fact that each cone has its own sensory neurone, these provide a visual picture of great acuity. Infact, when light falls directly on the fovea that it can be clearly focused.

The way retina is arranged is quite interesting; it is, what is called "back to front" arrangement. The outer segments are actually next to the choroids, and the neurons are at the interior edge of the eye ball. That is the reason why there is a blind spot where all the neurons pass through the layers of the eye to ho into the brain. The light has to go through the synapses and the inner segments before reaching the outer segment containing the photo-pigments. The reason for this bizarre arrangement has the basis in its origin during the embryonic development. Besides, the confusion is further confounded in interchanges of left and right images. The optic nerve, conveying the visual information cross over on their way to the visual cortex in the brain, so that, the information seen with right eye is taken to the left side of the brain for processing and vice-versa.

4.10.5 The Working of Rods and Cones

The working of rods and cones are identical; it is based on the reactions of the visual pigments with light. The visual pigment, rhodopsin is made up of two components, the opsin and retinene. Opsin is a lipoprotein and retinene is a light-absorbing derivative of vitamin A, that exists is two isomeric forms; Cis-retinene and trans-retinene. In the dark is all in the Cis-retinene from when a photon his a molecule of rhodospin, it converts the Cis-retinene into trans-retinene by breaking up of rhodospin into opsin and retinene. This process is referred as bleaching. This sets up a generator potential in the rod. In case this is large enough alternatively, several rods are fixed at once, and, an action potential is set up in the receptor neurone. Once the bleaching of visual pigment takes place, the rods cannot be stimulated till the rhodospin is reformed. The requisite energy (ATP) is produced by the innumerable mitochondria present in the inner segment to convert and reconstitute with opsin to become rhodopsin. The eye has a capacity to adapt to both bright light and dim light. In a day light when the rods do not respond to the dim-light because the rods are in light adapted condition and are completely in bleached condition. If eyes are exposed to complete darkness for thirty minutes, the rhodospin will be fully formed and the eye become sensitive to dim light. The process is called dark adaptation of the eye.

4.10.6 The Perception of Colour: The Role of Cones and Colour Vision

Cones working is similar to that of rods with one exception. The visual pigment is iodospin, unlike rhodospin in the rods. There is a theory called, trichromatic theory, which envisages the presence of three types of iosospin, each sensitive to one of the primary colours of the light. Idospin is not sensitive to low light intensities, since it needs more light energy that rhodospin to disjoin the pigment. However, the cones provides colours (Table 4.9).

Table 4.9 Brain interpretation of different colour as per the trichromatic theory of colour vision.

Stimulation of light by the type of cones			
Red cones	**Green cones**	**Blue cones**	**Perception of type of colour**
+	–	–	Red
–	+	–	Green
–	–	+	Blue
+	+	–	Orange/Yellow
–	+	+	Cyan
+	–	+	Magenta
+	+	+	White

4.11 The Sense of Smell and Taste

Humans, like any other vertebrate animal have developed acute sense of smell and taste, which is provided by specialized cells called as chemo receptors. These identify the chemical nature by the sensory receptors present in tongue nasal cavity. The taste and embedded in tongue have small hair like receptor cells which perceive the taste and discriminates four basic tastes: Sweet, salty, bitter and sour. The hair cells present in nasal cavity perceive various odar and pungent smell as well as different variations of smell.

The process of chemo reception is an universal sense. Cells that are preferentially respond to ions or molecules in food may be designated at taste receptor cells. Sensory modality is called as taste or gestation.

4.11.1 Human Nose: Anatomy and Physiology

Olfactory system in humans is constituted by a nose and its associated structure

Human nose consists of the following structures :

- The external nose, made up of cartilage
- Two nasal cavities made up of nasal bones separated by the bony and cartilaginous septum
- Two posterior choanae, leading to nasopharynax

Air flow in and out of the lung through the nose and nasopharynx. The nose effectively does three functions: Wards, moistens and filters incoming air. The nasal mucosa goes through a reflex nasal cycle, in which every 2-6 hours, one lumen of the nose widens due to vasodilation is mucosa, while the other narrows, giving a constant air way resistance. There is what is called Eccles' reflex when there is a pressure in the crook of the arm or lying on the one side of the body causing the contralateral air way to open. The purpose of these reflexes is not fully known.

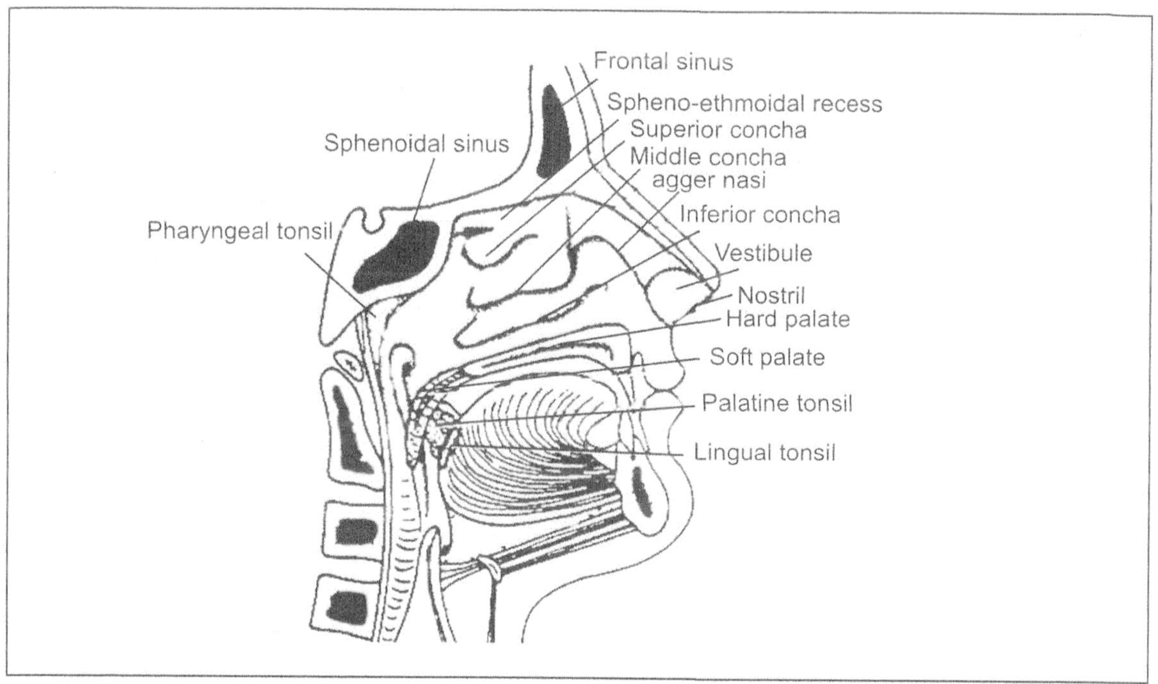

Fig. 4.47 Human nose, its anatomy and associated structures.

Most of the nasal cavity is lined with columnar ciliated respiratory epithelium, apart from the area of the olfactory cleft and cribriform plate known as the olfactory region. It is lined with olfactory mucosa and is innervated by bipolar fibres of the olfactory nerve. About 20 fibres run to the primary olfactory centre of the olfactory bulb. The nerve fibres travel through the olfactory tract to the secondary olfactory centre, eventually reaching the denate the semilunargyri in the cerebrd cortex. Olfactory nerve fibres are surrounded by supporting cells. A lipid secretion in this area helps in differentiation of odours.

4.11.2 The Taste System

Many foods are tasted by the olfactory nerve, but the sensation of sweet, salty, bitter and soul are exclusive to the system. The various sensations are picked up on the tongue (Fig. 4.48) by the taste buds. These lien in the vallte papillae in front of the sulus and fungi form papillae, peripherally. Saliva is required as a transport medium to facilitate tasting. The fili form papille do not contain taste receptors. The sensory supply of the anterior two thirds of the tongue comes form the lingual nerve a branch of the facial nerve. In the posterior two thirds of the tongue, both taste and sensation are supplied by the gasopharyngeal nerve along. The sulcus terminalis divides the tongue into two sections. The inferior surface of the tongue has the frenum in th midline, with the lingual artery nerve and vein present more laterally. The sublingual glands and submandubular duct open here as well.

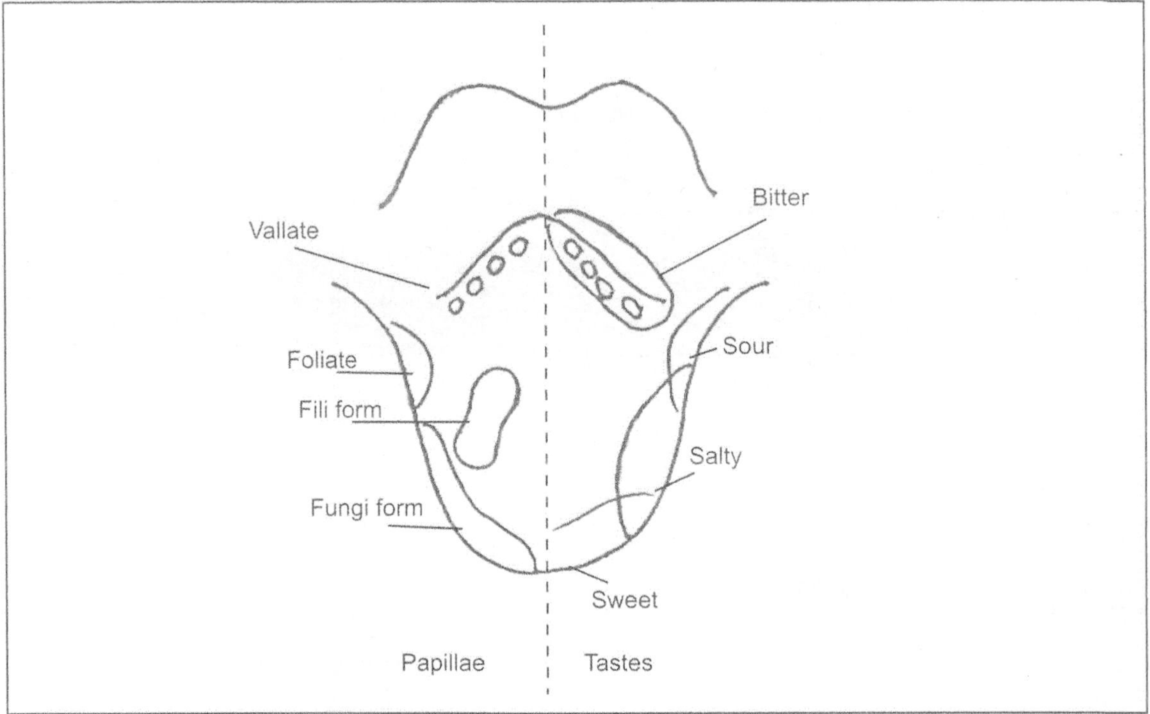

Fig. 4.48 (a) Human tongue showing taste areas and papillae region.

The tongue muscles are of two types

- ***Intrinsic :*** superior and inferior longitudinal, transverse, vertical – all are supplied by hyupoglossall nerve.
- ***Extrinsic :*** genionglossus, hyglossus, styloglossus – all are supplied by hypoglossal nerve. Platroglossue is supplied by cranial root of an accessory nerve.

4.11.3 Taste Receptors

Cells, which preferentially sensitive to ions or molecules in food are called as taste receptor cells. The sensory modality is called as taste or gestation. Taste receptors are found on the tongue, back of the mouth, pharynx, epigothics, and upper esophagus. As the cellular level, taste receptor cells are grouped together in taste buds. Taste buds are arranged on papillae, appearing as blunt pegs on the surface of the tongue. Several types of papillae exist with different degree of distribution.

4.11.4 Taste Transduction

All the taste sensations can be grouped into four basic qualities: sweet, salt, sour, and bitter. These qualities have basis for the behaviour of organism. Sweet in generally associated with nutrients food, salts is essential for fluid balance, sour is harmful in excess, and bitter is reliable caution that are injurious and harmful.

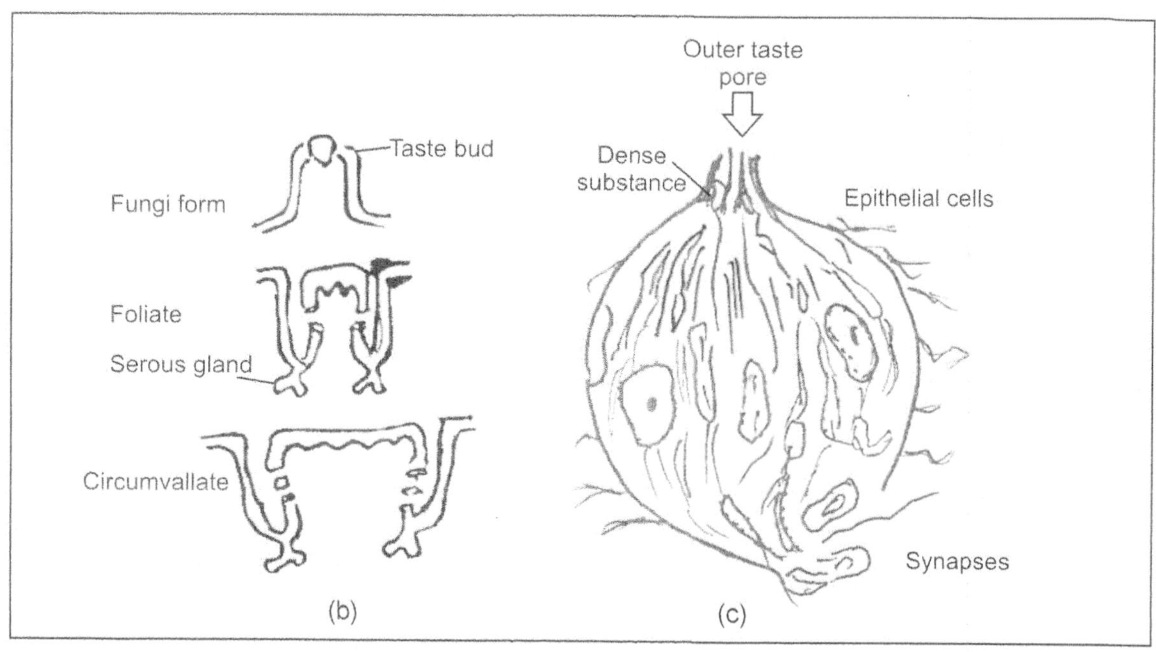

Fig. 4.48 (b) Types of papillae (c) taste buds of human tongue.

Essay Type Questions

1. Outline the organization of human body and mention respective functions.
2. List different tissues of human body and explain their function.
3. What is the composition of body fluids ?
4. Outline the composition of blood and list functions of each fraction.
5. Give an account of structure and function of blood cells.
6. How does blood carry oxygen ?
7. Mention the role of platelets and different pathways in blood clotting.
8. Give briefly the structure and function of WBC.
9. Describe human digistive system.
10. List different secretions of liver and pancres with their functions.
11. Outline briefly the respiration system.
12. Sketch briefly the human endocrine system.
13. Outline the organization of nervous system listing three basic functions.
14. What is resulting or membrane potential. Explain physiological and molecular basis of membrane potential.
15. Define action potential. Elaborate the physiological and molecular events involved in origin of action potential.
16. Write an account on origin and propagation of nerve impulse.
17. What is the synapse; give an account on types of synaptic transmission.
18. Discuss the molecular and physiological events involved in chemical synapse transmission.
19. Write an account on neurotransmitters.
20. Explain the physiological and molecular events involved in chemoreception of taste.
21. Describe the structure of olfactory receptor cell. Add a note on the mechanism involved in olfaction.
22. Describe the nerve physiology of the photoreception.
23. Illustrate the structural basis of photoreception.
24. Discuss the physiological and molecular events involved in photoreception.

Short Answer Type Questions

1. Write a brief notes on the following.

   (a) Neuron

   (b) Glial cell

   (c) Sensory receptors

   (d) Resting potential

   (e) Action potential

   (f) Na^+ K^+ pump

   (g) Local circuit mechanism

   (h) Saltatory conduction

   (i) Ion channel

   (j) Electrical synapse

   (k) Post synaptic receptors

   (l) Neuropeptides

   (m) Biogenic amines

   (n) EPSP and IPSP

   (o) Taste cell

   (p) Olfactory cell

   (q) Hair cell

   (r) Organ of corti

   (s) Photoreceptors

   (t) Rhodopsin

   (u) Singnaling in photoreceptors

   (v) Color vision

5 Introduction to Parasites and Immune Responses

5.1 Introduction to Parasitology and Biology of Parasites

Parasitology, the study of parasites and their relationship to their host is one of the most significant and fascinating area of biology. It encompasses the contributions from phylogenic, ecologic, morphologic, physiologic, chemotherapeutic, nutritional, immunologic and with great advances of cellular as well as molecular biology approaches indicating interdisplinary role. However, the parasitism is a function of whole organism, though the manifestations of parasitic way of life are evident at molecular, cellular, tissue and population level of organization. Because of its importance in public health and economic well being, socio-economic and socio-geographic approach is also part of parasitology. It is assumed that out of unconquored human diseases like schistosomiasis, trypanosomiasis, filariasis, leishmaniasis, malaria etc., the extent of human infections are roughly about by 700 million by hook worms, 180 million by schistosomiasis, 25 million by malaria and 1800 millions by all forms of intestinal parasites. Thus, medical parasitology is of immense importance.

Also because of the economic implications of parasitic infections among live stock and domestic animals, the importance of veterinary parasitology may not be over emphasized. With significant contributions made from aquaculture, mariculture and fisheries, it is pertinent to understand and control destructive parasites of fish and non-fish fishery organisms. Above all, the role of pathogenic nematode and protozoan parasites of certain crops need to be looked into because of the immense agricultural importance. Thus, the understanding of biology of parasites and parasitism is very important not only from the view point of advancement of knowledge but also from practical viewpoint, particularly, in relation to emerging field of biotechnology.

Parasitism

Parasitism is defined as an intimate and obligatory relationship between two heterospecific organisms during which the *parasite* usually the smaller of the two partners is metabolically dependent on the other, usually bigger one, the *host*. The relationship may be permanent, as found in case of intestinal helminth parasites of man or temporary, as found in feeding of mosquitos, leeches and ticks on their host's blood. Parasitism is obligatory because parasite usually cannot survive without host and *metabolic dependency* of parasites on hosts make the relationship an obligatory. In addition to metabolic dependency of parasites, these are also generally involves in *immunologic response* on the part of host.

Several types of parasitism are recognized. A parasite does not absolutely depend on the parasitic mode of life, but is capable of adapting to it, if placed in such a relationship, it is called as *facultative parasite* (eg. *Mycobacterium tuberculosis* in man). On the other hand, if parasite is completely dependent on the host during its life cycle, it is known as *obligate parasite* (eg. *Taenia solium*). Occasionally, some organisms acquire an unnatural host and survives, it is known as an *incidental parasite*. The parasites that live within the body of their host in locations such as gut, liver, lung, and muscle are called as *endoparasites* (e.g., *Ascaris*) and those attached to the outer surface of the host are known as *ectoparasites* (e.g., *Argulus*).

The host is commonly larger than two species and it may be categorized as,

(i) *definite host* : if parasite attains sexual maturity in it,

(ii) *intermediate host* : it serves as a temporary but essential environment for completion of parasite's life cycle. For example, snails serve as intermediate hosts for trematodes to complete life cycle.

(iii) *transfer* or *paratenic host* : if it is a temporary refuge and a vehicle for reaching definite host. For example, arthropods (e.g., Mosquito) serve as hosts as well as carriers for protozoan (e.g., *Plasmodium*) parasite which are referred as *vectors*.

(iv) Animals that become infected and serve as a source from which other animals can be infected are known as *reservoir* hosts. For example, filarial worm, *Brugia malayi*, causes filariasis in humans, is transmitted to humans through the bites of mosquitoes, the vector. The *B. malayi* can also be transmitted from cats and monkeys, which serve as reservoir hosts, to humans via mosquitoes. A *hyperparasite* is an organism which parasitize another parasite. For example, a protozoan *Nosema dolfusi* is a hyperparasite on larval stage of a flatworm, *Bucephalus cuculus* which inturn is a parasite of American oyster. Recently, the hyperparasites are being used as biological control agents. A number of parasites can cause diseases in both animals and humans and such phenomenon is referred as *zoonosis*.

Some Definitions and Terms Used in Parasitology

Parasite : An organism which is dependent on another organism for its survival

Host : An organism which harbours the parasite and is usually larger than the parasite

Ectoparasite	:	A parasite which lives on the outside of the host. e.g., head louse on hair of man
Endoparasite	:	A parasite which lives within the host.
Facultative parasite	:	A parasite which is capable of living both freely and as a parasite
Obligate parasite	:	A parasite which is completely dependent upon the host.
Pathogen	:	A parasite which is able to produce disease.
Reservoir host	:	Where the parasite is maintained in nature and which acts as a source for individual new cases.
Intermediate host	:	A host in which the intermediate stages of the parasite develop.
Definite host	:	A host in which the definitive or the final stages of the parasite develop.
Paratenic host	:	A host which acts as transporting agent for the parasite and in which the parasite does not undergo any development.
Disease	:	If parasite has a deleterious effect on the host
Epidemic	:	The sudden appearance of infection which spreads rapidly and involve a large population.
Endemic	:	An infection which was always existed in region.
Pandemic	:	A widespread epidemic, usually of world wide proportion.
Zoonosis	:	Diseases of animals which are transmissable to man.
Morbidity	:	Incidence of disease in a given population.
Infection	:	The presence of a parasite in or on the tissues of host
Infestation	:	The presence of parasite on the skin of host.

Biology of parasites : Protozoans, Helminths and Arthropods

Among Animalia, many organisms have parasitic mode of life on varied organisms with characteristic features of host parasite relationship. The pathogencity of parasite, which depends on specific host parasite relations results in a number of diseases in human, livestock, domestic animals, such as chicken, cattle, sheep, pig, cat, dog etc., aquatic organisms of fisheries, and some as parasitic forms of crop having influence on human health and economic well being. The main pathogenic parasites belong to animal groups are *protozoa, helminths* and *arthropods*. This chapter briefly presents the overview of parasites belong to protozoa, helminths and arthropods with resultant important diseases. The biology and diseases of some important parasites are also described along with brief measures of disease control.

5.2 Parasitic Protozoa, Diseases and their Control

The protozoans, unicellular organisms, are free living in aquatic and moist environments and also many species are true parasites. These parasites include amoebae, flagellates, gregarines, coccidians, microsporians, ciliates etc.

Table 5.1 gives the overview of different parasitic protozoans and the pathological diseases they cause in man, live stock, domestic and economically important animals.

Table 5.1 *Some important parasitic protozoans of human, livestock, domestic and other economically important animals.*

Parasite	Principal Hosts	Habitat	Main Characteristics	Disease
Entamoeba				
E. histolytica	Humans, other mammals	Large intestine	15–60mm in diameter, distinct ecto - and endoplasms, vacuoles enclose host cells, centric endosome	Amoebiasis
E. coli	Humans, monkeys pigs	Large intestine	15–50mm in diameter, sluggish, vacuoles do not enclose host cells, eccentric endosome	Nonpathogenic
E. gingivalis	Humans, dogs, cats, monkeys	Mouth	5–35mm in diameter, distinct ecto- - and endoplasms, vacuoles rarely enclose erythrocytes	Periodontitis
E. bovis	Cattle	Rumen	Approximately 20mm in diameter	Nonpathogenic
E. gallinarum	Chickens, turkeys	Caecum	9–25mm in diameter, centric endosome	Nonpathogenic
E. anatis	Ducks	Intestine	Similar to *E. histolytica*	Enteritis
E. cuniculi	Rabbits	Intestine	Similar to *E. coli*, 10–20mm in diameter	Nonpathogenic
E. polecki	Pigs	Intestine	5–15mm in diameter, resembles precystic stage of *E histolytica*	Nonpathogenic
Iodamoeba *butschlii*	Humans, pigs	Intestine	8–20mm in diameter, broad pseudo -podia sluggish, large glycogen mass	Usually nonpathogenic
Ciliates				
Balantidium *coli*	Pigs, monkeys, humans	Large intestine	50–100mm long, large ovoid body, anterior end more pointed, peristome leading into distinct cleft, two contractile vacuoles, kidney - shaped nucleus	Balantidiasis (in humans)
Balantidium sp.	Sheep	Intestine	45mm long, 33 wide	Nonpathogenic
Ichthyophthirius multifilis	Fish	Primarily integument	100–1000mm long, ovoid, with large cytosome measuring 30–40mm in diameter	*Ichthyophthiriosis*
Enchelys *parasitica*	Trout	Integument	Flask -shaped, cytosome slitlike, about 12mm long	*Enchelysiosis*
Flagellates				
Leishmania				
L. donovani	Human dogs, cats, horses, sheep, cattle	Spleen, bone liver, monocytes	2–4mm in diameter, eccentric rounded nucleus	*Kala-azar*

Contd.. Table 5.1

Parasite Amoeba	Principal Hosts	Habitat	Main Characteristics	Disease
L. tropica	Humans, dogs	Dermal sores	Indistinguishable from *L. donovani*	*Oriental sore*
L. braziliensis	Humans, dogs, monkeys	Dermal sores especially mucous membranes	Indistinguishable from *L. donovani*	*Mucocutaneous leishmaniasis*
Trypanosoma				
T. equiperdum	Horses, cattle, donkeys	Genitalia and internal reproductive organs	25–28mm long, 1–2mm	*Dourine*
T. theileri	Cattle	Blood	60–70mm long, 4–5mm wide, myonemes	Nonpathogenic
T. melophagium	Sheep	Blood	50–60mm long	Nonpathogenic
T. evansi	Horses, mules, donkeys, cattle, dogs, camels, elephants	Blood	25mm long	*Surra*
T. equinum	Horses	Blood	20–25mm long, no blepharoplast	Mal deCaderas
T. hippicum	Horses mules	Blood	16–18mm long	*Murrina*
T. brucei	Horses, donkeys, mules, camels, cattle swine, dogs	Blood	Pleomorphic, 15–30mm long	*Fatal nagana*
T. simiae	Pigs, monkeys, sheep goats	Blood	Seldom with free flagellum	Virulent
T. congolense	Cows, other domestic animals	Blood	Monomorphic, 819mm long, 3mm wide; on free flagella	*Bovine - trypanosomiasis*
T. vivax	Ruminants, equines	Blood	15.5–30.5mm long, free flagellum	Virulent
T. gambiense	Humans, monkey, antelopes, dogs	Blood, lymph	15–30mm long, 1–3mm wide, spiral undulating membrane	*Gambian typanosomiasis (sleeping sickness)*
T. rhodesiense	Humans, cats, dogs, pigs, opposums	Blood, lymph	Usually indistinguishable from *T. gambiense*	*Rhodesian trypanosomiasis (sleeping sickness)*
T. cruzi	Humans, cats, dogs, monkeys, squirrels	Blood	C- or U-shaped 20mm long	*Chagas' disease*
T. americanum	Cattle	Blood	17-25mm long or longer	Nonpathogenic
Trichomonas				
T. canistomae	Dogs	Mouth	Four anterior flagella, one trailing, 9mm 3.4mm wide	Nonpathogenic
T. felistomae	Cats	Mouth	8.3mm long, 3.3mm wide	Nonpathogenic
T. gallinae	Turkeys, chickens, pigeons	Upper digestive tract, liver	6.2–18.9mm long, 2.3-8.5mm wide 2.3–8.5mm wide	*Avian tricho-*moniasis
T. gallinarum	Turkeys, chickens, pigeons	Caecum	Pear-shaped, 9–12mm long, 6-8mm wide	Nonpathogenic

Contd., Table 5.1

Parasite Amoeba	Principal Hosts	Habitat	Main Characteristics	Disease
T. anseri	Geese	Caecum	Oval body, 7.9mm long, 4.7mm wide large cytosome	Nonpathogenic
T. vaginalis	Women	Vagina	10–30mm long, 10–12mm wide, cystome inconspicuous	*Vaginitis*
T. hominis	Humans	Intestine	5–20mm long	Nonpathogenic
T. tenax	Humans	Mouth	10–30mm long	Nonpathogenic
Tritrichomonas				
T. equi	Horses	Colon, caecum	Three anterior flagella, undulating membrane, slender axostyle, 4-6.5mm long	Nonpathogenic
T. foetus	Cattle	Genital track	Pear-shaped, three anterior flagella, undulating membrane, axostyle, 10–25mm long, 3-5mm wide	Tritrichomonas *abortion*
T. suis	Pigs	Intestine	Three anterior flagella, undulating membrane, axostyle, 8–10mm long	Nonpathogenic
T. eberthi	Chickens	Caecum	9mm long, 4 – 6 mm wide	Nonpathogenic
Retortamonas				
R. ovis	Sheep	Intestine	Pear-shaped, two flagella of length of body, 5.2mm long, 3 – 3.7mm wide	Nonpathogenic
R. cuniculi	Rabbits	Caecum	Posterior flagellum thick, one-half as long as body; 7.5-13 mm long; 5.5-9.5mm wide	Nonpathogenic
R. intestinalis	Humans	Intestine	Pleomorphic, 4-9mm long, 3-4mm wide, cytostome one -third length of body	*Diarrhoea*
Pleuromonas	Chickens	Caesum	One short anterior flagellum, one long flagellum, 5-12mm long, 5mm wide	Nonpathogenic
Pentatrichomonas				
P. hominis	Humans, other primates, dogs, cats	Instestine	Three to five anterior flagella	Nonpathogenic
P. gallinarum	Turkeys, chickens, guinea-fowls	Caecum	Pear-shaped, five anterior flagella, one flagellum, axostyle, 6 – 8mm long	*Caecal lesions*
Histomonas meleagridis	Chickens, turkeys, ducks, geese	Caecum, liver, other tissues	Pleomorphic four flagella	*Blackhead histomoniasis*
Monocercomonas				
M. ruminantium	Sheep, cattle	Rumen, prepuce	Three anterior flagella, 12-14mm long, 8 – 10mm wide	Nonpathogenic
M. gallinarum	Chickens	Caecum	5-8mm long, 3-4mm, pear-shaped, three anterior flagella, one longer trailing flagellum	Nonpathogenic

Contd., Table 5.1

Parasite Amoeba	Principal Hosts	Habitat	Main Characteristics	Disease
Chilomastix mesnili	Humans	Large intestine	Pear -shaped, three free anterior flagella, one flagellum, in cytostome	Nonpathogenic
Embadomonas intestinalis	Humans	Large intestine	Slipper-shaped, two unequal anteriorly directed flagella, large cytostome	Nonpathogenic
Tricercomonas intestinalis	Humans	Large intestine	Pear-shaped, three anterior flagella, one trailing flagellum	Nonpathogenic
Giardia lamblia	Humans	Intestine	Pear-shaped, bilaterally symmetrical, two nuclei, two axostyles, four pairs of flagella	*Diarrhoea*
Hexamita spp.	Turkeys, pigeons, chickens	Small intestine	Elongate, two anterior nuclei, two pairs anterior flagella, two trailing flagella	*Severe enteritis*
Callimastix				
C. equi	Horses	Colon, caecum	Kidney-shaped, 12-15 flagella at hilus, 12-18mm long; 7 – 10mm wide	Nonpathogenic
C. frontalis	Cows	Rumen	Anterior end disc-shaped, 12 flagella, 30mm long	Nonpathogenic
Sporozoans				
Cryptosporadium				
C. parvum	human, cattle	Intestine	–	Highly pathogenic
C. baileyi	chicken	Trachea, bursa cloaca	–	Pathogenic
C. nasorum	fishes	Stomach, intestine	–	Pathogenic
E meria				
E. necatrix	chicken	Small intestine, ceca	–	Highly pathogenic
E. ducephalae	ducks	Small intestine	–	Highly pathogenic
E. bovis	cattle	Intestine	–	Moderate pathogenic
E. arloingi	sheep	Small intestine	–	Moderate pathogenic
E. deblecki	pigs	Intestine	–	Moderate pathogenic
E. stiedai	rabbits	Intestine	–	Moderate pathogenic

Contd.. Table 5.1

Parasite Amoeba	Principal Hosts	Habitat	Main Characteristics	Disease
E. canis	dogs	Intestine	–	Moderate pathogenic
Taxoplasma gondii	human domestic cats	Reticuloendotholial system and CNS		Toxoplasmosis
Isospora suis	pigs	Small intestine	–	–
I. bigemina	dogs	Small intestine	–	
Sarcocystis tenella	cat, dog, sheep, goats	Intestine Muscle		Sarcocystiosis
S. lindermanni	cat, dog, humans	Intestine Muscle		
S. gigantea	sheep	Muscle		
S. suihominis	primate, pig	in habit host RBC		
Plasmodium				*Malaria*
P. ovale	human	Compact in apperance		*Ovale tertain*
P. vivax	human	amoeboid in appearence, cytoplasmic pigment in RBC		*benign tertain*
P. malariae	human	1/3 of RBC, compact somet-times band shaped.		*Quartain*
P. falciparum	human	Small, rarely observed		*Malignent subtertain (cerebral malaria)*
P. gallinareum	domestic hens			
P. lophure	chickens, ducks turkeys			
Babesia				
B. bigemia *B. bovis* *B. major* *B. divergens*		inhabit host RBC		*red water fever*
B. equi	horses	inhabit host RBC		*bilary fever*
B. ovis	sheep, goat	inhabit host RBC		low pathology
B. trautmani	pig	inhabit host RBC		low pathology
B. canis	dogs	inhabit host RBC		*ticks fever*
Theileria		RBC		High pathology;
T. annulate	cattle, buffaloes	Lymphocytes		*Medittarian*
T. parva	cattle, buffaloes	Lymphocytes		*coast fever*
T. ovis	sheep, goats	Lymphocytes		

As shown in Table some protozoans cause diseases in man as well as animals and these are termed as parasitic protozoa. They occur in all classes of protozoa. The biology of few representative of parasitic protozoa as well as important diseases of protozoans are briefly described.

A. *Biology of Protozoan Parasites*

1. *Entamoeba*

It is cosmopolitan in distribution but seen more often in tropical countries with poor sanitary conditions. The genus *Entamoeba* include numerous parasitic species in vertebrates. These amoeba possess a vesicular nucleus with a centrally or acentrally placed endosome and with chromatin granules along the inner surface of nuclear membrane. The best known species is *E. histolytica* which may live as commensal in the lumen of large intestine or it may invade the host tissues. It is causative agent of *amoebic dysentry* or *amoebias*is (Fig. 5.1). The uninucleate trophozoite of *E.histolytica* is very common in lumen of colon and rectum of humans and primates.

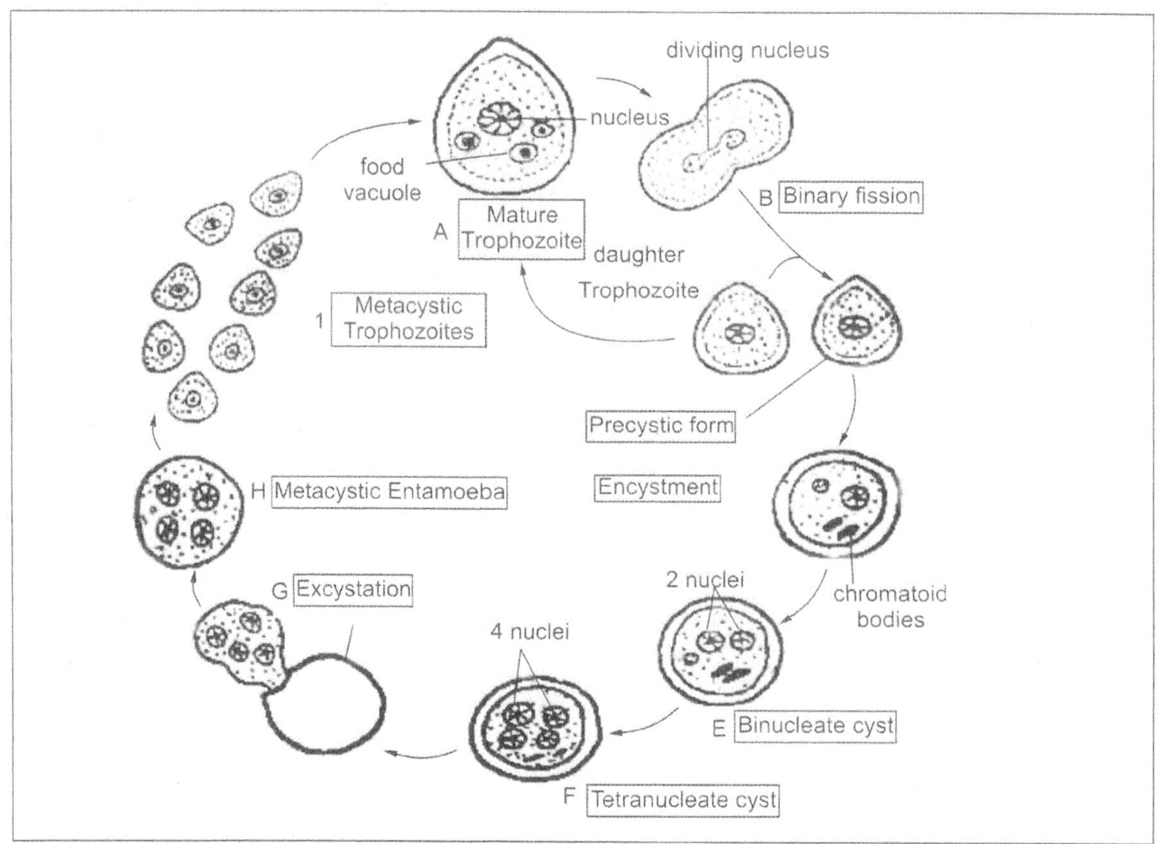

Fig. 5.1 *Entamoeba histolytica* life-cycle

It measures 15-60 mm in diameter, and typically monopodial i.e., with one podium. The cytoplasm indistinctly differentiated into ecto and endoplasms containing food vacuoles that enclose host erthryocytes, leucocytes, fragment of epithalial cells etc. Within host's gut, the trophozoite multiply asexually by binary fission. Some trophozoites invade the lining of large intestine with help of proteolytic enzymes causing amoebiasis. Few trophozoite may enter blood circulation and reach liver, lungs, brain and other organs producing abcesses. This is known as *secondary amoebiasis*. Among precystic form, few trophozoites of size 3.5 to 20 μm in diameter encyst. These cysts are infective forms and are spherical, smaller chromatidal bodies (stored nutrient, ribosome etc.) present. They are highly resistent and passed out in feaces. When food or water contaminated are the cysts pass along the alimentary tract to ileum, where excystation occurs and metcystic amoebae pass into large intestine and begin to reproduce by binary fission.

2. *Plasmodium*

Plasmodium, a protozoan intracellular parasite in red blood corpuscles (RBC) and liver cells of man. It causes dreadful disease called 'malaria'. About 60 species of plasmodium live in man and other animals like birds, bats, squirrels and reptiles. The medically most important species are *plasmodium vivax, P falciparum, P. ovale*, and *P malariae*. In 1897 *Sir Ronald Ross* a British Scientist while working in Secunderabad, India identified *Plasmodium* oocyst in stomach of female anopheles mosquito and described plasmodium life cycle for which he was awarded Nobel prize. Plasmodium

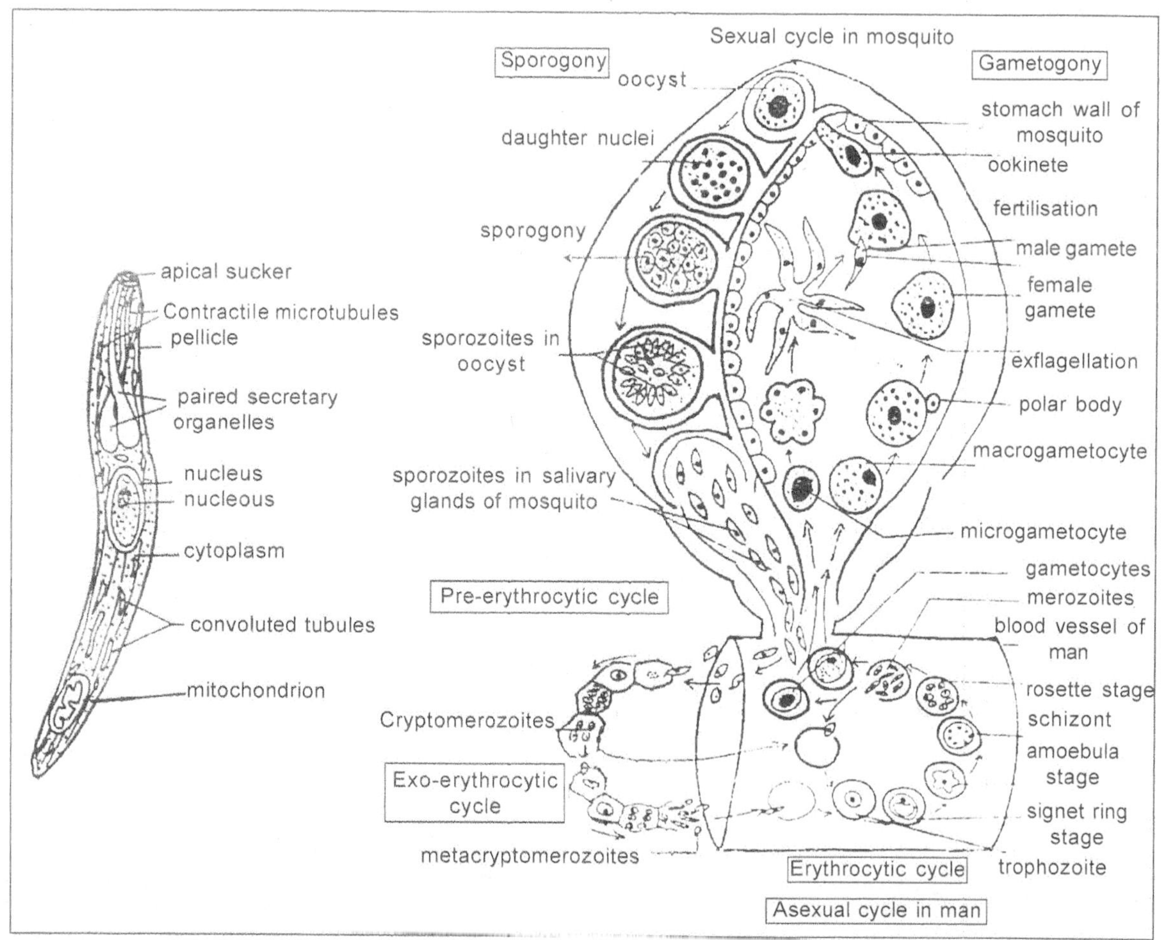

Fig. 5.2(a) Plasmodium - structure of sporozoite

Fig. 5.2(b) Plasmodium vivax - life cycle

passes through different stages. In liver cells of man it appears in the form of sporozoites and merozoites, in RBC it is in trophozoite and amoebula stage. In stomach of mosquito, it occurs in ookinete stage. It infects the man in sporozoite stage. The sporozoite is spindle like with a swollen middle part and pointed at two ends (Fig. 5.2(a)). It has dimensions of 14 μ in length and 1 μ in width. Body is covered with an elastic pellicle and consists of nucleus, golgi body, endoplasmic reticulum, mitochondria and an apical cup with secretary organelles containing cytolytic enzymes which helps in penetration at anterior end of the body. The life cycle of plasmodium is completed in two hosts, the primary host is female anopheles mosquito and secondary or intermediate host is man. Plasmodium reproduces asexually in man and sexually in mosquito. When, vector, female anopheles mosquito bites man the sporozoite of plasmodium enters human blood, reproduces asexually through *schizogyny* first in liver cell and later in RBCs producing *merozoites*. The Asexual cycle in man consists preerythrocytic (sporozoite → cryptozoite → cryptomerozoites (merozoites) cycle in liver cell in 8 days) exoerythrocytic cycle (some of merozoites of pre-erythrocytic cells enter liver cell and produce two types of merozoites, macro merozoites attack liver cells and micromerozoites attack fresh RBCs) and *erythrocytic* cycle - (merocytes enter RBCs and are multiplied in RBC by schizogyny released into blood along with yellowish brown pigment called *haemzoin* which produces malaria) completed in 48 h in *P. vivax*. *P. falciparum* and *P. ovale* and 72h in *P. malariae*. Owing repeated cycle of schizogyny man becomes weak, then merozoites transform into male and female gametocysts in RBC. These gametocyte continue the life cycle when they reach primary host female anopheles while they suck blood from malaria patient. The gametocytes enter sexual life cycle or *Cycle of Ross* which has the following stages : *gametogony*, fertilization, formation of ookinete, formation of oocysts and finally *sporogony* as shown in Fig. 5.2(b). At the end of cycle each oocyst produces about 10,000 sporozoites, which are released into haemocoel of mosquito and travel into salivery glands and ready for infecting man with mosquito bite.

3. *Haemoflagellates – Leishmania and Trypanosoma*

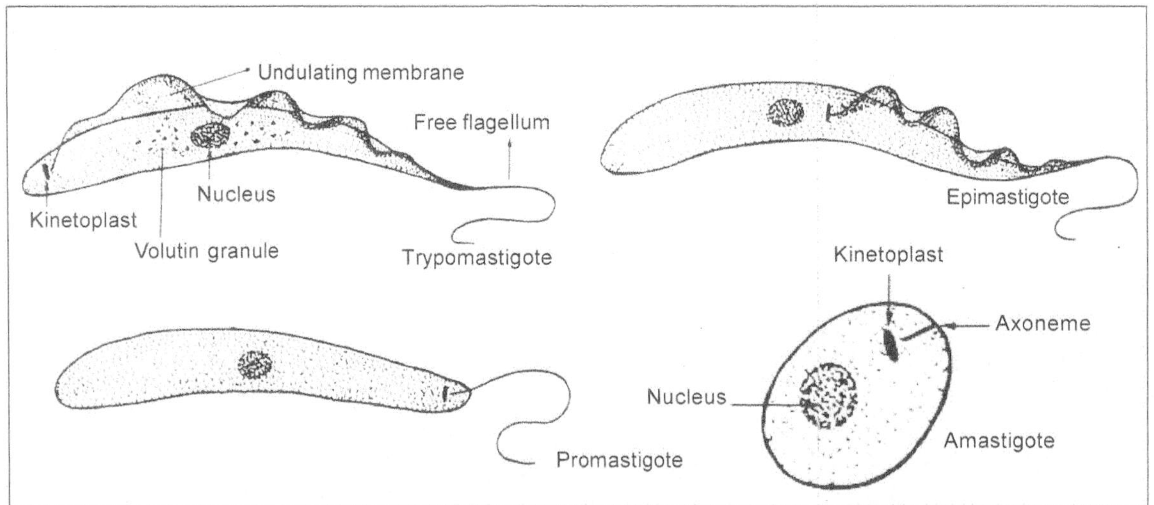

Fig. 5.3 (a) Morphological forms in haemoflagellates. *Trypomastigote* – Kinetoplast is at the extreme posterior end. *Epimastigote* – Kinetoplast is anterior to the nucleus. *Promastigote* – Kinetoplast is at the extreme anterior end. *Amastigote* – Free flagellum is absent and the cell has become ovoid.

(a) *Leishmania spp.*

About 7 species are clearly differentiated. These are *L.donovani, L.major, L.tropica, L. ethiopica, L.mexicana, L.brazilensis* and *L.percuiana*. The four morphological types- trypomastigate, epimastigote, promastigote and amastigote can occur in haemoflagellates (Fig. 5.3(a & b)). Morphologically, the *amastigote* (*Leishmania*) stage of various species of Leishmania are indistinguishable. Each is ovoid measuring about 2 to 3-4 µm and known as Leishman -Donavan body. Multiplication is by binary fission. The most characteristic feature of *trypomastigote* (*Trypanosoma*) is the locomotory flagellum, a part of which is attached to the body of parasite and the remaining is free. The *trypomastigote* has pellicle with microtubules and help in maintaining its shape and flexibility during movement. Kinetoplast is clearly visible and mitochondria is highly developed. The *promastigote* (leptomonad) form is elongated with a large nucleus and a short flagellum arising from a blepharoplast, located near the anterior end of the body. The *epimastigote* (crithidia) form has elongate body with blepharoplast and kinetoplast being immediately anterior to the nucleus. Undulating membrane is inconspicuous.

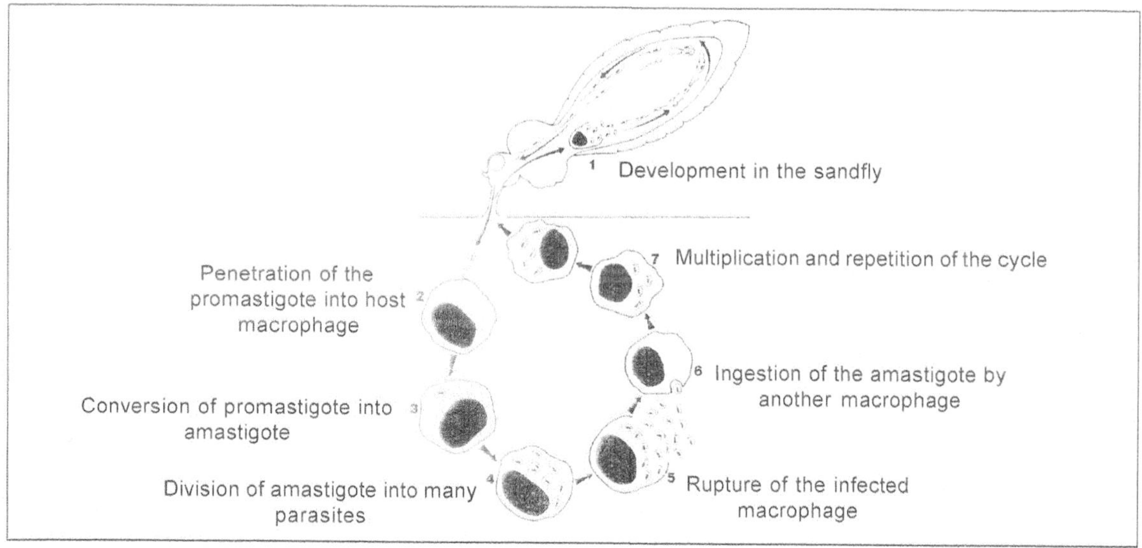

Fig. 5.3(b) Life cycle of Leishmania sp.

The vectors are female sand flies of genera *Phlebotomus*. The promastigote forms of Leishmania are injected into the tissues of host by sand fly. They enter the various phagocytic cells, loose their flagella and become amastigotes. The amastigotes multiply by binary fission inside host cell and liberated amastigotes invade new cells. The infection is transmitted to the vector when it ingests host cells containing the parasites. Inside the vector, the amastigote form which is infective. The major clinical manifestations are visceral, cutaneous and muco-cutaneous leishmaniasis.

(b) *Trypanosoma spp.*

Trypanosomes infecting man and domestic animals belong to the following species *T. rhodesiense, T. gambiense, T. cruzi, T. evansi, T. brucei, T. lewisi, T. suis, T. theileri* etc. These produce the characteristic diseases in man and domestic animals as shown in Table 5.1. The African trypanosomes (*T. gambiense* in equatorial Africa and *T. rhodesiense* in east Africa), causative agents of sleeping

flagellum
anterior end
pellicle
ectoplasm
endoplasm
undulating membrane
nucleus
endosome
volutin granules
basal granule
rhizoplast
parabasal body
posterior end

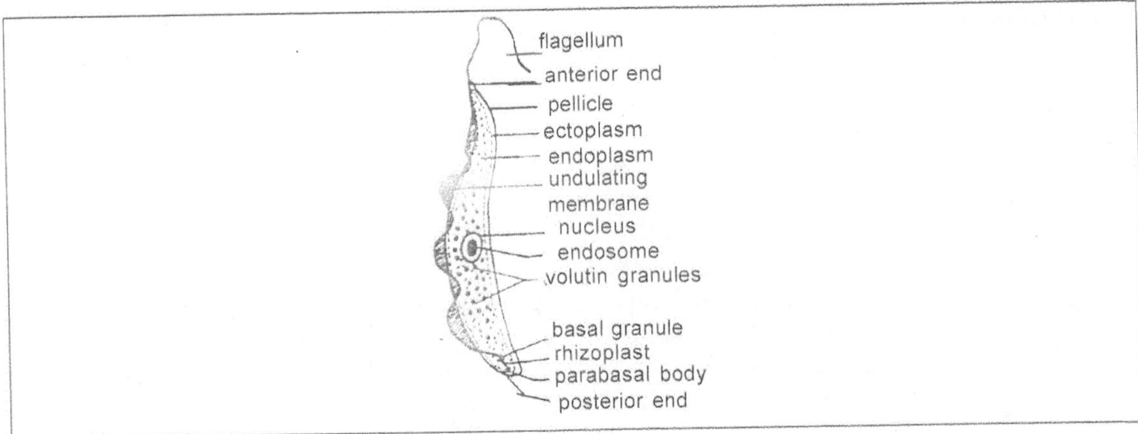

Fig. 5.4(a) Trypanosoma

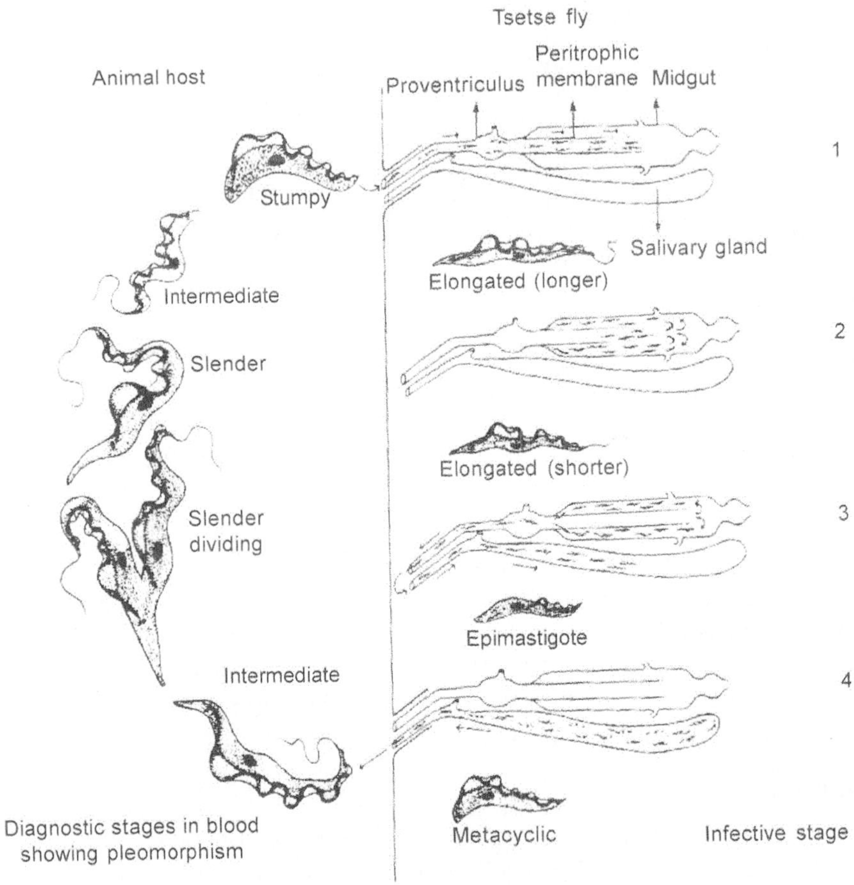

Tsetse fly
Animal host
Proventriculus
Peritrophic membrane
Midgut
1
Stumpy
Salivary gland
Elongated (longer)
Intermediate
2
Slender
Elongated (shorter)
3
Slender dividing
Epimastigote
Intermediate
4
Diagnostic stages in blood showing pleomorphism
Metacyclic
Infective stage

Fig. 5.4(b) Life cycle of *Trypanosoma rhodesiense and T. gambiense*

sickness are transmitted by tsetse fly, *Glossina*. All trypanosomes (Fig. 5.4(a)), are mononucleate, body bears a single flagellum from blepharoplast, and possesses kinetoplast and exhibit polymorphic (four) forms *viz.*, – *Leishmania* (amastigote), leptomonad (promastigote), crithidia (epimastigote) and Trypanosome (trypomastigote). Two or more of these forms occur in life cycle of Trypanosomes. The trypanosome in infected blood are extremely active, about 15 μm long with a single flagellum with undulating membrane. Trypanosomes undergo morphological changes during their life cycle. In the animal-man host blood stream, the slender intermediate and stumpy forms are seen (Fig. 5.4(b)). The stumpy forms infect the tsetse fly in its bite. Inside *Glossina*, stumpy forms changes into elongate form in peritrophic membrane of host. In the space of peritrophic membrane and gut wall, these become shorter, enter hypopharynx and salivary glands. At this stage, they change into epimastigote forms. In salivary glands, they develop further and finally change to metacyclic forms which are injected into the tissues with the saliva.

T.cruzi, causative agent of *chagas disease*, is common in south and central America. The vectors are blood sucking reduvid bugs of genera *Triatoma* and *Rhodnius*. It has large animal reservoirs and is found in armadillos, opossums, cats, dogs, and pigs in endemic areas. The infective metacyclic forms are passed in faeces of infected vector, which is usually deposited at the site of bite. The metacyclic forms enter blood stream via skin damaged into cells of mesenchymal origin, mainly cardiac muscle. Inside the muscle they change to amastigote, epimastigote and trypomastigote forms. Replication occurs at amastigote stage. Trypomastigote emerge from the muscle and enter the blood stream. They may reinvade the muscle or circulate without division, which enter the vector when they bite the host. These change to metacyclic forms through epimastigote form in hindgut of vector and which are passed off in feaces.

4. *Sarcocystis*

Sarcocystis, a coccidian parasite, form characteristic cysts in muscle of infected animals and humans. The cysts have thick straited wall and contain zoites which are slightly curved elongated bodies. The life cycle involves 2 hosts, a predator and prey. The sexual stage develop in the intestinal tract of predator like cat, dog etc. and sexual stages in various tissues of prey like sheep, cattle, pig etc. The life cycle, as shown in Fig. 5.5, the sporozoites liberated from the sporocysts of intestinal tract of prey, develop in the intestinal cells and blood vessels before entering the muscle cells. The muscle cells rupture in the intestinal tract of predator, liberating cystozoites, which enter the intestinal cells producing male and female gometocytes. Fertilization of female gametocytes result in the formation of oocytes and sporocysts. The sporocysts are infective to the prey and the cystozoites are infective to predator.

5. *Trichomonas*

It is a multiflagellate and most common endoparasite. The parasitic human species of *Trichomonas* are *T. vaginalis* (Fig. 5.6) of vagina, *T. hominis* of colon, and *T. tenax* of mouth. The other species include *T. foetus* of cattle and *T. gallionae* of doves. Trophozoite is actively motile and moves by gliding motion. *Trichomonas* body is pear or ovoid shaped with slightly tapering at posterior end. The anterior end bears basal body complex consisting of four flagella. The trailing flagellum border the

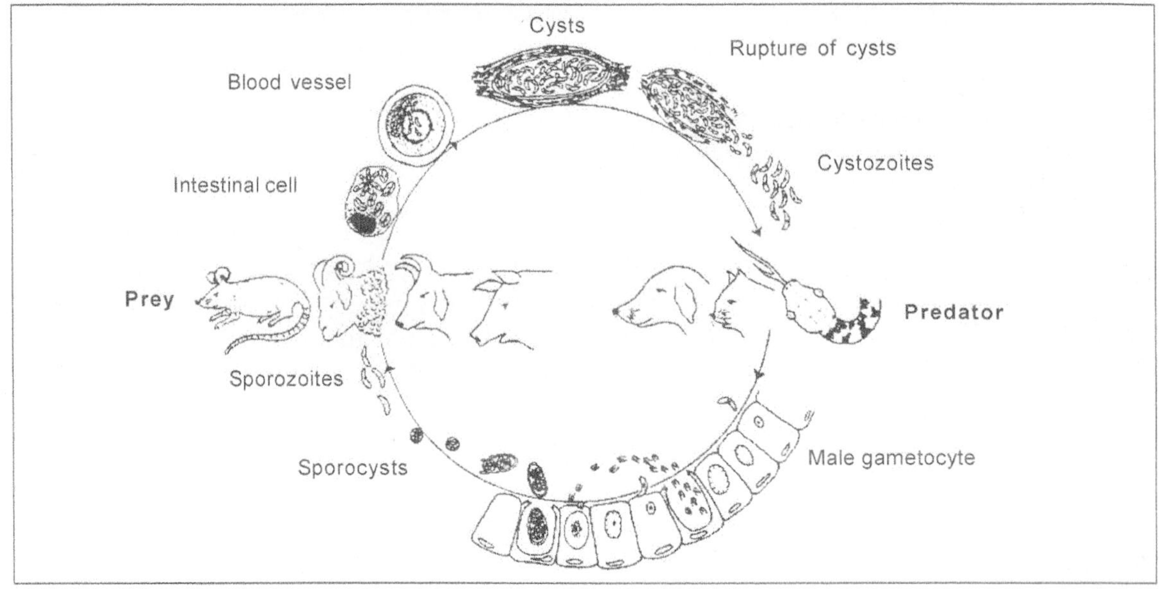

Fig. 5.5 Life cycle of *Sarcocystis*

undulating membrane. Two large straited parabasal fibres also arise from basal body complex. The costa, the contractile fibres, runs along the undulating membrane. Body is supported by central axostyle which also projects posteriorly serves to anchor the animal when it feeds. Broader anterior end bears ventral cytostome, which helps in ingestion of food. It is mainly holozoic and rarely saprozoic but rarely some time of saprozoic. Reproduction takes place exclusively by longitudinal binary fission. The disease caused by species of *Trichomonas* is called '*trichomoniasis*'.

6. *Giardia*

It is a diplomonadid parasitic flagellate occurring in intestine of humans and other animals and known as grand old man of intestine. *Giardia intestinalis*, intestinal or duodenal parasite of man, has bilateral symmetrical body with a length of 10-18 μ. Four flagella on each side of body are present. A single

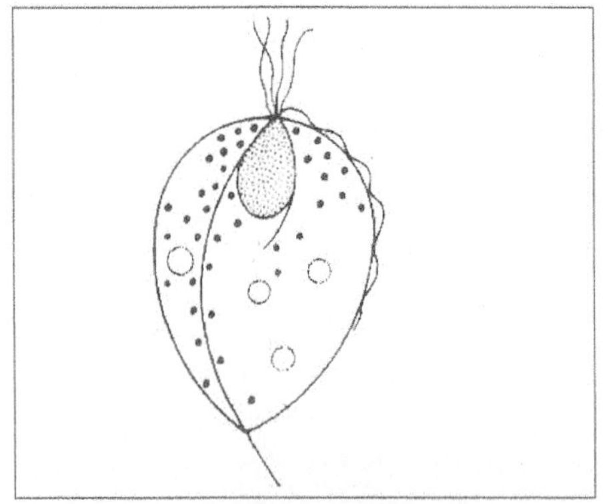

Fig. 5.6 Trichomonas vaginalis

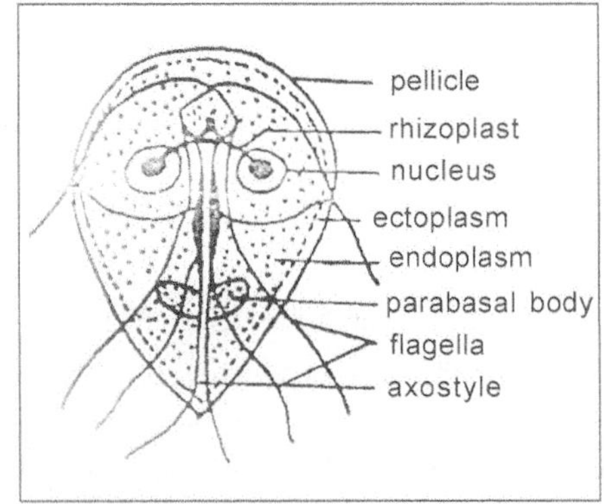

Fig. 5.7 Giardia

axostyle forms the longitudinal axis of body. Infection results from ingesting cysts via food and water. Excystation occurs in the upper region of small intestine where parasite multiplies by longitudinal binary fission (Fig. 5.7). Infection is by uninucleate cysts and infect others through contaminated food and water. Giardia interferes with normal digestive process. It causes diarrhoea, fever, anemia and allergic reactions.

7. *Balantidium*

It is a trichostomatid ciliate parasite inhabiting the intestine of man, pig, sheep, guinea pig, camel, ostrich, fish, frog etc. The *B. coli* infects the man. The parasite has two stages in its life cycle, the trophozoite stage and the cystic stage. Both these stages have large macronucleus and a small micronucleus may also present in concavity of macronucleus. The *tropozoite* is 30 to 300 μm in size and has two contractile vacuole. At the anterior end it has funnel shaped depression peristome, which leads to mouth. Nutrition is holozoic. The cilia, the locomotary organs are present on pellicle in longitudinal rows. The *cyst* of Balantidium is spherical or ovoid and is 40-60 μm in diameter. The cyst wall is thick and transparent and parasite is visible inside the cyst. Reproduction is by binary fission (Fig. 5.8). The phenomenon of conjugation is also seen. Human infection occurs from close association between human and pig. The direct infection to human to human transmission can also occur through contaminated food and water. The infective stage of parasite is the cyst. Excystation occurs in the small intestine and the parasite multiply in the large intestine by binary fission. The parasite may be excreted in the faeces as cysts or trophozoites. It causes the 'Balantidiasis'.

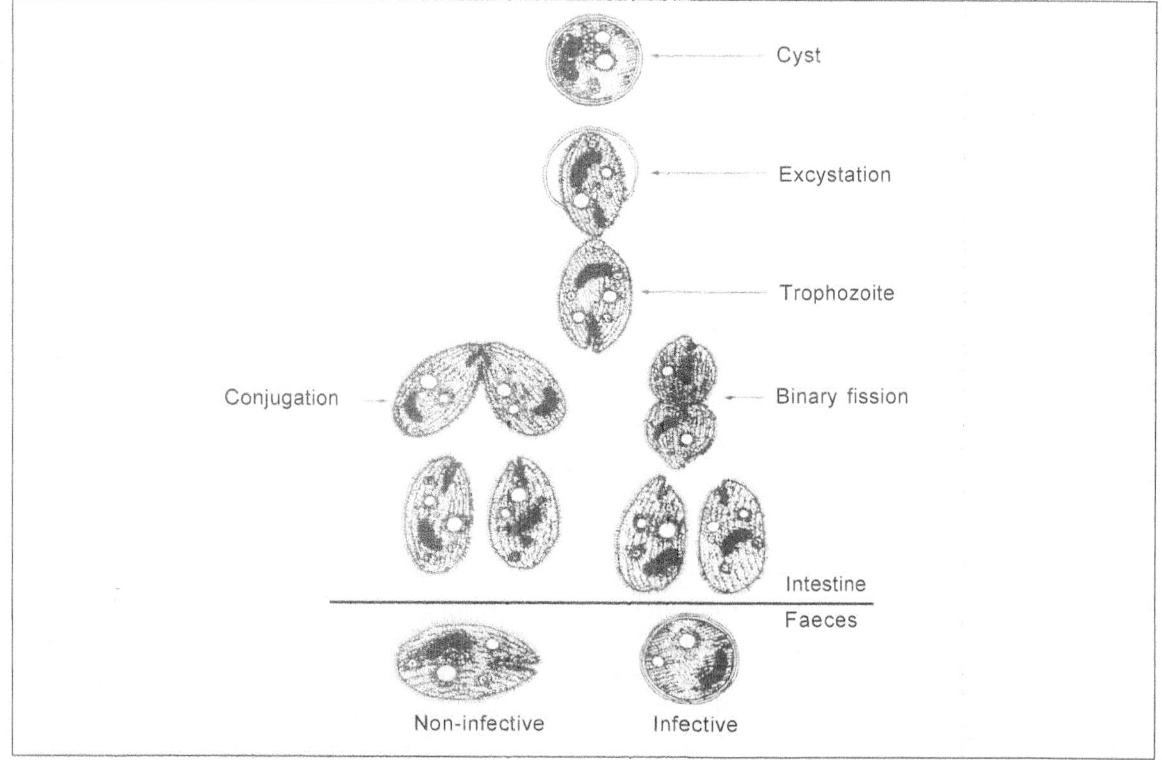

Fig. 5.8 Life cycle of Balantidium coli

8. *Toxoplasma*

Toxoplasma gondii is very common parasite of man and animals causing '*Toxoplasmosis*'. The range of animals involved is very large and include herbivores, carnivores and omnivores. However, definite hosts are members of cat family. The development pattern is like any other coccidian (Fig. 5.9). The infective stages which on ingestion penetrate the epithelial cells of intestine. They round up and grow within the host cells and asexual form of division occurs first leading to the formation of merozoites. Some merozoites enter extra intestinal tissues resulting in the formation of tissue cysts in other organs of the body. Other merozoites are transformed into sexual stages initiating gametogony. A macrogamete is fertilized by a motile microgamete resulting in the formation of an oocyst. The oocysts are found in faeces in an immature form. During sporogony, which occurs in the external environment, two sporoblast form initially and become sporocysts which ultimately form four sporozoites. This is mature oocyst and infective to other animals. The extra intestinal development can also simultaneously occur with intestinal phase.

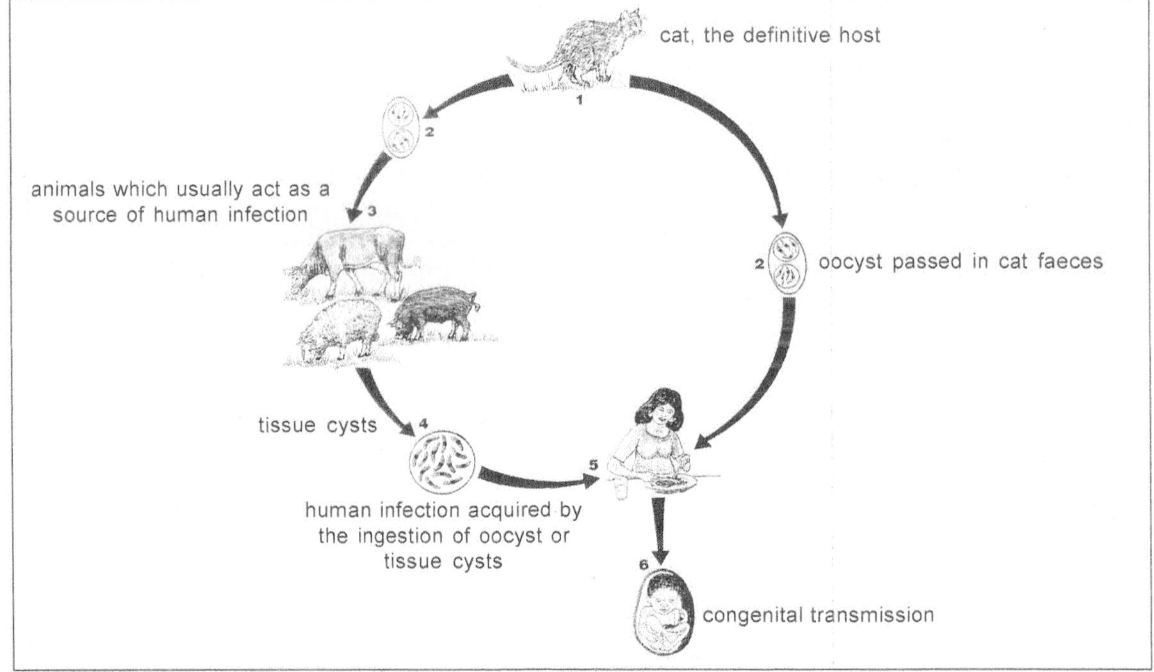

Fig. 5.9 Life cycle of *Toxoplasma*

The infection of Toxoplasma in man occurs after ingestion of oocysts from cat or by eating improperly cooked meat containing cysts or meat of various domestic animals, like pork, mutton, beef and poultry. Only asexual development occurs in man and oocysts are not formed in the intestine. Merozoites arising from the asexual development enter the lymphatics as well as the blood and from pseudocysts and cysts in various organs of body. Toxoplasmas are essentially intracellular parasites. They enter host cells either by rupturing the membrane or by invaginating it. Within the macrophage the parasite multiplies by repeated endodyogeny giving rise to large colonies. The macrophage finally rupture liberating the endozoites. The cyst formation occur mostly in central nervous system, eye, cardiac and skeletal muscle.

B. *Some Important Protozoan Diseases and their Control*

1. *Amoebiasis*

Amoebiasis or amoebic dysentry, endemic in warm countries, is caused by *Entamoeba histolytica.* Its trophozoites penatrate the walls of hosts intestine, secrete histolytic enzymes and feed upon its cells resulting in formation of ulcers. These ulcers rupture, discharge blood and mucus into the intestine which passes with stools. The trophozoites under certain circumstances reach liver, lung and brain where they cause abscess formation. Life cycle of E. histolytica is without intermediate host and infection takes place through tetranucleate cysts. The cysts through feacal matter reach environment, contaminate food and water and spread to new hosts. Some times house flies help in contamination.

The infection of *E. histolytica* results in amoebic dysentry and later as hepatic amoebiasis. The symptom of amoebiasis is diarrhoea which accompanied by blood, mucus, tenesmus with abdominal tenderness. In chronic amoebiasis diarrhoea alternate with constipation, abdominal pain, dyspepsia, and asthenia. Cysts are passed in stools. In hepatic Amoebiasis, hepatomegaly marked tenderness of liver, mild to high grade fever, leucocytosis and raised ESR are found.

Prevention and control is closely linked to food hygiene, fly-control, sewage disposal, proper water supply, identification and treatment of carriers. Water should be boiled or filtered to remove the parasite. Iodination of water can kill the parasite. Nitroimidazole derivatives (metronidazole, tinidazole, ornidazole) are used to eliminate the trophozoites. Emetin and its derivatives are used to control ulcerative amoebiasis.

2. *Trypanosomiasis*

Trypanosomiasis is caused by the species of Trypanosoma which are flagellate parasites of blood of vertebrate hosts and of gut in invertebrate host. The *sleeping sickness* (gambian form) a dangerous disease of man in tropical Africa, is caused by *T. gambiense* which is transmitted by TseTse fly, *Glossina palpalis*. With initial interstial inflammation of skin the parasite lodges itself in blood stream followed by infestation of lymphatic system leading to glandular swelling which is symptomatic for sleeping sickness. Later, the parasite penetrates into cerebrospinal fluid, cause damage to brain and results in lethargy, which ultimately leads to death if it is untreated. The pathology of CNS consists laptomeningitis, and cellular infiltration by mononuclear cells around blood vessels causing perivascular cutting. The symptoms of disease also include headache, arthritic pain, weakness of legs, precordial pain, disturbed vision, anaemia etc. *Rhodesian form* of sleeping sickness, caused by T. rhodesiense is mainly a zoonosis, while Gambian form is primarly a human infection. The trypanosomiasis may be prevented by protecting human being from bites of Tse Tse fly, isolating infected persons and treatment regimen include the drugs like Suramine , Diamidines, Germanin, Lamidine, Melarsoprol.

T. Cruzi is the causative agent of *American trypanosomiasis* or *chagas disease*. It is widespread in south and central America and is more common in children. It is transmitted by bug, Triatoma. The transmission is through bugs feacal material, via exposed skin or mucous membrane. The parasite entry into macrophages produce a small granuloma (chagona) which obstructs the flow of lymph and progressively results in systemic toxemia. Further, parasite entry into visceral organs, myocardium, endocrine glands and brain cells leads to severe pathological condition. Swelling of body parts, severe headache, continuous fever, Anaemia and later failure of cardiac muscle leads to death. Primaquine and pauromycin are used for temporary relief. Chagas disease can be prevented by checking the presence of Triatoma bug.

3. *Leishmaniasis*

The species of Leishmania, parasitic flagellates that inhabit phagocytic cells in body organs in vertebrates and gut in insect hosts cause Leishmaniasis. L.donovani causes *kala-azar or visceral leishmaniasis* in humans. The disease is widespread in India, south China and in mediterranian countries. The major characteristic feature of disease is a considerable enlargement of spleen due to blockage of reticulo endothelial system by parasites. The earliest symptoms are generally low grade fever with malaise and sweating. In later stages, the fever becomes intermitant and both liver and spleen grossly enlarged because of hyperplasia of lymphoid - macrophage system. Treatment with antimony compounds, Neostibosan, urea stibamin is useful. *L.tropica* is the causative agent of *cutaneous or skin leishmaniasis* (oriental sore). Infection is restricted to endothelium of skin capillaries and leads to lump like boils. Disease is endemic in tropical countries especially south west Asia, eastern Mediterranian and tropical American countries. Treatment include regular cleaning and dressing of boils with Atebrine and Berberine sulphate. L.brasillensis causes a disease called *Espurdia* producing multiple sores over large areas of body. Development of ulceration in nasal cavities, mouth and pharynx is quite frequent.

Leishmania species are transmitted from man to man by bites of sand flies belonging to genus *Phlebotomus*. The elimination of these through proper insecticidal practices can prevent the disease. The disease is also controlled by treating effected person with pentavalent antimony compounds. The cutaneous leishmaniasis can be treated by sodium stibogluconate or mepacrine.

4. *Malaria*

Malaria is one of the serious disease of human beings occuring in India, China, Phillipines, South Africa, Mexico, USA, Japan etc. and epidemic in form. It is caused by species of Plasmodium transmitted through the bites of female anopheles mosquito. Man is the principal host. Four species of plasmodium cause different type of malaria in human being viz, P. vivax, P. ovale, P. malariae and P.falciparum. The plasmodium attacks the RBC and liver cells and releases toxic substance *Haemozoin* which causes malaria. Malaria is characterized by periodic attack of fever. The plasmodium parasite makes the patient weak, causing anaemia ultimately resulting in death. The fever is repeated in *tertian malaria* caused by P. Vivax every third day, in *ovale malaria* caused by P. ovale every third day, in *quartan malaria* caused by P.malariae every fourth day and in *malignant tertian malaria* caused by P.falciparum every second day. In case of falciparum, the brain is infected with parasite resulting in congestion, swollen and haemorhage condition of cerebral blood vessel. Thus, this infection produce serious complications involving brain (cerebral malaria), massive haemoglobinuria (black water fever), acute respiratory distrome syndrome, hyperpyrexia, hypotension and shock (algid malaria). The cascade of pathophysiological effect of malaria is shown in Fig. 5.10.

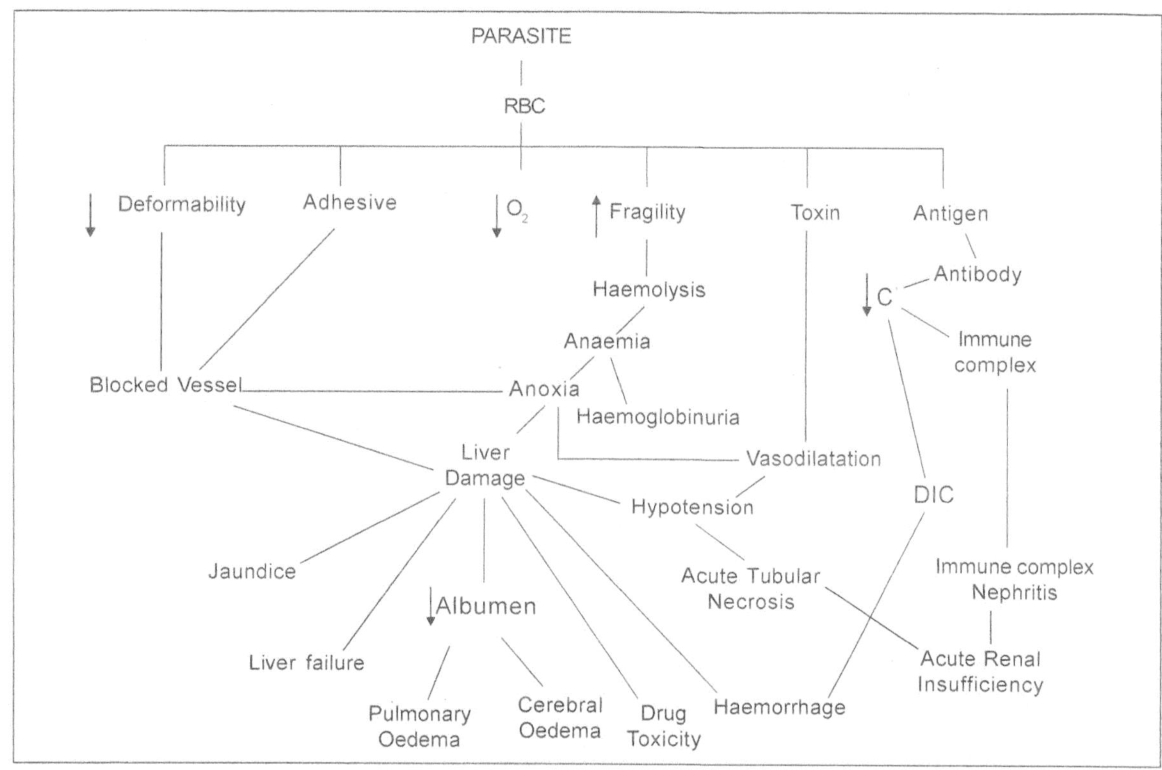

Fig. 5.10 The complex pathophysiology of P. falciparum infection

The prevention of malaria include the avoiding the bites of infected mosquito through physical and chemical measures. The various drugs which are now used in treatment of malaria include Quinine, Atebrin, Comoquine, Chloroquine, Paludrine, Daraprim, Primaquine, Proguaril etc.

5. *Trichomoniasis*

Trichomoniasis is caused by flagellate protozoan *T. vaginalis* that inhabits vagina of women and causes vaginitis. The parasite secrete toxic substance that injure the cells. Disease is characterized by inflammations, burning sensation, itch and frothy vaginal discharge. Although generally thought to be a parasite of women, it also inhabits the male urinogenital tract and prostrate gland. Transmission is always through sexual intercourse by male members who act as intermediaries. Arsenic and iodine drugs, antibiotics like Aureomycin and Tetramycin have proved to be helpful in control of the infection of the parasite.

6. *Toxoplasmosis.*

Toxoplasmosis is caused by *Toxoplasma gondii*, a sporozoan parasite of man and animals (Felidae) with world wide distribution. The infection to man occurs through ingestion of oocyts from cats or improper cooked meat of *domestic animals*, pork mutton, beef and poultry. The parasite occupies the cells of the raticulo endothelial and central nervous system. Usually the parasite remains in hosts body without causing any symptoms, but if an infected women conceives, abortion takes place. And also the nervous system of developing foetus is affected and infant dies after birth. The other symptoms include myocarditis or myositis. Pyrimethamine and sulphadiazine combined treatment is effective remedy. The disease can be prevented by isolating infected cats and proper cooking of infected meat.

7. *Balantidial Dysentry*

This disease is caused by *Balantidium coli*, an intestinal ciliate. It is characterized by diarrohea and ulceration of the large intestine. Transmission of parasite to a new host takes place through cysts in contaminated food or water. The infection is best treated with oxytetracycline or metranidazole.

8. *Giardiasis*

It is characterized by loose bowels, is caused by a flagellate parasite, *Giardia intestinalis*. It is present in the small intestine of man. The parasite may produce a wide range of gastrointestinal symptoms especially in children. These include vomiting, flatulence and diarrohea. This also indicated in the intestinal malabsorption syndrome and cholecystitis. The shortening of villi and cellular infiltration of lamina also occur. The symptoms include epigastric pain, abdominal discomfort, loss of apetite and headache. Transmission of parasite takes place through cysts which are passed along with faeces and enter new hosts in food or water. Mepacrine, nitroimidazole, Tinidazole are effective in treatment.

9. *Babesiosis*

The infection produced by piroplasmid protozoan *Babesia* spp is known as *Babesiasis* mimicking the malarial parasite. This is a very important and common parasite in live stock and other domestic animals throughout the world and is conveyed through tick bites. Sporadic cases in humans are also reported. The two species are found in humans are *B. bovis* and *B. microts*. The *B.bigemina*, is found in cattle, cause lethal haemoglobnuric fever, or red water fever. The trophozoite feed on erythrocytes and destroy, resulting in anaemia and give rise to toxins producing the diseases. The bite of tick *Boophilus annulatus* transmits the disease. The Deer, and water buffaloes are also effected by this disease. The symptoms of infection in humans include fever, anorexia, sweating, rigors and general myalgia. Treatment includes the aromatic diamidines.

10. *Theileriosis*

Several members of *Theileria* genus, parasitize cattle, sheep and goats resulting in a disease known as *Theileriosis*. This disease results in heavy losses of domestic animals in southern Europe, Asia and Africa. *T.parva* is causative agent of coast fever in cattle and buffalo. The parasite found in erythrocyte and transmitted by ticks. Life cycle is like Babesian sp. *T. parva* and *T. ovis* also infect sheep and goat causing theileriosis.

11. *Sarcocystiosis*

Sarcocystis, a coccidian parasite, forms characteristic cysts in the muscles of infected animals like cat, dog, pig, sheep, goat, horse and humans. Infection by this parasite known as Sarcocystiosis, is common in animals but relatively rare in humans. The parasite may give pain and swelling in the muscle. The herbivore is intermediate host and carnivore is definite host. *S.hominis, S.lindemanns* are present in human muscle. The symptoms of sarcocystiosis of animals are loss of apetite, fever, anemia, weight loss, lameness, abortion and even death.

Thus, these are the some of the common protozoan diseases of man, livestock, domestic and other important animals.

5.3 Helminth Parasites, Diseases and their Control

Helminths belong to platyhelminthes and nematodes. They are soft bodied, dorsoventrally flattened and bilaterally symmetrical having variety of adhesive organs with well developed parasitic adaptations. Many parasitic helminths (worms) are endoparasites of gut and blood of human and other animals causing a number of diseases which are called *Helminthiasis*. The life history is of complicated type passing through one to three hosts with a number of larval stages. The Helminth parasites belong to three groups, these are Cestoda, Trematoda and Nematoda.

A. *Cestodes*

The parasites of this group are known as *'Tape worms'*. They are endoparasites of man and other animals. The body is slender, elongated, flat, ribbon-like in shape and consists of a row of segments or *proglottids*. The *suckers* and *hooks* are found only on scolex (head region) which help in locomotion and attachment to host tissues. The nervous system consists longitude cords running laterally in each segment. Osmoregulation is by flame cells. The mouth and alimentary canal is absent and the food is absorbed directly from the body wall. Tape worms are hermaphrodite and each segment consists male and female reproductive organs. Almost all tapeworms have an indirect life cycle, the only exception is *Hymenolepis nana* which has direct life cycle and can complete its larval stage in the definite host. About 1500 species of tapeworms are known to be parasitizing the animals from fishes to mammals including man causing a number of troubles as well as diseases for their hosts. Some of the important cestode parasites of economic importance are shown in Table 5.2.

Table 5.2 *Cestode parasites in Humans, Live stock and Domestic Animals and the diseases*

Cestode	Principal Hosts	Habitat	Main Characteristics	Disease Caused
***Diphyllobathrium** latum*	Humans, dogs, cats, Minks, bears, other fish eating mammals	Intestine	Extremely large, 3-10 meters consisting of 3000 or more proglotidds, scolex 2.5mm long; 1mm wide, eggs 70 by 45 mm, operaculate	*Diphyllobothriasis* sometimes anemia
***Diplogonoporus** grandis*	Normally a parasite of whales, occasionally in humans	Intestine	1.4-5.9 meters long, two sets of reproductive organs per proglottid, eggs 63-68 by 50 mm operculate.	Diarrhoea, constipattion, secondary anemia
Spirometra erinaceieuropaei	Dogs, cats, humans	Intestine	85-100 cm long, multiple testes larger than those of *D. latum*, vitellaria numerous, eggs 57-60 by 33-37 mm, operculate.	Similar to *D. latum* infections
S. mansonoides	Dogs, cats, usually bobcats, sparaganum occasionally in humans	Small intestine	Rarely over 1 meter long; scolex 200-500 mm wide, bothria shallow, cirrus and vagina open indpendently on ventral surface, eggs	Diarrhoea and secondary anemia; larvae may cause sparganosis in humans
***Ligula** intestinalis*	Fish-eating birds occasionally in humans	Intestine	Specimens found in humans small, less than 80 cm long	Nonpathogenic
Bertiella studeri	Monkeys, apes, humans	Small intestine	275-300mm long, 10mm wide, scolex subglobose, set off from strobila, eggs irregular in outline, 45-46 by 50 mm, non-operculate.	No apparent symptoms

Contd.. Table 5.2

Cestode	Principal Hosts	Habitat	Main Characteristics	Disease Caused
***Anoplocephala** magna*	Horses	Intestine	350-800mm long, 10mm wide, 400-500 testes per mature progl-tidds, eggs 50-60 mm in diameter, non-operculate.	Anoplocephaliasis, secondary anemia.
A. perfoliata	Horses	Large and small intestine	10-80m long, 10-20mm wide, scolex 2-3mm in diameter with lappet behind each sucker.	Anoplocephaliasis, secondary anemia.
***Paranoplocephala** mamill*ana	*Horses*	Small intestine	6-50mm long, 4-6mm wide, suckers slitlike, eggs 50-88mm in diameter, non-operculate	Nonpathogenic
***Moniezia** expansa*	Sheep, goats, cattle	Small intestine	Up to 4 mtrs. long, 1 cm wide, two sets of reproductive organs per proglottid, 100-400 testes per segment, eggs 50-60mm in diameter, non-operculate	Nonpathogenic
M. benedeni	Sheep, goats, cattle	Small intestine	Up to 4 mtrs. long, larger scolex than *M. expansa*	Nonpathogenic
***Thysanosoma** acti*	Sheep, goats,	Small intestine	150-300mm long; large and proglottids broader margin, two sets of reproductive organs per proglittd, eggs expelled in capsules, each egg 19.25-26.95mm in diameter non-operculate.	Thysanosomiasis
***Taenia** solium*	Pigs, humans	Small intestine	Up to 2-7 mtrs. long, scolex quadrate with diameter of 1mm, rostellum armed with double row of hooks, eggs 31-43mm in diameter, non-operculate	Taeniasis solium
(Cysticercus cellulosae)	Various animals, including man	Various tissues	Oviod, withish, 6-18 mm long	Cysticercosis cellulosae
***Taenia** taeniaeformis*	Cats, Dogs, humans	Small intestine	15-60mm long, 5-6cm wide, armed rostellum, double row of usually 34 hooks	Taeniasis taeniaeformis
***Taenia** hydatigena*	Dogs	Small intestine	Up to 5 mtrs. long, 4-7cm wide, scolex armed with double row of 26-44 hooks, 600-700 testes per segment, gravid uterus with 5-10 lateral branches	Taeniasis hydatigenis
***Taenia** ovis*	Dogs	Small intestine	Approximately 1 mtr. long, scolex armed with two rows of 24-36 hooks, 300 testes per proglottid, gravid uterus with 20-25 lateral branches	Taeniasis ovis
T. tenuicollis	Minks	Small intestine	Up to 70mm long, large suckers, 237-303mm in diameter, two rows totalling 48 hooks, eggs 17-20mm in diameter	Taeniasis tenuicollis
T. pisiformis	Dogs, Cats, Rabbits	Small intestine	500mm long, 5mm wide, scolex with double row of 34-48 hooks, genital pores with 8-14 branches, eggs 36-40 long, 31-36mm wide, non-operculated	Taeniasis pisiformis

Contd., Table 5.2

Cestode	Principal Hosts	Habitat	Main Characteristics	Disease Caused
Taeniarhynchus *saginatus*	Cysticercus in cows, adults in humans	Small intestine	10-12 mtrs. long, no rostellum unarmed, 1000-2000 proglottids, eggs similar to those of *T. solium*	Taeniasis saginatus
Multiceps *multiceps*	Dogs, foxes, humans	Smal intestine	Up to 1 Mtr. long, 5mm wide, scolex armed with double row of 22-32 hooks, approximately 200 testes per proglottid, gravid uterus with 9-26 lateral branches, eggs 31-36mm in diameter	'Gid' in cysticerous infection; like taeniasis in adult infection
M. serialis	Dogs, occasionally humans	Small intestine	70 cm long, 3-5cm wide, rostellum with double rwo of 26-32 hooks, gravid uterus with 20-25 lateral branches	taeniasis
Echinococcus *granulosus*	Hydatid cysts in sheep, horses, deer pigs, humans	Liver and other organs	Hydatid cysts with thick two-layered wall, filled with fluid	Hydatid disease
Hymenolepis *nana*	Rats, mice, humans	Intestine	25-40 mm long, 1mm wide, short rostellum with 20-30 hooks in one ring, three testes, eggs 30-47 mm in diameter.	Hymenolepiasis
Hymenolepis *diminuta*	Rats, mice, humans, dogs	Intestine	200-600mm long, 1-4mm wide, rostellum unarmed, eggs 60-80 by 72-86 mm	Hymenolepiasis diminuta
H. carioca	Chickens, turkeys	Small intestine	300-800mm long, 500-700 mm wide, rostellum unarmed	Hymenolepiasis carioca
H. cantaniana	Chickens, turkeys, quails	Small intestine	2-12 mm long, rostellum shorter than that of *H.carioca*, otherwise the two species are similar	Hymenoplepiasis cantaniana
Fimbriaria *fasciolaris*	Chickens, ducks	Small intestine	14-85mm long, with pseudoscolex	Nonpathogenic
Dipylidium *caninum*	Dogs, cats, foxes occasionally	Small intestine	15-70cm long, 3mm wide, conical rostellum armed with 30-150 hooks, 200 testes per proglottid, eggs in capsules, each 35-60 mm in diameter	Chronic enteritis dipylidiasis
Raillietina *cesticillus*	Chickens, pheasants	Small intestine	100-130 mm long, 1.5-3mm wide, scolex broad and flat and about 100mm in diameter, rostellum armed with double row of 400-500 hooks	Enteritis and hemorrhage
R. tetragona	Chickens, turkeys	Small intestine	Up to 250mm long, 1-4mm wide, rostellum with double row of 90-130 hooks, suckers armed with 8-12 rows of hooklets, 6-12 eggs in single capsule	Enteritis, and hemorrhage
R. chinobothrida	Chickens, turkeys	Small intestine	Up to 250 mm long, 1-4mm wide, rostellum with double row of 200-250 hooks, suckers with 8-15 rows of hooklets	Enteritis and hemorrhage

Contd., Table 5.2

Cestode	Principal Hosts	Habitat	Main Characteristics	Disease Caused
Davainea *proglottina*	Chickens	Small intestine	Up to 4mm long, usually only 2-5 proglottids, rostellum with 66-100 small hooks, one egg per capsule	General physiologic retardation
D. meleagridis	Turkeys	Small intestine	Up to 5mm long, composed of 17-22 proglottids, rostellum with double row of 100-150 hooks, suckers armed with 4-6 rows of hooklets	Unknown
Mesocestoides *latus*	Cats, skunks.	Small intestine	12-30cm long, 2mm wide, scolex unarmed, vitellaria bilobed in posterior region of proglottid	Non-pathogenic
M. Lineatus	Dogs		30cm to 2 meters long, genital atrium midventral	Non-pathogenic

The biology of some cestode parasites of man, livestock and domestic animals is described below.

1. *Taenia saginata*

It is commonly known as *'Beef tapeworm'*. It is cosmopolitan in distribution and more common tapeworm of man in beef eating countries. It is an endoparasite found in posterior coils of intestine of man. *T. saginata* is comparatively bigger in size, about 5 m long and has about 2000 proglottids. The body is divided into scolex, neck and strobila. Four suckers are found on the scolex but hooks are absent. The eggs are spherical, with diameter of 31-43μm, covered with egg shell and outer envelope. Inside the egg is the hexacanth or *oncosphere* which bears 6 hooks. *T. saginata* is obligatory parasite of man and intermediate host is cattle (Fig. 5.11). The adults are located in small intestine of man and gravid proglottids migrate out of the anus or are discharged in faeces. Cattle become infected by ingesting eggs from the pasture. After ingestion the onchosphere hatches out and penetrates the intestine. Then enters the circulation and is transformed into the cysticercus stage in the muscle, known as *Cysticercus bovis*. Man is infected by eating muscles containing cysticercus bovis. After ingestion, the scolex of the cysticercus evaginates in the small intestine, attach itself to the mucous membrane and then develops into an adult worm. It causes *Taeniasis*.

2. *Taenia solium*

It is commonly called as *'Pork tapeworm'* and found in the small intestine of man. It is cosmopolitan in distribution and endemic in all parts of the world where pork or pork products are eaten like Indian sub-continent, Southeast Asia, Indonesia, Latin America, and South Africa. Its body is elongated, flattened and ribbon like with a maximum length up to 3m and differentiated into scolex, neck and strobila. The scolex has four suckers and two rows of retractile hooks. Each proglottid is hermaphrodite with the help of suckers and hooks worm attach with mucosa of intestine and absorbs digested food of host alimentary canal. The life cycle of *T. solium* is like *T. saginata* and involves, a primary host, man and intermediate host, pig, except that the intermediate host pig and larval stage (Fig. 5.12) which is known as *Cysticercus cellulosae* (instead of Cysticercus bovis) can also develop in humans. The adult T.solium infection results in a disease known as *Taeniasis*.

Scolex attach to small intestine
MAN
Gravid proglottid
Egg
Ingested
Cysticercus in muscle (infective stage)
(diagnostic stage)
in faeces
in faeces or environment
Circulation
CATTLE
Oncosphere hatches penetrates intestine
Embyronate egg or proglottids ingested

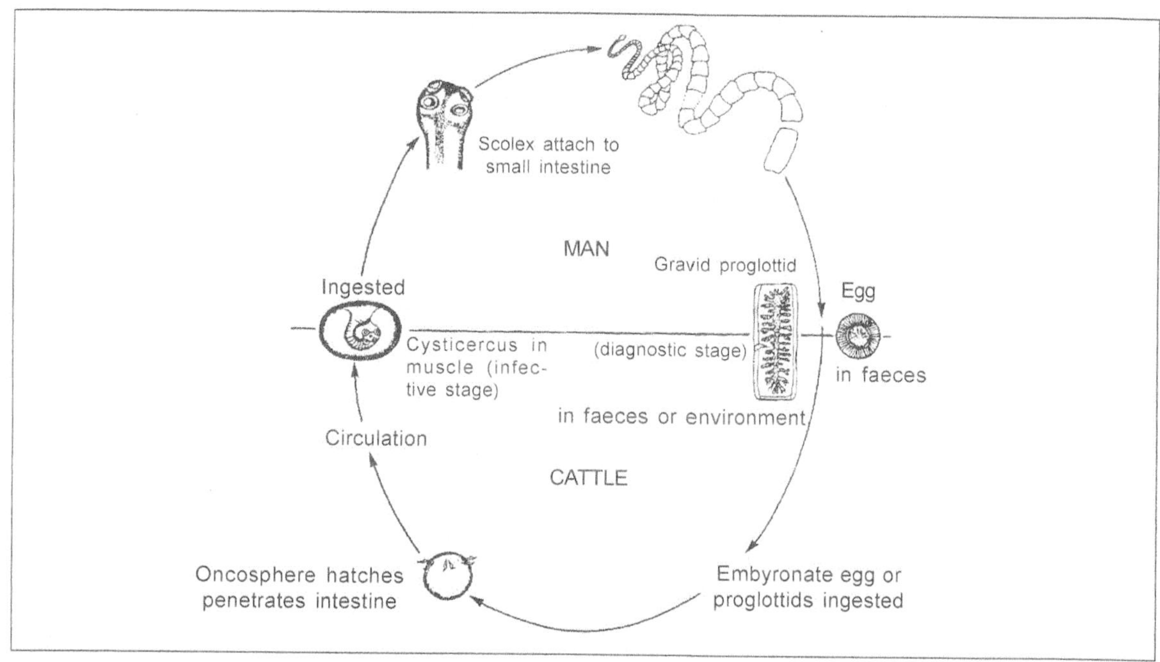

Fig. 5.11 Life cyle of Taenia saginata

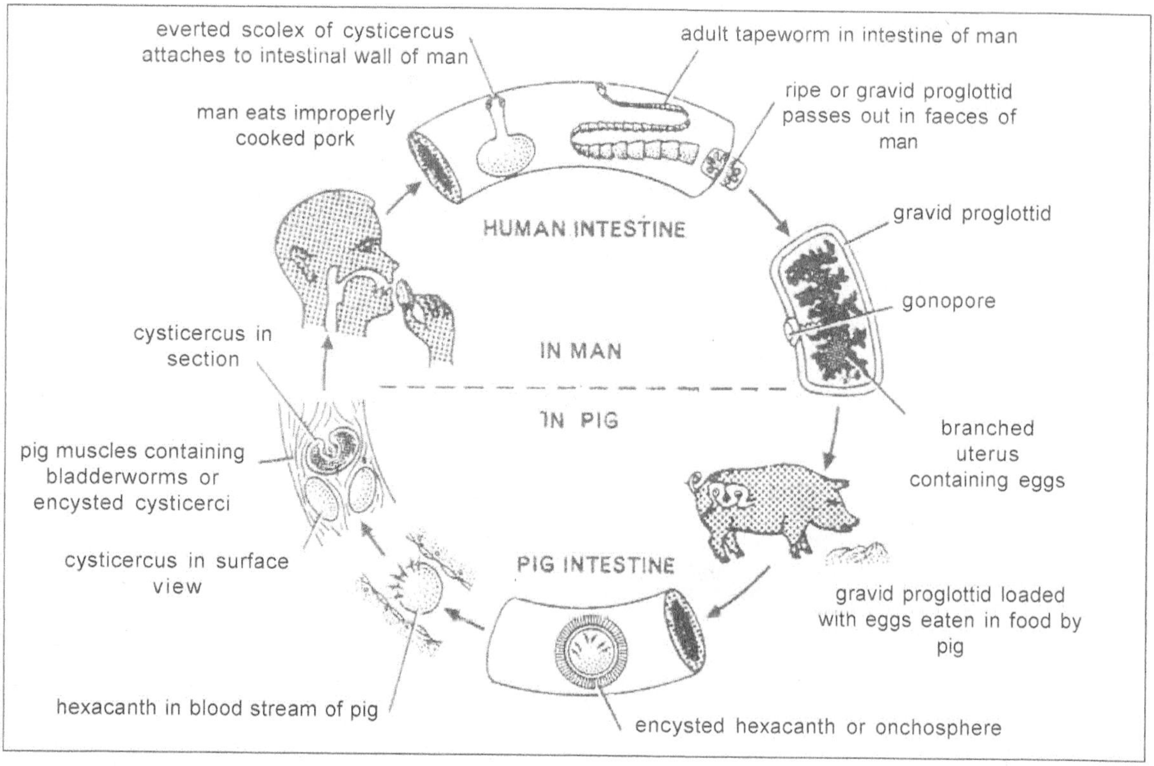

everted scolex of cysticercus attaches to intestinal wall of man
adult tapeworm in intestine of man
man eats improperly cooked pork
ripe or gravid proglottid passes out in faeces of man
HUMAN INTESTINE
gravid proglottid
gonopore
cysticercus in section
IN MAN
IN PIG
branched uterus containing eggs
pig muscles containing bladderworms or encysted cysticerci
cysticercus in surface view
PIG INTESTINE
gravid proglottid loaded with eggs eaten in food by pig
hexacanth in blood stream of pig
encysted hexacanth or onchosphere

Fig. 5.12 Life history of Taenia solium

Cysticercus cellulosae - Human infection with this stage is known as *'Cysticercosis'* and occurs in all areas where *T. solium* is present. *Cysticercus*, is typically a semitransparent fluid filled ovoid body with an average size of 10 × 5 mm. The scolex is invaginated bearing four suckers and a row of hooks. Man become infected by accidentally ingesting eggs of *T. solium* with contaminated food and water. The infection may also occur sometimes with retro-infection, the proglottids are regurginated into the stomach where they disintegrate and liberate a large number of eggs. The oncospheres from eggs then find their way into various parts of the body and develop into cysticerci infecting tissues like muscular and nervous system.

3. *Echinococcus spp.*

Cestodes of genus *Echinococcus* are distributed in many parts of the world. These are : *E. granulosus*, *E. multilocularis*, *E. oligarthus* and *E. vogeli*. *E. granulosus* is commonly known as *'Hydatid worm'* and it inhabits the small intestine of dogs, foxes, cats and other carnivorous animals. The intermediate host is man. The body is formed of scolex, neck and 3-4 proglottids, 3-6 mm in length. The scolex

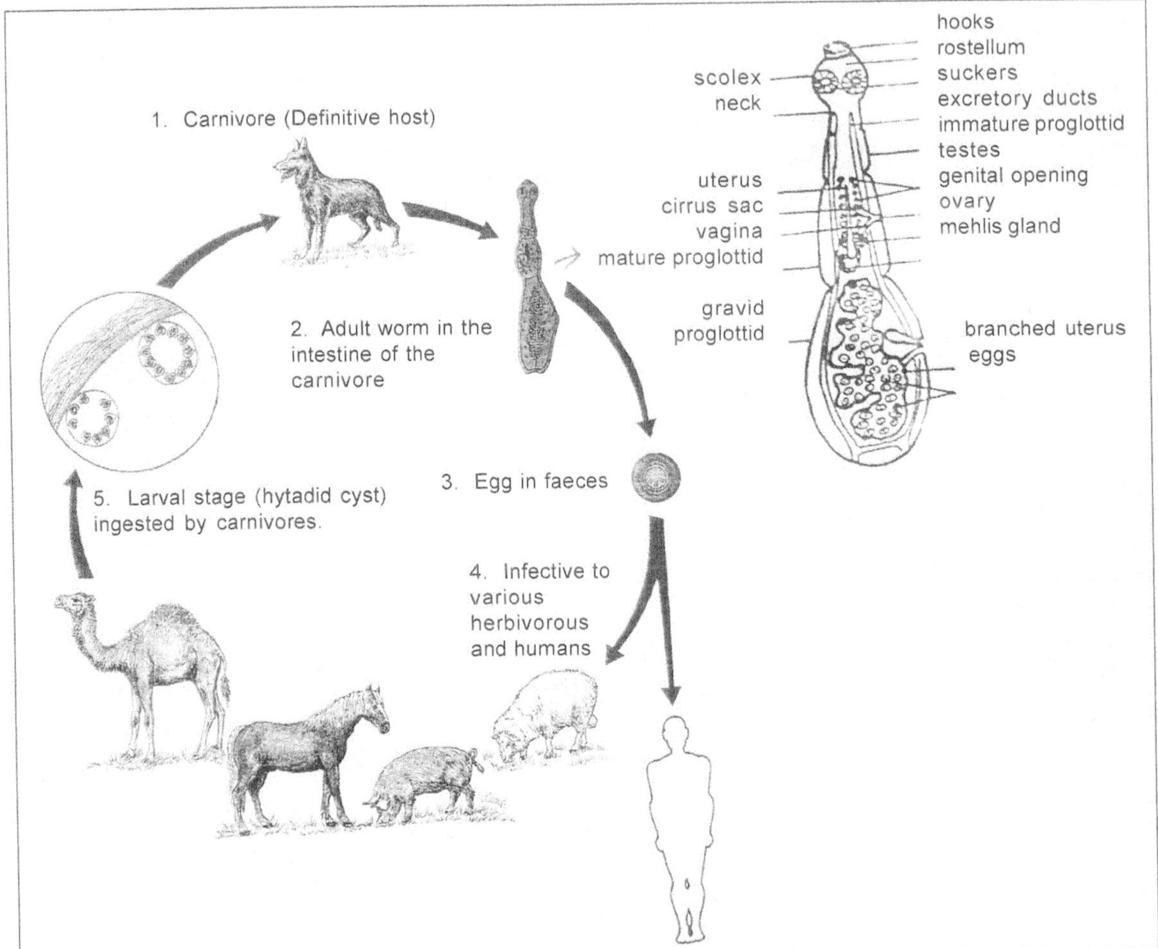

Fig. 5.13 Life cyle of Echinococcus granulosus

has 2 rows of 28-40 hooks and 4 suckers. The worms are firmly attached to the wall of small intestine of host. The terminal segment is gravid and contain 5000 eggs. The gravid proglottids disintegrate in the dog of intestine and eggs with onchospheres are then discharged in the faeces. If they are deposited on pasture, they are ingested by various grazing animals (sheep, camel, cattle, pig etc.). The oncosphere hatches in the duodenum, penetrate the intestinal wall and reaches various organs like liver, lung, brain, kidney, spleen etc. via blood stream. The hooks disappear and parasite begins to grow into '*Hydatid cyst*', hydatid fluid secreted by parasite fills it. The development of cyst is slow and protoscoleces are formed during 1 to 2 years. The carnivores which act as definitive host become infected by eating hydatid cysts from infected visceral organs of intermediate hosts. These scolex are liberated from cyst, attach themselves to the intestinal tissues where they develop into adults (Fig. 5.13). Man becomes accidently involved in the life cycle of parasite by ingesting the eggs and develop hydatid disease. Another hydatid worm, *E. multilocularis*, larval stage, occasionally produce multilocular cysts affecting liver and lung of man.

4. *Dipyhlidium caninum*

This is a common dog tape worm (Fig. 5.14) and on rare occasions infects humans. The adult worm is 200-400 mm long and has 60-175 proglottids. The proglottids are discharged in the faeces. The eggs are ingested by larvae of dog and cat fleas. The oncosphere hatches and penetrate the intestine and develops into larval stage, *cysticercoid*, in the body cavity of the insect. When the insect is ingested, the scolex emerges and attaches itself to the intestine of definite host where it matures into an adult worm. Man becomes infected by accidentally ingesting fleas. The infected child may show gastrointestinal disturbances with allergic manifestations such as urticaria and fever.

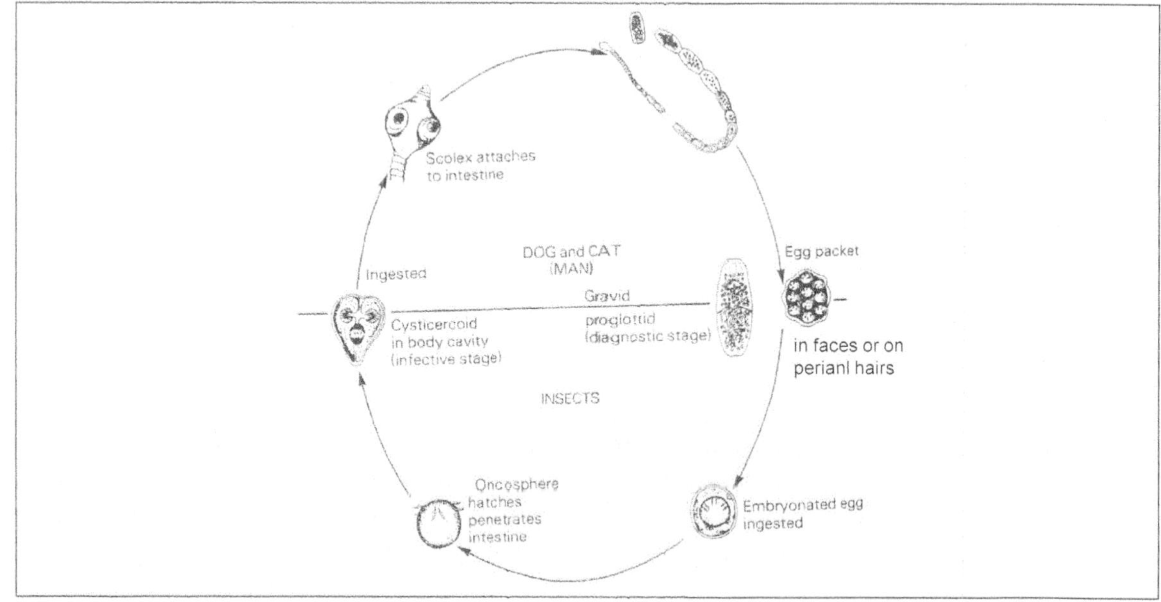

Fig. 5.14 Life cyle of Diphylidium caninum

5. *Hymenolepis nana*

It is cosmopolitan and is a common parasite in the Middle East, Central Asia, India, Pakistan Mediterranean regions, Africa, South and Central Americas. It is smallest tapeworm with length varying from 25 to 40mm and has about 200 segments. The eggs are ovoid with 20-47μm in diameter, and with two membranous shells. Inside the egg is the oncosphere or hexacanth embryo with 6 hooks. There is no intermediate host (Fig. 5.15). Infection occurs by swallowing the eggs as a result of faecal contamination of water and food. After ingestion oncospheres hatches and penetrates villus of small intestine where the cysticercus develops, emerges and later transforms into the adultworm. Gravid proglottids disintegrate in the lumen of intestine and eggs are discharged in the faeces. The infection results in gastrointestinal disturbances such as anorexia, vomiting and diarrhoea accompanied by weakness, insomnia, restlessness and dizziness.

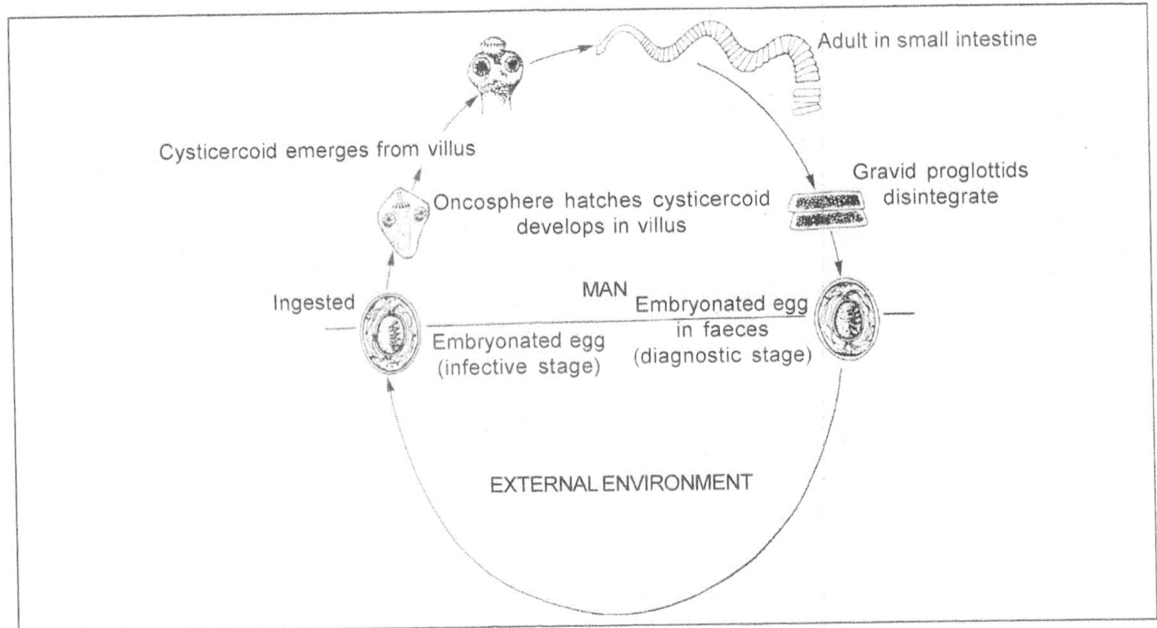

Fig. 5.15 Life cyle of *Hymenolepis nana*

6. *Diphyllobothrium latus*

This tapeworm infects many fish eating mammals like sea lions, cats, dogs and also man. It is a large tapeworm upto a length of 10m. The scolex is elongated and elliptical without any suckers, but possess groove. Each proglottid is hermaphrodite. The eggs are operculated and unembryonated at the time of ejection along with faeces. A ciliated embryo develops in 11-15 days, known as coracidium. The coracidium emerge from the egg and swims freely in water. When ingested by crustacean, it transforms into a Procercoid larva. If the infected crustacean is eaten by a fish, the procercoid larva changes into Plerocercoid larva. Humans are infected by eating raw or semi cooked infected fish. The released larva then enters the small intestine and develops into an adult worm. The infection in man shows mild gastrointestinal symptoms and perinicious anaemia.

B. *Trematodes*

The parasites of this group are known as *flukes*. The body is generally flattened, leaflike, except schistosomes which are elongated. They have organs of attachment, suckers, or hooks or both on anterior end of the body. Well developed branched alimentary canal is present. Trematodes are monoecious except for schistosomes. The male and female reproductive systems are well developed. The excretory system consists bilateral array of flame cells. The nervous system is with 2 ganglia near the pharynx from which transverse and longitudinal nerve fibre extends. The eggs of trematodes are operculated except schistosomes. The eggs develop into ciliated larva miracidium. All trematodes pass through a sexual development in the snail host. The miracidium after entry into snail transform into sporocyst, rediae and cercariae larval stages.

The animals of this class are ecto or endoparasites and they include fluke parasites in man, domestic, livestock and other economic important animals. Table 5.3 gives the overview of trematode parasites and the disease they cause in man and other animals.

Table 5.3 *Trematode parasites found in Human livestock, and Domestic Animals and the diseases*

Trematode	Principal Hosts	Habitat	Main Characteristics	Disease Caused
***Schistosoma** mansoni*	Humans	Blood	Male and female paired, male 6.4-9.9 mm long, female 7.2-14 mm long, conspicuous tuberculations on male, 6-9 testes, eggs 114-175 by 45-68mm	Schistosomiasis mansoni
S. japonicum	Humans, dogs, rats, mice, cattle,	Blood	Male and female paired, male 12-20 female 26 mm long, tegument spinous on pigs, males, 7 testes, eggs 70-100 by 50-65 mm	Schistosomiasis japonicum
S. *haematobium*	Humans, monkeys	Blood	Male and female paired, male 10-15mm long, female 20mm long, tegumental tuberculations on males, 4-5 testes, eggs 122-170 by 40-70mm	Schistosomiasis haematobium
***Fasciola** hepatica*	Humans, sheep,	Bile ducts	20-30mm long, 13mm wide, cone-shaped process at anterior end, eggs 130-150 by 63-90 mm	Fascioliasis (liver rot)
F. gigantica	Horses, cattle	Bile ducts	Similar to *F. hepatica*	Fascioliasisgigantica
***Fascioloides** magna*	Cattle, horses,		Only larger than *F. hepatica.* often over 200-300 mm long, eggs 109-168 by 75-100 mm	Fascioloidiasis
***Fasciolopsis** buski*	Humans, pigs	Duodenum,	Broadly ovate, 30-75mm long, 8-20mm wide, highly dendritic testes, eggs 130-140 by 80-85 mm	Fasciolopsiasis
***Dicrocoelium** dendriticum*	Sheep, cattle, Horses, pigs, rabbits, dogs humans	Liver, bile ducts	Slender, lancet-shaped, 5-12mm long, 1mm wide, extremities pointed, eggs 38-45 by 22-30 mm	Dicrocoeliasis
***Opisthorchis** felineus*	Cats, rarely humans	Biliary and pancreatic ducts	Lancet-shaped, rounded posteriorly, 7-12 mm long 2-3 rom wide, intestinal caeca along entire length of body, eggs 30 by 11mm	Opisthorchiasis

Contd., Table 5.3

Trematode	Principal Hosts	Habitat	Main Characteristics	Disease Caused
Clonorchis *sinensis*	Humans, dogs, cats	Bile ducts	10-25 mm long, 3-5 mm wide, testes, eggs 27.3-35 by 11.7-19.5 mm	Clonorchiasis
Metorchis *conjunctus*	Dogs, cats, foxes, humans	Gallbladder, bile ducts	1-6.6mm long, 590mm to 2.6mm wide, linguiform, testes slightly lobed, cirrus absent, eggs 22-32 by 11-18mm	Experimental hosts killed in heavy infection
Parametorchis complexus	Cats	Bile ducts	3-10mm long, 1.5-2mm wide, uterus rosette shaped and located in anterior half of body	Cirrhosis of liver
Amphimerus *pseudofelineus*	Cats.	Bile ducts	12-22mm long, 1-2.5mm wide, uterus with only ascending limb, eggs 25-35 by 12-15mm	Cirrhosis of liver
Heterophyes *heterophyes*	Humans, cats, foxes	Small intestine	Elongate, pyriform, 1-1.7 mm long, 0.3-0.4 mm wide, small oral sucker	Heterophyiasis
Metagonimus *yokogawai*	Humans, dogs	Small intestine	Similar to H. *heterophyes* but with acetabulum deflected to right of midline	Metagonimiasis
Postharmostomum *gallinum*	Chickens	Caecum	Elongate body, 3.5-7.5mm long, 1-2mm vitellaria well developed along lateral margins of body	Irritation in heavy infections
Sehaeridiotrema *globullus*	Ducks, swans	Small intestine	Body subspherical; 500-850mm long; short, in front of acetabulum, containing 4 to 5 eggs; eggs 90-105 by 60-75mm	Ulcerative enteritis
Riberioria *ondatrae*	Chickens, fish-eating birds, muskrats	Proventriculus	1.6-3mm long, testes at posterior end of body, ovary anterior to testes, eggs 82-90 by 45-48mm	Proventriculitis
Clinostomum *attenuatum*	Chickens	Trachea	5.7mm long, 1.6mm wide, dorsal body convex, ventral surface concave, oral sucker surrounded by collar	Nonpathogenic
Echinoparyphium *recurvatum*	Chickens, turkeys usually in water birds	Small intestine	4.5mm long, 500-600mm wide, collar of 45 spines around oral sucker, four large comer small spines, others arranged in two rows	Severe inflammation of small intestine
Echinostoma *ilocanum*	Humans, rats, dogs	Small intestine	2.5-6.5mm long, 1-1.35mm wide, circumoral disc with 49-51 spines, eggs 83-116 by 58-69mm	Colic, diarrohoea
E. evolutum	Chickens, usually in water birds	Caecum, rectum	10-22mm long, 2-3mm wide, collar water birds spines, five grouped together as corner spines, eggs 90-126 by 59-71 mm	Hemorrhagic diarrohoea in heavy infections
Himasthla *muehlensi*	Humans	Intestine	11-17.7mm long, 0.41-0.67mm wide, body thin, elongate, collar of 32 spines, two on each side, remaining 28 arranged in horseshoe pattern, eggs 114-149 by 62-85 mm	Gastroenteritis
Hypoderaeum *conoideum*	Ducks, chickens pigeons	Small intestine	6-12mm long, 1.3-2mm wide, collar spines in two rows, eggs 95-108 by 61-68mm	Non-pathogenic
Paragonimus *westermani*	Humans, cats	Encapsulated in lungs	Plump, ovoid, 7.5-12mm long, 4-6 mm scalelike spines, deeply lobed testes, no cirrus pouch or cirrus, ovary lobed, eggs 80-118 by 48-60mm	Paragonimiasis

Contd., Table 5.3

Trematode	Principal Hosts	Habitat	Main Characteristics	Disease Caused
P. kellicotti	Humans, dogs, cats, goats, rats, lions	Encapsulated in lungs	Plump, ovoid 9-16mm long, 4-8mm similar to P.*westermani*, eggs 78-96 by 48-60 mm	Kellicotti paragonimiasis.
Nanophyetus *salmincola*	Dogs,foxes,bobcats, raccoons, humans, Fishes	Intestine	Pyrifom, 0.8-1.1mm long, 0.3-0.5mm uterus simple with few eggs, vitellaria eggs 60-80 by 34-35 mm	Salmon poisoning
Collyriclum *faba*	Chickens, turkeys	Encysted in skin	Each cyst with two wonns unequal in size 4-5mm long, 3.5-4.5 mm wide, eggs 19-21 by 9-11mm	Emaciation and anaemia resulting in death
Watsonius *watsoni*	Humans	Intestine	Pear-shaped, 8-10mm long, 4-5mm wide, thick body, acetabulum near posterior end, eggs,122-130 by 75-80mm	Severe diarrohoea
Gastrodiscoides *hominis*	Humans	Caecum	Pyriform, 5-10mm long, 4-6mm wide, prominent genital cone, large acetabulum covering posterior half of body, eggs 150-152 by 60-72mm	Mucous diarrohoea
Cotylophoron (= *Paramphistomum*) *microbothrioides*	Cattle	Rumen	3-11mm long, 1-3mm wide, conical, convex dorsally, concave ventrally, acetabulum posterior end, testes large and lobate, eggs 132 by 68mm	Paramphistomiasis
Mesostephanus *appendiculatum*	Dogs, cats	Small intestine	900mm to 1.75mm long, 400-600mm large adhesive organ behind acetabulum, genital pore posterior, uterus short with 4-5 eggs, eggs 117 by 63-68mm	Non-pathogenic
Typhlocoelum	Ducks, geese	Trachea, bronchi and air sacs	6-12mm long, 3-6mm wide, caeca form complete ring, eggs 122-154 by 63-81mm	Suffocation in heavy infections

Biology of Some of Tremade Parasites of Man and Other Animals is Described Below

1. *Fasciola hepatica*

It is commonly known as *'liver fluke'* and is found commonly in bile ducts of sheep. Man is occasionally effected. It also occurs in bile ducts of cattle, elephant, horse, dog, and monkey. These are found in many parts of the world where cattle and sheep are found. The body is leaf like dorsoventrally flattened measuring 30 × 13mm and provided with both oral and ventral suckers. Mouth is located at anterior end surrounded by oral sucker. The body has extensive branching of vitelline glands, testes, intestinal caeca and uterus is coiled. The ova of Fasciola are operculated with thin egg shell.

The eggs after fertilization stay in uterus for some time which later reach the bile duct, then into the intestine of host which finally pass out with the faeces. Under suitable conditions, the eggs are hatched into ciliated *miracidium* larva (Fig. 5.16), which swim in water in search of intermediate host, snail, Lymnaea, where in pulmonary sac it develops into *sporocyst* consisting of 3-8 *rediae* larva. The rediae produce more rediae which are transformed into other larval form *Cercaria*, that has slender tail, sucker, and intestine comes out of snail. The free swimming larva attaches to grass

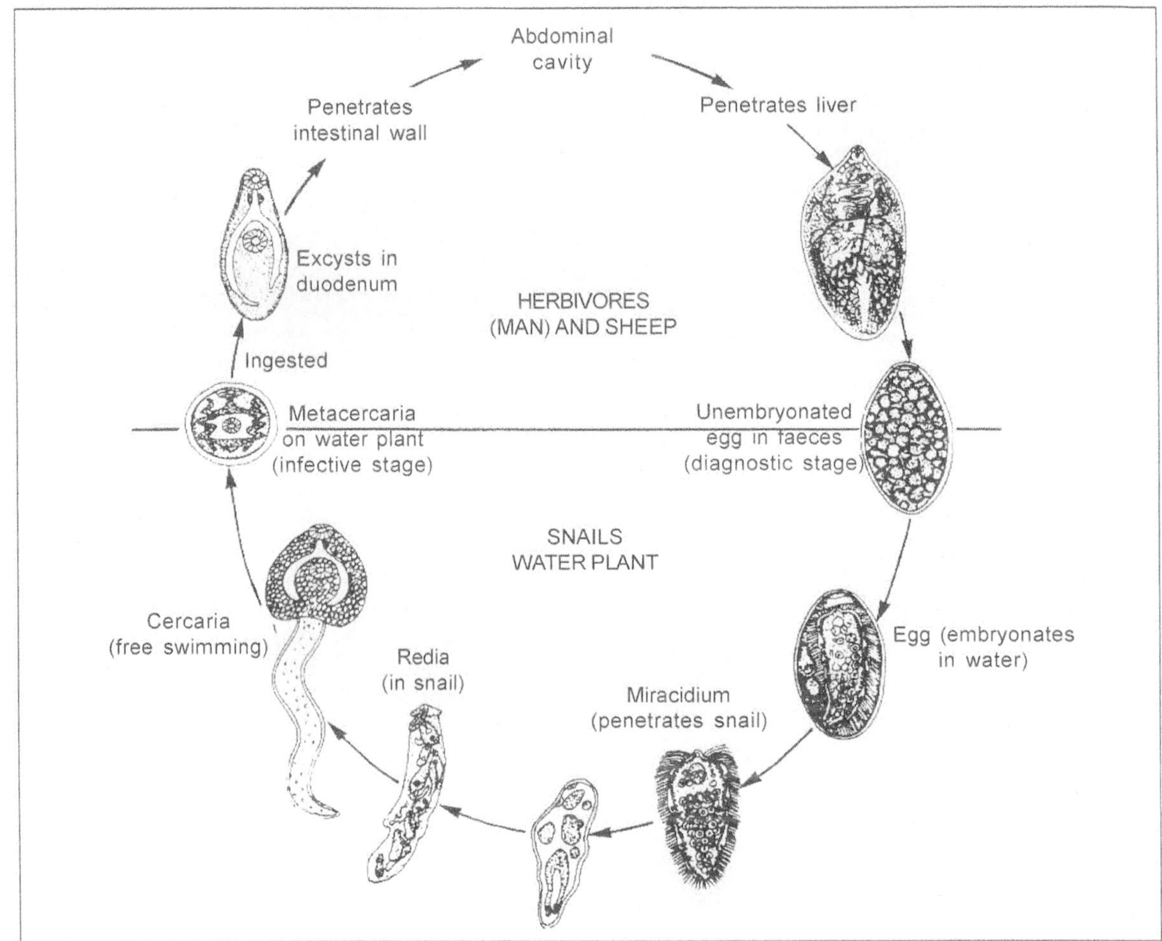

Fig. 5.16 Life cyle of Fasciola hepatica

available, looses its tail, encysts and forms *metacercaria* which is infective. During grazing, or drinking metacercaria enters the alimentary canal of sheep or cattle. The cyst is dissolved in alimentary canal, the metacercaria reach liver and later settle in bile duct. Man is occasionally infected by ingesting metacercaria. It causes '*fascioliasis*'.

2. *Fasciolopsis buski*

It is commonly called as '*intestinal fluke*' and found in the digestive tract of man and pig. This is mostly seen in far east and southeast Asia, particularly in India, China and Bangladesh. It is a large trematode, ovoid with oral sucker smaller than ventral sucker. The alimentary tract consists of a short oesophagus and 2 unbranched caeca. Characteristically testis are highly branched. The eggs are unembryonated when passed with an indistinct operculum. The intermediate hosts of this parasites are fresh water snails, Lymnaea. Man is infected by ingesting metacercariae on various water plants. The embryo, after egg reaching out with faeces, develops in water into miracidium and penetrate the snail, intermediate host (Fig. 5.17). The sporocyst develops in the snail and forms radiae. Radiae

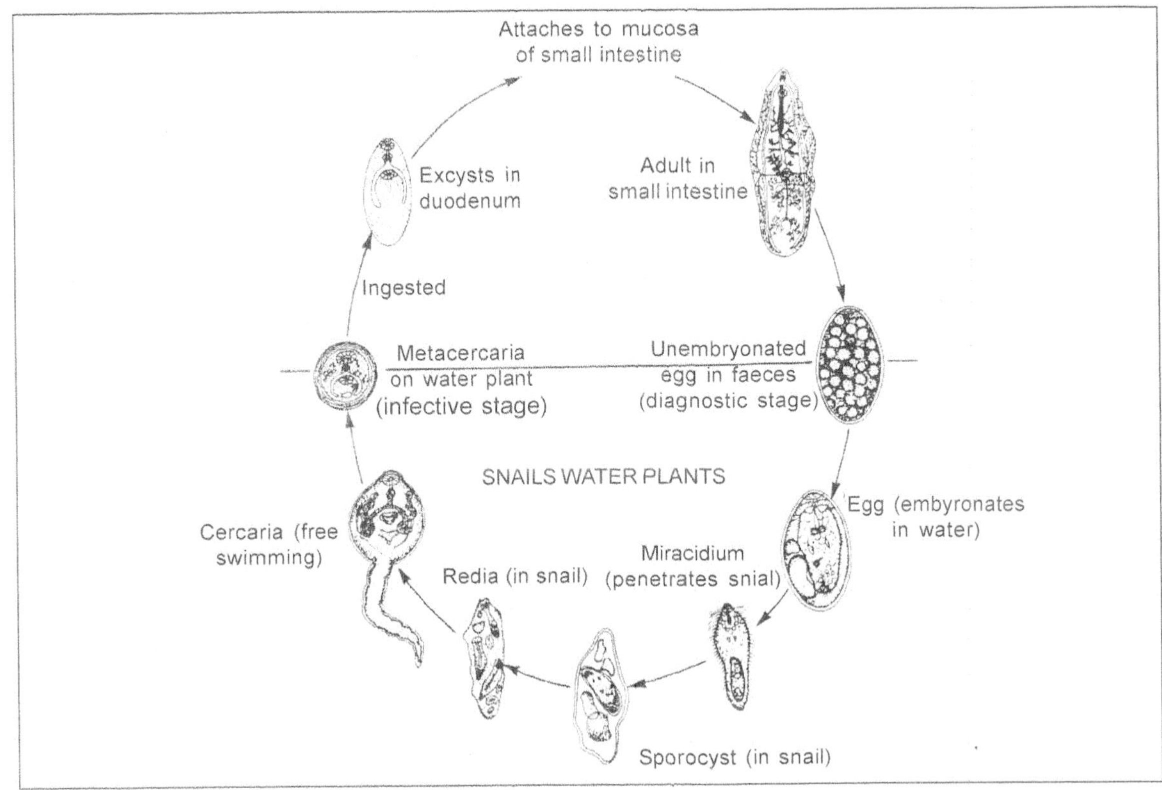

Fig. 5.17 Life Cycle of *Fasciolopsis buski*

produce cercariae which swim in water and have a single cercariae which settle down on aquatic plants and encyst to form metacercariae. Natural mammalian host other than man include pigs and dogs. The infection results in gastrointestinal symptoms.

3. *Schistosoma Spp.*

These are commonly known as *'blood flukes'* and found in blood vessels of humans. The three main species infecting human are *S. mansoni, S. japonicum* and *S. haematobium* and in addition to this the minor species are *S. intercalatum, S. bovis, S. mekongi* etc. *S. mansoni inhabits* the blood vessels of large intestine and is found in Africa, Egypt and tropical America. *S. japonica* inhabits the portal veins and mesenteric vessels and is distributed in Japan, Philippines, Formosa, China and other oriental regions. S.haematobium inhabits the blood vessels of urinary bladder and urinary tract and is distributed in Africa, India, South west Asia and Madagascar. The infection of blood fluke causes *Schistosomiasis* in human beings.

Schistosomes are dioecious and measure 10 to 20mm in length 0.5 to 1mm in width and feed on blood. The male has a deep ventral groove known as gynaecophoric canal in which the female lies during copulation (Fig. 5.18). Both sexes have two suckers, an anterior and ventral sucker. The life span is about 30 years. The mature female lays eggs in the form of chain of beeds. The eggs are carried out in faeces in case of *S. mansoni* and *S. japonium*, whereas in urine in case of

S. haematobium which hatches out with contact of water. Miracidum swims actively to penetrate an appropriate snail host. After entry into the snail intermediate host they develop through various stages (sporocyst, radiae) to become *cercariae* with forked tail, the infective stage of man. The cercariae emerge from the snail and infect animal or human host by penetrating the skin particularly the foot. Now larva looses tail, forms *metacercaria* and passes through heart, lung by systemic and portal circulation reaches liver and feeds on blood. After maturation it migrates to its final habitat according to the species.

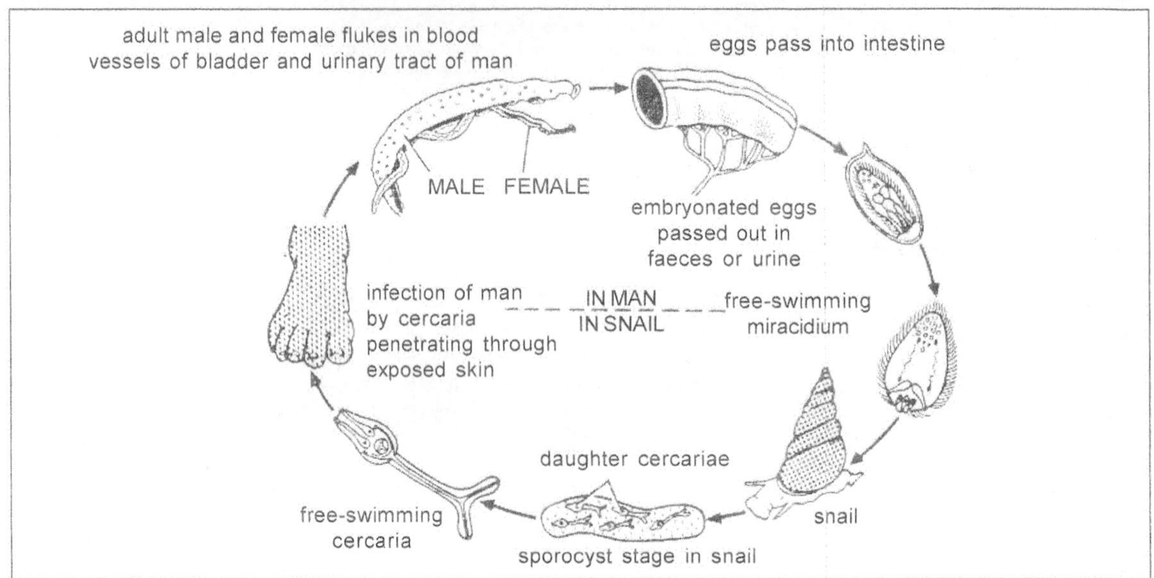

Fig. 5.18 Life cycle of Schistosoma

4. *Paragonimus westermani*

It is commonly known as *'lung fluke'* and is distributed in many areas of Africa, South Asia, Southeast Asia and South and Central America. Man is the primary host in addition to certain carnivores whereas intermediate hosts of parasites are snails, crabs and crayfishes. The adult is oval in outline (8-16mm × 4-8mm) and oral sucker is small. The worm may survive for 10 to 20 years. The eggs of this parasite come in contact with water through cough expelled out by infested persons. The eggs develop into miracidia which enter snail, first intermediate host, and develop into cercariae (Fig. 5.19). The cercariae emerge out from snail and enter into crab or crayfishes, the second intermediate host. The cercariae change themselves and encyst into metacercarial cysts. When uncooked cray fishes and crabs are eaten by man, the metacercarial cysts reach the intestine, then juvenile migrate into human lungs where they attain adult stage and further change into encapsulated form. From here the eggs are discharged and may appear in the sputum of faeces. Occasionally the eggs may enter the circulation and may be distributed to various organs of the body. Adult worms may occasionally enter ectopic sites such as the brain.

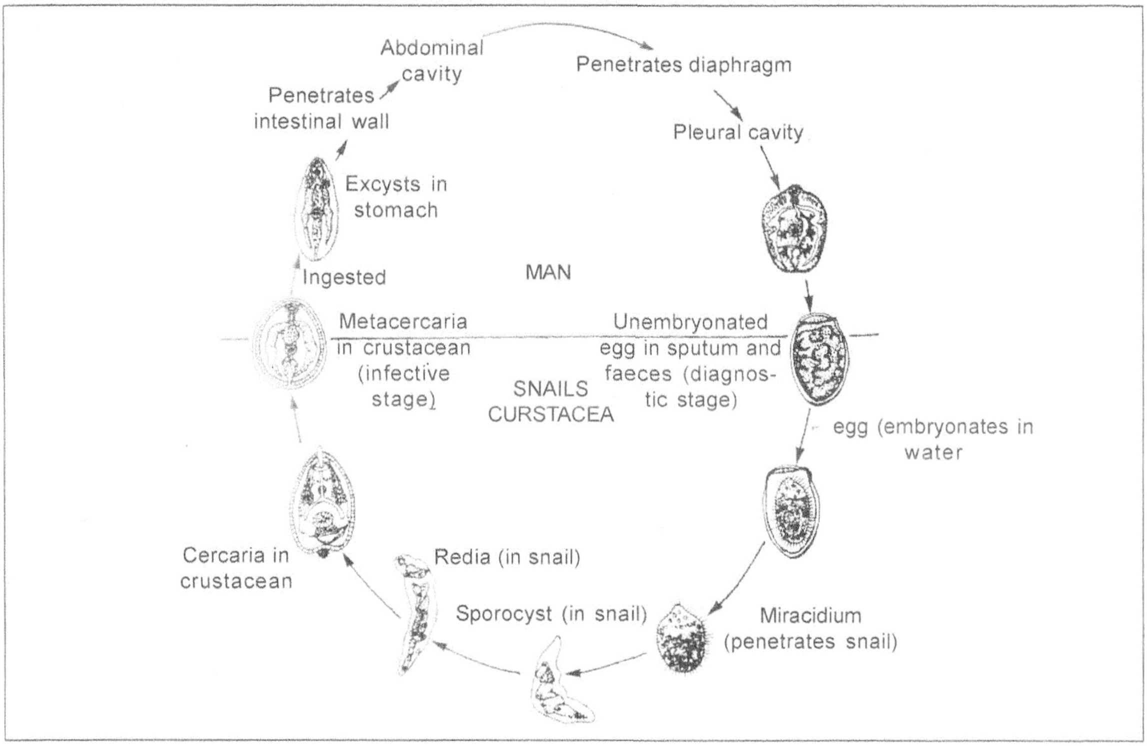

Abdominal cavity
Penetrates diaphragm
Penetrates intestinal wall
Pleural cavity
Excysts in stomach
Ingested
MAN
Metacercaria in crustacean (infective stage)
Unembryonated egg in sputum and faeces (diagnostic stage)
SNAILS CURSTACEA
egg (embryonates in water
Cercaria in crustacean
Redia (in snail)
Sporocyst (in snail)
Miracidium (penetrates snail)

Fig. 5.19 Life cycle of Paragonimus westermani

Migrates to bile ducts
Excysts in duodenum
Adult in bile duct
Metacercaria in fresh water fish
Embryonated egg in faeces (diagnostic stage)
Penetrates under scales of fresh water fish
Ingested by snail
Miracidium hatches
Cercaria (free swimming)
Redia (in snail)
Sporocyst (in snail)

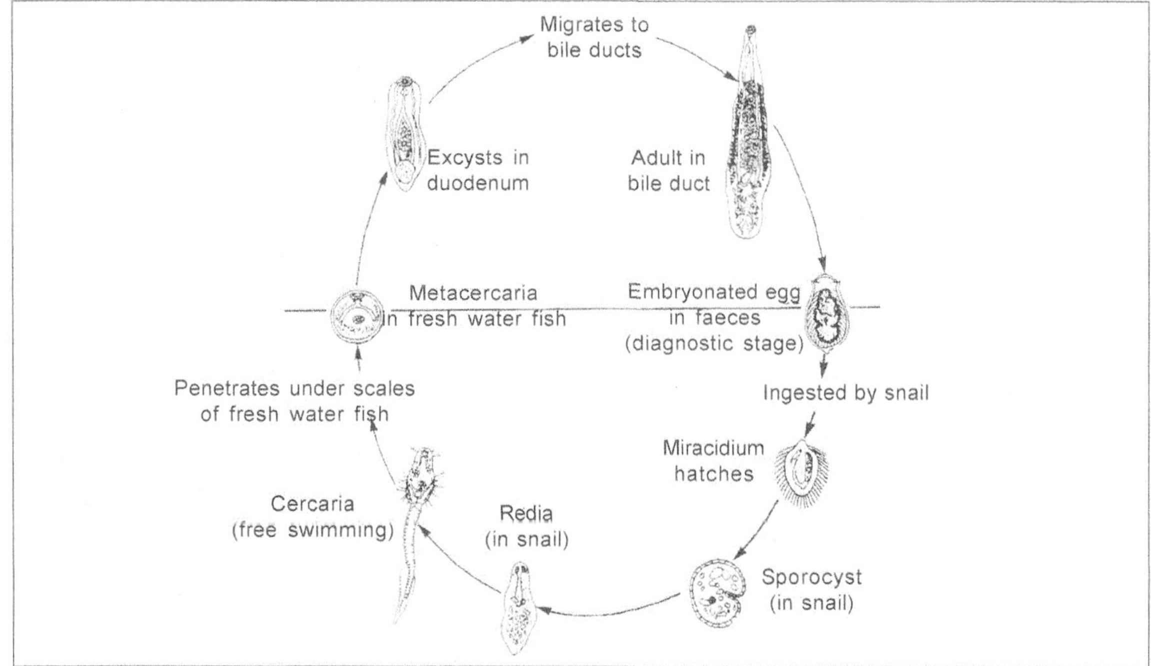

Fig. 5.20 Life cycle of Clonorchis sinensis

5. *Clonorchis sinensis*

This is most important '*liver fluke* infecting man, and is widespread in the far east Asia and highly endemic in China, Korea, Japan and India. This is moderate sized trematode measuring 10 to 25mm × 3 to 5mm. It has elongated and coiled uterus. The eggs of clonorchis have a distinct operculum. Infection results from eating raw fish. In addition to man, cats and dogs are also infected and these animals act as important reservoir hosts. The dumping of human night-soil into fish ponds as food for fish perpetuates infection. Embryonated eggs are passed in the faeces and are ingested by snails in which miracidium hatches out (Fig. 5.20). The sporocyst and redia are formed in snail tissues. The free swimming cercaria with a finned tail is liberated and penetrates under the skin of fresh water fish and is transformed into a metacercaria. On ingestion of metacercaria, the larva excysts in the duodenum and migrates to the bile ducts and adult develops in the biliary passages and gall bladder.

6. *Gastrodiscoides hominis*

This is mostly endemic in Assam in India, Bangladesh, and Indonesia. It has characteristically a large acetabulum at the posterior end of the worm. Eggs are spindle shaped. Man becomes infected by ingesting metacercariae from the skin of water plants. Life cycle is similar to Fasciolopsis buski. Pigs are the main animal reservoirs in the endemic areas. The adult worms are located in large intestine. In heavy infection, inflammation of mucosa of large intestine and caecum with diarrhoea is observed.

7. *Echinostome Spp*

A number of Echinostome spp are reported in man in Far east and south east Asia. These parasites are characterised by the presence of spines on their cuticle, circum oral disc on their anterior end, which is surrounded by spinous processes. The eggs are large, with thin shell and operculum and are immature when passed in faeces. Man is infected by ingesting metacercariae in tissues of edible snails. The adult worms are inhabited in the small intestine.

C. *Nematode Parasites*

Nematodes are commonly known as *round worms* and these are parasitic on man, animals as well as on plants. They are distributed in all ecological conditions. Nematodes are pseudocoelomate, elongated worm like cylindrical animals with unsegmented body. They range in length from a few mm to many cms. The body is covered by cuticle. Alimentary canal is tubular and simple. The excretory system consists of two tubes running inside as lateral cords. The nervous system consists of a nerve ring encircling the oesophagus and from which trunks radiate anteriorly and posteriorly. Except Strongyloides stercoralis, all are dioecious. All are unisexual, male is smaller with curved posterior end. The male genital system is tubular and can be differentiated into a small ejaculatory duct, a seminal vesicle, a vas deference and testis. The ejaculatory duct opens into cloaca. The female genital system is also tubular consists of an ovary, an oviduct, seminal recepticle, a uterus, vagina and vulva. Nematodes generally undergo 4 moults during their life. The first and occasionally second moult takes place in egg, which are discharged by definite host ; The 3rd stage larva after 2nd moult is *infective larva*. The 3rd moult occurs inside the host forming 4th stage larva. After 4th moult, adult formed which matures to produce eggs.

Nematode parasites of man and animals are shown in Table 5.4 and phytoparasites are discussed elsewhere.

Table 5.4 *Some of important nematode parasites of man, livestock as well as domestic animals with the diseases*

Order	Superfamily	Genus	Host	Diseases
Enoplida	Trichinelloidea	1. Trichinella	pig, man mice	– Trichinosis
		2. Trichuris	man, primates	– Trichuriasis
		3. Capillaria	man, fish	– Capillariasis
Rhabditida	Rhabdiasoidea	1. Strongyloides	man, cat, dog, horse, pigs	– Strongyloidiasis
Strongylida	Ancyclostomaoidea	1. Ancylostoma	man, dog, cat	– Ancylostomiosis
		2. Necator	man	– Ancylostomiosis
	Metastrongyloidea	1. Angiostrongylus	goat	– Meningocephalitis
		2. Metastrongylus	Pig, sheep, horses, rabbit, man cattle,	– lung pathogenesics
	Trichostrongyloidea	1. Trichostrogylus	ruminants, horses, cattle, sheep	– Gut, bile ducts, lung pathology
		2. Haemonchus	sheep, cattle	– pathology
Ascarida	Ascaridoidea	1. Ascaris	man, pig	– Ascariasis
		2. Toxocara	dog, cat, man	– Toxocariasis
		3. Anisakis	man, sea mammals	– gastric and intestinal granulomas, Anisakiasis
Oxyurida	Oxyuroidea	1. Enterobius	man	– Enterobiasis
	Spiruroidea	1. Gongylonema	deer	– Gnathostomiasis
	Thelazoidea	1. Thelazia	man, cattle	– Eye-filariasis
	Gnathostomatoidea	1. Gnathostoma		– Gnathostomiasis
	Filariodea	1. Wuchereria	man	– Elephantiasis
		2. Brugia	dogs, monkey, cats, man, dogs	– Malayan filariasis
		3. Onchocerca	man	– Onchocerciasis
		4. Loa loa	man	– Calabar swelling
		5. Dirofilaria	dogs, and cats	– dirofilariasis
	Dracunculoidea	1. Dracunculus	man, monkey	– dracontiasis or dracunuliasis

Biology of Some Nematodes of Man and Other Animals is given below

1. *Ancylostoma duodenale*

It is commonly known as '*hook worm*' and distributed in the tropical, subtropical and temperate regions of Asia, Africa, Europe, America, Japan, China and India. The adult sucks blood from the intestine of man and other domesticated animals. The adults are cylindrical with the head bend sharply backwards giving them a hooked appearance. The males are smaller than females and possess a bursa at posterior end. The sexes are separate and exhibit sexual dimorphism. The fertilization takes place in host's intestine and fertilized egg passout with faeces of host. The life cycle is simple without intermediate host (Fig. 5.21). The eggs are hatched into larva, *rhabditiform larva*, in soil actively feed on debris and organic matter, moult twice and attain third larval stage *filariform larva* which is infective. The larva penetrate the skin of the feet and through blood circulation reach the lungs, capillaries, trachae and intestine and attain their adult stage in intestine. The life span is 5-7 years. It causes the disease called as *Ancylostomiasis*.

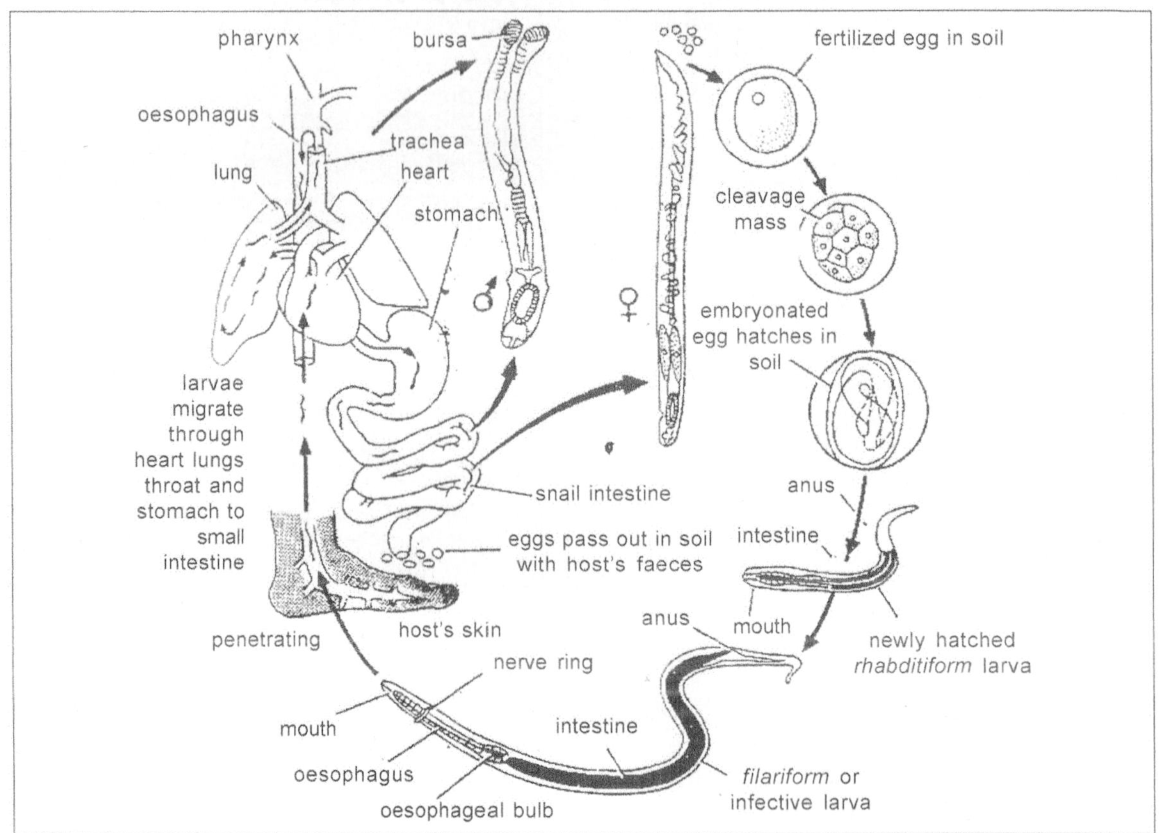

Fig. 5.21 *Ancylostoma duodenale* Life cycle

2. *Ascaris lumbricoides*

Ascaris is known as *round worm* and inhabits the small intestine of man and more common in children. It is cosmopolitan in distribution. The body is elongated, cylindrical with tapering ends. The female measure 20 to 35cm long and 3-6mm wide. The male is smaller measuring 12 to 31 cm and 4 mm wide and has a curved tail. The mouth is at anterior end and is guarded by 3 lips. The alimentary canal comprises of foregut, midgut and hindgut. It completes life cycle without intermediate host. Infection occurs from the ingestion of the embryonated food or water (Fig. 5.22). The infective larva hatches out in small intestine and penetrates the intestinal wall to enter the portal circulation. From liver it is carried to heart and via pulmonary artery to lungs where it undergoes another moult to become fourth stage larva. From lungs larva moves upto branchi and swallowed or crawled to enter digestive tract. In the intestine it moult again to become a sexually mature worm. The children are more effected. A gravid female produces 2,00,000 to 2,50,000 eggs daily which are expelled along faeces. The roundworm causes *ascariasis*.

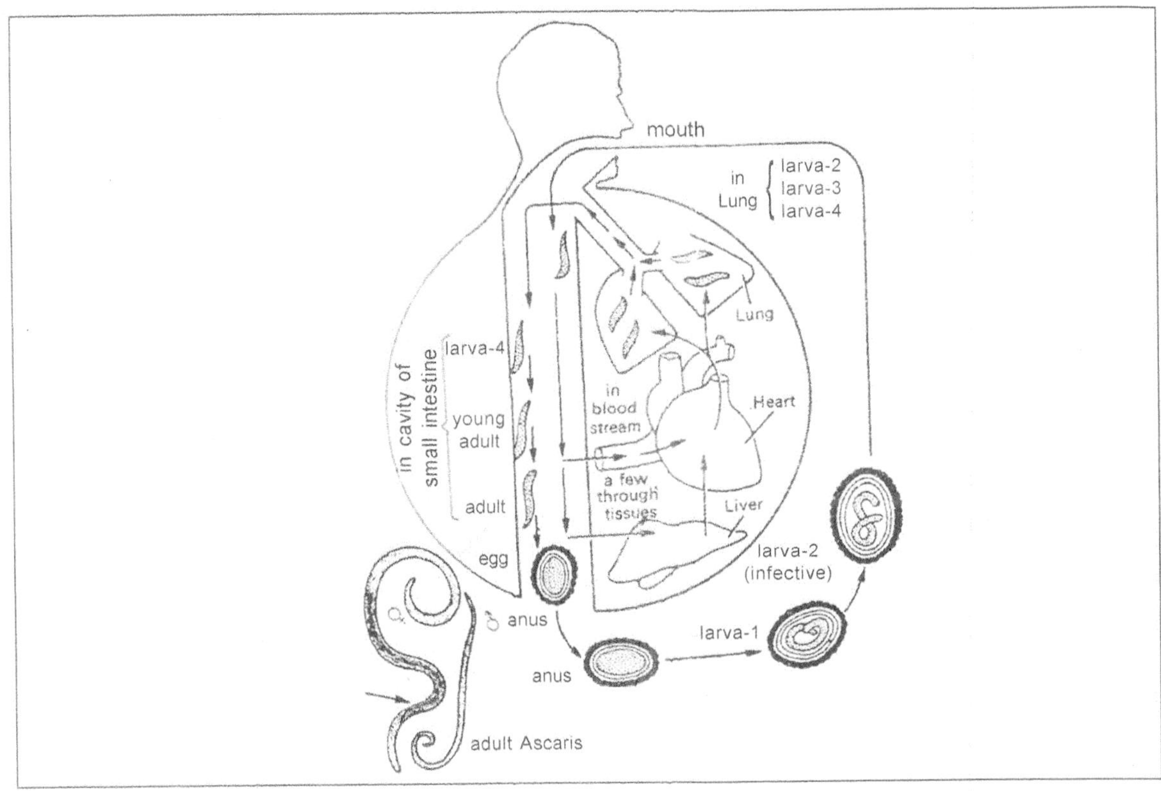

Fig. 5.22 Life history of Ascaris.

3. *Trichuris trichiura*

This has world wide distribution being the most common intestinal nematode in tropical areas such as southeast Asia. It is commonly known as *Whip worm* because the anterior 3/5th is elongated, thin and posterior 2/5th is fleshy and bulbous. It is endoparasite and is found in large intestine and caecum of human beings. The adult male is 30-45 mm long and female is 35-50 mm long. These worms have characteristic capillary like oesophagus surrounded by gland cells known as stichocytes. The infection results from the ingestion of embryonated eggs (Fig. 5.23). The larva does not undergo visceral migration but penetrates the gut wall for a short period before returning to the human to mature into adult stage. The fertilized egg pass out along with faeces and reach the soil.

4. *Enterobius vermicularis*

It has world wide distribution, and commonly known as *'pin worm'*. It is an intestinal parasite and more common in children. The female (12mm/0.3 to 0.5 mm) is larger than male (5mm/0.1 to 0.2 mm). The anterior end is provided with three lips and a pair of cephalic expansions. The posterior end of male is blunt and curved where as female is long and pointed. The life cycle is direct and without intermediate host. The adults are mainly located in caecal region and female deposits eggs in the anus and on peripheral skin and human infection occurs through inhalation and swallowing of eggs or

Penetrates and develops in mucosa
Adults in caecum
Larva hatches in intestine
MAN
Embryonated egg (infective stage)
Eggs in faeces (diagnostic stage)
EXTERNAL ENVIRONMENT
advanced cleavage
2-cell stage

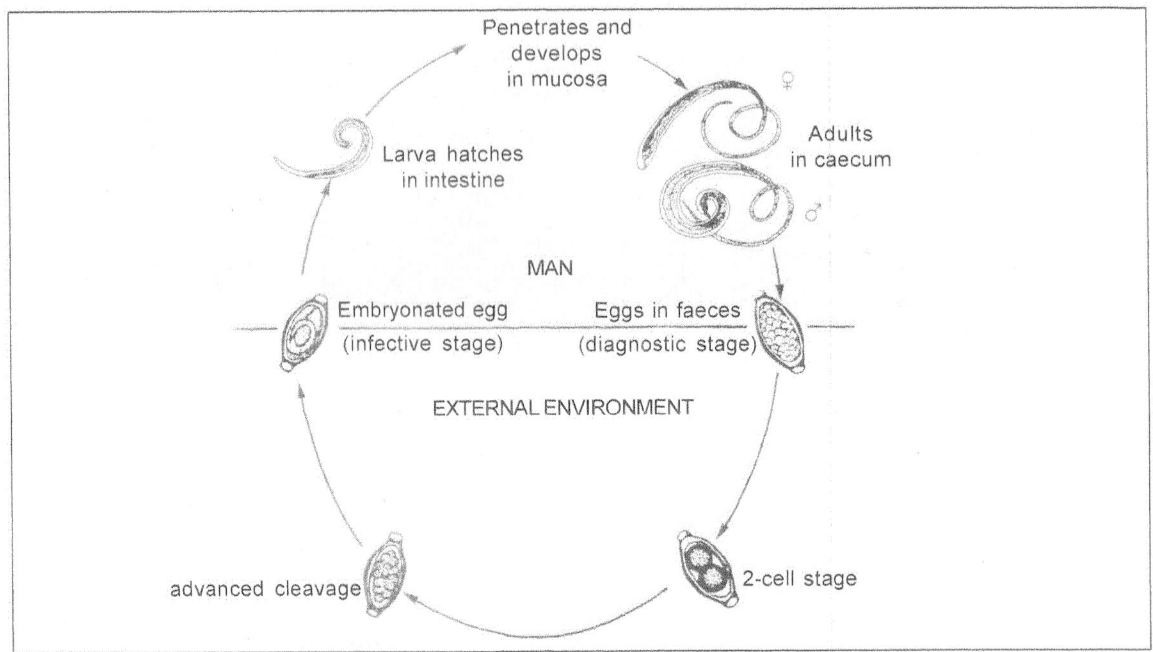

Fig. 5.23 Life cycle of Trichuris trichuria

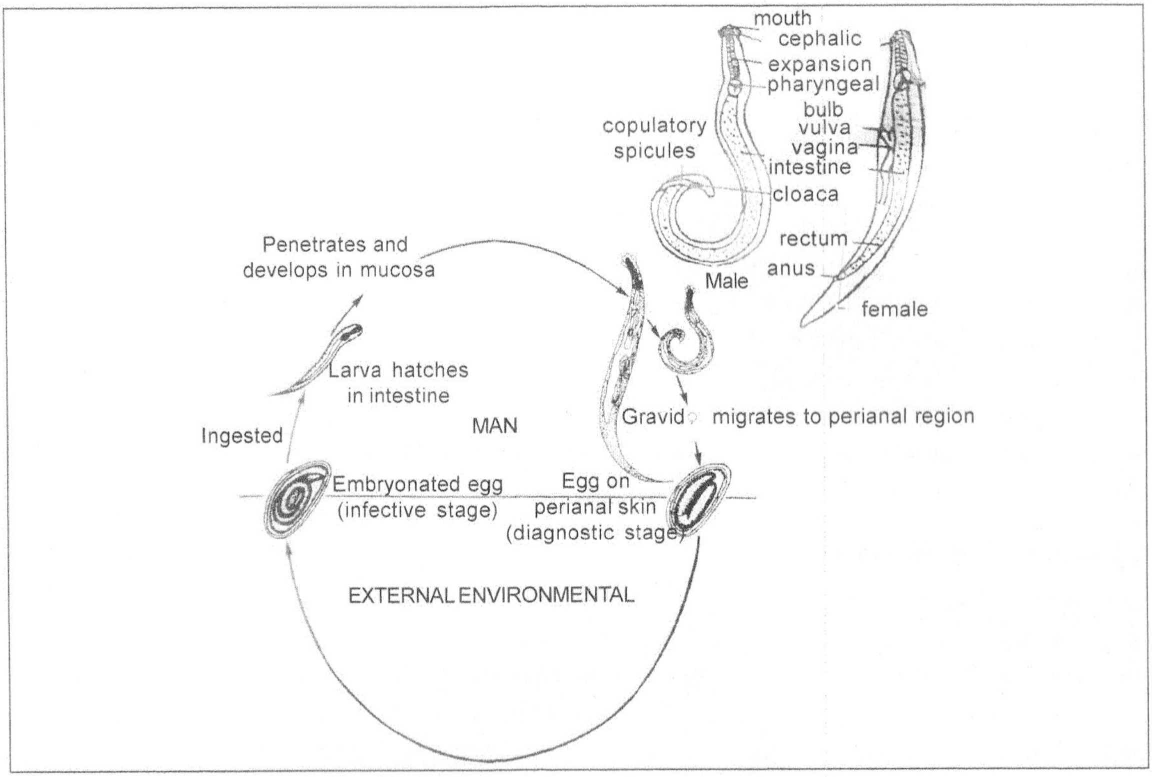

mouth
cephalic expansion
pharyngeal bulb
vulva
vagina
intestine
cloaca
copulatory spicules
rectum
anus
Male
female
Penetrates and develops in mucosa
Larva hatches in intestine
Ingested
MAN
Gravid ♀ migrates to perianal region
Embryonated egg (infective stage)
Egg on perianal skin (diagnostic stage)
EXTERNAL ENVIRONMENTAL

Fig. 5.24 Life cycle of *Enterobius vermicularis*

sometimes with faeces and reach the gut of host with contaminated food (Fig. 5.24). There is no visceral migration. The larva matures into adult with two moultings. The life cycles of the parasite is completed in about 6 weeks.

5. *Dracunculus medinensis*

It is commonly known as '*Guinea worm*' and it infects humans, dogs, cats, horses, cattle, wolves, monkeys, baboons etc. It's adults inhabit the subcutaneous tissues, particularly the arms and legs (Fig. 5.25). The cyclops, water fleas are the intermediate host which carry infective larval stage of parasite. The male measuring 12-40 mm in length and 0.4 mm in diameter, is much smaller than the female which measure 70-120 cm in length and 0.9-1.7 mm in diameter. The anterior end is provided with mouth. The female containing eggs migrate underneath the skin of feet, legs and hands which are regularly in touch with water. Parasites secrete toxic substance producing blisters on skin with ulcers. When these come in contact with water, the parasite lay eggs and these egg liberate free living larva, enter the body of cyclops through penetration. Within cyclops it moults twice and attain the infective stage after 20 days. When infected cyclops are swallowed, the larva gets free in duodenum and burrows the mucosa, undergoes two moults and lodged in liver, or subcutaneous tissue where it matures in 8 to 12 months.

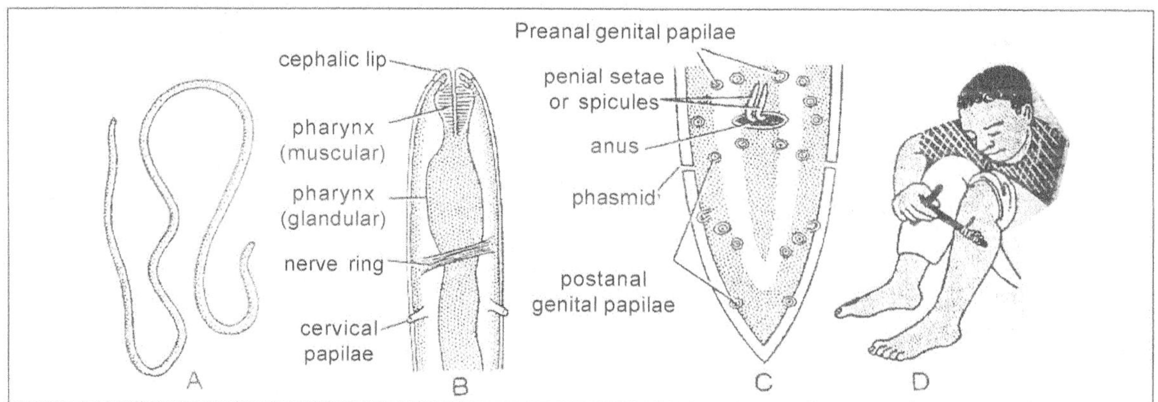

Fig. 5.25 Dracunculus medinensis – A. Entire worm, B. Anterior end of male, C. Posterior end of male in ventral view and D. Guinea worm being wound on a stick

6. *Wucheraria bancrofti*

It is commonly known as *Filarial worm* and found in lymphatic vessels, muscular tissues and glands of man. The worm is found in tropical and temperate countries and causes filariasis or elephantiasis. The adults are elongated, thread like worms, male measures upto 4 cm while female up to 10 cm in length. The female is viviparous and liberates sheathed microfilariae into lymph which find their way into blood stream and migrate into blood capillaries underneath the skin. The intermediate host is mosquito, *Culex*. During the night, if mosquito bites the man, the microfilariae enters the stomach of mosquito from where they migrate to the thorasic muscles, grow and attain length of about 1.5 mm in few weeks (Fig. 5.26). Then larvae migrates to the proboscis and when an infected mosquito bites a man, the larvae are transferred to the blood stream of man from where they go to the lymph glands and passages and grow into adult worms.

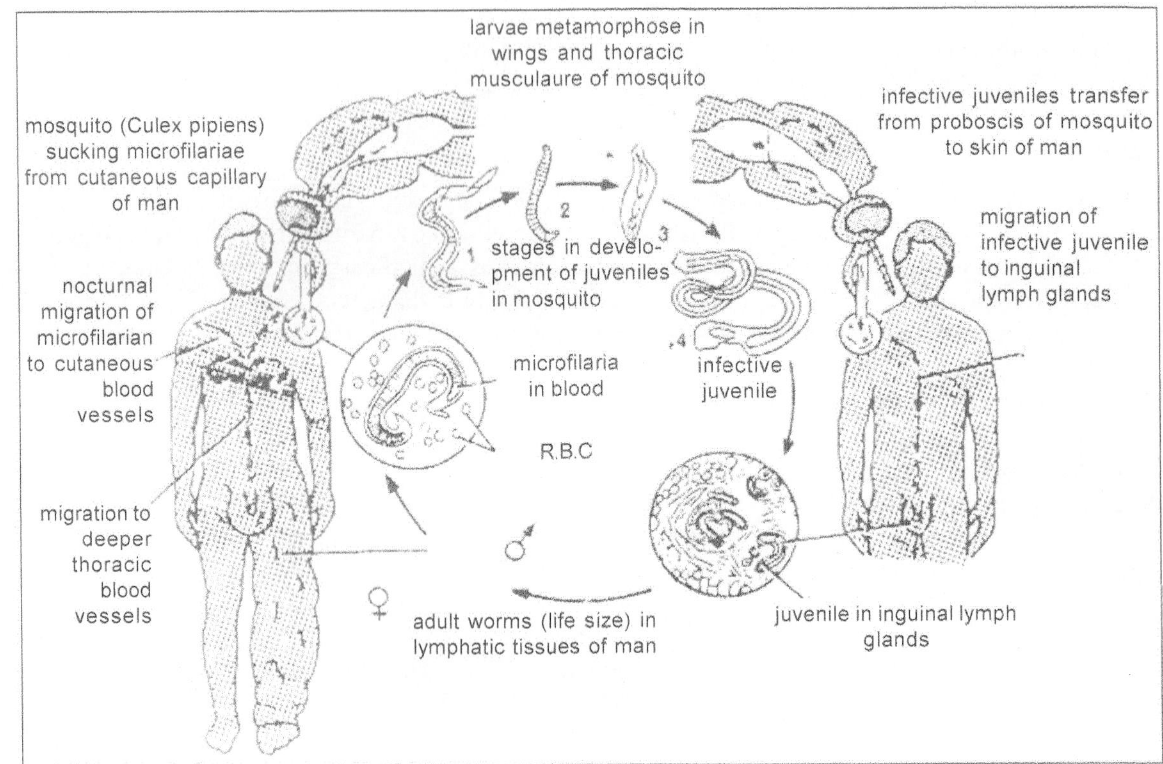

Fig. 5.26 Life cycle of Wuchereria bancrofti.

7. *Brugia malayi*

It is found in eastern regions of Indian subcontinent and south east Asia. It has wild animal reservoirs which include monkeys, dogs, cats, insectivores and pangolins. It is similar to *W.bancrofti* but smaller in size, and life cycle is also similar to W.bancrofti. The infection is not associated with elephantiasis.

8. *Onchocerca volvulus*

It has found in tropical Africa. Female worm measures 23-50 cm and male 16-42 cm. The parasite is transmitted by black flies, *Simulium*. Infective larvae enter human tissues during feeding and develop into adult worms in the deep facial planes and dermis. Microfilariae are discharged by gravid female in subcutaneous tissue from where they are picked up by black flies. After ingestion, they enter stomach and then thorasic muscles where they moult and form infective larvae. They enter proboscis and transmitted to human during bite. The microfilariae are main cause of the disease. It affects skin, eyes, lymph nodes and various internal organs of body. It results in itching, scratching and dermatitis (Sowda). It may cause disease of bony prominences known as *'Onchocercomas'*.

9. *Loa loa*

It is known as *Eye worm*, mainly found in tropical Africa. In addition to man, it also found in monkeys and baboons. They commonly invade subcutaneous tissue and during their migration may pass across the eye ball. Male and female measure 20 to 35 mm and 20 to 70 mm in length respectively (Fig. 5.27). Body is covered by numerous worts. Gravid female produce ensheathed juvenile, or

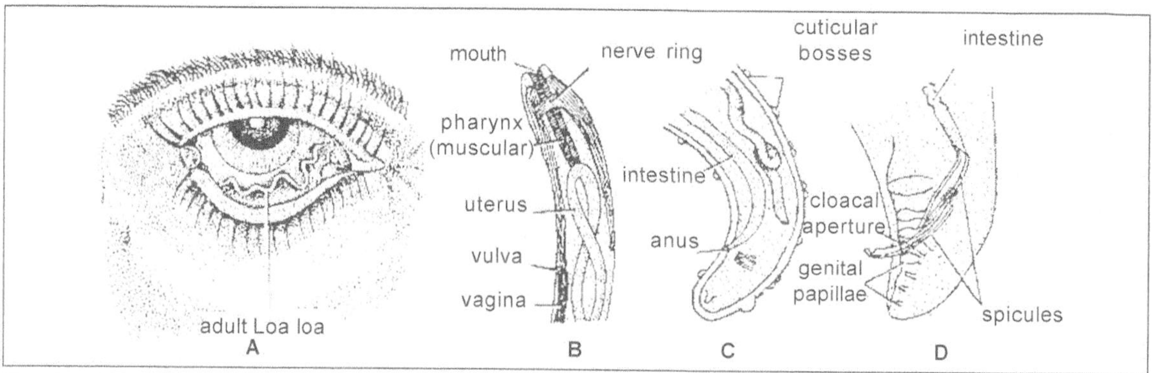

Fig. 5.27 Loa loa – A. adult across the bulbar conjuctive of human eye, B. Anterior end of female, C. Posterior end of female and D. Posterior end of male

microfilariae. The intermediate hosts are mango flies, *Chrysops*. Larvae moult and develop up to infective stage in fly and again infect the man through proboscis during bite. During migration, as adult worms develop in connective tissue, they cause itching and swelling. They also cause swelling and pain in eyes known as '*Calabar swellings*'. Loa microfilariae cause very injurious and fatal cases when they penetrate brain and spinal cord.

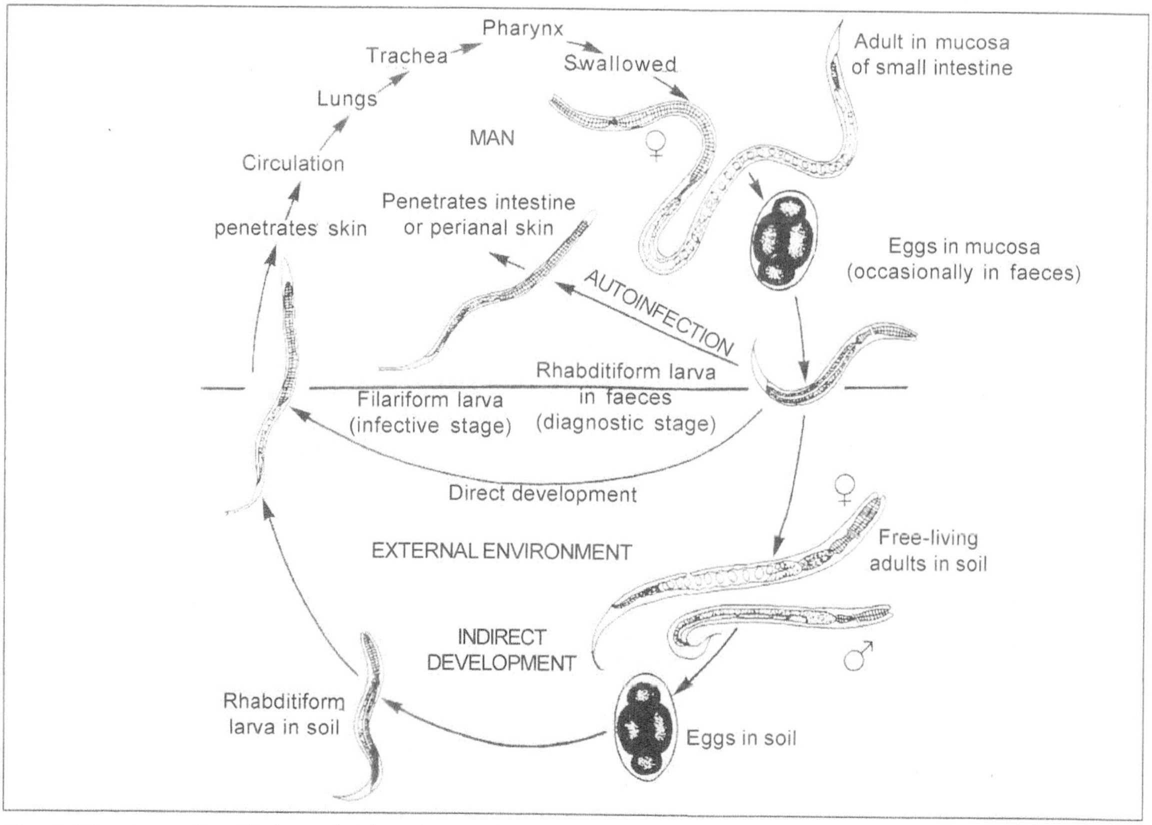

Fig. 5.28 Life cycle of *Strongyloides stercoralis*

10. *Strongyloides stercoralis*

This is widely distributed in tropical areas of Asia, Africa and South America. It persists for many years in man by autoinfection. The adult parasitic female resides in the mucosa of small intestine and is about 2.7 mm long and 30-40 μm wide and reproduces by parthenogenesis. The eggs are thin shelled, ovoid and transparent and larva usually hatches in the intestinal tissues and is passed in the faeces. The parasites has 3 different modes of development, indirect development, direct development and autoinfection (Fig. 5.28). Indirect development occurs in the external environment where the free living adult produces egg, rahbditiform larva and filariform larva. The filariform larva penetrates the skin, enter the circulation, pass through the lungs and can be swallowed to become adult females in the intestinal tissues. The rahbditiform larva passed in the faeces can undergo direct development to become filariform larvae without developing into adults. Autoinfection occur when rahbditiform larva undergo same cycle of development, which leads to continuous buildup of worm burden. The infection results in strongyloidiasis.

11. *Trichinella spiralis*

It is commonly known as *Trichina worm*, is an important nematode parasite of man and other mammals like pig, dog, cat, rat and bear (Fig. 5.29). It causes the disease known as *Trichinosis*. It is endemic in many areas where pork is consumed and prevalent in central and eastern Europe, America, parts of

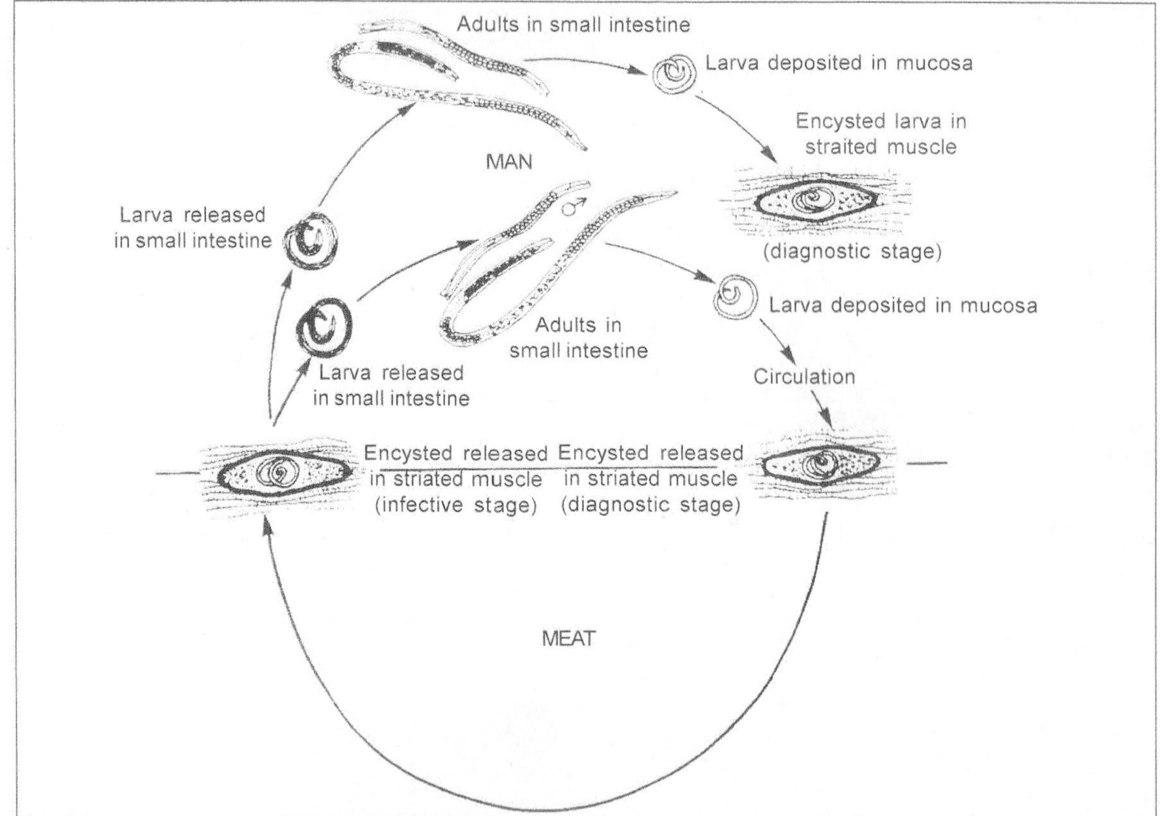

Fig. 5.29 Life cycle of *Trichinella spiralis*

Africa and Asia. The adult males are small nematodes 1.4 to 1.6 mm long. The female is viviparous and about twice as long as the male. The anterior part of body contains a row of glandular cells known as stichocytes. Humans infected mainly by eating improperly cooked pork or pork products (Fig. 5.29). After ingestion, the larvae are liberated in small intestine, and mature into adults. The female being viviparous deposits larvae in the intestinal tissues from where they find their way into blood circulation and these into striated muscle of body. The most heavily parasitised muscles are diaphragm, tongue, abdominal and other striated muscle. After penetration, larva coil into spiral shaped body which eventually becomes in thick walled cyst, and this may remain viable for longer periods (10-20 years). Pig become infected by eating infected scraps and garbage from slaughter house or forms. Occasionally they are infected by eating carcases of infected rats.

D. *Helminth Diseases of Man and other Animals and Disease Control*

In the aforesaid description, the important helminth parasites and the diseases caused by them are indicated in Table 5.1, 5.2, 5.3. The helminths-cestodes, trematodes and nematodes cause disease in humans, livestock animals like cattle, sheep, goat, horse etc, and other economically important animals like chicken, turky, many fishes etc. This section describes some of the most common diseases of helminths, called *Helminthiasis*, with possible disease control measures.

1. *Ascariasis*

Ascariasis is a highly prevalent disease of man and mostly seen in children caused by round worm. *Ascaris lumbricoides* with adult residing in intestine, the patient complaints of fever, utricaria, abdominal pains, vomiting, diarrhoea, nausea, headache, irritability, dizziness and night terrors. When adult worm migrates through intestinal wall, they cause severe peritonitis. These may enter other organs causing appendicitis, gall bladder and liver diseases. In children, juveniles may enter lungs and cause branchiasis. This may also lead to varying degree of pneumonia and bronchospasm (known as *Loeffler's syndrome*). The larvae of *A. suum* (pig Ascaris) may produce severe pneumonitis and branchospasm.

As ascaris eggs survive in the soil for many years, prevention and control in endemic areas is difficult. Mass chemotherapy at regular intervals along with environmental sanitation, care in taking food particularly pig meat and clean vegetable and salads may reduce the incidence of infection. Treatment of human ascariasis is fairly successful with administration of Pipereazine citrate, Hexylresercinol, Santonin, Levamisole, Pyrantel, Mebendazole or Albendazole at appropriate treatment regimen.

2. *Ancylostomiasis*

It is caused by the intestinal hook worms *Ancylostoma duodenale* and *Necator americanus*. It is most prevalent and important of the parasitic diseases. Man acquires infection when parasite juvenile penetrate through soft skin of hands and feet. During the stage of skin penetration larvae can produce intense itching and dermatitis at the site of entry (known as *ground itch*). The larval migration through lungs may have symptoms of pneumonitis and branchitis. The characteristic symptoms of the ancylostomiasis are gastrointestinal disturbances, anaemia, nausea, dizziness, headache. The diseases can be prevented to some extent by taking care of faecal soil contact, health education and mass chemotherapy and sanitation of environment. The effective treatments include the antihelminthic drugs, safe as well as effective drugs for hookworm are bephenium hydroxynapthate, pyrantel pamoate, mebendazole, tetrachloroethane etc.

3. *Enterobiasis*

This disease is caused by pin worm or seat worm, *Enterobius vermicularis*. It is more frequent in children than adults. Symptoms of disease include severe itching around anus (Pruruitusani) loss of apetite, sleeplessness, bed wetting, grinding of teeth, nausea and vomiting. Occassionally, these worm with ectopic migration into female genital tracts causing fallopian tube pain and symptoms of salpingitis. Sometime this parasite infection leads to appendicitis. Improved personal and group hygiene is important for prevention of the disease. The treatment of drugs include with piperazine, pyrantel, mebendazole of pyrvinium.

4. *Trichuriasis*

It is caused by whipworms, *Trichuris trichiura*, which inhabits large intestine, caecum or vermiform appendix. Patients suffering from this disease complain of nausea, vomiting, constipation, headache, fever and pains resembling appendicitis. In severe cases anaemia and eosinophilia may develop. In children infection may lead to dysponea, oedema and sometimes cardiac failure. Drugs most commonly used for expulsion of whip worms are oscarsol, and dithiazanine. The other drugs include combination of oxantel plus pyrantal and also mebendazole therapy is useful.

5. *Trichinosis*

This disease is caused by trichina worm, *Trichinella spiralis*. It is transmitted when man eats raw meat especially of pork. The patient at invasion stage complains of abdominal pain, nausea, vomiting and diarrhoea of varying intensity. There may be fever, profuse perspiration and tachycardia. During migration stage, symptoms include hypersensitive reaction to parasitic antigens. There may be oedema of face, periorbital tissues, muscular tenderness, fever and hyper eosinophilia. At encystment stage, the symptoms worsen and may have myocardial failure, respiratory and central nervous system involvement. The disease can be prevented and controlled by screening of meat of pork, destruction of rats which are source of infection to pigs and proper education about disease. Use of mebendazole is highly active against parasite.

6. *Strongyloidosis*

It is caused by thread worms, *Strongyloides stercoralis*. These invade the lining of alimentary canal. Skin reactions often seen at the site of entry of larvae, which occurs as rash and pruritis. There may be pneumonitis during larval migration and intestinal symptoms on maturation. The patient with strongyloidosis complains of nausea, dizziness, blood diarrhoea, vomiting, cough and fever. Prevention and control measures are like ancylostomiasis. Most effective drugs for treatment of this disease is gentian violet, Thiabendazole, Albendazole, Mebendazole, Dithiazamine etc.

7. *Elephantiasis or Filariasis*

The filarial worm, *Wuchararia bancrofti* causes this disease. These worms live in lymphatic system and connective tissue of body and in circulating blood during night. The acute stage in disease manifests itself in lymphangitis in the extremities, transient skin swellings and recurrent inguinal, auxiliary or epitrochlear lymphadenitis. In chronic stage the infection results in elephantiasis, hydrocoele and chyluria. The disease is characterised by enlargement of limbs, scrotum and mammae. The antivector measures against culex, and mass chemotherapy for Wucheraria can prevent and control the disease. In the treatment of *Wuchereria* infections the drug of choice is diethylcarbamazine. Levamisole and cyanine dyes are also used.

8. *Taeniasis*

It is caused by *Taenia* spp. (*T. solium*-pork tapworm and *T. saginata*-beef tape worm). Man acquires infection by eating raw or undercooked pork or beef that contains cysticerci. Presence of tapeworms in intestine causes gastrointestinal disorders-like vague intestinal discomfort, vomiting, diarrhoea, intestinal obstruction, sometimes appendicites and anaemic conditions etc. The human infection with cysticercus develop the disease called *Cysticercosis*, mainly involving central nervous system, the symptoms vary from mild nervous manifestations to severe epileptiform attacks. Other symptoms include headache, dizziness, eye complaints like blurred vision, photophobia and diplopia, aphasia and amnesia. Sometimes cysticerci block flow of cerebrospinal fluid pathway causing hydrocephalus. The prevention and control of *Taeniasis* and *Cysticercosis* should involve early treatment of cases, improvement of sanitation, prophylactic chemotherapy of workers involved in with pig rearing and avoiding improperly cooked meat. Treatment of Taeniasis uses the drugs atabri and quinacrine. The cysticercosis involves the surgical removal and przaizquantal treatment is choice of therapy.

9. *Hydatid disease*

This disease is caused by hydatid worm, *Echinococcus granulosum*. Primary host of this worm is dog in whose intestine the eggs are released. They pass out in faeces and develop into onchosperes. Humans acquire infection on eating food or drinking water contaminated with onchosphere containing eggs. The hydatid cysts develop in liver, lungs and rarely in brain, eye, kidney, muscles, bone and heart. As the cyst enlarges, symptoms of a space occupying lesion develop. Allergic reactions and sometimes anaphylactic shock may occur if the cyst ruptures. The sheep are also affected by these cysts. The control measures of this disease include prevention of dog eating meat of infected animals, reducing dog populations, avoidance of contact with infected dogs. The chaemotherapeutic agents include treatment with mebendazole or albendazole.

10. *Schistosomiasis*

This disease is caused by blood flukes-*Schistosoma manosoni*, *S. haemotabium* and *S. japonicum*. The disease is widespread throughout the world. These parasite live in the blood stream, and feed on it. The eggs penetration into blood vessels and tissue by secreting enzymes results in pathological damage. The tissue reaction may be in form of granulomas to extensive fibrosis. Cercarial dermatitis (swimmer itch) is a pruritic rash resulting from circaria contact. At acute stage with blood fluke infection, the patient may have cough, fever, asthama like symptoms. Lymphadenopathy, hepatic and splenic enlargement may be present. During chronic stage of infection, the disease is characterised by diarrhoea, abdominal pain and headache, enlargement of liver, spleen, anemia, haematomesis and ascites. The infection may also have immune and neurological consequence. In *S. haematobium* the symptoms are haematuria, polyuria and dysuria as well as papilloma formation which may lead to hydronephrosis, kidney failure and uraemia. The control measures of the disease include proper sanitation and water supply, elimination of intermediate hosts by physical and chemical means and treating all cases. Chemotherapy of schistosomiasis is to kill the worm so that egg laying stops or reduced as even a reduction in egg laying is of potential value to the patient. Praziquantol is now the drug of choice for the treatment.

11. *Paragonimiasis*

This disease is caused by the lung fluke, *Paragonimus westermani* and disease is widespread in Asia, Africa, South and Central America. These are found in lungs in encapsulated form. Lung fluke cause chronic cough with emission of bloody sputum. Heavy infections cause chest pain with pleurisy, shortness of breath, fever and anemia. Haemoptysis is common. The clinical picture often simulates pulmonary tuberculosis. Abscess formation is quite common in ectopic sites. Proper cooking of fresh water crabs and cray fishes is completely preventive. Emetine hydrochloride, sulpha drugs and praziquantel are used for treatment of paragonimiasis.

12. *Clonorchiasis*

It is caused by *Clonorchis sinensis*, which inhabits the bile duct. The disease is widespread in China, Japan, Korea, Vietnam and India. The adult flukes cause thickening of bilary ducts wall. Severe cases usually lead to cirrhosis, cholangitis, obstructive jaundice and ultimate death. Malignant changes of the liver and pancreas have also been attributed to the parasite. The prevention and control include health education recommending cooking of fish before eating, better sanitation, and improved methods for storage and treatment of night soil. Treatment with praziquantel, gelation violet and chloroquine are useful in curing infections.

13. *Fasciolopsiasis*

This disease in man is caused by the intestinal fluke *Fasciolopsis buski*. Light infections are generally asymtomatic. In heavy infections various gastorintestinal symptoms such as colic, diarrhoea and vomiting may occur. There is also marked eosinophilia, facial oedema and utricaria symptoms. With large number of parasites, the patient may develop acute intestinal obstruction. The disease can be controlled by cooking water plants or growing them in water uncontaminated with infected human and pig faeces. Hexylresorcinal and crystaloid anthelminthic drugs are useful in eradication of disease.

14. *Fascioliasis*

Fascioliasis or liver rot of sheep and goat are caused by liver fluke. In sheep the bile ducts as well as liver are damaged. It causes inflammation and hepatitis resulting in loss of its epithelium, thickening of wall, calcification and formation of gall stones. Heavy infection may interfere with metabolism of liver, haemorrhage and irritation resulting the disease, liver rot or Fascioliasis. Death of sheep may also result due to cerebral apoplexy. It may also results in anaemia, and oedemas. The liver rot may result in huge toll of sheep. Man is also infected by Fascioliasis. Prophylaxis is by control of intermediate host snails. The treatment of disease is with anti-helminth drugs such as hexachloroethane, filicin, emetine hydrochloride, phenothiazine etc.

5.4 Parasitic Arthropods - Diseases in Man and Animals

Many arthropods are free living and are found in an array of aquatic and terrestrial habitats. However, some members of crustacea (copepoda, isopoda, cirripedia, amphipoda), arachnida and insecta are parasitic. The arthropodan parasites are of considerable interest to biologists, specifically parasitologists, from the biological standpoint. Some of the parasitic arthropods like *ticks, mites, mosquitoes, fleas, bugs, flies* etc, are of considerable medical, veterinary, fishery and other economic importance not only because they cause direct injury to their hosts but also because many serve as vectors for various pathogens including numerous microorganisms and viruses. The arthropods may transmit diseases by two methods : 1. *Mechanical transmission* - the disease producing agent does not multiply in the arthropod. For example, *Salmonella* transmission by the house fly. The housefly is not an integral part of life cycle of the microorganism. The disease producing agent may develop in the arthropod 2. *Biological transmission* - the arthropod is an integral part of the life cycle. The disease is produced either by (i) multiplication, (e.g., *Yersinia pestis* in fleas). or (ii) by development (e.g., *Wuchereria bancrofti* in mosquitoes) or (iii) both by multiplication and development (e.g., *Plasmodium* sp. in mosquitoes).

The important parasitic arthropods are given in Table 5.5(a) and 5.5(b) and the arthropod born diseases are given in Table 5.5(c).

Table 5.5(a) *Classification of some Parasitic Importanct Arthropods*

Class	Order	Family	Common name
Insecta	*Diptera*	Simuliidae	Black flies
		Psychodidae	Sand flies
		Culicidae	Mosquitoes
		Ceratopogonidae	Biting midges
		Tabanidae	Horse flies Deer flies
		Gasterophilidae	Bot flies
		Oestridae	Warble flies
		Muscidae	House flies Tsetse flies etc.
		Calliphoridae	Flesh flies Blow flies
		Hippoboscidae	Keds Louse flies
	Hemiptera	Reduviidae	Assassin (conenose) bug
		Cimicidae	Bed bugs
	Siphonaptera	Pulicidae	Fleas
	Anoplura	Pediculidae	Sucking lice
	Hymenoptera	Apidae	Honey bee
		Vespidae (many other families)	Wasps
	Orthoptera	Blattidae	Cockroaches

Table 5.5(a) Contd...

Class	Order	Family	Common name
	Coleoptera	Staphylinidae Cantharidae Scarabaeidae	Beetles Beetles Beetles
Arachnida	*Acarina*	Ixodidae Argasidae Sarcoptidae Trombiculidae Demodicidae	Hard ticks Soft ticks Itch mites Chiggers Follicle mites
	Araneida	Many families	Spiders
	Scorpionida	Buthidae Scorpionidae Centuridae	Scorpions
	Pentastomida	Linguatulidae	Tongue worms
	Subclass	Porocephalidae	
Myriapoda	Diplopoda Chilopoda		Millipedes Centipedes
Crustacea			Includes lobsters, crabs and water fleas which may act as intermediate hosts of some helminths

Table 5.5(b) *Important arthropodan parasites of livestock and domestic animals*

I. Blood-sucking flies :

Phlebotomus spp. (sandfly, Psycondidae)
Tabanids (horse fly, Tabanidae)
Stomoxys calcitrans (stablefly, Muscidae)
Haematobia (=Siphona) irritans (hornfly, Muscidae)
Glossina palpalis (tsetsefly, Muscidae)
Simulium aureohirtum (blackfly, Sirnuliidae)
Hippobasca maculata Leach (lousefly, Hippoboscidae)
Siphona exigua (buffalo-fly, Dip. : Muscidae)

II. Myiasis flies : (a) Botflies

Gastrophilus intestinalis (horse bot, Oestridae)
Oestrus ovis (sheep bot, Oestridae)
Hypoderma lineatum (warble fly, Oestridae)

(b) Blow flies

Chrysomyia bezziana (Indian screwwonn, Calliphoridae)
Cochliomyia (=Callitroga) hominivorax (American screwworm, Calliphoridae)
Calliphora erythrocephala (European blue bottles, Calliphoridae)
Lucilia cuprina (green fly, Calliphoridae)
Phormia regina (blue fly,) Calliphoridae)

Table 5.5(b) Contd...

	(c) Flesh flies
	Sarcophaga spp. (flesh fly, Sarcophagidae)
III. Lice	
(a) Sucking lice :	*Haematopinus* spp. (hog louse, Haemampinidae) *Linognathus vituli* (blue lice of cattle, Linognathidae)
(b) Biting lice :	*Menopon gallinae* (shaft louse, Menoponidae) *Columbicola columbae* (pigeon louse, Philopteridae) *Eomenocanthus (= Menacanthus) stramineus*, (chicken body louse, Monoponidae) *Bevicola* spp. (animal lice, Trichodectidae) *S. trichodectis canis* (dog louse, Trichodectidae)
IV. Fleas :	
	Echidnophaga gailinacea (stickfast flea, Hectosyllidae) *Ctenocephalides canis* (dog flea, Pulicidae)
V. Ticks :	*Boophilus microplus* (cattle tick, Ixodidae) *Argas persicus* (fowl tick, Argasidae) *Rhipicephalus sanguineus* (dog tick, Ixodidae)
VI. Mites :	
	Sarcoptes scabiei (manage mite, Sarcoptidae) *Demodex canis* (dog mite, Demodicidae)

Table 5.5(c) *Arthropod-Borne Disease of Man*

Class of pathogen	Pathogen vector	Arthropod	Disease	Mode of transmission
I. Bartonellacea	*Bartonella bacilliformis*	*Phlebotomus nogouchi*	Carrion's disease	Bite
II. Rickettsia	1. *Rickettsia typhi*	*Pediculus*	Epidemic typhus	Bite
	2. *R. prowazeki*	*Xenopsylla*	Endemic *typhus*	Bite
	3. *R. tsutsu-gamushi*	Mite	Scrub typhus	Bite
	4. *R. ricketlsi*	Tick	Spotted fever	Bite
	5. *R. quintona*	*Pediculus*	Trench fever	Bite
	6. *R. diaporica*	Tick	Q fever	Bite
III. Virus	1. Virus	*Aedes*, Tick	Yellow fever	Bite
	2. Virus	*Aedes*, *Stegomyia*, Tick	Dengue fever	Bite
	3. Virus	*Aedes*, *Culex*	Encephalitis	Bite
	4. Virus	*Stomoxys*, *Musca*	Poliomyelitis	Bite
	5. Virus	*Phlebotomus*	Papataci fever	Bite
IV. Spirochaetes	1. *Treponema partenue*	*Hippelates*	Yaws	Bite
V. Bacteria	1. *Pasteurella tularensis*	Tick	Tularemia	Bite
	2. *Yersinia pestis*	*Xenopsylla*	Bubonic plague	Bite
	3. Shigella *sonnei*	*Musca*	Bacillary dysentery	Food contamination
	4. *Salmonella typhi*	*Musca*	Typhoid	Food contamination

Table 5.5(c) Contd...

Class of pathogen	Pathogen vector	Arthropod	Disease	Mode of transmission
	5. *Vibrio cholerae*	*Musca*	Cholera	Food contamination
	6. *Myobacterium tuberculosis*	*Musca*	T.B.	Food contamination
	7. *Myobacterium leprae*	*Musca*	Leprosy	Contact
VI. Protozoa	1. *Entamoeba histolytica*	*Musca*	Amoebiasis	Food contamination
	2. *Giardia lamblia*	*Musca*	Giardiasis	Food contamination
	3. *Leishmania donovani*	*Phlebotomus*	Visceral leishmaniasis	Bite
	4. L. *tropica*	*Phlebotomus*	Oriental sore	Bite
	5. L. *brasiliensis*	*Phlebotomus*	American leishmaniasis	Bite
	6. *Tryparwsoma gambiens*	*Glossina palpalis*	Gambien sleeping sickness	Bite
	7. *T. rhodesciense*	G. *morsitans*	*Rhodesian sleeping sickness*	Bite
	8. *T. cruzi Triatoma*	*Rhodnius.*	American sleeping sickness	Bite
	9. *Plasmodium spp.*	*Anopheles spp.*	Malaria	Bite
VII. Helminths Nematoda	1. *Wuchereria bancrofti* culex.	*Aedes Mansonia*	Filaria	Bite
	2. *Loa loa*	*Chrysops spp*	Loasis	Bite
	3. *Onchocerca volvolus*	*Simulium*	Onchocercosis	Bite
Cestoda	1. *Diphyllobothrium latum*	*Cyclops sp.*	Diphyllobothriasis	Fish
	2. *Hymenolepis dimunita*	*Blatta germanica*	Hymenolepisiasis	Food contamination
Trematoda	1. *Paragonimus westermani*	snail to crab man	Paragonimiasis	Oral
VIII. Acanthocephala	*Monoliformis monoliformis*	Cockroaches	Monoliformiasis	Food contamination

The important parasitic arthropods mosquitoes, flies including house fly and related species, fleas, bugs, lice, ticks, mites, etc., are briefly described along with their parasitological importance.

1. *Mosquitoes*

These belong to the family culicidae. This is a large family containing approximately 31 genera and many hundreds of species. The genera of greatest medical importance are *Anopheles, Culex, Aedes,* and also of lesser importance are Coquillettida, Culiseta and Manosonia. Mosquitoes have elongated mouth parts and only female is capable of taking a blood meal. The proboscis is a complex structure and is made up of a pair of mandibles, a pair of maxillae and a hypopharynx, which are all enclosed in the labium, which forms a protective sheath. The mosquito head is globular in shape and bears two large compound eyes. Eggs are laid on water or on moist substrate and are characterized by the presence of floats in case of Anopheles. The larvae and pupae are both aquatic (Fig. 5.30). The adult mosquitoes emerge from the pupae and mate when they are 1 to 2 days old. The female generally

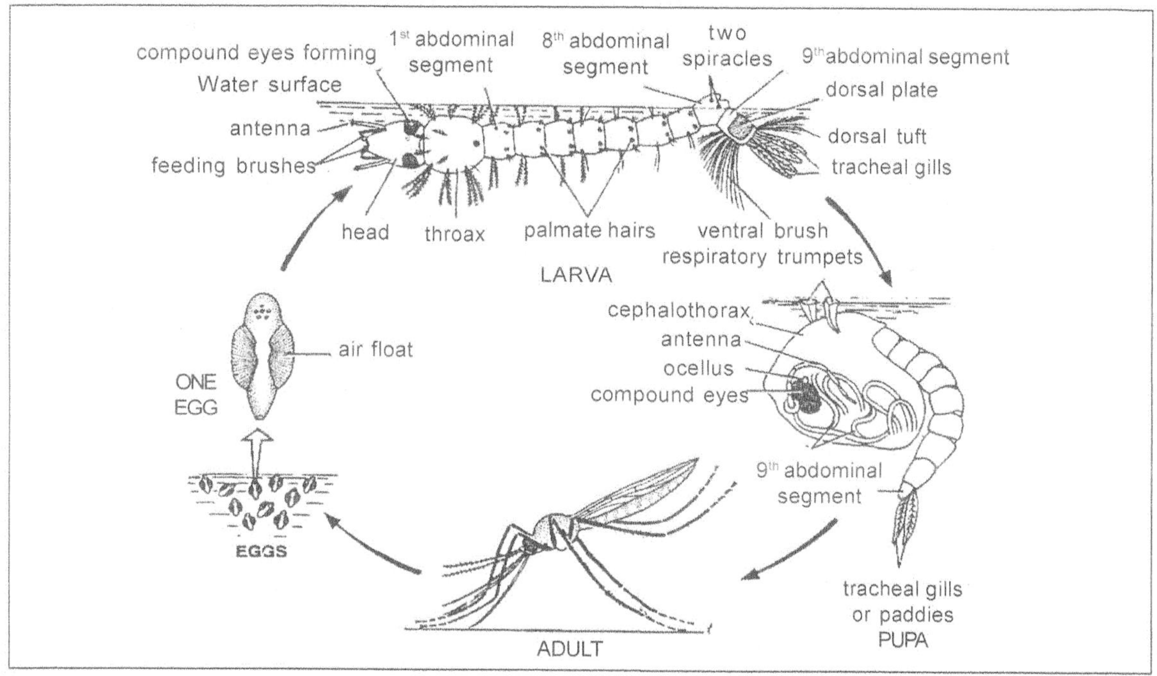

Fig. 5.30 Anopheles – stages of life cycle

takes a blood meal every 4 to 5 days, after which she lays eggs. Human plasmodium is transmitted only by genus *Anopheles*, which acts as malarial vector. The *Culex pipens* is major vector of filarial worms Wucheraria boncrofti and Dirofilaria immitis. They also transmit avian malaria, avian pox, encephalotides caused by arboviruses. *Aedes* genus include about 800 species, out of which *Aedes aegypti* is principle yellow fever virus, and *Aedes dorsalis* known for its bites on humans, horses, cows, sheep and even birds. *Mansonia* is an important vector for microfilariae (*Wuchereria bancrofti, Brugia malayi, Dirofilaria immitis*).

Mosquitoes are not only serve as vicious pests but also serve as vectors for various pathogenic organisms. The major mosquito transmitted diseases are Malaria, Yellow fever, Dengue fever, equine encephalitis, filariasis etc.

2. *Flies*

Flies with parasitic importance in man and animals include housefly, blood sucking flies, sand flies-tabanids, tsetse flies, black flies etc., myiasis flies, bot flies, warble flies, blow flies, flesh flies etc.

(a) *House fly*

There are many genera in the family muscidae. The common house fly, *Musca domestica* is one of the most important ones. These are medium sized flies with a mouse-grey coloration. The head has a pair of large compound eyes. The terminal labella on the proboscis is covered with parallel transverse ridges through which the food is sucked in. The egg laying starts from 4 to 8 days after copulation and is done on substratum like putrefying organic matter and manure which is suitable for larval development

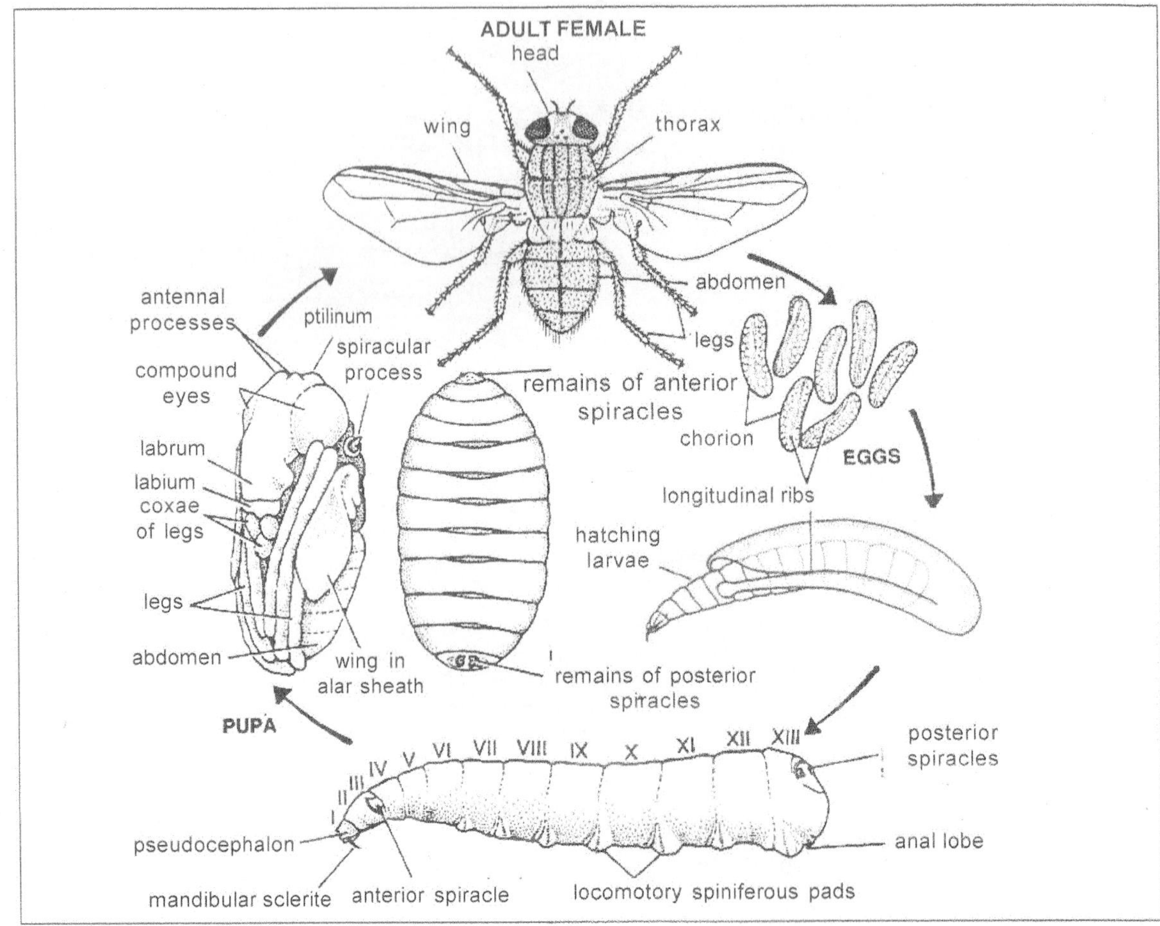

Fig. 5.31 Musca domestica – life history

(Fig. 5.31). Adult are non parasitic but are household pests. However, they are mechanical transmitters, by their contaminated feet, of such diseases as *typhoid fever* (by *Salmonella typhi*), *cholera* (*Vibrio cholerae*) and *yaws* (*Treponema pertenue*). Interestingly, larvae of *Musca domestica* have been recovered from human vomitus, faeces and urine. Thus, *Musca domestica*, sometimes results in *accidental myiasis* through ingestion of its egg or young larvae causes gastrointestinal as well as urinary myiasis or infestation on uncerated wounds resulting in dermal myiasis. In tropics the important myiasis producing genera are *Dermatobia, Callitroga, Chrysomyia, Lucilia, Sarcophaga* and *Cardylobian.*

(b) *Blood Sucking Flies*

(i) *Phlebotomus spp (Sand flies)*

These hematophagus insects are small, about 3-4 mm long, slender and yellowish buff coloured with a hairy body. They live in damp and dark places, crevices and breed in sandy soil; (therefore known as *sand flies*). The egg laying starts after 30-36 hours of blood meal. A female lays 30-35 eggs which hatch in 6-12 days, undergo 4 moults to become pupa and which moult in 6-14 days into adult. They suck blood preferably of cattle but can also feed on horses, birds, man, dogs, and hogs in that order.

The three species found in India are *P. papatasii*, (Fig. 5.32(a)) P. argentipes, and P. minutes. The insect bite is a sharp painless incision with oozing followed by itching and formation of red pustule. In addition to debilitating effect and reduction of milk yield, these insects transmit pathogens of diseases such as papatasi fever (by virus), visceral (kalazar), cutaneous leishmaniasis and Bartonellosis (carrion's disease) in humans.

(ii) *Tabanids*

These are commonly known as Horse flies, gad flies, deer flies, green headed flies etc. These insects are larger than the common house fly with a thick-set body, large head and big brilliantly coloured eyes. The genus *Taba*nus (Fig. 5.32(b)) has several species such as *T. rubidus, T. viva, T. striatus* and *T. macer* all of which live in marshy places and prefer sunlight. A female deposits 200-500 egg masses, enclosed in transparent and water proof material, on aquatic and subaquatic vegetation. Hatching takes place in 5-8 days and liberates carnivorous larvae, which moult about 6 times in 5-6 months before pupation. The image emerges, 10-18 days. The male feed on nectar while female feed on blood of human and animals. They feed on horse, cattle, camel, elephant and occasionally on man. The loss of blood (100-200ml) by cattle resulting by tabanid flies is a serious problem. Weight loss and decline in milk production commonly result from the bites. In addition to inflicted injuries, discomforts, loss of blood, tabanids serves as vectors for bacterial, protozoan and helminth parasites. The *Trypanosoma evansi*, the causative agent of "*surra*" disease in cattle, horses and dogs is transmitted by *Tabanus striatus*. *Trypanosoma equiperdium* causative agent of '*daldecaderas*' in horses and other animals and *Trypanosoma theleri* of cattle are transmitted by these flies. Besides, the tabanids transmit pathogens of bacterial diseases like anthrax and Tularemia in humans and animals, equine anemia and surra in hoofed animals and Trypanosomal infections in animals of Africa and filarial disease, loaris (causative agent Loa loa) is also transmitted.

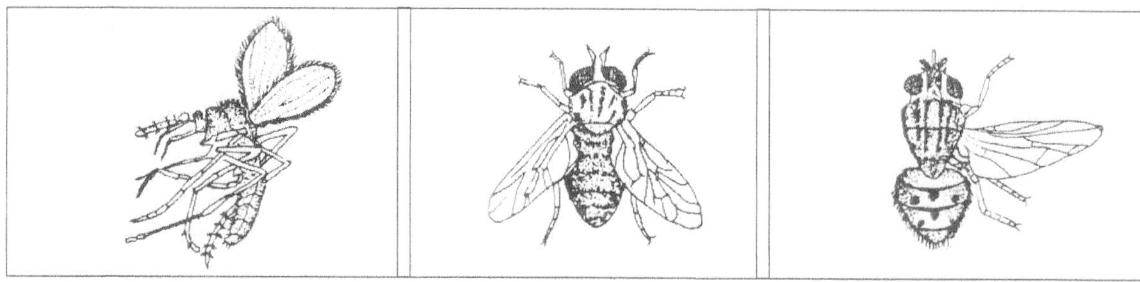

Fig. 5.32(a) Phlebotomus papatasii Fig. 5.32(b) Tabanus viva Fig. 5.32(c) Stomoxys calcitrans

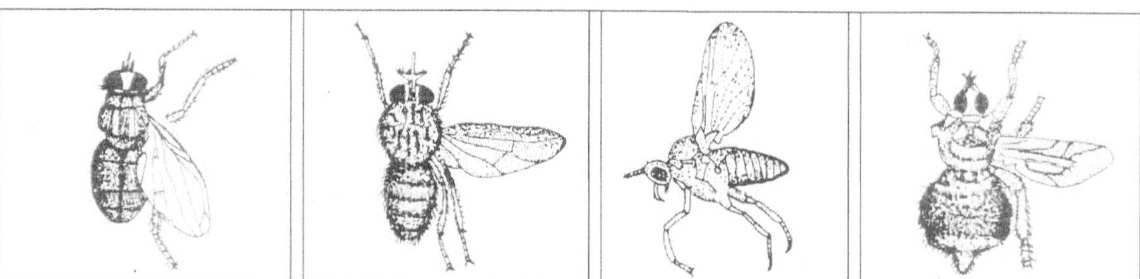

Fig. 5.32(d) Haematobia (=Siphona) irritans Fig. 5.32(e) Glossina palpalis Fig. 5.32(f) Simulium Fig. 5.32(g) Hippobosca maculata

(iii) *Stomoxys* :

These are cosmopolitan in distribution and are called as *stable flies*, storm flies and stinging flies. These are oval, 6-8mm long greyish insects, resembling house flies, with longer proboscis (Fig. 5.32(c)). They suck blood from animals and also feed on fruits, straw and manure. The female lay about 20-50 eggs after three blood meals, egg hatch in 1-3 days, mature in 7-21 days, pupate for 6-26 days and develop into adults. Both sexes bite to suck blood from dairy cattle, horses and man resulting in lose of weight and milk yield in cattle. The flies also transmit virus causing poliomyeletis, leprosy in man, anthrax, surra, swamp fever, trypanosomiasis and leishmaniasis in animals.

***(iv) Haematobia irritans* :**

These are called as *horn flies*. They are is dark or yellowish grey in colour, half the size of the house flies and stable flies (Fig. 5.32(d)) and remain on their hosts all the time. A female lays 300-400 eggs in fresh manure and spreading among herd. The egg hatches into maggot in 2 days, pupate and pupa hatch in 6-8 days. Both sexes suck blood from cattle, goats, sheep, horses and dogs. They cause great annoyance and interference with the feeding of animals at they parasitise.

(v) *Tsetse flies* :

These are found in Africa; of them there 22 are well known species. They are robust yellowish to brownish black flies measuring 6 to 15mm. Both sexes suck blood. They have conspicuous proboscis which projects anteriorly. The larvae is fully grown within mother, (viviparous), and delivered upto 10 in number, larviposition occurs during vegetation, in tree holes or caves. The larva pupate and adult emerges in about 1 month. The medically important species are *Glossina pal*palis (Fig. 5.32(e)) *G. trachinoides, G. morsitans, G. swynnertoni*, and *G. pallidipes*. These are capable of transmitting trypanosomiasis in animals, Nagna disease of cattle and horse, and G. palpalis and G.morsitons spread African sleeping sickness in man.

(vi) *Black flies* :

These are small flies measuring 1 to 5 mm and they generally have a strong humped appearance. Only the female sucks blood and it lays eggs below the surface of streams and rivers. The larvae hatch out and pupae are always found in water. The fly emerges under water and then comes to surface. They belong to genera simulium (Fig. 5.32(f)). The insects are vicious biters causing severe irritation and suck blood from cattle, horses and other domestic animals, transmit with debilitating effect. They transmit filarial disease called *Onchocerciasis* in man living in vicinity of wooded streams in Africa and Central American countries.

(vii) *Hippobosca maculata* :

These louse flies are wingless, flattened leathery looking more like louse than a fly (Fig. 5.32(g)). They have world wide distribution living on birds and mammals as ectoparasites on whom they pass their life cycle. Females are viviparous depositing 10-20 full grown larvae on the ground, pupate and adult with the help of claws cling to host body and become permanent ectoparasite of cattle, goat, sheep, horse and dog.

3. *Myiasis flies*

Myiasis refers to an infestation of living tissues or organs of man, the livestock as well as domestic animals by maggots or larvae of non-sucking, dipterian flies. These include bot flies, warble flies and blow flies. Myiasis may be of atrial, cutaneous, intestinal, urinary, nosopharyngal myiasis and wound type depending on nature of site of infestation.

(a) *Bot flies :*

Gastrophilus (Fig. 5.33(a)) commonly known as horse bot fly, has superficial resemblance to honey bee. The flies appear during November-December, buzz around the animals, horses and lay the eggs on body of animal. Eggs are licked inside the mouth and hatched by friction and moisture of tongue, burrow into tongue, moult into second instar then pass into stomach, attach the wall and feeds, moult into third instar and become full grown in 9-10 months. Then pupate and pass in faeces, hatch into adult flies. Horses, mules, donkeys, rarely dogs, rabbits and man are attacked. The infestation interfere with glandular activity of stomach, cause inflammations, ulcers, secrete toxins and also obstruct passage to the intestine. If not treated the host dies.

Oestrus ovis (Fig. 5.33(b)) is sheep bot fly and are viviparous depositing their larvae in the nostrils which after 2 or 3 weeks invade the frontal sinuses and other parts of head. In 2-3 weeks, these are full grown, drop out pupate on ground, emerging in 4-6 weeks as adult flies. The infestation results in profuse discharge of mucus and cause distress to the host. Heavy infestation may cause extensive mortality in sheep.

(b) *Warble flies :*

*Hypoderma lineatu*m (Fig. 5.33(c)), these are hairy yellow or black flies popularly called as *Warble flies* or *heel flies*. A female fly lay about 100 eggs in hairs around hoofs of cattle or some times back and belly. In 3 days maggots hatchout, penetrate the skin, migrate through the cavities to underskin of back. The 2-5th instar larvae encapsulate beneath the skin of host causing subcutaneous swelling, *warbles*, and perforation in hide. The larva moult and mature inside warbles in 6-9 months, comes out and pupate in dry soil where infection results in irritation, pain, lower milk yield, and hide quality and may also lead to eye myiasis.

(c) *Blow flies :*

The blow flies are medium to large sized insects of metallic colour, oviparous depositing eggs in decaying organic matter or in cutaneous sores and abrasions causing cutaneous myiasis. They infect man, sheep and cattle. The important species are *Chryomyia beeziana, Cochliomyia hominivorax* (Fig. 5.33(d)), *Calliphora erythrocephala* (Fig. 5.33(e)), *Lucilia cuprina, Phormia regina* etc. These flies have thoracic stripes and transverse abdominal bands. It is attracted to wound in which it deposits its eggs, the maggots hatch in a day, look like a screw, burrow the tissue, feed on it develop into third instar in 7 days and leave the host to pupate on ground. The larvae produce foul smelling lesions that even erode bone. *C. bezziana* causes human myiasis in India. *C. homnivorax* infests wounds of domestic and wild animals, particularly cattle, even damaging cartilage, brain and sinuses. *C. erythrocephala* causes myiasis in man and other animals causing extensive tissue destruction. L. serenissima causes cutaneous myiasis in cattle. Phormia regina, its larva are common maggots in wool of sheep.

(d) *Flesh flies :*

The flesh flies are parasites during their larval stages, medium to large sized, thick set, greyish and larviparous flies that are usually found in vicinity of latrines or near decaying animals and vegetable matter or sometimes on healthy animals also. These be family sarcophagidae. The *Sarcophaga carinaria* (Fig. 5.33(f)) and S. haemorrhoidalis, are two important species, which produce cutaneous and nasal myiasis in mammals including human. Sometimes the intestinal and rectal myiasis are also found on man due to deposition of eggs at anus.

(e) *Biting midges :*

These are smallest of biting flies measuring between 1 and 4 mm. The female is blood feeder and deposits her eggs on plants or objects close to water. The larva is worm-like with a distinct pigmented head. Medically the most important genus is *Culicoides*.

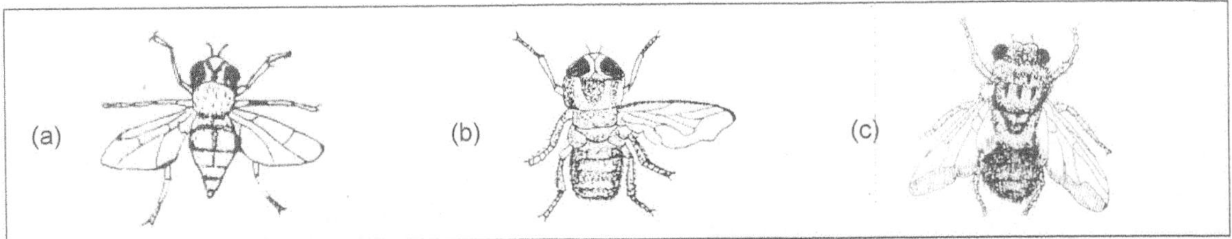

Fig. 5.33(a-c) (a) Gastrophilus intestinalis (b) Oestrus ovis (c) Hypoderma lineatum

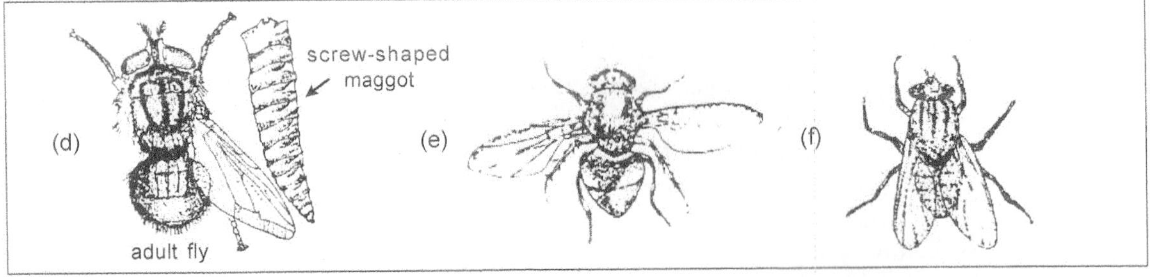

Fig. 5.33(d-f) (d) Cochliomyia hominovorax (e) Calliphora erythrocephala (f) Sarcophaga carinaria

4. *Fleas*

Fleas are small, apterous and ectoparasitic insects, laterally compressed, and without wings. The antennae lie in a groove in head region. The legs are long and muscular, allowing the fleas to jump. The mouth parts are of piercing and sucking type and they suck blood from warm blood animals (Fig. 5.34). The adult feed periodically and do not stay permanently on their hosts. The eggs are laid in burrows and larvae are maggot like. The flea of medical importance are Pulex irritans (common human flea), *Ctenocephalides canis* (Fig. 5.34), *C. felis* (dog and cat flea), *Xenopsylla cheopis* (oriental rat flea) which is vector of plague. On ingestion of *yersinia* by the flea, the bacteria multiply in stomach and oesophagus becomes blocked, such a flea moves from person to person to obtain satisfying blood meal and in this process rapidly transmit infection of plague. The flea *Tunga penetrans*, of 1mm in length burrows the skin, cause extreme irritation and requires surgical removal.

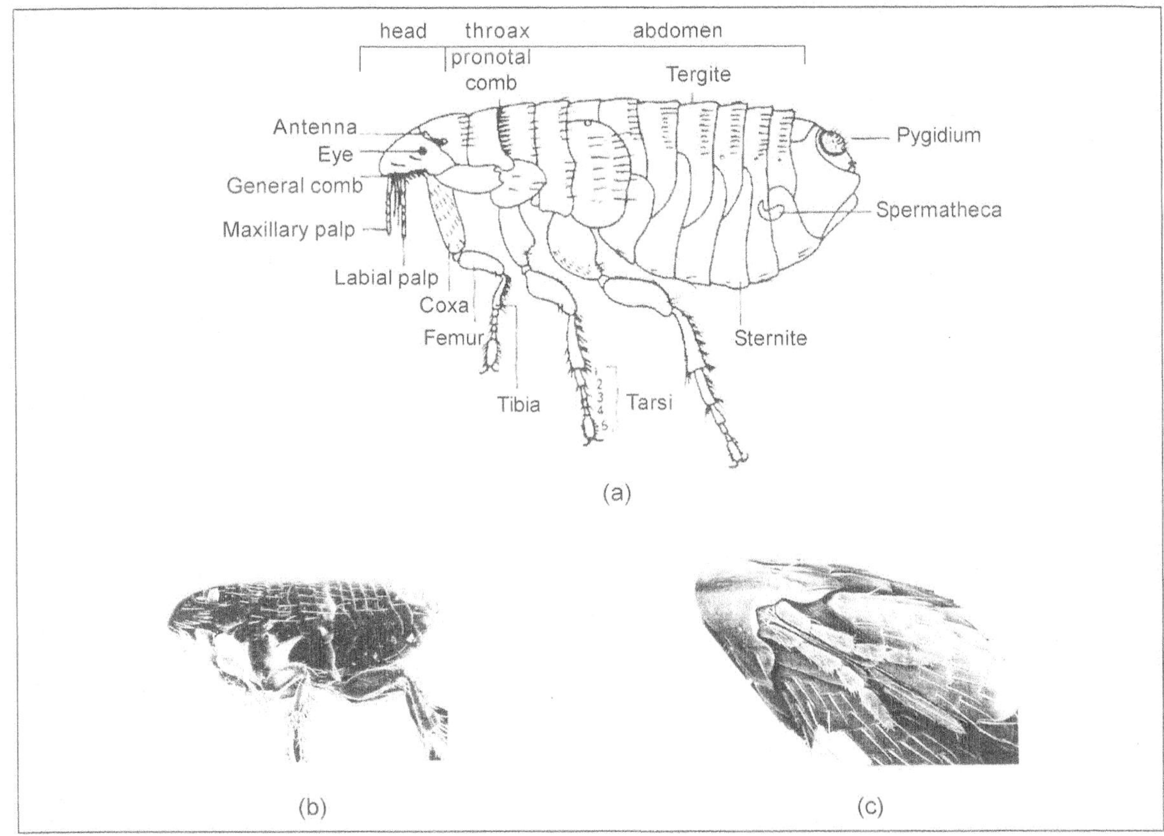

Fig. 5.34 Fleas – (a) adult flea (b) Ctenocephalides (c) Xenospsylla

Ctenocephalides canis, found in dogs and cats, have two rows of combs of spines which help to move through hairs on skin. The larva feed on blood, moult twice, and pupate in cocoons. The infection cause itching, irritation and annoyance. The stickfast flea, *Echidnophaga gallinacea*, infects poultry, resulting in anaemic with decreased egg production or may even lead to death in birds.

5. *Bugs*

The arthropods belonging to families cimicidae, reduvidae and polytenidae include parasitic species. These are cimax, triatoma, rhodinus, panstronglus etc. The common bed bugs are ectoparasite on humans, *Cimex lectularis* (temperate regions) and *Cimex hemipterus* (Fig. 5.35) (tropical regions). The body is flattened dorsoventrally and are covered by bristles. The head bears 4-jointed antennae and a 3 jointed proboscis, which is short and lies folded on the ventral surface of head. The legs have distinct claws. The female lays eggs after bloody meal. The eggs take 7-9 days to hatch, pass through 5 ecdyses and all nymphal stages are haematophagus. The bite of bed bug can be a source of severe skin irritation and discomfort.

The Triatomine bugs, are naturally infected with the haemoflagellates, *Trypanosoma cruzi*, are important as vectors of chaga's disease. The bug, *Triatoma*, measures 18-20 mm in length and is characterised by a flattened body that is dark brown and splattered with orange or dark brown spots.

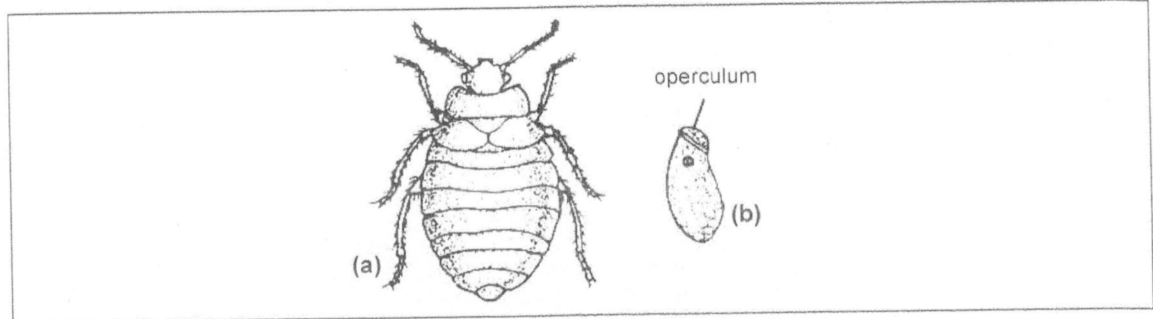

Fig. 5.35 Cimex hemipterus – (a) adult bug, (b) a nit or egg

These species are vicious biters of various mammals including humans. Their bite result in very painful and itchy swelling due to toxin injected during feeding. The genus *Rhodinus*, also infected with T.cruzi, is common in brazil. The other genus like Panstrongylus, Melanolestes and Rasahus bites human and hosts reaction to toxin is severe.

6. *Lice*

(a) ***Human lice*** (Fig. 5.36(a)) are *Pediculus humanus capitis* (head louse), *P.h. humanus* (body louse) and *Pthirus pubis* (pubic or crab louse). Adult lice are 3 to 4 mm long, varying from white to grey black in colour. They attach to the host tissue by strong crab-like claws and suck blood through a stylet. In case of head lice, the eggs are firmly cemented to the base of the hair. The nymphs hatch in about 5 days and adult develop in about 16 days. The body louse oviposits on clothing. The lice migrate from one person to the other and more so in hyperpyrexia conditions.

(b) ***Sucking lice*** are all apterous insects living as blood sucking endoparasites of mammals. They are small flat bodied insects with piercing and sucking mouthparts remaining hidden inside the head when not in use. Their thoracic segments are fused and legs are clawed. *Haematopinus eurystermus* (Fig. 5.36(b)) known as Cattle louse, about 3mm long flat bodied and wingless. *H. quadripertusus* found in tail, anus, vulva and eyes of cattle. *H. suis* occurs on both domestic and wild pigs. They thrust their mouth parts into skin, suck the blood and simultaneously inject saliva which causes irritation and itching. *Linognathus vituli, blue lice*, found in cattle in India particularly on sick and emaciated ones. The hides of these animals get damaged.

(c) ***Biting lice*** resemble sucking lice in their shape, wingless, flattened, reduced eyes and clawed legs. They have biting mouth parts and are ectoparasites mainly on birds. *Menopon gallinae* (Fig. 5.36(c)) shaft louse, of birds, cause abrasions, itching annoyance, the young birds growth is retarded and older birds become weak and lay fewer eggs. The pigeon louse, *Columbicola columbae*, similar to shaft louse causes damage. *Menaconthus stramineus* is chickens body louse of fowl and turkey causing serious injury. *Bevicola* found on animals (not on birds) B. caprae (goat), B. ovis (sheep) B. bovis (cattle) and B. equi (horses and mules). *Trichodectes canis* found on the dogs and acts as a vector of dog cestode.

Ticks and Mites

Ticks and Mites belong to family Acarina of Arachinida, with characteristic mouth parts consisting pedipalps and chelicerae, they have cephalothorax and abdomen with four pairs of appendages. All

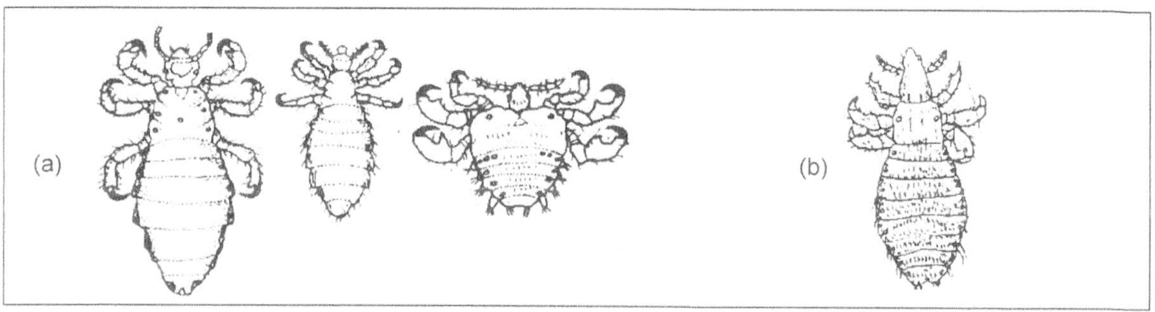

Fig. 5.36 (a) Pediculus humanus (body louse); Pediculus humanus capitis (headlouse); Pthirus pubis (pubic louse) (b) Haematopinus sp.

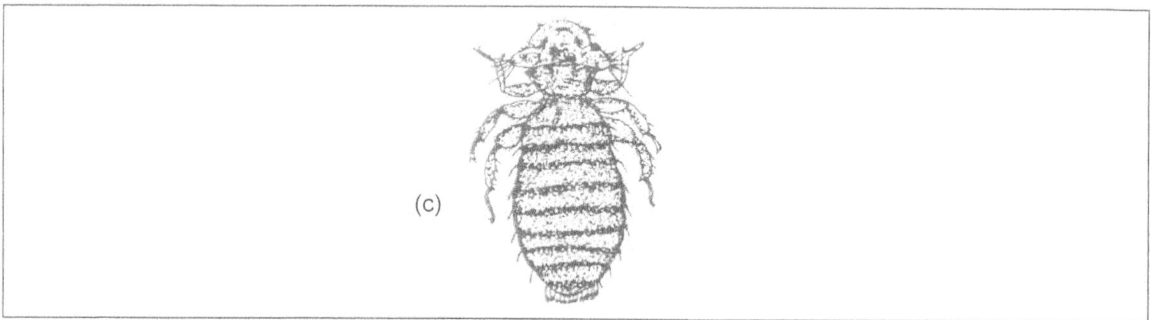

Fig. 5.36 (c) Menopon gallinae

ticks and many mites are parasitic. Ticks differ from mites it being larger, having short hairs, a leathery body-texture, exposed and toothed hypostome (mouth parts). In contrast mites are small, with longer hairs, membranous body texture, hidden unarmed hypostome.

7. *Ticks*

These are of two kinds – the Argasidae or soft ticks and Ixodidae or hard ticks. The ticks usually hide in cracks and crevices during the day. The female produces eggs in very large number, the larva which emerges has three legs and this becomes a nymph with 4 legs after moulting. The nymph then moult to become an adult. All stages are blood sucking. In addition to being a vector of certain infections, ticks also produce local and systemic damage. The local inflammatory lesion results from trauma caused by hypostome. The systemic damage is produced by saliva which blocks the release of acetylcholine. This leads to ascending flacid paralysis or *tick paralysis*. *Boophilus microples* (Fig. 5.37(a)) and *Hyalomma anatolieum* are common cattle ticks of India. The female measures about 2-5 mm and lays about 2-3 thousands eggs which hatch in 2-6 weeks. The larvae feed on host's blood and moults into nymphs. H. anatoleium parasites on goat, horse, camel etc. The ticks cause effect on host through mechanical injury, causing inflammation, abrasion, haemorrhage and produce tick paralysis through their secretions in sheep, cattle, dog, less frequently in man and also transmit diseases like tick fever, texas fever, tularemia, American spotted fever. *Argus persicus* (Fig. 5.37(a)), the fowl tick attack birds like fowl, duck, turkey, goose, pigeon and apart from sucking blood, cause weakness, reduction in eggs, and transmit fowl spirochaetosis. The dog tick, *Rhipicephalus sanguineus* (Fig. 5.37(a)), beside blood sucking, transmit canine piroplasmosis caused by the protozoan *Babesia canis*.

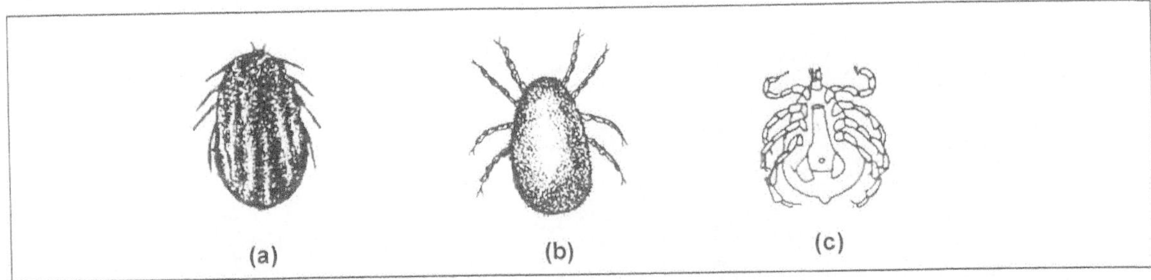

Fig. 5.37 (a) Boophilus microplus, (b) Argas persicus (c) Rhiphicephalus sanguineus

8. *Mites*

These are usually small to microscopic. The cephalothorax and the abdomen are fused without any line of demarcation. The mouthparts have chelicerae. Three major groups are of medical importance. (i) trombiculid mites (ii) sarcoptic mites and (iii) house dust mites.

Trombiculid mites: These live in soil and vegetation. The larval stages attack animals or humans. The larvae feed on tissue fluids and lymph by embedding their mouth parts in the skin (Fig. 5.38(a)). After feeding they drop off to moult into nymphs and then into adults. The ricketessial organisms are carried by these mites and are passed through the egg from one generation to the another.

Sarcoptic mites: These belongs to *Sarcoptes scabiei* (Fig. 5.38(b)) and produce contagious disease scabies. The disease is characterised by a popular rash which appears over the body particularly in interdigital areas, the auxilla, the genitalia, the buttocks, ulnar surface of fore arms and the wrist. Itching occurs and the secondary infection may result in ulceration.

Dust mites: A number of species of mites are found in dust (Fig. 5.38(c)). These are microscopic and release antigenic material in the respiratory passages when they are inhaled. These mites are an important cause of asthma in man.

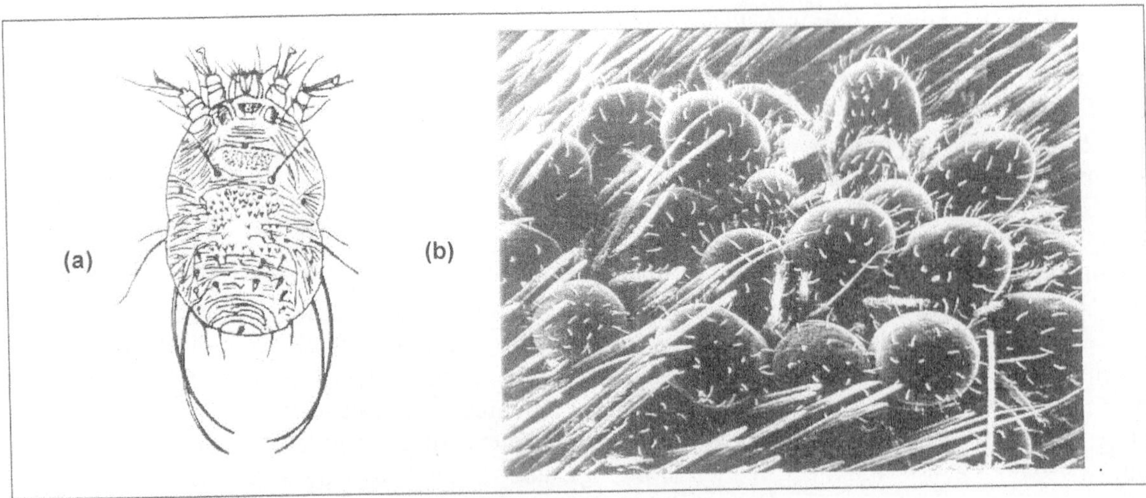

Fig. 5.38 (a) Sarcoptes scabiei (b) Larval mites taking a blood meal on a mouse. Mites are distended with blood, therefore, appear as oval bodies.

Fig. 5.38 (c) An adult house dust mite standing on a dust particle

Some representative Etiological Agents of Diseases transmitted by Mites

Disease	Etiological agent	Mite vector
Scrub typhus	*Ricketsia trustsugamushi*	*Trombika akamushi* *T. delienris* *T. pallida* *T. scutellaris*
Rickatesial pox	*R. akori*	*Allodermanyssus sanguineus*
Q. fever	*Coxialla burnetti*	*Orinthonyssus bacoti*
Nonpathogenic	*Moniezia expansa, M. benedeni* (cattle, sheep, lambs)	*Sheeloribates* spp. Galumna sp.

Disease control of parasitic arthropods

As seen in earlier discussion, parasitic arthropods like ticks, mites, flies, fleas, bug and mosquitoes are mainly external parasites and act as vectors or disease carrying or hosting animals. Hence, the prevention or removal of these parasites from reaching the host is the best choice in disease control. Pulling off ticks, mechanical control of parasites like fleas, mosquito or removal of myiasis larvae etc., are some of the physical methods. Using of chemicals like sulphur ointment, benzylbanzoate, tetmosol and lindane for control of scabies, DDT and other insecticides like pyrethrum, organic phosphorus as well as organochlorine compound, and repellents for lice, flies and mosquitoes control are some of the measures. Myiasis can also be controlled by using antibiotics or surgical removal. The utilization of biological methods like sterilants (azidirne and phosphine oxide), use of predators, attractants, pathogenic microorganisms and genetic measures are of much significance in control of arthropod parasites particularly which act as vector or intermediate hosts. Thus, the disease control programme of parasitic arthropod must include prevention of contact with host, destruction of parasities, their larva and vectors with chemical and biological methods and use of proper therapeutics whereever necessary.

5.5 Parasitic Diseases of Fish and Shell Fish

Fish and shell fish organisms are affected by several diseases due to environmental stress as well as by parasitic afflictions. Diseases of aquatic fishery organisms at times assume the magnitude of epidemics. From causative point of view, fish diseases may be classified as

(i) *Parasitic diseases caused by*, (a) protozoa, (b) worms, (c) crustacea, (d) bacteria and (e) fungi and

(ii) Non-parasitic diseases due to physical as well as environmental factors.

This section presents a brief description of parasitic diseases, caused by Protozoans, helminths and crustaceans parasites.

A. *Protozoan diseases*

Numerous protozoan parasites live on fish and other aquaculture species and cause both external and internal diseases with serious mortalities. Even moderate numbers of these organisms on small fish may prove fatal because the infections cause the fish stop feeding. The principle protozoan diseases include those caused by *Icthyiopthirius, Costia, Myxobolus* and *Trypanosoma.*

1. *Icthyopthiriasis*

Icthyopthiriasis or white spot disease is caused by the protozoan parasite *Icthyopthirius multifilis* a ciliate, (Fig. 5.39) is considered to be one of the detrimental diseases in fresh and brackish water fish culture. It is a fairly widespread disease. The common carp, Chinese carp, trout and cat fishes are susceptible to the disease. *I. multifilis* has a round or ovoid body 0.5 – 1.0 mm long with longitudinal rows of cilia on surface of body and with a small rounded mouth. This species multiplies by binary fission. The mature parasite (trophants) enters the water and on attachment parasite becomes enclosed in a gelatinous cyst and multiplies. After emerging from cyst, they swim free for 2 to 3 days and if host is found they penetrate under skin, grow and mature.

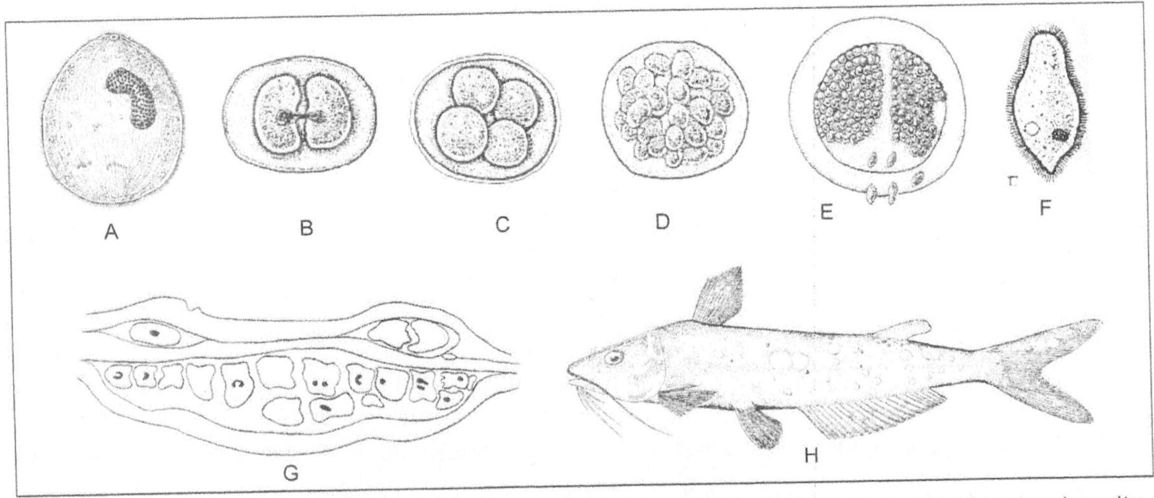

Fig. 5.39 Ichthyophthirius multifilis, a parasite in the skin of fish. A. Mature trophant or trophozoite, B to E. stages in the development, F. Young theront that has escaped from cyst, G. Section through fin of infected carp showing numerous parasites, and H. Catifish heavily infected with and I. multifilis.

The symptoms include appearance of numerous small white specks (early infection) or white spots (advanced stage of infection) over the entire skin, gills and fins. The spots range between 0.1-1 mm in diameter. The spot is infected in a small chamber in which the parasites are accumulated. In heavy infections the severe symptoms are restlessness, fish rubbing against bottom, sides and collecting in masses at inlet, small tubercles occur on body, the fish stop feeding and cease reacting to stimuli.

The destruction of carrier fish is an essential aspect of prevention. Disinfection of contaminated water is recommended. The dilute salt of 0.5 % kill encysted parasites and the ciliated bodies. The ciliated bodies can also be killed by drying ponds. The treatment of ponds with 0.1 ppm malachite green or 15 ppm formalin, and 1 to 2 ppm quinine solution has been found effective.

2. *Ichtyobodosis*

Ichtyobodosis, previously known as costiasis, is caused by the flagellate *Icthyobodo necator* (Costia necatrix) (Fig. 5.40) affects many species of fish (carps and trouts) particularly at younger ages. The parasite attaches itself to the skin of host and anterior end penetrate into the cells of host and suck its contents. Parasite multiplies by longitudinal division and transmission takes place through water. Indications of the Ichtyobodosis are the presence of a bluish coating on the skin of the fish and presence of large amount of mucus.

The parasite causes irritation and disturbs respiration. In very advanced stages red patches may appear on affected parts. Fins may be folded up and gills partially destroyed. To prevent the diseases, the young one should not be kept too long in spawning and rearing pond. Ponds should be disinfected with quick lime. This disease can effectively be treated with 15-50 ppm of formalin, 2-3 ppm potassium permanganate or 0.1 ppm malachite green.

3. *Whirling disease*

It is caused by protozoan *Mysosoma cerebralis* (Fig. 5.41) in salmon and trouts. A common symptom is rapid tail chasing (whirling) behavior when the fish are frightened or trying to feed. At advanced stages of disease, skeletal deformation, including deforming head, jaws, gill covers as well as spinal curvature can be observed. The spores of this parasite may survive for 10-15 years in contaminate mud. The prevention of disease is not to allow contact of susceptible fish with the parasite. There is no effective therapy for the disease.

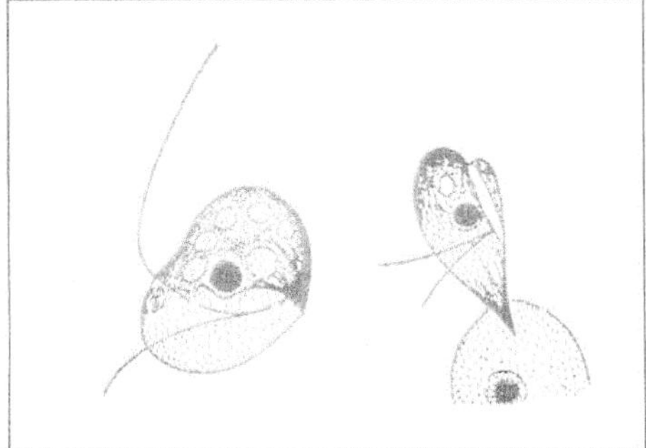

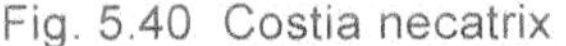

Fig. 5.40 Costia necatrix

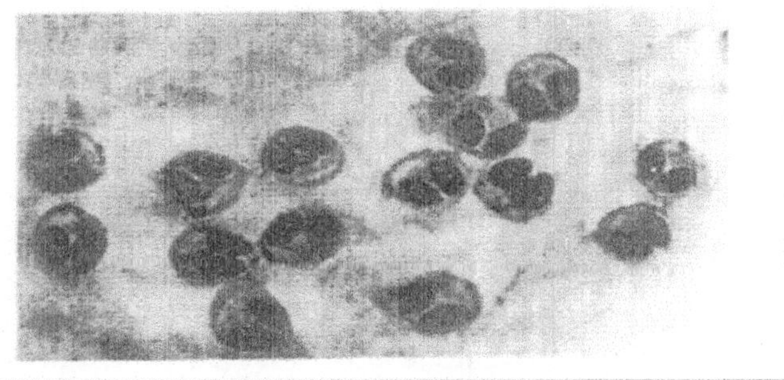

Fig. 5.41 Stained spores of Myxosoma cerebralis under microscope

4. *Coccidiasis*

Cocidia or *Eimeria carpella* and *E. epithiliasis* are cytozoic parasites in the intestine of freshwater and brackish water fishes. The white tumours are formed in fishes. The disease can be controlled by using furazolidin at 100 gm/1 kg of fish.

5. *Other protozoan diseases*

The other protozoan diseases include chilodonelliasis and trichodiniasis. *Chilodonelliasis* is caused by cilliate, Chilodonella cyprini (Fig. 5.42) affects many fish including carp, trouts etc. causing heavy mortality. The parasite feeds on epithelial cells, with its protrusible pharynx. Multiplication is by transverse binary fission. Transmission is through contaminated water. In severe infection, body is covered with bluish – grey film, fish appear restless, lose weight and become lethargic.

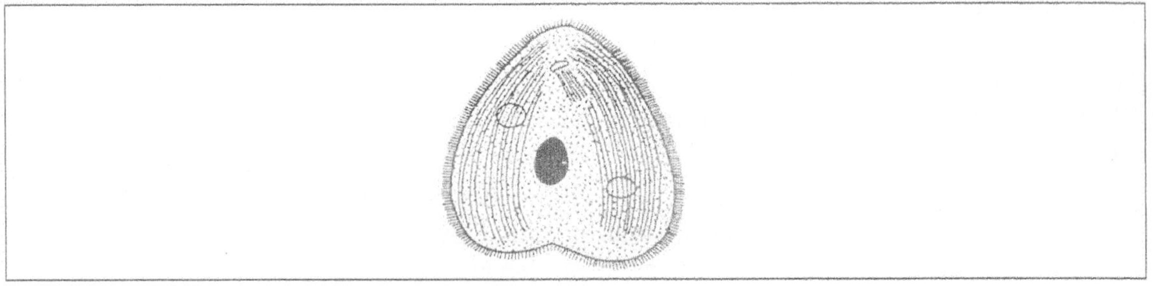

Fig. 5.42 Chilodonella cyprini, causative agent of chilodonelliasis

Trichodiniasis

It is caused by *Tricodina domerguei, T. pediculus, T. nigra, T. reticulata T. epizootica* and *T. bulbosa* and results in mortality in number of fish including carps. The disease affects fry, fingerlings and adults act as carrier of disease. The disease is transmitted with contact. The other ciliate infections observed in India are caused by *Scyphidia pyriformis* affecting rohu, catla, mrigal, reba and *Cyclochaeta* sp infecting pond fishes. The mastigophoran parasites are *Bodomonas rebae* in mrigal, rohu and catla, *Trypanosoma* clariae on *Clarias batrachus*, *Trypanosoma punctati* in *Channa punctatus*. The Myxosporidians constitute typical fish parasites known to produce cysts on different regions of body and internal tissues and organs. The common myxosporadion genera are Leptotheca, Chloromyxum, Myxobolous, Henneguya, Thelohanellus, Myxidium, Lentospera etc.

B. *Worm diseases*

Worm diseases are caused by flatworms (trematodes) , tapeworms (cestodes), roundworms (nematodes), thorny headed worms (Acanthocephala) and Hirudineans.

1. *Dactylogyrosis*

Dactylogyrosis is a common disease found in fishes particularly the carps. Dactylogyrus (Fig. 5.43) is small in size, not more than 1 mm in length and occurs on gill filament, fins and body. The most common disease causing species are *D. vastator, D. extensus, D. lamellatus, D. ctenopharyngodonis, D. hypophysalichthys* and *D. nobilis*. They are hermaphrodites and lay eggs which fall to bottom, hatch into free swimming larva and attack the gill or body surface of fish. It feeds on blood in thin

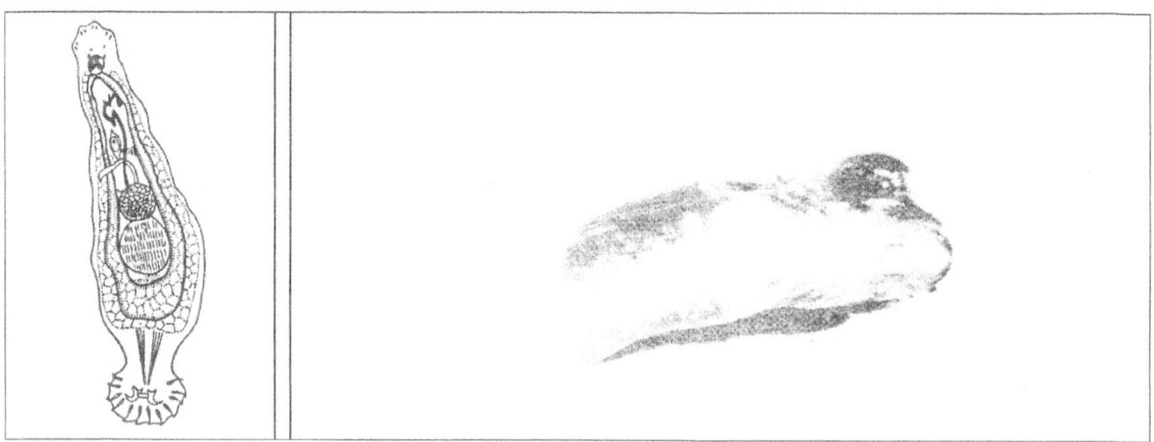

Fig. 5.43 Dactylogyrus

Fig. 5.44 Nanophyetiasis in fish. Photograph showing exophthalmia in an Atlantic salmon with infection of Nanophyetus salmincola cercariae.

capillaries. The infected fish become restless and collect near outlets. The gills are damaged and become covered with mucus inhibiting normal respiration. The usual sanitary measures, the bath of ammonia and use of Bromex-50 and Dipterex are successfully used in disease control.

2. *Gyrodactylosis*

It is caused with infection of monogenatic trematode *Gyrodactylus* in major carps and found on gills, peduncle of fins and skin. The symptoms include fading of colours, drooping of scales, occurrence of excessive mucus on gill surface and body, peeling of skin etc. Infected fish may be seen rubbing their body against hard surface in order to get rid of the parasites. Alternate day bath in 1 : 2000 acetic acid and sodium chloride solution have been found effective.

3. *Diplostomiasis*

Diplostomum spp. causes the black spot disease or Diplostomiasis. This disease is prevalent in rearing and stocking ponds. The disease include the appearence of numerous small black nodules or cysts of about 1.3 mm diameter all over the body of fish. Isolation of infected fish and their treatment with picric acid (3 ppm) for 1 hour has been found effective.

Nanophyetiasis is found in (Fig. 5.44) Atlantic salmon is caused by *Nanophyetus salmincola*, metacercaria are pathogenic. The symptoms of disease include exoopthalmia, damage to intestine, gills, eyes and almost all internal organs.

4. *Tape worm*

Tape worm infection is more serious than the flatworm infection, and relatively common in wild fish than in cultivated fish. The commonest parasite is Ligula intestinalis attaining a length of 15 to 40 cm. Infected fish shows a swelling of belly, depending on the number as well as size of parasites. Infected fish appear dull, sickly, with parts of alimentary canal swollen or completely choked by cestodes cysts. Gall bladder is also affected in some cases. The gonads are adversely affected and finally leads to death.

5. *The nematodes*

The nematodes are also found as parasites in fishes and these are Anisakis, (Fig. 5.45). *Camallanus, Neocamallanus, Para camallanus, Procamallanus, Cucullanus, Heliconema, Proleptus, Rhaldochona Philometra, Spiroptera, Agamofilaria, Eustrongylides, Zealanema* etc. Little is known about the pathological effects of nematode infections and their remedies.

6. *Acanthocephala*

Of genera *Acanthogyrus, Pallisentis, Acanthocephalus, Neoechinorhynla-chus, Rhadinorhynchus, Centrorhynchus* have been found to infect fishes.

7. *Hirudinea*

Leeches belonging to gnanthobdella and rhynchobdella infect fishes. The species of *Hemiclepsis, Myzobdella, Piscicola* hold the skin of fish and suck fish blood. After feeding, when leeches detach themselves from fish, the wound is open for secondary infection. This causes irritation, abnormal movement of fish and loss of growth of fishes. The dip treatment method in 2.5 % sodium chloride is effective. Alternatively, treatment with 1 ppm dylox for 5 days is also effective.

C. *Crustacean diseases*

Copepod parasites of the families Arguilidae, Learnaeidae and Ergasilidae of crustacea are important fish ectoparasites. These parasitic crustacea are world wide distribution. And there is no involvement of intermediate hosts in development. All other parasites multiply if the fish are overcrowded and temperature is favorable. Argulosis and Learnaeosis are best known *crustacean fish diseases*.

1. *Argulosis*

Argulosis is caused by species of Argulus (fish lice) namely *A. foliaceous, A. japonicus* and *A. giordani* and most common external infection of several species of fish *viz* trout, common carp, grass carp, and eel. Argulosis is an epizootic fish disease causing mass mortality. Argulus is large parasite with a wide oval, flattened body with organs of attachment having curved hooks at the end (Fig. 5.46). Argulus has a suctoral probocis and suckers on the ventral side and four pairs of swimming legs. They lay egg in batches of 250-300, hatch into swimming larva and within 2-3h attach to the host, otherwise it die. Argulus attaches itself to the fish, pierces the skin with its probocis, injects cytotoxic secretion and sucks blood from the host. The feeding site becomes wound, haemorrhagic, inflammatory with profuse secretion of mucus and oedema and become secondarily infected with

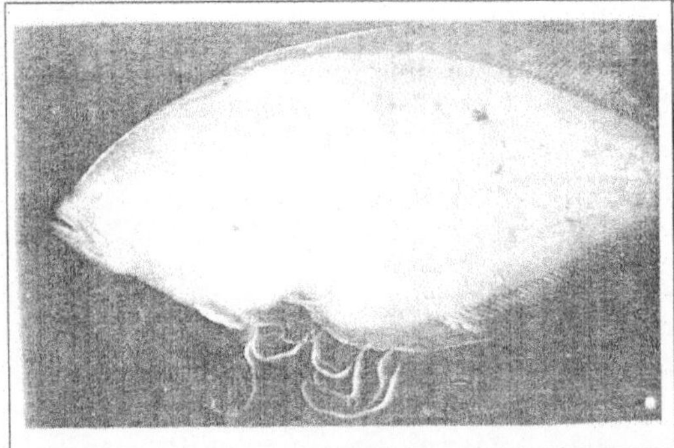

Fig. 5.45 Anisakid nematodes in fish. Specimens, of Porrocaecum sp. migrating from the digestive tract of a summer flounder, Paralichthys dentatus.

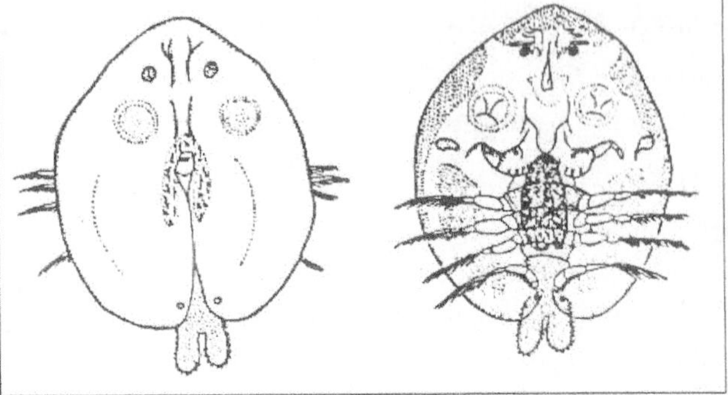

Fig. 5.46 Argulus foliaceous

other pathogens. Younger fish appear more susceptible to infection and because lack of host specificity Argulosis is generally considered a greater risk for fish health. Prevention of argulosis consists of isolation of infected fish, removal of physical base used by as artificial substratum to deposit eggs. For curative treatment several chemicals have been suggested including malathion and dipterix at 0.25 ppm or bromex at 0.12 ppm, 0.25 ppm of dylox or 2000 ppm of lysol etc.

2. *Lernaeosis*

Lernaeosis or anchor worm disease is caused by *Lernaea cyprinacea* (Fig. 5.47) and *L. ctenopharyngodonis* in a number of fresh water fishes. The female parasite has long unsegmented body, the head has branched processes which help parasite to penetrate the body of host. There are no intermediate host and free swimming larvae parasitize the skin or gills of fish. These are temporary parasite on fish feeding on mucus and blood into the muscle and causes deep ulcers, abscesses or fistulas at the point of attachment leading to formation of parasitic fibrous nodules. The parasite some time causes focal traumatic hepatitis. Prevention is better than cure. The chemical treatment includes potassium permanganate at 2 ppm, or Boromex at 0.12-0.15 ppm with treatments repeated three times at weekly intervals.

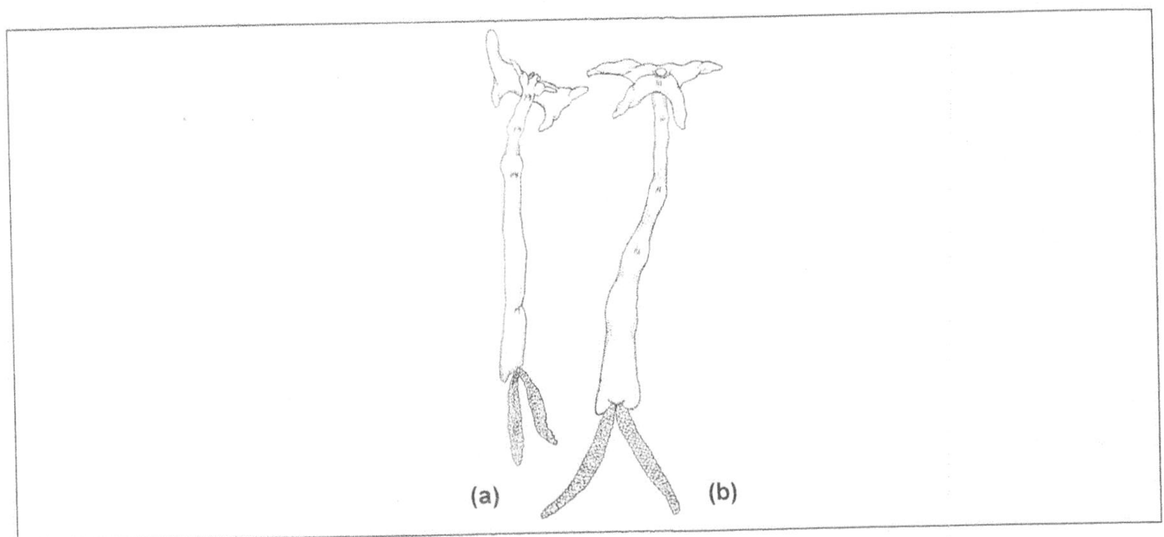

Fig. 5.47 (a) Lernaea cyprinacea from common carp, (b) Lernaea ctenopharyngodonis from grass carp

3. Other crustaceans diseases of fishes are caused by *Ergasillus, Salmincola* and *Actheres* which are attached to gills and feed on blood and epithelium or on sometimes on fins and body also. The infection by these parasites results in impaired respiration, anaemia, epithelial hypertrophy, retarded growth. This infection facilitates the fishes for secondary infection with other pathogens. These diseases can be treated with combination of 0.5 ppm copper sulphate and 0.2 ppm ferric sulphate.

D. *Fish Diseases caused by other Pathogens*

In addition to protozoan, helminth and crustaceans causing fish diseases, the fish are affected with diseases caused by bacteria, virus and fungus. The viral diseases include Lymphocytosis, viral hemorrhagic septicemia, infectious pancreatic necrosis, infectious haemopoetic necrosis, chinook disease, spring viremia of carps, carp pox, channel cat fish viral disease etc. The bacterial diseases in fishes include Furuculosis, colummar disease, vibriosis, Dropsy, Gill or Fin or tail rot, cotton mouth disease, Fish tuberculosis, Bacterial gill diseases, Enteric red mouth disease, Edwardsiellosis, etc. The main fungal diseases found in fishes are Branchiomychosis, Icthyophonsis, saproligniasis etc.

E. *Parasitic Disease on Crustaceans (Prawn / shrimp, crabs)*

The protozoan parasites have ability to grow on body including on gills, appendages, eye etc. These protozoan attach on host or some penetrate deep into the body proper. The resultant injuries facilitate further infection with bacteria, virus, fungi etc. Protozoans also cause stress, irritation, sometimes block respiration in prawn. Some of the examples of protozoan parasites on prawn causing diseases are *Acinata, Epistylis, Norima, Loginofress, Pleisthosphora, Zoothamnium, Thelohania Sp.* The prolific and colonial growth of these protozoans on prawn body gives an appearence of cotton wool, hence it is called as "*Cotton shrimp disease*". The protozoon infection changes the colour of body, formation, feeding and growth, moulting of prawn etc. There are no proven methods to cure for this disease.

The ciliate gill diseases are caused by protozoans of genera *Zoothaminium, Epistylis* and *verticella.* When the surface of gills are covered by these organisms, hypoxia and death occurs. The disease can be controlled with formalin treatment. The '*Gray crab disease*' in crabs result in mortalities is caused by *Paramoebas periniciosa.* The affected crab portray a greyish discolouration of the ventral body surface (Fig. 5.48).

The other diseases of prawn/shrimps include bacterial diseases like baculovirus diseases, shell disease, *Leucothrix* disease, viral diseases like hypodermal and hematopoietic necrosis, hepatopancreas parvo-like virus diseases, the fungal disease of non–inflammatory mycosis, *Fusarium* fungal diseases and other non-parasitic environmental diseases like tail rot, Gas bubble disease, soft shell syndrome. *Paragonimus westermani*, a trematode, tend to encyst in intermediate host, crab and form metacercarial cysts in gills and muscles of body and legs and sometimes in hepatopancreas of crab (Fig. 5.49).

F. *Parasitic Disease of Moluscans – Oysters and Mussles*

The epizootics are believed to have been caused by three or four protistan parasites, starting with *Marteilia refringens* and then *Bonomia ostreae.* Shell diseases caused by the fungus *Ostracoblabe implexa* has caused serious loses of young oysters. Another reported disease is the *Pit* disease which is described as a congestion caused by the rapid multiplication of the flagellate *Hexamita.* It generally occurs when oysters are kept too long in a storage basin at low temperature. In oysters, the serious diseases are caused by protozoans, for example *Minchinia nelsonii* (Haplosporadian) causes the salinity dependent *Delaware Bay* disease and M. Costalis causes the '*Seaside disease or winter mortality*'. Viral infections and mycosis of larvae have been reported to cause major losses in hatcheries. The other parasites which cause damage to oysters are *Polydra ciliata, Ancistrocama, Trichodina* and copepod *Mytilicola.*

The most disastrous losses of mussel have been caused by the intestinal copepod, *Mytilicola intestinalis* and the infection is more pronounced in areas with dense populations. Dinoflagellates of genus *Boniaulax* are more dangerous as they cause paralytic shell fish poisoning. The trematode parasite *Gymnophallus* infects the mussel. The metacestode larva of *Tetragonocephalum* in oyster, Crassostera virginica, develops a capsule around it comprised of fibroblasts, leucocytes and connective tissue fibres (Fig. 5.50).

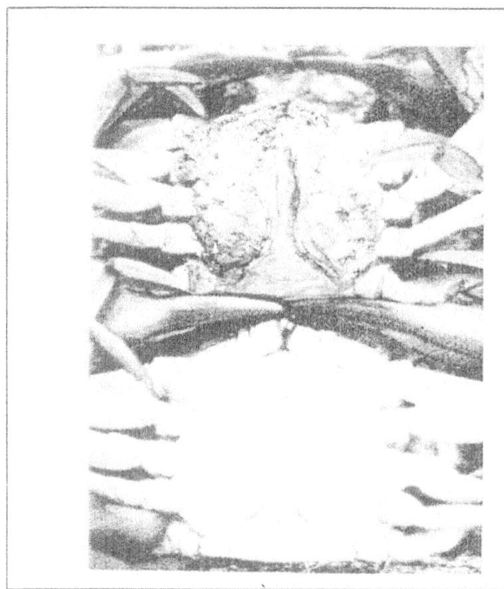

Fig. 5.48 Gray crab disease, Infected crab showing gray ventral body surface (above) and uninfected crab (below).

Fig. 5.49 Several metacercariae of Paragonimus westermani encysted in gill filament of crab host.

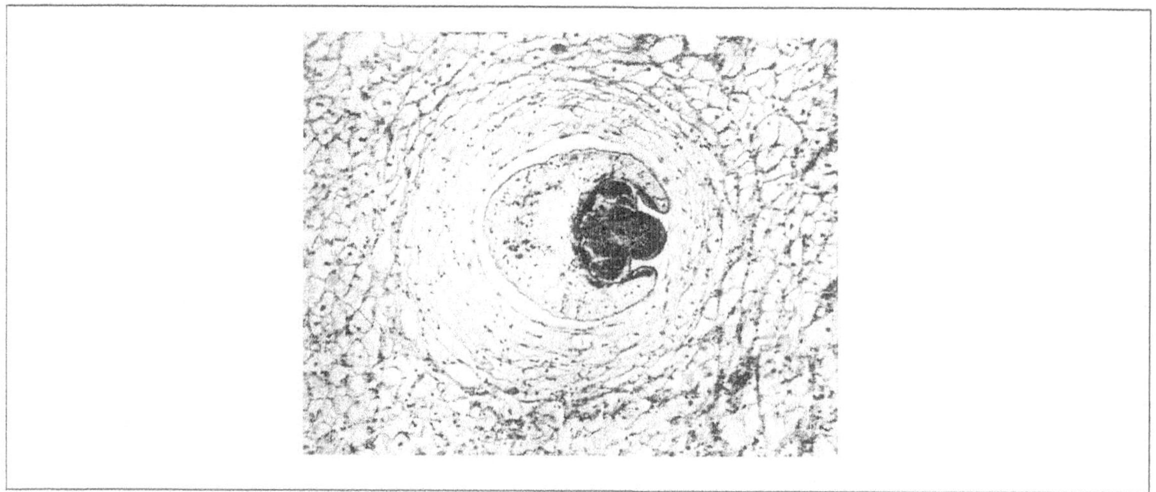

Fig. 5.50 Histologic section showing metacestode (larva) of Tetragonocephalum in connective tissue of the American oyster, Crassostrea virginica.

Relatively fewer clam or cockle diseases have been reported. Many of the known diseases of Juveniles and adult clams are caused by the haplosporadian, *Perkinsus marinus*, the coccidia *Hyaloklossia* and *Pseudoklossia*, the gregarine Nematopsis and ciliate like Tricotina and Ancistrocoma.

5.6 Parasites of Plants

As mentioned earlier, an organism that lives on or in some other organism and obtain food from the latter is called a parasite. A plant parasite is an organism that becomes intimately associated with a plant and multiplies or grows at the expense of plant. The removal by the parasite of nutrients and water from the host plant usually reduces efficiency in the normal growth of the plant and becomes detrimental to the further development and reproduction of the plant. In many cases, parasitism is intimately associated with pathogenicity, since the ability of parasite to invade and become established in the host generally results in the development of disease condition in the host.

In most plant diseases, the substances secreted by parasite or produced by host in response to stimuli originating in parasite produces changes in tissues and metabolism. Then, this ability of the parasite to interfere with one or more of the essential functions of the plant and thereby cause diseases, has role in pathogenicity. Of the large numbers of groups of organisms, only a few members of a few groups can parasitize plants and act as pathogens which includes bacteria, mollicutes, viruses and viroids, parasitic higher plants, nematodes and protozoans. Depending on the parasitic relationship, they may be obligatory, and facultative parasites. Parasitism in cultivated crops is a common phenomenon. Among animalia, the most common parasites of crops belong to *nematodes* and few to *protozoa*.

A. Nematode Parasites of Crops

The nematodes with worm like appearance are responsible for a number of plant diseases. There are about 197 genera, and 4300 species of phytoparasites. Nematodes are elongated worms generally about 0.3 mm to 4.0 mm in length by 15 - 35 μm wide. The anterior end is hounded by lips and posteriorly it tapers. In the male anus serves as reproductive aperture. The female reproductive aperture is distinct and ventral in position. The development of these parasites is very simple type. The eggs are oval in shape. They undergo a series of divisions, from first stage larvae, moult within eggs, second stage larvae emerge out from the egg, and which resemble the adults except for their small body size. The second stage larvae starts feeding in living plants and undergo 3 moults and attains mature adult stage.

Phytoparasitic nematodes are obligate parasites and feed only on living plants. The nematodes which move through tissues are called as *migratory endoparasites*. The females which attach to the root permanently are sedentary endoparasites which include root knot and cyst nematodes. Migratory endoparasites are lesion, stem, bulb, lance and spiral nematodes. The nematodes feeding on plants from outside are ectoparasites which include the ring nematode (sedentary), the dagger, stubby root and sting nematodes (migratory).

Nematode bears a protrusable hollow stylet which helps it to puncture protective wall of host plant. These nematodes release secretions containing enzymes which dissolve the cell wall, digest the plant material and prepare cell contents for ingestion. The nature of damage varies from species to species. Some of nematodes like *Tylenchus, Pratylenchus* feed on host plant causing necrosis at site of formation, *Augunia* causes gall formation, *Trichodorus* produces stubby roots, thus, affecting growth, structure and yield of crops. The most damaging plant parasitic Nematodes are root knot nematodes (*Meloidogyne*), lesion nematode (*Pratylenchus*), cyst nematode (*Heterodera, Blobodera*), stem and bulb nematode (*Ditylenchus*), citrus nematode (*Tylenchulus*), dagger nematode (*Xiphinema*), burrowing nematode (*Radopholus*), reniform nematode (*Rotylencholus*), spiral nematode (*Helicotylenchus*) and sting nematode (*Belonolaimus*).

All plant parasitic nematodes belong to the phylum nematoda, Most of the important parasitic genera belong to the order Tylenchida but a few belong to the order Dorylaimida. The following list provides all phytoparasitic nematodes.

PHYLUM : NEMATODA

Order : *Tylenchida*

Family : Anguinidae

Genus :
1. ***Anguina***, wheat or seed–gall nematode
2. ***Ditylenchus***, stem or bulb nematode of alfalfa , onion, narcissus, etc

Family : Belonolaimidae

Genus :
3. ***Belonolaimus***, sting nematode of cereals , legumes, cucurbits, etc.
4. ***Tylenchorlaynchus***, stunt nematode of tobacco , corn , cotton , etc.

Family : Pratylenchidae

Genus :
5. ***Pratylenchus***, lesion nematode of almost all crop plants and trees
6. ***Radopholus***, burrowing nematode of banana, citrus, coffee, sugarcane, etc.
7. ***Nacobbus***, false root–knot nematode

Family : Hoplolaimidae

Genus :
8. ***Hoplolaimus***, lance nematode of corn, sugarcane, cotton, alfalfa, etc.
9. ***Rotylenchus***, spiral nematode of various plants
10. ***Heliocotylenchus***, spiral nematode of various plants
11. ***Rotylenchulus***, reniform nematode of cotton, papaya, tea, tomato , etc .

Family : Heteroderidae

Genus :
12. ***Globodera***, round cyst nematode of potato
13. ***Heterodera***, cyst nematode of tobacco, soybean , sugar beets, cereals
14. ***Meloidogyne***, root – knot nematode of almost all crop plants

Family : Criconematoidea

Genus :
15. ***Criconemella***, formerly Criconema and Criconemoides, ring nematode of woody plants, cause of peach tree short life
16. ***Hemicycliophora***, sheath nematode of various plants

Family : Paratylenchidae

Genus : **17. *Paratylenchus***, pin nematode of various plants

Family : Tylenchulidae

Genus : **18. *Tylenchulus***, citrus nematode of citrus, grapes , lilac , etc.

Suborder : Aphelenchina

Family : Aphelenchoidadae

Genus : **19. *Aphelenchoides***, foliar nematode of chrysanthemum , strawberry, begonia,rice, coconut,etc.

20. *Bursaphelenchus*, the pine wilt and the coconut palm or red ring nematodes Oder : Dorylamida

Family : Longidoridate

Genus : **21. *Longidorus***, needle nematode of some plants

22. *Xiphinema*, dagger nematode of trees, woody vines, and many annuals

Family : Trichodoridae

Genus : **23. *Paratrichodorus***, stubby – roor nematode of cereals , vegetables , cranberry, and apple

24. *Trichodorus*, stubby – root nematode of sugar beet, potato, cereals, and apple.

Some Phytoparasitic Nematodes are briefly described below

1. *Meloidogyne spp. (Root-Knot Nematodes)*

These are commonly known as root – knot nematodes. They are found in light sandy soils in warmer regions. They cause severe damage to potato, papaya, brinjal, tomato, jute, groundnut etc. The symptoms of infested plants include yellowing of leaves and stunning of shoots. The most distinguishing symptom is the formation of knots in the roots. The shape and size of root knots vary according to host plant and environmental factor. Some times infested host plants even die in severe infection. The female lay about 500 – 1000 eggs. The eggs hatch into larvae which are found free living in soil. The second stage larva attack the new host (Fig. 5.51). The larvae which develop into females establish feeding zone in pericycle and moult three times and develop into female. Gall formation is visible sign of root knot nematode infection resulting from the hypertrophy and hyperplasia within the roots. For copulation male leaves the host and enters the sac like matrices of the female.

2. *Radopholus spp. (Burrowing Nematodes)*

These are distributed in tropical and subtropical regions. It causes severe damage to a number of economically important plants like banana, sugarcane, citrus, tea, coffee etc. It damages the cortex of roots of host plants. All stages of nematode are found inside the roots and also facilitates the penetration of pathogenic fungus leading to severe damage to host plant. The affected plants have little growth, foliage loss and bear smaller fruits. The life cycle is of duration of 2-3 weeks (Fig. 5.52).

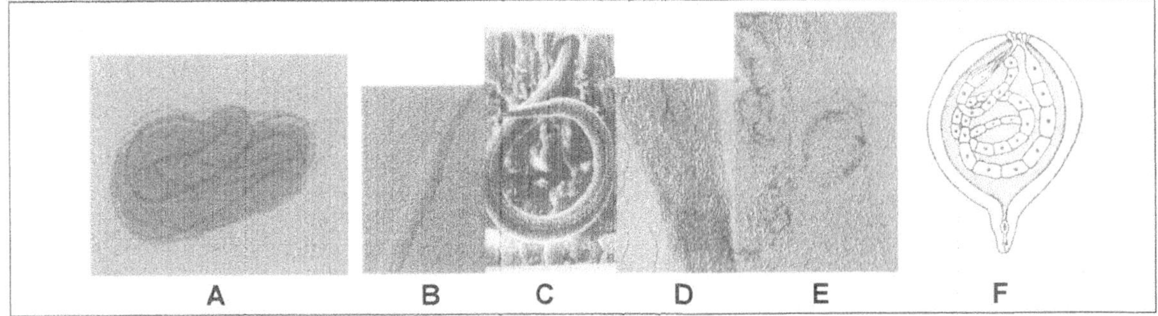

Fig. 5.51 Infectious cycle of root knot nematode, Meloidogyne – A. first juvenile moult inside egg, B. Juvenile 2 in free soil, C. J-2 penatrating root, D. J-2 migrating to vascular cylinder, E. adult and F. sketch of adult.

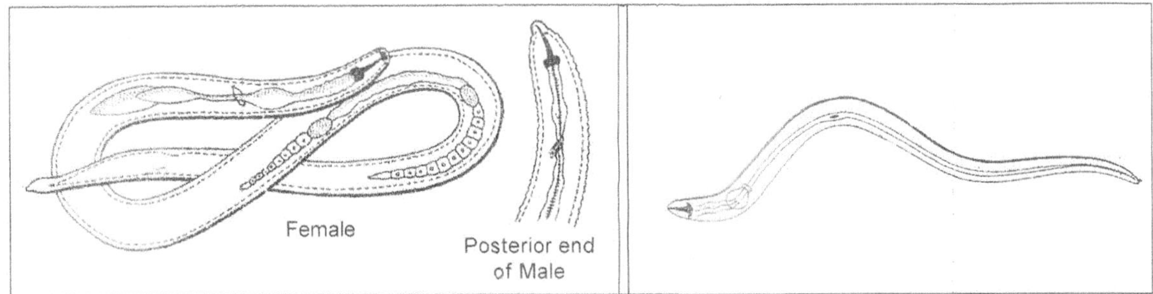

Fig. 5.52 Radopholus

Fig. 5.53 Heterodera

3. *Heterodera spp. (Cyst nematodes)*

The formation of cysts containing eggs is the characteristic feature of this nematode. They infect soyabean, cereals, cloves, sugar beet, potato, oats, woody perineal plants. In certain cases of severe infestation, almost total loss of crop is reported. The yellowing of foliage and stunting of shoots are major symptoms. In severe infestation the roots of host plants become hairy which causes complete failure of the crop. The fertilized female lays eggs inside the cyst. The hatching of first instar larva takes place within the eggs. The second stage larva emerge from the eggs, move freely in soil and attach with host roots. The larvae penetrate at root tip. The second stage larva undergoes second, third and fourth moults and leave the host in the form of adult male (Fig. 5.53). In case of development of female the second stage larva remain sedentary and attains adult stage after its fourth moult. The complete life cycle takes about one to two months.

4. *Ditylenchus*

Ditylenchus (the stem and bulb nematode) is migratory endoparasite mainly in shoots with diverse feeding habits.

Important species are :

1. *D. dipsaci* : Can infect over 500 plant species
2. *D. destructor* : Potato dry rot (endoparasite of underground plant parts)
3. *D. myceliophagus* : Fungivore, pest for mushroom industry

Life Cycle and Host – Parasite Interactions

- reproduction by amphimixis (needs males)
- females lay 200 - 500 eggs which usually hatch in the plant tissue
- all stages are infective, penetration at shoot apical meristem
- migratory feeding leads to cavities in the parenchymous tissue
- pectinase production is speculated in degradation of the middle lamella that results in cell separation and cavity formation during infection
- swelling, galling, stunting, distortion, and necrosis result from serious infection
- 4th stage juvenile survival up to 23 years.
- cause galling and distortion to leaves, brown rings in bulbs.
- life cycle is 19 - 23 days.

5. *Xiphinema spp.* (Dagger nematode)

- Major hosts : woody plants, tree fruits, soyabean, corn. sugar beet , turnip, spinach, cucumber, potato and some cereals
- sedentary ectoparasite feeds at or near root tips, may cause slight galling and necrosis
- X index is a problem on grapes and also vectors grapevine fanleaf (GVFL) virus
- large nematode, long stylet, feeds deeply (Fig. 5.54).
- very long life cycle: several months to 2 years
- causes "giant cells and galls" like root knot nematode
- giant cells caused by karyokinesis without cytokinesis yielding multinucleate cells
- cessation of root growth caused by less than 12 h of feeding
- root swelling may continue after feeding has stopped

6. *Trichodorus spp.* (Stubby root nematode)

This is ectoparasite on vegetables, grains, fruit crops and fodders. Due to the attack of the nematode, the growth of root tip is checked resulting in formation of stubby root. Thus, the attached plants have very poor root system with short root lets . It complete its life cycle in about 16 days (Fig. 5.55).

7. *Rhadinaphelenchus spp.*

This causes 'red ring disease' on coconut and oil palm. In primary stages of infection is yellowing and drooping of lower leaves followed by the formation of brown or red discolouration of stem tissue in ring form. The life cycle is completed in about 7 days (Fig. 5.56).

8. *Aphelenchoides spp.*

These are called as 'leaf and bud nematodes' because they damage leaves and buds of plants. They are both ecto and endoparasites. They infest rice, straw berries and vada orchids. They start attacking from the base of plants and reach upwards towards flowers. In severe infestation the plants may become completely defoliated before flowering (Fig. 5.57).

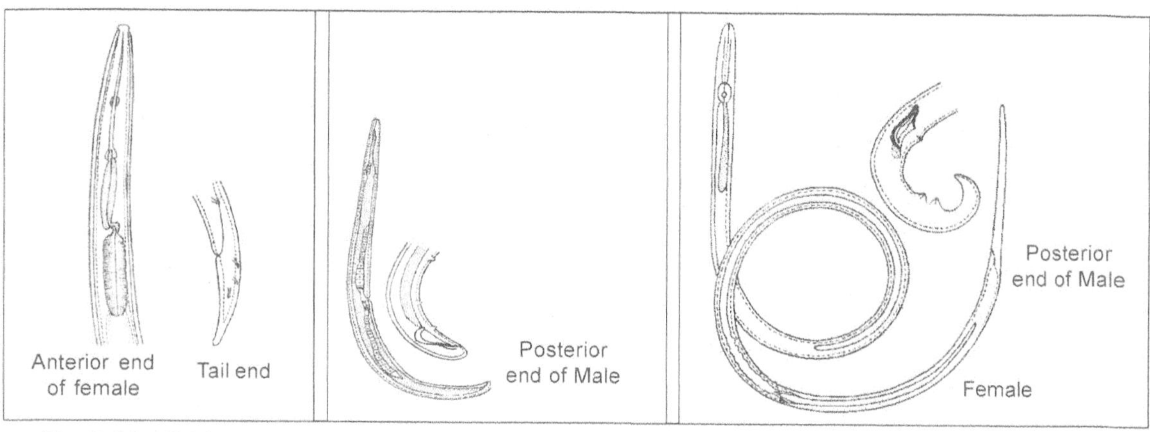

Fig. 5.54 Xiphinema Fig. 5.55 Trichodorus Fig. 5.56 Rhadinaphelenchus

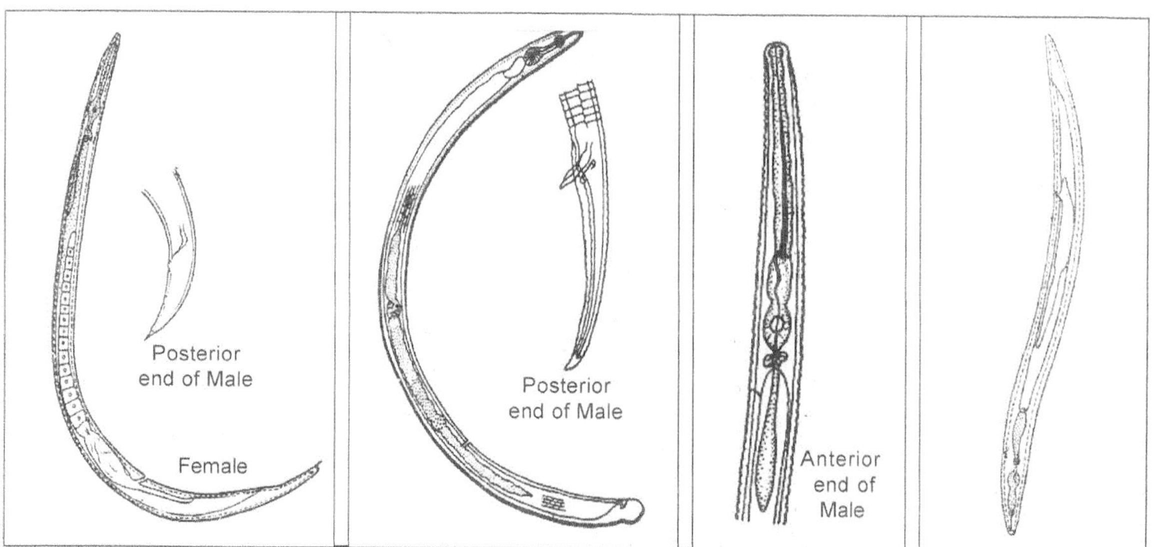

Fig. 5.57 Aphelenchoides Fig. 5.58 Hoplolaimus Fig. 5.59 Belonolaimus Fig. 5.60 Pratylenchus

9. *Hoplolaimus spp.*

These are called as '*lance nematodes*' and are world wide distribution. They are ectoparasite and migratory endoparasite. The major host plants are cotton, millets, maize, sugarcane, onions, roses, apples. They cause stunting and slow development of root system (Fig. 5.58)

10. *Belonolaimus spp.*

These are called as '*sting nematodes*' and are migratory parasites in nature. The plants attacked are maize, cotton, groundnut, soyabean, strawberries and cowpea. The sunken dark lesions appear on the attached root and also on root tips are swollen (Fig. 5.59).

11. *Pratylenchus spp.*

These are called as 'root lesion nematodes'. They are endoparasites of maize, cotton, wheat, sugar-cane, cowpea, tomato, banana, pine apple, apple, grape, strawberry etc. They are cosmopolitian in distribution. The stunting and malnutrition is main symptom of affected plants. The root system of plants is poorly developed. Brown to dark lesions in the cortex have also been reported. The life cycle is completed within 30 to 70 days depending on host plant and environmental conditions (Fig. 5.60).

The phytoparasitic nematodes are not as host specific as zoo–parasites. They have a wide host range and each parasitic nematode may infest almost all host species. The pattern of life history is less variable in comparison of zoo–parasites. The infective stages are found living freely in soil. Therefore, the control measures of phytoparasitic nematodes is to stop life cycle at a possible stage. The most commonly used control measures are chemotherapy using of nematicides like parathion, methylbromide, Nemagon etc.; thermotherapy–steam treatment of plants to control nematodes, crop rotation and soil treatment by fumigants along with physical means like heat, radiation are some of measures of disease control.

B. Protozoan parasites of crops

Certain Trypanosomatid flagellates, belonging to (order–kinotoplastida) protozoa have been known to parasitize plants. The flagellate protozoa parasitize the latex bearing cells of the laticiferous plant *Euphorbia* by *Phytomonas davidi. Phytomonas* has been reported from plants belonging to families of Asclepialdacae (*P. elmaissiani*), Moraceae (*P. bancrotti*), Rubiaceae (*P. leptovasorum*), Euphorbiacae (*P. francai*). The genus *Phytomonas* completes its life cycle in two hosts, a plant and an insect. *Phytomonas* species have been isolated from fruits of tomato and in phloem of coconut, oil palms and coffee. These parasites produce enzyme degrading pectin and cellulose causing damage. These are transmitted by root grafts, and by insect genera *Lincus* and *Ochlerus.* These parasites cause *heartrot* disease of coconut palms, *sudden wilt* of oil palm, *empty root disease* of cassava are some of examples.

5.7 Pathological Effects of Parasites

This chapter describes pathology with reference to parasites. The first part describes the various pathological mechanisms with examples, while the second part gives an account on pathological effects of parasites on tissue system and cellular level of hosts.

A. Pathological Mechanisms

Parasitic organisms can produce pathology in their hosts by a number of different mechanisms. These include mechanical insult, biochemical intervention and immunological factors among others . Some of the general observations made on pathological mechanism in parasitic diseases are shown in Table 5.6. These may vary from direct penetration of intestinal wall by adult *Ascaris* which produces mechanical damage and potential to secondary bacterial invasion to immune suppression caused by African trypanosome that also lead to secondary infections.

Pathology may result because of *Mechanical Insult.* For example, the growth of larval stages of Taenia and cysticercosis and migration of larval helminths (eg. *Fasciola* larvae) through host tissue produce mechanical tissue damage. The pathology may also occur by *consumption* of host tissue. Hook worms, for example, feed on host blood and this can lead to severe anemia. The pathology may also result indirectly through *Biochemical Interventions* of various kinds. For instance, Entamoeba histolytica releases hydrolytic enzymes that are capable of destroying host tissue. These enzymes split macromolecules of host and use them for parasitic nutritional needs, in this process thereby causing utmost damage to host. Many parasites compete with host for *consumptions of nutrients* and severely deplete the host essential nutrients required for its survival. Some parasites produce *toxic molecules* which alter host physiological functions, or host's immunological responses or both. It is examplified in trypanosome, plasmodium and schistosome infection. Parasites can also alter the host's *hormonal* activity pattern by which host's physiological, immunological and behavior responses to environmental stimuli get effected. The parasite may also cause pathology due to direct *host parasitic interaction* but mediated by host's immune responsiveness resulting in inflammatory reactions. The host's inflammatory reaction to foreign antigens, parasites, may be due to both non-specific and specific defense mechanisms including of host immunopathology in chronic schistosomiasis, in which host responds immunologically to the eggs that are trapped in liver. This include the B–cell and T–cell responses, migration of macrophages and eosinophils. The continuous influx of these, lead to formation of a ring of host cells surrounding the egg, called as *granuloma*. Eventually, granulomas with scarring and calcified area's join together, resulting in severe liver damage. Similar tissue destruction may occur in any host in which larval stages or eggs persist.

Table 5.6 *Some examples of Pathological mechanisms in parasitic diseases.*

Organism	Observed pathology
	Direct mechanical factors
Plasmodium	Lysed RBCs, clog CNS microvascular bed
Hookworm	Feed on blood from intestinal wall
Ascaris	Penetrate intestinal wall
Taenia solium larvae	Cysticercosis (larval growth in confined space)
	Biochemical factors
	1. *Nutritional depletion*
African trypanosomes	Hypoglycemia
Plasmodium	Anaemia, Fe^{2+} deficiency
Hook worm	Anaemia, Fe^{2+} deficiency
Diphyllobothrium latum	Vitamin B 12 deficiency
	2. *Toxic parasite products*
	(a) *Small molecular weight*
African trypanosomes	Aromatic amino acid catabolites
Fasciola hepatica	Proline excess, bile duct hyperplasia
	(b) *Large molecular weight*
Entamoeba histolytica	Secreted proteases, pore-forming peptides, lectin-mediated adherence
African trypanosomes	Hydrolytic enzymes (proteases), B – cell mitogen, T – lymphocyte triggering factor, endotoxin
	Immunological factors
	1. *Immediate hypersensitivity*
African trypanosomes	Glomerular nephritis

Contd.. Table 5.6

Organism	Observed pathology
Plasmodium	Glomerular nephritis
Schistosomes	Granuloma formation
Taenia solium larvae	Cysticercosis, allergic reaction
	2. *Delayed hypersensitivity*
Leishmaniasis	Mucocutaneous, cutaneous pathology
Schistosomiasis	Egg granulomas, pipe – stem fibrosis
Taxoplasmosis	Brain pathology
	3. *Immunosuppression*
Plasmodium, African trypanosome, schistosomes, Possibly most parasitic infections	Decreased ability to respond immunologically; increased secondary infections
	4. *Autoimmunity*
Plasmodium	Anti – RBC , anemia
Trypanosoma cruzi	Anti – nerve and / or heart tissue; nerve and heart damage

In some instances, the pathology may result due to not only with immediate hypersensitivity but also with release of *endotoxin* like substances that can activate fever response and induce other physiological and mechanical changes (Table 5.7). For example, this type of pathological mechanism is evident in African trypanosomiasis. In addition, animals chronically affected with trypanosomes become immunosuppressed and secondary infections are common cause of morbidity and mortality. Similarly, in chronic schistosomiasis, although the delayed type of hypersensitivity is the major cause of pathology, the other causes include parasite toxins, mechanical destruction, activation of *complement, clotting and kinin cascades*. For example, the inflammatory reaction initiates the raised red, hard painful pustule following a splinter penetrating the skin. First there will be vascularization followed by activation of immune system. If the microorganism on splinter is eliminated, the reaction quickly subsides, otherwise, if they persist, the host continues to respond immunologically, and the small pustule may enlarge into a boil, tissue damage will follow and scarring may result.

Table 5.7 *Combination of Pathological mechanisms suggested in African Trypanosomiasis and Schistosomiasis*

Mechanism	African trypanosomes	Schistosomes
Toxin (s)	+	+
Immunosuppression	+	+
Immediate hypersensitivity	++	±
Delayed hypersensitivity	±	++
Mechanical damage	–	+ (pipe – stem fibrosis)
Metabolic rate and catabolite production	+ (hypoglycemia)	+ (malignancy)

A much more *coupled pathological syndrome* is also seen in parasitism. In African trypanosomiasis, following entry of parasite in host, parasite number increase stimulating both on acute phase response, fever, malaise, cachexia followed by a humoral B cell response. The antigen – antibody complex through active complement attracts phagocytes to site of deposition. This accumulation leads to a decrease in p^H and oxygen tension at the site and release a variety of hydrolytic enzymes that cause host tissue destruction. The antigen-antibody complex also activate kinin cascade producing changes in vascular bed and resulting in edema and immediate hypersensitivity. Another example due to fibrosis and parasite induced immunosuppression. The schistosome catabolites – carcinogens can induce bladder cancer. A similar mechanism of pathology may also exist for liver fluke *Opisthorchis viverrini*.

The pathological changes induced by parasite appear to evolved to *assist parasite in completion of its life cycle*. For example, the granuloma formation in schistosome infection around eggs allow safe passage across tissue. The parasite induced host immunosuppression also permit the parasite to survive within host for an extended period of time increasing the opportunity for transmission to occur. In contrast, some mechanisms that are *involved in the defense of host* are also involved in causing host pathology. For instance, both humoral and cellular reactions of the defense system can lead to severe disease if the parasite is able to persist within the host. The host remains healthy as long as its immune system is intact and able to rapidly eliminate the parasite. However, if the parasite can persist even at low levels the immune mechanism, which actually maintain low parasitemia, continues to respond and there by lead to both local tissue damage, like schistosome – egg granuloma or to a systemic inflammatory response, like African trypanosome-glomerular nephritis.

B. Pathological Effects of Parasites on Host

The pathological effects of parasite on host are specifically described under (1) destruction of hosts tissue (2) depletion of hosts nutrients (3) utilization of host's non-nutritional material (4) mechanical interference (5) effects of toxins and secretions and (6) other effects.

1. Destruction of Host's Tissue

The degree of destruction of host's tissue varies greatly. Some parasites injure the host's tissue during the process of entry, other inflict tissue damage after they have entered, still others induce histopathological changes by eliciting cellular immunological response to their presence or a combination of these three types of injury may also occur. The hook worms, *Ancylostoma duodenale* and *Necator americanus*, infective larvae inflict extensive damage to cells and underlying connective tissue during penetrations of host's skin . The various helminths, acanthocephalans, flukes and tape worms, armed with attachment hooks and spines, irritate the cells lining the lumen of their host's intestine while they are holding on. The repeated irritations over long periods can result in appreciable damage and also becomes sites for secondary infection by microorganisms. The amoebic dysentery caused by protozoan *Entamoeba histolytica*, actively lyses the epithelial cells lining of host's large intestine causing ulcerations, that are not only damaging themselves but are also sites for secondary infection (Fig. 5.61(a)). The ulcerations result from a combination of mechanical destruction and action of secreted enzymes. The amoeba cause large abscesses in host's liver (Fig. 5.61(b)). The larvae of large nematode, *Ascaris lumbricoides*, during migration within human host cause physical damage to lung tissue and induce a profound immunological reaction, the ascaris pneumonia (Fig. 5.61(c)). The above example of histopathological alterations reflect the involvement of one or a combination of all categories of damage.

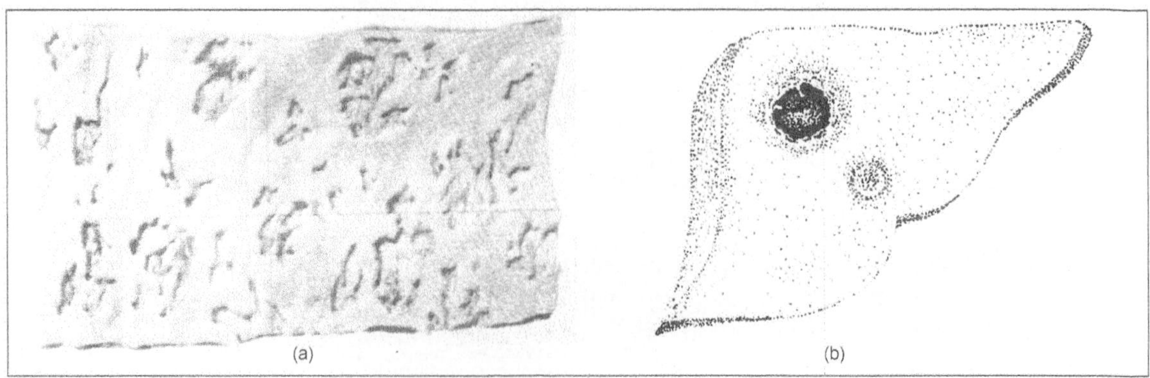

Fig. 5.61 (a) Amoebiasis – Large ulcerated areas on mucosal surface of human colon resulting from confluence of smaller ulcers caused by Entamoeba histolytica, (b) Amoebiasis – Large lesion in human liver due to invasion by Entamoeba histolytica

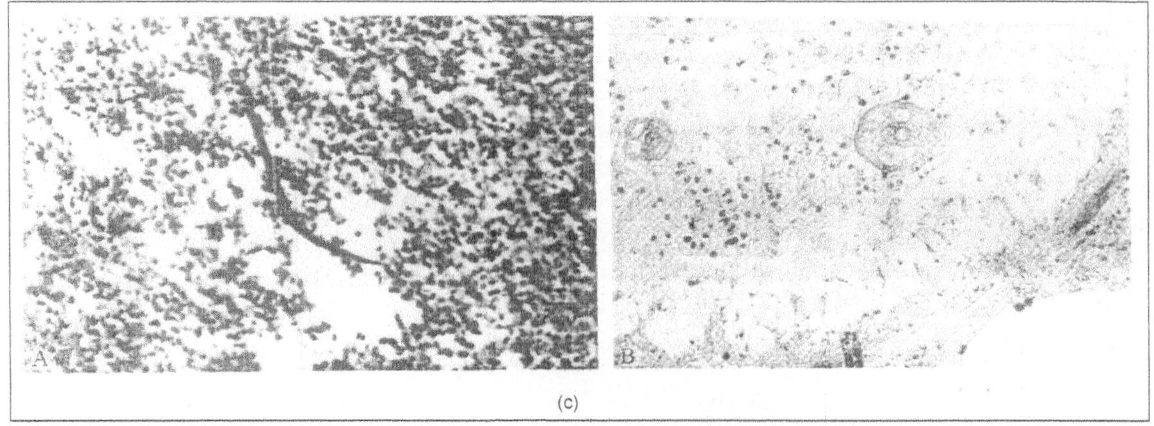

Fig. 5.61 (c) Larva migrans – A. Larva of Ascaris lumbricoides during migration through lung of human host. Notice displacement of host cells,
Section of human lung showing bronchiole filled with mucopurulent material containing a larva of A. lumbriocoides cut transversely through the esophagus and midintestine.

Histopathological studies of parasite damaged tissue show that cell damage other than removal by ingestion or mechanical disruption is of three categories. (i) *albuminous* or *parenchymatus degeneration*, in which cells become swollen and packed with albuminous or fatty granule, the nuclei become indistinct and cytoplasm looks pale. This is found in cells of liver, kidney and cardiac muscle cells. (ii) *Fatty degeneration* refers to the process in which cells become filled with an abnormal amount of fat deposits giving them a yellow appearance. Example is liver cell in parasitic infection. (iii) *Necrosis* refers to the persistence of cell degenerations in any type of cell. The cells finally die and tissue becoming opaque in appearances. The best example is necrosis of tissues followed by calcification as a result of encystment of Trichinella spiralis in mammalian skeletal muscle cells.

The possible consequences of parasitism associated with cell and tissue parasites is a change in growth pattern of affected tissue and these *tissue changes* can be divided into four major types. (i) Hyperplasia (ii) Hypertrophy (iii) Metaplasia and (iv) Neoplasia.

Hyperplasia : Hyperplasia refers to accelerated rate of cell division associated with increased level of cell metabolism. It commonly follows inflammation and is consequence of an excessive level of tissue repair. For example, the Fasciola, liver flukes, infection resulting in thickening of bile duct of their host. This shows the hyperplastic condition resulting from excessive division of the epithelial lining of duct stimulated by the presence of parasite.

Hypertrophy : Hypertrophy refers to an increase in cell size and commonly found with intracellular parasites. For instance, the red blood cells are commonly enlarged during the erythrocytic phase of *Plasmodium vivax* life cycle.

Metaplasia : Metaplasia refers to the changing of one type of tissue into another with parasitic infection. For example, the infection of human lungs with fluke, *Paragonimus westermani*, it is surrounded by a wall of host tissue composed of epithelial cells and elongate fibroblasts, which are transformed cells from other type of cells in the lungs.

Neoplasia : Neoplasia is the growth of cells in a tissue to form a new structure like a tumor. The neoplastic tumor is not inflammatory and not confirm to normal growth pattern. The Neoplasms may remain localized or invade adjacent tissue or other parts of body through the blood or lymph. Several species are associated with tumors including cancers (Table 5.8).

Table 5.8 *Parasites associated with Tumor Formation (Neoplasia) in Mammals*

Group	Parasite	Host	Site of Tumor
Protozoa	*Eimeria stiedae*	Rabbit	Liver
Trematoda	*Schistosoma mansoni*	Human	Intestine and liver
	Schistosoma haematobium	Human	Bladder
	Schistosoma japonicum	Human	Intestine
	Paragonimus westermani	Tiger	Lung
	Clonorchis sinensis	Human	Liver
Cestoda	*Cysticercus fasciolaris*	Rats	Liver
	Echinococcus granulosus	Human	Lung
Nematoda	*Gongylonema neoplasticum*	Rat	Tongue
	Spirocerca lupi	Dog	Esophagus

2. *Depletion of Host's Nutrients*

Competition for the host's nutrients by parasites, especially endoparasites, resulting in depletion of host's nutrients leading to serious consequences particularly when the parasite density is sufficiently great. *Diphyllobothrium latum* infection in humans causes anaemia because it absorbs 10 to 50 times of vitamin B_{12} compared to other tapeworms, thus interfering the blood formation in host. The parasites absorb not only simple sugars but also certain amino acids, other essential nutrients. This drainage results under nourishment of host having considerable effect.

3. *Utilization of Host's Non-Nutritional Material*

The parasites, in some cases, also feed on host substances other than stored or recently acquired nutrients. The best example are the ecto and endo parasites that feed on host's blood (Table 5.9).

Table 5.9 *Pattern of blood intake in some parasites*

	Host	Number of parasites	Amount of Blood Lost from Host
Ticks			
Ixodes ricinus (larvae)	Sheep	1000	5 ml
Ixodes ricinus (adult female)	Sheep	1	1 ml
Leeches			
Haemadipsa zeylannica	Humans and animals	–	Sufficient in heavy infections to cause anaemia
Limnatis nuotica	Humans and animals	–	Sufficient in heavy infections to cause anaemia
Nematodes			
Ancylostoma caninum	Dogs and humans	1	0.5 ml each day
Ancylostoma duodenale	Humans	500	250 ml each day
Necator americanus	Humans	500	250 ml each day
Haemonchus contortus	Sheep	4000	60 ml each day

Thus, the blood lost through parasitic infections can constitute an appreciable amount over a period of time. The loss of even 50 ml of blood per day results in serious loss of blood cells, haemoglobin, serum and other essential constituents.

4. *Mechanical interference*

There are significant effects to host resulting from mechanical interference by parasites. The elephantiasis is best example. The filarial nematode, *Wucheraria bancrofti*, infection of human lymphatic ducts, with increase of their number coupled with aggregation of connective tissue may result in complete blockage of lymph flow and consequent sepage of lymph to surrounding tissue causing edema. With time, the build up of scar tissue and fluid leads to enlargement of limbs, breasts and scrotum as found in elephantiasis. The shear occupation of a large portion of liver and other human organs with hydatid cysts of tape worm *Echinococus granulosus* is another example of mechanical interference. The cysticercus larva of dog tape worm, *Multiceps multiceps* infect sheep brain, an intermediate host, exert pressure on brain and spinal cord that results in neurological damage to brain. The parasitization of monogenetic trematodes and dinoflagellates on gills of fishes suffocating them is yet another example of mechanical damage to hosts.

5. *Effect of toxins, poisons and secretions*

Many parasites release specific toxins or poisons which cause irritation and damage to host . A best example of an irritation of parasite secretion that elicits an allergic reaction in the host is schistosome cercarial dermatitis. During blood sucking in mosquito bite, the swelling represents the host's resp-

-sponse to the irritating salivery secretions of the insect. A known parasite toxin is body fluid of the nematode, *Parascars equorum*, which irritate hosts cornea and mucous membrane of nasophary ngeal cavity. The intestinal amoeba, *Entamoeba* produces a potent toxin that not only kill cells maintained in culture but also produces toxic symptoms in parasitized mammalian hosts.

6. *Other Parasite Induced Alterations*

Beside above effects, there are some interesting effects which are peculiar to parasitism. These include *sex reversal* and *parasitic castration* (sacculina on crab, carcinus) and *enhanced growth* (increased size of ant, *Pheidole commutata* parasitized by nematode Mermis).

5.8 Elements of Immunology

The immunobiology of parasites or immunoparasitology is the major field of study of parasitology. It has contributed to a greater extent in understanding of host parasite interactions. This chapter provides brief introduction of the basic concepts of immunology with emphasis on immunoparasitology. It also illustrates the production and application of antibodies in diagnosis of parasitic diseases.

A. *Immunity and Immune Response*

All organisms including human being are continuously exposed to the invasion of a wide variety of infectious organisms like virus, bacteria, protozoan, helminth, and arthropod parasites. Defence mechanisms are therefore essential if hosts are to survive and to lead successful life. Immune responses are most effective form of defence once the pathogen gains entry into the host. But before that many fixed structural, biochemical and physiological characteristics of host provide formidable barriers to prevent the initial process of infection which is known as *natural resistance* or *innate immunity* that protects against many of parasites with which host comes into contact. During invasion and initial establishment the parasite is subjected to further innate defences. All organisms show *cellular responses* to foreign organisms entering their bodies. This is mostly through capacity of phagocytic and inflammatory cells to bind via membrane receptors to molecules on the surface of the invader and these cells can distinguish between self and non-self. In invertebrates, recognition by phagocytic cells result in phagocytosis and digestion, in encapsulation or in cytotoxicity depending on the nature and size of foreign target. Whereas, in vertebrates, the cells can act as phagocytic or cytotoxic cells and also contribute to complex inflammatory response.

In addition to cellular defences, both invertebrates and vertebrates possess *humoral recognition factors*. The humoral factors in invertebrates include the molecules like *agglutinins* and *lectins* which bind to glycoproteins on the surface of pathogens resulting in agglutination and facilitation of phagocytosis. Vertebrates however use a fundamentally different system of humoral process involving specific *immunoglobulin molecules* or antibodies, which form part of defence system, known as *adaptive immune* response, is usually restricted to vertebrates.

The internal defence mechanisms of animals, both invertebrates and vertebrates, are of two types; *innate* and *acquired*. The *innate mechanisms* represent the native capability of the host to act against an invader. These are assumed to be genetically mediated rather than expressions of previous experiences of some foreign material. The *acquired mechanisms* are those that develop in the host in response to previous exposure to foreign material. As shown below, both innate and acquired internal defence mechanisms are may be of two types : *cellular* and *humoral*.

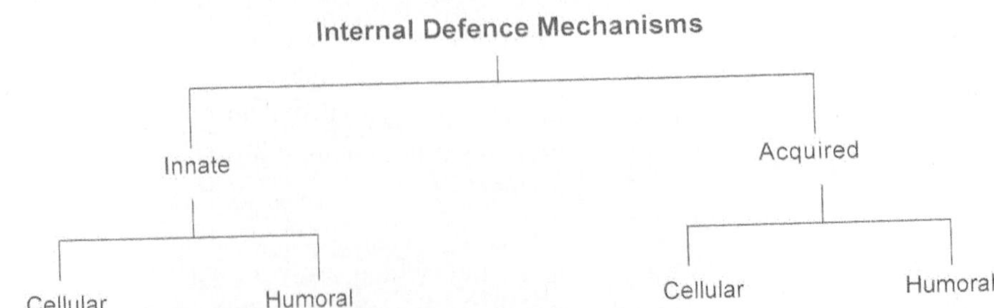

The central theme in immunology is the recognition of self from non-self by an animal. In immunoparasitology, the animal is the host and parasite is either self or non-self. The parasite is recognized as non-self (foreign). Parasite in some way camouflages itself as part of self and therefore is not reacted against. Thus, a host reacts to the parasite in two ways (1) *Cellular or Cell mediated reactions* and (2) *Humoral Mediated reactions*. In the former, the specialized cells become mobilized to check and eventually destroy the parasite whereas in latter, specialized molecules, antibodies or immunoglobulins are released into circulation resulting in immobilization and destruction.

An brief presentation on internal defence mechanisms of invertebrate hosts followed by a discussion on molecular processes involved in immune response of vertebrate hosts is described

B. *Internal Defence Mechanisms of Invertebrates*

The invertebrates generally do not synthesize immunoglobulins against foreign substances including zoo-parasites. However, the types of reactions elicited in invertebrates by invasion of parasites and other foreign materials are referred as immune responses. The mechanisms of immune responses are of prime importance in host parasite relationship in invertebrate hosts. Recently much attention has been directed towards these mechanisms particularly, the role of humoral immunity. The known type of phenotypic manifestation of innate internal defence mechanisms in invertebrates can be categorized as (a) phagocytosis (b) encapsulation (c) nacrezation (d) melanization and (e) humoral factors.

(a) *Phagocytosis*

In most of invertebrates (insects and molluscs) when a small foreign parasite invades, it is usually phagocytosed by host's leucocytes. There is an increase in number of phagocytic cells in presence of parasites. This phenomenon is called as *leucocytosis*. The process of phagocytosis consists of three phases as shown in Fig. 5.62. (i) attraction of phagocytes to foreign material by chemotaxis (ii) attachment of foreign material to the surface of phagocyte through chemical binding site and (iii) internalization of foreign substance or engulfment by the phagocyte. The phagocytosed foreign material depending on the nature may be degraded intracellularly, transported by phagocytes across epithelial borders to exterior or sometimes remain undamaged within phagocytes. However, all small parasites are not phagocytosed. For example, the haplosporadian, *Haplosporidium nelsoni* in oysters rarely phagocytosed (Fig. 5.63). The oyster phagocytes recognise the parasite as self and consequently do not attack it. On the other hand, when the parasite surface become altered then it is recognized as non-self and phagocytosed.

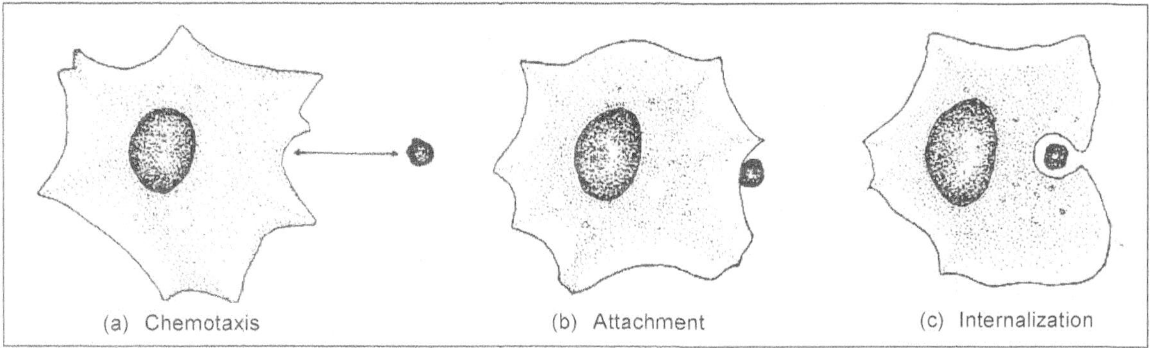

Fig. 5.62 Three phases of phagocytosis – A. Chemotactic attractions between phagocyte and foreign particle, B. Attachment of foreign particle to surface of phagocyte and C. Internalization of foreign particle.

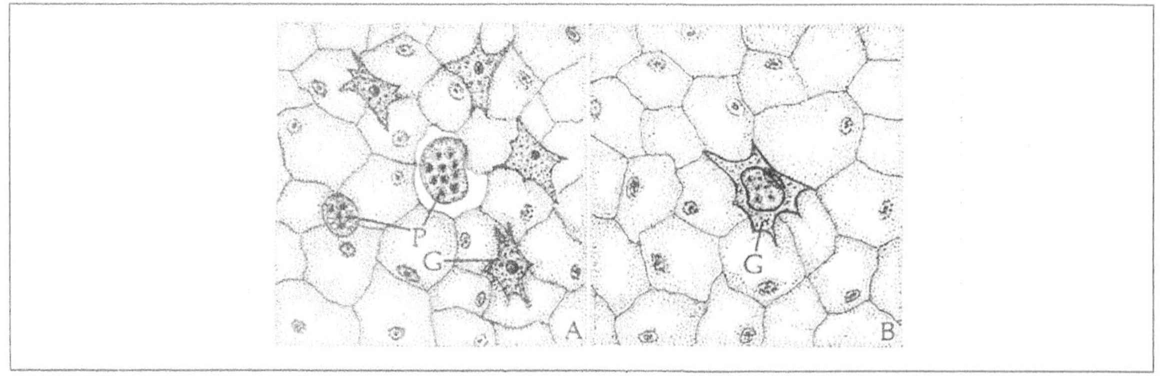

Fig. 5.63 Haplosporidium nelsoni in oyster (connective) tissue – A. Nonphagocytosed multinucleated plasmodia, B. Plasmodium that had become phagocytosed after the molecular structure of its surface had been altered G. Oyster granulocyte, P. Plasmodia of H. nelsoni.

(b) *Encapsulation*

All invertebrate hosts capable of encapsulation of foreign material that are too large to be phagocytosed. The encapsulation involves the enveloping of an invading non-self mass by cells and fibres of host origin. When a parasite enters the host and recognized as non-self it results in leucocytosis. Many of these cells migrate toward the parasite and form a capsule of discrete cells around it. The host cells are chemically attracted to the invading parasites like trematodes and cestodes that have mucopolysacharides on the body surface, while the chemo-attractants are secreted by protozoans and nematodes.

Thus, the initial stage of encapsulation resembles the phagocytosis and with time the number of host cells surrounding parasite increases. For example, the leucocyte of snails transform into epitheloid cells which become intimately abutted to form a wall surroundings the parasite. In some insect hosts, the host's leucocytes may become fused to form a syncytial tunic. Mostly the encapsulation is cellular type, but in some cases like in molluscans, consists of fibres of reticulin rather than collagen, synthesized and secreted around parasite by host's cells. This type of encapsulation is seen in American oyster, *Crassostrea virginica* in response to the larva of tape worm, *Tylocephalum*. The phagocyte cell then migrates into parasites and eventually degrades it.

(c) *Nacrezation*

Nacrezation is another type of cellular defense mechanism known in molluscans. For example, certain helminth parasite like metacercariae of trematode (*Meiogymnophallus minutus*) occur between the nacreous layer of shell and mantle of marine bivalves, the mantle secretes the nacre around the invaded parasites and enclosed parasite is killed.

(d) *Melanization*

Yet another type of defence mechanism seen in insects against helminth and arthropod parasites is melanization. The melanization is characterized by deposition of black-brown pigment melanin around invading parasite and may lead to death by interfering with vital activation like hatching, moulting or feeding. Melanization is the result of enzymatic oxidation of polyphenols primarily by tyrosinase. The melanization is found in nematode *Heteratylenchus autumnalis*, the parasite in the haemocoel of a larva of house fly, *Musca domestica*.

(e) *Humoral factors*

Though the role of innate humoral factors is evident, it requires extensive studies to elaborate. The innate humoral factors in invertebrates are of two functional categories. Those are directly *parasitocidal* and those that *enhance cellular reactions*. For instance, the tissue extract of marine molluscan contain active molecules that are lethal to cercariae of nematode *Himasthla quissetensis* showing the parasitocidal activity of humoral factors. The example of invertebrate humoral factors that enhance cellular reactions are naturally occurring agglutinins or lectins. These glycoprotein molecules enhance phagocytosis of foreign material and analogous to vertebrate opsonins. Recently, the additional information on invertebrate humoral factors has revealed two other categories of humoral factors (i) secreted lysosomal enzymes and (ii) synthesized antimicrobial molecules. Both of these are acquired *lysosomal* enzymes, found in a membrane bound organelle in cells, occur abundently in invertebrate heamocyte, the granulocytes. The lysosymal enzymes include peroxidases, phospholipases A_2, acid phosphatases, ribonuclease, sulfatase, glucosidase, cathepsin B and E, etc. For instance, when an invertebrate host like mollusca or insect are challenged with foreign material including parasite, there is hypersynthesis of some specific lysosomal enzymes which are released into serum. Consequently, when parasite makes contact with these enzymes it get destroyed. The destruction mechanism may be direct, where the enzymes attack and destroy the parasite or indirect, when one or more enzymes cause chemical alteration of parasite body surface and parasite is then attacked by host phagocytes (Fig. 5.65).

Antimicrobial molecules

It has been demonstrated that when invertebrate hosts like insects4 are challenged with microorganisms, they respond by synthesizing antimicrobial molecules, quite different from classical vertebrate immunoglobulins. For example, the cecropia moth, *Hylophora cecropia*, is challenged with *E. Coli*, the insect synthesizes two small basic proteins P 9 A and P 9 B which kill the *E. Coli*.

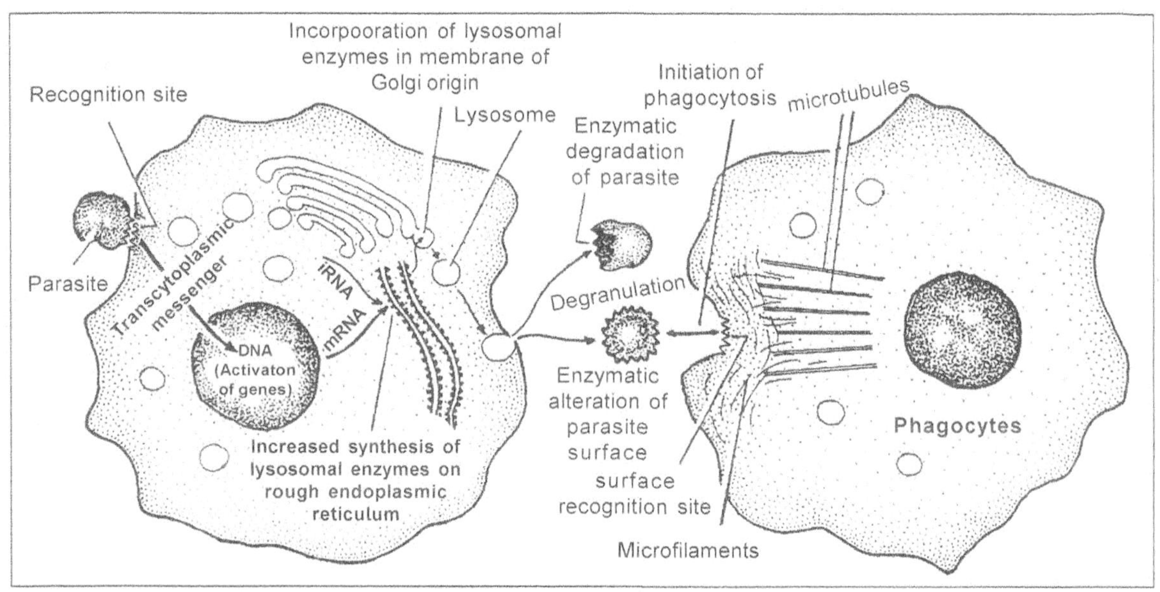

Fig. 5.64 Cellular events leading to hypersynthesis of lysosomal enzymes and their subsequent discharge into serum (degranulation).

C. *Vertebrate Immunology*

Although Eli Metchnikoff had studied internal defence mechanisms of invertebrates in 1882, the credit of initiation of immunology of vertebrates especially of mammals as discipline goes to Charles Ricket. Because of rapid development of vertebrate immunology and its obvious practical implications, all fundamental principles of immunology have been based on vertebrate system. Here, it is appropriate to review briefly basic concepts, components and mechanisms of immunity.

The protozoan or metazoan parasite comprises a number of different molecules some of which are antigenic to host. An *antigen* is any substance that is capable of inducing the synthesis of antibodies and of reacting specifically with such antibodies. The term *antibody* refers to proteins synthesized in response to the presented antigen and which react specifically with that antigen and to a variable extent of similar structure. Based on development of circulating antibodies and their reaction with antigen it is referred as *humoral immunity* or *B cell immunity*.

The second type of acquired immunity is achieved through the formation of large number of activated lymphocytes that are specifically designed to destroy the foreign agent. This type of immunity is called *Cell mediated* or *T-cell immunity* (Fig. 5.65). Thus, there are B and T type of lymphocytes, in addition to macrophages, or granulocytes, plasma cells, cytotoxic cells and helper cells.

1. *Antigens*

A large variety of macromolecules like proteins and synthetic polypeptides, polysacharides, nucleoproteins, glycoproteins, lipoproteins, and small molecules which suitably linked to peptides serve as antigen. For a substance to be antigenic, it usually must have a high molecular weight of 8000 daltons or greater. Furthermore, the process of antigenicity depends on regularly recurring

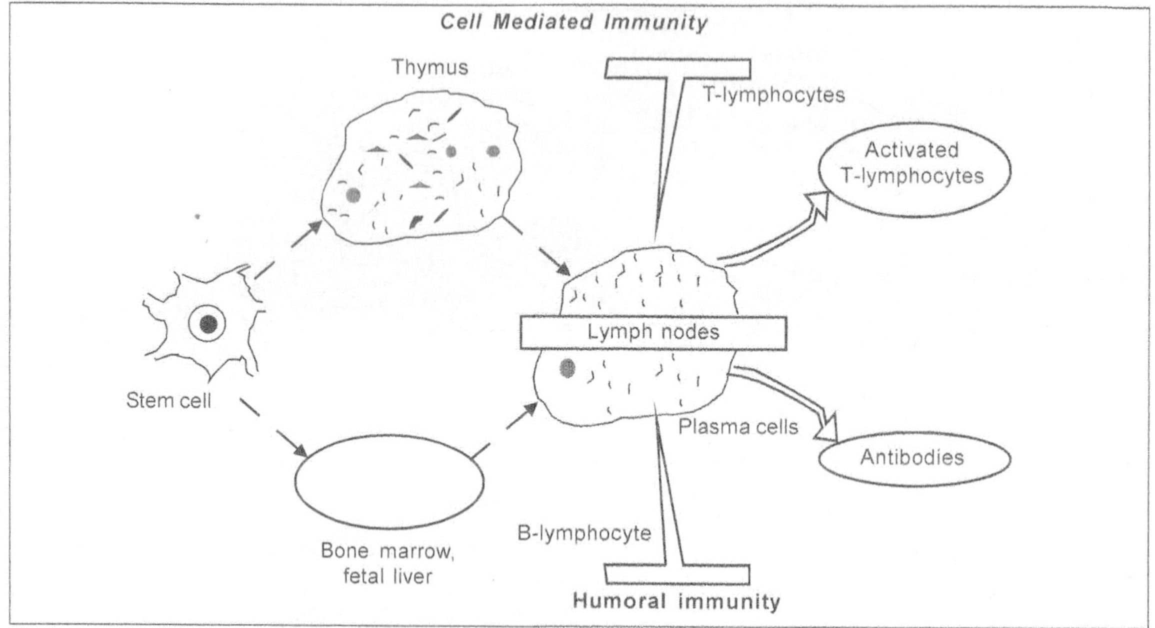

Fig. 5.65 Formation of antibodies and sensitized T-lymphocytes for cell mediated and humoral immune processes of lymph nodes.

molecular groups called antigenic determinants or '*epitopes*' on the surface of large molecules. These epitopes determine the specificity of antigen-antibody reaction. Antibody molecule do not bind to whole of an infectious agent. Because of antibody specificity, each one bind to specific epitope of antigens. A particular antigen can have several different epitopes or repeated epitopes (Fig. 5.66(a)). Antibodies are specific for epitopes rather than the whole antigen molecule. The substances with low molecular weights of less than 8000 daltons seldom act as antigen but when combined with antigenic substance like protein and then in combination elicits immune response. Such low molecular weight substances are called *hapten*, these generally include low molecular weight drugs, chemical constituents of dust, breakdown products of dandruff of animals, industrial chemicals, toxin so forth. For example, if NO_2 groups are attached to rabbit serum proteins and injected to rabbits, the antisera produced will react with those of horse, chicken or other animals . The antisera however will not react with non-nitrated serum proteins. Thus, antibodies are capable of recognizing the nitro groups or some other uniquely altered antigens. Here, the NO_2 group is epitope. In another instance, rabbits injected with p–azobenzone arsonate–globulin form antibodies that react with them, but p–azobenzone arsonate alone does not elicit antibody production. In this case, p–azobenzene arsonate is example of hapten (Fig. 5.66(b)).

Antigens of Parasites

All parasites have characteristically multiple antigenicity Two categories of parasite antigens exist. These are *somatic* antigens and *metabilic antigens*. The former are those associated with molecules comprising the soma of parasite whereas latter are those associated with secretions and excretions of parasites. For example, during moulting process of nematode, *Haemonchus contortus*, the ensheating fluid is highly antigenic and employed with considerable success in the vaccination of hosts. Further

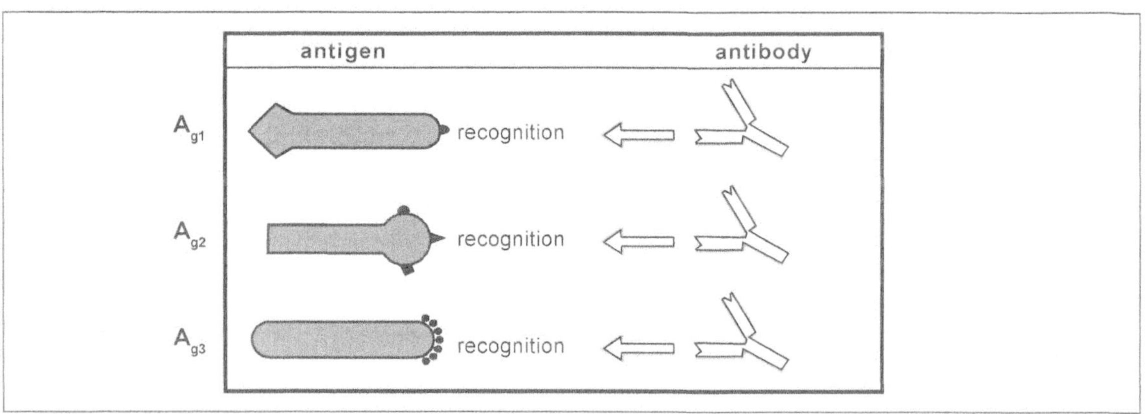

Fig. 5.66 (a) Molecules that generate antibodies are called antigens. Antigens molecules each have a set of antigenic determinants, also called epitopes. The epitopes on one antigen (Ag_1) are usually different from those on another (Ag_2). Some antigens (Ag_3) have repeated epitopes. Epitopes are molecular shapes recognized by antibodies and T-cell receptors. Each antibody receptor recognises one epitope rather than the whole antigen. Even simple microorganism have many different antigens which may be protein, lipid or carbohydrate.

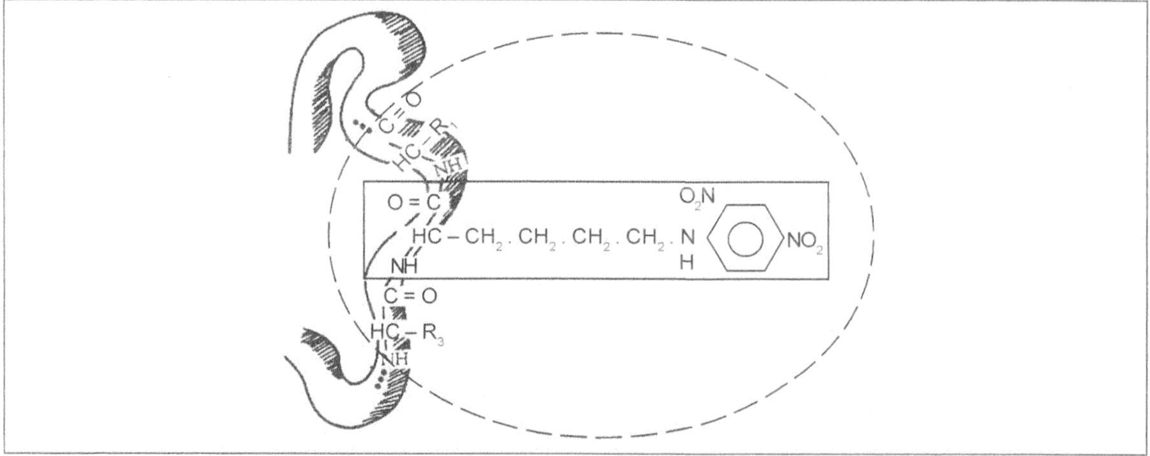

Fig. 5.66 (b) A representative haptenic group. 2,5 dinitrophenyl (DNP) group substituted in the e-NH_A group of a lysine residue. The haptenic group is outlined by the solid line and the antigenic determinant is visualized as the area outlined by the broken line. Amino acid residues contributing to the antigenic determinant need not be the nearest covalently linked neighbors of the e-DNP-lysine residue as shown; they could be parts of distant segments of the polypeptide chain hooked back to become contiguous with the DNP-lysyl residue.

more, antibodies produced as a result of antigenic stimulation by one species or one stage in life cycle of parasite may react with certain antigens of another species or an other development stage. Thus parasites show cross reactivity indicating multi antigenic parasites have common antigens.

2. *Antibodies*

Antibodies synthesized by vertebrates are gamma globulins called immunoglobulins and they have molecular weights between 160000 and 970000. They usually constitute about 20% of all plasma proteins. All immunoglobulin are composed of combinations of light and heavy polypeptide chains, most are combination of 2 light and 2 heavy chain (Fig. 5.67(a)). Some of immunoglobulins have combinations of as many as 10 heavy and 10 light chains, which gives rise to the high molecular weight immunoglobulins.

In brief, each immunoglobulin is made up of four polypeptide chains held together by sulfide bonds. Two of them smaller (each with molecular weight of 22,000) known as light chains and the other two each with a molecular weight of 55,000 are called heavy chains. The antibody has two parts, the *variable portion*, designated end of each light and heavy chain and the remainder of each chain is called as *constant portion*. The variable portion is different for each specificity of antibody and this part attaches specifically to a particular antigen. The constant portion of antibody determines the other properties of antibody establishing such as factors as diffusability of antibody in the tissues, adherence of antibody to specific structures in tissue, attachment to the complement complex, transport through membranes etc.

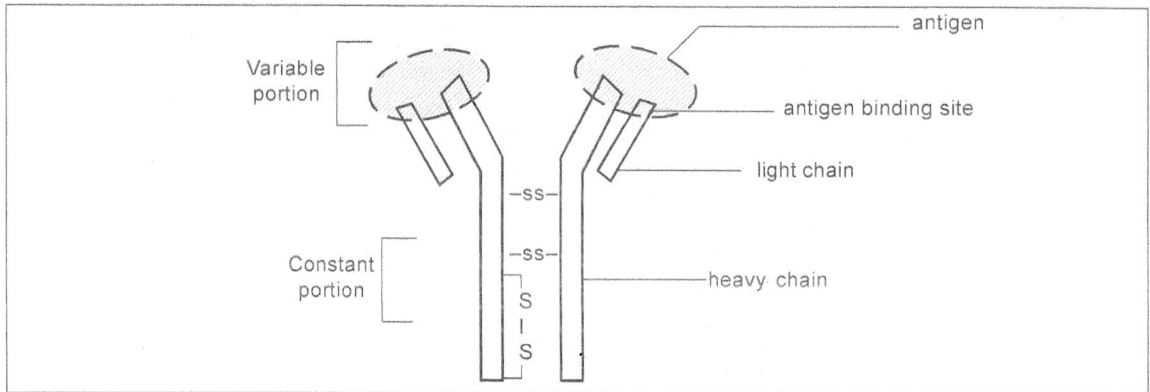

Fig. 5.67(a) Structure of typical antibody-consisting of two heavy polypoptide chains and two light polypoptide chains. The antigen binds at two different sites (shaded) on the variable portion of the chains.

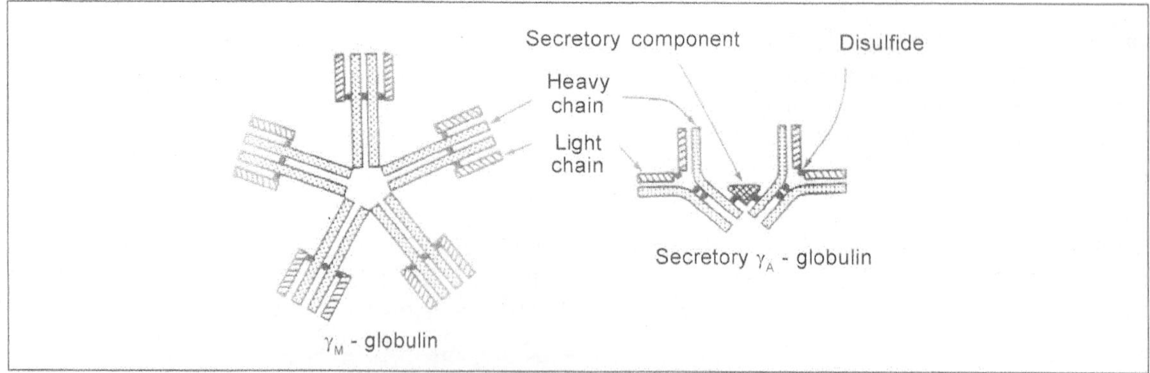

Fig. 5.67(b) Models of γ_M and secretory γ_A. The former is shown in its usual pentameric form while the latter is shown attached to a secretory component.

Specificity of Antibody

Each antibody is specific for particular antigen, this is caused by its unique structural organization of amino acids in the variable portion of both heavy and light chains. The amino acid organization has a different stearic shape for each antigen specificity and allow rapid bonding between antigen and antibody (Fig. 5.67(b)). The bonding is non-covalent and in specific case the coupling is exceedingly strong with help of hydrophobic bonding, hydrogen bonding, ionic attractions and Vander wall forces.

Classes of antibodies

There are five different classes of antibodies, respectively named I_g M, I_g G, I_g A, I_g D and I_g E, each with a distinct chemical structure and specific biological role. These classes are shown in the Table. 5.10 along with some of their properties and structure of I_gM and I_gA are shown in Fig. 5.67(b) for illustration.

Table 5.10 *Classes and Characteristics of Immunoglobulins*

Isotype	Structure	Molecular mass (kDa)	Mean survival (Days)	% Total serum Ig	Characteristics
IgM	Pentamer	900	5	6	First isotype in response; good agglutinator, fixes complement
IgG	Monomer	150	23	80	Major isotype in body fluids, fixes complement, binds to macrophages and polymorphs, facilitates ADCC, crosses placenta, several subclasses
IgA	Monomer Dimer (+ secretory piece)	160 400	6	13	Major isotype at mucosal surfaces, crosses epithelial cells, secreted in milk
IgE	Monomer	200	1.5	0.002	Binds to mast cells and *basophils*, involved in immediate hypersensitivity, binds to eosinophils, facilitates ADCC
IgD	–	150	2.8	0.2	–

Of the five classes of immunoglobulins, IgG is most abundantly present in vascular and extra vascular tissues, has long half life of about 23 days and provide the bulk of immunity against invading organisms. Alpha globulin or Ig A is the second most abundant type, present in lymphoid tissue lining gastrointestinal, respiratory and urinogenital tracts. IgM is the largest in terms of size, limited to vascular system, efficient aglutinator of antigens and predominant in primary response. The IgE is predominant in helminth infection. The Ig D role in parasitism is not yet established.

3. *Cells of Immune System*

The reticuloendothelial system provides an array of cells which elaborates cell products associated with immune mechanism. These cells are strategically distributed throughout the body including the lining of lymphatic and vascular system. The specific cell types, the mechanism which trigger them and their cell products are shown in Table 5.11.

Table 5.11 *Cell Types and Effector Mechanisms in Immune Reactions*

Cell type	Agent responsible for mobilization of cell	Cell product
1. Macrophages (monocytes)	Chemotactic factors, migration inhibitory factor	Processed immunogen.
2. Granulocytes		
– Neurtrophils	Chemotactic factor complement associated factor	Kinins, SRS.A, basic peptides
– Eosinophils	chemotactic factor complement associated factor	Kinins, SRS.A, basic peptides
– Basophils	–	Vasoactive amines
3. Platelets	Factors for platelet aggregates	Vasoactive amines
4. Lymphocytes	–	Anti body, interferon, cytotoxin, transfer factor etc.
5. Plasma cells.	–	Antibody

Macrophages

These occur in various tissues including blood where they are known as *monocytes* (Fig. 5.68(a)). Unlike granulocytes, these are capable of dividing in tissues. These are capable of phagocytosis and important in cellular defense mechanisms in arrest as well as removal of foreign material including parasites (known as *clearance*). The phagocytosis of foreign material is sometimes facilitated by opsonin (which is elaborated elsewhere). Macrophages are also involved in hypersensitivity reactions and are attracted to an area of injury by chemotactic factors.

Granulocytes

These cells have their origin in bone marrow and are released into circulatory system at a sufficient rate to replace dying cells. They are three types, namely, neutrophil, basophil and eosinophil (a) *Neutrophils* : (Fig. 5.68(b)) polymorphonuclear leucocyte, comprises about 60 to 70 % of total leucocytes and they do not divide. Its primary function is to phagocytose and digest particulate foreign material. Further, during inflammation, neutrophils migrate to and congregate at the site of injury through mediation of complement factors. The neutrophils also have a role in production of humoral factor known as slow reactive substance (SRS - A). (b) *Basophils* : Constitute about 0.5% of circulating leucocytes. Mast cell, indistinguishable from basophil, not found in circulation, has similar properties of basophil. These cells, along with platelets, secrete vasoactive substances like histamine and serotinin. The secretion of these substances is signalled by contact with antigen-antibody complexes through complement dependent and independent mechanisms (Fig. 5.68(c)). (c) *The eosinophils* make up from 1 to 3% of circulating leucocytes. They are capable of phagocytosis. They are attracted by antigen–antibody complexes and respond to complement derived chemotactic factors. They increase in number during infections (Fig. 5.68(d)).

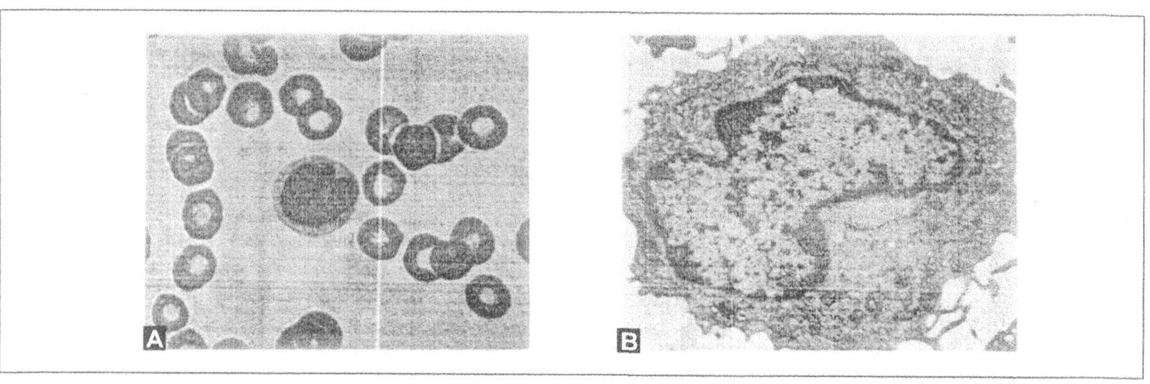

Fig. 5.68(a) Human macrophages (monocytes) A. Macrophage from peripheral blood stain, x 1400, B. Electron micrograph of macrophage, x 11, 000

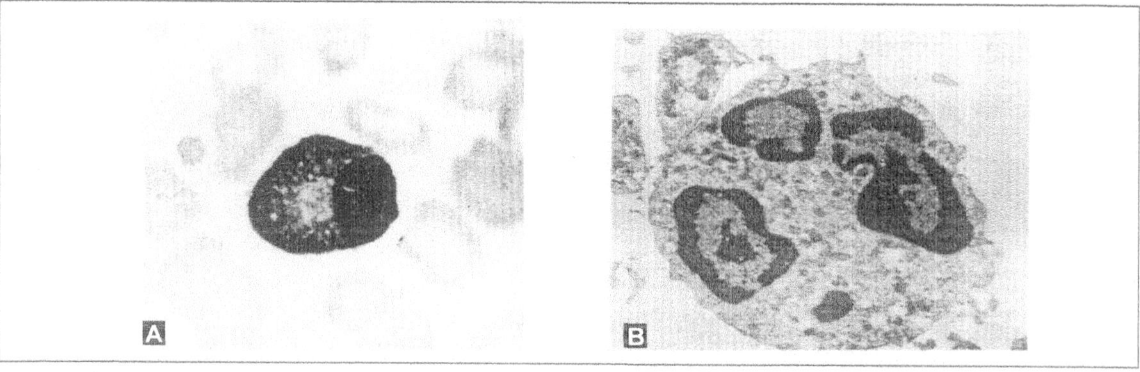

Fig. 5.68(b) Human neutrophils or polymorphonuclear leucocytes from peripheral blood –
A. Neutroophil showing segmented nucleus x 1350.
B. Electron micrograph of neutrophil showing cytoplasmic granules, x 10,500

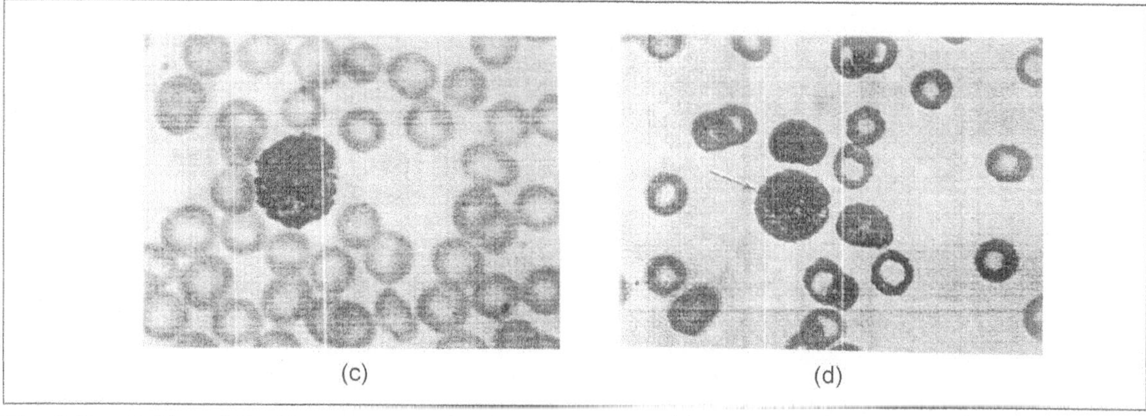

Fig. 5.68 (c) Basophil – Human basophilic leucocyte (basophil) from peripheral blood showing relatively large basophilic granules in the cytoplasm, Wright-Giemsa stain, x 1400.

(d) Eosinophil – Human eosinophilic leucocyte (eosinophil) (arrow) and a small lymphocyte from peripheral blood x 1400.

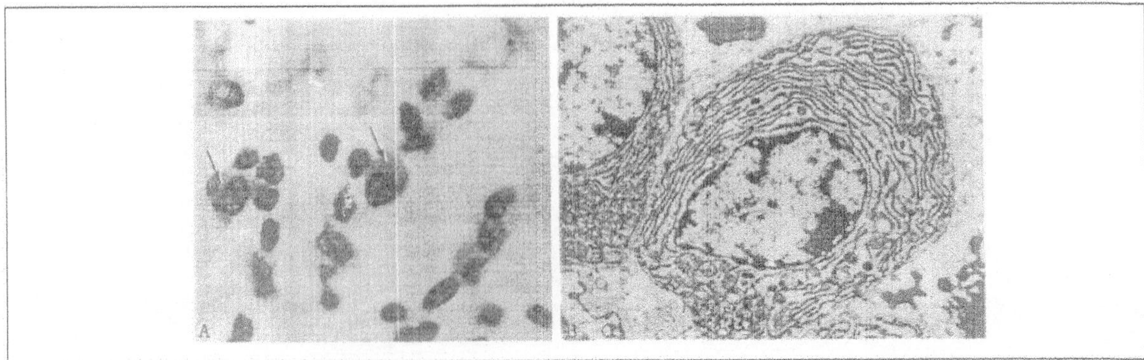

Fig. 5.68(e) Human Plasma cells – Plasma cells (arrows) from section of inflamed aorta, x 1000. Electron micrograph of plasma cell showing well-developed endoplasmic reticulum, x 11,000

Platelets : In addition to their role in clotting, these are involved in immune response and especially in inflammation. They express MHC products, receptors for Ig G and Ig E. Following injury, the aggregated platelets release substances that increase permeability as well as factors that activate complement and attract leucocytes.

Plasma cells : These cells (Fig. 5.68(e)) upon antigenic stimulation, elaborate antibodies. These cells are characterized by their RNA-rich cytoplasm, indicating active production of antibodies.

Lymphocytes

There are two major populations of lymphocytes, B and T cells. *B-cells* formed in bone marrow and subsequently processed through the bursa of Fabricus of birds, a lymphoid organ attached to intestine near cloaca, through liver and spleen of mammals and possess receptors comprised of antibodies at a concentration of 10^5 molecules per cell. Whereas *T-cells*, formed in thymus, have characteristic cell surface receptors. Binding of antigen to a receptor initiates either humoral or cell mediated immune response, depending on whether B cell or T cell receptor is stimulated. Each lymphocyte carries a single type of receptor. Mammals have about 10^8 to 10^{12} lymphocytes to respond to enormous variety of immune system. The lymphocyte on immunogenic stimulation proliferate, differentiate and a clone of progeny of lymphocytes are formed. Each cell of this clone has same idiotypic surface receptors of original stimulated cell. In the process of proliferation, some cells differentiate into effector cells, the functional product of immune response (Fig. 5.69).

B cell effector cells are called as *plasma* cells, which secretes antibodies of same idiotype and antigen recognition specificity as their cell surface receptors. On antigenic stimulation, T-lymphocytes differentiate into several type of effector cells with different functions. These are *Cytotoxic* or *Killer T cells* (T_C cells), *helper cell* (T_H *cell*), *suppressor cell* (T_S *cells*) and T_D, T_A cells. The cytotoxic (T_C) cells eliminate foreign cells directly through direct contact with surface membrane of target cells or through secretion of non-specific messenger molecules *lymphokines* (interleukins, interferon). The lymphokines act as lymphotoxin to foreign cell tissue or stimulate phagocytosis by macrophages or attract inflammatory cells to the site of injury. The other type of effect T cells are responsible for coordination of B cell differentiation and proliferation-T_H cells, for delayed hypersensitivity-T_D cells, for amplifying cytotoxic cell differentiation and proliferation-T_A cells and suppressing immune response-T_S cells.

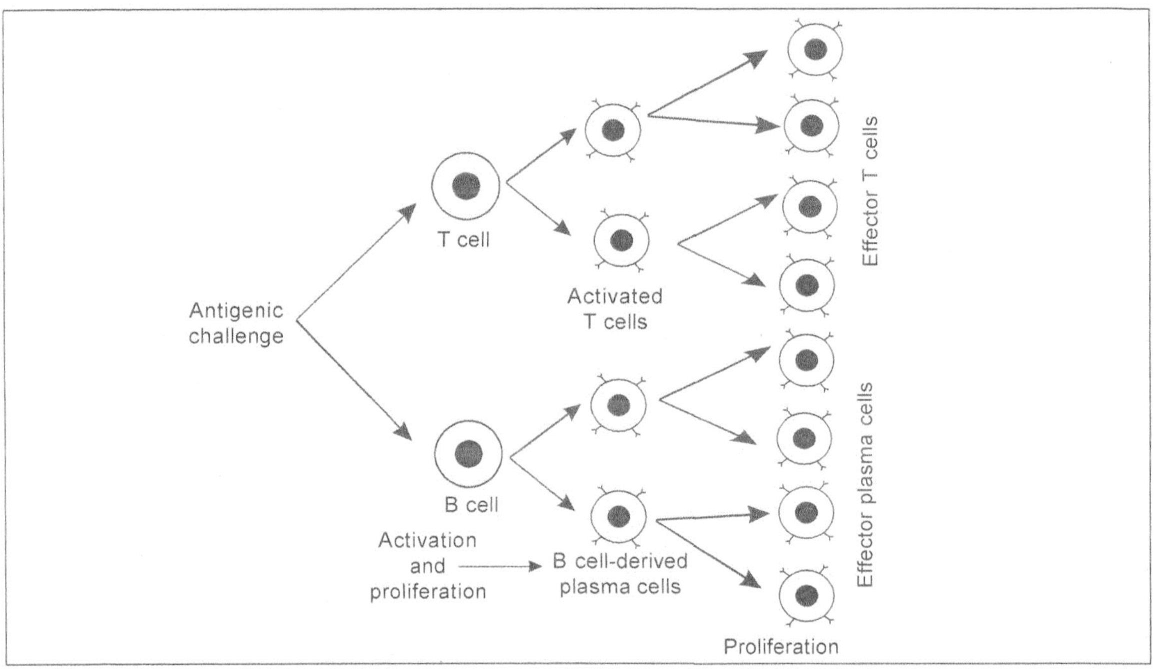

Fig. 5.69 Schematic diagram showing activation and proliferation of T and B lymphocytes of vertebrates immune system upon antigenic challenge

5. *Antigen - Antibody interaction*

Antigen - antibody interactions can be categorized into three : Primary, secondary and tertiary. (*a*) *Primary interaction* is the basic event during which the antigen bound to two or more available sites on antibody molecule (Fig. 5.70(a)). This type of interaction is rarely directly visible and more common through the occurrence of secondary and tertiary events, the primary interaction is ascertained. (*b*) *Secondary interaction* include direct action on invader and by activation of complement for destroying the invader. The secondary manifestation include agglutination, precipitation, neutralization, lysis, complement dependent reactions.

1. *Agglutination* – the multiple large particles with antigens on their surface, such as bacteria or blood cell are bound together into a clump.
2. *Precipitation* – in which the molecular complex of soluble antigen, such as tetanus toxin, and antibody becomes so large that it is rendered insoluble and precipitate.

In case of both agglutination and precipitation a two stage reversible chemical reaction occurs. In the first stage, serum antibodies in the host react with specific antigens on the foreign molecule and second stage, the lattice formation (Fig. 5.70(b)) the unbound receptor sites on antibody molecule become attached to suitable receptors on additional antigen molecules forming a lattice.

3. *Neutralization* in which the antibodies cover the toxic site of the antigenic agent
4. *Lysis*, in which some potent antibodies are capable of directly attacking membranes of cellular agents and thereby causing ruptures of cell.

The direct action of antibodies on antigen are not strong enough in normal conditions in protection of body. The total protection comes through the amplifying effect of complement system.

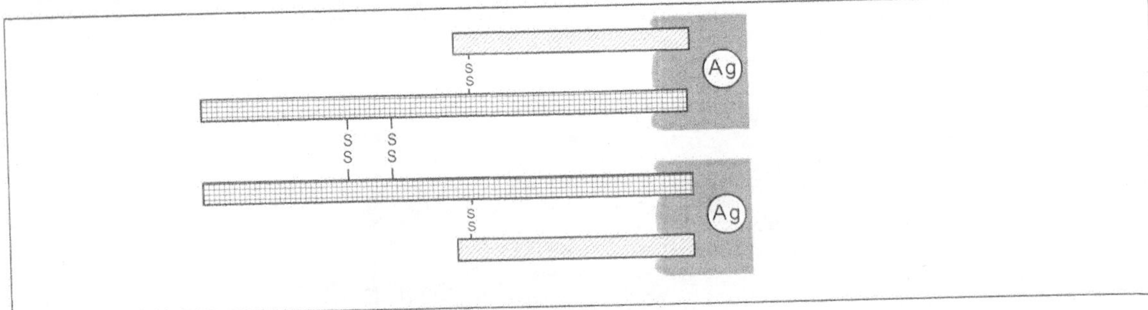

Fig. 5.70(a) Schematic representation of the binding of antigen to antibody sites

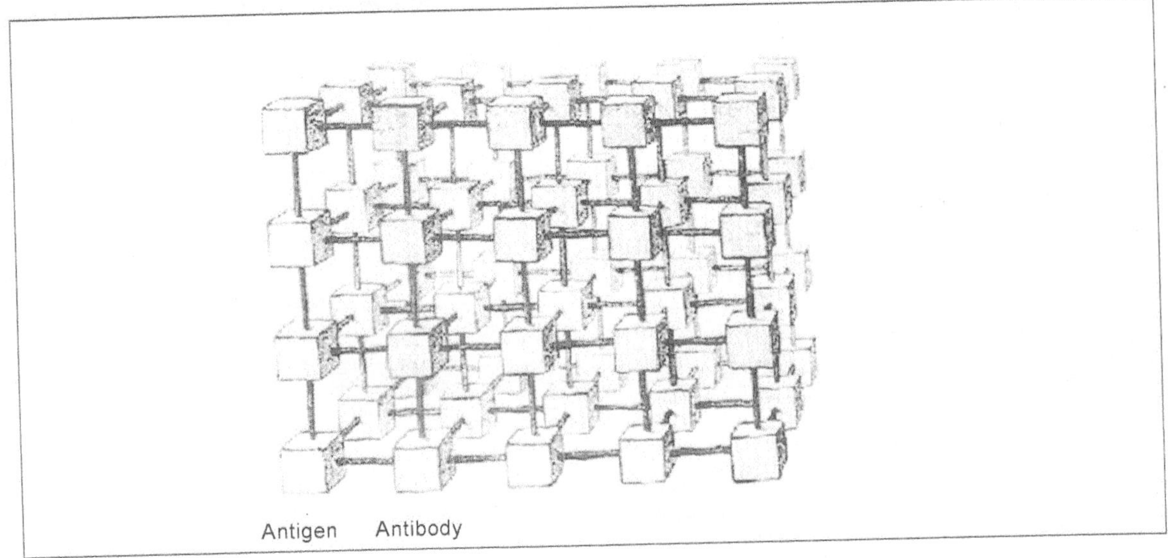

Fig. 5.70(b) Schematic representation of antigen-antibody lattice formation

5. *Complement System-Antibody Action*

Complement system refers to cascading action of group of 20 proteins, many of which are enzyme precursors. The important ones are 11 proteins designated as C_1 to C_9, B and D, as shown below.

The enzyme precursors are normally inactive, and they can be activated in two ways 1. *Classical pathway* and 2. *Alternate pathway.*

Classical pathway

The antigen-antibody reaction activates the complement pathway as shown in Fig. 5.71, through binding and activation of C1 molecule of complement which then activate successively increasing quantities (amplification) of enzymes in later stage resulting in formation of multiple end products. Several of these products cause important effects that prevent damage by the invading organism or toxin and more important effects are

(a) *Opsonization and phagocytosis* : The C3b complex activate phagocytosis through opsonization by both neutrophils and macrophages and maximize the effect.

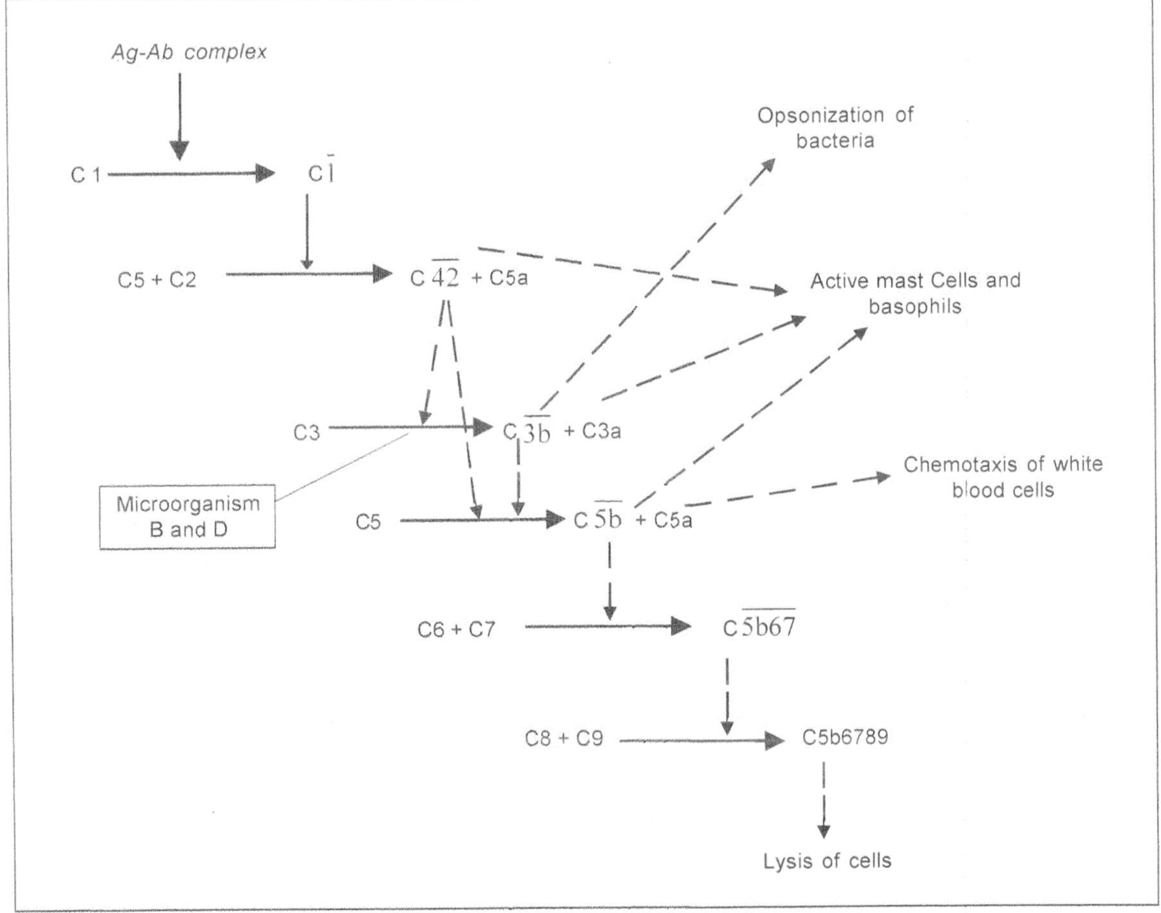

Fig. 5.71 Cascade of reactions of complement during activation of classical pathway.

(b) *Lysis* : The lytic complex (C5 b 678) has direct effect of rupturing the cell membrane of invading organism.

(c) *Agglutination* : The complement product by altering surface of invading organism promotes the agglutination.

(d) *Neutralization* : Some enzymes and products of complement can attack structures of organisms and render them as non-virulent.

(e) *Chemotaxis* : The C5a causes chemotaxis of neutrophils and macrophages, causing migration of these cells to local region of antigenic agent.

(f) *Activation of basophils and mast cells as well as inflammation* : The C3a, C5a and C5a activate mast cells and basophils resulting in liberation of histamine, heparin etc. which help in inactivation or immobilization of antigenic agent. These factors also have major role in inflammation.

Alternate pathway

The complement system sometimes is activated without antigen and antibody reaction, in response to large polysaccharide molecules of cell membrane of invading organism. These substances react with factor B and D, as shown in Fig. 5.71 forming an activation product that activates factor C3 and setting off cascade beyond C3 level of complement system. Thus, without involvement of ag–ab reaction, it is one of first line defense mechanism against invading microorganisms.

6. *Role of T-lymphocytes in cell mediated immunity*

On exposure to antigen, the specific T-lymphocyte clone proliferate and release number of activated T-lymphocytes in ways that parallel antibody release by activated B cells. These cells distribute throughout the body through circulation. Some of the activated T-cells, preserved in lymphoid tissue, known as Memory T cells producing more T-cell rapidly on subsequent exposure. There are 3 major type of T-cells – helper cells, cytotoxic T-cells and suppressor T-cells, as shown in Fig. 5.72. Helper T-cells, through formation of lymphokines, regulator T-cells a series of protein mediators, (interleukin 2 – 6, interferon), serve as major regulator of virtually all immune functions, like activation of immune system, differentiation to form plasma cells and growth and proliferation of cytotoxic T and helper T cell, stimulation of B cells and antibodies, activation of macrophage system and feedback stimulatory effect on helper cells themselves.

7. *Tertiary interactions*

These are *in vivo* expressions of antigen-antibody interactions. They may be of survival value, the antibodies produced serve protective function forming the basis for total or partial parasitic resistance. On the other hand, the deleterious effects can occur leading to disease through immunologic injury. The pathological effects of *immunological process* or *hypersensitivity* are classified into four types (Table 5.12).

Table 5.12 *Immunological mechanisms of tissue injury or Hypersensitivity*

Type	Manifestation	Mechanism
I	Immediate hypersensitivity	γ_E, other I_g.
II	Cytotoxic antibody	γ_G and γ_M
III	Antigen – antibody complex	γ_G only
IV	Delayed hypersensitivity	Senistized lymphocytes.

Type I or *immediate hypersensitivity* result in a reaction within minutes after introduction of soluble antigen into a previously sensitized host. This type of hypersensitivity is also called as anaphylaxis and is manifested in a variety of ways. For example, gunea pig in anaphylaxis manifests as scratch, sneeze, and cough and finally convulse, collapse and die. In humans, the sign of amaphylactic shock are itching, erythema, vomiting, abdominal cramps, diarrhoea and respiratory distress.

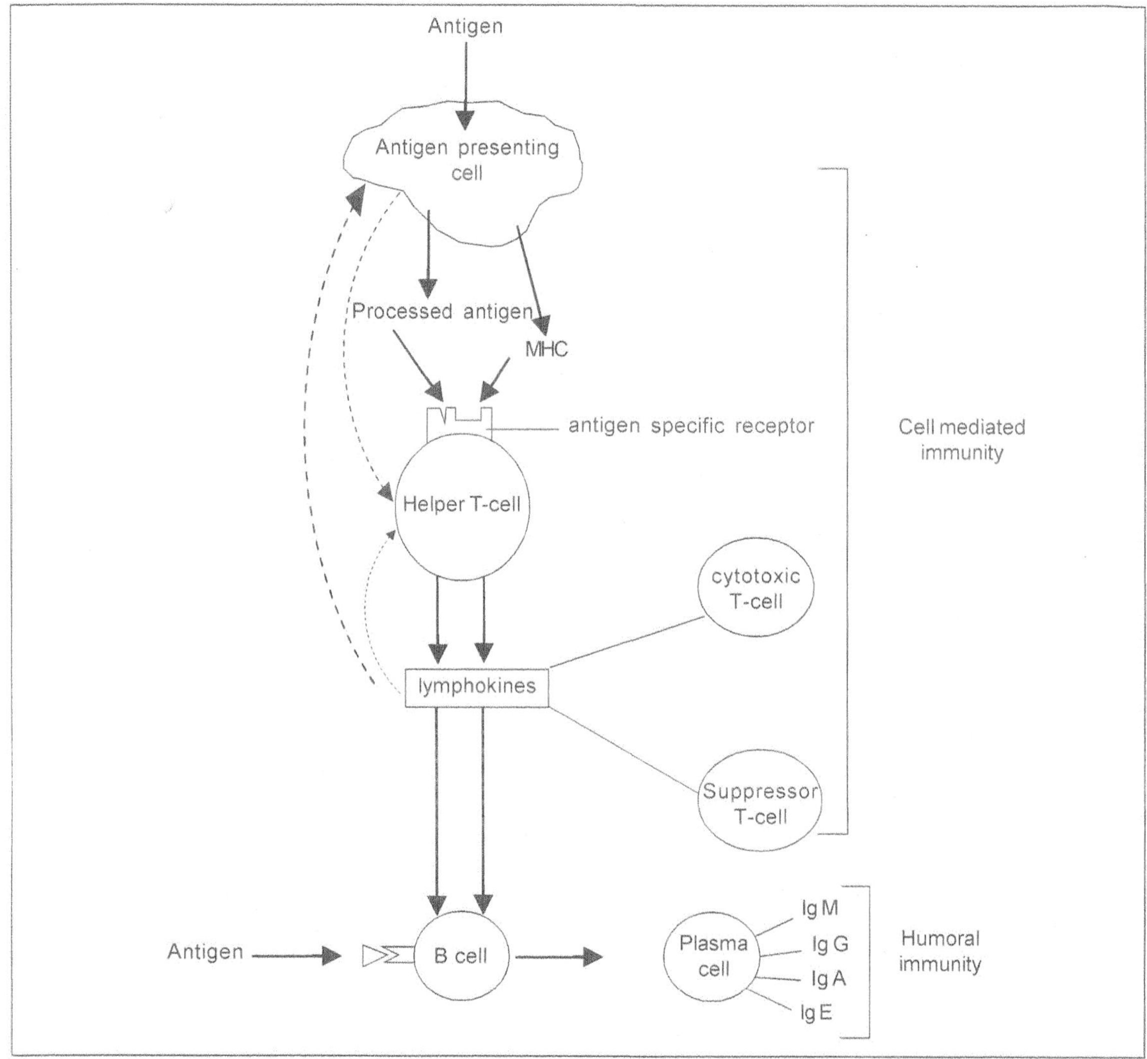

Fig. 5. 72 Regulation of Immune system, role of T-cells

Type II or *cytotoxic antibody* occurs when the antibodies are directed to tissue antigens. There are two main mechanisms. The antibody react with host own tissue cell causing cytolysis and killing through complement system and in second one the antitissue antibody reacts with its antigen, the complex interacts with host's phagocytic cells which acts on host cells. Host erythrocytes are not only lysed but also phagocytosed by other cells of reticula endothelial system. Because host's antibodies affect its own tissues in type II reactions, this is referred as *autoimmunity*.

Type III reaction is manifested as tissue injury produced by ag–ab complex. During immune responses, the phagocytosis, elimination of ag–ab complex, a variety of inflammatory reactions occur which result in tissue injury.

Type IV reaction is a manifestation of cell mediated injury called as *delayed hypersensitivity*. It is defined as an increased reactivity to specific antigens mediated not only by antibodies but by cells. It has slow onset and take 25h to reach maximal intensity. In this, there is accumulation of neutrophils, macrophages and lymphocytes at the site of antigen.

8. *Primary, challenge infections and Anamnesis*

A *primary infection* is the one during which the host experiences its initial exposure to the antigen while a *challenge infection* refers to all subsequent exposures. A host exposed to a primary infection rapidly eliminates antigen and this must precede immune elimination which is combination of antigen with antibody. This is known as *primary response*. The primary response is associated with a latent

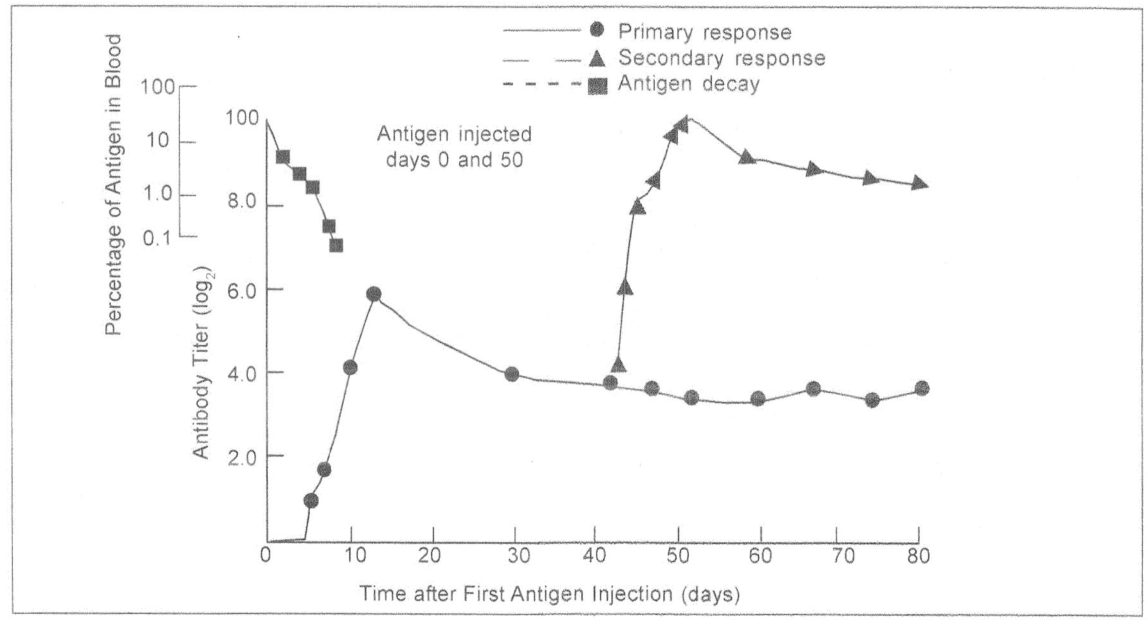

Fig. 5.73 Antigen decay (elimination) and primary and secondary immunoglobulin synthesis curves. The antigen elimination curves shows the three phases of equilibration, metabolic elimination, and immune elimination, the latter beginning at about the fifth day. Circulating antibodies are not detectable until about the fifth day. Notice how the secondary immunoglobulin response following the readministration of antigen at the fortieth day reaches a very high titer compared to the primary response

phase as shown in Fig. 5.73 and during this period antibody is not traced. As the latent period ends, the primary antibody response is demonstratable and antibody level increase for a few weeks, plateus and then begins to drop. The host is subjected to a challenge by combining with antigen and fall in antibody level in blood. Almost immediately thereafter, there is a spectacular rise in antibody level, which may be 10 to 50 times higher than the primary response. This accelerated enhanced response to the challenge infection is called as *Memory response* or *Anamnesis*. Thus, the host immune system remembers the early exposure of same antigens and remains primed for a second exposure. Anamnesis is characteristic of vertebrate hosts.

5.9 Opsonization

Opsonization is a process by which phagocytosis is faciliated by the deposition of opsonins - antibody and C3b complex - on the antigen or foreign material and enhances the phagocytosis. The opsonin activities are mediated by activation of the complement sequence. The role of antibody and C3b complex in opsonization process is explained below.

Antibody as Opsonin

Antibodies are a group of serum immunoglobulin molecules produced by B-lymphocytes and act as soluble form of B cell's antigen receptors. Normally in immune response, while one part of antibody molecule, Fab portion, binds to antigen, the other parts interact with other elements of immune system such as phagocytosis or one of the complement molecules (Fig. 5.74). For example, when a microorganism lacks the inherent ability to activate complement or bind to phagocytes, the body provides antibodies as flexible adaptor molecules. The body can make several million different antibodies that able to recognize a wide variety of infectious agents. Thus the antibody, as shown in Fig. 5.74, binds microbe 1 but not mocrobe 2, by its antigen binding portion (Fab). The Fc portion may activate complement or bind to Fc receptors on host cells, particularly phagocytes. In effect, antibodies act as *Flexible adaptors*, linking various elements of the immune system to recognize specific pathogens and their products.

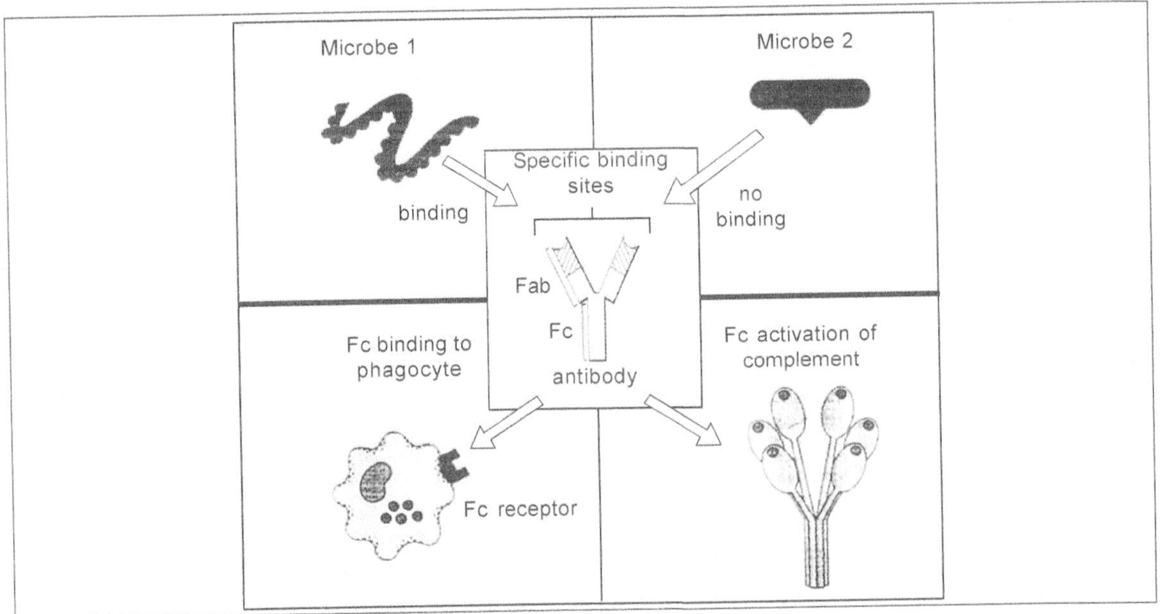

Fig. 5.74 Antibody as a flexible adaptor

The other part of antibody, Fc portion, interacts with the cells of immune system. Neutrophils, macrophages and mononuclear phagocytes have Fc receptors on their surface. If an antibody binds to pathogen, it can link to a phagocyte through the Fc portion. This allows phagocytosis, i.e., the pathogen to be ingested and destroyed by the phagocyte. Here, the antibody acts as opsonin. As illustrated in Fig. 5.75, the phagocyte can recognize material using either activated complement (C3b) or antibody as the opsonin, but phagocytosis is most effective when both participate.

Phagocyte	opsonin	binding
1.	–	±
2.	complement C3b	+ +
3.	antibody	+ +
4.	antibody and complement C3b	+ + + +

Fig. 5.75 Opsonization process, role of antibody as well as C3b complex – 1. The intrinsic ability of phagocytes to bind directly with microorganisms is much enhanced if complement is activated, 2. This binding is through C3b via C3b receptors, 3. Organisms without activated complement are opsonized by antibody which bind to F_c receptors on phagocyte and 4. Antibody can also activate complement and if both antibody and C3b opsonize organism, binding is greatly enhanced.

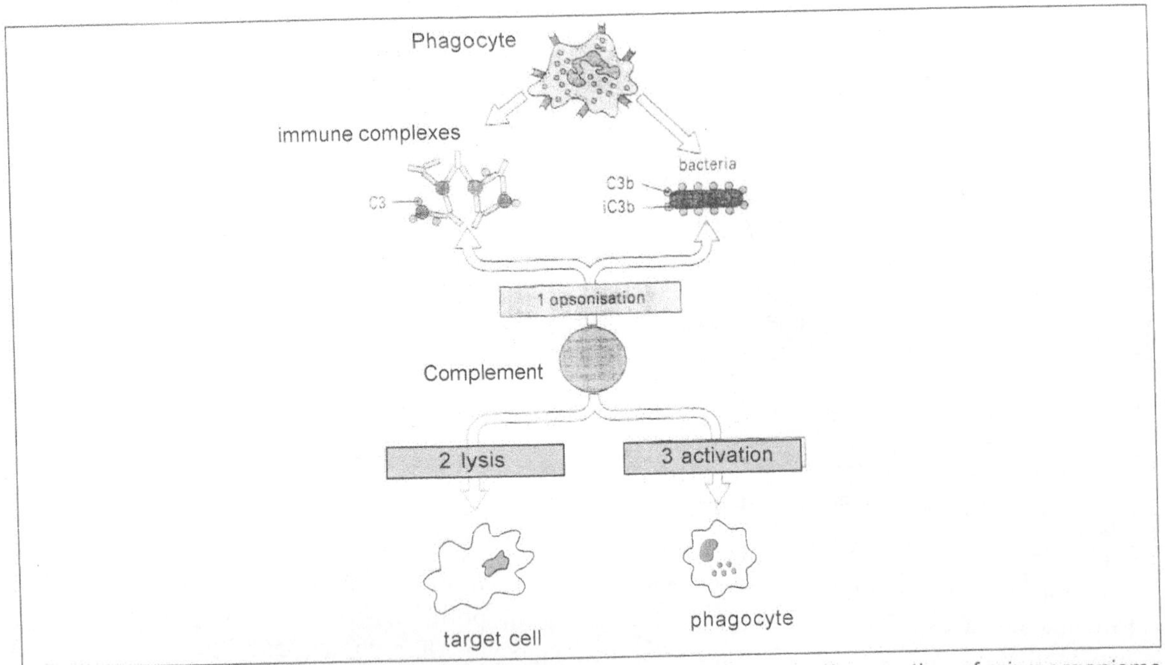

Fig. 5.76 Three major biological activities of complements – 1. Opsonization coating of microorganisms of immune complexes, 2. Lysis of target cells and 3. Activation of phagocytes including macrophages and neutrophils.

Complement system : Role of C3b Complex and Receptors in Opsonization in Phagocytosis -

Complement system is part of the innate immune system and consists of many proteins that act as a cascade, where each enzyme acts as a catalyst for the next. The most important component is C3. The two main pathways for complement activation are :

(a) *Classical pathway* links to adaptive immune system through the binding of immune complexes of *Clq* and

(b) *Alternate pathway* is activated by chance binding of C3b to the surface of microorganisms. The consequences of complement activation are : *Opsonization, Activation of leucocytes* and *Lysis of target cells*, as shown in Fig. 5.76.

Opsonization involves complement proteins coating the surface of target. Phagocytic cells carrying receptors for these complement components are then able to bind the target leading to cell activation and endocytosis of target. *Activation of leucocytes*, stimulate directed cellular movements and activation. *Lysis of target cells* involves insertion of hydrophobic plug in membrane and disruption.

In complement activation, both classical as well as alternate pathway lead to formation of a convertase that cleaves C3 to C3a and C3b. This is very important step in the process of complement activation (Fig. 5.77). The C3b in turn activates the terminal lytic complement sequence C5-C9. The C5 convertase catalyses the first step that leads to the production of membrane attack complexes. The first stage leading to C3 fixation by classical sequence is the binding of antigen to its antibody. The alternate pathway does not require antibody and is initiated by the covalent binding of C3b to hydroxyl groups on the microorganisms cell membrane. The alternate pathway provides non-specific innate immunity, whereas classical pathway represents more recently evolved link with adaptive immunity.

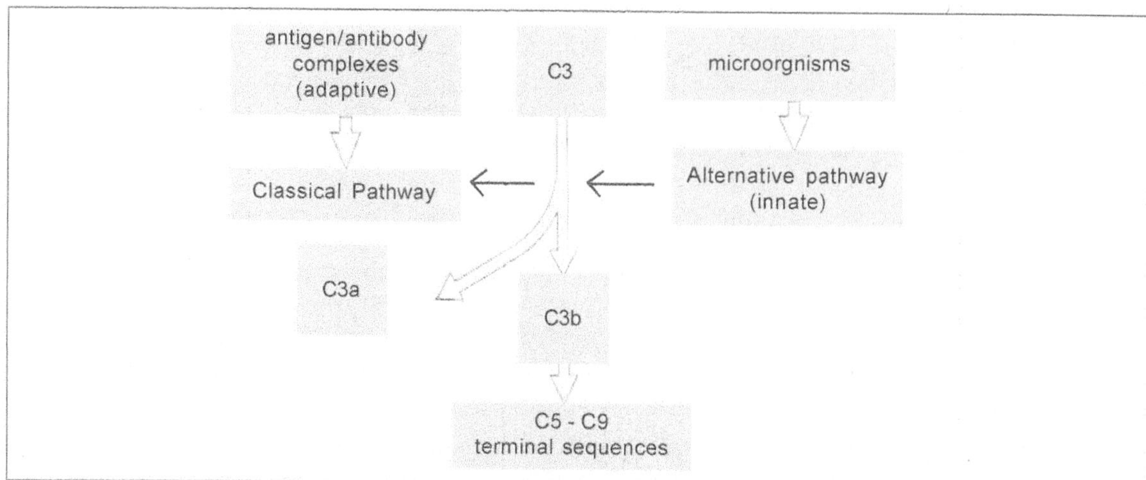

Fig. 5.77 Comparison of the classical and alternative complement pathways

Complement receptors. Many of fragments of complement proteins produced during activation, bind to specific receptors on the surface of immune cells. This is an important mechanism of mediation of physiological effects of complement including uptake of particles opsonized by complement and activation of cell bearing receptors. The opsonic fragments of C3 products - C3b, iC3b and C3dg - bind to the membranes of target cells. Four different receptors (CR1, CR2, CR3 and CR5) for these opsonic fragments are known and their potential ligands and their cellular distribution are given in Table 5.13.

Table 5.13 *The Complement receptors for fragments of C3*

Receptor	Ligands	Cellular distribution
CR1 (CD35)	C3b > iC3b C5b	B cells, neutrophils, monocytes, macrophages, erythrocytes, follicular dendritic cells, glomerular epithelial cells
CR2 (CD21)	iC3b, C3dg Epstein - Barr virus interferon-α	B cells, follicular dendritic cells, epithelial cells of cervix and nasopharynx
CR3 (CD18/CD11b)	iC3b zymosan certain bacteria fibrinogen factor X ICAM-1	monocytes, macrophages, neutrophils, NK cells, follicular dendritic cells
CR5 (p 150-95) (CD18/CD11c)	iC3b fibrinogen	neutrophils, monocytes, tissue macrophages

The CR1 (CD35) has multiple physiological activities as a receptor for C3b and iC3b. The cellular binding reaction or immune adherence mediated by CR1 dependent cellular binding. The trypanosomes or microbes opsonized with antibody and complement adhere to platelets of rodents and to erythrocytes of primates. The CR1, the C3b/C5b receptor or CD35 have four important physiological activities.

(i) It is an opsonic receptor on neutrophils, monocytes and macrophages, mediating phagocytosis by appropriately primed cells (Fig. 5.78).

(ii) With CR1 role cleavage of iC3b (into C3c and C3dg), protect self cells from attack by complement.

(iii) CR1 help to pick up opsonized immune complexes and transport them to the cells of the fixed mononuclear phagocytic system and

(iv) CR1, along with CR2, on lymphocyte, act as receptor mediating lymphocyte activation. The *CR2* or CD21, is located on B-lymphocytes, folliculor dendritic cells, epithelial cells, EBV (Epistein Barr Virus) etc., and with its ligands iC3b, C3dg, interferon α

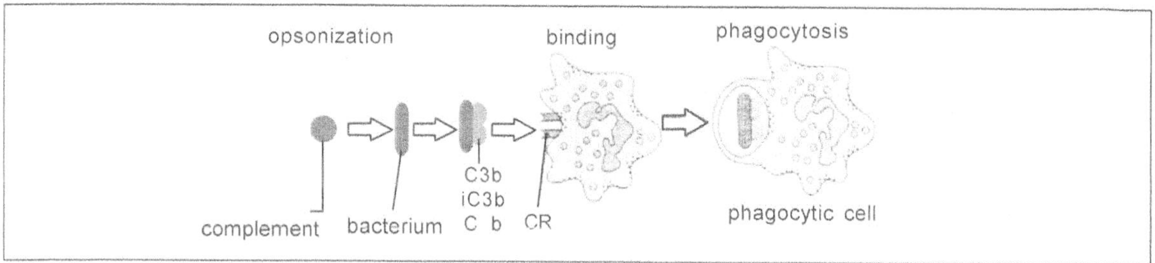

Fig. 5.78 Steps in the uptake of a particle such as a bacterium opsonized by C3b

involved in B-lymphocytes activation. The main pathophysiological effect of *CR*3 and *CR*5 are leucocyte integrin family cell surface receptors. The CR3 mediates phagocytosis of particles opsonized with iC3b where as CR5 also binds with iC3b and involves in adhesion of monocytes and neutrophils.

Thus, biological activity of complement system among others by opsonizing the microorganisms enhances phagocytosis process.

5.10 Monoclonal Antibodies and Diagnostics for Parasites

Antibodies are routinely obtained by injecting an animal or man with the antigens for which an immune response is sought. The immune system then reacts by generating a variety of antibodies, each specific to a different part of injected antigen molecule. Blood serum taken from such an animal contain this antibody mixture. Thus, obtained antiserum is being a mixture of antibodies with varying affinity, cross reactivities and effecter functions. Indeed, classical serology is like a jungle as it showed endless cross reactions. Although specific antisera can be generated with purification methods, it has problems of heterogenecity and intrinsic unpredictability. A way out of foregoing limitations of classical serology was found by George Kohler and Cesar Milstein (1975), who described the technique (called *Hybridoma technique*) for monoclonal antibodies. This has revolutionized classical immunoassay and realized the potential of antibodies as reagents not only in research and diagnostic laboratories but also as therapeutic agents. The modern approach permits the production of nearly pure and homogeneous antibodies by fusion of an antibody forming cell with a cell capable of perpetual proliferation or division.

Production of Monoclonal Antibodies

Monoclonal antibody is a specific type of antibody derived or produced from a single clone and are homogeneous. These are developed in the laboratory to recognise only one small part of the specific protein in a foreign substance that allows researcher to be very specific when they want to seek out a target such as a piroplasmosis causing parasite. Although monoclonal antibody is a well defined reagent it does not have a greater specificity than a polyclonal antiserum which recognizes the antigen by means of different epitopes.

Immunologists often need to isolate pure antibodies, which may be either antigen specific or non-specific immunoglobulin. *Isolation* of antigen specific immunoglobulin is carried out by affinity chromotography using antigen coupled immunosorbant with chaotropic agents such as sodium thiocyanate, or *glycine - HCL* buffer or diethylemine buffer. Thus using affinity chromatography pure antibody or antigen is obtained.

Another way of obtaining, most-common, pure antibody of a defined specificity is to produce monoclonal antibody from cells in culture by hybridoma technique. By creating an immortal clone of cells which manufacture a single antibody of defined specificity and production can be maintained indefinitely, avoiding vagaries and limitations of antiserum production.

Principle and Procedure

The essential features of monoclonal antibody production is depicted in Fig. 5.79. The animals (usually mice or rats) are immunized with antigen (antigen of corresponding desired antibody). Once the animals are making good antibodies response, their spleens or lymph nodes are removed and a cell suspension is prepared. These cells are fused with a myeloma cell line by addition of polyethylene glycol (PEG) which promotes membrane fusion. A small proportion of cells fuse successfully. The fusion mixture is then setup in culture with medium containing a mixture of hypoxanthine, aminopterin and thymidine (HAT). Aminopterin, is inhibitor, which blocks metabolic pathway and this pathway can be bypassed if the cell is provided with medium containing hypoxanthine and thymidine. The spleen cells grow in HAT medium but myeloma cells die in HAT medium because they have a metabolic defect and cannot use the bypass pathway.

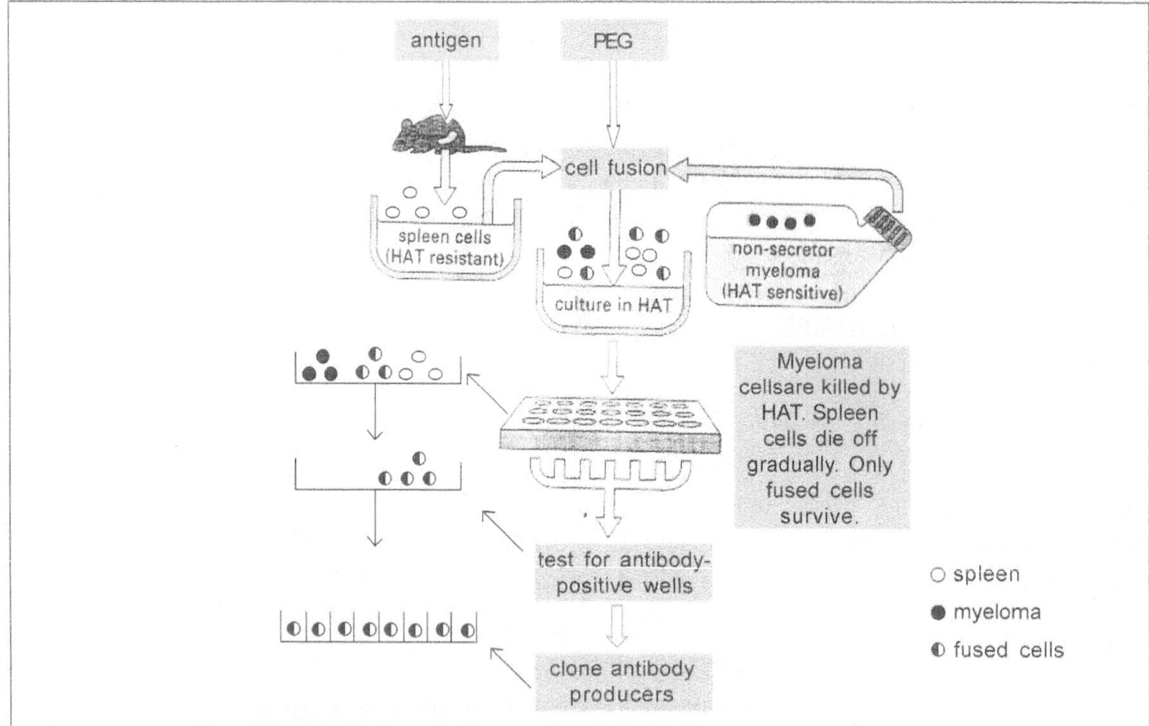

Fig. 5.79 Monoclonal antibody production

When the culture is set up in HAT medium it contain spleen cells, myeloma cells, and fused hybrid cells. The spleen cells die naturally after 1-2 weeks and myeloma cells are killed by HAT medium. Fused hybrid cells, as they have immortality of myeloma and metabolic bypass of spleen cells, only survive. These cells have the antibody producing capacity of the spleen cells. The cells containing growing cells are tested for the production of the desired antibody and the positive cells, the culture are cloned by plating out so that there is only one cell in each well. This produces a clone of cells derived from a single progenitor which is both immortal and producer of monoclonal antibody. The culture of these cells continued for the mass production of characteristic and desired monoclonal antibody.

Monoclonal antibodies have found widespread use in many biological applications, where the antibody is used as a highly specific probe. Most monoclonal antibodies are generated by fusion of mouse splenocytes with a B-cell myeloma (hybridoma) from the same strain which does not secretes its own antibody. An alternative method is to transform B-cell for example, human B-cell may be infected with Epstein-Barr virus.

A new method of generating antibodies is by *involve phage display*. In this, it is possible to express antibody variable region (V_H and V_L) as port molecules (F_V) defined antigen binding specificity and affinity on the surface of M13 filamentous phage so that they can be selected by antigen.

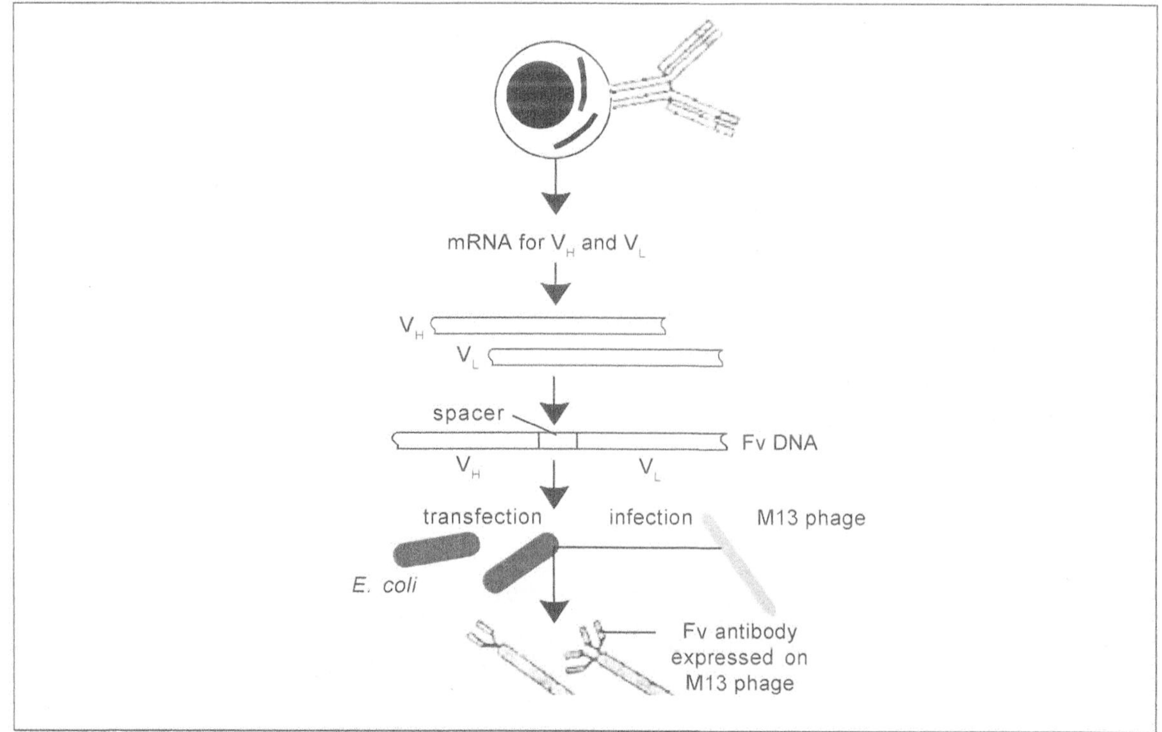

Fig. 5.80 Production of Fv antibodies by phage display

To produce F_V antibodies by phage display, antibody V_H and V_L genes are first amplified from B-cell mRNA by polymerase chain reaction. As shown in Fig. 5.80 the genes are joined with the gene in a phagemid vector containing a leader sequence, a fragment of gene expressing phage coat-protein 3 and an M13 origin of replication and then infected with M13 phage. The phages replicate and express F_V. The phages with desired specificity is amplified. These antigen specific phages used to infect known bacteria and F_V protein is secreted in large amounts into culture medium. This approach does not require the deliberate immunization of animals or humans.

Characterization or Screening of Parasites Using Monoclonal Antibodies

The landmark application of monoclonal antibodies in parasitology has revolutionized not only research in parasitology but also diagnosis, prevention and treatment. The production and screening of monoclonal antibodies against parasites is illustrated with following representative examples.

1. *Characterisation of malarial parasite, Plasmodium berghei :* During progress of utilization of monoclonal antibodies, at the early period, Potocnjak and others (1982) have reported the production of monoclonal mouse antibody directed against the circumsporozoite protein of murine malarial parasite, *Plasmodium berghei* infection. By injecting mice with this antibody against idiotype of the anticircumsporozoite protein immunoglobulins used as immunogen. This anti-idiotype antibody reacted with the antigen-binding antibody. The inhibition of interaction of these two antibodies by *P. berghei* sporozoites constitutes the basis of radio immunoassay which permits quantitative determination of small amount of relevant circumsporozoite antigen, even when present in a crude extract.

2. *Characterisation of Endotrypanum parasites using specific monoclonal antibodies :* The screening of monoclonal antibodies (MAbs) against parasites is further elaborated in detail with example of *Endotrypanum*. Endotrpanum parasites are unique among kinetoplastida protozoans in that they infect erythrocytes of mammalian host. Inside erythrocyte the parasite assumes an eptmastigate or trypomastigote form, while in sandfly (vector) and during *in vitro* culture the parasite assumes promastigote morphology. These parasites are alike with several species of *Leishmania* sp. and *Trypanosoma* sp. Besides morphological and biochemical indicators, the more precise taxonomic markers have resulted from application of molecular technique, particularly the serotype analysis using specific MAbs. The parasitic differentiation of *Endotrypanum* is characterized by their reactivities with MAbs and using distinct assay systems.

The MAbs used were produced against membrane components of *Endotrypanum* or *Leishmania*. The following species specific or group specific clones among many others were employed for typing parasites

1. E – 24, CXXX – 3G5 – F 12 (*Endotrypanum* sp.)
2. E – 2, CXIV – 3C7 – F 5 (*E. schaudinni*)
3. T – 12 –IS1 2G7 – F 1 (*Leishmania tropica*)
5. D – 2, LXX VIII 2D5 – A 8 (*L. donovani*) etc.,

Typing of *Endotrypanum* with MAbs was performed with an indirect immunofluorescence assay using whole fixed promastigote as antigen or by distinct immunobinding assay systems, such as indirect radio immunoassay, the ELISA technique or dot-blot ELISA using whole parasite lysate as antigen. Characterisation of molecules associated with the specific antigenic determinant can also be analysed by western blot analysis.

Using different immunological assay systems, MAbs considered specific for genus *Endotrypanum* (E25 – C XXX – 3 G5 – F12) or strain of *E. shaudinni* (E – 2, CXIV - 3C7 – F5) reacted variably according to test used but in the ELISA or immunofluorescence assay both reacted with all strains tested. Analyses of these MAb's showed antigenic diversity occurring among the *Endotrypanum* strains. Western blot analyses of parasites further showed that MAb's recognized multiple components. Differences existed either in epitope density or molecular forms associated with the antigenic determinants and therefore allowed the assignment of strains to specific antigenic group. These strains formed a polymorphic population according to their enzyme profiles and were grouped into 12[12] zymodemes. About 17 *Endotrypanum* isolates were classified through application of technique employing specific MAbs in comparison with standard reference strains. The characterisation of parasites with specific MAbs is quite useful in identification of stock, and group/ differentiate further among these parasites.

In a related study on cutaneous Leishmaniasis, direct smears and cultures were made from cases suspected of disease. Out of 166 cases, 163 (98%) were smear positive, while 133 (80%) were culture positive. Using standard monoclonal antibodies on culture isolates and IFAT and ELISA techniques characterised as 80% as *L. tropica* and 9% as *L. major*. The use of PCR and biochemical characterisation using isozymes method further confirmed, these assay.

3. *Monoclonal Antibody Capture ELISA for Immuno Diagnosis of Paragonimus heterotremus*

A monoclonal antibody (MAb) capture enzyme-linked immunosorbent assay (MAb capture enzyme-linked immunosorbent assay - MAb capture ELISA) was developed and evaluated for its application in the detection of circulating antibodies to *Paragonimus heterotremus* in infected human sera. Mouse IgG monoclonal antibodies derived from a hybridoma clone 10F2 which reacted to 31.5 and 22 kDa components of the excretory-secretory (ES) antigen of the parasite were used to sensitize a microtiter plate; the bound antibodies captured their respective epitopes in the subsequent added ES antigen. Individual sera of patients with either parasitologically confirmed Paragonimus heterotremus, other parasitic infections or pulmonary tuberculosis, and parasite-free healthy controls were incubated to the appropriate antigen containing wells. Finally, the bound human antibodies were detected by the anti-human immunoglobulin-horseradish peroxidase conjugate and substrate. The Mab capture ELISA had 100%, 97%, 91.6% and 100% diagnostic sensitivity, diagnostic specificity, and positive and negative predictive values, respectively, when tested on sera 33 patients with parasitologically confirmed Paragonimiasis heterotremus, 68 patients with other parasitic infections, eight patients with pulmonary tuberculosis and 29 normal controls. The MAb capture ELISA offers an alternative for sensitive and specific diagnosis of paragonimiasis by means of antibody detection without the requirement of purified, specific antigen of the parasites.

These are the few examples showing the underlying principles and procedures for monoclonal antibody production and screening against parasites. With ever increasing development, this area has immense potential to play in diagnosis, prevention and disease control in parasitic infections.

5.11 Experimental Immunization and ELISA for Diagnosis

ELISA for Detection of Antibodies Produced by Experimental Immunization

Diagnosis of parasitic infections is based on parasitological findings like parasites in blood smears or biopsies, egg or cysts in faecal samples etc. These provide unequivocal evidence of infection which is limited especially applied to low-level infections or to infections where the parasites occur in deeper tissues. The immunological diagnostic tests have indispensable role in detection of low level of infections as well as large scale screening of populations in endemic areas. These tests based on immunological principles of antigens-antibody reactions, used successfully are beset with considerable problems like provision of appropriate antigens and antibody, achievement of specificity and sensitivity etc. However, the advance in knowledge of parasitic antigens and ability to synthesize and clone suitable molecules, antibodies, have considerably solved these problems and improved application of the tests. It is now possible to select antibodies with high specificity and use these in a variety of sensitive tests, of which the *enzyme linked immunosorbant assay* (ELISA) and its derivatives are among the most widely used tests.

ELISA tests are used to measure antibody responses to defined antigens, the pattern and level of isotype response can be used to provide complete analysis of parasitic infection. ELISA can be used to detect the presence of parasitic antigens in body fluids or in faecal material which not only indicate certain parasitic infection but also can be adapted to provide quantitative data on the level of infection present. The efficiency of ELISA has been enhanced by availability of monoclonal antibody reagents which give a greater specificity and allow improved degree of quality control and giving better reproduciability. ELISA testing is also amenable to automation and also for rapid colorimetric applications like dipstick tests. The ELISA test is further modified as in immunoblotting test where parasitic antigens on prepared nitrocellulose papers are used in reactions with sera. This helps in recognition of pattern of infection in field conditions or detection of antibodies from unseparated parasitic antigens (dot-blot test).

Principle

The techniques of immunoassay using labelled reagents for detecting antigen and antibodies are exquisitely sensitive and extremely economical in the use of reagents. Immunologists first employ a number of techniques which are common to other biological sciences. For example, the method used to isolate antigens and antibodies are those from biochemistry and protein fractionation. Immunologists however, developed number of techniques based on antigen – antibody reactions. For example, any molecule that acts as an antigen can be identified in tissues by immunocytochemical methods. Very low concentrations of such molecules can be quantified by radioimmunoassay (RIA) and enzyme linked immunosorbant-Assay (ELISA). Solid phase assays for antibodies employing ligands travelled with radio isotopes or enzymes (ELISA) are probably the most widely used of all immunological assays because large numbers can be performed in a relatively short time.

ELISA involves an antigen and antibody linked to an enzyme. Degradation of the enzyme substrate by the enzyme-linked antigen-antibody complex can be measured photometrically and is proportional to the concentration of antigens or antibody in the test material. This technique has been applied with success for the detection of parasite or antibodies to parasite. The advantages of ELISA are

(a) it is more quantitative and not subjective (b) more tests can be performed in a given time and (c) it is more sensitive

Procedure of ELISA

The general protocol of ELISA Test is as follows : (Fig. 5.81).

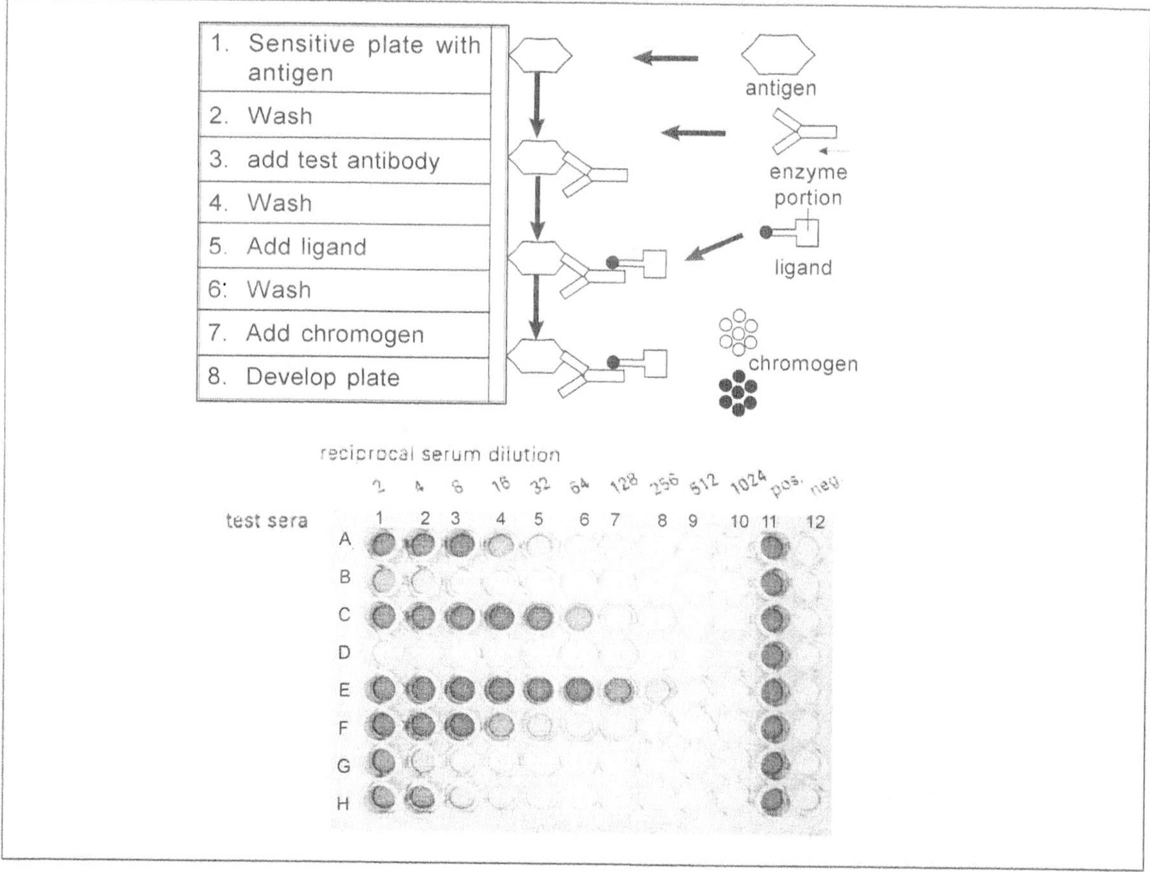

Fig. 5.81 Enzyme-linked immunosorbent assay (ELISA)

1. The antigen in saline is incubated on plastic plate and small quantities become absorbed onto the plastic surface.
2. Free antigen is washed away with wash buffer. (This is to remove excess protein, by which the plate may be blocked with excess of an irrelevant non-specific binding of protein)
3. Test antibody is added which binds to the antigen
4. Unbound protein are washed away.
5. The antibody is detected by labelled ligand. In this system, ligand is a molecule which can detect the antibody and is covalently coupled to an enzyme such as peroxidase, Alkaline phosphatase or cholinesterase. This binds the test antibody and after that free ligand is washed away.

6. The bound ligand is visualized by the addition of chromogen – a colourless substrate which acts on the enzyme portion of ligand to produce a coloured end product. The plate is developed.
7. The amount of test antibody is measured by assessing the amount of coloured end product by optical density scanning of the plate with help of microplate reader.
8. Specificity of test will range due to nature and relation of antibody to antigen like polyclonal antibodies, monoclonal antibodies and nature of antigen. The purification of antigen is accomplished by isolation from either hosts or vectors, with sufficient titer of particles and ability to produce secondary antibodies to primary antibodies.

More recently ELISA has been applied for serological diagnosis of a number of parasitic infections of medical and veterinary importance throughout the world. These tests have been shown to be more specific than those used previously.

The utilization of ELISA is illustrated with the following examples of experimental immunization.

1. *Use of Monoclonocal Antibodies and ELISA for Detection of Leishmania Infection*

Cutaneous leishmaniasis is a complex disease with a wide spectrum of clinical manifestations determined by causative species. This is essentially a disease caused by *Leishmania* sp. transmitted to humans mainly by sand fly. Routine diagnosis of leishmania is based on clinical presentation together with conformation of parasite by examination of direct smear, culture or biopsy. The reported sensitivity of these procedures is 50–70%. In a few patents, the parasites cannot be detected even when all conventional procedures are combined. Thus, the identification of *Leishmania* is based on extrinsic factors like clinical manifestation, epidemilaogical parameters and geographical distribution and established methods like isozyme analysis based on intrinsic features. However, use of monoclonal antibodies in diagnosis is much superior and rapid. In an experimental study, parasitic identification in 166 isolates were identified by their reactivities with a panel of specific monoclonal antibodies using ELISA, which were further confirmed by immuno-fluorescent antibody test, biochemical technique of isoenzyme electrophoresis and polymerase chain reaction.

ELISA Test :

Promastigotes were cultured in RPMI and then washed with Phosphalic buffer saline (PBS) by centrifugation. The pellet was resuspended in buffer (1 ml of lysis buffer per 8 ml of parasite culture) and maintained on ice. The antigen protein concentration was adjusted to 100 μg/ m/ with PBS and dispensed 50 μl/ well on polystyrene plate. The plates are incubated overnight at 4 ^{0}C, washed with wash buffer. The specific monoclonal antibodies (50 μ l / well) at concentration of 1000 PBS was added to each well and incubated at 37^0 C for 2h. The plates washed with wash buffer. Fifty μl of rabbit anti–mouse Ig G conjugated to horse radish peroxidase diluted to 1/20,000 in PBS added to each well. Incubation was carried out for one hour at 37^0 C and plates were washed with wash buffer. The substrate containing 100 μl of orthophenylenediamine was added to each well, left at room temperature for1/h and absorbance, was read on plate reader at wave length of 590 nm.

2. *Detection of Antibodies in Experimental Amoebiasis by ELISA*

In humans, the protozoan, Entamoeba histolytica infection results in intestinal amoebiasis. Experimental amoebiasis can be produced in mice. The mice were inoculated with monoxemic E. histolytic culture. Briefly, mice underwent laprotomy and then abut 5×10^5 trophozoites in 0.15 ml culture medium were inoculated intracaecally. Control mice were inoculated in the same manner with culture medium contains *E. coli* but free of *E. histolytica*. Mice were sacrificed at regular intervals and positive diagnosis for caecal infection was done.

ELISA Test :

The infected and non–infected control mice were bled and antibody titres were determined by ELISA. The 96 well plates were coated with 5 µg amoebic protein per well in 100 µL carbonate and bicarbonate buffer and incubated at 5^0 C. The plates were washed with PBS–Tween and blocked with BSA. The plates were incubated with 100 µl of serum samples at a dilution of 1 : 100 for one hour. Plates were thoroughly washed and 100 µL of peroxidase labelled antibodies of goat anti mouse Ig M, Ig G or Ig A were added to each well. Colour was developed by adding to each well O – phenylenediamine and H_2O_2 in 0.1 M citrate butter. The reaction was stopped by adding 100 µl of 1M H_2SO_5 to each well. The plates were then read at 495 nm on ELISA processor and the expected result are shown in Table 5.14 for illustration. The infected animals showed high Ig A, Ig M and Ig G compared to non–infected controls.

Table 5.14 *Determination of anti-amoebic antibody response in the sera of infected mice by amoebiasis*

Days after infection	Mean OD value at 595 nm		
	IgM (SD)	**IgG (SD)**	**IgA (SD)**
5	1.66 (0.07)	0.18 (0.06)	0.08 (0.03)
10	1.74 (0.06)	0.55 (0.25)	0.21 (0.11)
15	1.23 (0.29)	0.85 (0.23)	0.15 (0.07)
20	1.63 (0.02)	1.67 (0.36)	0.21 (0.03)
25	1.22 (0.13)	1.68 (0.41)	0.55 (0.23)
30	1.66 (0.17)	2.53 (0.42)	1.39 (0.08)
50	1.52 (0.15)	2.58 (0.43)	1.28 (0.33)
50	1.55 (0.32)	2.09 (0.31)	1.51 (0.05)
60	1.55 (0.31)	2.53 (0.46)	1.50 (0.07)
Non- infected controls	0.17 (0.03)	0.15 (0.05)	0.12 (0.04)

3. *Hypoderma Infection of Cattle and Determination by ELISA*

The first description of the use of ELISA for detection of antibodies in cattle affected by Hypoderma infestation and subsequent studies demonstrated that ELISA was an extremely useful tool for the detection of antibodies in serum (Fig. 5.82).

The antigens prepared from first instar Hypoderma larvae and most commonly used for ELISA testing. The secretary antigens may be prepared by incubation of Hypoderma larva in culture media at 37^{0}C and purified. Further, the recombinant antigens have been cloned from cDNA of hypodermis and produced from a gt11 library expressed in E . Coli host cells are used in diagnostic ELISA, with 100 % sensitivity and 97 % specificity.

Likewise, ELISA can also be used in specific cases like serodiagnosis of onchocerciasis, antibody detection of Babesia, detection of antibodies of T.cruze in chagas disease, antibodies to Schistosoma, Trichinella etc. and many other parasitic infections.

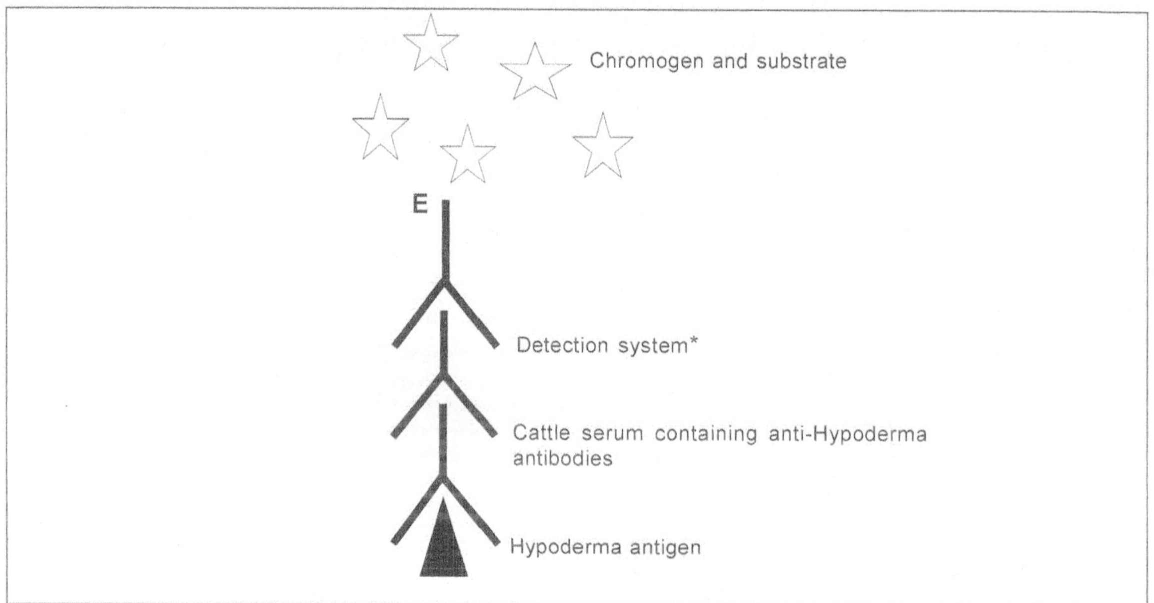

Fig. 5.82 Diagramatic representation of the standard ELISA for hypodermosis,

*either an anti-bovine peroxidase conjugated antibody or a two step system using rabbit anti-bovine immunoglobulin followed by a goat anti-rabbit peroxidase conjugated antibody.
If all components are in place a colour reaction results and can be seen or optical density calculated spectrophotometrically.

Essay Type Questions

1. Enumerate the various kinds of protozoan parasites with suitable diagrams and diseases they cause to man and animals.
2. Give an overveiw of helminth parasites with an emphasis on their pathological importance.
3. Descirbe briefly the biology of any four nematode parasites of man.
4. Write an account on the protozoan diseases of man and animals.
5. Give an account on helminth diseases of man and animals caused by Helminth parasites.
6. Discuss the importance of parasitic arthropods to man and economically important animals.
7. Give an account of parasitic arthropods.
8. Describe the parasitic disease of Fin fish and Shell fish suggesting measures of disease control.
9. Discuss about the phytonemodes and their importance as parasites of crops.
10. Discuss about the pathological effects of parasites in Host-parasite interactions.
11. Explain the process of Immune response with suitable examples.
12. Discuss about the internal defense mechanisms of invertebrates.
13. Describe the role of cellular systems involved in immunity.
14. Explain the mechanism of cell mediated immunity and humoral immunity.
15. Illustrate the role of complement system in immunity.
16. What is Opsonization ? Illustrate the mechanisms of opsonization and its importance.
17. Describe the immunological methods for detection of diseases in animals.
18. Explain how the monoclonal antibodies are produced against parasites with examples.
19. What are monoclonal antibodies ? How are they used in detection of parasitic diseases ?
20. Illustrate application of monoclonal antibodies in the screening of parasites with any two examples.
21. Describe the principle and procedure of ELISA technique.
22. Illustrate the application of ELISA technique for detection of antibodies produced in experimental immunization.

Short Answer Type Questions

1. Write the brief notes on biology and parasitological significance of following in 200 words.

   1. Entamoeba
   2. Plasmodium
   3. Trypanosoma
   4. Sarcocystis
   5. Giardia
   6. Toxoplasma
   7. Taenia
   8. Echinococus
   9. Dipylidum
   10. Fasciola
   11. Paragonimus
   12. Schistosoma
   13. Ancylostoma
   14. Ascaris
   15. Dracunculus
   16. Wucheraria
   17. Trichinella
   18. Blood sucking flies
   19. Mosquitoes
   20. Strongyloides
   21. Myiasis flies
   22. Fleas
   23. Lice
   24. Mites
   25. Ticks
   26. Meloidogyne
   27. Ditylenchus

2. Describe briefly the following diseases and the control measures in 100 words.

   1. Malaria
   2. Amoebiasis
   3. Sleeping sickness
   4. Leishmaniasis
   5. Sarcocystiosis
   6. Ascariasis
   7. Ancylostomiasis
   8. Elephantiasis
   9. Hydatid disease
   10. Taeniasis
   11. Shistosomiasis
   12. Whirling diseases
   13. Icthyopthioriasis
   14. Argulosis
   15. Lernaeosis

3. Write briefly about
   (i) Pathological mechanisms in parasitism
   (ii) Pathological effects of parasites

4. Write brief notes on

   1. Immune response
   2. Phagocytosis
   3. Antigens
   4. Antibodies
   5. Complement
   6. Lymphocytes
   7. Antigen-antibody interaction
   8. Teritiary interactions
   9. Primary and challenge infection
   10. Monoclonal antibodies
   11. ELISA

5. Write brief notes on any four of following.

   1. Schistosoma
   2. Trypanosomiasis
   3. Phytonematodes
   4. Monoclonal antibodies
   5. Parasitic arthropods.

6. Explain any two of the following with illustrations.

   1. Monoclonal antibodies in screening of parasites.
   2. Oposonization in immunity
   3. Parasitic diseases of fishes and shell-fish.

6 Introduction to Genetics and Molecular Biology

6.1 Genetics – The Science of Life

The science of genetics deals essentially with the study of heredity and variation. It has passed through several major phases of development and each stage has contributed, directly or indirectly, to some of our present insights into the nature of genetic material, a gene. In this chapter, we will briefly examine major insights into genetics and trace the roots of molecular biology.

6.1.1 Mendelian View of the Biological World

Man has been considered as unique among all life forms, since he alone has developed complicated languages and processing capacity that let meaningful and complex interplay of ideas and emotions. In a fraction of time, compared to the evolution of life (3.5×10^{10} years ago), man has evolved from apes, made possible the emergence of great civilizations, transformed the world's environment in ways quite inconceivable for any other forms of life. Biologists and philosophers have always thought that there is something special that differentiates man from rest of the organisms. This notion has found expression in man's religious beliefs, by which he attempts to find an origin for his existence and in that endeavor, self-limiting but workable rules for his own conduct of life were provided. It is quite natural to think that man was not always present but descended at a fixed moment through evolutionary processes, perhaps the same way as rest of the life forms.

This belief was first seriously questioned just more than 150 years ago, when Charles Darwin and Alfred Russell proposed their theory of organic evolution based on natural selection of the most fit. In nut shell their theory is that the various forms of life are not constant but are continuously giving rise to slightly modified versions of animal and plant forms, some of which are adapted to survive and reproduce more effectively. When they proposed this theory they could not envisage the origins of this continuous or discontinuous variation but they did correctly perceived that these new modified traits must persist and passed on to the progeny, if such modifications were to form the basis of evolution

An immediate consequence of Darwinian concept is the realization of the fact that life evolved on this space-ship, Earth, some 2-3 billions years ago as a pro-genote in a simple form, possibly resembling bacteria (prokaryote), the simplest form of prokaryote that exists today. The very presence of such microbes suggests that the essence of living state is found in every minute organisms. Besides, the evolutionary theory further suggests that the same basic principles of life exist in all living forms.

6.1.2 The fundamental unit of Life is a Cell

The second great principle of nineteenth century biology is the cell theory, put forth by Schleiden and Schwann, independently in 1839 that all plants and animals are constructed by a small fundamental units of life called cells. All living forms are made of them and governed by the same basic principles. All cells are enclosed by a membrane, and contain an inner body the nucleus, which is also bound by a membrane, called nuclear membrane (Fig. 6.1). Most cells are capable of growing and dividing, to give daughter cells with the same information content.

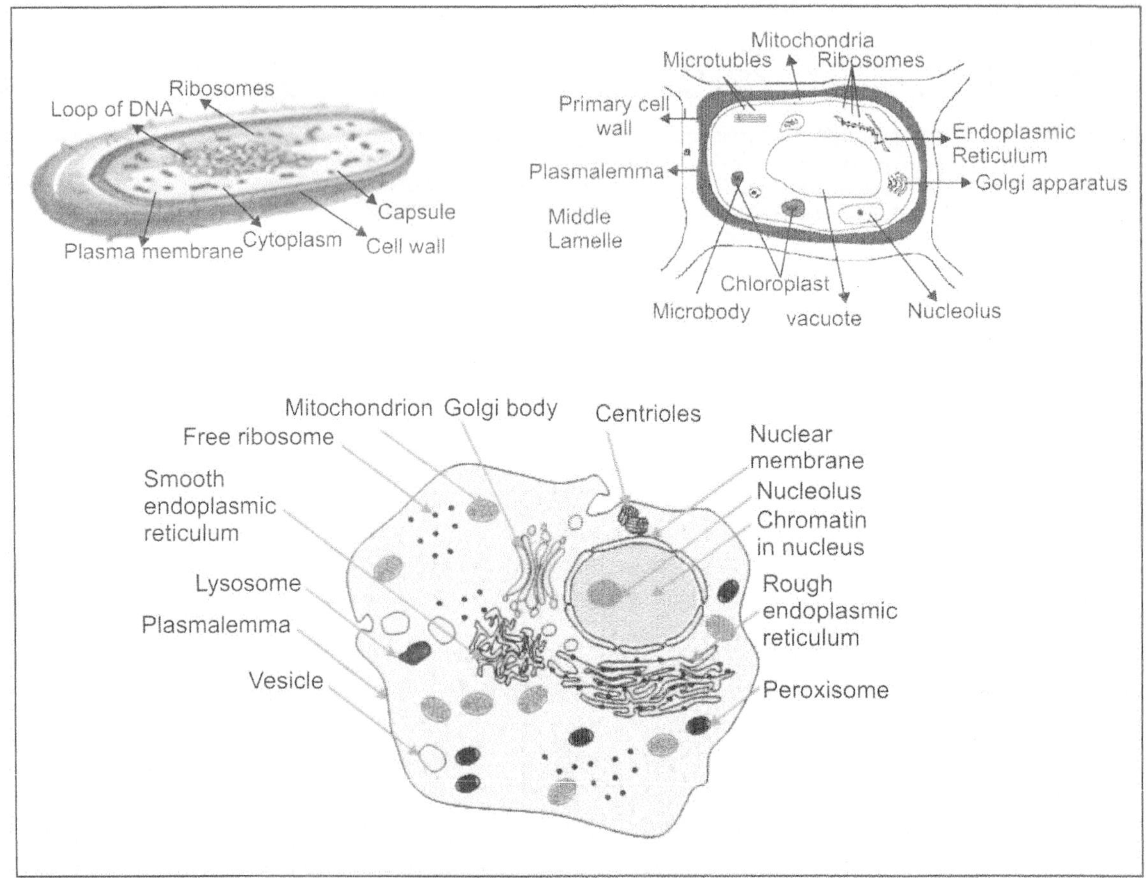

Fig. 6.1 Cell as a primary unit of life: Typical Bacterial, Plant and Animal cell.

6.1.3 Universality of Cell theory

Although the cell theory developed from observations on eukaryotes, it also holds good to prokaryotes and other microbes like protozoa, etc., Each bacterium or protozoa is a single celled organisms, whose division normally produces new daughter cells identical to parent from which it soon separates. In eukaryotes, on the other hand, the daughter cells not only remain together, but also differentiates into radical cell types to form tissues and organs such as nerve and muscle tissues, each made up of component cells, while maintaining the genome integrity (Chromosome complement) of the zygote. Furthermore, the zygote itself arises by the union of highly differentiated specialized cells, sperm and egg and a new cycle of differentiation takes place to develop organism. Thus the most complex organism like man (*Homo sapiens*) is built by a large number of cells ($5x10^{12}$), all from a single zygote cell. The fertilized egg comes to possess all the information necessary for the growth, differentiation and development of an adult plant or animal. The message is that the living state *per se* does not demand complicated interaction that occurs in complex organisms. The essential property of life can be found in single growing cells.

6.2 Mendelian Genetics

6.2.1 Mendel's perception of the unit of inheritance

In 1854, at about the same time Darwin was penning his thoughts on the organic evolution, Gregor Mendel (1822-1884) was doing his experiments on garden pea to investigate the basis of heredity. He showed convincingly to the world that simple mathematical treatment (algebra) could describe the nature of inheritance. Mendel clearly established that the two essential attributes of the basic unit of inheritance is that 1. the gene is transmissible to offspring and 2) the gene exerts its influence on the organism in which it occurs, although the term gene was not coined until 1911 by Johnnsen. Today it is well known that gene is a smallest basic unit of an organism that is capable of transmitting and expressing genetic information from generation to generation.

Mendel studied the hereditary characteristics of a garden pea (*Pisum sativum*). He chose seven contrasting characters of the peas for the genetic analysis (Table.1) and looked for simple mathematical relationship (quantification of data) to describe the behaviour of the characters in genetic crosses (hybrids and their progeny). For each factor (now called as gene) he studied two different alternatively expressed forms of the gene, which we call as alleles or allelic forms. In modern jargon alleles are alternative forms of a gene. The different forms are homologues having a nearly identical gene product and acting equivalent as functional entities.

Table 6.1 *Mendel's group of seven genes of the* Pisum sativum *(Garden pea)*

Site of effect of the gene on the character	Phenotype generated by the alternative (allelic) forms of the gene
Seed Shape	*Round*, wrinkled
Colour of seed	*Yellow*, green
Colour of seed coat	White, *grey*
Shape of seedpod	*Loosely fitted* seed Tightly fitted
Pod colour	*Green*, yellow
Flower position	*Bunched*, distributed over stem
Stem length	*Tall* (6-7ft) short (3/4 – 1 ½) unit

* The dominant trait is italicized, its alternative recessive trait is indicated in normal font.

Mendel had no clue about either the physical or chemical nature of the gene. His conceptualization of gene is only through its influence or expression on the organism containing it. The organism's external manifestation due to gene is called the *Phenotype*. The genetic information contained in a gene of the organism for the phenotype is called *genotype*. Mendel thus needed to figure out the genotype from the morphological appearance and subsequent genetic behaviour of each plant. The gene per se remained a mathematical abstraction to Mendel. In his observation, transition forms between different alleles were never noticed. Each gene could exists two distinct allelic forms but each allelic unit was indivisible. For instance, he never obtained an allele for 3ft. high plant from a cross between short (1ft. high) and tall plants (6ft. high).

6.2.2 Basic tenets of inheritance : Mendel's Law of segregation of Alleles

Mendel's contribution to understanding of basis of inheritance began with the idea that in peas, genes occur in pairs. It should be remembered of that peas are diploid organisms. In the first instance of Mendel's mathematical analysis of behaviour of genes governing characters, he analyzed what happened to pair of alleles of a single gene in a *monohybrid cross* when he followed their distribution to offspring for three consecutive generations, F_1, F_2 and F_3 (first, second and third filial). The outline of the experiment is diagrammed in *Punnet square* mode of analysis of gametes and progeny (Fig. 6.2).

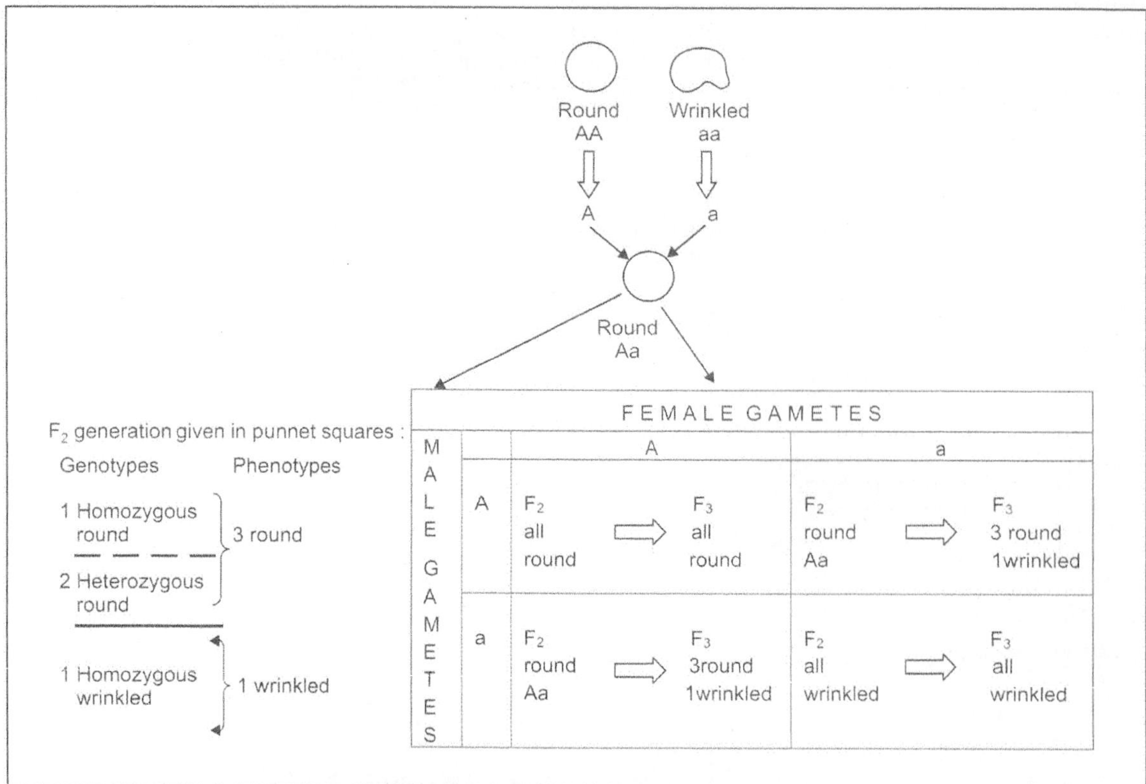

Fig. 6.2 A monohybrid cross showing the law of segregation of Mendel.

Mendel made a cross between homozygous pure breeding (having identical alleles) plant producing round pea seeds (two copies of allele A, genotype AA) with another plant producing wrinkled seeds (two copies of allele 'a', genotype 'aa') by depositing pollen of one plant to the ovule of another plant. Since each plant is true breeding by virtue of homozygosity, forms only one type of gametes in sperms (pollen) and eggs. In a monohybrid cross all of the seeds resulting in F_1 (first filial) generation were round. These were *heterozygous* (Aa) for seed shape, having a round allele (A) from one parent and wrinkled allele (a) from the other parent. Since the allele round expressed, it is called dominant and the other allele, 'a' is masked, it is called *recessive*.

Peas are autogamous plants, left to their own nature, self fertilize. That is to say that plant reproduce themselves by their own pollen fertilizing their own ovules. Mendel thus obtained a second filial generation of pea plants by sowing the seeds of F_1 (Monohybrids) and letting the plants to self fertilize. In each of the F_1 plants, *meiosis* takes place. Meiosis is a kind of cell division limited to the germline and at the end of the division haploid gametes (sperms and eggs) are formed. Thus, the two alleles present in a diploid individual would be separated and pass onto products (gametes) of meiosis.

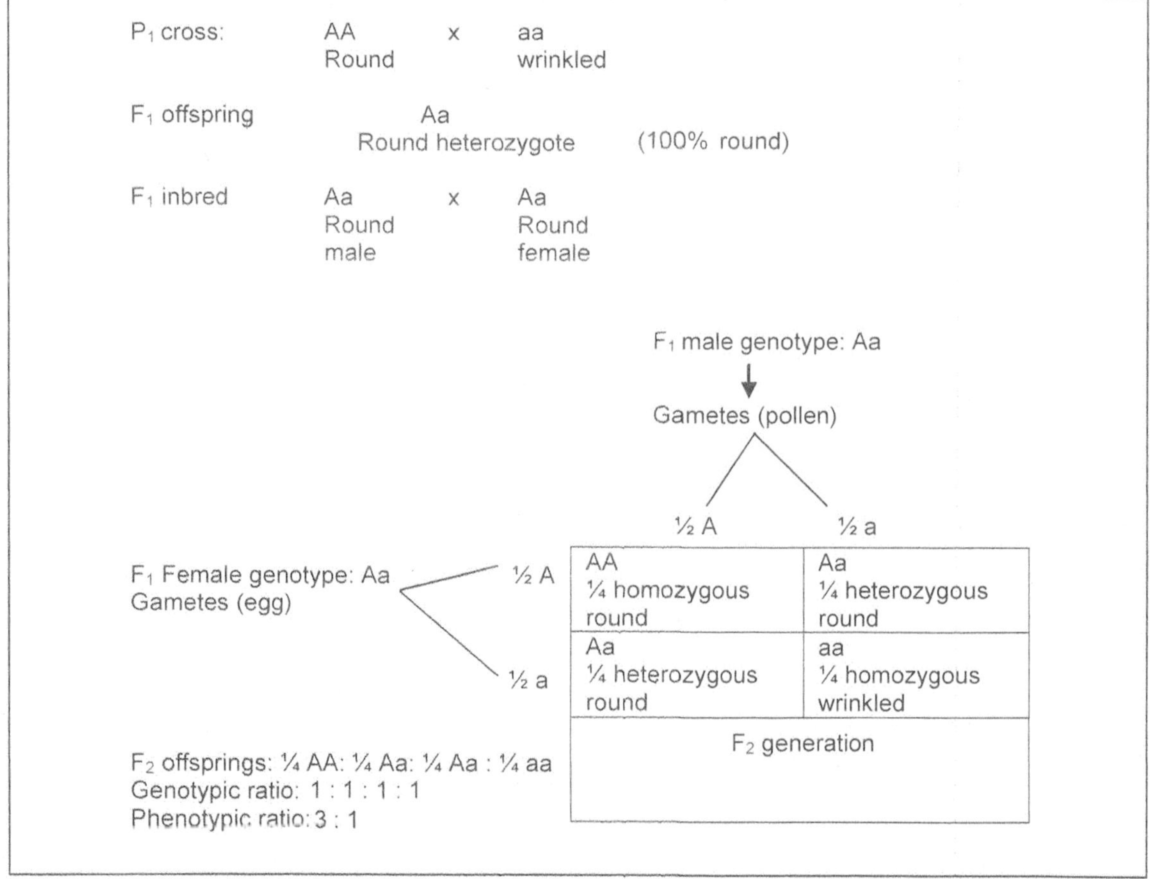

Fig. 6.3(a) Genotypes and phenotypic expectations in F_2.

In this case A and a separate to from different gametes. It may be noted that in the Punnet square (checker board), in the case of either male (?) or female (?) gametes, the probability of occurrence of A would be one half and a one half (1:1 ratio). The F_2 generation results from the fusion of these gametes. Hence, in F_2 generation the resultant progeny would be the product of respective probabilities of A and a. Infact Mendel has precisely found three round seeds for every one-wrinkled seeds in his experiments. The phenotypic (3:1) ratio of segregation of Round and wrinkled seeds in F_2 and F_3 along with genotypic ratio is presented in the Fig. 6.3(a).

Mendel further planted all the F_2 generation seed and raised F_3 plants. The results of each $F_2 \times F_2$ crosses to obtain F_3 plants are shown in Fig. 6.3(b). All the wrinkled seeds produce only wrinkled seeds. Of the F_2 plants, from round seed, one third of them produce only round seeds. The remaining F_3 plants (two third of them) like F_2 that is three round seed to one wrinkled seed. The conclusion that Mendel made is that one fourth of the F_2 were purely homozygous wrinkled (aa), one fourth of them purely homozygous round (AA), but rest of the half of the F_2 appeared to be round seeded externally, but still possessed within them the allele to be wrinkled and hence heterozygous (*Aa*). These masked wrinkled alleles could be uncovered by self-fertilizing the F_2 plants to form another generation (F_3). For a monohybrid cross of two allelic forms of a gene, the genotypic ratio in F_2 was thus found to be 1:2:1; that is one half of each parental types (homozygous) and rest of the half (two individuals) would be still having hybrid constitution (Aa) having two alternative alleles together.

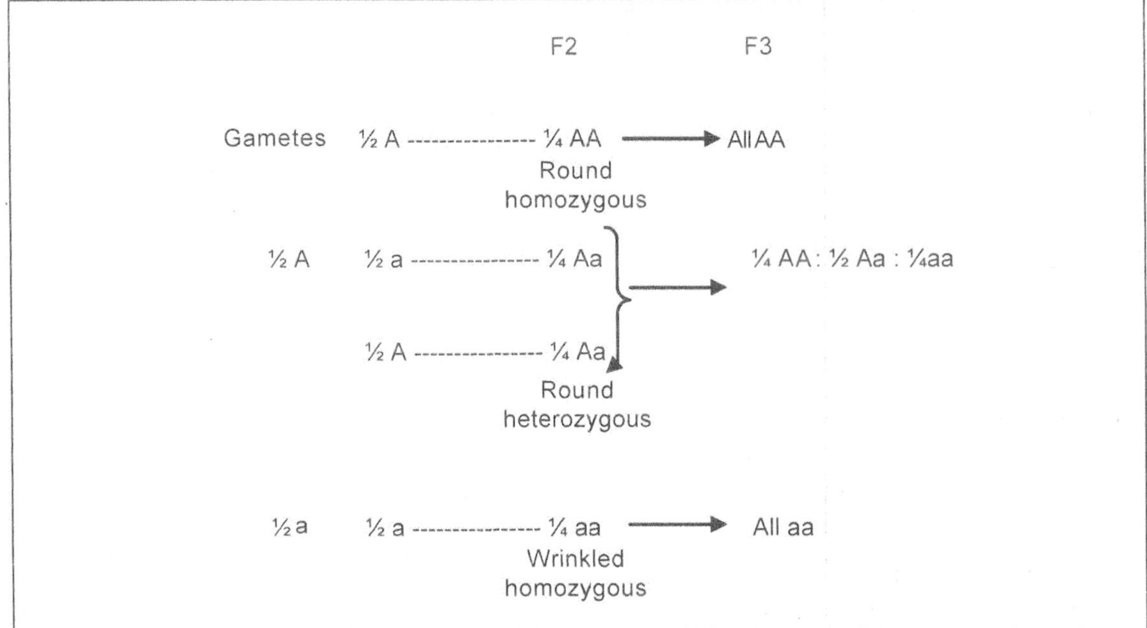

Fig. 6.3(b) Results of $F_2 \times F_2$ crosses to obtain F_3.

This typical ratio was expected if each of the resultant progeny got one allele of this gene from the pollen (male) parent and from the ovule (female parent) and the selection of gametes for fusion is a random event. Mendel with his background of studying physical science noted quite correctly, which his predecessors (biologists) failed to note, that the genotypic ratio would perfectly fit to bionomial expression

$A^2 + 2Aa + a^2$, where A^2 symbolizes the homozygous dominant round seeded plant (AA), Aa symbolizes the round seeded heterozygous seed plant still containing the allele for wrinkled seed (Aa), a^2 symbolizes the homozygous wrinkled seed plant (aa)

From Monohybrid experiments involving the seven characters, Mendel deduced the ***law of Segregation of alleles (Genes)): members of a pair of alleles separate from each other in genetic crosses***

We now are aware that the separation of allelic pairs occur regularly, as a rule, in all diploid organisms during the gametogenesis. Nature invented a mechanism with which this is accomplished by a cell division called *meiosis*, where the genome (chromosome complement) becomes reduced from diploid (2n) to haploid (n) in order to maintain the species integrity, when zygotes are formed with the union of gametes during the reproduction. Mendel envisaged the need for separation to take place during germ cell formation (spermatogenesis and oogenesis), even though nothing is known about chromosomes, meiotic process and significance of the reduction of chromosome number to a haploid state in the gametes. Exception to Mendel's Law of Segregation of alleles seldom occurs, and very rare.

6.2.3 Mendel's Law of independent assortment of Genes

Mendel's approach to genetic analysis to gain insight into the inheritance process was very methodical and meticulous. In the second phase, he considered what happened when a dihybrid cross involving two genes each with an alternative form (heterozygous for two genes). Mendel chose two plants, one with round seeded with yellow coloured endosperm and the second one with a homozygous having wrinkled seed and green endosperm, and obtained a dihybrid. The results are diagrammed in Fig. 6.4. Round and wrinkled are represented by symbols, 'A' for round seeded and 'a', for wrinkle seeded, and yellow and green by 'B' and 'b', respectively. The actual data obtained by Mendel as well as the resultant F_2 and F_3 progeny is shown in Table 6.2.

The genotypes of the F_2 individual plants have been deciphered by Mendel based on their phenotype and genetic behaviour in F_2 and F_3 generations. For instance, the data in the first line indicates that in the F_2 generation there are 38 plants that are round with yellow colour seed which when self fertilized to raise F_3 generation, all of them produce true breeding round yellow coloured seed. That implies that these are homozygous for another gene Yellow colour endosperm (BB not Bb) since no green or wrinkled seed are produced. In the second line 65 plants also gave round and yellow seeds in the F_2. However in F_3, seeds were all round implying that F_2 was round homozygous. As per the colour, half of them yellow and rest of the half were green, which implies that F_2 was heterozygous for yellow and green i.e. Bb. The logic of inference is same for rest of them.

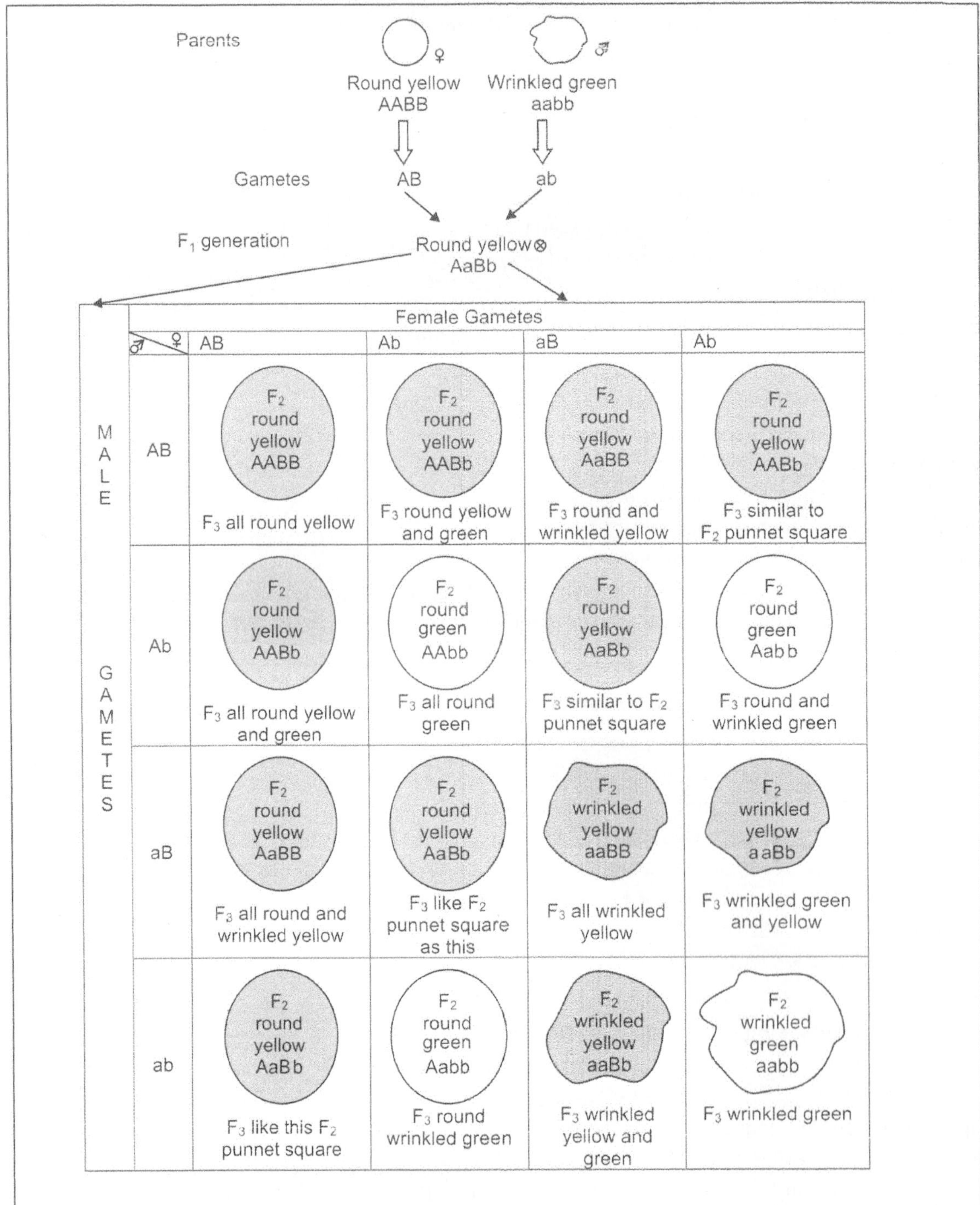

Fig. 6.4 A dihybrid cross showing Mendel's law of independent assortment.

Table 6.2 *Mendel's data an a dihybrid cross.*

Phenotype ratio	No. of plants observed	F2 seed phenotype	Genpotypes inferred	No. of F2 with genotypes	Seed phenotypes found among F3 after slefing
9	315	Round yellow	AABB	38	Round yellow
			AABb	65	Round yellow, green
			AaBB	60	Round, wrinkled, yellow
			AaBb	138	Round, wrinkled, yelloow and green
3	101	Wrkinled yellow	aaBB	28	Wrinkled, yellow
			aaBb	68	Wrinkled, yellow and green
3	108	Round green	AAbb	35	Round green
			Aabb	67	Round and wrinkled
1	32	Wrkinled	aabb	32	Wrkinkled green

The F_2 phenotypic ratio of 9:3:3:1 was arrived at by dividing all of the numbers observed in each class by the number of progeny of the smallest class. The ratio can also be obtained by counting each phenotype class in the punnet square (Fig. 6.4).

The 9:3:3:1, ratio can be fragmented as Mendel has demonstrated to one part each possible pure type (AABB, aaBB, AAbb, aabb), two part each homozygote for one gene but heterozygote for the other (AABb, aaBb, AaBB, Aabb) and four parts of the mixture at both of the genes (AaBb). One can verify this analysis by dividing the number in fifth column of the Table by 32, the least number of progeny represented by aabb. The results stem from the random combination of (A^2 + 2Aa + a^2) and (B^2 + 2Bb + b^2) to generate $A^2B^2 + A^2b^2 + a^2B^2 + a^2b^2 + 2A^2Bb + 2a^2Bb + 2AaB^2 + 2Aab^2 + 4AaBb$. Such a random combination could be formed only if each gene acted independently during a cross without being influenced in the sighted by what another gene was acting in the same dihybrid cross. As Mendel described "The relation of each pair of different characters in hybrid union is independent of the other differences in the two parental stocks". Roughly equal number of all possible combinations of the A and a alleles with the B and b allele will be found as has been assumed in generating the punnet square. Thus second basic tenet, the ***Mendel's law of independent assortment of genes*** was arrived at, which states that, ***different genes assort independently of each other in genetic cross.***

This generalization holds good for genes that are on separate chromosomes or situated far apart from each other if they are on the same chromosome.

6.2.4 Gist of Mendel's view of the Gene

- A gene is a basic unit of inheritance responsible for the transmission and expression of a hereditary character
- A gene exist in two alternative forms called alleles
- The two alleles of a gene can be present in an individual. When such situation arises, the dominant allele hides the recessive allele
- Genetic crosses revealed that each of the allele remain uncontaminated or unchanged in the hybrid

- When two or more genetic differences are tracked through a cross, it is obvious that the different genes do not influence each other or interact but instead are distributed to the offspring's completely and independently
- Random distribution of alleles into a gametes does occur as a rule

Elaboration of Mendel's view on the gene as basic unit of inheritance.

In the beginning of 20th century, three biologists, Hugo de Vries, Correns and Tschermark rediscovered Mendel's work, which remained latent and unappreciated for 34years. Once it is done, for next three to four decades, several elaborations were made on his theory and the science of genetics developed enormously as a modern subject. ***Multiple allelism*** phenomenon was discovered, basis for quantitative inheritance was established and it was demonstrated that genes can be mutated alerting their function or impaired to generate new alleles.

6.3 Physical Basis of Inheritance

6.3.1 Gene is a segment of Chromosome

Early 20th century, with rapid developments in Genetics, soon the scientists were addressing the problem of physical basis of hereditary. The perception that genes are carried on chromosomes arose in the same period as the one in which Mendel's ideas were rediscovered and extended to all other organisms, both animals and plants. The concept is a sequel to the grand convergence of ***cytology*** (the study of cells) and genetic analysis, leading to the development of another specialty in biology called ***Cytogenetics***. The first attempt to connect genetics and cytology was made early in the renaissance of Mendelian or ***transmission genetics***. Mendel had demonstrated that the hereditary material consists of unit factors (Genes) that segregate instead of bloods that mix (Paint pot theory). The understanding that genes are carried in chromosomes "physically" or a segment of chromosomes came from the initiative observation of nuclei and chromosomes at cell division, fertilization and subsequent development and differentiation of embryo (zygote). This problem faced by the then biologists was solved simultaneously by ***W.S. Sutton*** a graduate student at Columbia university (US) and by the eminent cytologist, ***"Theodor Boveri"***. Both reached the conclusion that the Mendel's factors (Genes) of inheritance are contained in the Chromosome.

6.3.2 Parallelism between Genes and chromosome behaviour

Sutton and Boveri based on their independent observations on the behaviour of chromosomes during meiosis, fertilization and differentiation and the genes have concluded that (1902) Mendelian genes must be carried on the chromosomes. The behaviour of chromosomes during cell division resemble very strikingly with the way the behaviour of genes as observed in genetic analysis through breeding experiments. One should understand clearly that they were sure of the existence of genes as a part of chromosome not because they knew the presence of genes through physical and chemical analysis (the way we know now) but because Mendel's Laws can be satisfactorily explained and understood only on the assumption that genes exist and are segment of chromosomes.

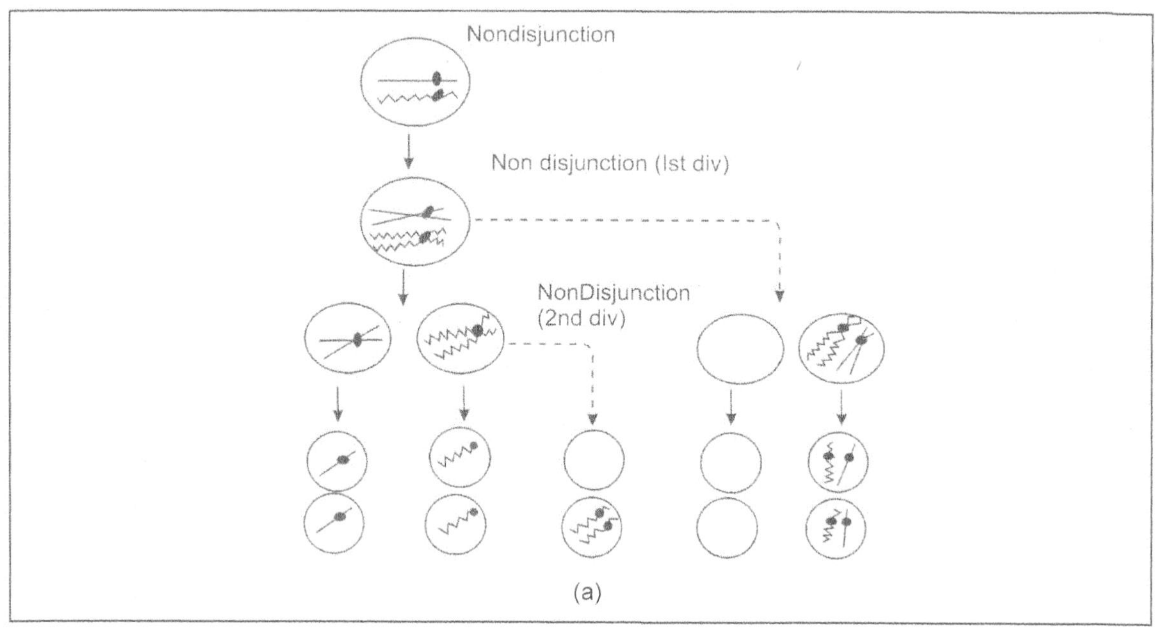

Fig. 6.6 (a) Non-disjunction.

The sex-determination in Drosophila is due the balance of genes present in X- chromosomes and autosomes. The constitution of normal females is XX and normal male is XY. In male Y chromosome is needed for fertility of sperms but its presence do not make necessarily a male. The differences between the constitutions of zygote is responsible for sex differentiation. The gene for eye colour is present on the non homologous (to Y) region of X chromosome. The dominant allele W^+ causes a red eye colour and its recessive allele W in homozygous condition makes the eye colour white. Males carry only one X chromosome, hence one of the allele (W^+ or W) where as females have two X chromosomes, they carry two alleles (W^+ W^+/W^+ W/WW). When a cross is made between the Red-eyed male and white eyed female, the resultant F_1would have white eyed male and Red-eyed female offsprings in 1:1 ratio. When reciprocal cross in made (female having Red eye allele and (W^+ W)male with white eye) the F_1 progeny was all Red, both male and female. This discrepancy in the reciprocal cross results is due to the fact that the gene is X-linked. In other words it is situated on X-chromosome. (See Fig. 6.6(b))

When non-disjunction takes place occasionally (1/2000) during meiosis the two X-chromosomes do not segregate and go to one gamete together, while the other gametes will not have any sex chromosome included in it. The resultant abnormal eggs (gametes) are called 'diplo-*X* and nullo-*X* eggs. When such eggs are fertilized by the normal male sperms (gametes) having either one 'X' chromosome or one Y chromosome both red eyed and white eyed females manifest instead of having only red eyed females. On cytological observation of rare white eyed female revealed that it contained non-disjoined X-chromosomes + Y chromosomes, triploid X female or XO female. Thus the abnormal behaviour of genes increase is correlated with abnormal behaviour of the X chromosomes, since the X chromosome carried the gene. All the result were consistent with the concept that these eye colour gene was a segment of X chromosomes (See Fig. 6.6(b)).

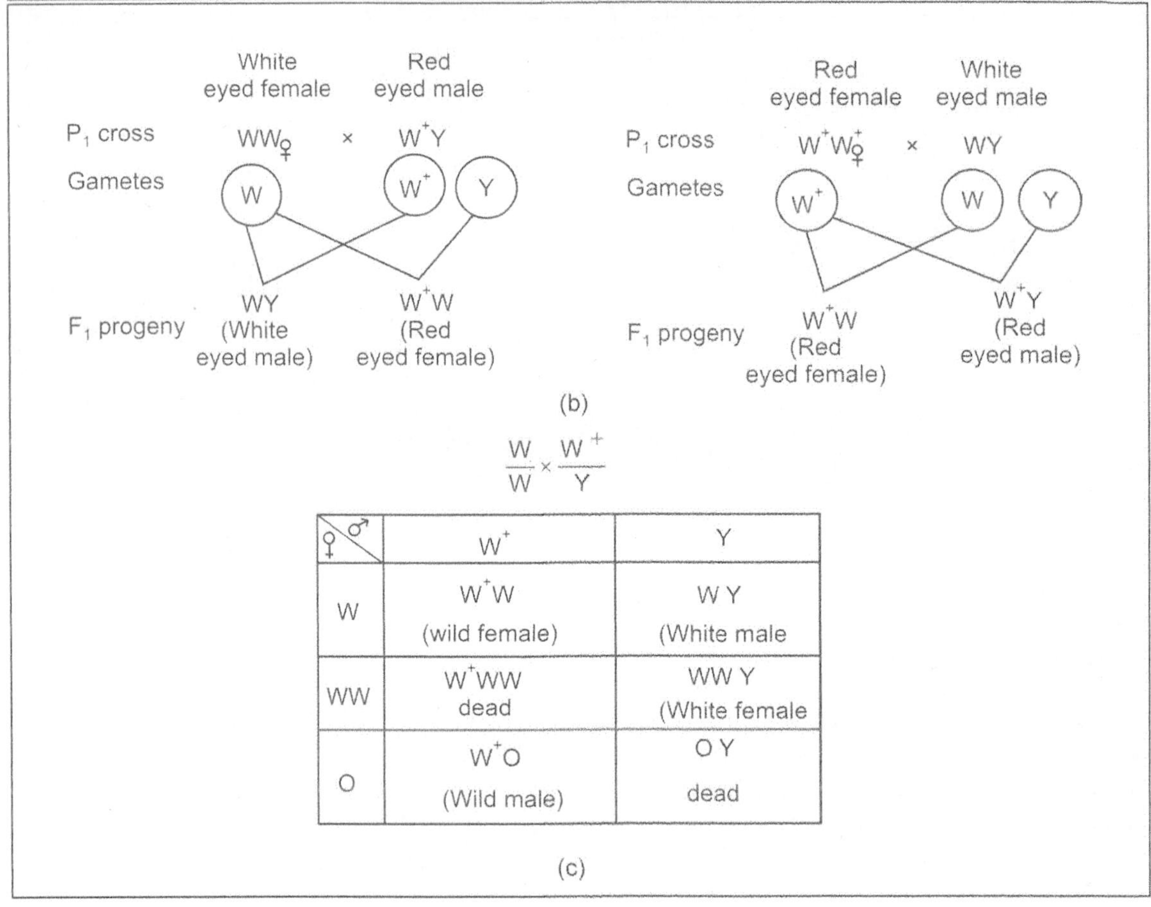

Fig. 6.6(b)&(c) Sex determination in drosophila.

6.3.4 Linkage and crossing over

Genetic analysis of many genes suggested that they are carried on chromosomes. All the genes of ***Drosophila*** formed four ***linkage groups***. Morgan and his students categorized these groups which assorted independently from rest of the groups. On the other hand when two genes within the same linkage group were being analyzed in a cross, quite often the genes did not assort independently, instead tend to remain together in the parental combinations. The fact that ***Drosophila*** has four types of chromosomes (n = 4) in their cells strengthened the inescapable idea that all the genes must be carried on these chromosomes to form independent linkage groups: each should correspond to one of four types of chromosomes

Geneticist soon became interested to see how the new allelic ***recombinations*** is accomplished in nature. It became clear that the recombination is achieved through the process of crossing over involving the exchange of segments of chromosomes by breaking and reassociation between nonsister chromatids of homologous chromosomes through chiasma formation. It was realized that farther apart two genes are on the chromosome, it was more likely that the two genes would be separated by recombination. Closely linked genes always tend to stay together while distantly placed genes would be separated quite frequently.

6.3.5 Cytogenetic detection of crossing over – proof that Genes are carried on Chromosomes

One of the most elegant experimental evidence for the correlation of gene with the chromosome segments was the cytological proof of crossing over by Harriet Creighton and Barbara McClintok, the genetic analysis of a specially designed cross of two corn strains was published in the year 1931. One of the corn strain was marked by a peculiar chromosome 9, possessing a heteropycnotic (stains dark) knob on one end and intercharged piece of another chromosomes (chromosome No. 8) attached to the other end. Besides they have used gene markers ('C/c Wx/wx'), and behaviour of the alleles C (colour of the kernel) and interchanged chromosome tag (chromosome No. 8) behaviour was correlated with behaviour of the allele wx (waxy kernel).

The experimental outline is presented graphically in Fig. 6.7. The genotypes indicated with land marks on chromosome shapes were crossed and the progeny was analyzed. Genotypes were inferred and any unknown allele was symbolized (-). The offsprings were classified into groups based on phenotypes: 1) Coloured aleurone with waxy kernel. The inferred genotype was C__ wx wx; 2) Coloured aleurone with non waxy kernel (C__ Wx__); 3) White coloured aleurone with non waxy kernel (cc Wx__); and 4) White aleurone and waxy kernel cc wx wx. Specific predictions were made about which of these phenotype classes should contain a chromosome with a knob or with an interchange all of these expectations were corroborated cytologically by examining the chromosomal constitution of offsprings. It has been established that the behaviour of two genes and two land marks on chromosomes correlated perfectly. The inference made was that genes are carried on chromosomes and when chromosomes exchange segments of chromosomes due to crossing over, alleles also long in regular predictable way.

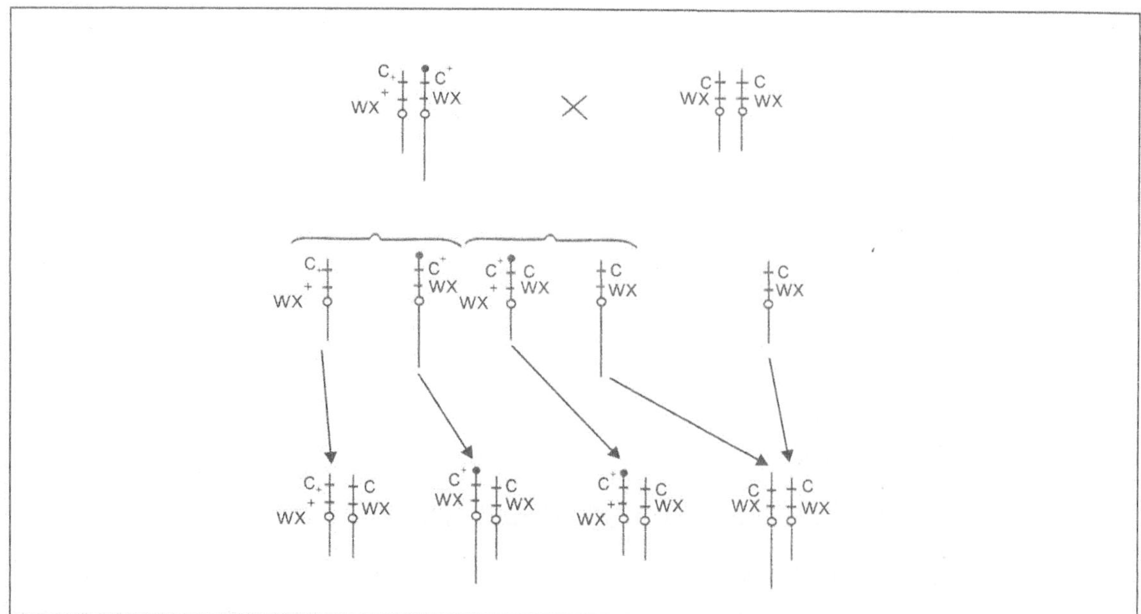

Fig. 6.7 Experimental evidence for the crossing over in maize

6.3.6 The bands in Salivary Gland Chromosome represents Genes

The discovery of polytene chromosomes in salivary glands of Drosophila and numerous bands in them (Painter, 1933) correlate with the mutations like duplications or deletions of genes in terms of the differences in the band number became an additional evidence for the generalization that gene are carried on the chromosomes. Many known genes were assigned to a region of one or few band in a specific chromosome. Besides, the loss or gain of bands due to chromosomes mutations like deletions or duplications and change of positions of arrangement of genes within a chromosome or between chromosomes reflected in the alteration in the band position or order (Fig. 6.8).

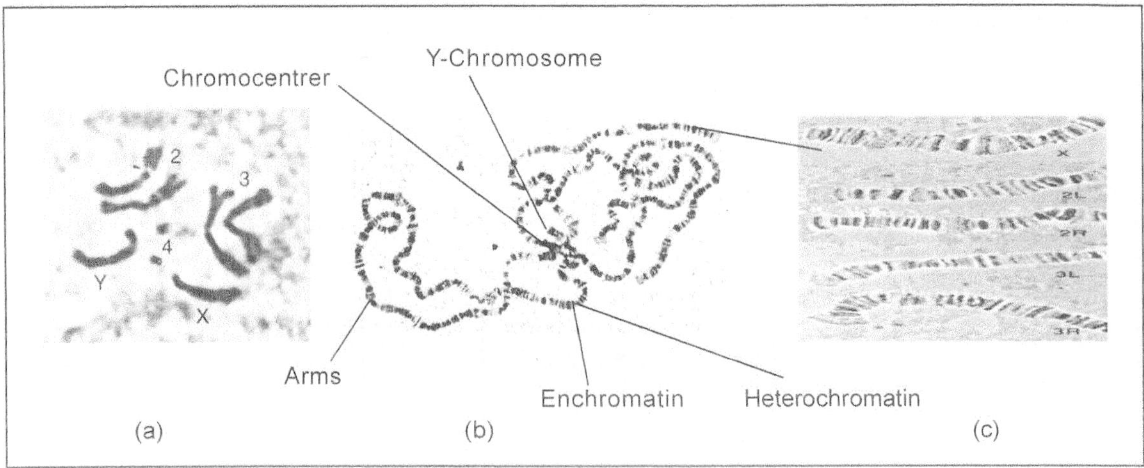

Fig. 6.8 Polytene chromosomes in Drosophila: (a) banding pattern in X, 2L,2R,3L,3R (b) arrangement of chromosome in a polytene chromosome (c) well spread polytene chromosome attached to chromocentre

Physically a gene is a segment of chromosome

- A gene is a unit factor responsible for transmission and expression of a hereditary trait.
- A gene may have a number of alleles (Multiple alleles) that arise from the wild type allele by mutation.
- Any diploid organism might contain at a given locus two such alleles.
- A dominant allele may mask expression of its alternative recessive allele.
- Allelic forms are constant not affected or charged by genetic crosses.

6.4 Gene from the Biochemical and functional point of view

6.4.1 Gene as a Nucleic Acid

As the physical basis was being established, many a biologists were addressing the problem of gene function; what gene does? And how it does? The next major revision in the concept of gene was concerned about gene function. Archibald Garrod (1902), while studying the ***metabolic disorders of***

man postulated that the phenotype ***alkaptonuria***, a condition in humans causing blackened urine, resulted from inactivity of an enzyme that, in normal individuals metabolizes the benzene ring of ***alkaptan***. Many eminent geneticist of the time like Sewal Wright, J.B.S. Haldane had opined that the Mendelian genes in expressing phenotypes acted by controlling the synthesis of enzymes. The enzymes are proteins and are vital to a cell metabolism and essentially all reactions in living cell (anabolic / catabolic) are catalyzed by them.

6.4.2 Chemistry of Gene action

A quantum forward leap in providing that genes control enzymes production resulted from the work of Beadle and Tatum (1941). It was shown by them that when haploid bread mold (a fungus), *Neurospora* was exposed to X-rays, *mutant* strain could be obtained. *The individual carrying altered genotype is called mutant. Mutation is a sudden heritable change in the gene.* Earlier, H.J. Muller in *Drosophila* (1927) and L.J. Stadler in plants have demonstrated that the artificial transmutations from one allelic state to another was possible by irradiating with physical rays (X -rays, α-rays, β-rays, UV etc.). Auerbach (1944) has shown that the chemicals also induce mutations. Beadle and Tatums work with *Neurospora* had shown that the wild type carryout normal functions, which a mutant could not do. The first three mutants they have isolated were unable to synthesize Vitamin, B_6, Vitamin B_1 and P-amino benzoic acid (PABA). The haploid nature of *Neurospora* was useful in unfolding the expression of one allele of each gene whenever a mutation occurs. These *nutritional mutants* requiring their supplementation for normal growth in culture medium are called *auxotrophic mutation*, while those have normal capability to produce them from the minimal medium (carbon as energy source) are called *prototrophs*.

The concept of a one-to-one relationship between mutations and blocks in metabolic pathways was extended further, in the biosynthesis of arginine in *Neurospora* (Srb and Horowitz, 1994). A mutant defective for arginine production thus has a predictable phenotype. Without supplementation of arginine, the fungus cannot grow. There are many steps in the synthesis of arginine from precursor molecules. Each step is catalyzed by an enzyme. If each enzyme is determined by a separate gene, it was predicted (Srb and Horowitz) that diverse classes of arginine requiring mutants should exist and each group of mutants should have a block in a different step. The blocking of metabolite is due to the inactive or altered or absence of enzyme conditioned by gene and hence unable to catalyze the normal reaction. Srb and Horowitz successfully isolated mutants for three of the known steps in the synthesis of arginine (Fig. 6.9(a)). Thus, correlating genes and enzymes in terms of their functions. It was elaborated subsequently by *One gene-one enzyme* hypothesis which states that ***A large class of genes exist in which each gene controls the synthesis of or the activity of but a single enzyme (Horowitz)***.

Various other genes were immediately discovered to be associated with enzymes, especially notable was the gene for the enzyme tryptophan synthetase whose genetics was quite extensively studied in several organisms (David Bonner, Yanofsky, Lacy, Suyama and others.)

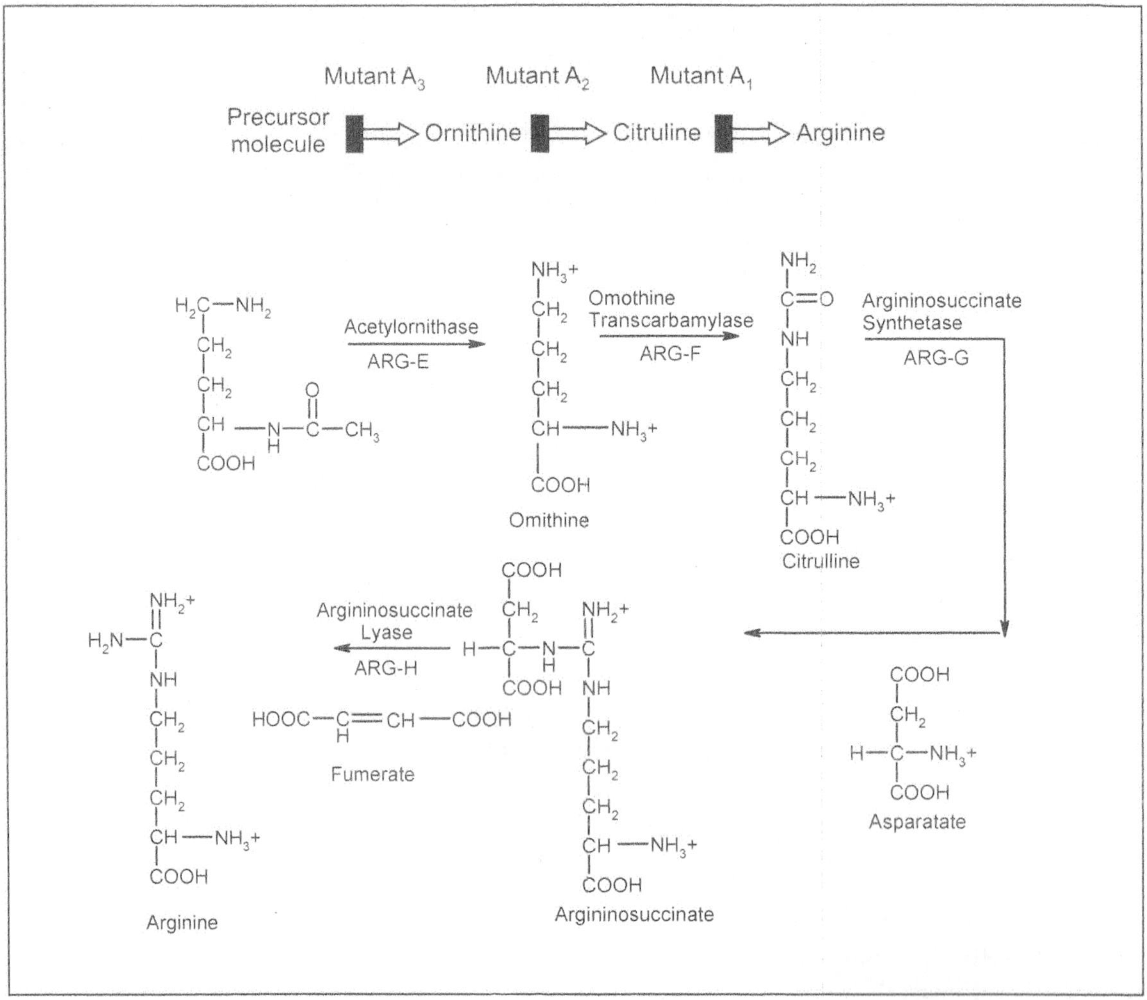

Fig. 6.9(a&b) (a) Arginine biosynthetic pathway (b) Mutant blocks in the three mutants of Neurospora failing to synthesize Arginine.

6.4.3 Genetic maps and changes in protein: Deciphering tryptophan biosynthetic pathway.

Many of the reaction that occur in cells have to do with synthesis of building blocks needed to make nucleic acids (DNA & RNA) and proteins from nutrient sources. Such chemical reactions includes reduction of the precursor compounds, oxidation of the part of the compounds, transferring carbons onto the compound, introducing nitrogen into specific locations and transferring phosphates and ultimately to form the building blocks. Analogous to chemists strategy to set reactions in the synthetic laboratory to achieve the goal, cells transform one particular sort of molecule into another sort, goes through a strategy of a set of reactions termed biochemically as ***biosynthetic pathways.***

Genetic analysis has been of paramount importance in deciphering biosynthetic pathways. Yanofsky was one of the pioneers in this regard. He had demonstrated in *E. coli,* that mutation sites within a gene for tryptophan synthetase had an exact point by point correspondence to the site of amino acid changes in the enzyme. The experimental evidence was provided by the comparison of a map of various mutational sites having a close similarity to the genetic map of the same mutation site. Thus, the ***co linearity principle that the sequence in gene is co linear to the sequence in amino acids in the protein (enzyme)*** was established, paving the way for understanding *genetic code* and the process of information transfer from gene to protein. Horowitz idea was that the gene contained a linear sequence of information (genetic code) needed for the synthesis of another linear macromolecule a protein. Yanofsky's extensive genetic studies and data allowed geneticist to gain an insight into the process by which new alleles of a gene are created by normal recombination. This is a novel evidence to show, how Mendel's Law of segregation of alleles can be violated and recombination within a gene could create a new allele.

6.4.4 One Gene One Polypeptide Chain

Further understanding of relationship between genetic and macromolecule synthesis (protein, enzymes, hormones, other RNA etc.) was the refinement of one gene-one enzyme hypothesis to ***one-gene one-polypeptide chain hypothesis***. The gist of this hypothesis is that proteins (enzymes) are made of a chain of amino acids linked to each other by a ***peptide bond*** and may be composed of one or more such chains and genes function was to determine a single polypeptide chain. The earliest evidence in favour of this came from the Linus Pauling's work on Sickle cell type of haemoglobin. Hemoglobin is a protein made up of two copies each of two different kinds of polypeptide chains. The polypeptide chains are called, alpha (Hbα) and beta chains (Hbβ). Each chain is specified by separate gene (α & β). The sickle cell type of Haemoglobin (Hbs) was shown by Linus Pauling and his collaborators

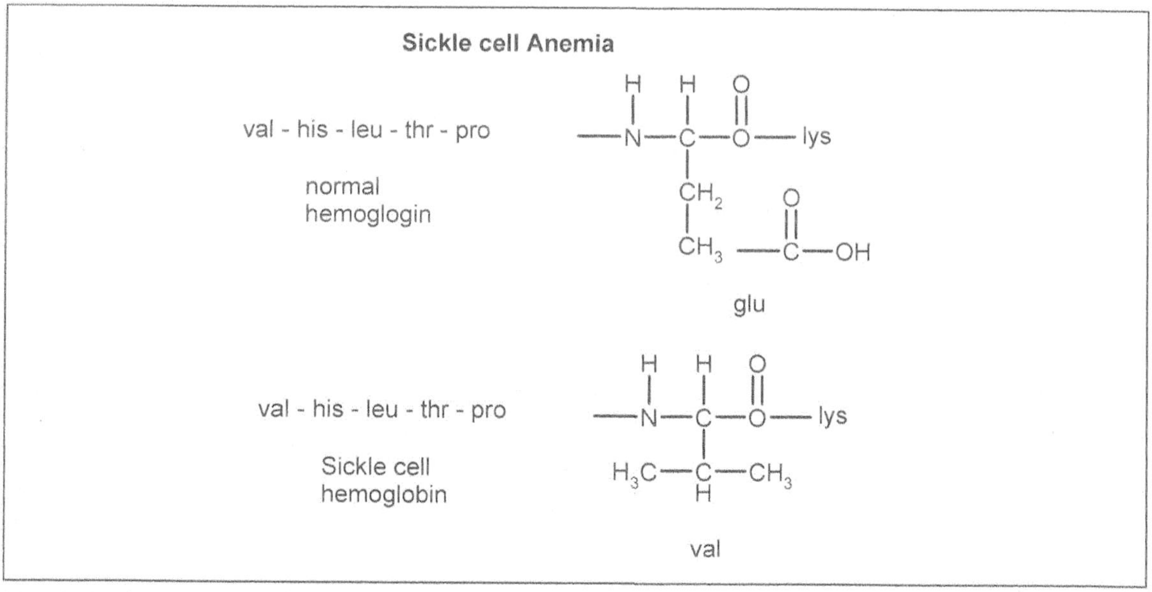

Sickle cell Anemia

val - his - leu - thr - pro —N(H)—C(H)—C(=O)— lys

normal hemoglogin

CH_2 — CH_3 — C(=O)—OH

glu

val - his - leu - thr - pro —N(H)—C(H)—C(=O)— lys

Sickle cell hemoglobin

H_3C—C(H)—CH_3

val

Fig. 6.10 Amino acid difference between normal and sickle cell hemoglobin leading to anemia.

(1949) that the abnormal haemoglobin have replacements of an amino acid at certain position in α chain. The two β chains are normal and identical to each other in Hbs, but the α chain (polypeptide) was different from the normal (Ingram, 1950) in one of the amino acid subunits (Fig. 6.10). The small change resulted from the ***point mutation*** in the HB α gene leading to ***Sickling phenotype*** and anaemia. The gene-polypeptide system for haemoglobin thus was analogous to the gene-enzyme systems studied in ***Neurospora***. The one gene – one polypeptide chain generalization still holds good for most of the gene. Further refinement is one cistron – one polypeptide which came after the elaboration of fine structure analysis of rII region of bacteriophage (Benzer).

To sum up, from the biochemical point of view, what a gene is:

- A gene is a unit character that is responsible for transmission of hereditary character.
- A gene might have a multiple number of alleles that arise either by mutation or intraallelic recombination of preexisting alleles.
- Each of the alleles produce a different polypeptides chains conditioned by the gene.
- Gene is collinear to amino acids in a polypeptide chain.
- Each of such gene is at one locus on the chromosome.
- Each such gene in diploid organisms have two alleles, but meiosis segregate them to form haploid gametes.
- Linked genes assort together but may be frequently separated by crossing over.
- Unlinked genes assort independently.

6.5 The Gene as a Nucleic Acid

During the period when genes were being discovered to contain information for making proteins and enzymes another sort of investigations were taking place to determine the chemical nature of gene itself. The idea that nucleic acid is the basis of inheritance is not new. Even prior to 1900, as a result of work of Miescher (1896), Kossel (1896) and Mathews (1897), it was shown that the nuclei in puss cells of humans, chromatin of sperm in sea urchins and fishes is composed of a salt of nucleic acid with histones or protamines. The Genetic analysis of prokaryotes in the early half of 20th century by several microbial geneticists made great contribution towards the understanding of chemistry of genes and DNA as a genetic material.

6.5.1 Genetic material – blue print of life

Biologists have attempted by chemical means to induce in higher organisms predictable and specific changes which there after could be transmitted as stable hereditary traits. Transformation of specific types of ***pneumococcus*** was first described by Griffith (1928), who succeded in transforming an attenuated and non-encapsulated R-variant derived from one specific type into fully encapsulated and virulent cells of heterologous specific type(S). Griffith found that Type IIR bacteria, which have a rough outer cell walls when injected into mice, there will not be any disease or death and hence are non-virulent (Type IIIS). On the other hand when another strain of bacteria, Type III S (smooth) are injected into the mice there was disease and death in the mice and hence the bacteria are highly infectious and virulent. When a small number of living Type RII bacteria together with a large inoculum of heat killed Type III S were injected into the mice, there was infection and mice succumbed to the disease leading ultimately to their death. Griffith has logically concluded that Type II R strain was avirulent and incapable of by itself causing disease, the ***bacteremia*** and additional fact that the suspension of heat killed Type III S contained no viable bacteria and hence will not be in a position to cause disease or come back alive. It was very convincingly argued that the R forms growing under those conditions had newly acquired the capsular structural characteristics and transformed to biological specificity of Type III ***Pneumococci***, by the ***transforming principle*** contained in the suspension of heat killed Type III cultures (Fig. 6.11).

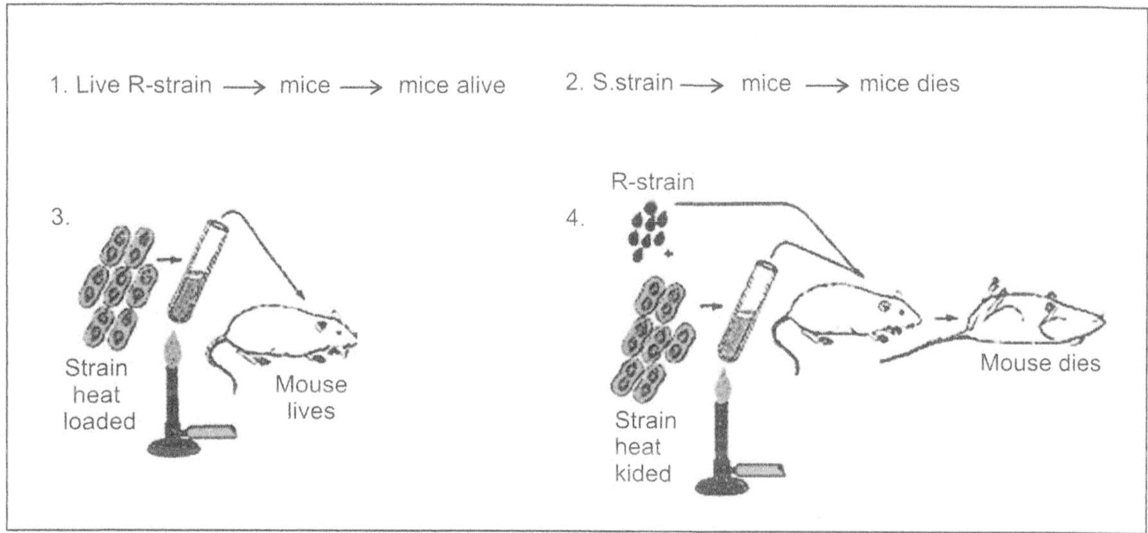

Fig. 6.11 Griffiths experiment on bacterial transformation

The original observations of Griffith were later confirmed by many biologists subsequently (Neufeld and Levinthal, 1928; Baurhenn, 1932; Dawson, 1930). Dawson and Sia (1931) even succeded in achieving transformation ***in vitro***. This was accomplished by growing R cells in a liquid medium containing anti R serum and heat killed Type III S cells. It was shown that in the test tube as in the animal body transformation can be selectively induced, depending upon the type specificity of the S cells used in the reaction systems. Later Alloway (1932) was able to achieve specific transformation ***in vitro*** using aseptic filtrate of S cells removing cellular debris by Berkfefeld filtration. He thus demonstrated that the crude extract containing ***transforming principle*** in soluble form are as effective in transformation as are the intact cells from which the extract is prepared.

6.5.2 Transforming principle is Deoxyribonucleic Acid (DNA)

Experiments using prokaryotes were pointing out clearly that the Deoxyribonucleic Acid (DNA) could be the transforming principle in bringing about stable, heritable changes in bacteria. However, not until a series of experiments published in 1944 by Avery, Mac Leod and MecCarty demonstrated that the transforming principles consists of deoxyribonucleic acid (DNA). They isolated from type

IIIS pneumococcus cultures, a biologically active fraction in highly purified form and mixed it with cells of Type IIR, non-virulent bacteria. They found that under appropriate conditions of culturing, the transformation was achieved in unencapsulated R variants of pneumococcus Type II into fully encapsulated cells of the same specific type as that of heat killed microorganisms from which the transforming principle was recovered. The chemical nature of the purified substance, DNA was ascertained and established by qualitative tests (positive to DNA), elemental analysis that corresponded closely to the predicted composition for DNA, sensitivity to DNase, an enzyme that destroys the chemical and the transforming biological properties of the DNA. Enzymes that destroy other cellular macromolecules (protein, carbohydrates etc.) did not affect the transforming property. Further the data obtained by chemical, enzymatic and serological analysis together with the result of electrophoresis, ultracentrifugation and UV-spectroscopy indicated that the active fraction comprised principally of a highly polymerized viruses of DNA. The evidence presented supported the belief that the deoxyribonucleic acid (DNA) is the fundamental unit of the transforming principle of pneumococcus Type IIIS (Fig. 6.12).

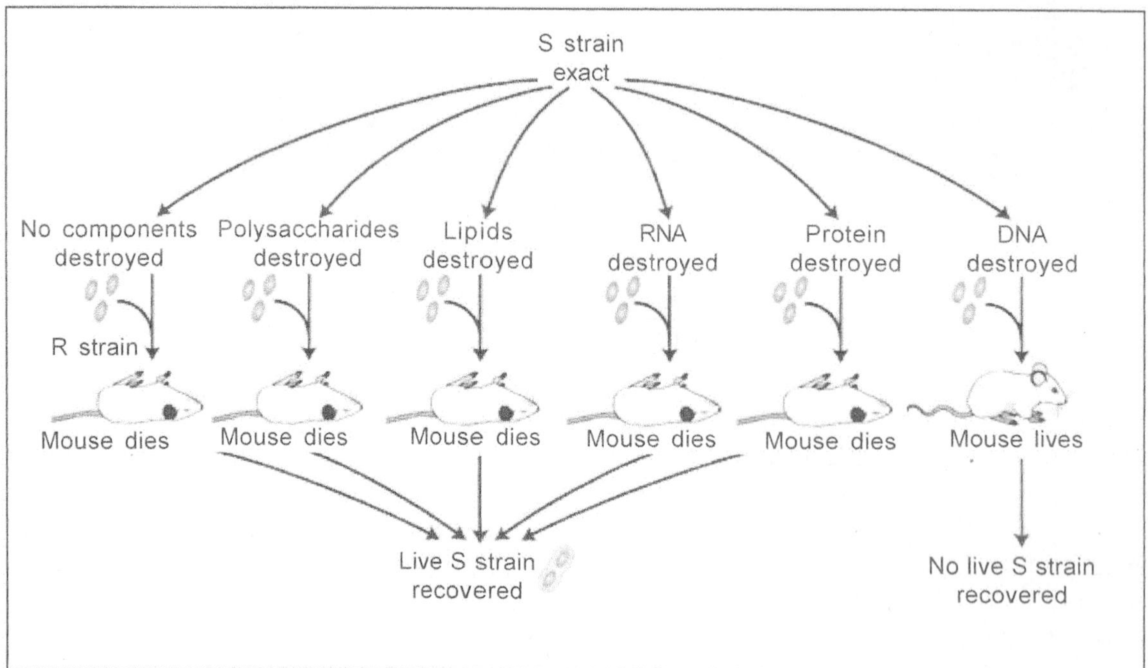

Fig. 6.12 Experimental proof for DNA as a Transforming Principle.

6.5.3 Viral Genomes are Nucleic Acid

Another important piece of evidence for the DNA / RNA (nucleic acids) as genetic material has come from the elegant work of Hershey and Chase (1952) and Frankel-Konrat (1955) respectively. Outline of their experimental approach is presented in Fig 6.13.

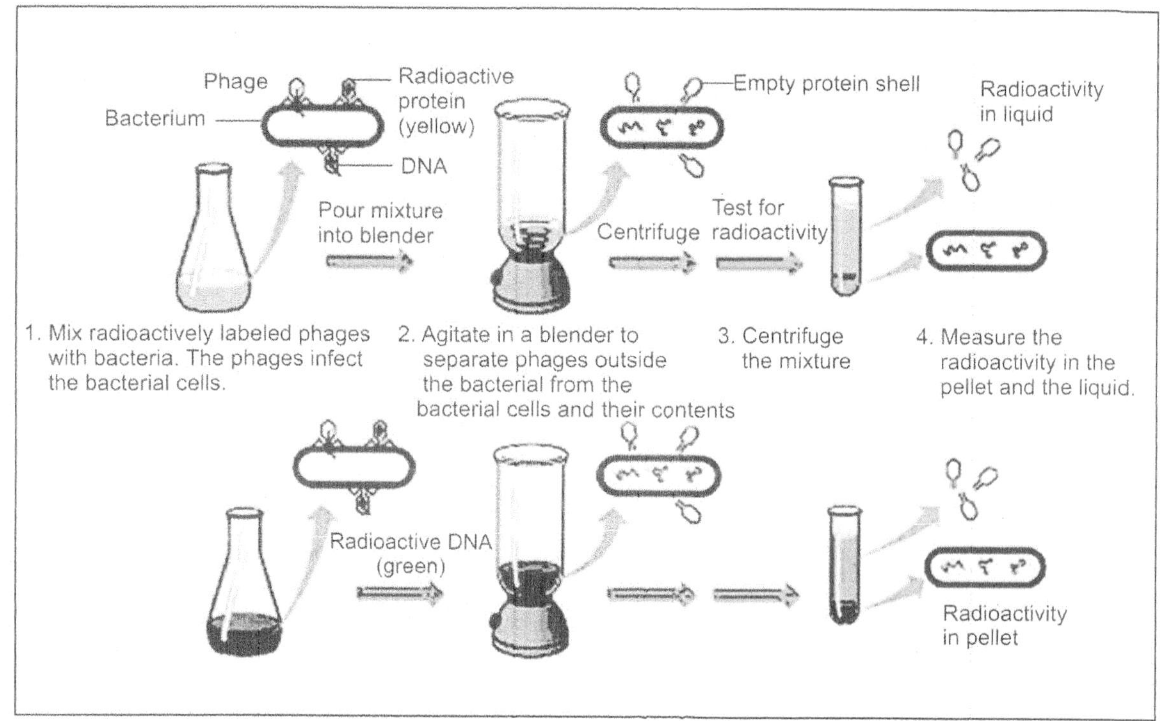

Fig. 6.13 Hershey and Chase experiment demonstrating DNA as genetic material

Bacteriophages are a class of viruses that infect bacteria and multiply in them and kill the bacteria by lysis and release new progeny virus particles. Bacterial viruses (phage for short) can be made radioactive by lablelling with either S^{35} or P^{32} in order to detect their presence or absence by chasing the radioactivity.

Hershy and Chase labelled T_2 phage with different radioactive compounds so that they could follow the fate of the two different components of the T_2 virus. The protein coat of the phage was labelled with S^{35} and the DNA core with P^{32}. The mode of infection of virus is by attaching itself to a bacterium and subsequently infect its DNA into the bacterium, where in it multiplies. A rapid mixing of attached bacteria in the process of injecting its core consisting of DNA in warning blender chapped off most of external coat protein which remained outside of the infected bacterial cells. The DNA injected into bacteria, however goes through the production of many copies of its DNA and assembles the protein coat required to make progeny virus particle.

The original DNA strands, labelled with P^{32} were conversed in the process and packaged a new coat proteins. Once the bacterial cell is filled to the brim with particles the cell wall is lysed and new generation virus of phages are released. When the new progeny particles were examined by Hershey and Chase, all the P^{32} label could be recovered while S^{35}label was not found in them. They logically concluded that the DNA labelled with P^{32} was being transmitted to the progeny generation of bacteriophage. In other words the genes must be made of Deoxyribonucleic acid, since it provided the genetic continuity from one generation of phage to another.

6.5.4 Tobacco Mosaic Viruses contains RNA Genomes

Frankel-Conrat and his collaborators were researching on the virus reconstitution by combining proteins and nucleic acid from different strains of Tobacco mosaic virus (TMV), that parasitize plants and cause disease. These viruses have another type of nucleic acid, Ribonucleic acid (RNA for short) as their genomes. They have employed commonly occurring strains TMV, and the masked (M), yellow aucuba (YA) and Holmes ribgrass (HR). Active viruses were reconstituted from 1) TMV nucleic acid and M-protein, 2) M-nucleic acid and TMV protein, 3) YA-nucleic acid and TMV- protein, 4) YA-nucleic acid and M-protein, 5) HR-nucleic acid and TMV protein, 6) TMV-nucleic acid and

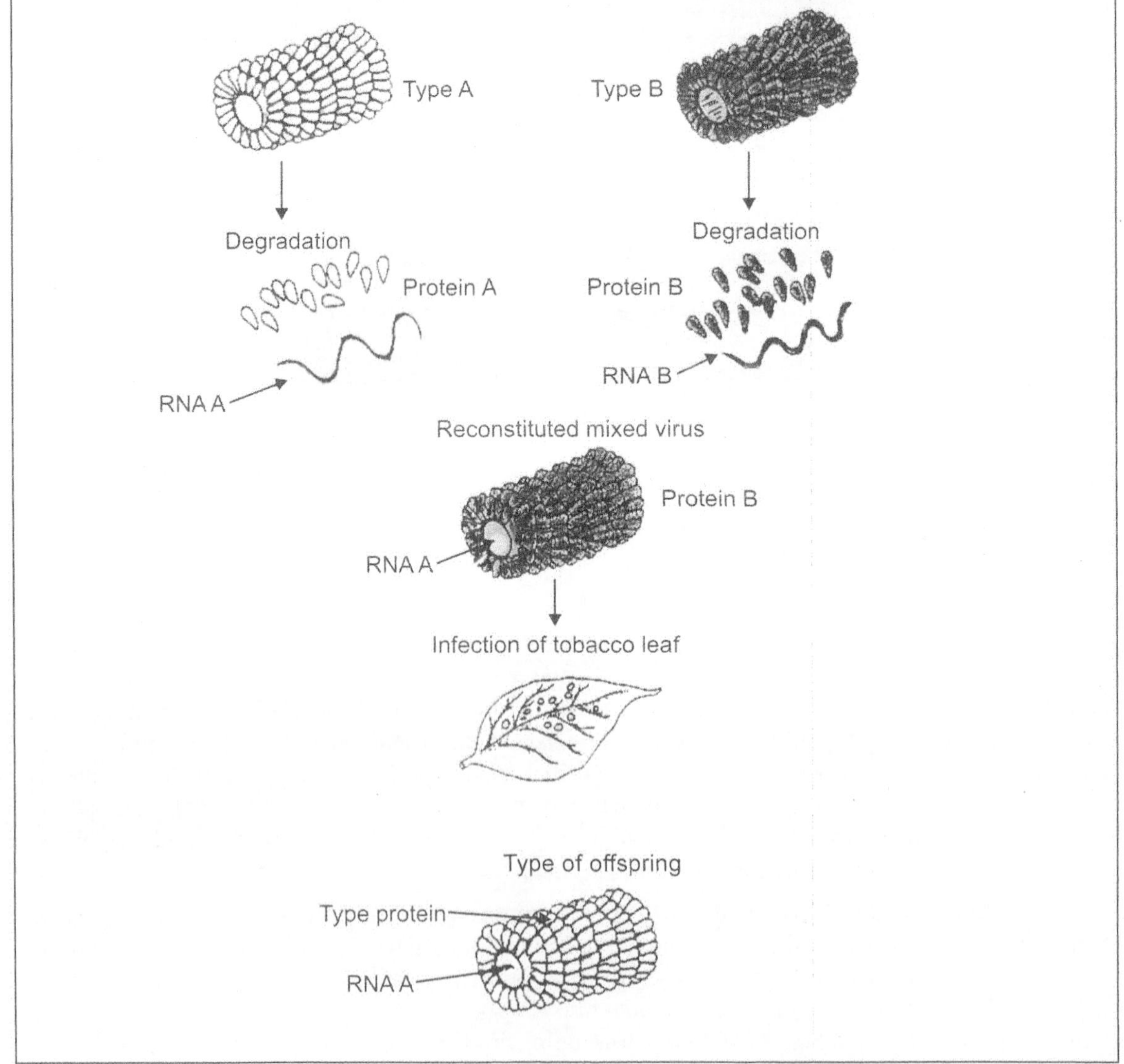

Fig. 6.14 Frankel-Conrat experiment demonstrating RNA as genetic material in virus

HR-protein and 7) HR-nucleic acid and M-protein. The reconstituted virus particles by mixing different types of nucleic acid and protein yielded a very infective preparations like original (native) strains. The nature of disease and the chemical nature of the progeny in terms of amino acid composition of mixed viruses reassembled closely to that of the RNA viruses. On the other hand the serological properties of mixed virus were those of the virus supplying the protein. The dual nature of activity of reconstituted virus has clearly demonstrated that the infectivity of virus and serological characteristics is actually a property of newly formed virus and the nucleic acid, RNA is a genetic material for transmitting the traits (see Fig. 6.14).

6.5.5 The evidence that DNA is Genetic material in Eukaryotes too

Prokaryotic genetic analysis established that bacteria and their viruses clearly have DNA as their genetic material. But what about eukaryotes? For a long time the evidence was deductive and inferential. DNA is present in chromosomes and quantitatively it is constant in all the somatic cells, but half the quantity is found in gametes correlating with the haploid chromosome complement (genome) in them. This can be expected of genetic materials behaviour, but it is not a direct proof. Avery, while discussing results of bacterial transformation noted that *"If we are right it means that nucleic acids are not merely structural important but functionally active substances in determining the biochemical activities and specific characteristics of cells and that by means of known chemical substance it is possible to induce predictable and hereditary changes in cell.* This comment has been made in the context of transformation in bacteria but has a wide ranging implications. Similar results have been obtained with the somatic cells of higher organism in culture.

The first discovery that heralded these possibilities was that chromosomes could be prepared from one cell and added to another set of cells by mixing. At a very low frequency, the recipient cells expressed new function. Such experiments subsequently were performed with the purified DNA, and attained refinement to the point that even only a single gene containing pure preparations of DNA can be made to transform the cells that lack the gene. Transformations can then be selected on a selective medium by their expression. This can be done for any gene that has a robust protocol for the assay. Even though it is analogous to bacterial transformation, for historical reasons, it is termed as ***transfection***. As an example of one of the standard system it is described (Fig. 6.15).

In a procedure quite akin to transformation of bacteria the eukaryotic cells growing in culture can take up exogenous DNA and get transformed displaying new phenotypes. Mammalian cells that lack ***Thymidine Kinase gene*** (Tk^-) cannot produce the enzyme thymidine kiinase and therefore are auxotrophic and cannot grow and subsequently die in the absence of thymidine in the medium, whereas its allelic form Tk^+ can grow, survive and produce clones in the medium lacking thymidine. A pure DNA preparation was made from Tk^+ cells and added to the Tk^- cells growing in a culture. After transfection it was found that while most of the cells were dying, a colony of Tk^+ cells were seen. Such results invocably demonstrated that not only DNA is genetic material in eukaryotes too, but also that gene can be transferred between different species and yet remain functional. At first these could be performed only with cells adapted to growth in defined medium. Subsequent developments in cell culture methods and molecular biology made it possible to introduce exogenous genes into mouse eggs by microinjection and are made to become stable components of the adult mouse genome. The genesis of genetically modified organism has roots in this kind of experimentation!

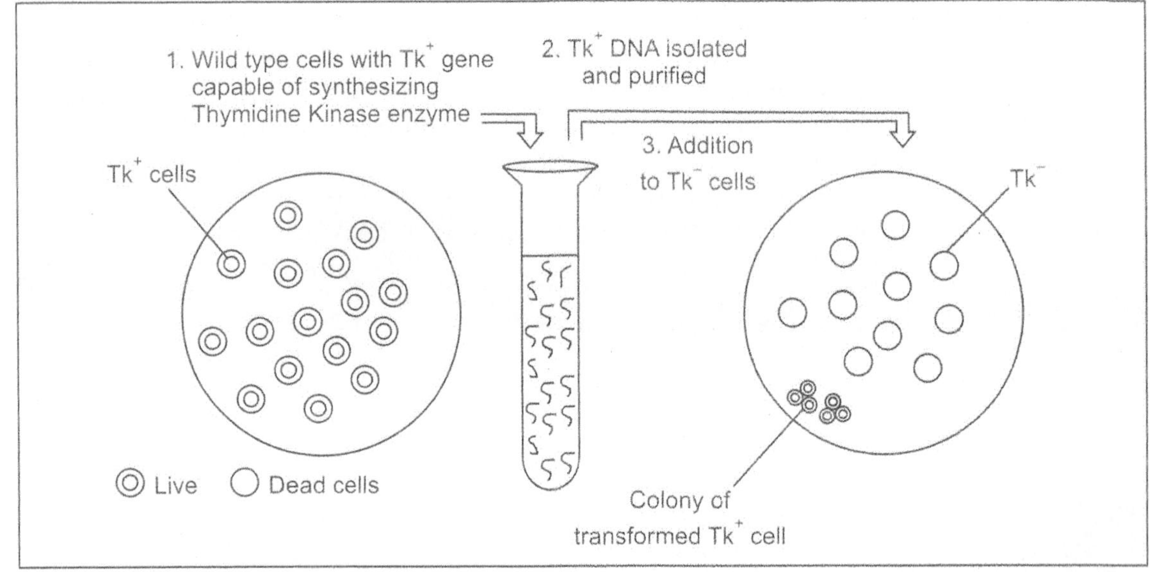

Fig. 6.15 Transfection of eukaryote cells – Acquisition of a new phenotypic trait by added DNA similar to bacterial transformation.

6.6 Structure of Deoxyribonucleic Acid (DNA)

Once it was established that the DNA is genetic material, biologists were addressing the problem of its structure and were considering how the structure relates to the functional requirements of the genetic material. A great impetus in the study of the nature of gene was provided by Schrodinger in 1945 in his book 'What is life'. He opined about the nature of gene from the point of view of physicist. He wrote 'Incredibly small group of atoms, much too small to display exact statistical laws, do play a dominant role in the very orderly and lawful events within a living organism. The gene is much too small to entail an orderly and lawful behaviour according to physics". He went on developing from the views of Delbruck, the concept that the laws of physics might not be enough to account for the properties of the genetic material, especially its transmission either from generation to generation at organism level or from cell to cell at cellular level growth, development and differentiation. It was clear to contemporary scientist that the structure of gene not only be expected to obey the laws of physics, but also the two fundamental biological properties of life reproduction (duplication) and variation (mutation). The thought and prospect that the study of properties of gene might unravel the new laws of physics attracted many physicists into biology.

Schrodinger summarized the properties of the gene which remarkably coincide with the present view: *"We shall assume the structure of gene to be that of a huge molecule, capable only of discontinuous change which consists in a rearrangements of the atoms and leads to an isomeric molecule. The rearrangement (mutation) may affect only a small region of the gene, and a vast number of different rearrangements may be possible"*. In his view, though the gene was conceived to be a periodic crystal under the constraint of accommodating demands placed on the genetic material, the notion that it might be a continuous part of the chromosome was evident.

Presently, of course it is known that the gene consists of DNA and that its structure is not maintained and perpetuated in isolation. It depends on enzymes that influence the accuracy of the process besides needed catalytic functions. All of them obey the known laws of physics and chemistry.

6.6.1 DNA is a Polymer

The fact that both DNA and protein were long polymer were understood by early 1930 and led to the concept that both have potentiality required to be genetic material. One of the reason for doubting about the genetic functions of DNA in early phase was the misconception about its structure. Although it was known to consists of four kinds of nucleotides it was through to comprise a regular repeating unit of tetranucleotide. By the 1930's Casperson had demonstrated that DNA comprises of extremely large molecule – infact much larger than proteins, but the unappealing monotony of its structure seemed to preclude any possible genetic role. On the other hand proteins consisting of 20 biological amino acids appeared much attractive for this role. The extensive work of Chargaff carried out on diverse group of organism demonstrated that the four bases found in DNA are quite variable quantitatively differing from species to species and characteristic of each species. This led to the concept that the sequence of bases of DNA might be the mode in which ***genetic information*** is encrypted and this sequence information is some way decoded in terms of the sequence of amino acids in a protein. By the 1950s the concept of genetic information was commonly understood. The twin problem of accommodating in a structure, the biological duplication and provision for information transfer (translation) to make protein was working out in a DNA structure.

6.6.2 Components of DNA

The fundamental unit of DNA molecule is a nucleotide. Nucleic acids consist of a chemically linked sequence of nucleotides. Each nucleotide comprises of a heterocyclic ring of carbon and nitrogen atoms (the nitrogenous base), a five carbon sugar in ring form (Haworth structure), and a phosphate group.

6.6.2.1 Sugar Components

Two types of pentoses are found in nucleic acids. These differentiate DNA and RNA and impart general name nucleic acids. In DNA the pentose sugar, 2-deoxyribose is present and hence the name deoxyribonucleic acid and in RNA the pentose sugar is ribose and hence the name ribonucleic acid. The difference in sugar manifests in the absence or presence of the hydroxy group at a position 2 of the pentose sugar ring shown in Fig. 6.16.

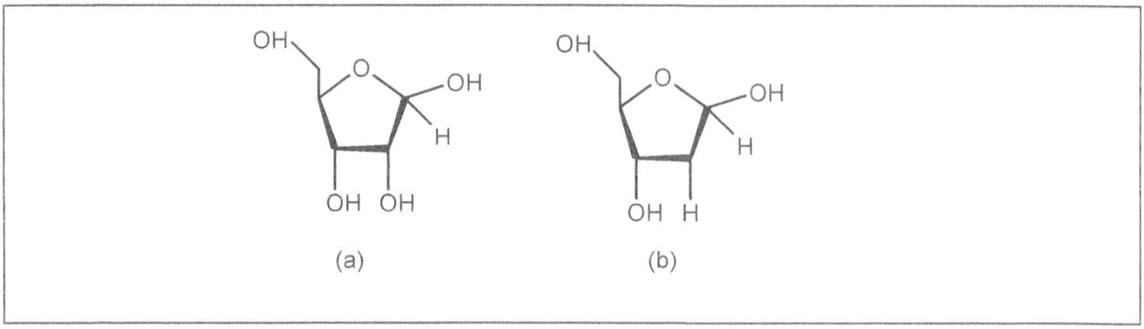

OH O OH H OH OH (a)

OH O OH H OH H (b)

Fig. 6.16 (a) Ribose and (b) Deoxyribose sugars of the Nucleic Acids

6.6.2.2 Nitrogeneous Bases

The nitrogenous bases can be categorized into two types: ***purines*** and ***pyrimidines.*** Purines have fused five-six member rings, where as pyrimidines have a six member ring. The same purines, Adenine and guanine are present in both the nucleic acids, DNA and RNA. The two pyrimidines present in DNA are cytosine and thymine whereas; in RNA thymine is replaced by uracil. The only difference between thymine and uracil is that in thymine the C_5 position has addition methyl group. The bases are usually referred to by their initial letter, A,G,C,T/U. So DNA contains A, T, G, and C. while RNA consists of A,U,G and C. The components structures are illustrated in Fig 6.17.

Base formula	Base X = H	Nucleosides X = Ribose or Deoxyribose	Nucleotide, Where X = Ribose Phosphate
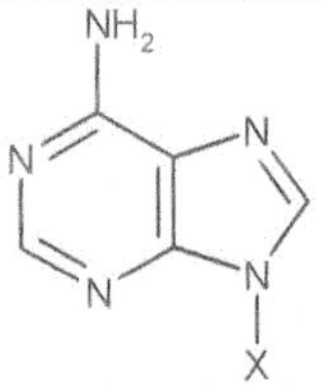 NH$_2$, N, N, N, N, X	Adenine A	Adenosine A	Adenosine monophosphate AMP
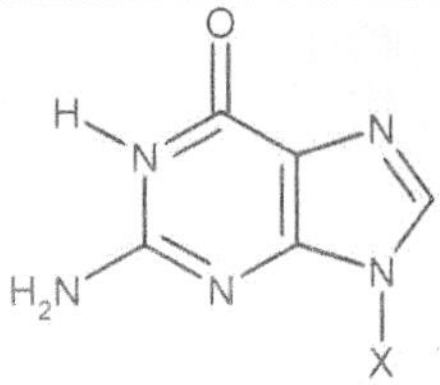 O, H, N, N, H$_2$N, N, N, X	Guanine G	Guanosine G	Guanosine monophosphate
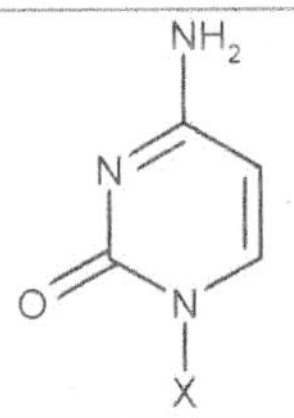 NH$_2$, N, O, N, X	Cytosine C	Cytidine C	Cytidine monophosphate CMP
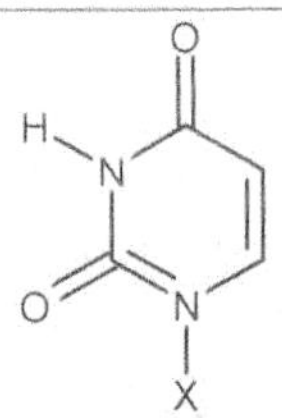 O, H, N, O, N, X	Uracil U	Uridine U	Uridine monophosphate UMP
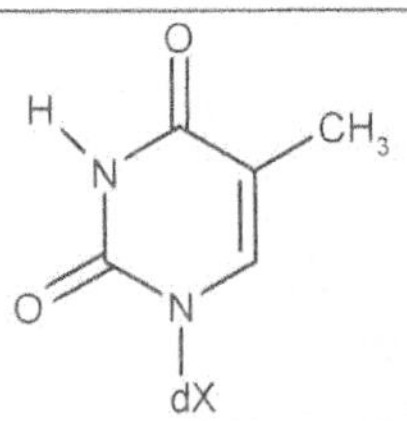 O, H, N, CH$_3$, O, N, dX	Thymine T	Thymidine T	Thymidine monophosphate TMP

Fig. 6.17 Components of a DNA/RNA nitrogenous base

6.6.2.3 The Phosphoric Acid Component

When a nitrogenous base is linked to sugar, such molecules are called ***nucleosides.*** When a phosphoric acid group attaches to 5-carbon of sugar, the same is called ***nucleotide.*** Up to three individual phosphate groups can be added in a series giving a nucleoside monophosphate (NMP), nucleoside diphosphate (NDP) and nucleoside triphosphate (NTP). The individual phosphate groups are designed α, β and γ depending upon their position in the compound. The first phosphate molecule to attach sugar is called γ, later ones as α and β. The nitrogenous base is linked to position 1 on the pentose sugar ring by a glycosidic bond from N_3 to pyrimidnes or N_9 of purines. In order to minimize the ambiguity between numbering systems of the heterocyclic rings and the sugar, the position on sugar are given a prime (') (Fig. 6.18).

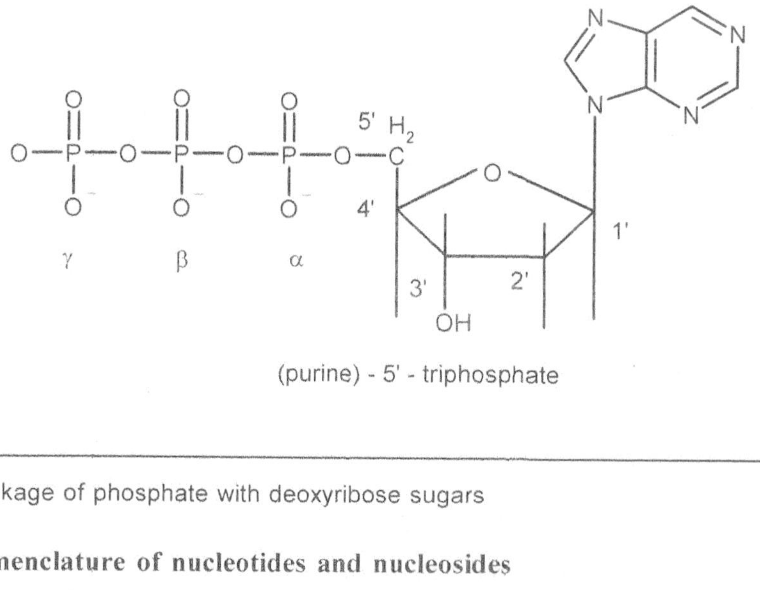

(purine) - 5' - triphosphate

Fig. 6.18 Linkage of phosphate with deoxyribose sugars

6.6.2.4 Nomenclature of nucleotides and nucleosides

Different nucleosides and nucleotide structure along with the nomenclature used is given Fig. 6.19 and Table 6.4.

Table 6.4 *The nomenclature of the components of Nucleic acid, DNA & RNA*

Base	Nucleoside (Base + Sugar)	Nucleotide (Base + Sugar + Phosphoric acid)
Adenine	Adenosine	Adenylic acid (AMP or d AMP)
Guanine	Guanosine	Guanilic acid (GMP or d GMP)
Cytosine	Cytidine	Cytidylic acid (CMP or d CMP)
** Thymine	Thymidine	**Thymidilic acid (dTMP)
* Uracil	Uridine	Uridylic acid (UMP)

** Present in DNA only, * Present in RNA only, rest of the three are present in RNA as well as DNA

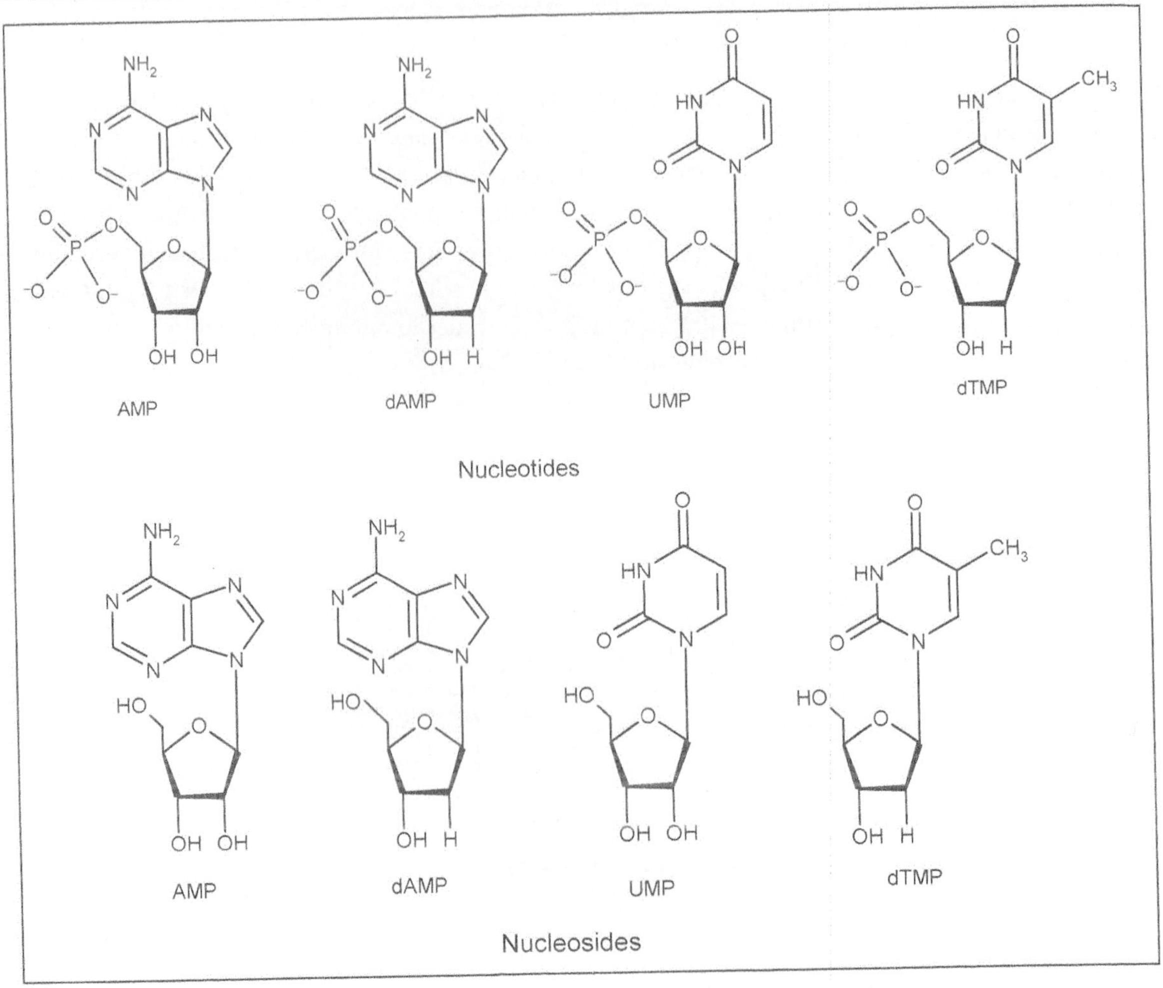

AMP dAMP UMP dTMP

Nucleotides

AMP dAMP UMP dTMP

Nucleosides

Fig. 6.19 Structure of Nucleotides and Nucleosides

6.6.3 Polynucleotide and their Linkage

Nucleotides are linked together into a polynucleotide chain in the process of polymerization by a backbone containing an alternating series of sugars and phosphate residues. The 5' position of one pentose ring is connected to the 3′ position of the next pentose ring through a phosphate group (Fig 6.20). Thus the ***nucleotides are joined by phosphodiester bonds*** hence the back bone is said to be consisting of 5′ – 3′ linkages. The nucleotide monomers are linked together by joining the α-phosphate group, attached to the 5-carbons of one nucleotide to the 3′-carbon of the next nucleotide in the chain (providing a polarity). The terminal nucleotide at one end of the chain has a free 5′ groups. The terminal nucleotide at the other end has a free 3′ group. It is conventional to indicate nucleic acid sequences from 5′ terminus towards 3′ terminus.

When nucleic acids are digested into their constituent nucleotides the cleavage may takes place on either side of the phosphodiester bonds. Depending on the factors, nucleotide might have their phosphate group attached to either 5′ or 3′ positions of the pentose as shown in Fig. 6.20.

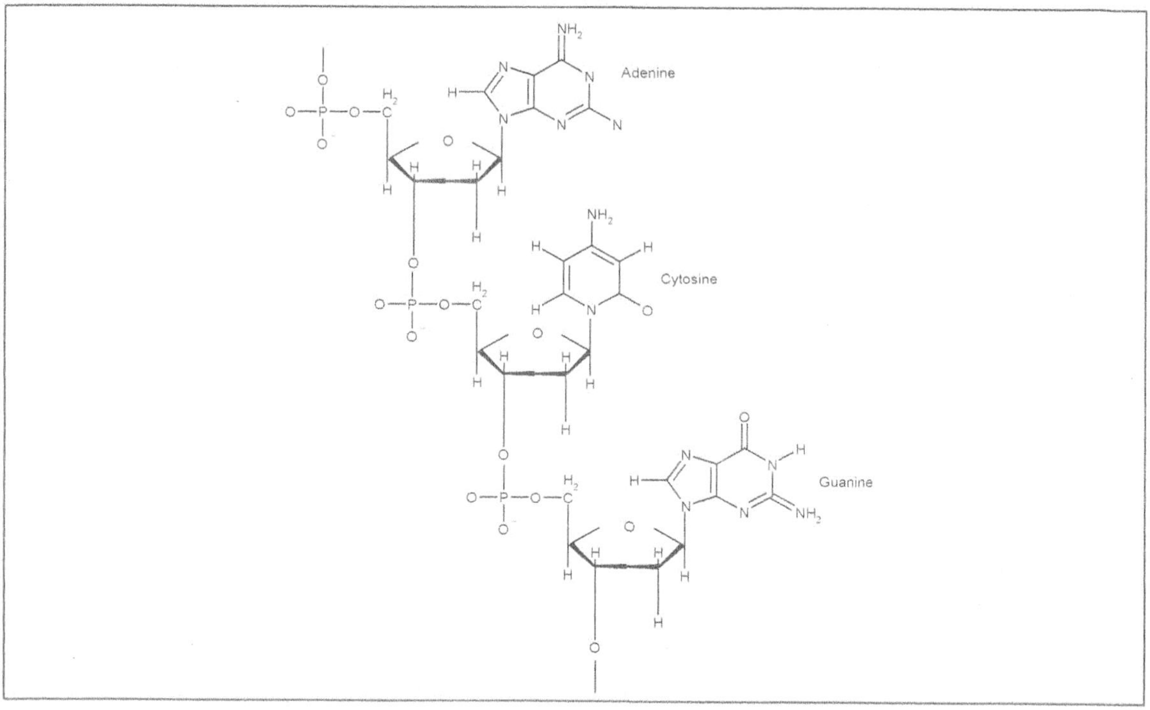

Fig. 6.20 Polynucleotide

Since the sugar base moiety by itself is a nucleoside two types of nucleotides released from nucleic acids: nucleoside 3′- monophosphates and nucleoside 5′-monophosphate.

All the nucleotides can occur in such form, that more than one phosphate group linked to the 5′ position (Fig 6.21).

The bonds between the first (α) and second (β) and between (β) and third (γ) phosphate groups are energy rich and are exploited to provide an energy source for various cellular activates. The nucleic acids triphosphates are the forms from which nucleic acids are synthesized.

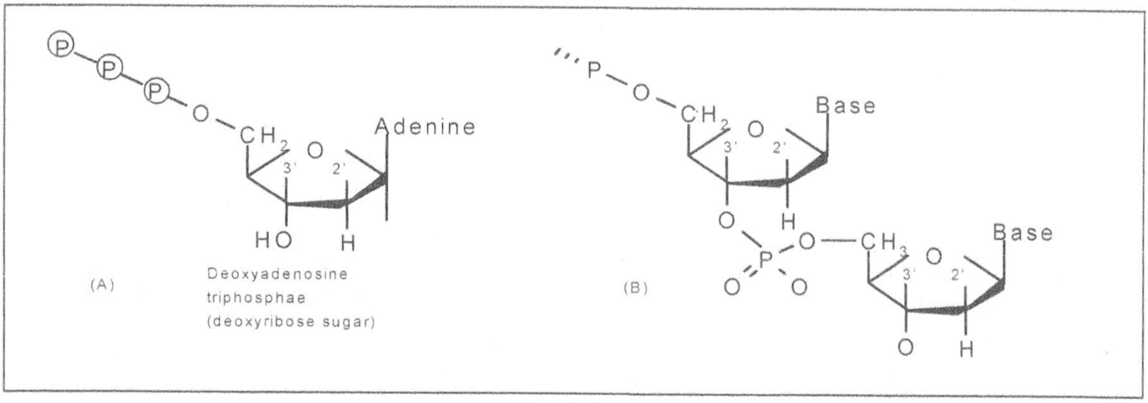

Fig. 6.21 Formation of Phosphodiester bond.

6.7 Structural Components of RNA

Besides DNA, another type of nucleic acid RNA is present in living cells and organisms. This plays a vital but innumerable roles in molecular biology of cell during gene expression, transcription, translation and regulation. Like DNA, it is also a polynucleotide but with two major differences:

- The sugar is ribose not 2′-deoxyribose as in the case of DNA
- RNA do not have thymine, instead it contains uracil in its place (Fig 6.22)

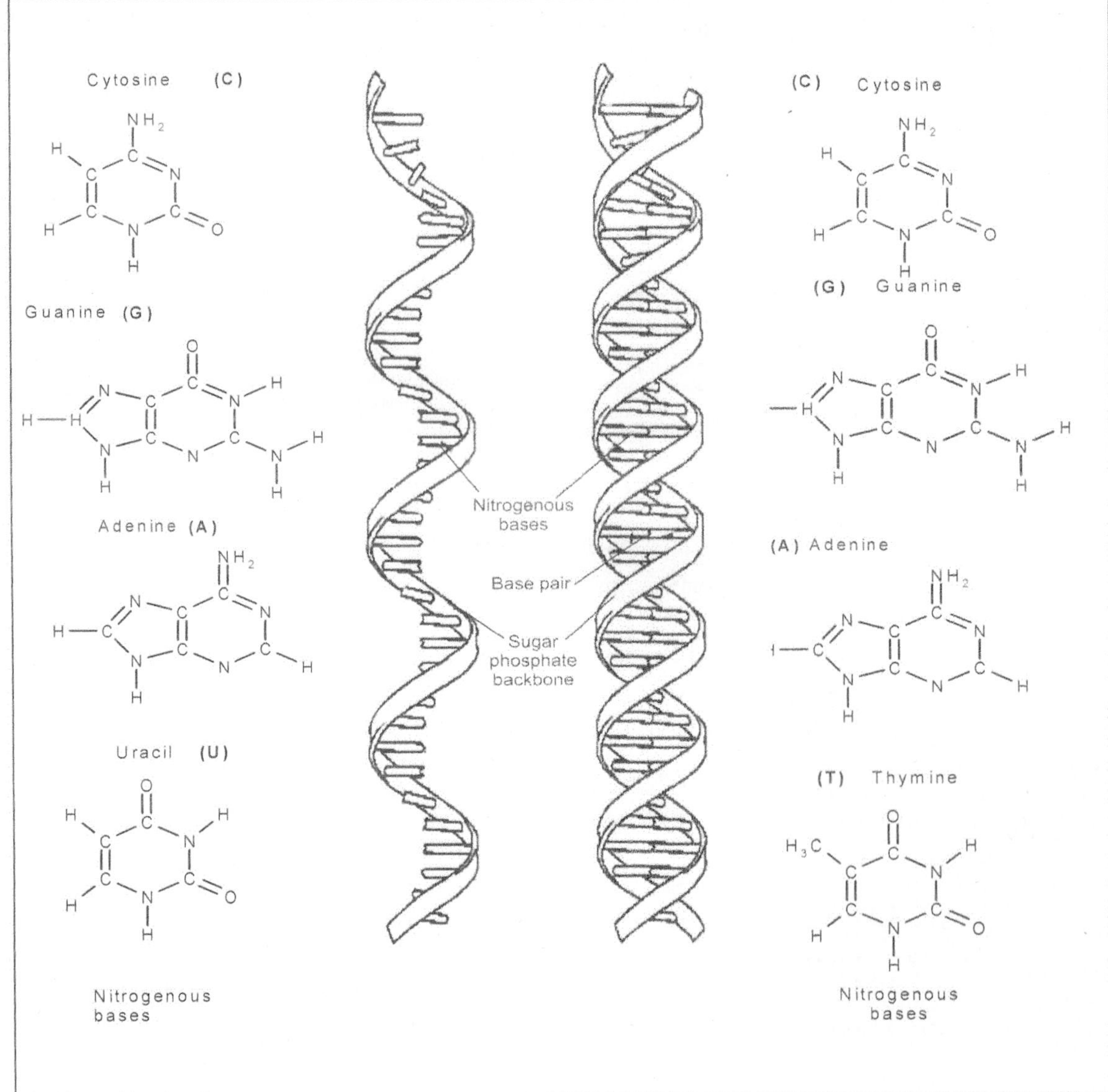

Fig. 6.22 Differences between RNA and DNA

Hence, the nomenclature of four nucleotides that makes a RNA polymer changes as given below along with their abbreviation

Adenosine 5´-triposhpate (ATP)
Guanosine 5´-triphosphate (GTP)
Cytidine 5´-triposhapte (CTP)
Uridine 5´-triphosphate (UTP)

The polynucleotide structure of RNA is similar to that of DNA, with 3´-5´ phosphodiester bond linkage between individual nucleotides in the molecule. There are no restrictions on the sequence of component nucleotides within the RNA molecule. One important difference between RNA and DNA is that RNA is a single stranded molecule, while DNA is helical, double stranded molecule invariably in all the cell.

6.8 The Watson and Crick Model of DNA: the Double Helix

The proposal of ***double helix*** molecular model for DNA structure by Watson and Crick and its quite dramatic announcement and its publications in Nature (as a letter to the editor) in 1953, was one of the greatest intellectual triumphs of the deductive logic in the history of natural science. It has profoundly influenced the next course of molecular genetics in particular and biology as a whole in general. At that time there were many known biological to chemical and physical facts about DNA. Many lines of evidences indicated that the DNA is carrier of the genetic specificity of the chromosomes and thus of the gene ***per se***. However, no evidence has been forthcoming to demonstrate, how it might fullfill the essential criterion required of a genetic material that of self duplication and manifestation of variations.

6.8.1 Nitrogenous base composition and the Analysis of Chemical specificity of Nucleic Acids

The composition of nitrogenous base and their ratios in Nucleic acid (DNA) by Chargaff paved the way for the prediction of accurate structure. Chargaff and his collaborators had earlier established that there were unities among the bases of native DNA molecule. The amount of Adenine (A) was equal to the amount of cytosine (C). The amounts of AT and/or GC content of genomes expressed as percentage (GC or AT% of total) varies widely among species but was constant for a species and organisms, and initially for prokaryotes, the parameter was used for classification purposes.

6.8.2 Physical and molecular structure studies by X-ray diffraction suggested a Helical conformation

Quite parallel to the chemical work on the constituents of nucleic acids and the nature of chemical bonds involved. Progress was being made on the macromolecular structure through crystallography and by means of X-ray diffraction, which was brought to the notice of Astbury (1945). Very accurate but elegant X-ray diffraction microphotograph and data was already in existence in the laboratory of Maurice Wilkinson at King's college which came handy for Watson and Crick to predict the model.

When a DNA crystal was bombarded with X-rays, atypical diffraction pattern was obtained. The theory of X-ray crystallography is quite complex, but is based on the angle at which X-rays are deflected on passing through the crystal and captured on a photographic film. By analyzing the pattern of spots, their position and intensities, the deduction of structure, the three dimensional structure was perceived on a photoplate.

X-ray diffraction methods were the standard analytical procedures to determine the structure of molecular crystal. It has been successfully used to determine the structure of important compounds. By 1950's several natural proteins, like keratinin, myoglobin and haemoglobin etc., have been analysed with an increasing sophistication to elucidate their complex three dimensional structures. Linus Pauling, one of the architect of haemoglobin structure was also interested in solving the structure of DNA and was already in the process of building a model. Watson and Crick had an access to the most elegant

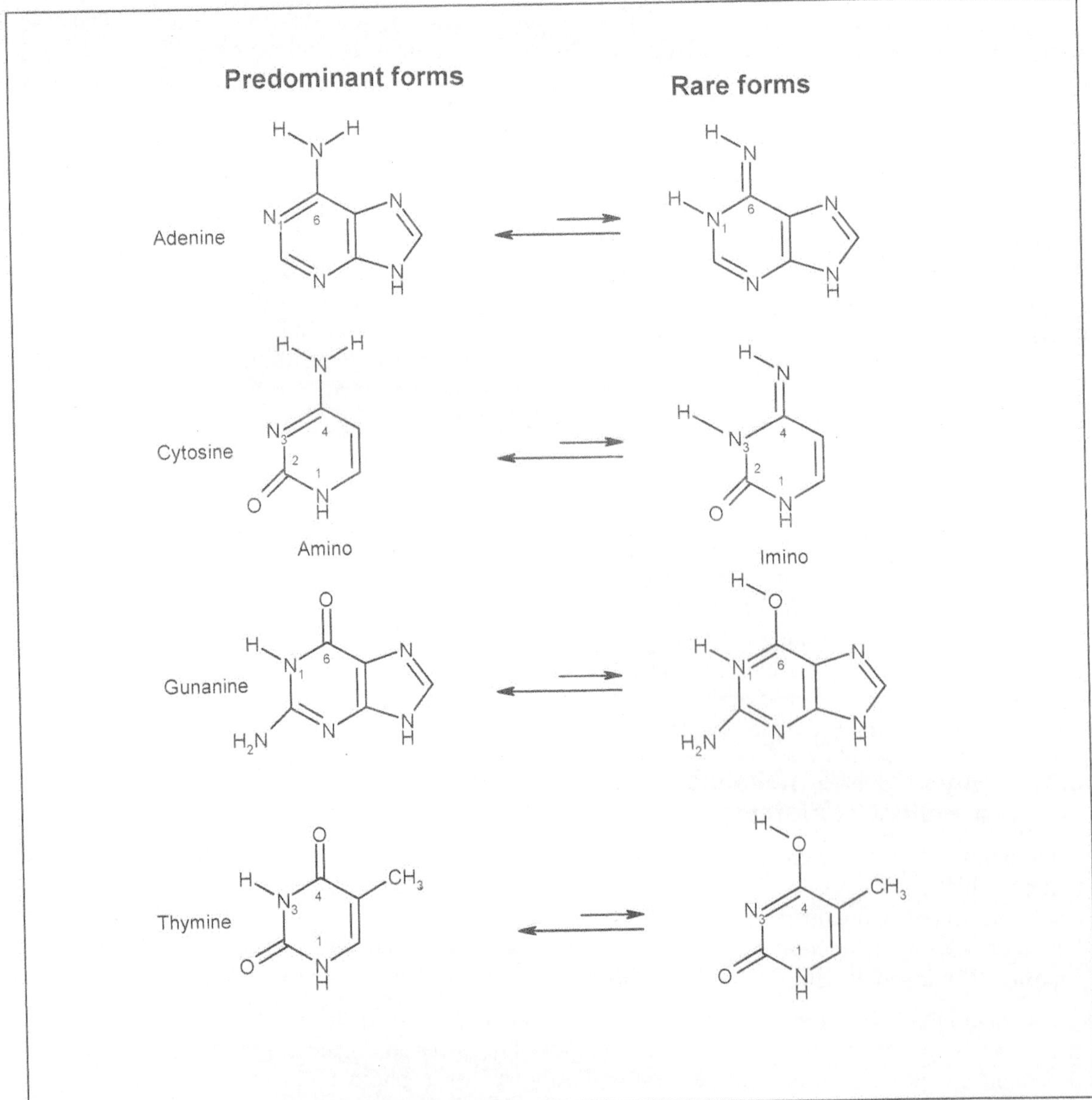

Fig. 6.23 Keto and enol forms of nitrogenous bases

X-ray diffraction pictures taken by Rosalind Franklin with crystals of DNA employing the technique earlier. These photographs clearly indicated that DNA is a double helix with two regular periodicities of 3.4 A^0 and 3.4 A^0 along the axis of the molecule. The actual riddle was how does the X-ray diffraction pattern relates to Chargaffs base pair rule and the actual structure of DNA. At Kings College, London during May 1952 using the database developed by Maurice Wilkins, Watson and Crick solve the riddle and predicted the structure for DNA. As described in J.D. Watson's book, ***The double helix***, Watson and Crick had found out from Linus Pauling's son, that the most prevalent tautomeric forms (isomers or different arrangement of atoms which are in equilibrium) of the bases were the keto forms, not the enol forms as had been assumed by the biologists. The bases are drawn in the keto forms (Fig.6.23) enol forms have – OH rather than = O groups attached to their rings. This single fact enabled them to figure out correctly, using chemical models how DNA bases could be bonded through hydrogen bonds into two, and that too only with specific pairs, A with T and G with C (Fig 6.24). A hydrogen bond is the sharing of one hydrogen atoms with an oxygen or a nitrogen, which has slightly +ve charges, between two nitrogen atoms or an oxygen and nitrogen atoms which have slightly –ve charges on complementary bases opposite each other.

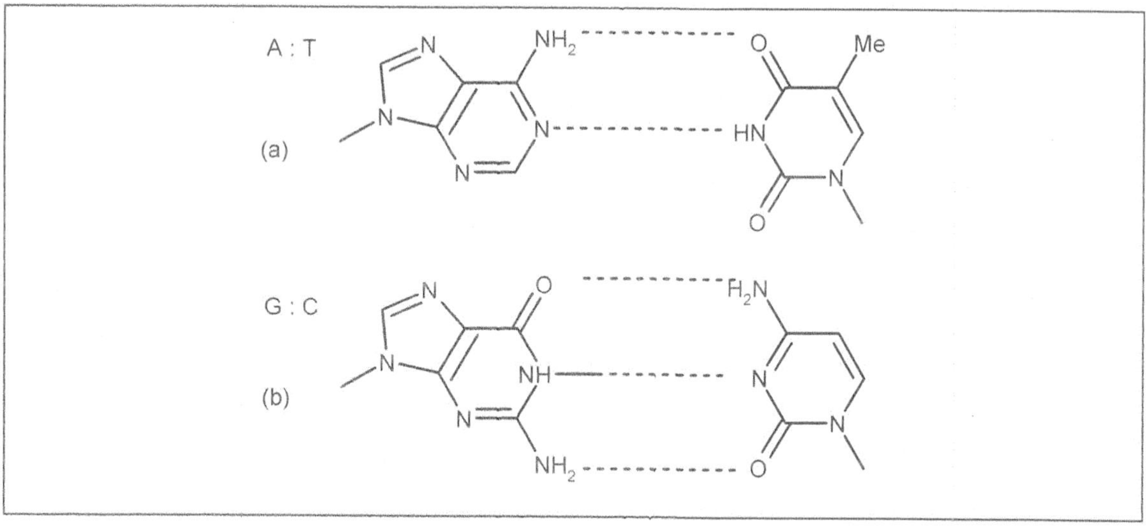

A : T

(a)

G : C

(b)

Fig. 6.24 (a) A•T and (b) G•C base pairs, showing Watson–Crick hydrogen bonding.

Watson and Crick in 1953 put together all the prevailing data related to DNA very intelligently and logically and deduced the ***double helix*** structure for DNA (Fig. 6.25). Some of the important features of their model is given below:

- ***The DNA is made up of a double helix comprising of two independent polynucleotide chains***

  To arrive at the number of polynucleotide in the helix was quite problematic and contentious issue for quite some years in the beginning. Several experimental evidences were demonstrating that the number was variable: two, three or four. Linus Pauling, an American Scientist, who already had a Nobel prize in physics was toying with an idea that DNA helix is made up of three helical chains for which he had proposed a model just few months before the Watson and Crick structure came up.

- ***The nitrogenous bases are stacked very snugly inside the helix, while sugar-phosphate backbone is outside***

  The experimental data indicated that the sugar-phosphate constituted the ***back bone*** of the DNA molecule and is externally situated with the bases stacked on top of each other inside the helix.

- ***The connectivity of the two polynucleotide chains is through hydrogen bonding***

  In order to provide allowance to the Chargaff's base pair ratio rule, it is conceived that an adenine residue in one of the polynucleotide chains is always bonded to a thymine in the other strand. Similarly, the Guanine always pairs with the cytosine in opposite strand (Fig. 6.26). Two hydrogen bonds between adenine and thymine (A = T) and three hydrogen bonds between guanine and cytosine are the only possibilities for hydrogen bonding. These hydrogen bonds are the only attracting forces between the two polynucleotide chains of the double helix that hold the structure together (vanderwal's force)

- ***For every turn of the double helix, there are ten base pairs.***

  The double helix takes one turn every ten base pairs. The pitch of the helix is 34A^0 implying the space between each successive base pairs is 3.4.A^0. Such a periodicities are apparent from the X-ray diffraction pattern. The diameter of helix is 20A^0 (Fig. 6.26).

- ***The two polynucleotide strands in the double helix are antiparall oriented.***

  The stable double helix has an antiparallel orientation. One polynucleotide chain orients in the 5′-3′ direction, the other in 3′-5′ direction. No other arrangement of polynucleotide chains confers a stable conformation (Fig. 6.25).

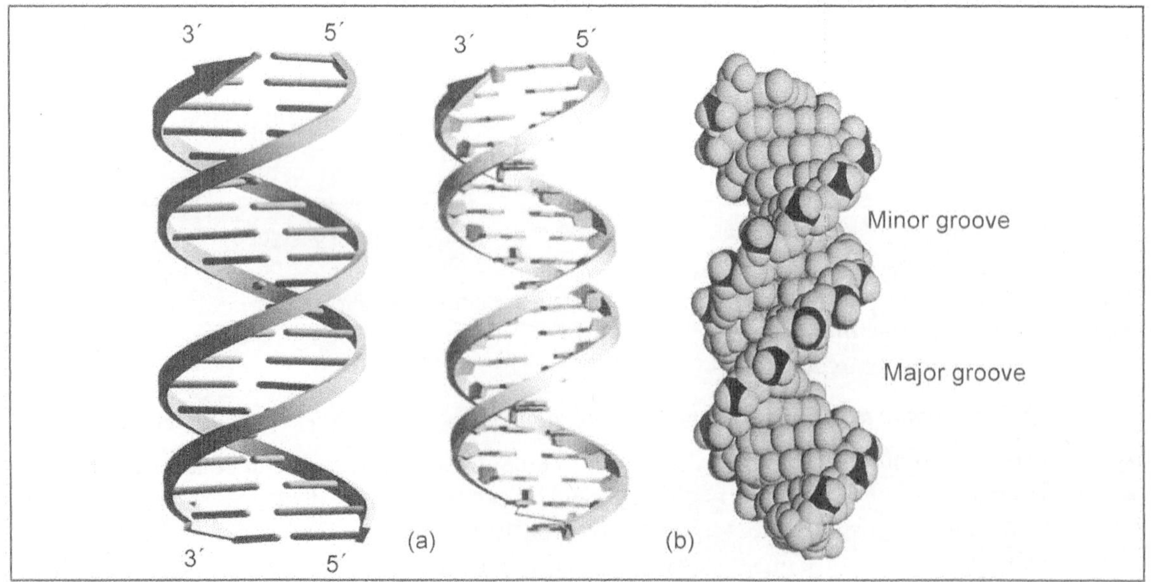

Fig 6.25 DNA double helix

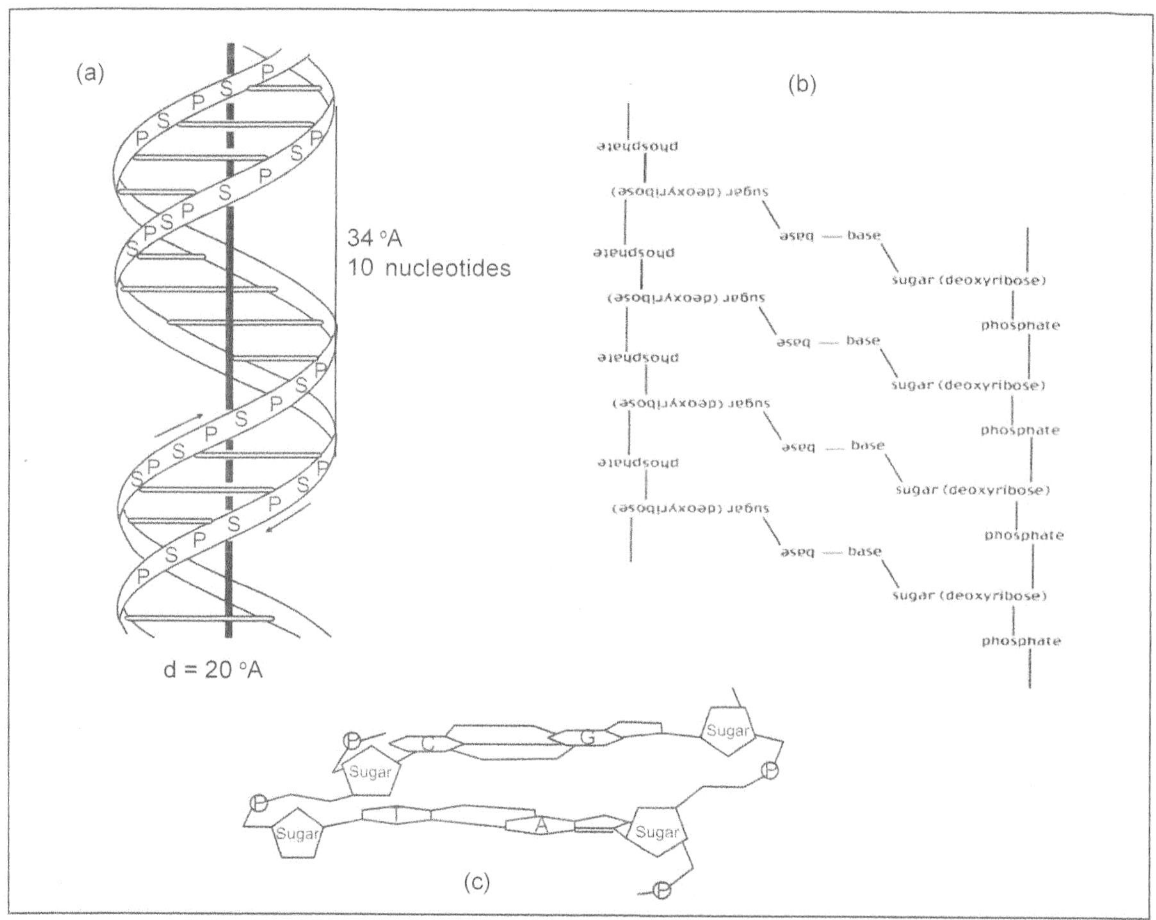

Fig. 6.26 The four nuclotides linked to the sugar and phosphate.

- ***The double helix has major and minor grooves in its structure***

The double helix is not an absolutely regular in appearance. It has variation in terms of two different grooves in its structural conformation: minor groove and major groove. This aspect is very important from the point of view of gene expression and regulation of information content, since these provide surfaces (docking space) for interaction between the double helix and proteins, involved during replication and expression of genes (Fig. 6.25).

- ***Major form of DNA helix is right handed one.***

The double helical structure that Watson and Crick described is right handed one and is called B-form of DNA. If one imagines DNA helix to be a spiral stairase, the sugar phosphate back bone is like a banister positioned towards right to support climbing. In cells, two other forms of DNA called A, and Z – DNA are also available under ***in vitro*** conditions; additional helical structures such as C, D and E – DNA have been found (Table 6.5) and (Fig. 6.27).

Table 6.5

Directionality of the helix	No. of base pairs/turn	Distance between base pairs(A^0)	Diameter of the helix (A^0)	Forms known as
Right handed	11	2.6	23	A-form
Right-handed	10	3.4	19	B-form
Right-handed	9.3	3.3	19	C-form
Left-handed	12	3.7	18	Z-form

A, B and C- DNA forms are formed under different relative humidities (75% for A, 92% for B and 66% for C); Z form when helix is made up of alternating purines and pyrimidines

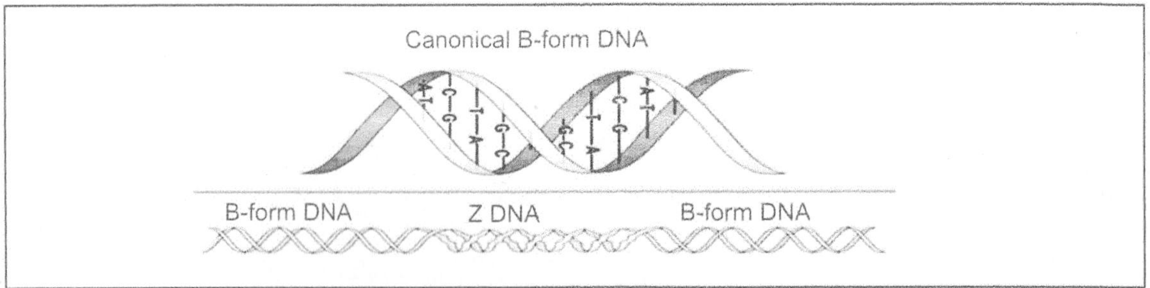

Fig. 6.27 Different Conformations of DNA

6.8.3 Complementary (base pairing) nature of the model implies self – replication

Of all the redeeming features of the double helix structure is its complementary nature of the base pairing which is important and suggestive of the most fundamental fact of molecular genetics, self duplication or replication. In fact Watson and Crick in their preliminary paper wrote, "It has not escaped our notice that the specific pairing we have postulated immediately suggests a possible copying mechanism for the genetic material. The base pair rule, a pair with T and G with C is implied and pairing between other base pairs is not possible.

6.9 Duplication of Genetic Material

6.9.1 Replication of DNA

One of the most fundamental properties of life is self-duplication, which when translated at cellular level means that every time a cell divides as a part of growth development and differentiation it must duplicate making a complete copy of all its genetic information. This is quite an imperative need to ensure that the products of cell division, the two daughter cells, must receive a complete copy of the genetic information contained in the parental cell. A dividing cell hence has to go through extensive duplication. How does at molecular level, the DNA replication occurs? How does the accuracy needed while replicating the information is achieved and maintained throughout the lifetime of organisms? The process of replication or information transfer from one cell generation to another has to be quite precise, maintaining a high degree of accuracy. Even an error rate of 1×10^{-4}–1×10^{-5} magnitude will result in a catastrophy and cause a significant level of alterations in the meaning of genes of a rapidly dividing prokaryote organisms or even eukaryote cells, making the gene sequence meaningless.

The pioneers of molecular genetics had the option of three aspects of DNA replication to be addressed: 1) Tenability of overall pattern of replication in terms of how each of the strands of the original parental double helix acts as a template for the synthesis of new polynucleotides; 2) Working out the biochemistry and enzymology of the replication process *viz.*, which kinds of proteins and enzymes are involved? What reactions do they accomplish during DNA replication? 3) How does DNA replication cellular machinery handles in diverse organisms with varying genomic conformations, circular, linear, that are prevalent in pro-and-eukaryotes?

The early period of developments in molecular genetics were quite so very exciting and intellectually challenging. The experimental strategies designed by the scientists in order to get solutions were so ingenious and bright that, they could serve as a model for the students of science!

6.9.2 An Over-view of DNA replication

In their publication of " ***Molecular structure of nucleic acids***" in Nature (1953), one of the last but one paragraphs, they stated that " ***It has not escaped our notice that the specific pairing, A pair with T and G with C we have postulated the base pair rule that immediately suggests a possible copying mechanism for the genetic material***". Indeed such a statement is an example of under statement. They were dropping clues quite humbly about the possible mode of replication of DNA. Their reference was to the fact that the complementary base pairing between the two polynucleotide chains of the double helix would let the each strand to act as a template surface for the synthesis (or copying) of its complement with cent percent accuracy. Thus, the double helix structure of DNA model proposed implies self-replication (Fig 6.26(a))

Despite such clear leads, molecular biologists of the time (early 1950s) were quite, unsure of about overall process of the replication. Three different plausible biological strategies for the self-duplication of DNA were being investigated (Fig. 6.28).

- ***Dispersive replication,*** in which each chain of every daughter molecule is made up of new and old bits of polynucleotide.
- ***Conservative replication,*** in which the original duplex is left as it is and a new chains are synthesized afresh to form a new duplex
- ***Semi-Conservative replication,*** in which the daughter molecule is formed from a one newly synthesized strand retaining one old parental strand to form a double helix.

Even though the semi-conservative mode of replication appears to be most likely way of biological replication, the scheme favoured by Watson and Crick, apart from intuition it was necessary to have an experimental proof which would distinguish between the three possible ways the DNA could replicate and conform to both prokaryotic as well as eukarytoic organisms.

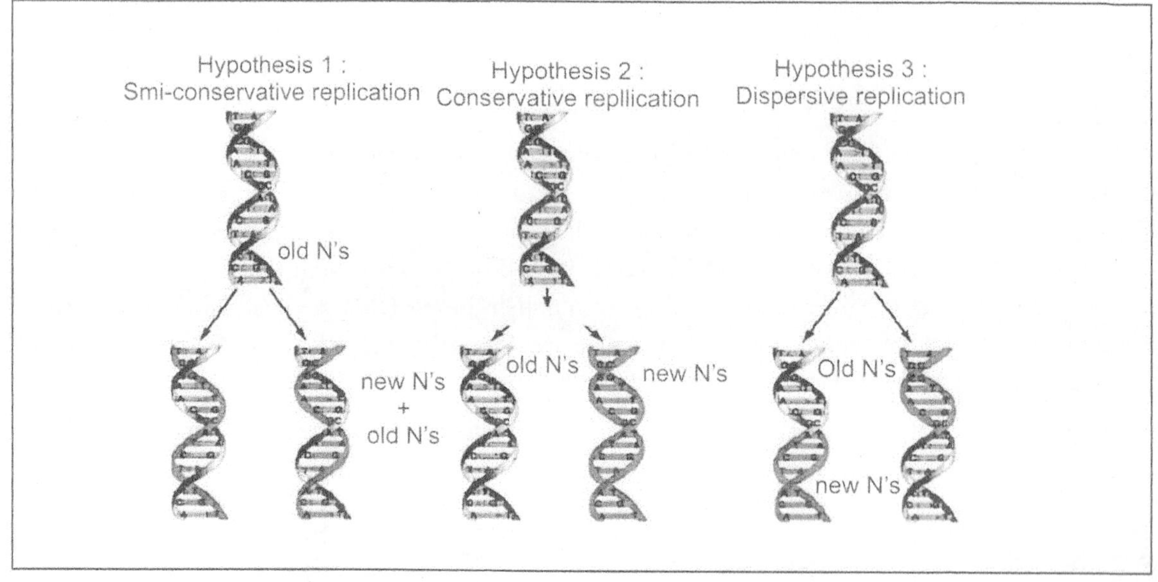

Fig. 6.28 Replication schemes, three possible ways DNA can replicate

6.9.3 The replication of DNA is semi conservative

The Meselson-Stahl Experiment

In order to test the hypotheses for the mechanism of DNA replication, Meselson-Stahl utilized the strategy of radioisotope labeling and tracking the distribution of parental atoms among progeny molecules. Furthermore, they believed that a label, which imparts increased density to DNA molecule, might additionally permit an analysis of the distribution by density gradient centrifugation technique. To start with **E.coli** were cultured on a simple medium at 36^0C with glucose and labelled NH_4Cl salt as the sole nitrogen source. Subsequently, the bacteria uniformly labeled with N^{15} were prepared by growing washed cells for fourteen generations in medium containing $N^{15}H_4Cl$ of 96.5% isotopic purity. Such bacteria, completely labeled with heavy nitrogen (N^{15}) were abruptly shifted to another media, containing glucose and NH_4Cl, so that new polynucleotide synthesized after resuspension, will contain only the normal isotope of nitrogen (N^{14}). The bacterial culture were then allowed to undergo one round of cell division, which takes about twenty minutes for E.coli, during which time each bacterial DNA molecule replicates just once. Some cells were taken from each culture; DNA was separated and analyzed by buoyant density gradient centrifugation. The result depicted a single band of DNA at a position corresponding to buoyant intermediate between the value expected from DNA labelled with N^{15} and N^{14} – DNA. This is suggestive of the fact that at the end of one round of replication, each DNA double helix is made of roughly equal amount of N^{15} – polynucleotide chains, an evidence that is in favour of semi conservative scheme of replication. If it were to be conservative mode, that DNA had replicated after one generation, only two bands, completely heavy (N^{15}/N^{15}) and completely light (N^{14}/N^{14}) should have been present.

In order to draw a distinction between semi-conservative and Dispersive mode of replication and the nature of distribution of the heavy (N^{15}) and light (N^{14}) label, the experimenters permitted **E. coli** to undergo another round of multiplication. At the end of the cell division, DNA was extracted and analyzed for the density gradient. Two bands one heavy (N^{15}/N^{14}) representing the same hybrid molecule (N^{15}/N^{14}) consisting of one old and another newly synthesized chain), another band, solely corresponding to (N^{14}/N^{14}). The data once again suggested, that the replication is semi-conservative. The Dispersive scheme could be discarded (rejected) because if it were to be so, the resultant progeny generation, should still have only one band made of hybrid composition (N^{15}/N^{14}) (Fig. 6.29). Meselson and Stahl, based on the data concluded that:

- *Nitrogen of a DNA molecule is divided equally between two subunits which remain intact through many generations*

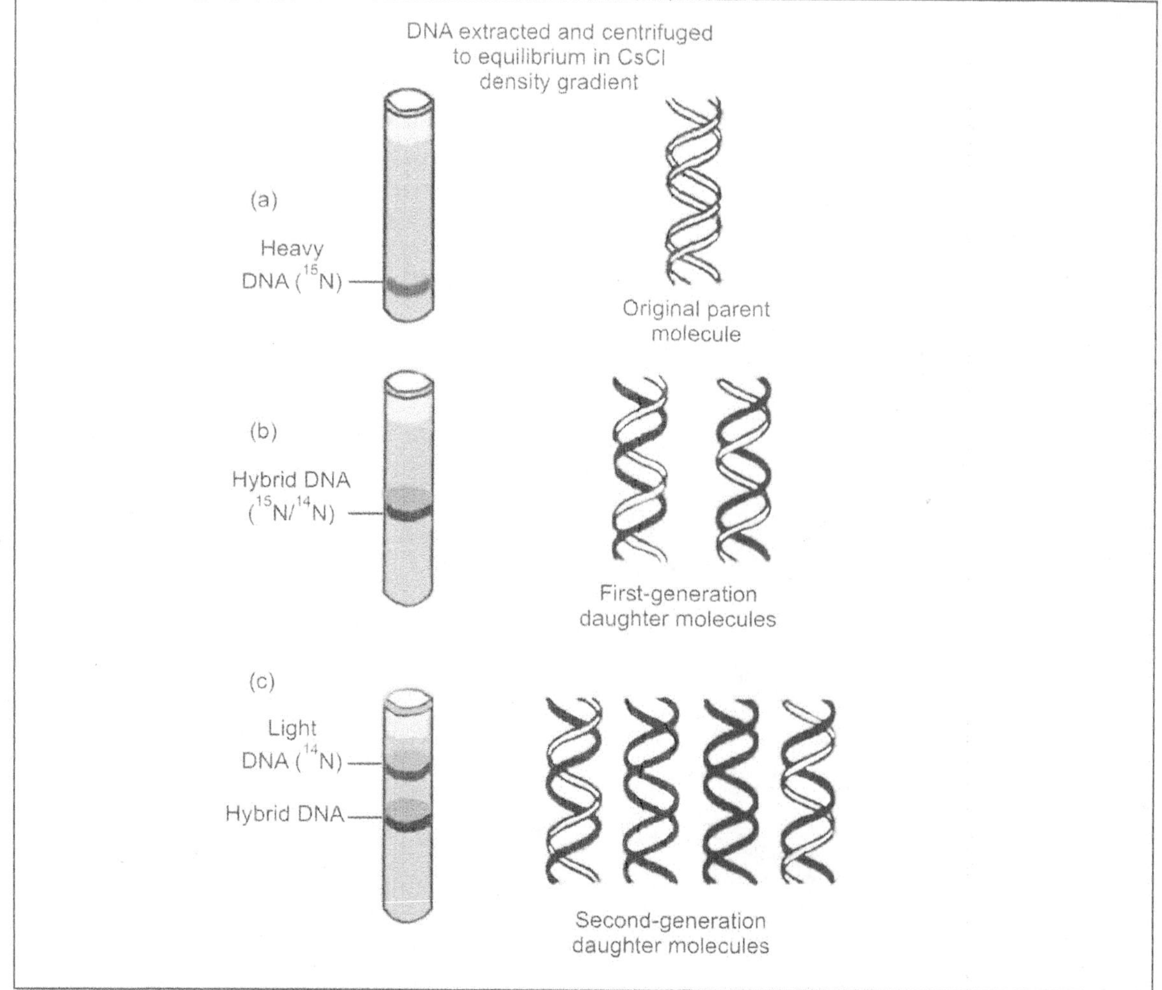

Fig. 6.29 Meselson and Stahl experiment

- *Following replication each daughter molecules has received one parental subunit.*
- *The replication results in a molecular doubling*
- *The result of the experiment are in exact accord with the expectations of the Watson-Crick model for DNA duplication*
- *DNA molecule is divided equally between two physically continuous subunits. Following duplication each daughter molecule receives one of these and conserved through many replication.*

The experiment of Meselson and Stahl strongly suggested the overall view of DNA replication. However, the details of replication of DNA are far more complex when looked into the diverse organisms, virus, bacteria and higher organisms.

6.9.4 The mechanism of DNA Replication

Tests for the replication scheme proposed were not immediately developed, until Meselson and Stahl (1958) made it almost certain that, in *E. coli* the replication follows the semi-conservative manner; (See Fig. 6.30) each of the strands is retained and made to act as template to synthesize new DNA strand, making use of complementary principle. Thus DNA is replicated and doubled. Even though results appeared difficult to reconcile with any other mode of replication, experiments dealing with eukaryotes were designed to test the idea that a chromosome might consists of a single DNA duplex, employing the titrated thymidine, a label which incorporates wherever thymine is by Herbert Taylor (1958), which supported the concept of replication of DNA. At the same time biochemists were developing strategies to synthesize DNA *in vitro*. Major impetus in this direction came from the

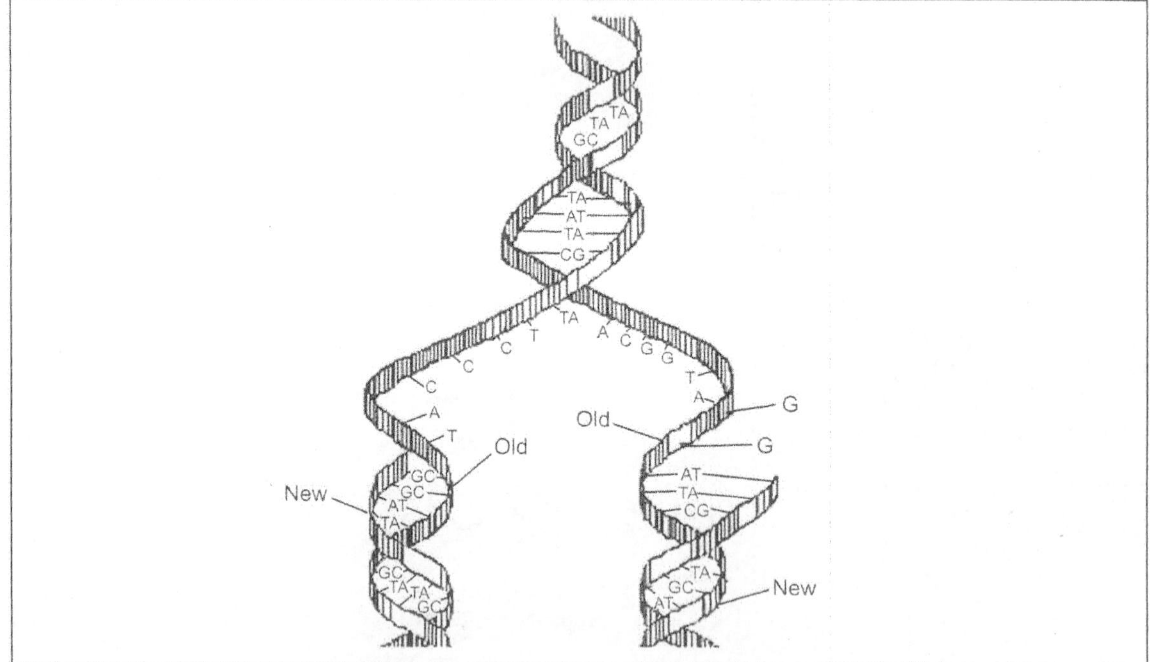

Fig. 6.30 Semiconservative Mode of Replication.

discovery of Grun berg-Manago and Ochoa (1955) that, an enzyme extract of **Azotobacter agilis** catalyzed the polymerization of nucleoside diphosphate into polyribonucleotide, paved the way to investigate the synthesis of DNA *in vitro*. Unfortunately, this system did not take direction from a template. In science, new discoveries are made with a combination of genius and luck. Arthur Kornberg, a former colleague of Ochoa, soon reported success with an extract from *E. coli*, which synthesized DNA polymer. The *in vitro* synthesis of DNA required a primer to start the reaction. The early evidence, by comparing the base pair composition and ratio of primer and newly polymerized polynucleotide product, which were identical, indicated that the primer is acting as a template surface for the synthesis. This discovery paved the way for understanding, not only the biochemical mechanism of DNA replication, but also played an important role in supplying the proof for Watson and Crick's scheme of proposed replication of DNA.

Most of the Prokaryotes have a single chromosome, ***nucleoid*** consisting of circular double stranded DNA. The size may vary considerably. ***E. coli*** for example has the genome size of 3.8×10^6 bp, whereas, ***Bacillus subtilis*** has 2×10^6 bp and ***Salmonella typhimurium*** 10.5×10^6 bp. If one extends to a linear form, it would be around 10mm long but has a compact structure because of supercoiling of the DNA. During the replication this must be opened up to let the access to enzymes involved in replication. When a DNA molecule is being replicated a small region opens up to assume non-base – paired form. The breakage in base pairing begins at a specific region, called the ***replication origin*** (Fig. 6.31). It gradually progresses like unzipping with simultaneous synthesis possibly in both the directions. The region at which the base pairs of the parental DNA double helix molecule are broken and the new polynucleotides are synthesized is referred to as a ***replication fork.***

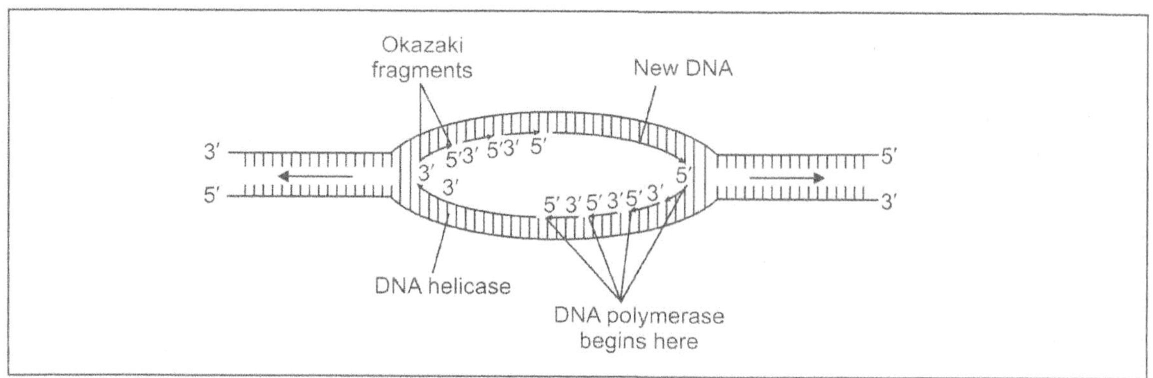

Fig. 6.31 Origin of Replication and its progression

6.9.4.1 DNA polymerase is the enzyme that catalyzes polynucleotide synthesis (Replication)

An enzyme that takes part in the synthesis of new ploynucleotide chain is known as ***DNA polymerase***. It requires deoxyribonucleotide for assembling the new chain of polynucleotides on the parental molecule acting as a template following the complementary base pairing principle. DNA polymerization takes place in 5' - 3' direction (Fig. 6.32).

In general outline, the mechanism of DNA replication is more or less similar in most of the pro- and-eukaryotic organisms. Differences exist only with respect to the enzymes and proteins that are needed for the initiation and regulation of the replication processes.

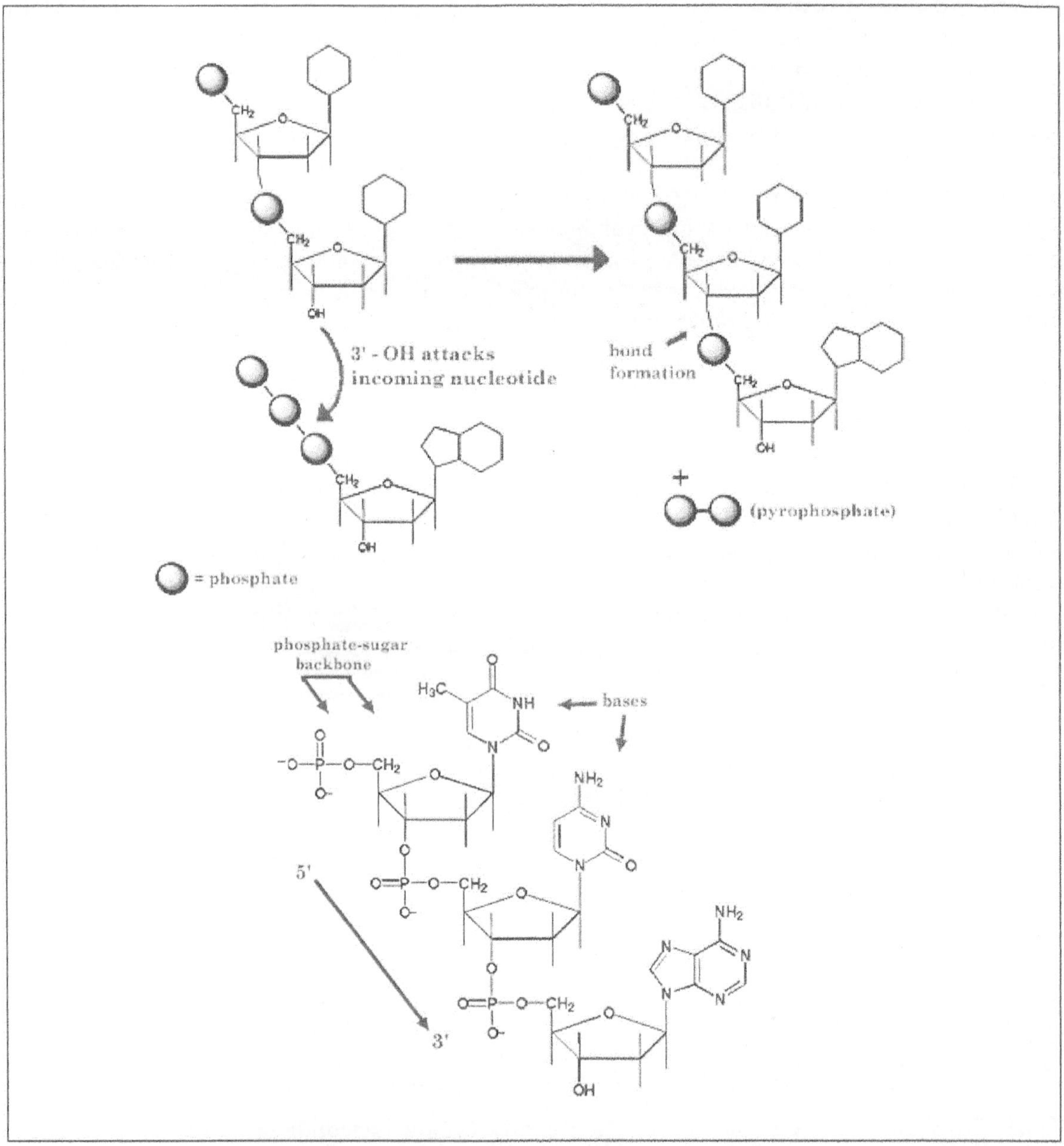

Fig. 6.32 Addition of nucleotide bases in 5´ to 3´ direction.

The cell free system developed by Kornberg for the synthesis of DNA ***in vitro*** employing an enzyme isolated from ***E.coli*** was quite competent to polymerise DNA on a polynucleotide template in the presence of (ATP, TTP, GTP, CTP) and Mg^{++} ions. It was believed that this enzyme was the DNA polymerase that catalyses the DNA replication, in the bacterial cells ***in vivo*** too. As the enzyme was studied intensively it became known that the gene that codes for ***DNA polymerase I*** in ***E.coli,*** even after mutational inactivation, the DNA replication takes place. It was then realized that even though DNA polymerase I is involved in the replication, it is not the primarily important and responsible for the replication in ***E. coli.*** Another gene coded product, ***DNA polymerase* III** is responsible primarily for the replication. It is a massive enzyme with a mass exceeding > 250,000 daltons made up of several subunits.

6.9.4.2 Replication origin and the events that ensure at the replication fork of DNA

The insight and experimental evidences of what exactly happens and how DNA is replicated right from its origin to termination has accrued over decades of hard and intelligent experimental work of biochemists and molecular geneticists all over the world. Fairly a good picture and a consensus has emerged about events and enzymatic process that takes place during replication. The starting point of replication has to be the **breakage of the parental double helix molecule,** in order to act as templates to provide information for the synthesis of daughter molecule. This necessitates the provision for breakage of the base pairing between two strands of the parental double helix. Two enzymes, known as **helicases** in conjunction with **Single Strand Binding Proteins** (SSBP) are known to play the role. The job of helicase is to unwind the two parental strands and permit the SSBPs to attach the single stranded DNA, so that the separated strands do not anneal in the creation of **replication fork.** The template surface on which nucleotides enable assembly by the action of **DNA polymerase III**

Since the DNA polymerase III is capable of synthesizing DNA molecule only in 5' to 3' orientation, the problem arises to the cell machinery in treating the two separated strands, having two different orientation, one in the direction of 5' - 3' and the second one in the opposite 3' - 5' in identical way (Fig 6.33). The parental strand that is in 3' - 5' direction there is no problem, since the synthesis can start from the opposite end in 5' - 3' direction continuously to make an exact replica. This strand is called **leading strand** in technical jargon. However the second strand known as **lagging strand,**

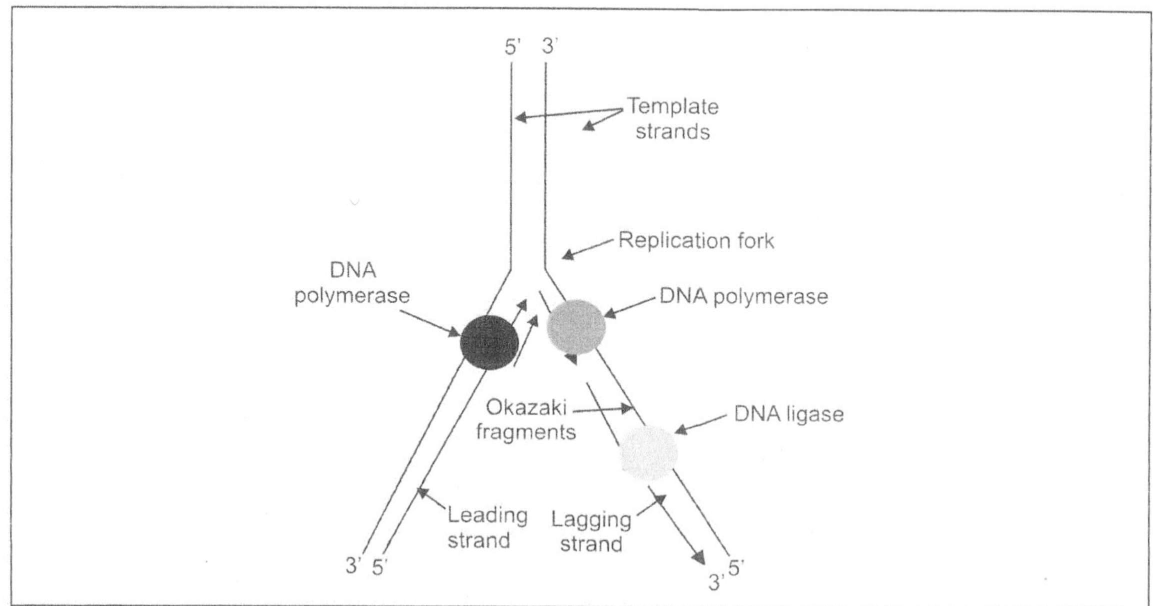

Fig. 6.33 DNA replication on leading and lagging strand

which is in 5' - 3' direction, cannot be copied continuously, because this needs orientation of synthesis in 3' - 5' which the enzyme DNA polymerase cannot perform. Only option left to the organism is to have a strategy of fragmentation of the replication process and to go through the synthesis part by part (Fig 6.33) as the strands are dissociating. This ingenous method adopted by the cell was hypothetically proposed and discovered by Reiji Okazaki (1968). A short varying lengths of 100-1000 nucleotides long being polymerized was confirmed by others and the Okazaki **fragments** infact exist at the time of replication, which are sealed enzymatically to get a completed strand.

6.9.4.3 Major classes of proteins used in the DNA Replication Machinery

Topoisomerases: Starts the process of DNA unwinding by nicking (breaking) DNA. As a consequence, the torsion (tension) force holding the helix and its supercoiled is released

Helicases: Unwinds the original double stranded DNA, once super coiling has been eliminated by Topoisomerase. DNA polymerases complex synthesize new strands. Complex aggregates of different protein subunits often includes proteins that perform **proof reading**. If any error is crept locally, the same will be identified and excised and in its place the correct pairs are aligned thus effecting the repair. In cells, DNA directed DNA polymerase participates in the replication by new DNA synthesis from an existing DNA template. A wide variety of polymerases exist in mammalian cells (α, β, γ, δ and ε). (see Table 6.6).

Major categories of DNA polymerases

[A] Classical DNA directed DNA polymerases with high fidelity:

	DNA polymerase type				
	α (Alpha)	**β (Beta)**	**γ (Gamma)**	**δ (delta)**	**ε (Epsilon)**
• Location	Nucleus	Nucleus	Mitochondria	Nucleus	Nucleus
• DNA replication	Synthesis and priming of Lagging strand	replication	mt DNA of leading strand	synthesis	—
• Exonuclease, 3'-5' with proof reading potential	No (-)	No (-)	Yes(+)	Yes(+)	Yes(+)
• DNA repair function	–	Base excision	mtDNArepair	base excision	base excision

Primases: Mere availability of single strand DNA is not enough to kick-start the replication. A primer is required with 3' OH group on to which the polymerase can attach a dNTP. The specific role of primase is to attach small RNA primer to the single stranded DNA to act as a substitute, 3' OH for DNA polymerase to initiate synthesis. Once the replication starts, the RNA primer is eventually removed by a nuclease and the gap generated is filled in using a DNA polymerase, then sealed off using another enzyme ligase

Ligases catalyze the formation of a phosphodiester bond given an unattached but adjacent 3' OH and 5' phosphate. Single strand binding proteins (SSBPs) are important to maintain the stability of the replication, as fork single strand DNA is very labile or unstable, so the proteins bind to it while it remains single stranded and keep it from being degraded.

[B] DNA directed DNA polymerase with low fidelity and error-prone

DNA polymerase eg [Zeta]: Involved in hypermutation; expressed in Immune system (B and T cells)

DNA polymerase n [eta] : Mutates A and T nucleotides during hyper mutation

DNA polymerase i [iota] : Involved in hypermutation involving very low of fidelity of replication thought to be involved in Hypermutation.

DNA polymerase ì(mu): highly expressed in B and T cells

[C] RNA directed DNA polymerases (reverse transcriptases)

- Telomerase reverse transcriptase (Tert) : Replication of DNA at the end of linear chromosomes
- LINE-1/ endogenous rectrovirus reverse transcriptase (L-1 Ert): occasionally converts in mRNA and other RNA into cDNA which integrates elsewhere into genome.

6.9.4.4 Kick-starting of Replication

The problem of priming and the Joining of Okazaki fragments

DNA polymerase III by itself cannot start the DNA synthesis, the way RNA polymerase does in transcription and synthesis of RNAs. To prime DNA synthesis, a short stretch of double stranded region has to act as a ***primer*** (Fig. 6.34). This is accomplished by the RNA polymerase enzyme by aligning a short stretch of ribonucleotides on the single stranded DNA to make initially a priming stretch of double stranded-DNA and the enzyme withdraws. Subsequently, DNA polymerase takes over and start synthesizing the vest of the strand on the template by adding deoxynucleotide base pairs. Primase is the name given to the enzyme RNA polymerase that initially starts polymerizing RNA on DNA. In *E. coli* the *primase* is a single polypeptide with a molecular weight of 60,000 daltons. This is different one from that of transcribing RNA polymerases. It acts along with six other proteins, which constitute a complex known as *primosome.*

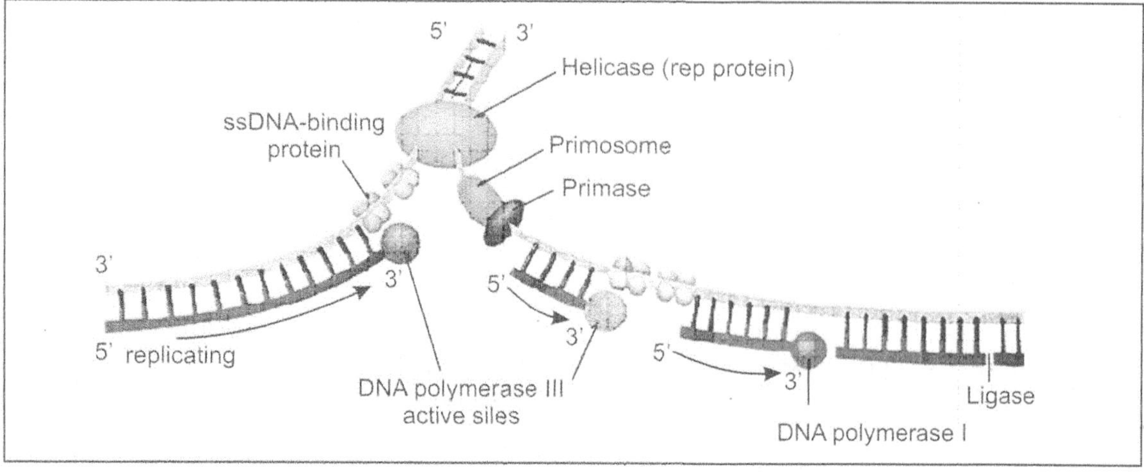

Fig. 6.34 Priming of DNA synthesis

Since, the synthesis of DNA on lagging strand is discontinuous, the DNA polymerase III can align bases (deoxy trinucleotides) only for a short distance till it reaches the RNA primer at the 5' end of the subsequent ***Okazaki*** fragments. DNA polymerase I takes over at this stage by relieving the DNA polymerase III and continuing not only the synthesis of the primer of the adjacent ***Okazaki*** fragments and replaces the same with deoxyribonucleotides. When the job is over, either it stops or continues to a short distance into the (DNA) region of ***Okazaki*** fragment replacing nucleotides before it dissociates from the double stranded DNA.

Finally, to complete the replication of lagging strand, there is a need to join adjacent Okazaki fragments, which have a gap between neighboring nucleotides. A phosphodiester bond has to be synthesized at this position joining the bases. DNA ligase does this job (Fig 6.35).

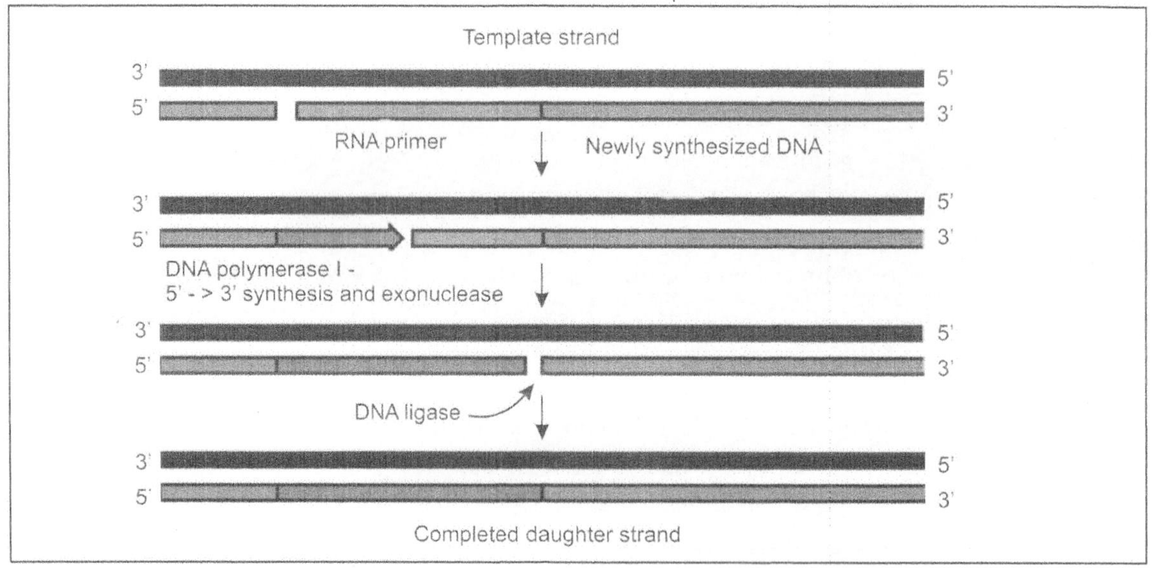

Fig 6.35 Completion of DNA synthesis on lagging strand

6.9.4.5 Unwinding of DNA to facilitate Replication–The Topological Problem

In understanding DNA replication, the major problem for a long time has been how the unwinding of double helix is handled by the cell, since the two polynucleotides chains of parental helix are wound round each other. This implies that the progression of the replication fork along the parent molecule is not just unzipping by snapping the hydrogen bonds, but it is more complex than that. This is a crucial problem because the genome sizes vary from 0.58 Mb to 3300Mb (0.58×10^6 in Mycobacterium to 3.3×10^9bp in humans). For instance the ***E.coli*** with a genome size of 4.6×10^6 bp, it has to go through 4.6×10^5 i.e., nearly half-a-million turns in the process of unwinding during the replication. This has to be done with in twenty minutes. The attendant implication is that the double helix should rotate at a speed of 6500 rpm! The topological conundrum is solved by a battery of enzymes.

6.9.4.6 DNA topoisomerase and its role in DNA Replication

The impossible feat of unwinding needed at the time of replication is accomplished by a battery of enzymes called ***DNA topoisomerase.*** There are two categories of them: Type I and Type II. Both of them unwind DNA molecule without rotating the double helix! They cause transient breaks in the backbone of DNA and achive the task. Type I topoisomerase break one of the polynucleotide chains and pass the other strand through the gap before reconstituting the backbone (Fig. 6.36(a)) this is accomplished by forming a covalent bond to one of the broken ends, moving one strand around the other and subsequently transferring bound end to the other broken end. Since, bonds are conserved, no input of energy is needed.

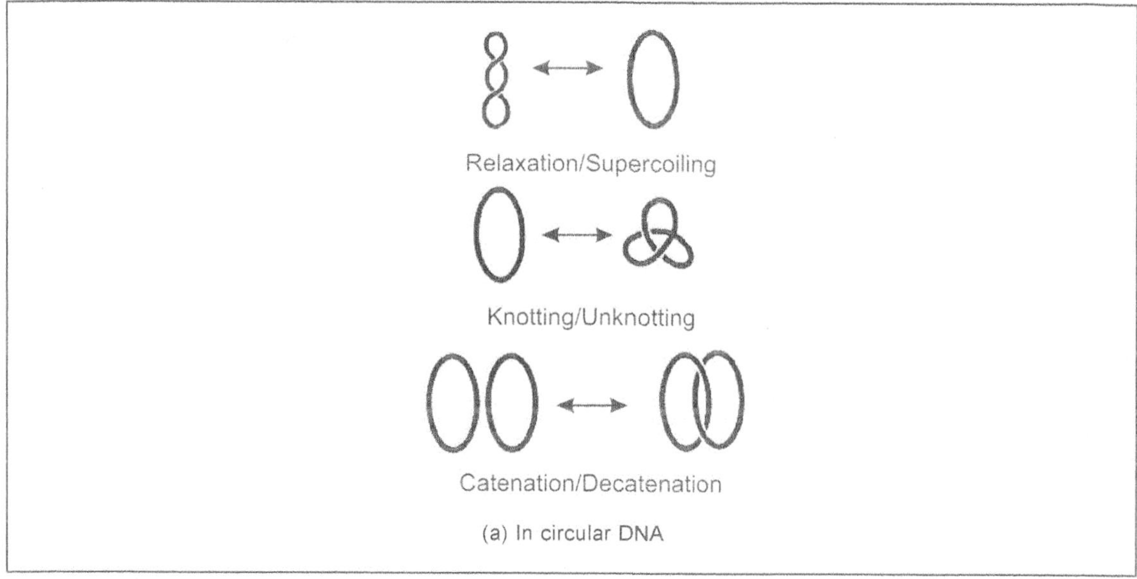

Fig. 6.36 (a) Action of topoisomerase in reducing the supercoiling

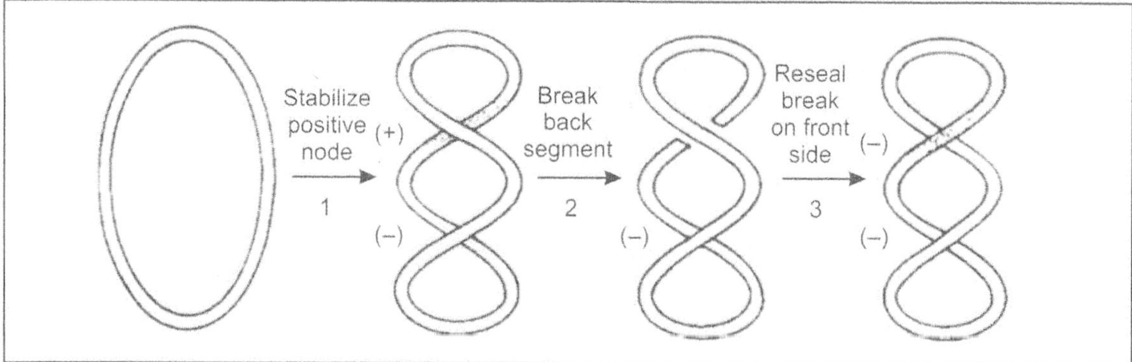

Fig. 6.36 (b) Topoisomerase II in circular DNA

Topoisomerase Type II is a well characterized which include the ***DNA gyrase*** in ***E. coli*** which carryout operations similar to Type I with one exception; it breaks both the chains at adjacent positions. Both the enzymes achieve the same end results i.e., positions of double helix just in advance of the replication fork are regularly unwind permitting the replication fork to progress without any hindrance along the parental molecule (Fig. 6.36 (b)).

DNA gyrase may introduce negative super coiling in duplex DNA, a reverse process to the earlier one and induct extra turns into a DNA molecule as opposite to unwinding (Fig 6.36(b)).

6.9.5 Replicon: Replication of Genomes

From the earlier presentation of DNA replication it appears that everything is known about duplication of DNA molecule contained in the genome. It is not true. There are several difficulties in explaining minor but significant events. The process of replications in organisms present a variety of modes depending on genome conformation, some are circular with only one DNA molecule as a genome, some are linear and yet some have additional genomes organellar, as well as plasmids. Even if an organisms has linear or circular DNA as their genome presents problems in explaining replication of its respective terminal region during replication. Origins of replicons can be mapped by autoradiography or electrophoresis. Prokaryotes like bacteria have a single genome as a replicon, whereas eukaryotic chromosomes have many replicons. Even though bi-directional replications form one or more origins will make copies of the bulk of linear molecule, the ends of linear DNA molecule pose a problem for replication. This is mainly due to the presence of RNA primer at 5' region. All known nucleic acid polymerases, DNA or RNA, proceed synthesis in 5' – 3' direction only. This poses a problem to replicate ends of the linear molecules. It is not known how the primer RNA is removed and the ends get replicated. Different organisms especially virus evolved ingenious method to stem the problem but the basic question as to how terminal RNA primer sequences are converted to DNA sequences remains. The strand which is in 3'-5' orientation there is no problem since the new strand synthesizes in 5'-3' orientation. The problem comes to the strand which is in 5'-3' position as the synthesis must start from the last base or if deleted it becomes shorter. This problem by changing linear to circular or multimeric replicon or forming unusual hairpin bends as in the case of mtDNA in ***Paramecium*** or by enzymatic activity of ***Telomerase*** or by a protein intervention as in the case of viruses.

Circular molecules are much facile to replicate perhaps this is the reason, why nature designed all the rapidly dividing genomes such as bacteria, viruses etc. are circular. Two main strategies have been adopted. Most of the bacterial or viral genomic DNA is replicated bidirectionally comprising of two forks around the circle from a single origin, forming an intermediate from prior to the joining of forks. Phages employ a different strategy known as rolling circle mode of replication. This begins with an inclusion in one strand of the parental molecule and extension of the free 3'-OH end by DNA polymerase (Fig. 6.37). The original parental strand is displaced and the daughter molecule is rolled off. The DNA synthesis may cease after one round of revolution, resulting in two daughter molecules (Fig. 6.37) or may continue around the circle several rounds, very rapidly producing a series of ***concatamers***. When single stranded copies are linked end to end are called as ***Concatamers***. These are subsequently cut into individual genomes and the ends are ligated to produce daughter molecules.

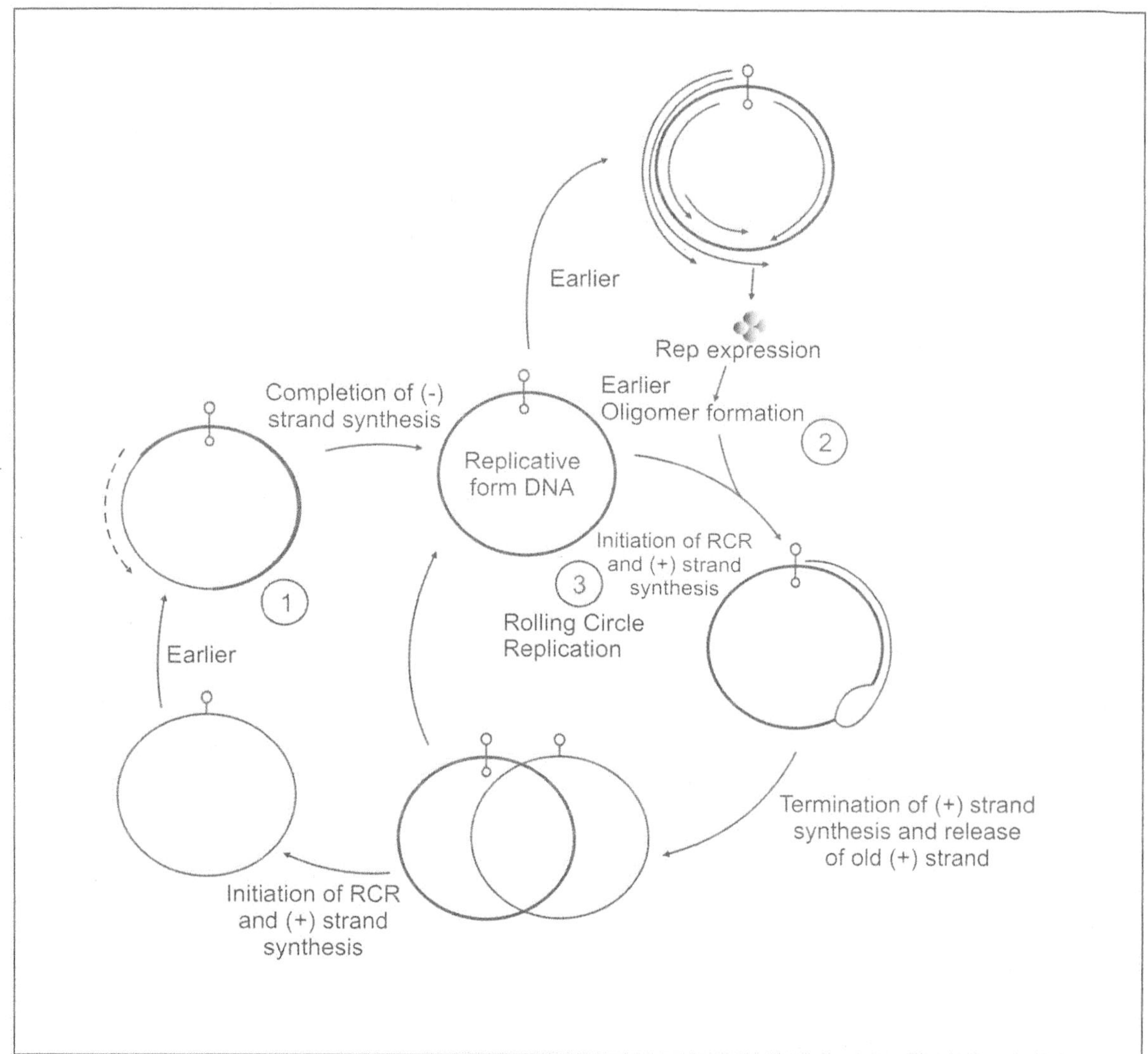

Fig. 6.37 Replication of circular genomes–Rolling circle model

6.10 Gene Expression

Gene expression is the process whereby the biological information encoded in genes as a sequence of nucleotides of DNA is made available to cell. It is a complex process but in outline it is relatively simple and straightforward. Before one sets forth to study the transcription, translation relating genes to protein synthesis and their function and regulation it is necessary to understand the overall process and what it accomplished to provide a frame work of *modus operandi* of the whole gamut of gene expression.

6.10.1 The Central Dogma of Molecular Biology

Francis Crick in 1958, while lecturing on protein synthesis at the society for Experimental Biology, propounded a theory that the biological information possessed in a gene encoded as a sequence of nucleotides is transferred in the first place to RNA and subsequently to protein. All experimental evidences so far suggest that in simplest form this is what happens during gene expression of any gene. Crick has also postulated that the information flow is a one-way traffic in that the proteins by themselves are incapable of directing the synthesis of RNA and RNA in turn cannot direct the synthesis of the DNA. An exception to this was soon discovered in 1970 by Howard Temin & David Baltimore independently in certain viruses, where the transfer of biological information takes place from RNA to DNA (Fig. 6.38). However, *the central Dogma* still remains conceptually as one of the bedrocks of molecular genetics.

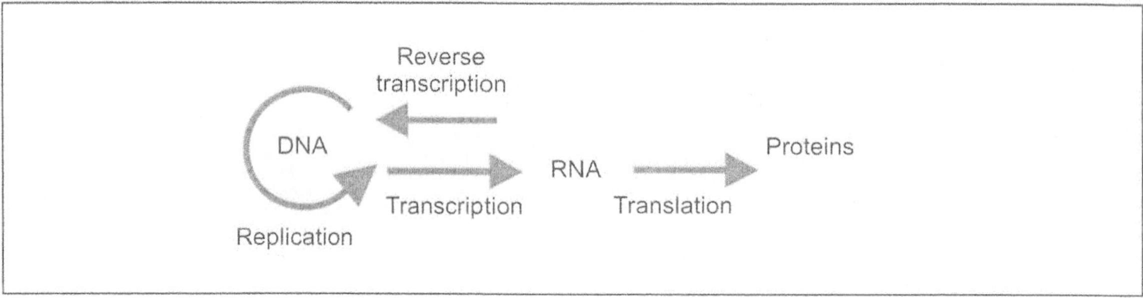

Fig 6.38 Central dogma of molecular biology: information transfer that occurs universally

6.10.1.1 The first stage of gene expression is transcription

The process by which one of the DNA (Gene) strand acts as a template to direct synthesis of RNA using base complementary principle is called *transcription*. The newly synthesized RNA is called *transcript*. The chemical reactions involved in RNA synthesis, the enzymes that drive the reactions and over all process of *transcription* differs between pro-and-eukarytoic organisms. The basic features of genes were established prior to their descent and divergence from *progenotes*. Pioneering work in molecular biology was carried out using prokaryotic organism, *Escherichia coli.* However, since early 1980's molecular biologists over came the difficulties involved in the study of eukaryotes by turning their attention to a single celled eukaryote, *Saccharomyces cereviciea.*

6.10.1.2 RNA Synthesis using the DNA template strand and the base complementary rule

In molecular biology all aspects of gene structure and expression is referred in terms of *nucleotide sequences*. Usually it is a convention that the sequences presented is always that of *non-template strand* of the gene (DNA) in the 5'– 3' direction. Such a convention may look odd, since the biological information is carried by the template strand not by its complementary strand. When RNA is synthesized as a transcript using the template it is done so by using the base complementation rule, and hence the sequence is same that of non-template strand, excepting the base thiamine, which is replaced by another base uracil. Hence, *when writing a nucleotide sequence of a gene, the non-template strand is given as this has the same sequence as that of RNA transcript* (Fig.6.39).

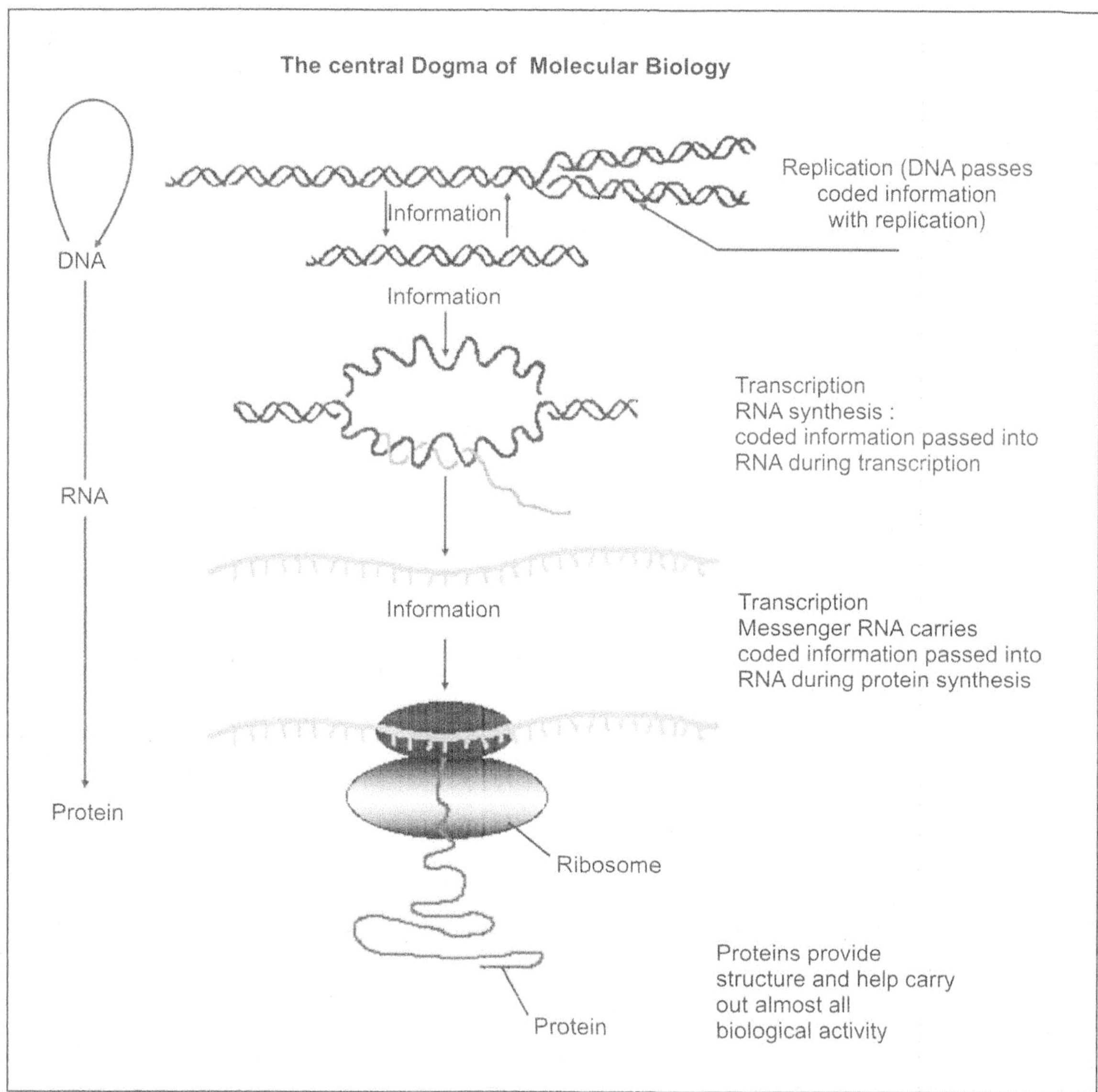

Fig. 6.39 Information flow during gene expression

6.10.1.3 Types of Major classes of RNA Molecules

One of the functions of DNA as the genetic material is to provide information (acting as template) to synthesize RNAs. There are three major categories of RNA molecules produced by transcription, apart from innumerable kinds of small RNAs. These are *ribosomal RNA (rRNA), transfer RNA (tRNA) and messenger RNA (mRNA).* Ribosomal RNA and tRNA are end results of gene expression and as such does the pivotal role in the cell needed for gene expression while, messenger RNA goes through the second stage of gene expression, namely translation and has no other function beyond playing the role of intermediary between gene and its expressed product protein. Ribosomal and transfer RNA are considered stable becoming part of protein synthesizing machinery and hence are long-lived when compared to messenger RNA, which is short-lived and has quick *turnover rate* (Fig 6.40).

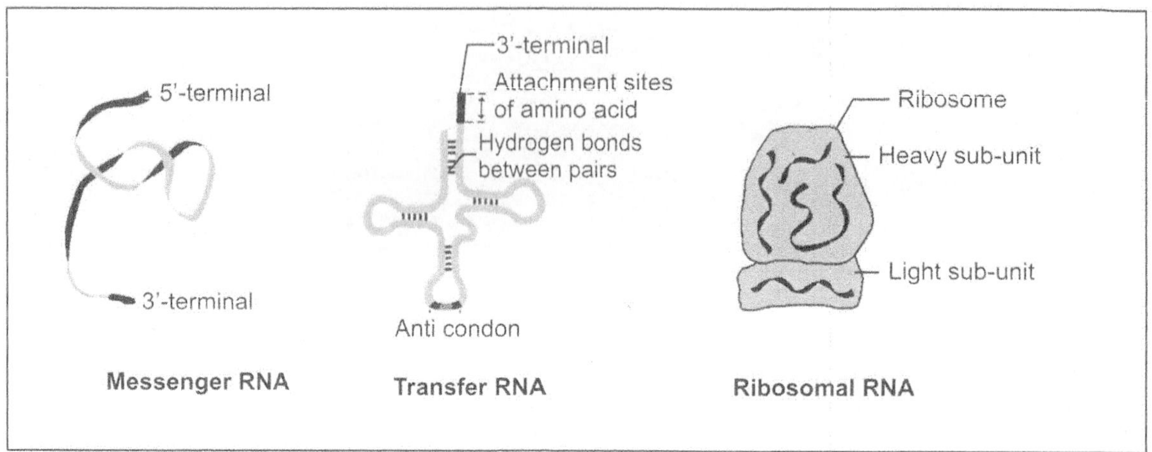

Fig. 6.40 Major Classes of RNA

6.10.1.4 RNA Synthesis

To view the process of transcription from chemical reaction point of view it is nothing but synthesis of an RNA molecule, be it stable rRNA and tRNA or transient mRNAs. This takes place by polymerization of ribonucleotides, which can be shown empirically as:

$$n(NTP) + \text{template (DNA)} + Mg^{++} \rightarrow \text{RNA of n nucleotide length} + n-1 \text{ PPi}$$

Essentially during polymerization the 3'-OH group of one ribonucleotide engages with the 5' phosphate group of the second one to form a *Phosphodiester bond* resulting in the removal of one inorganic pyrophosphate molecule (PPi) for each bond accomplished. The chemical reaction is guided by the presence of DNA template to direct the sequence or ribonucleotide to be polymerized. The RNA transcript formed is in 5'-3' direction. This means the template strand (DNA) must be read in the 3'-5' direction.

6.10.1.5 RNA Polymerase is the primary enzyme that catalyzes RNA synthesis during transcription

The enzyme that catalyses RNA synthesis during transcription is DNA-dependent *RNA polymerase* (RNA polymerase), which was first discovered in 1958. Prokaryotes and eukaryotes differ with respect to structure and type of polymerase. Polymerase performs several tasks during transcription. RNA polymerase is a massive protein made up of several polypeptide subunits.

6.10.1.6 The RNA polymerase of Prokaryotes typified by E.coli is made up of five subunits

In E.coli at any given time, the cell contains nearly seven thousand RNA polymerase molecules of which nearly 2000 - 5000 may be actively engaged in transcription. Each molecule is made up of two alpha(α) polypeptides, one each of Beta(β) and homologue Beta prime (β') and sigma (σ). Generally the structure is indicated as α_2 β β' σ (Fig. 6.41(a)). Such a structure is called as ***holoenzyme*** and has molecular weight of 48KD (48,000/daltons). There is another version of this called as a core enzyme with a molecular mass of 395,000 daltons (39.5kD). Typically this is not associated with sigma (σ) and is described as α_2 β β'. These two versions play a different role during transcription. (Fig.6.41(b))

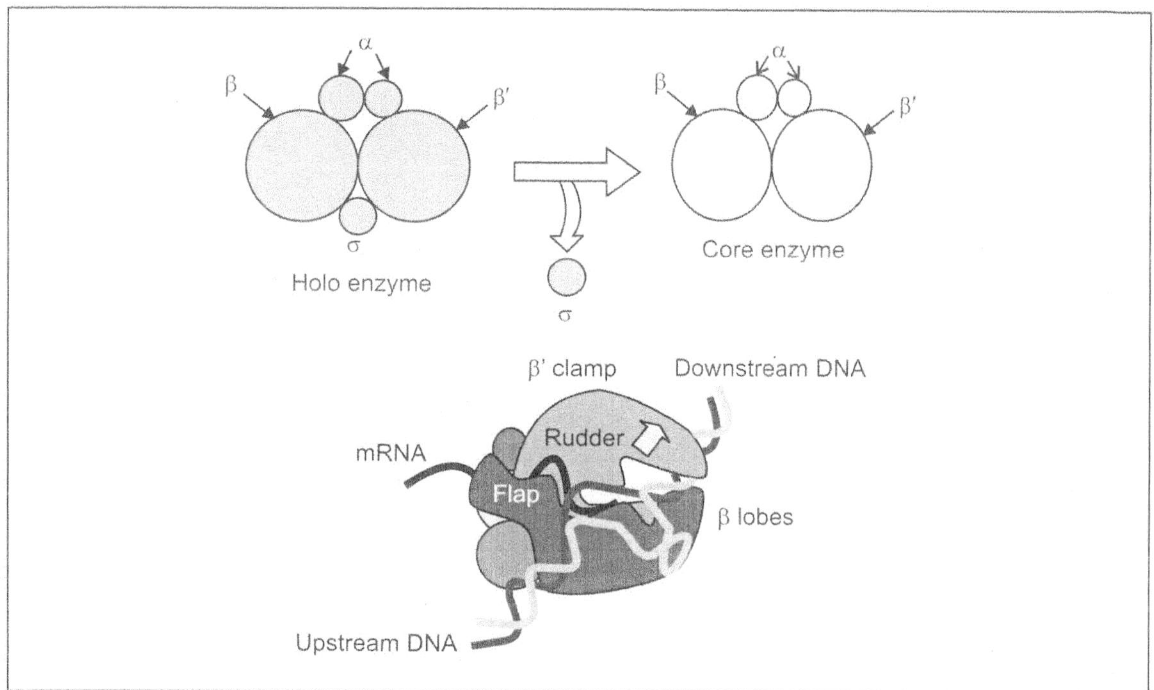

Fig. 6.41(a) Structure of RNA polymerase (Prokaryotic)

6.10.1.7 RNA Polymerases of Eukaryotes are more complex and of three different types

In contrast to prokaryote (***E.coli***), where all genes rRNA, tRNA and mRNA are transcribed by the same polymerase enzyme, in eukaryotes there are three different kinds of polymerase, one each for the transcription rRNA, tRNA and mRNA. Besides, the enzymes are also large when compared to ***E.coli.***, each composed of 8-14 subunits and having a total molecular weight far exceeding 500 KD. The yeast (***S.cerevisiae***) polymerase are well characterized and studied among eukaryotes. However, there is great deal of homology between the components (subunits) of polymerase between pro-and-eukarytic organisms. For instance the largest subunits of RNA polymerase II, that transcribe genes for protein designated as RPB_1, RPB_2 and RPB_3 are homologous to α, β and β^1 subunits of E.coli. This was made possible by disaggregating the polymerase enzyme subunits and determinig the amino acid sequence of each polypeptide, besides studying their role in transcription. However, the role of each and every other subunits remains unknown and infact certain experiments suggested that they are redundant as the transcription goes on even if they are completely dispensed with. (Fig. 6.41(b) Table 6.6)

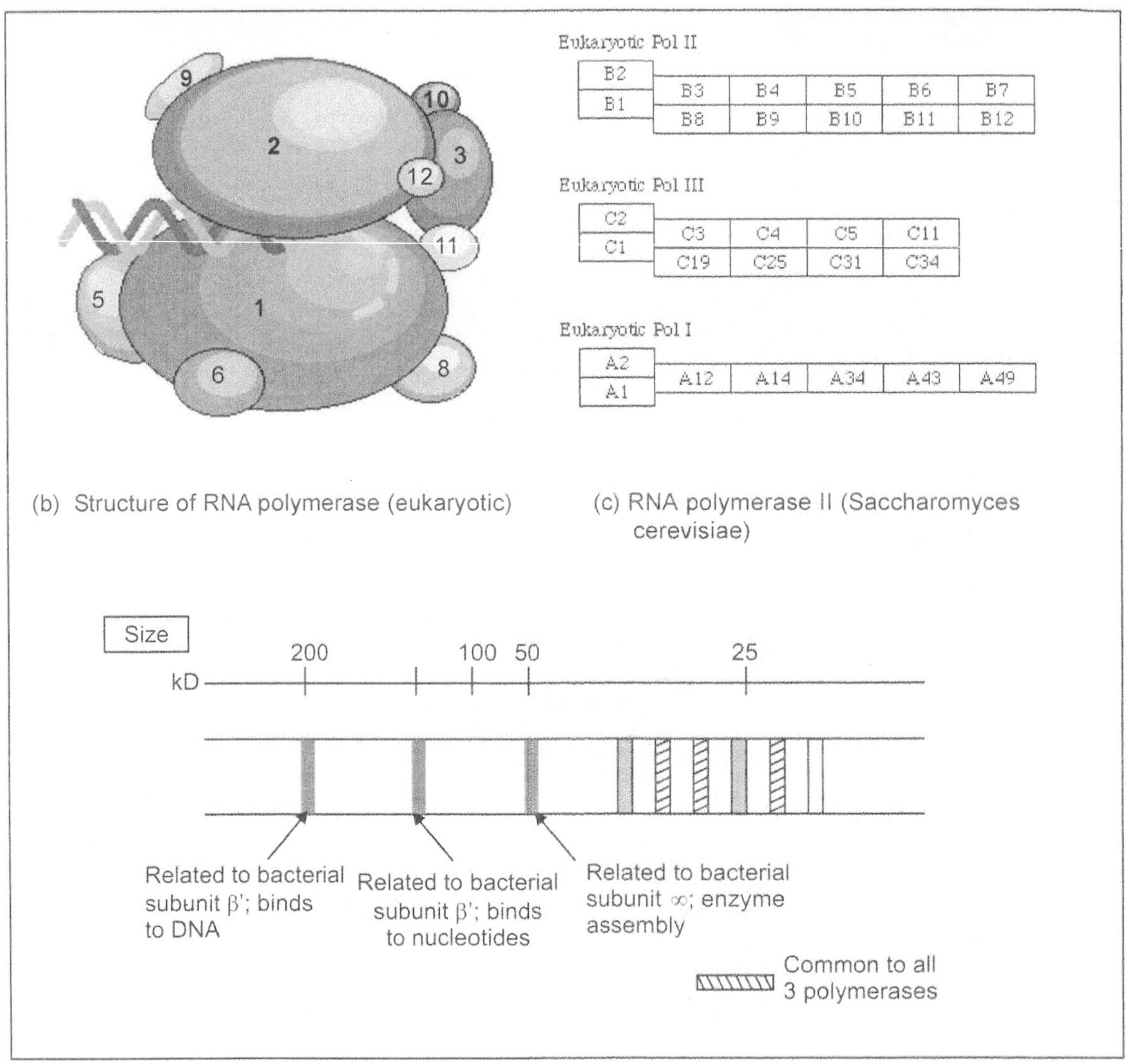

Fig. 6.41(b) Structure of eukaryotic polymerase.

Table 6.6 *Three different RNA polymerase are involved in eukarytoic gene transcription.*

Type of Polymerase	**location in Nucleus**	**Responsible for transcription of**	**Relative activity (%)**	**Sensitivity to α – amanitin**
RNA Polymerase I	Nucleolous	rRNA	50-70	–
RNA Polymerase II	Nucleoplasm	Nuclear RNA	20-40	++
RNA Polymerase III	Nucleoplasm	tRNA	10	species specific

6.10.2 The process of transcription

The actual process of transcription varies between prokaryotes and eukaryotes. However, it can be divided into three distinct stages: (a) Initiation (b) elongation and (c) termination.

(a) ***Initiation :*** The critical point in the transcription is its accurate initiation of the gene sequences, and not random sequences of any regions of DNA. This implies that the responsibility of polymerase is to bind at specific *start point* or position just upfront of the gene (upstream) to be transcribed. Specific binding sites are quite essential because in all organisms a large proportion of the total DNA is not codogenic. The efficiency of transcription of genes is well regulated and does not happen as a chance event.

Promoter signal heralds the initiation of transcription

The initiation of transcription firstly involves the binding of polymerase to the double stranded DNA (dsDNA). The binding of polymerase takes place at the specific sites designated as *Promoters. Promoters are the nucleotide sequences in DNA, upstream of gene (5' side) to which RNA polymerase binds to initiate transcription.* The promoters sequences are often conserved between genes, but differences do exist. Sequence differences between the promoters of different genes may confer differential efficiency for transcription initiation and are involved in regulation of gene expression. Besides RNA polymerases there are a number of DNA binding proteins that bind to short conserved sequences within the promoter region involved in initiation or regulation of transcription.

The DNA helix locally unwinds to facilitate the accessibility of the template strand to be used for aligning nucleotides for the synthesis of RNA. The unwinding starts at the promoter region to which RNA polymerase binds prior to the initiation of synthesis of RNA, which starts from a specific nucleotide called as the initiation site or start point (Fig.6.42). The RNA polymerase and its co-factors, when congregated on the DNA template are referred to as the *transcription complex or transcriptosome*. (Fig.6.43).

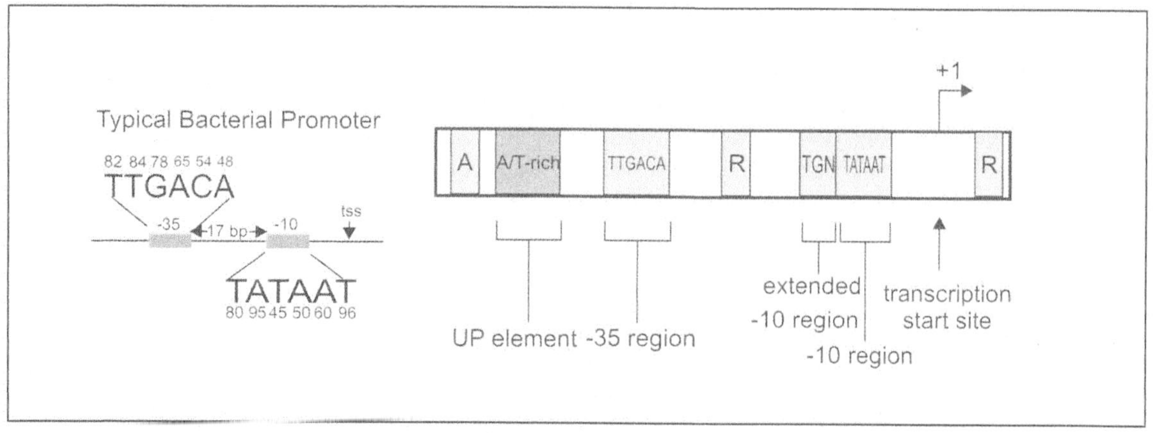

Fig. 6.42 Promoter sequences (a) prokaryotes (b) eukaryotes

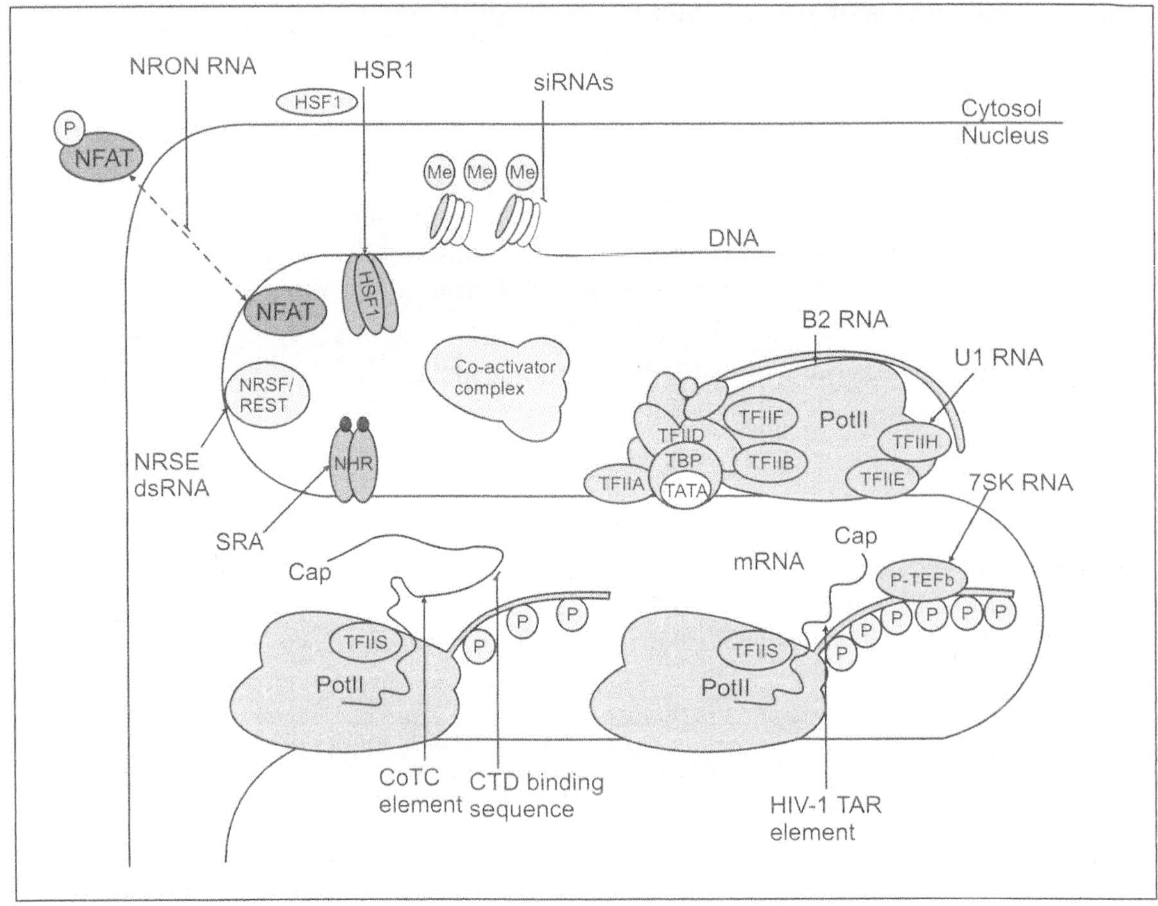

Fig 6.43 An eukaryote transcriptosome

Since, only one types of RNA polymerase is involved in prokaryotes for the transcription of genes (rRNA, tRNA and mRNA), the promoter sequences in DNA must have identical or near identical nucleotide sequence in order to be recognized by the enzyme. In *E.coli,* the promoter sequences have been well analyzed and found to consist of two distinct boxes known as *Pribnow* box, named after its discoverer, at the position of –35 and –10. These are also designated as –35 box with a sequence of 5' -TTGACA -3' and – 10 box with a sequence 5' TATAAT – 3'. The actual sequence of the promoter may vary from gene to gene. However, all of them are related and recognizable with the *Consensus Sequence*. It is a nucleotide sequence employed to describe a large number of related but not identical sequences. Each point of the consensus sequences represents the most probable nucleotide present at that position in majority of cases analyzed (Table 6.7). The real significance of slight variations found in different promoters positioned up stream of different genes is not understood. It has been conjectured that the real sequence variation found may affect the efficiency of RNA polymerase in locating and binding to promoter region. Clearly this is a strategy employed by nature in regulating and influencing the extent to which genes have to be expressed.

Table 6.7 *Certain Promoter sequences recognized by E.coli RNA polymerase*

Gene	– 35 box	– 10 box
lac operon	TTTACA	TATGTT
trp operon	TTGACA	TTAACT
tRNA genes	TTTACA	TATGAT
Consensus Sequence	TTGAC	TATAAT

Certain antibiotic molecules inhibit transcription

Inhibition of transcription initiation occurs in prokaryotes when treated with certain antibiotics like Rifamycin B and Rifampicin etc., by binding to β subunit of polymerase enzyme.

(b) **Chain elongation**

Once the transcription is initiated, the elongation of transcript (RNA) starts. During this phase of transcription, the RNA polymerase enzymes slides along the DNA molecule, melting down and unwinding the double helix as the transcription process progresses, at the same time adding sequentially ribonucleotides to the 3' end of the growing RNA molecule. The recognition of ribonucleotides to be added is solely determined by the base pairing to the template polynucleotide of the gene. However, three basic points one should note about the outcome of transcription.

- **The transcript (RNA) is longer than the gene.** This is mainly due to the fact that the actual transcription takes place even before start point and beyond the end point of gene sequence itself. Almost all genes have a ***leader segment*** that is transcribed by the polymerase prior to reaching the start point of genes. The actual length may vary from gene to gene. For instance in *E.coli* it may vary 20-600 nucleotides long. Similarly, ***trailer segment*** is transcribed even after reaching the end of gene (Fig. 6.44)

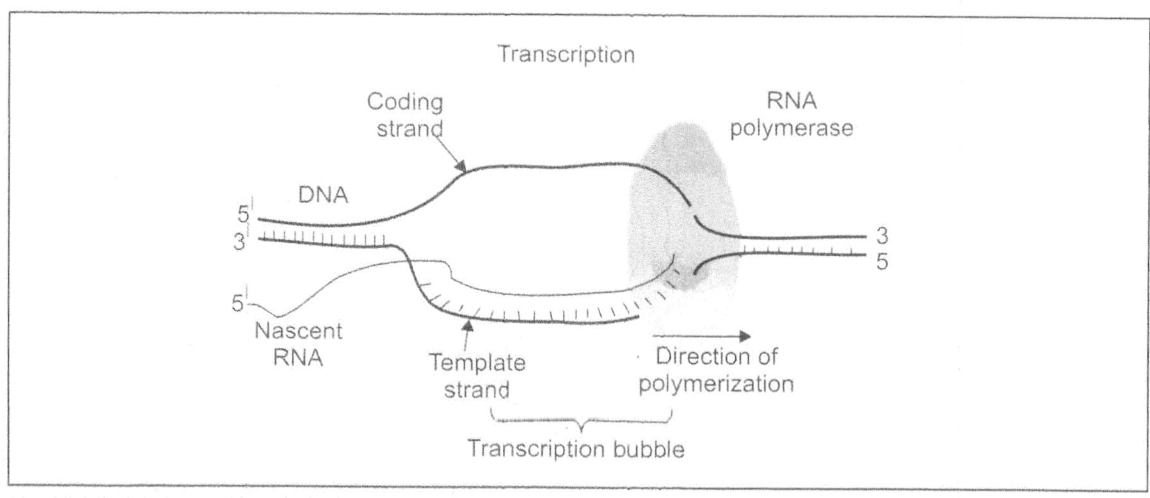

Fig. 6.44 Transcription bubble

- **At any point of time only a small region of the DNA helix is unwound.** Only limited region of the DNA molecule is melted to facilitate the addition of ribonucleotides at any one time. Once the nascent RNA chain gets elongated, the polymerized portion dissociates from the template (DNA) strand permitting the double helix to restitute to its original state. The unwound region is called as '***transcription bubble***' contains 12-17 RNA – DNA base pairs and constitutes only a short stretch of DNA as a whole. From physical point of view this is important because unwinding a region of DNA helix needs overwinding the neighbouring areas and as such there is limited freedom for this to take place.
- **The elongation speed is not constant.** Though the direction of chain elongation is universally 5' $\rightarrow$ 3' some time polymerase enzyme decelerates, stutters and then reaccelerates or even reverses to a short stretch removing nucleotides and then resumes. The rate of elongation is not constant. About 20-50 nucleotides per second at 37°C with an error frequency off 1 per every 10^4 nucleotides has been noted. Certain molecules added like actinomycin D, ethidium bromide proflavin etc., inhibit chain elongation.

(c) **Termination of transcription**

As is the case with initiation the termination of transcription also has to take place at a precise position at the end of the gene sequence and hence is not a random event. Polymerase enzymes reads a complex signal in the form of sequences and attendant conformational changes. There are no analogous sequences specific to the promoter to mediate the termination.

Terminator signals emanate from complementary palindrome sequence.

(a) In the case of prokaryotic organism like E.coli, the terminatory sequences show a wide variation, but all have a common characteristic feature viz., they are complementary palindromes. This implies that the base pairing can take place not only between the two strands of the DNA helix but also intrastand by folding back as well as within RNA chain synthesized as a transcript (Fig. 6.45). These results in the formation of ***Cruciform*** conformation with DNA or a ***stem loop*** configuration in a single stranded RNA transcript. Since, bioenergetically DNA is stable, because of the presence of much large number of stable base pairs it is unlikely to form cruciform structure within the DNA helix, but more likely loops are formed in growing chain of RNA due to complementary palindrome sequences, thus exerting influence on transcription to terminate. The DNA sequences at termination site that are ***Rho**-independent* is 5' - CCC - GGG - TTTTT - 3' that constitute GC rich stem loop RNA structure. Formation of the stem loop is an essential part of termination.

(b) Some of the terminators are ***Rho**-dependent* (*trp* operon) and function only in presence of a protein called Rho, which is an hexamer with molecular weight of 27.6 kD.

(c) The end of transcription

Termination of transcription ensues with the dissociation of RNA polymerase from the DNA helix and release of the RNA transcript. The core enzyme may once again reassociates with σ subunit and commence a fresh round of transcription of the same or different gene sequence. The RNA molecule formed at the end of transcription may be ready to play its role either as end product of gene expression or the information (mRNA) to synthesize protein polypeptides by translation processes.

In eukaryotes, the termination of transcription is due to multiple sites beyond poly- A addition site that are recognized by the RNA polymerase II as termination, as does it takes place in HIV *Tat*, *C-myc*, Heat shock genes etc.

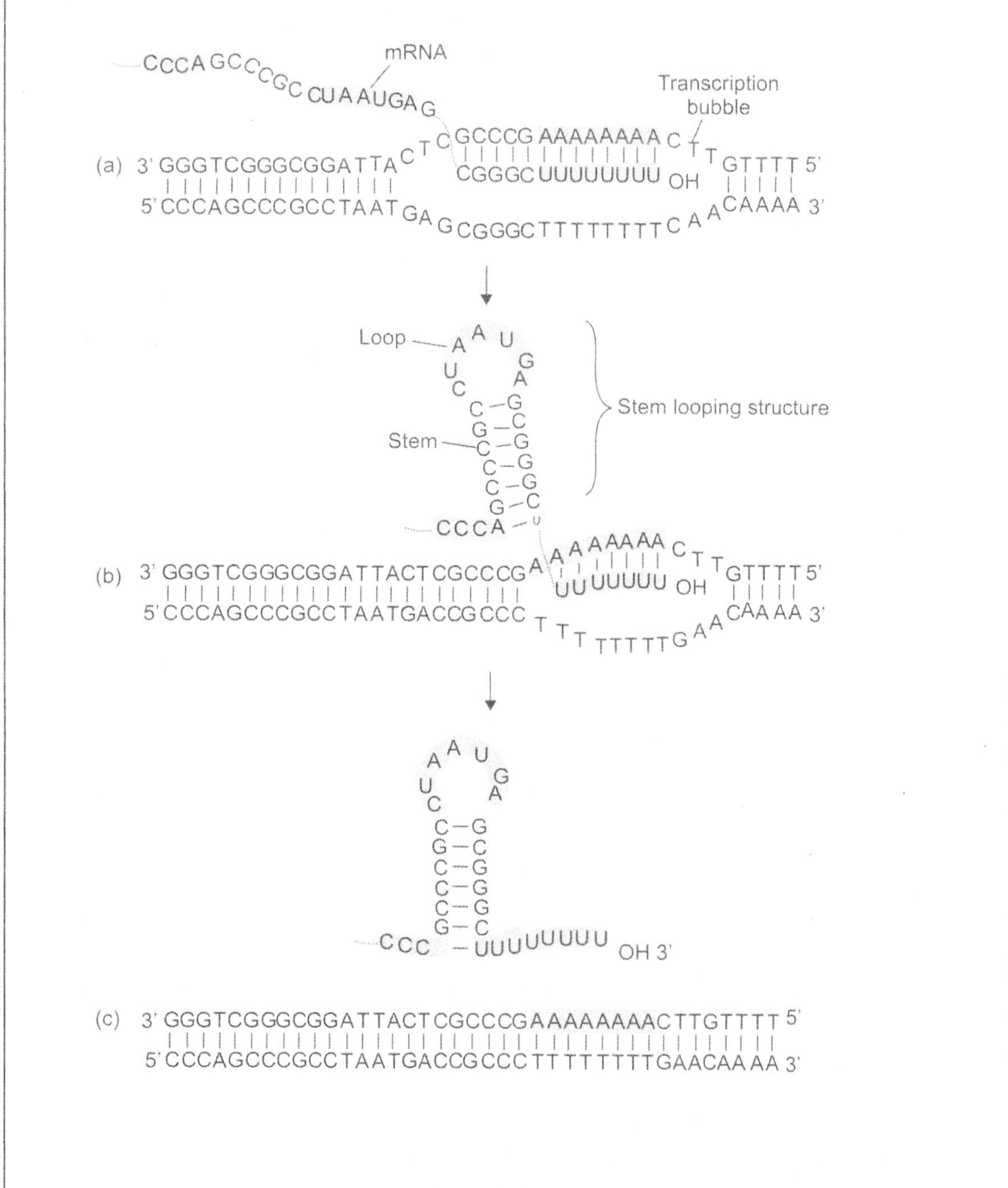

Fig. 6.45 Termination of transcription by stem and loop formation

6.11 Flow of Genetic Information: Translation of Genetic Code

Introduction

With precise replication and high degree of fidelity as well as its transmission, DNA serves to carry genetic information from cell to cell and generation to generation. What is important is, to understand the way genetic information is translated and expressed as phenotypes. Without any exception, all phenotypic effects of all living organisms at cellular or subcellular, tissue and organ level are end results of a network of biochemical reactions taking place all the time at subcellular level. All metabolic reactions need enzymes, and all enzymes are proteins either wholly, or partially. Proteins are synthesized in cytoplasm, outside the nucleus though genetic information is housed in the Nucleus. The biological problem of protein synthesis basically reduces to addressing the question of how DNA information contained in the nucleus mediates the protein synthesis in the cytoplasm.

6.11.1 Information Transfer: Genes to Protein

Even much before the unraveling of molecular events in protein synthesis, many genetic experiments laid foundations and furnished clear proof of gene to protein relationship and protein-to-protein interactions. The groundwork for a functional relationship between genes and enzymes was laid after the re-discovery of Mendelism in 1902 when **Bateson** reported that a rare human defect, *alkaptonuria*, was inherited. In 1909, **Archibald Garrod** an English Physician, published a book, *Inborn errors of metabolism* wherein he suggested a relationship between genes and specific biochemical reactions mediated by enzymes. Several metabolic disorders like *Alkaptonuria, Phenylketonuria, albinism, cretinism* etc occur in humans; each one governed by a recessive gene. **Beedle** and **Tatum** in 1941, working with *Neurospora* mutants, finally connected the relationship between genes to protein by developing a concept of one gene-one enzyme one phenotype in his famous *one gene– one enzyme hypothesis.*

6.11.2 Protein Synthesis: Genetic Code

The genetic information contained in DNA (gene) is copied into a new nucleic acid (mRNA) by transcription in a single biological language consisting of nucleotide bases A, U (T), G, C in a sequence, whereas biological proteins are made up of amino acids polymerized together by a peptide bond. Translation of mRNA occurs at the ribosomes by tRNA's carrying amino acids in different combination thus making seemingly endless variety of orders to make an almost infinite variety of polypeptides. The amino acids are put together to make a protein as a result of translation of the genetic information contained in the gene (DNA). In a DNA double helix, only variation along the structure is its component bases. Hence it was deduced that the sequence arrangement of bases constitutes the genetic information. Infact, starting from 1953, soon after unraveling of double helical structure for DNA by **Watson & Crick**, major intellectual predilection of molecular biologists was in favour of a simplistic code for information. The most enlightened scientists of the day flocked around **Francis Crick** to address the problem of genetic code. Gradually it was realized that any features of code should be compatible with the established facts of inheritance. Working hypothesis for delineating code was evolved gradually. The working model was based on two basic assumptions that genes and proteins are collinear and each codon is a triplet thus the genetic code forming first principle in information transfer was conceived.

6.11.2.1 The Principle of Colinearity between Nucleotide Sequences of DNA (gene) and Amino Acid Sequences of Protein

Right from the day, double helical structure for DNA was proposed in 1953, it was assumed that, the genetic information stored in an array of sequence of nucleotides in DNA is collinear to that of amino acid sequences in proteins (Fig. 6.46). The colinearity principle made the problem of protein synthesis simple and straightforward in that, the genes could code for proteins. However, the experimental proof for the same was not available till 1964. This is partly because, at that time the type of experiments needed to insert a nucleotide into the specific site of the gene, (like present day *in situ* mutagenesis), affecting the gene sequence was not readily available or protocols perfected to determine whether the resulting change in amino acid sequences of the protein occurred at the same relative position. **Charles Yanofsky** along with his colleagues provided the much needed analysis with *E. coli* gene mutations that affected tryptophan. **Tryptophan synthase** is the enzyme that catalyses the final step in pathway, but impaired due to change in the sequence. The result demonstrated clearly the principle of colinearity between gene and protein. Subsequently, **Sydney Brenner** and his group presented evidence with a similar analysis of genes and its proteins from T_4 phage thus, corroborating the **Yonofsky**'s results. Further, it was also established that the N(amino) terminus of the protein corresponds to the 5' end of the gene.

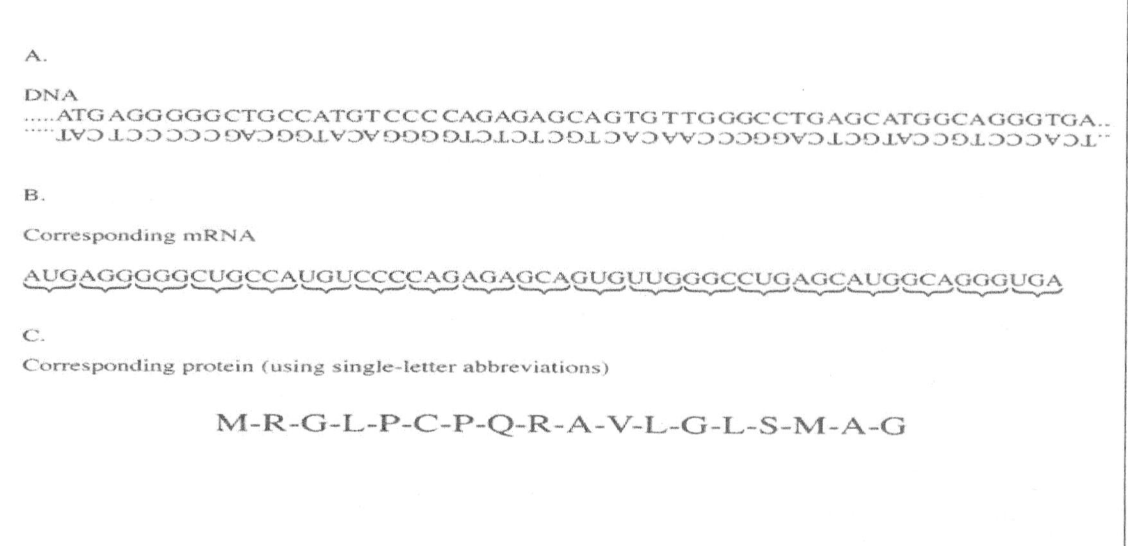

Fig. 6.46 mRNA transcript synthesized form sense strand coding for amino acid sequence of a protein

6.11.2.2 Each Codon specifying amino acid sequence in polypeptide is a triplet of nucleotides in the gene

The second basic assumption about the genetic code is the number of nucleotides that makes codon (code word) for a single amino acid in a protein, whose information is present in a gene as an array of nucleotides. It is obvious that a codon cannot be just one nucleotide (A,T,G,C) because it can only code for four amino acids. Similarly, duplet codons (AT, TA, TT, GC etc;) also will not serve the purpose because such a system can only create 16 codons capable of coding for 16 amino acids only (4^2=16) . However, the triplet nature of code (codon size is three; AAA, TGT, GGG, CAC etc;) would be more than enough to code for 20 amino acids since 64 codons are feasible (4^3=64). To provide biological proof for the triplet nature of code is not so simple as in the case of colinearity. However, using the chemical mutagens that cause specific base pair changes including deletions or addition in double stranded DNA, it was established that reading frame can be shifted by insertion or deletion of one or two bases which impair the genetic information, but by inserting or deleting three bases, down the change meaning of the gene in terms of amino acids sequence can be restored. Most elegant experiments were conducted by **Francis Crick**, **Brenner** and others employing IIb gene of T_4 phage, to see whether or not the ability to infect *E. Coli* is restored after inducing the changes by proflavin. The results of such work established the triplet nature of genetic code. However, direct evidence came forth by 1961, even before the results appeared in journals.

6.11.2.3 Protein synthesis out side the cell employing synthetic homo and random heteropolymers

Marshall Nierenberg and **Heinrich Mathaei** working at the National Institutes of Health Laboratories, USA, eventually cracked the genetic code in 1961. They have perfected a cell-free preparation from *E. coli* capable of synthesizing polypeptide chains specific to introduced synthetic RNA message. Since, it was easy to make a simple homopolymeric RNA, they have used Poly(U)- homopolymer in the cell free system. The result was the synthesis of polypeptides made up of polyphenylalanine. It was obvious that the code signifying amino acid is UUU triplet. Thus the first codon assigned is 5'-UUU-3' for phenylalanine. With similar logic, three other RNA homopolymers, CCC, AAA and GGG were made and evaluated for their codogenic property. The codon AAA was assigned to lysine, CCC was assigned for proline. However, GGG product could not be identified at that time.

6.11.2.4 Unravelling of the Genetic Code

The elucidation of genetic code was made possible because of gradual developments in the field of genetics coupled with technological advances in analytical tools biochemistry and cell biology that permitted the question which triplet codon specified which amino acid. Two major breakthroughs in this direction were:

- Synthesis of RNA molecules *in vitro*, and
- Cell-free *in vitro* protein synthesis.

In vitro synthesis of RNA

Severo Ochoa (1955), discovered **Polynucleotide Phosphorylase**. This enzyme *in vitro* degrades RNA but the same in test tube catalyzes the synthesis of RNA. This step does not require a DNA template and is nothing to do with transcription. These proteins were made use in synthesizing artificially a RNA molecule of known or predictable nucleotide sequence. If such an artificially made RNA is translated *in vitro*, by looking at the sequence of amino acids in the resultant protein one could assign the individual codons.

In vitro synthesis of protein

To employ an artificially synthesized RNA molecule to act as a messenger for translation, a cell extract capable of synthesizing proteins is needed. Such *in vitro* system comprises of essential components of cells, ribosomes, tRNAs, Amino Acids, etc, minus endogenous messenger RNA, so that protein synthesis takes place when synthetic RNA is provided with.

The next thing the researchers have tried was to construct artificial RNA heteropolymers consisting of more than one nucleotide. This was accomplished by using mixture of nucleotides (A,T,G,C) in different proportion with **polynucleotide phosphorylase**. This approach faced a problem. A mixture of mere two nucleotides, for example A and C generates eight possible codons (2^3=8) but polypeptide made out of them have six amino acids, proline, histidine, threonine, aspargine glutamine and lysine in any order. The problem was how to determine which codon is codon for which amino acids. The answer sought was partially provided by using a pre-determined quantity of each nucleotide in the reaction soup for RNA synthesis. Say for example the ratio of C to A is 10:2, then the probability of codon AAA occurance is much higher that the chances of CCC codon. The frequency of each of eight possible codons can be computed and then compared with the composition of each amino acid in the resultant protein. This kind of allocation of codons is a statistical concept and can be evaluated by changing the components of heteropolymers. These kind of logical approach let the **Nirenberg**, **Mathaei** and **Ochoa** to propose a coding dictionary comprising of 64 codon words of the Genetic code. However, the completion of genetic code has to wait for additional experimental approach for unambiguous determination of each codon despite the unraveling of genetic code employing homopolymers and random heteropolymers.

H. Gobindkhorana's approach of synthesizing ordered heteropolymers, consisting of known dinucleotides, like poly (AC), producing ACA CA CA CA CA CA, which has two codons ACA and CAC, or trinucleotides, consisting of UGU giving rise to UGU UGU UGU, with codons UGU, GUU, UUG, instead of polymerizing with mononucleotides, paid a great dividend in cracking the code of several difficult codons. Further more, *the triplet binding assay* developed by **Nirenberg** and **Leaderberg** (1964) as a modification of cell free system of protein synthesis demonstrated that the purified ribosomes attach to an mRNA molecule of only three nucleotides (a single codon) and bind to the correct amino acid apply to tRNA molecule. Since it was possible by then to synthesize a triplet of known sequence, the newly discovered binding assay enabled to cross-check the previously assigned codons. Besides, it was easy to allocate rest of the codons. By 1966, exactly five years after Mathaei's experiment with cell free system using homo polymers, the meaning of every remaining codons was confirmed and the riddle of genetic code was complete (Table 6.8) & (6.9).

6.11.3 General Characteristics of Genetic Code

The Genetic Code dictionary is given in Table 6.9. Some of the main features are:

6.11.3.1 The code is degenerate

Barring methonine and tryptophan, all of the amino acids have more than one Codon. Most of the amino acids are coded for by more than one triplet codon. That is to say, that certain codons are synonymous, and code for the same amino acids. Hence they are grouped into a sort of families for instance codons GGA, GGU, GGG, GGC all code for glycine. This feature is called *degeneracy of code*. This arises because, despite 64 codons, each coding for an amino acid, there are only 20 amino acids needed to be coded.

Table 6.8 *Dictionary of genetic code.*

Fisrt position				*Second position*									*Third position*
	U			C			A			G			
U	UUU	Phe	F	UCU	Ser	S	UAU	Tyr	Y	UGU	Cys	C	U
	UUC	Phe		UCC	Ser		UAC	Tyr		UGC	Cys		C
	UUA	Leu	L	UCA	Ser		UAA	Stop		UGA	Stop		A
	UUG	Leu		UCG	Ser		UAG	Stop		UGG	Trp	W	G
C	CUU	Leu	L	CCU	Ser	P	CAU	His	H	CGU	Arg	R	U
	CUC	Leu		CCC	Ser		CaC	His		CGC	Arg		C
	CUA	Leu		CCA	Ser		CAA	Gln	Q	CGA	Arg		A
	CUG	Leu		CCG	Ser		CAG	Gln		CGG	Arg		G
A	AUU	Ile	I	ACU	Thr	T	AAU	Asp	N	AGU	Ser	S	U
	AUC	Ile		ACC	Thr		AAC	Asp		AGU	Ser		C
	AUA	Ile		ACA	Thr		AAA	Lys	K	AGA	Arg	R	A
	AUG	Met	M	ACG	Thr		AAG	Lys		AGG	Arg		G
G	GUU	Val	V	GCU	Ala	A	GAU	Asp	D	GGU	Gly	G	U
	GUC	Val		GCC	Ala		GAC	Asp		GGC	Gly		C
	GUA	Val		GCA	Ala		GAA	Glu	E	GGA	Gly		A
	GUG	Val		GCG	Ala		GAG	Glu		GGG	Gly		G

- **The Code has punctuation to**

  Three codons, UAA, UGA and UGA, instead of coding for amino acids, act like a punctuation marks to stop the protein synthesis if present in the midst of a heteropolymer, resulting in a shorter version of the polypeptide than anticipated. These three triplets are known as *termination codons;* one of them is sure to be found at the end of a genes information where translation must stop. Similarly, there is another codon AUG that virtually present at the beginning of a gene and marks the point where translation should start. This codon is termed as chain *initiation codon*. It codes for amino acid methionine and every nascent polypeptide has in the beginning methionine at the amino terminus. However, it is removed subsequently due to post-translational

Table 6.9 *Amino acids structures, three and one letter symbols.*

Name	Symbol	Structural Formula
With Aliphatic side chains		
Glycine	Gly [G]	$H—CH(NH_3^+)—COO^-$
Valine	Val [V]	$CH_3—CH(NH_3^+)—COO^-$
Leucine	Leu [L]	$(H_3C)_2CH—CH(NH_3^+)—COO^-$
Isoleucine	Ile [I]	$CH_2—CH_2—CH(CH_3)—CH(NH_3^+)—COO^-$
With side chains containing Hydroxylic (OH) Groups		
Serine	Ser [S]	$CH_2(OH)—CH(NH_3^+)—COO^-$
Threonine	Thr [T]	$CH_3—CH(OH)—CH(NH_3^+)—COO^-$
Tyrosine	Tyr [Y]	See below
With side chains containing Sulfur Atoms		
Cysteine	Cys [C]	$CH_2(SH)—CH(NH_3^+)—COO^-$
Methionine	Met [M]	$CH_3—S—CH_2—CH_2—CH(NH_3^+)—COO^-$
With side chains containing acidic groups of their amides		
Aspartic acid	Asp [D]	$^-OOC—CH_2—CH(NH_3^+)—COO^-$
Asparagine	Asn [N]	$H_2N—C(=O)—CH_2—CH(NH_3^+)—COO^-$
Glutamic acid	Glu [E]	$^-OOC—CH_2—CH_2—CH(NH_3^+)—COO^-$

Name	Symbol	Structural Formula
Glutamine	Gln [Q]	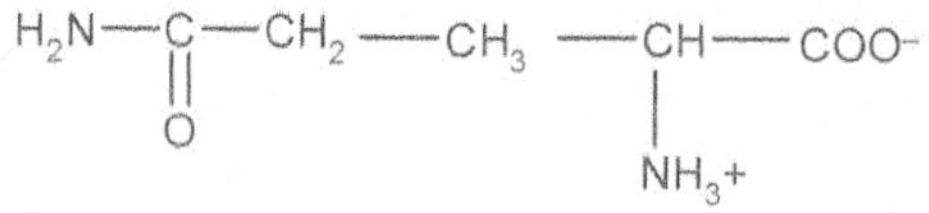 $H_2N—C(=O)—CH_2—CH_3—CH(NH_3^+)—COO^-$
With side chains containing basic groups		
Arginine	Arg [R]	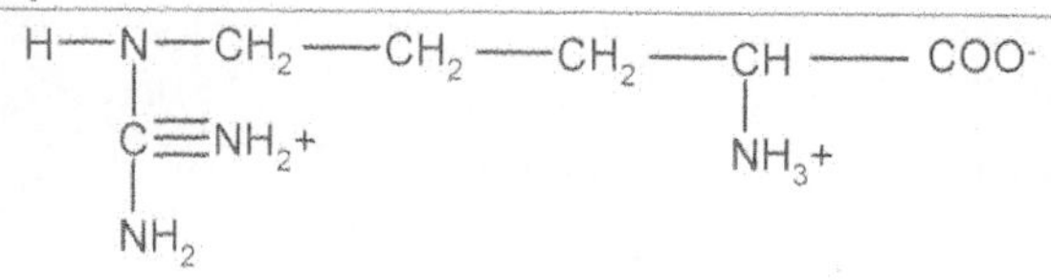 $H—N(—C(\equiv NH_2^+)—NH_2)—CH_2—CH_2—CH_2—CH(NH_3^+)—COO^-$
Lysine	Lys [K]	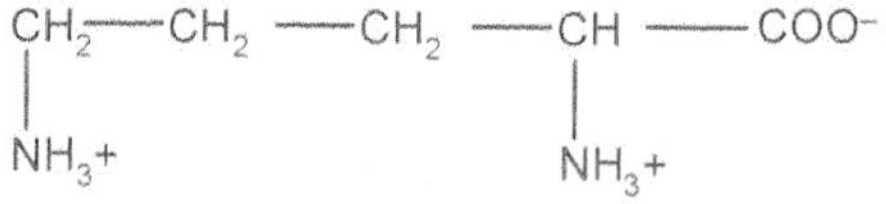 $CH_2(NH_3^+)—CH_2—CH_2—CH(NH_3^+)—COO^-$
Histidine	His [H]	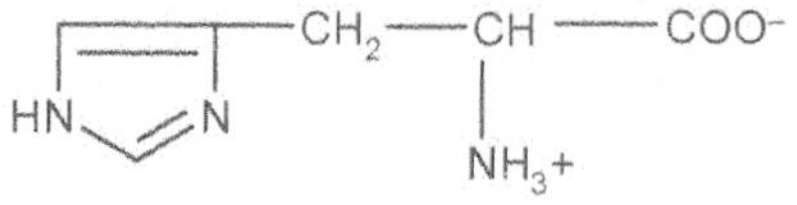 (imidazole ring: HN, N)$—CH_2—CH(NH_3^+)—COO^-$
Containing Aromatic Rings		
Histidine	His [H]	See below
Phenylalanine	Phe [F]	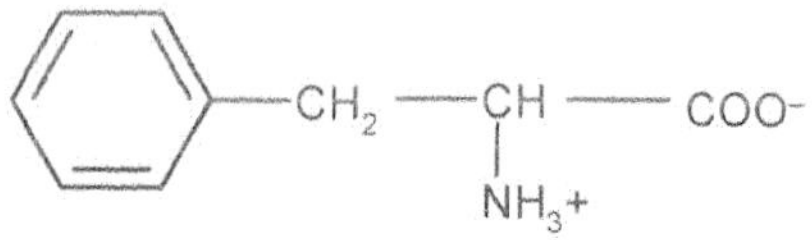 (benzene ring)$—CH_2—CH(NH_3^+)—COO^-$
Tyrosine	Tyr [F]	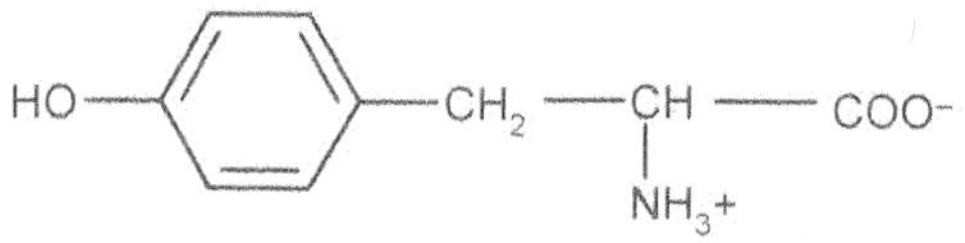 $HO—$(benzene ring)$—CH_2—CH(NH_3^+)—COO^-$
Tryptophan	Trp [W]	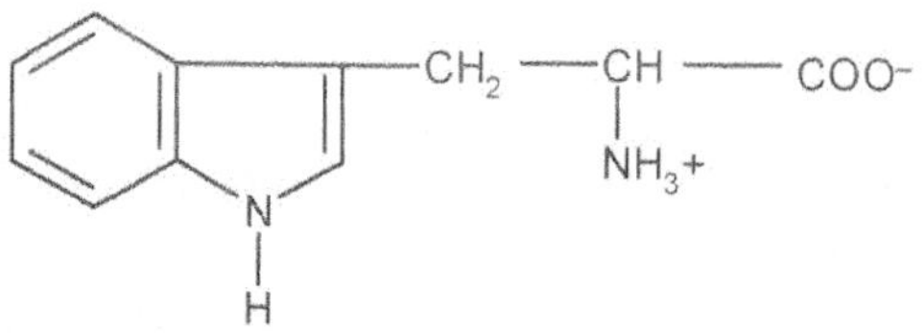 (indole ring, N—H)$—CH_2—CH(NH_3^+)—COO^-$
Imino Acid		
Proline	Pro [P]	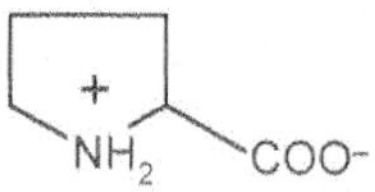 (pyrrolidine ring, $\overset{+}{N}H_2$)$—COO^-$

modification of the protein. It should also be noted that AUG is the only codon for amino acid methionine, and as such it may also occur in between the sequence where it needs to code for methionine, not for chain initiation.

6.11.3.2 Is the genetic code Universal?

As soon the problem of genetic code stood resolved by 1966, it was taken for granted and believed that genetic code is universal. The reason for such a generalization was that the frozen genetic code that does not evolve has an advantage because the constancy of genetic code in biological world imply that the meaning of codons do not impair the proteins of the cell. Thus, consolidating the process of evolution. The cell free system employed originally obtained from *E. coli* when replaced with mammalian system also worked without any change exactly reproducing the same results. Hence, the universality of genetic code became another dogma. Soon after, the scientific community was shocked with the discovery of Fredric Sangers group at Cambridge in 1979 that, mitochondrial genome employs different genetic code (Table 6.10). The terminator codon UGA in mitochondrial genome codes for tryptophan AGA and AGG codons that code for arginine normally codes instead for chain termination. Similarly AUA that normally codes for isoleucine codes for methionine. The universality of genetic code is violated in number of organism including other mammalian species, Drosphila, Fungi, Yeast, Protozoa etc.

Table 6.10 *Altered Genetic code in Mitochondria.*

Code	Amino acid coded in mtDNA	Amino acid coded in nuclear DNA (Universal code)
UGG	Trp	Trp
UGA	Trp	Stop
AGG	Stop	Arg
AGA	Stop	Arg
AUG	Met	Met
AVA	Met	Ile

6.12 Translation

The genetic code manifests during the final stage of gene expression, when transcribed genetic information mRNA gets translated into proteins. Translation is thus a very complex phenomenon and involves many of cell components. However, genetic code is vital deterministic central factor around which the entire process revolves. It is the tRNA molecule that acts as an adaptor, thus becoming a link physically between the nucleotide sequence of mRNA and amino acid sequence of the proteins. All this takes place with the strict compliance of genetic code.

6.12.1 The tRNA and its role

Each cell possess a number of diverse kinds of tRNA molecules. They are distinguished by their differences in sequences, part of the sequence is invariant and part of the sequence may be variant.

Functionally, each tRNA is distinguished by its specificity for one of the 20 biological amino acid related to protein synthesis. For instance the $tRNA^{Phe}$ is specific to phenylalanine, $tRNA^{Pro}$ proline. Each tRNA molecules forms a covalent linkage with respective designated amino acid by recognizing as well as attaching to a codon with anti codon, specifying the amino acid. As the genetic code is degenerate there may be more than one kind of tRNA molecules for a single amino acid. The tRNA that bind the same amino acid are termed as *isoacceptors*.

6.12.1.2 Interactions of Codon and Anti-Codons

Amino acylation of tRNA is the crucial first step in which each tRNA molecule forms a covalent bond with specific amino acid. This is called as charging. Every amino acid is recognized by a specific tRNA attached and to the CCA end of acceptor arm of the tRNA clover leaf (Fig. 6.47). The enzyme that controls the charging is *Aminoacyl-tRNA* synthetase which ensures so that right amino acid is attached to the right tRNA. The function of the enzyme is to recognize both the correct amino acid and appropriate tRNA. However, how it accomplishes this is not known. The anticodon (three-base sequence) on tRNA, is complementary to specific codon found in the mRNA and attaches after recognition following complementary base pairing rules. Since each tRNA carries a specific activated amino acid, the binding of tRNA to the mRNA by *base* pairing between codon-anticodon provides the

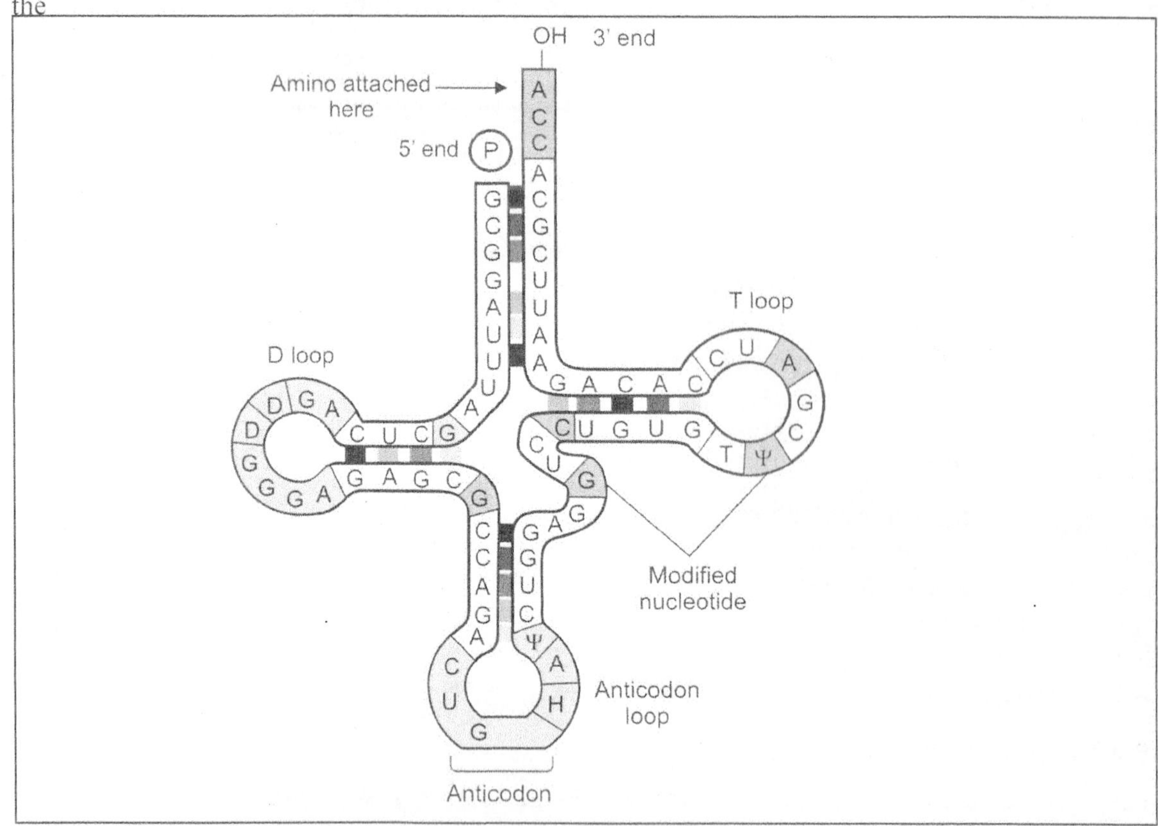

Fig. 6.47 Structure of tRNA.

specificity for placing and linking a particular amino acid to the growing polypeptide chain. The chain termination codons, UAA, UAG, and UGA do not have a tRNA with a complementary anticodon. Hence, these are not recognized by any tRNA and no amino acid is added to the growing polypeptide chain. However, their presence in mRNA results in chain termination and signals the end of the protein synthesis.

6.12.1.3 Changes in the DNA base sequences results in mutation

Once the reading frame of the mRNA commences at AUG, each triplet is read in linear order with no interruptions or gaps. If the sequence changes even by a single base, the amino acid specified in the protein will be different or sometimes, the protein synthesis will be terminated. Accordingly such changed codons are referred to as sense, missense or nonsense codons. Sometimes the entire reading frame changes due to frame shift mutation.The process of translation must be extremely accurate to avoid the synthesis impaired proteins, changes occurring in the sequence (DNA), either missense or sense or frame shift mutations, are permanent and are heritable. Such changes may be leading cause of heritable human diseases.

6.12.2 Mechanism of Protein synthesis

Once the adaptor hypothesis was confirmed and the role of tRNA has been established in protein synthesis, the exact *modus operandi* by which translation takes place has been followed and established. The main components of the protein synthesis apparatus consists of mRNA, tRNA, amino acids and ribosomes. mRNA provides the information in the form of sequence of codons that are translated into protein; tRNA bind specific amino acids and then transfer that amino acid into polypeptide using the specific recognition of anticodon-codons in the mRNA. Ribosomes, tRNA and proteins act at the site of protein synthesis. Traditionally, the entire process of protein synthesis can be divided into three phases-initiation, elongation and termination, although the entire process is continuous and merge with each other imperceptibly.

6.12.2.1 Initiation of translation

The translation, information transfer *from a gene to protein* takes place in the cytoplasm. The first step is the interaction of mRNA with a specialized complex called ribosomes which enables to read the information present in base sequences of mRNA. Ribosomes are the most abundant fraction of the cells made of ribonucleoproteins. There are two subunits, smaller 30s and a larger one 50s size in prokaryotes (30s + 50s); the small subunit 40s and a large one 60s in eukaryotes. The mRNA attaches to smaller subunit (30s or 40s); when not involved in protein synthesis these dissociate and form the ribosome pool in the cytoplasm.

6.12.2.2 Translation Commences precisely at a position binding site of the smaller subunit of ribosome

Translation commences accurately by binding to smaller subunit of ribosome called, *ribosome binding site* (Fig. 6.48). The aptness of attachment point is ensured by a sequence known as the *Shine-Dalgarno Sequence*. It is believed that a transient base pair attachment to 16s rRNA occurs thus enabling the 30s subunit to bind to mRNA. A Consensus sequence of 5'-A GGA GGU-3' is present in *E. coli* (Table 6.10) and (Fig. 6.48).

Sequence of the ribosome binding sites of *E. coli* gene

Table 6.10

nucleotides Sequence of genes	Concerned gene	Product of the gene	Distance upstream of AUG (initiation) codon
5'-AGGA-3'	*lac Z*	β-galactosidase	7
5'- GGAG-3'	*gal E*	Hexose-1-P-uridyl transferase	6
5'-AGGA-3'	*gal T*	Galactose-1-P-uridyl transferase	6
5'-AGGAG-3'	*r PIJ*	Ribosomal protein L10	8
5'-AGGU	*cro*	phage λ regulatory protein	3
5'-AGGAG-3'	*B*	Phage 174 O_X coat protein	7
5'-AGGAGGU-3'		Consensus sequence in *E.coli*	

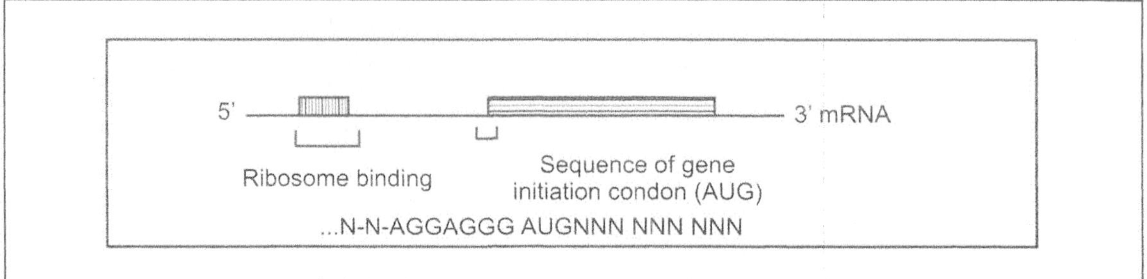

Fig 6.48 The Relative position of ribosome binding site in *E.coli* from initiation codon

6.12.2.3 Initiation Complex

The translation process *per se* starts with the binding of aminoacylated (charged) tRNA to the initiation codon of mRNA that has been already positioned by the 30s subunit of the ribosome. The initiator tRNA carries charged amino acid, methionine since the codon AUG signifies it. In bacteria it is formylmethionine because methionine as soon as it is charged modifies by substituting group (-COH) with one of the hydrogen atom of the amino group (NH_2) thus becoming N-formyl methionine (*f met*). Nature seems to have contrived a mechanism to block peptide bond formation with modifying the amino group. This step thus directs the polymerization of amino acids into polypeptide along one way path i.e., amino (NH_2) to Carboxyl (COOH) direction. The formation of complex comprising primarily of ribosome mRNA, 30S subunit, and charged (aminoacylated) tRNA *fmet* is known as *initiation complex*.

6.12.2.4 Factors of Initiation

Finer details of the role of different factors in translation process especially non-ribosomal proteins is not clearly understood. However, it is well known that three proteins IF_1, IF_2, IF_3 are quite essential for protein synthesis to take place. IF_1 and IF_2 appears to play a role in the dissociation of 30s and 50s subunits of ribosome as the holistic ribosome is incapable of initiation. Probably, IF_3 plays a role in the recognition of ribosome binding site. IF_2 also seems to take part in the attachment of charged tRNA *fmet* and may also mediate binding to the initiation complex. The GTP molecule of the complex provides energy for the subsequent step initiation (Fig 6.49)

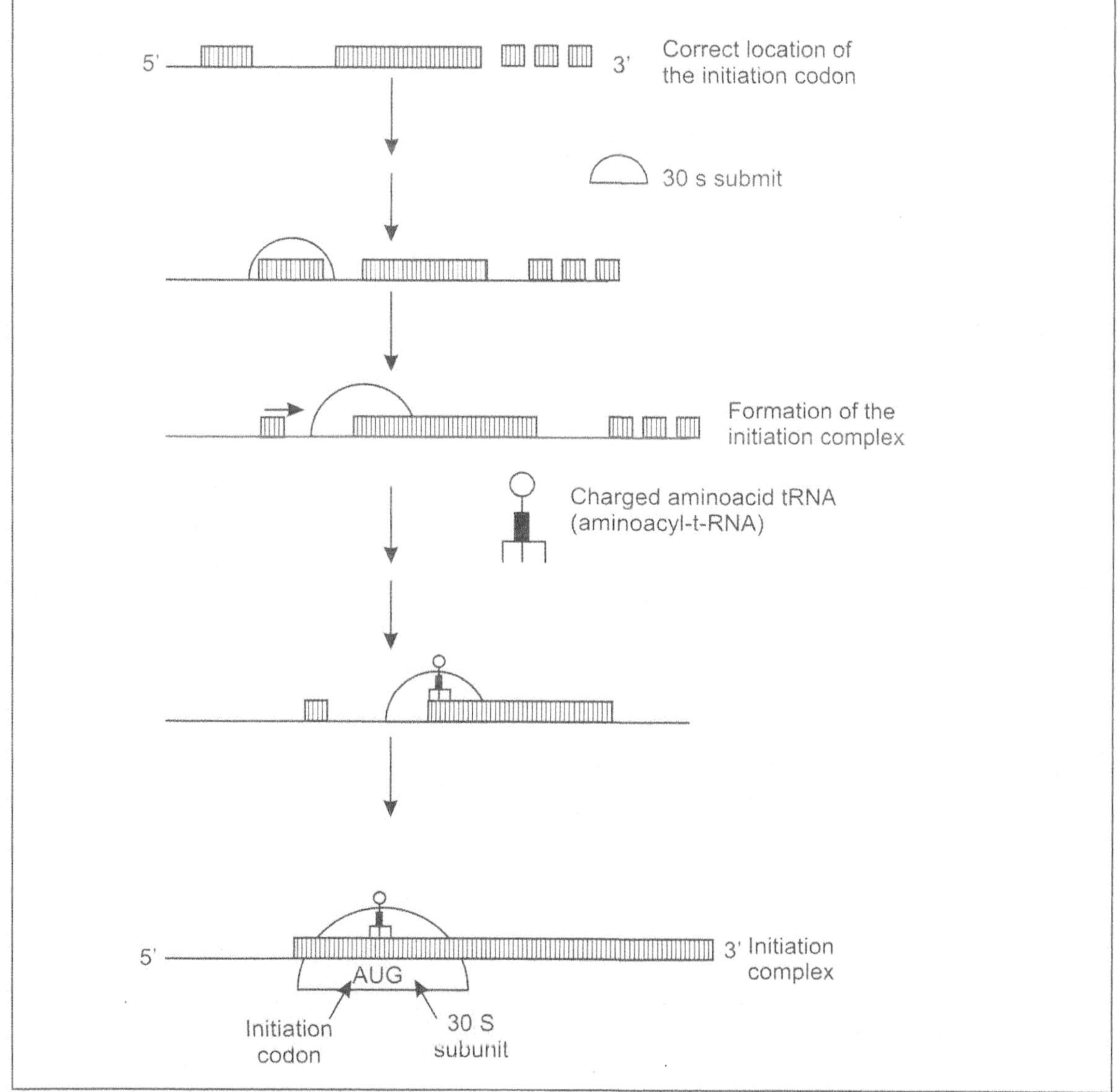

Fig. 6.49 Initiation of translation (Gene to protein) in prokaryotes (representative sp, E. coli)

6.12.2.5 Growth and Elongation of the polypeptide chain

Initiation complex once formed, it signals and reads the large ribosomal subunit to attach to it. The energy is derived by hydrolysis of the GTP molecule associated with the initiation complex. This creates two distinct but individual sites for the binding of tRNAs: 1. *Peptidyl* – or *P-Site*, 2. The aminacyl – or A-site. The *P* site to begin with is engaged to tRNA fmet base paired with initiation codon. The *A*-site is positioned over the next codon of the gene and to begin with it is empty. The peptide chain elongation starts once the correct aminoacylated tRNA occupies the *A-site* and establishes base pairing with second codon in the sequence of information contained in the mRNA (gene). The two elongation factors EF-T_4 and EF-Ts are needed to continue chain elongation. EF-Tu is directly committed for the entry of the charged tRNA into A-site and the second molecule of GTP is general requisite energy. Thus deactivated EF-Tu is again made active by another factor EF-Ts enabling its participation is further elongation of the chain. This goes on cyclically in the cytoplasm till the entire information of the gene in the from of mRNA gets translated to protein (Fig. 6.50).

6.12.2.6 Formation of a peptide bond and translocation.

Once both sites of the ribosomes engaged by the charged tRNAs, the two amino acids are juxtaposed. Subsequently, these closely positioned two amino acid forms a peptide bond between the carboxyl group of *fmet* and the amino group of the second amino acid. The reaction is catalyzed by an enzyme called peptidyl transferase, in conjunction with another ribosomal enzyme, *tRNA deacylase*. This enzymes responsibility is to undo the fmet tRNA link. Thus at the end of the enzymatic reaction a dipeptide left attached to tRNA positioned in the *A-site* (Eg) Translocation is the process where by the ribosome moves along the mRNA by a codon (three nucleotides) so that another charged tRNA (aa-tRNA) enters P-site replacing the now discharged tRNA *fmet* thus making the *A-site* vacant so that the next (third aminoacylated) tRNA enters and the cycle of elongation is repeated. Fact of these cycle needs hydrolyses of a GTP molecule with the help of a third elongation factor EF-G.

6.12.2.7 Several ribosomes simultaneously translate each mRNA

Electron micrography revealed a presence of a polysomes both in prokaryotic and eukarytoic cells. After several cycles of elongation the start point of the mRNA molecule is not associated with the ribosome. A second round of translation can begin. Another 30S subunit can attach to the ribosome binding site and form a new initiation complex, thus leading to polysome.

6.12.2.8 Termination of protein synthesis

Sensation of translation takes place when a termination codon (UAA,UAG or UGA) of mRNA enters the *A-site*. Since there are no tRNA molecules with anticodons that can base pair with any of the termination codons one of the two *release factors* RF_1 or RF_2 enters the *A-site* to cleave the already synthesized polypeptide from the last tRNA. Another release factor, RF_3 plays a supportive role in this process. The ribosome releases the polypeptide and mRNA disintegrates into 30s and 50s components which enters into the cellular pool ready to be engaged in another round of translation. The protein polypeptide chain bonds to assume its tertiary conformation and commences its function in the cell. (Fig. 6.51).

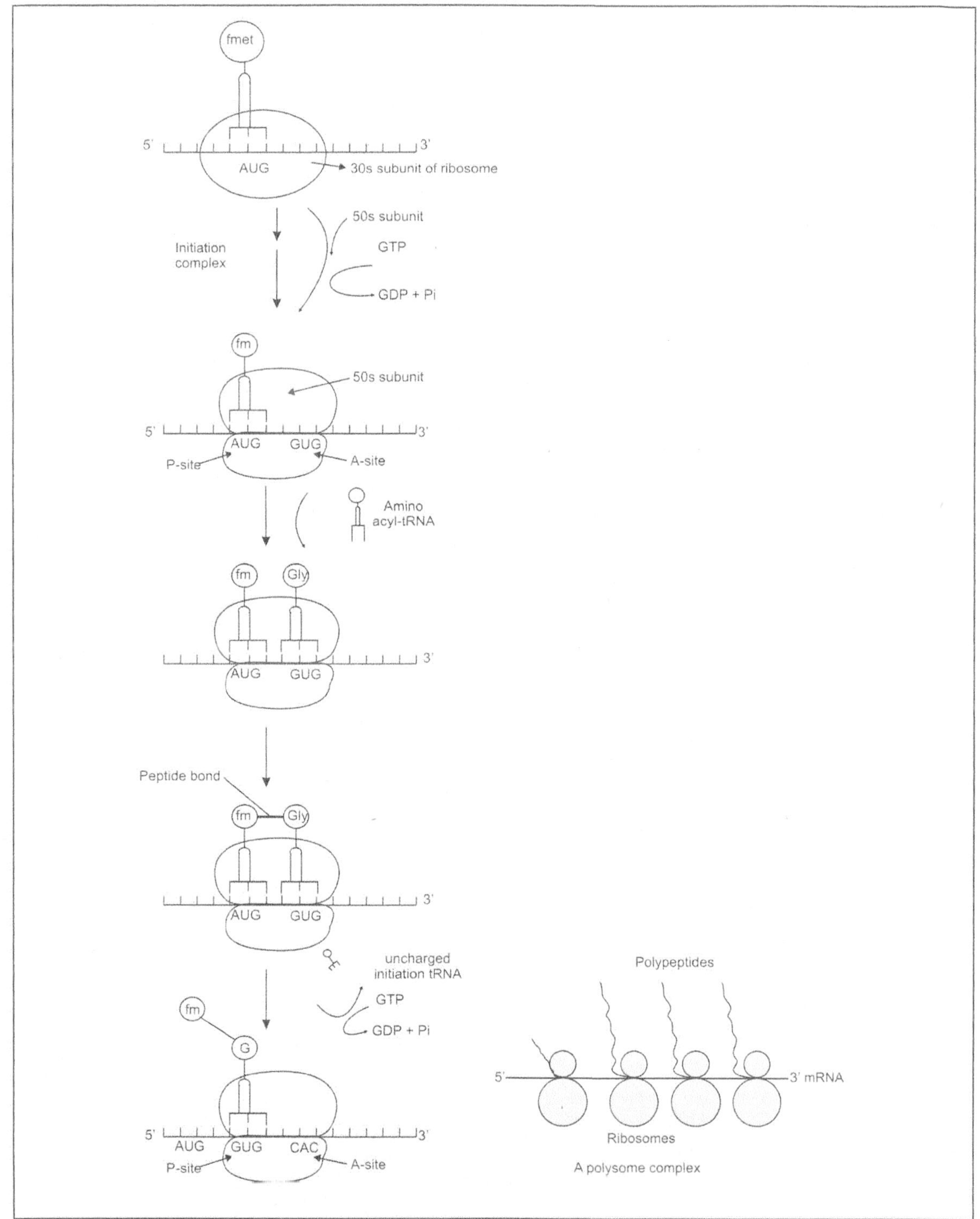

Fig. 6.50 Translation process of Prokaryotes–Elongation of polypeptide chain and polysome

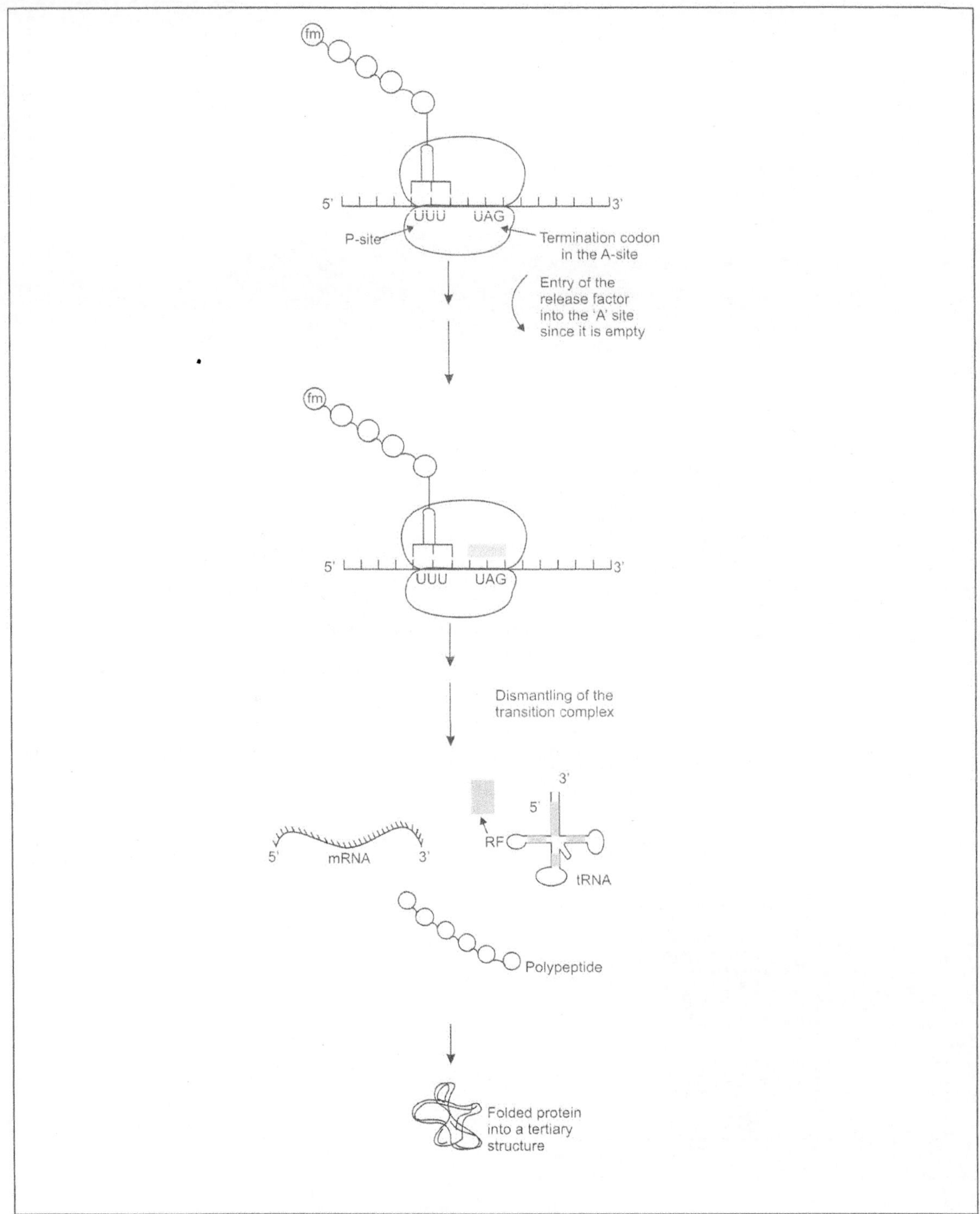

Fig. 6.51 Termination of translation (in prokaryotes)

A list of bacterial translation factors is given in Table 6.11.

Table 6.11

Protein factors and stage of participation and stage at which they take part	Molecular weight	The role of each factor
1. *Initiation* of polypeptide chain *step*		
I F_1	9,000	Dismantles the ribosomes subunits to basic units (30s + 50s).
I F_2	97,000	Helps in attachment of met tRNA to the initiation complex.
I F_3	23,000	Dissociation of ribosome, and binding of mRNA to the initiation complex.
2. Elongation of polypeptide chain		
EF-Tu	43,000	Helps in binding of charged tRNA (a-tRNA) to the A-site
EF-Ts	14,000	Helps in activation of EF-Tu
EF-G	77,000	Helps in the process of translocation
3. *Termination and release of polypeptide chain*		
RF_1	36,000	Releases the chain by occupying UAA, UAG codons
RF_2	41,000	Releases the chain by occupying UGA, UAA condon
RF_3	46,000	Cooperates with RF_1 & RF_2

6.12.3 Translation in Eukaryotes in principle is the same as that of prokaryotes

Translation process – from gene to protein, is essentially follows the same fashion as has been shown in bacteria (e.g., *E.coli*). Minor differences are : firstly the size of subunits of ribosomes which are 40s and 60s in eukaryotes. Secondly the way the small ribosomal subunit (i.e., 40s) binds to the mRNA and locates the start codon, AUG at which translation begins. An internal ribosome binding site equivalent to Shine-Dalgarno sequence of *E.coli* is present in very few eukaryotic mRNA. In its place the small subunit of ribosome is believed to recognize the cap structure as its binding position. Hence, it attaches to extreme 5' end of mRNA. The subunit subsequently slips along the mRNA till it reaches an AUG codon for initiation. This is not necessarily same in all cases. Some cases the translation begines with later AUG triplet. It appears that region surrounding AUG triplet is crucial whether such initiation codon is made use of or not. Some of the distinct attributes of translation in eukarytes is listed below:

- The initiator tRNA carries an unmodified methionine in eukaryotes unlike in prokaryotes where *fmet* amino acid is employed.
- A large number of initiation factors are employed in eukaryotes, unlike prokaryotes, where only three factors take part.

- Unlike prokaryotes, the initiation complex in eukaryotes needs hydrolysis of ATP molecules to provide energy for their build up as well as to scan and drive along the mRNA.
- Termination of translation needs hydrolysis of GTP while it is not needed in prokaryotes (e.g., *E.coli*)

Table 6.12 *A conspectus of Factors of Eukarytoic translation*

Factor & State of Participation	Responsible
A. Initiation of translation	
elF_3	Attach the ribosomal smaller subunit (40s) prior to
$elF4_C$	mRNA binding
$elF4_A$	Attach to the cap structure and help in binding
$elF4_E$	of the 40S ribosome subunit
$elF4_G$	
elf 4B	During the scanning process it probably breaks the stem loops (of mRNA)
elf2	Binds to the initiator tRNA
elf2B	Regenerates elF_2 to make it active again
elF5	Helps in release of elF_2, elF_3, and $elF4_C$ from the growing initiation complex
elF6	Assists in the dissociation of ribosome subunits
elF1	Not known definitely
B. Elongation of the Polypeptide Chain	
eEF_1	Helps in Binding of aminoacyl tRNA to the A-site
eEF_2	Assists in translocation.
C. Termination of translation	
eRF_1	Participate in the recognition of chain termination
eRF_3	codons. AAU. AAG. AUG.

6.13 Regulation of Gene Expression

Introduction

The genetic information contained in genomes, whether in unicellular or multicellular organisms is very massive and comprises of many thousands of genes. Some of the information is needed by the cells all the time. Such genes that are active all the time are called *house keeping* genes. Examples are rRNA genes, ribosomal protein genes, RNA polymerase genes, and genes of metabolic pathways etc., Whereas, many genes have to perform specialized roles and their expression is needed only in certain special circumstances. Furthermore, the complexities of growth, differentiation, development and senescence requires precise control of gene expression in terms of qualitative as well as for quantitative production of proteins. Therefore, there should be mechanism to regulate the differential gene expression so that, the genes whose RNA or protein products are not needed at that specific point of time to be switched off. Though it seems too simplistic and straight forward, it is very complex process with several implications. The regulation of gene expression varies in pro-and-eukarytoic organisms in certain basic ways.

6.13.1 Levels of Gene regulation

The gene expression can be regulated and controlled at different levels as listed below:

- Transcriptional Control
- Translation Control
- Posttranslation Control
- Development Control

Most of the control takes place through network of regulatory genes. Some of these genetic elements have evolved precisely for the regulation of transcription of genes encoding proteins and enzymes. Gene expression also can be controlled at other stage. The Pro-and-eukaryotes evolved a great flexibility in responding to intrinsic and extrinsic environmental changes.

6.13.2 Gene regulation in Prokaryotes

Prokaryotic organisms like bacteria (e.g., *E.coli*) inhabit widely varying ecological environments often not congenial or in a state of flux. Gene regulation is quite essential for the survival of bacteria to respond to changes in environments. Genes can be switched *off* or *on* depending on the types of sugars available in the medium (Fig. 6.52).

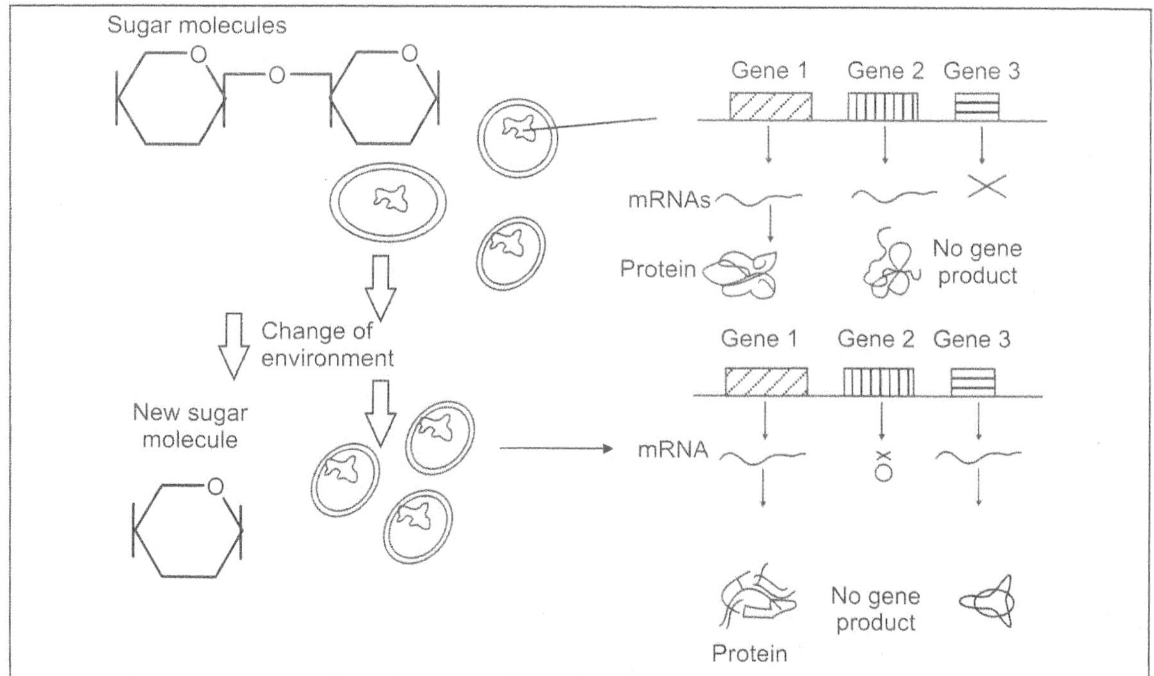

Fig. 6.52 Regulation of expression of gene – Gene-1 expresses always, while gene-2 and 3 expression is regulated depending on type of sugars present in the medium.

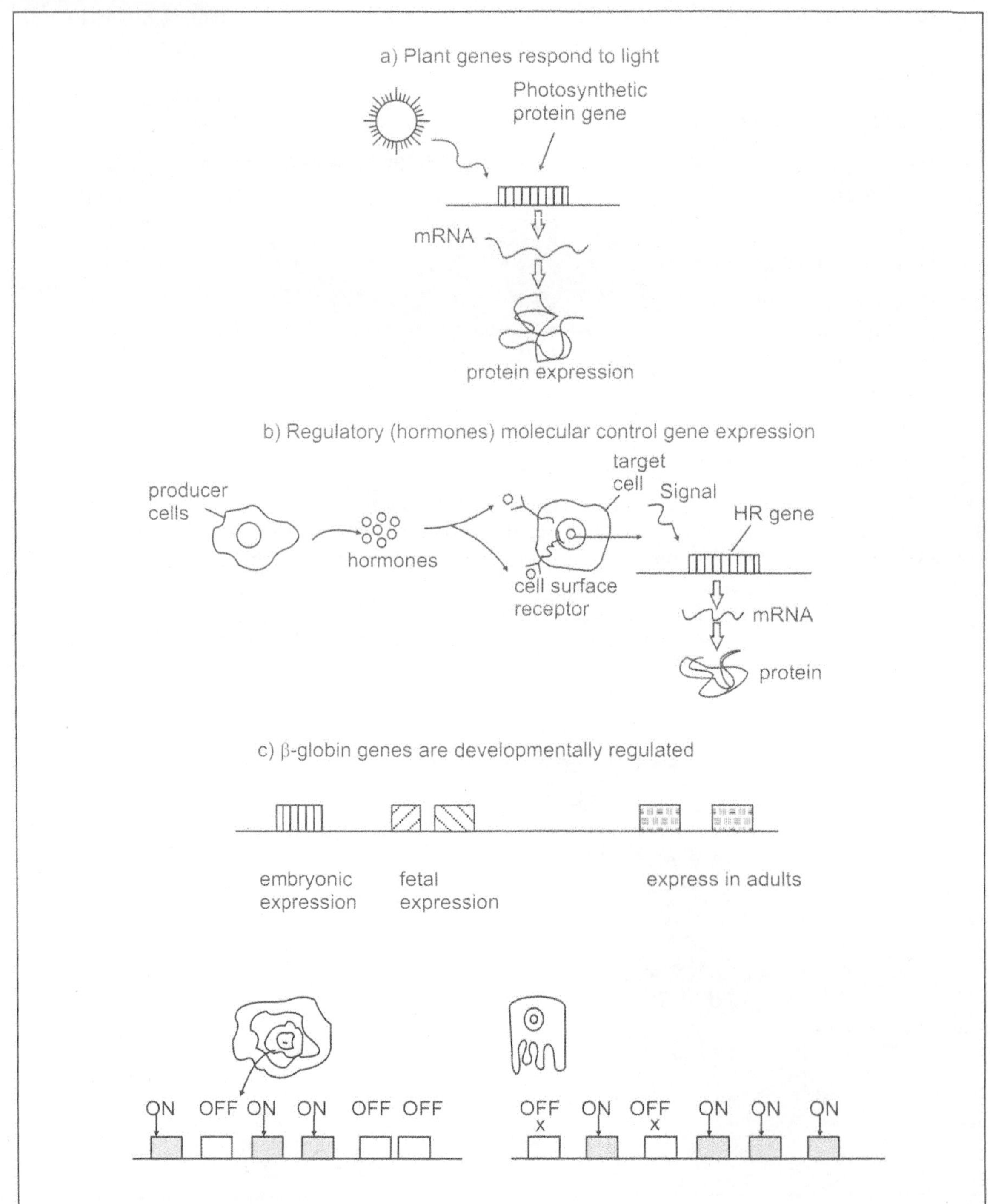

Fig. 6.53 Different aspects of regulation in multicellular eukaryotes.

6.13.3 Eukaryotic Gene Regulation

Basically, regulation of gene expression accomplishes the same thing in both pro-and-eukaryotes. Such process enables a cell to change its biochemical capabilities. However; a great degree of sophistication with respect to signals as well as the impact of that on the organism has evolved in eukaryotes. To summarize some of the regulatory process:

- The range of environments that stimulate response of eukarytoic cells is very wide

  As in bacteria, eukarytoic cells also respond to changes in the environment by changing the pattern of gene expression. For instance, a single celled eukaryote, yeast (*Sacharomyces cervisiae*), regulates its genes for sugar utilization similar to *E.coli*. Plant cells switch on genes responsible for photosynthetic protein by the light. In multicellular organisms certain individual cells or group of cells may respond to internal situation like hormones.

- Genes of Eukaryotes are developmentally regulated

  Most of the eukaryotes (some prokaryotes too), switch on or off depending upon the developmental stage in which they are placed or committed. Human hemoglobin gene expression varies depending on the developmental stage. The protein varies, though perform the functions.

- Regulation of genes in eukaryote organisms results in cell specialization. As soon as a zygote is formed in eukaryotes, the differentiation takes place leading to specialized cells with their set of genes expressing in certain specific ways. The differentiation is irrevocable. In humans, there are more than 250 cell types, each performing a specialized function with distinct morphology, biochemistry, and the physiological role. Liver cell is quite different from that of nerve cell or muscle cell because it expresses different genes. The gene expression pattern attendant on differentiation during early development (Fig. 6.53).

  (a) Plant genes respond to light

  (b) Regulatory (hormones) molecules control gene expression

  (c) β- globin genes are developmentally regulated.

6.14 Recombinant DNA Technology

Introduction

One of the most potential practical applications of molecular genetics is the emergence of the field *of recombinant DNA (rDNA) technology*. It is also popularly referred to as *Genetic Engineering* or *Genetic modification of organisms* (GMO). In simple terms the technique involved is nothing but the manipulation of genes in such ways bypasses the formal transmission and expression as sexual or asexual modes of reproduction, as has been successfully done in crop and animal improvement. The rDNA technology enables the process of moving genes from one genome (organism) to, another despite the barriers of reproductive isolation mechanisms (RIMs) that safe guard the integrity of biological species and the flow of genes. Conventional breeding techniques of plant and animals (the recombination and selection of recombinants) is limited to exploitation of intraspecific genetic variations only. The technique of rDNA not only moves genes from one genome to another with felicity, but also lets their expression and biological functioning of the transformed organism (GMO) with its altered genetic material (DNA). Thus the rDNA technology have made possible to create new genomes that can express human or any genes *and make* human or any organisms proteins in alien surroundings.

6.14.1 Strategies for making Recombinant Organisms

Presently there are three strategies for genetic manipulation that have been developed to different degrees:

- Phenotype level
- Somatic cell level
- Germ cell level

Whatever may be the level and complexity of the system, the technique involves first locating the desirable donor gene , second moving such a desired gene into another organisms genome where it will be expressed. The proteins produced in response to the newly added piece of DNA (rDNA) might have useful effect in the host (e.g., gene therapy or correction and addition of metabolic step etc.); alternatively, the rDNA may be harnessed for use elsewhere.

The real protocols used are too complex to be covered here. More details are provided elsewhere in the text of other chapter. The main outline of inserting a new gene into a bacterium and hypothetical correction of insulin deficiency (gene therapy) in diabetic humans is given graphically in (Fig. 6.54).

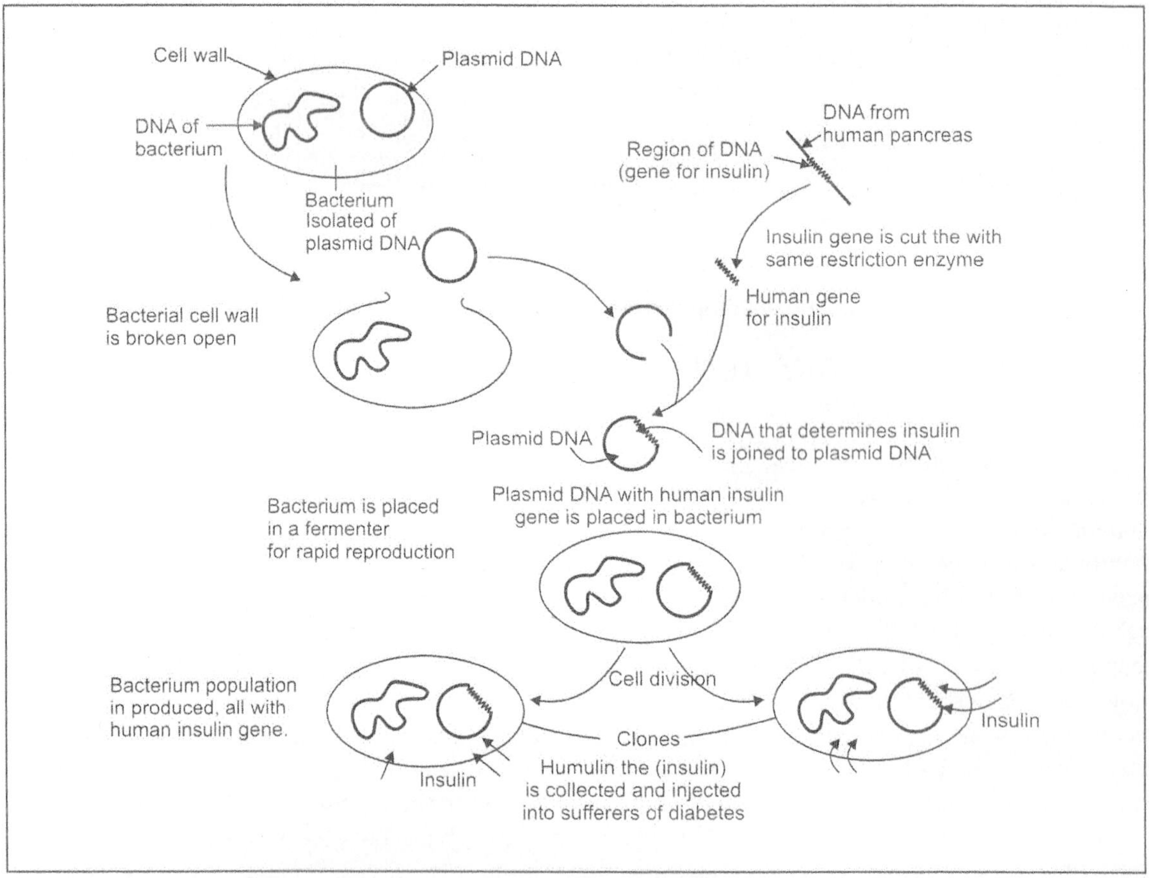

Fig. 6.54 Illustrate example of Gene cloning for gene therapy.

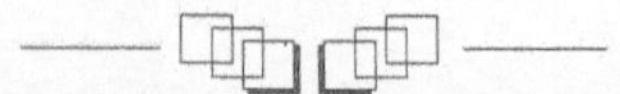

Essay Type Questions

2. Describe the basic tenets of mendelism.
3. Outline and explain the Mendelis law of Independent assortment.
4. What is the physical basis of Inheritance.
5. Give an experimental proof that genes are carried on chromosomes.
6. Correlate the genes and enzymes in lieu of a suitable example.
7. One-gene-one enzyme hypothesis.
8. One-gene-one polypeptide hypothesis.
9. Outline the experimental proof that genetic material is DNA
10. Outline the experimental proof that genetic material is RNA in TMV.
11. Outline the experimental protocol of Hershery and Chase experiment demonstrating DNA as genetic material.
12. How do you prove experimentally that DNA is genetic material in eukaryotes.
13. Explain briefly the structure of Deoxyribo nuclic acid.
14. Explain Walson & Crick model of DNA.
15. Give an outline of DNA replication
16. Provide experimental proof for semi-conservative replication of DNA.
18. List the major classes of proteins employed by the DNA replicating machinery.
19. What is the role of topoisomerase in replication?
21. Outline the central Dogma of Molecular Biology and list various stages of gene expression.
22. List and explain about major classes of RNA.
23. Write a note on pro and eukaryotes RNA polymerases.
24. Describe the process of transcription.
25. Write a brief account of genetic code.
26. Describe the process of translation.
27. List and explain briefly the factors of eukaryotic translation.
28. Mention briefly the expression of gene regulation in pro and eukaryotes.
29. What to you understand by recombinant DNA technology?
30. Mention strategies for the production of GM organism.

Short Answer Type Questions

1. Briefly outline cell theory.
2. Briefly describe mechanism of DNA replication.
3. Briefly describe replication of Genomes.
4. Define the following
   - (i) Genotype
   - (ii) Phenotype
   - (iii) Chromosome division
   - (iv) Linkage
   - (v) Crossing over
   - (vi) One gene - one enzyme hypothesis
   - (vii) Semiconservative replication
   - (viii) Promoters
   - (ix) Polymerases
   - (x) Endonucleases
   - (xi) rDNA
   - (xii) Vector

Part - B

Elements of Biotechnology

7 Introduction to Biotechnology

7.1 An Introduction to Biotechnology

There is little doubt that biotechnology is leading new biological revolution. Undoubtedly, the 21st centaury will witness an unprecedented economic growth like never before. The advances made in modern biology and bewildering array of sub-disciplines besides other physical and material sciences has given birth to a new economic growth engine – biotechnology, generating new materials (bioterials) that will fuel and accelerate economic growth dramatically in this century. If one goes by the number of Nobel Prizes awarded during the last two and half decades, it is clear that much emphasis has been placed on description of life processes at the cellular and molecular level, thus paving the way for biotechnology enabling to embrace the changes in bio-economy based on genetically modified (engineered) products of the natural world rather than on Chemical synthesis and industrial process.

The term biotechnology was originally coined by KARL EREKY, an Hungarian engineer in 1919 to encompass all the lines of work by which products are made from raw materials with the aid of living organisms. He has envisaged and envisioned a biochemical (biotechnological) age similar to that of Stone and Iron ages (BUD, 1989)

What is biotechnology?

During 1970's biotechnology has emerged as a novel discipline as a result of fusion of developments in biology and technology. There are widespread unnecessary controversies on the basic definition of biotechnology, mainly due to one's own prejudices and emphasis on specializations. Consequently, different organization defines biotechnology in diverse ways. *The biotechnology has been broadly defined as "the development and utilization of biological processes, forms and systems for obtaining maximum benefit to man and other forms of life. In other words it is the science of applied biological process"* (Biotechnology, Dutch Perspective, 1981)

The organization for Economic Co-operation and Development (OECD) defines the biotechnology as *the application of Scientific and Engineering principles to the processing of materials by biological agents to provide goods and services* (0ECD, 1981).

The International Union of the Pure and Applied Chemistry (IUPAC) defines it as *the application of biochemistry, biology, microbiology and chemical engineering to industrial process and products and environment* (IUPAC, 1981)

The office of Technology Assessment of US congress has accepted a broad definition of biotechnology as *"any technique that use living organisms, or substances from those organisms to make or modify a product to improve plants or animals or to develop microorganisms for specific use"*

With the broad definition, one can include old world practices in agriculture (Animal and Crop improvement) the production of diary products, the production of beverages by fermentation as well as antibiotics and vaccines. The real cutting edge of biotechnology — the modern biotechnology has emerged from the developments in modern genetics especially the spectacular developments made in molecular biology and genetic engineering. The biotechnology has moved from Genetic engineering to discovery science with the innovations in manufacturing the health care products especially the drug design and discovery.

An unified definition of current biotechnology has been provided by smith (1996) as *"the formation of new combination of heritable material by the insertion of nucleic acid molecules produced by what ever means outside the cell into any varies bacterial Plasmid or any other vector system so as to allow their incorporation into a host organism in which they do not naturally occur but in which they are capable of continued propagation and expression and can offer services and goods for the economic growth and development of mankind* (P.H-SMITH, 1996. Developments in Industrial microbiology).

7.2 Nature and Scope

If all the hopes and aspirations of biotechnology can be accomplished, all the development that has gone before pale into insignificance. The newly acquired biological knowledge has already made vastly important contributions to the health care and welfare of man. Broadly speaking all the applications of biological organisms, systems or process to the manufacturing and service industries constitute the core nature of biotechnology. Thus biotechnology is the science that deals with the integrated approach on the application of biology, biochemistry and process technologies on biological systems for their use in the industry. It creates industries requiring little fossil energy and fuel the world economy in 21st century. The essential nature of biotechnological processes is in most instances has three essential features:

- Functions at low temperature or room temperature
- Will consume little energy inputs and
- Rely mainly on inexpensive inputs for biosynthesis

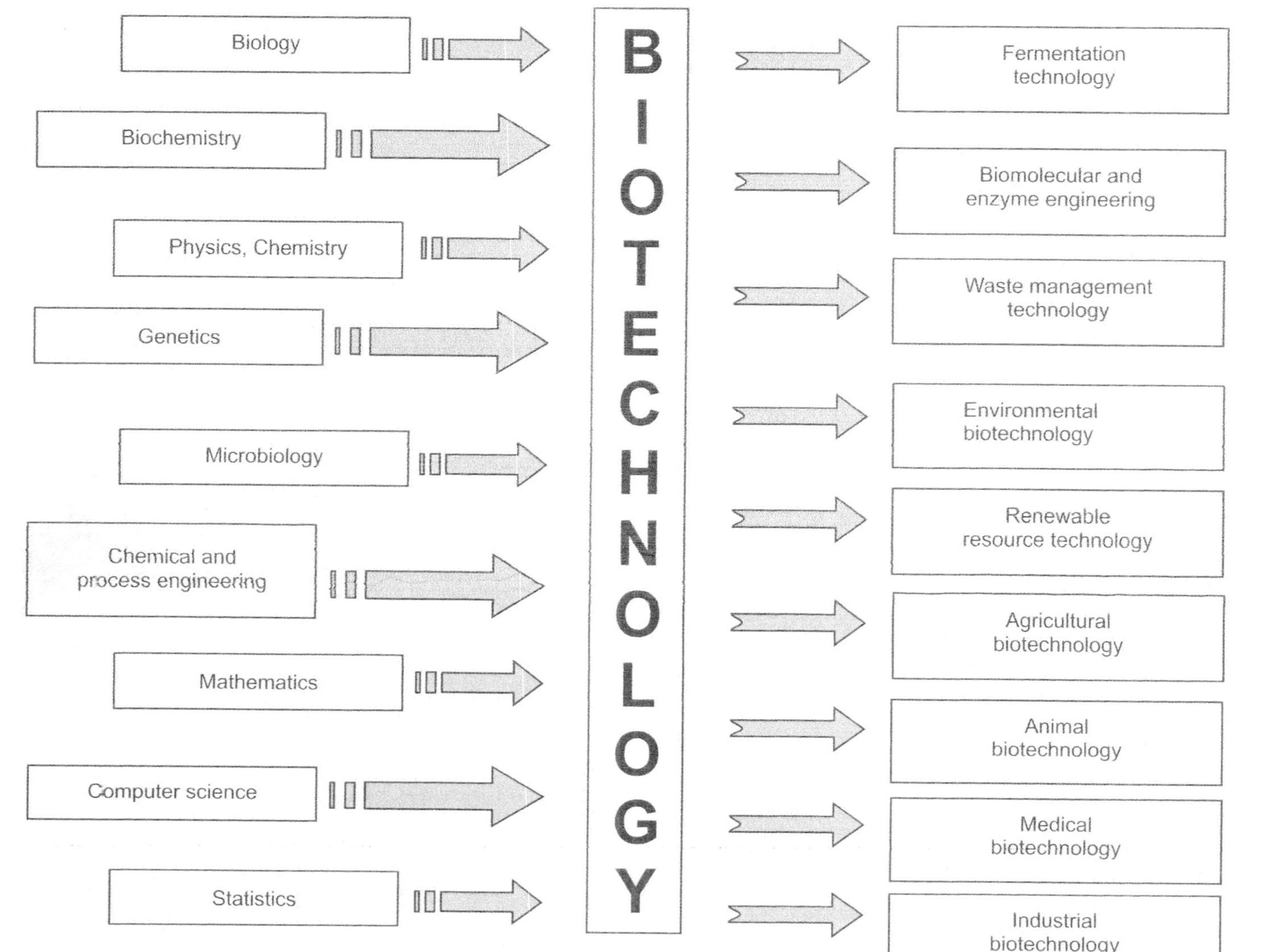

Fig. 7.1 Biotechnology and its interrelation hip with other sciences.

Almost all the present day industrial activities will be influenced and affected by the developments in biotechnology. Some of them are listed bellows:

- Human and Animal food production
- Provides chemical feed stocks to replace petrochemical sources
- Provides alternative energy sources
- Waste recycling
- Pollution control and improvement of the environment
- Agriculture and Animal improvement
- Health care products including pharmaceuticals and neutraceuticals

Biotechnology is internationally considered as most potential technology for generating wealth and new economic order. Interacting with each other the six new technologies – microelectronics computers, telecommunication, new man-made materials, robotics and biotechnology, have emerged at the end of 20th century and beginning of 21st century creating break through technologies that have allowed a whole set of new big industries. A global economy is possible for the first time in human history. It is going to be an era of man-made brainpower industries. The old foundation of success has gone; knowledge is the new basis for wealth. In future capitalist talk about control of knowledge. Biotechnology is one such industry with a vast potential and driving the global economy.

7.3 Relationship with other Sciences

Biotechnology essentially is an interdisciplinary pursuit. In solving the problems of Science and technology often-multidisciplinary strategies are resorted to. Biotechnology will employ technical inputs derived from physical sciences and process engineering, mathematics and computer science. The main objective would be the innovation development and optimal operation of process in which biochemical catalysis has a fundamental irreplaceable role. It also has to draw inputs from medicine, nutrition, pharmacy, agriculture, industrial chemistry, environmental protection, biodiversity, waste process technology etc. (Fig. 7.1).

Essay Type Questions

1. Ennumerate different definitions of biotechnology.
2. What is biotechnology ? How does it relate to other sciences ?
3. Define biotechnology and trace its developments and interrelationship with other basic sciences.

8 Microbial Biotechnology

8.1 Introduction

The microbial and human world are interwined interwound very intimately. The relationship is often helpful and positive in many ways, but also harmful and negative in certain other ways, especially when when cause miner/and devastation through disease and death. Despite the fact that microbes cause diseases, human spirit conquered and exploited them beneficially for the human welfare in diverse ways the newly developed rDNA technology enabled biotechnologists to harmers the potentials of microbes the genetically modified organisms for the benefit of man-kind. Some of the key developments are : microbial biofertilizers, biofuels, biopesticides, single cell proteins biosensors, sewage treatment technologies microbial food and food

8.2 Biofertilizers : Nitrogen Fixation in Nature

Nitrogen, phosphorus and potassium are the important nutrients that influence the plant growth and productivity. Nitrogen, though abundant in the aerial and soil environment cannot be directly used by plants. Nitrogen fixation occurs in free living and symbiotic microorganisms. Legumes are the largest group of plants having symbiotic association in their roots with *Rhizobia* that fix atmospheric nitrogen. Thus nitrogen fixing bacteria *Rhizobium* and *Bradyrhizobium* are considered to be important biofertilizers as they fix the nitrogen and enrich the soils in association with many agricultural crops. The water fern, *Azolla* that forms a symbiotic association with the Cyanobacterium, *Anabaena,* is used to fertilize the paddy fields by supplementing the nitrogen fertilizer.

The free-living bacteria capable of fixing molecular nitrogen are of three types :

1. Obligate aerobic
2. Facultative aerobic and
3. Anaerobic organisms.

Obligate aerobic bacteria of the family *Azotobacteraceae* constitute the majority of heterotrophic free–living bacteria. e.g. *Azotobacter*, *Beijerinckia*, *Derxia*. Facultative aerobes include the genera of *Aerobacter*, *Pseudomonas* and *Klebsiella*. Anaerobic nitrogen fixing bacteria belong to *Clostridium*, *Chlorobium*, *Rhodopseudomonas*, *Rhodospirillum* and others.

Nitrogenase is the key enzyme involved in the fixation of molecular nitrogen and it has been isolated from many genera of free-living nitrogen fixing organisms such as *Clostridium*, *Bacillus*, *Klebsiella*, *Rhodospirillum*, *Anabaena*, *Gloeocapsa*, *Plectonema*, *Azotobacter* and *Mycobacterium*.

The cyanobacteria or blue–green algae (BGA) are an important source of biofertilizers produced at practically no energy cost to mankind. These are photoautotrophic diazotrophs which can utilize solar energy to reduce atmospheric dinitrogen to ammonia. The mass production of BGA is cheaper and easier than the chemical fertilizers. BGA fertilizers have gained importance in the tropical paddy fields as they can fix about 25–30 Kg nitrogen per hectare per season. Besides nitrogen fixation, BGA have the ability to ameliorate problem soils and to remove the pesticide residues from the soils. Thus the use of BGA fertilizers has considerably reduced the input costs and dependence on costly chemical fertilizers.

Blue Green Algal (BGA) Biofertilizers

A large number of BGA are known to fix nitrogen which is partly released during the growth phase of the cyanobacteria while the root is released after the death and autolysis of the algae. BGA facilitate the solubilization of inorganic insoluble phosphorus into soluble phosphorus making it available to crop plants. Cyanobacteria also excrete plant growth promoting substances like vitamins, auxins, amino acids, and sugars (polysaccharide). The polysaccharides improve soil structure by binding soil particles and improving soil aggregation. Improved soil structure results in soil permeability, infiltration rate and water holding capacity. Cyanobacteria help to reclaim saline and alkaline soils as they change the soil p^H towards neutrality.

Beneficial effects of BGA technology are :

1. Enhanced crop yield by 5-30% in different agroclimatic zones in the absence of chemical N-fertilizers.
2. BGA application increases crop yield by 10% when it is used in addition to chemical N-fertilizers; BGA application result in net saving of 20-30 Kg N/ha in rice fields. Among several nitrogen fixing BGA, *Anabaena variabilis*, *Nostoc muscorum*, *Aulosira fertilissima* and *Tolypothrix tenuis* are the common strains used for biofertilizers.

Production

Algal biofertilizers may be produced at individual farmer level or on commercial basis. General production methods include : 1. Trough method 2. Pit method and 3. Field method.

Shallow troughs of approximately 2m × 1m × 23cm size are used in BGA production. Similarly pits are dug and lined with polythene to prevent water seapage. Algal production should be done in an open area so as to receive sunlight throughout the day. A light intensity of 1000 to 4500 lux is good for BGA growth. Spread about 3 Kg soil per unit and add 200g single super phoshpate. Soil p^H should be between 7.0 to 7.5. Soil p^H can be adjusted with either gypsum or lime. Multiplication units (Trough/pit) have to be filled with water (up to 3-4") and mixed. Starter cultures containing *Aulosira* and *Tolypothrix* or *Anabaena* and *Nostoc* are sprinkled in different troughs so that they grow equally avoiding competition. Carbofuran (3%-20g/unit) granules helps to prevent insect breeding in multiplication units. Hot summer months with temperature over 30C is more suitable for BGA growth. It takes about 7-10 days for the formation of a thick algal mat. Dried algal flakes are collected, mixed, packed in polybags and stored in a cool dry place.

Field Scale Production

It is a scaled up process for large scale production which is essentially similar to trough/pit method. In this method about $40M^2$ plots with strong 15cm earth embankments. About 12 Kg of single super phosphate is required. Application of insecticides (carbofuran or ekalux) prevents predators and algal grazer. Inoculate the plot with 5 Kg of composite and good algal mat is formed in about 2-4 weeks depending on the type of soil. About 16-30 Kg of BGA is produced from a 40 M^2 plot.

BGA can be sun dried and stored for a long time in dry state at normal room temperature in shade.

Field Application

BGA is broadcast as a dry inoculum over the standing water in the rice field at the rate of 10 Kg/ha. BGA fertilizers should be used in every rice crop as a kind of insurance to the crop yield. It also prevent the deterioration of soil physico-chemical properties.

Mycorrhizae as Biofertilizers

Some fungi enter into a mutualistic association with plant roots to form '*mycorrhizae*'. The fungus derives nutritional benefits from the plant roots and help the plant to obtain certain nutrients from soil without causing any disease. The mycorrhizal associations are widespread and unique in forming integrated morphological units. Enhanced up take of water and nutrients such as phosphorus and nitrogen have been demonstrated in many mycorrhizal associations. There are two basic types of mycorrhizal associations 1. Ectomycorrhizae and 2. Endomycorrhizae.

Ectomycorrhizae

Ectomycorrhizae associations are common in gymnosperms and angiosperms. Ectomycorrhizal fungus may be an ascomycete or a basidiomycete. The fungus forms an external pseudoparenchymatous sheath constituting up to 40 % of the dry weight of the combined root–fungus structure (Fig. 8.1).

The fungal hyphae penetrate the intercellular spaces of the epidermis and cortex forming a thick mantle. The morphology of the root gets altered by mycorrhizal association as the roots form shorter, dichotomously branched clusters. The common ectomycorrhizal fungi belonging to ascomycetes are truffles while the basidiomycetes are *Amanita* and *Boletus*. Ectomycorrhizal fungi are common in most trees of temperate forests with optimal growth temperatures of 15–30C and p^{H} of 4.0 – 6.0. Most ectomycorrhizal fungi utilize complex organic sources of nitrogen, simple carbohydrates and produce a variety of metabolites including auxins, gibberellin cytokinins, vitamins, antibiotics and fatty acids.

The benefits the plant may derive from ectomycorrhizal association include longevity of feeder roots, increased nutrient absorption from soil, mobilization of selected ions (phosphate and potassium) from soil, disease resistance, increase tolerance to drought, p^{H} and other toxins. The ectomycorrhizal sheath provide an effective physical barrier to penetration by plant root pathogens. Ectomycorrhizal roots also produce volatile organic acids with fungistatic properties.

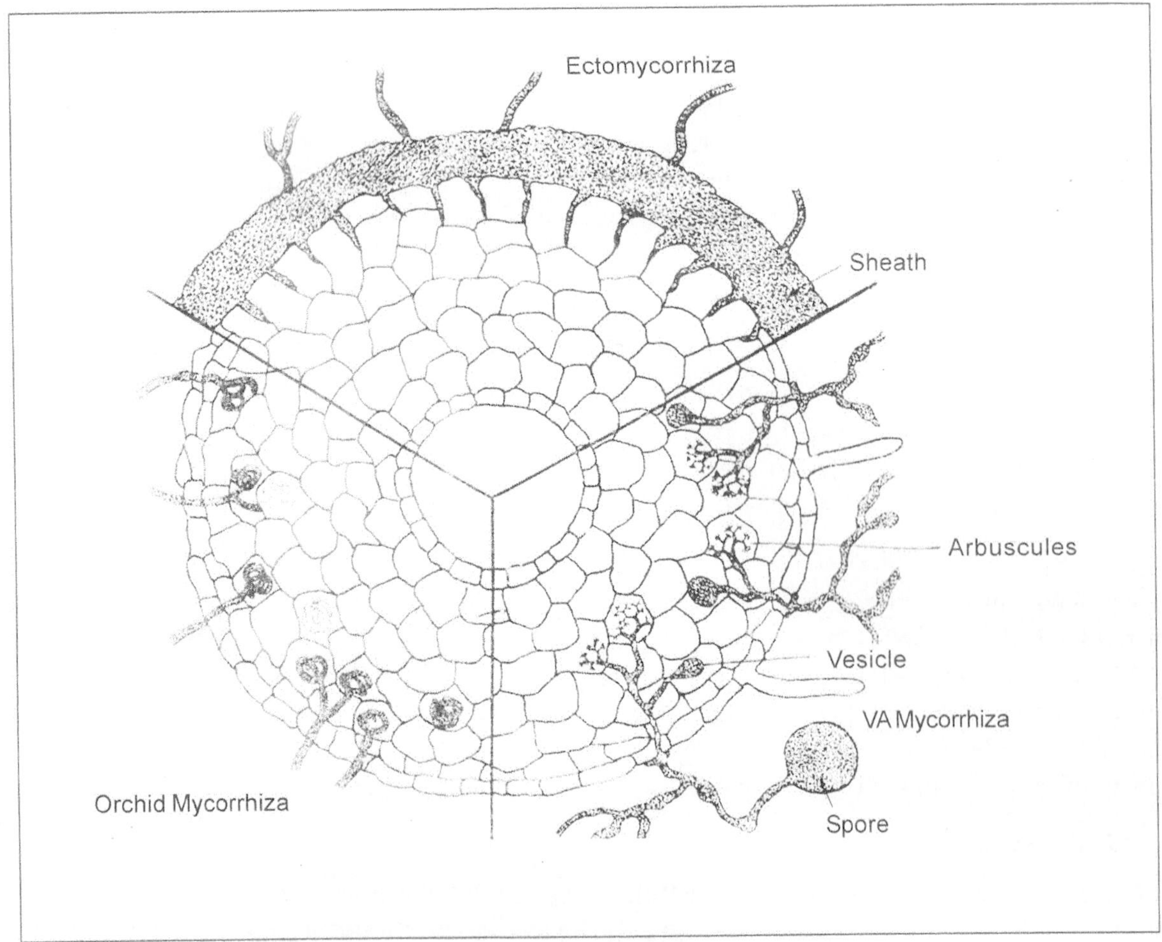

Fig. 8.1 Three types of Mycorrhizal associations in Plant roots (diagrammatic)

Endomycorrhizae

Endomycorrhizal associations are characterized by the penetration of fungi into the plant root cells as seen in members of Ericales (e.g., *Heath*, azal and *Rhododendron*) and orchids. Vesicular arbuscular mycorrhizal fungi form endomycorrhizal association in most of the angiosperms, gymnosperms, ferns and bryophytes. The endomycorrhizal fungi of the plant genera belonging to the ericales are characterized by non-pathogenic penetration of root cortex by septate fungal hyphae forming intracellular coils. The ericoid mycorrhizae are involved in better nitrogen nutrition and greater phosphatase activity in mycorrhizal roots that helps the transfer of phosphorus from external sources to the host plant.

Orchids are obligately mycorrhizal under natural condition. All orchid roots are internally infected with fungal hyphae to form mycorrhizal association. The fungi form coils within the cells of the outer cortex which are digested, utilized by the host cells later. Orchid mycorrhizal fungi include *Rhizoctonia solani* and *Armillaria mellea* which help to enhance the orchid seed germination.

Vesicular Arbuscular Mycorrhizae (VAM Fungi)

VA mycorrhizae are the most predominant and widespread endomycorrhizal association occurring in most of the 2.6 million known plant species of angiosperms, gymnosperms, pteridophytes and bryophytes. VA mycorrhizae are known to enhance nutrient uptake, plant growth and productivity in many economically important crop plants like wheat, maize, potato, soybean, tomato, groundnut, cotton, tobacco, apple, orange, grape, sugarcane, rubber etc. VA mycorrhizae are aseptate, zygomycetous fungi belonging to the families Glomaceae, Acaulosporaceae and Gigasporaceae. The characteristic feature of VAM fungi is the presence of swollen vesicles and intricately branched arbuscules in the root cortex (Fig. 8.12). The fungus is both inter and intracellular in the root cortex and produce external mycelium that spreads into the soil. The external mycelium forms a loose net work in soil around the mycorrhizal root. These fungi produce largest known resting spores of any fungi with diameters of 20–400 μm. Bacterial endo symbionts are known to inhabit these fungal spores. The polymerase chain reaction (PCR) analysis of these bacterial endosymbionts revealed that these are species of *Burkholderia*. The VAM fungi increase plant growth through improved uptake of nutrients like phosphorus by the external hyphae beyond root hair and phosphorus depletion zones. The beneficial effect on plant growth is prominent in phosphorus deficient soils. The potential of VAM fungi in tropical agriculture is much greater as the tropical soils contain low levels of phosphorus.

Species of *Glomus* are used in many crops as a biofertilizer in the form of soil-root bit inoculum. VA mycorrhizal fungi could not be cultured due to their obligate nature of their association. VAM inoculum is multiplied by growing grasses, such as *Cenchrus celiaris* and the roots colonized by the fungus along with soil is used as inoculum for the crops.

Biofertilizers are widely accepted as less expensive, ecofriendly supplements to chemical fertilizers which provide nitrogen and phosphorus to plants. Biofertilizers preparations contain microorganisms in sufficient numbers that help plant growth and nutrition. Several microorganisms, particularly bacteria fix nitrogen which is abundant (78%) in the atmosphere. Rhizobia have the ability to fix atmosphere nitrogen symbiotically in the root nodules of leguminous plants. Besides Rhizobia, diazotrophs like *Azotobacter, Azospirillum* and blue-green algae (Cyanobacteria) are capable of fixing atmospheric nitrogen. Cyanobacterium, *Anabaena* fixes nitrogen in water fern called *Azolla.*

Phosphorus is the important nutrient next to nitrogen required for the growth of plants. Most of the soil phosphorus is in unavailable form and only 1-2% of it is incorporated into aerial parts of the plants. Phosphorus solubilizing microorganisms consists of various bacterial, fungal and actinomycetes forms that help to convert insoluble inorganic phosphate into simple soluble forms species of *Pseudomona, Micrococcus, Bacillus, Penicillium, Fusarium, Sclerotium* and *Aspergillus* are some of the important phosphate solubilizing microorganisms. These organisms solubilize tricalcium phosphates in a medium producing clear zones around the growing colonies of the organisms. Several rock phosphate dissolving bacteria, fungi, yeasts and actinomycetes were isolated from soils from rock phosphate deposits and rhizosphere soils of different leguminous crops.

The most efficient bacterial isolates include *Pseudomonas striata, P. rathonis, B. polymyxa* and fungal isolates such as *A. niger, A. awamori, P. digitatum* and *Schwanniomyces occidentalis. P. striata, B. polymyxa* and *A. awamori* have been selected at IARI for carrier based inoculant and is known as IARI Microphos Culture. Bacteria can be used in neutral to alkaline soils and fungi are better suited for acidic soils. These cultures have the capacity to solubilize insoluble inorganic phosphate such as rock phosphate, tri-calcium phosphate, iron and aluminium phosphates by producing organic acids. They also mineralize organic phosphates present in organic manures and soils.

Microbial Biofertilizers

Microbial biofertilizers are the preparation containing primarily active strains of microorganisms in sufficient numbers. Among the microbial fertilizers, *Rhizobium* inoculant is specific for different legume crops. Besides *Rhizobium* other bacterial fertilizers produced in India are *Azotabacter, Azospirillum* and phosphate solubilizers. In India, pulse crops are grown in about 24 million hectares with groundnut occupying almost 1/4 of the area. Pulses have a special role to play as a source of high quality protein in the poverty stricken developing world. Rhizobia are widely distributed in the soils of the tropics and it forms a good candidate for a biofertilizer in India.

Isolation of Rhizobium

Clean nodules are individually crushed in sterile water and the suspension is streaked on yeast extract mannitol agar (YEMA) media in Petri plates. The bacteria appear within 3-10 days as white, translucent, glistering and elevated colonies.

Rhizobia have been divided into two groups based on their growth. The fast growing acid producers on YEMA are *Rhizobium* species (*R. phascoli, R. trifolii, R. leguminosarum, R. meliloti*) while the slow growing strains producing alkali on YEMA are *Bradyrhizobium* species (*B. japonicum*). DNA-base ratios of the fast growing species of Rhizobium are less closely related to the slow growing species.

Nodulating bacteria have developed certain definite host preference with the leguminous species and vice versa. Each of the groups of legumes and their specific rhizobia constitute a cross-inoculation group. *Rhizobium* isolated from one of the legume members of the cross-inoculation group would nodulate all other members of that group. Seven cross-inoculation groups have been recognised on the basics of ability of rhizobia to form nodules with legume plants (Table 8.1).

Table 8.1 *Cross-Inoculation groups of Rhizobium*

Rhizobium	Cross-inoculation group	Legume host
R. trifolii	Clover group	*Trifolium*
R. meliloti	Alfalfa group	*Melilotus, Mediacarp, Trigonella*
R. phaseoli	Bean group	*Phaseolus*
R. lupini	Lupine group	*Lupinus*
R. leguminosarum	Pea group	*Pisum, Vicea*
R. japonicum	Soy bean group	*Glycine*
Rhizobium sp.	Cow pea group	*Vigna, Arachis,*
Bradyrhizobium japonicum	–	*Glycine*
Bradyrhizobium sp.	–	*Cajanus, Cicer, Vigna*

Production of Rhizobium Inoculants

Rhizobial strains should be selected based on the criteria to form association with the host, nodulation ability, and fix adequate amount of nitrogen in many host species under the same cross-inoculation group. Inoculum is prepared either in flasks on a shaker or in a fermentor. The inoculum from the log phase of growth be suspended in liquid. Air flow rates of 5 l/h has been found satisfactory for small scale units. Incubation temperature ranges from 26-30C for most of the strains. Lignite and charcoal are universally adopted carriers for rhizobial cultures in India. Sterilization of the carrier is necessary before blending of the broth culture to suppress the proliferation of other microbes. The rhizobial number in broth must be about 10^9 cells/ml at the time of mixing with sterile and neutralized carrier. Method of mass production of *Rhizobium* is shown in Fig. 8.2.

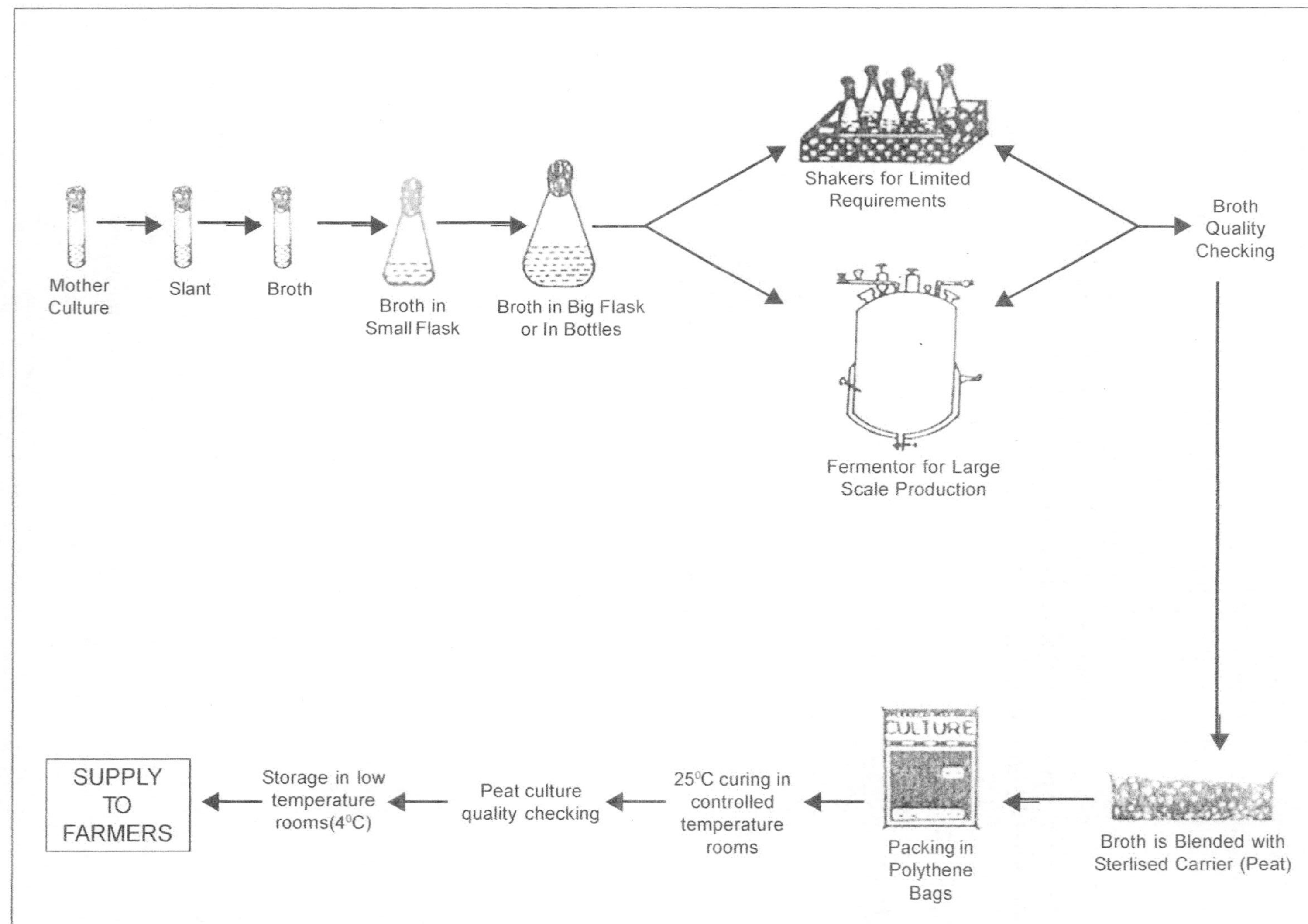

Fig. 8.2 Steps involved in the mass production of *Rhizobium*

Azotobacter and Azospirillum Inoculants

The producers for mass multiplication of *Azotobactor* and *Azospirillum* are similar to that of *Rhizobium* culture production but for the media used. The Jensen's medium is used for *Azotobacter* production and the Okon's medium is used for production of *Azospirillum*.

Seed Bacterization

The method of seed inoculation for all the three microbial inoculants is similar. The microbial inoculant is mixed in clean water or in a 5% adhesive to make a thick slurry. The slurry is then mixed with required quantity of seeds for one acre of land. The seeds are dried in shade and stored in cool dark place before use. Immediate use of such seed gives optimum benefit. Seeds may also be inoculated by sprinkle method. In this method the seeds are mixed with adhesive in water and the inoculants is sprinkled over the seeds subsequently. For transplanted crops, the seedling are dipped in the slurry of the carrier based inoculum for 10-50 minutes before transplantation. The inoculant may also be mixed in farm yard manure and broadcast near the root zone.

8.3 Biofuels

Plants efficiently utilize energy and convert CO_2 into biomass. Biomass is the total cellular dry weight or organic matter produced by the living organisms that can be used as a source of energy. In fact, biomass is the only available energy source to man until the development of fossil fuels like coal and oil. Rapid decline of fossil fuel resources necessitated the search for alternative sources of energy. Biological agents including microorganisms have the ability to convert biomass into fuels such as methane, ethanol, butanol, biodiesel and hydrogen. The fuels produced through biological means are referred as 'biofuels'. Biofuels are less expensive, non-polluting (ecofriendly) renewable resources generally produced by utilizing waste materials (municipal waste).

Bioethanol is widely used for transport purpose is Brazil and USA. The use of bioethanol replacing petrol has begun during 1980s. Biofuel used for transport purpose should have certain desirable features like (i) portability of sufficient quantities in the vehicle, (ii) burning in the internal combustion engines and (iii) it should be roughly equivalent to petrol in energy content. Ethanol also has certain advantages over petrol. Such as :

(i) higher latent heat of vaporization (855 MJ/Kg) than petrol (293 × 5 Kg)
(ii) higher octane number (99) than petrol (80 – 100)
(iii) lower hydrocarbon omission compared to petrol
(iv) ethanol has much higher flash point (45C) than petrol (ethanol is much less likely to catch fire and explode in case of fuel leakage)

Ethanol can be mixed with petrol which increases the octane rating of petrol. In USA, ethanol is blended with petrol in 20:80 ratio and marketed as 'gasohol'. Recently, ethanol mixed petrol is marketed in several states of India.

Ethanol has certain disadvantages as it is highly hydrophilic in nature, starting problems when the air is cool and consumption of 10% more fuel by ethanol run engines. Downstream processing for ethanol is expensive as it requires energy.

Bioethanol is produced mainly from three important sources

(i) Starch or sugar crops

(ii) Cellulose (following enzymatic hydrolysis)

(iii) Cellulose (following chemical hydrolysis)

Several microorganisms including bacteria and fungi are involved in the production of various biofuels utilizing different substrates. Commercial production of ethanol is based on starch and sugar crops, cellulose materials are first converted into glucose and other fermentable sugars which are subsequently fermented to alcohol by yeast. *Trichoderma reisei* (convert cellulose to glucose) *Sacharomyces cerevisiae* (hexoses) and *Escherichia coli* (pentose) are the three important microorganisms involved in the process of bioethanol production.

Conversion of cellulose to ethanol involves four important steps : (i) production of the enzyme-cellulase, (ii) hydrolysis of cellulose (iii) fermentation of hexose sugars and (iv) pentose fermentation. Cellulose hydrolysis yields hexoses like glucose and cellobiose and hemicellulose digestion releases pentoses. Biotechnology can help to develop more efficient strains for alcohol production as well as devising efficient use of these microbes in cell immobilized systems.

Biobutanol is produced from strictly anaerobic fermentation by *Clostridium acetobutylicum* using molasses/cornmeal as substrate. Diesel like liquid obtained from materials of biological origin is known as 'biodiesel'. Plants and algae produce lipids and hydrocarbon which serve as a source of biodiesel. Several oilseeds such as sunflower, rape seed, linseed, soybean, olive, groundnut, safflower etc., produce high energy value lipids that can be used as diesel engine fuels. Lipids are esterified to reduce the viscosity of the oil and these esterified fatty are referred as "biodiesel". Algae are known to accumulate over 60% of their biomass as lipids. Inspite of the growing interest in oilseed based biodiesel production, it has not picked up in developing countries due to higher costs involved.

The plant species accumulate hydrocarbons in the form of latex that can be used a fuel. Such plants include members of Euphorbiacae (*Euphorbia latyris*), milkweeds (*Asclepias* spp.) and a leguminous tropical tree (*Capaifera multijuga*) some fresh water and marine algae are known to accumulate hydrocarbons as much as 75% of its biomass (*Botryococcus braunii*). Lipid based biodiesel production is now at an experimental scale and it may be scaled up for commercial feasibility in future.

Hydrogen

Hydrogen is a promising fuel resource of the future in view of the depleting fossil fuels. Hydrogen is biologically produced as an end product from the carbohydrate fermentations by bacteria. But methane formation from organic compounds is preferred over hydrogen due to high convention efficiency of carbohydrates to energy. However, phototrophic microorganisms are capable of producing significant amounts of molecular hydrogen. Various photosynthetic bacteria are known to produce hydrogen using solar energy. The photosynthetic bacteria include both purple and green sulfur bacteria. Some cyanobacteria are also capable of producing hydrogen. The enzyme hydrogenase is crucial in the photoproduction of hydrogen. Photosynthetic microorganisms which can simultaneously produce molecular hydrogen and remove excess CO_2 from the atmosphere may reverse the phenomenon of global warming, combustion of hydrogen does not produce gases that alter the atmosphere and hence it is a clean and safe fuel. Photoheterotrophic bacteria, such as *Rhodospirillum* or *Rhodopseudomonas* are capable of utilizing waste materials for the production of hydrogen.

Single-Cell Proteins

Microbial biomass may be used either as food for humans or as feed for animal consumption. It is used as a food additive as it contain a large percentage of high-grade protein . The microbial biomass production for food or feed is referred as single-cell protein production. Human diet in the developing world is mostly grain-based and deficient in proteins as well as some essential amino acids. Microbial biomass can augment the deficiency and help to overcome the problems of malnutrition. Microorganisms also have an additional advantage of converting relatively in expensive materials such as cellulosic wastes and molasses.

Yeasts are the most extensively grown microbes for single cell protein due to their high nutritional value (50-60% protein), easy cultivation and lack of toxic by-products. *Saccharomyces* species are important microorganisms as they can grow on a common substrate like molasses. *Candida lipolytica*, a feed yeast can be grown on hydrocarbon like alkanes. In view of the escalating prices of petroleum products, use of waste materials for single cell protein has become attractive in recent times. Animal waste from cattle feed lots and waste sulfite liquor or from wood-pulp processing have been successful utilized for the commercial production of yeasts for animal feed.

Algae and cyanobacteria have the advantages as a source of single cell protein as they require only inorganic nutrients and sunlight. *Spirulina* has been traditionally used as a food supplement by the natives in Africa and Mexico. This cyanobacterial bloom that floats on the surface of the Lake Chad, Africa is simply scooped up, dried and used. *Spirulina maxima* is commercially cultivated at Lake Texcoco, Mexico at about 2 metric tons per day. *Spirulina* contains 60-70 % high grade protein and the total nucleic acids are less than 5% which is less than the most bacteria and yeasts.

Similarly Sea weeds like *Ulva*, *Caulerpa*, *Laminaria* and *Porphyra* are also used in Pacific islands. They are eaten fresh as salad, dried as a snack or are added as supplements. They provide important vitamins and minerals to the people dependant on rice and fish-based diets. The unicellular fresh water green alga, *Chlorella* is cultivated on simple media and used as a protein and vitamin supplement in yogurt, ice cream, bread and other food products.

8.4 Microbial Production of Plastics

The commercial manufacturing of polymers by microbial system has been well established. For instance, one of the triumphant examples is the production of xanthan, a unique substance, which find manifold uses in making food-stuffs with a range of reological properties, besides having its own uniqueness. Xanthan gum, an extracellular polysaccharide was manufactured by employing *Xanthomonas campestris*.

However, the creation of new structural polymers analogous to the common plastics like polyethylene, polypropylene and polyvinyl chloride etc. through microbes is an exciting proposition since such a new material will not only be an alternative to petrochemical based plastics (which are not degradable) but also will be biodegradable obliterating the environmental pollution and concomitant biological hazards due to plastic that we use today.

PHB Polymer and its Structure

A wide variety of bacteria have poly –3 –hydroxy butyrate or simply PHB an energy storage medium. It is an aliphatic polyster with repeating unit as a primary structure (Fig. 8.3). The molecule adopts two helical conformation i.e., two repeat units for a complete twist of the helix and crystalises in orthorhombic system with two antiparallel helices, and a total of four repeats in the crystallographic unit cell. The space group describing the crystal symmetry is $P_{21,}\ 2_1\ 2_1$. The transmission electron microscopy of bacteria followed by embedding, sectioning and replication revealed that, within the cell, the PHB is accumulated as discrete granules that can reach a size of 200. Each granule is bound by a membrane, which hold the necessary enzymes for polymerization or depolymerisation.

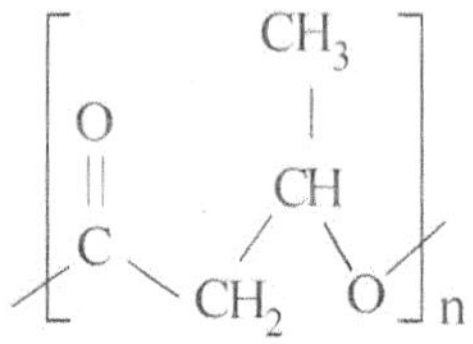

$$\left[-\overset{\overset{\large O}{\|}}{C} - CH_2 - \overset{\overset{\large CH_3}{|}}{CH} - O - \right]_n$$

Fig. 8.3 The Chemical Structure of poly 3-hydroxy butyrate - a repeating primary unit

Biosynthesis

PHB synthesis in bacteria follows relatively a simple metabolic pathway starting with the conversion of any wide range of carbon sources to acetate and thence to acetyl coenzyme A. Condensation of two acetyle co-enzyme. A molecules forms acetoactyl co A and subsequently 3-hydroybutyryl co A by a process of reduction. The linking of the 3-hydroxybutyrate unit by a polymerase regenerates the enzyme and generates PHB micro molecule (Fig. 8.4).

Polymers of other 3-hydroxyalkanoates can be synthesised by similar mechanisms, some alternative alkyl group replaces the methyl side group of PHB. Many poly 3-hydroxyalkanoates have been reported to occur in bacteria, but less common when compared to PHB; side chains groups may be as large as pentyl. However, both homo and co-polymers have been reported.

Fermentation

PHB production is carried out in a two stage batch culture fermentation. The first stage encompasses cell growth typically with glucose as a principal source of carbon. In the second stage the bacterial cells are starved for essential nutrients like oxygen, nitrogen, phosphorous while providing only carbon source i.e, glucose.

Unable to grow and divide, the bacterial cell force to convert the carbon source into PHB till 80% of total dry mass of bacterial cell is nothing but synthesised PHB. The cell literally bursting at seam full of polymer granules. The cell wall becomes so taut that it will not be possible to hold more. The product to reach about 100g/litre concentration takes several days of fermentation.

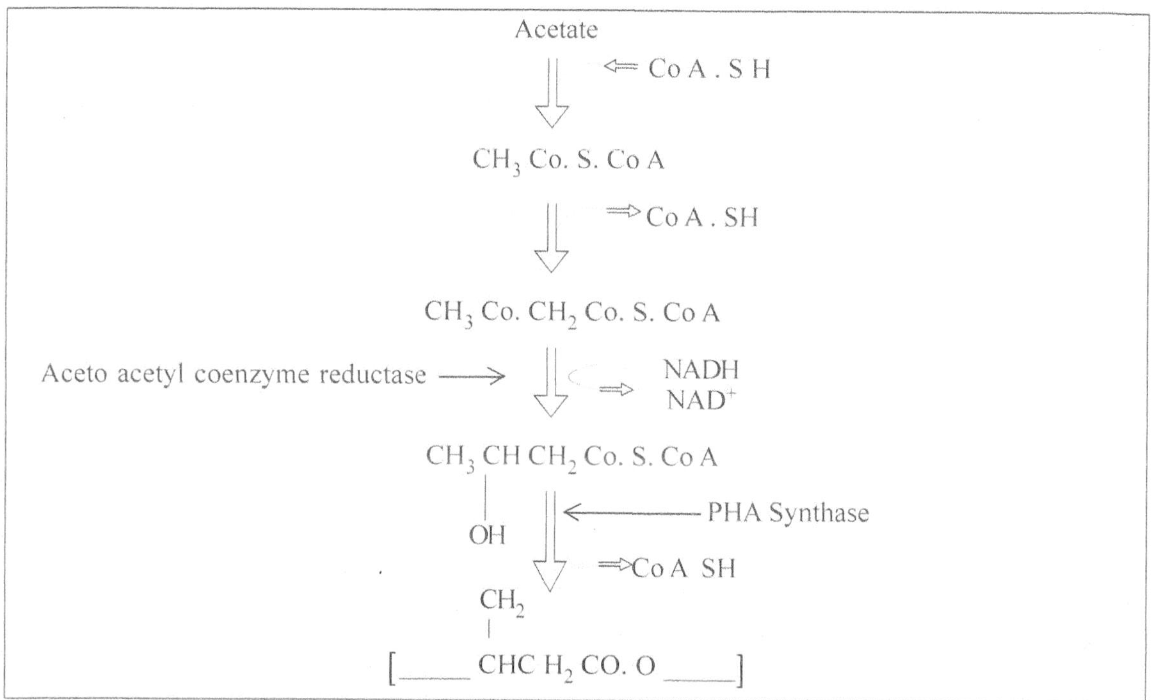

Fig. 8.4 PHB biosynthesis

Product Recovery

There are several methods of recovery of the product after fermentation is terminated. First important step is to break the cells so that the cell contents become accessible to the next stage of recovery. The breaking of cells can be conveniently done by subjecting the fermented broth to increased pressure and temperature and relax the pressure suddenly. This causes the cells to rupture and the contents are liberated. Subsequently, the PHB can be recovered by solvent extraction with chloroform or methylene chloride. Alternative procedure is the remnant cellular material is digested using protease and lipase enzyme and the PHB is recovered as a powder after flocculation. The selection of extraction method is crucial step, since along with choice of ideal organism and growth conditions, it exert influence on the molecular weight of the recovered product.

Finally, the product form is very important so that the polymer is to be readily processable by the users. For most of the applications the PHB powder is dried, extruded and granulated into a '*chip*' form suitable to standard plastic equipment. Additives like segments or plasticizers to impart distinctiveness are incorporated in the beginning itself, before the extrusion and granulation step.

Polymer Properties

The PHB polymer have excellent properties when compared to the those conventionality made synthetically. The former are free from catalyst residues typical of synthetic ones. The repeat unit has a chiral center D(–) (or R absolute) configuration as dictated by enzymatic synthesis. Most significant property of bioplastics is that these are biodegradable by the microorganisms present in the environment. A comparison of properties of conventionally made plastics such as polypropylene (pp) and biopolymer (plastics) is given in Table 8.2. One can see that they share the same properties.

Table 8.2 *A comparison of bioplastics (PHB) of microbial origin and conventionally made plastic polypropylene (pp)*

Physical/chemical attribute	PHB	PP
Tmelt (°C)	180	180
Tg (°C)	15	–10
Crystallinity (%)	80	70
M_W	5,00,000	20,00,000
Density	1.25	0.9
Tensile Strength (MP_a)	40	38
Flexural Modules (GP_a)	4.0	1.7
Extension to Break (%)	8	400
UV – resistance	good	poor
Solvent resistance	poor	good
O_2 permeability, 25μ film (Cm^3/ m^2 / atm / day)	45	1,700

As a thermoplastic PHB can be melt processed and can be used in injection moulding, blow moulding, and the production of films and fibres, as well as being incorporated into polymer blends and composite materials.

Applications : A changing perception

Ever since the isolation and identification of PHB from microbial source by Lemoigne (1925), its potential was well known. However, successful exploitation on industrial scale of production of PHB was not taken up immediately. In early 1960s, W. R. Grace, an American company even though took patent for the production soon it was discontinued due to some constraints in its production and extraction, though the patent literature indicated some possible medical application.

However, the exceptionally superior qualities of biopolysters over weighed the cost factors and interest in them was renewed. The sky-rocketing of oil prices due to middle East war in 1973, spurred the development of biopolysters. It became imperative necessity to develop new materials that can replace the commodity polymers derived from petro-chemicals. The similarity of PHB to conventional plastic made it an unenviable candidate of choice and intensive research took place for its development and characterization. Subsequently, there was a set back with the charged political climate in development of PHB with the reversal of oil prices because of cost factor. The production of PHB became costlier than the conventional plastics. However, bioplastics possess two superior qualities when compared to the conventional plastics. High degree of toleration to mammalian tissue with a minimal inflammation and biocompatibility leading to its conferring high added value due to potential medical and healthcare applications (Fig. 8.5).

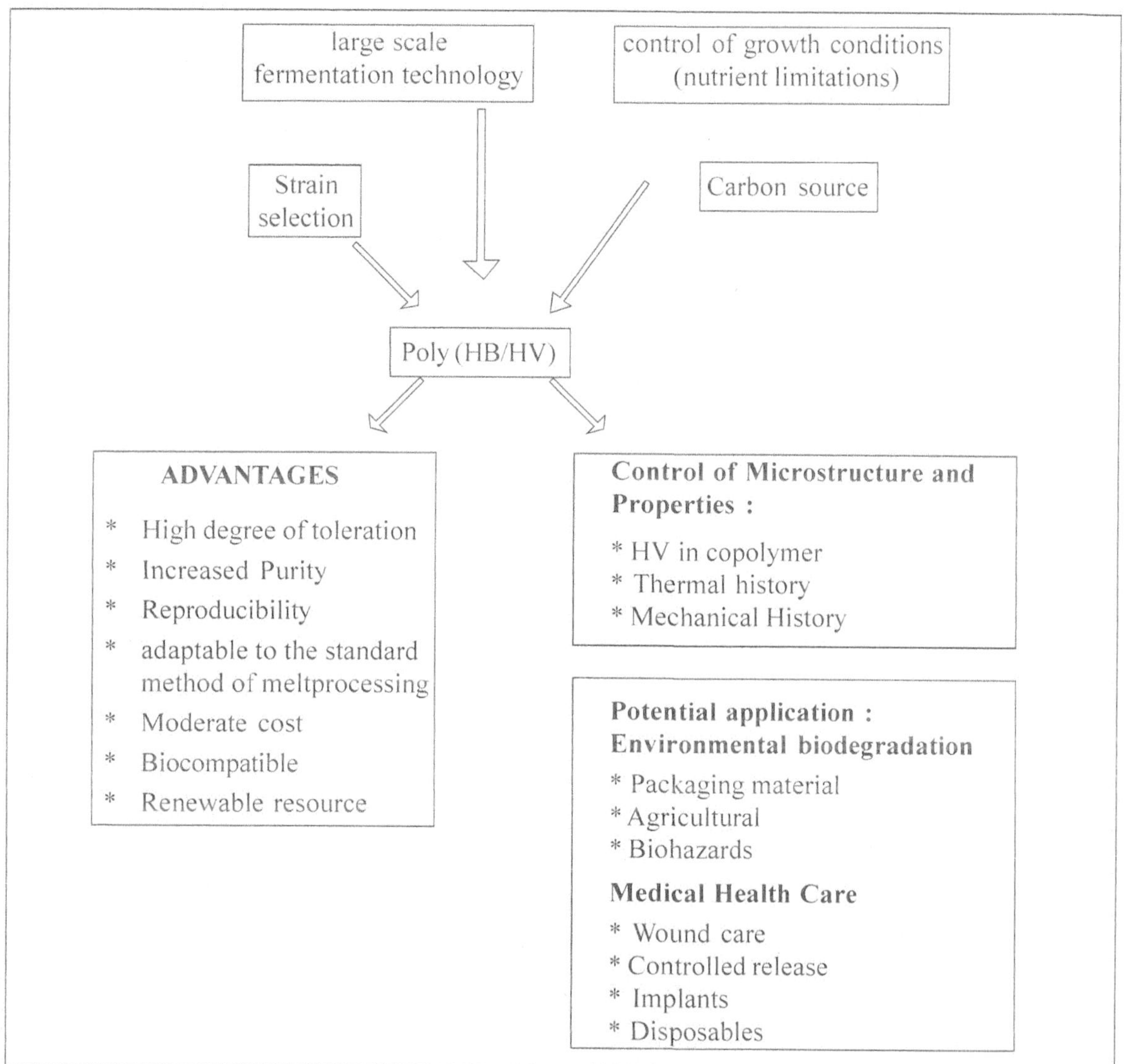

Fig. 8.5 Factor controlling (PHB / HV) fermentations product properties and potential application of PHB/HV polymers.

PHB Possessed Certain Special Characteristics

- There is an exceptionally high degree of toleration by mammalian tissue , with a least inflammatory reaction i.e., the polymer is unusually biocompatible and hence has a great potential for its application in medical and health care produces, which is definitely an added value
- Another of its speciality property is its biodegradability and hence it can be used in producing a range of environmental friendly biodegradable products for use in disposable products and as packaging material.

Another important development was the finding that, by altering carbon source appropriately during the fermentation just when polymer is being accumulated, the bacterial culture of *Alcaligenes eutrophus* could be coaxed to synthesize not only PHB polymer, but also other copolymers of 3-hydroxy butyrate with regulated quantities of 3-hydroxy valerate (HV). The monomers of PHB and PHV is presented in Fig. 8.6.

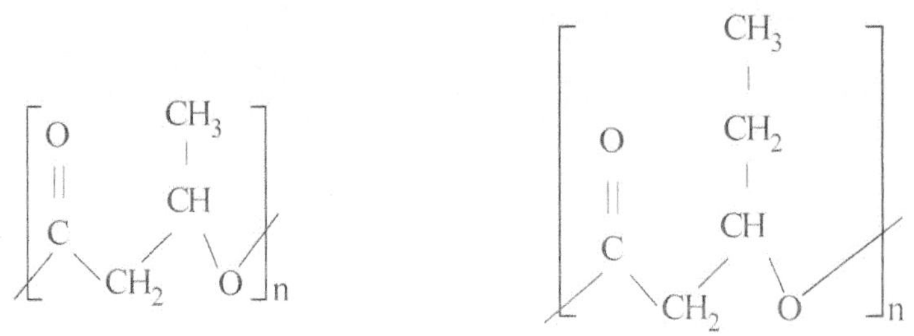

Hydroxy butyrate (HB) | Hydroxy valerate (HV)

Fig. 8.6 Polyhydroxy butyrate/3-hydroxy valerate co-polymers, the inclusion of HV units brings about improved toughness and resilience and thus overcomes the limitation that use of homopolymer PHB has.

The exploitation of the microbial (bacterial) polymers continued because of myriads of uses. Parallel with the development efforts for finding many potential applications, the scale up of fermentation process and methods of extraction were pursued actively. In the initial stages, low–volume, high value products were made with PHB/HV, as the cost of production was high and the polymer was available in limited quantities. The '*speciality*' business attracted a small organization, Malborough Biopolymers limited, a subsidiary of ICIPLC and MTMPLC based in UK, a country which has pioneered and built up expertise in biotechnology. The product was marketed outside the USA under trade name '*BIOPOL*'. The BHP/PV polymers are made in tonnage quantities.

Once large quantities were being manufactured, there was a distinct shift towards producing environmentally acceptable plastic. This trend was mainly diverted by pressure of public opinion backup by the legislation in major industrial countries. The falling costs and increasing of production enable the PHB/HV material keep up with market pressure from the environmental trend. Nevertheless speciality uses in medical/healthcare sector remained significant. Widespread use in surgery, especially in developing orthopaedic device, wound care and controlled release of implants/ drugs was found. The chiral purity of PHB repeat unit has led to the use of monomer resulting from the hydrolysis of PH/HV polymer became a valuable intermediate in the synthesis of fine chemicals where only a single optical isomer is necessitated .

A summary of factors controlling PHB/HV fermentations, product properties and some potential areas of application is presented in Fig. 8.6.

8.5 Microorganisms in Waste-treatment and Bioremediation

Microorganisms have an extensive ability to recycle synthetic organic molecules. Indiscriminate use of industrial chemicals, pesticides and herbicides has resulted in the accumulation of hazardous wastes on to space-ship, the earth, leading to the ground degradation and pollution of the environment. Technologies employing microorganisms with extensive biodegradative capabilities have been developed for prevention of pollution and site remediation making the same clean and green. The method of using the innate or cloned ability of microbes in the removal of pollutants from the environment is known as *bioremediation*. Bioremediation also involves the use of plants to extract heavy metals from contaminated soils and water.

The success of bioremediation technology for the clean up of a specific pollutant depends upon the biodegradability of the chemical. The biodegradation of pollutants in the environment is a complex process depends on the nature and amount of pollutant, environmental conditions and the indigenous microbial community etc. Bioremediation can be applied to clean up a variety of chemical pollutants such as aromatic compounds (chlorobenzene, benzene, toluene, xylene creosote), fuel oil (alkanes, alkenes), polychlorinated biphenyls, pyridine, chloroform and trichloro ethylene etc.

Bioremediation can be achieved by environment modification through nutrient application, aeration, and by the addition of appropriate microorganisms. The end products of effective bioremediation are water and carbondioxide which are non-toxic, harmless to the living organisms. Bioremediation to remove pollutants is inexpensive compared to physical methods for decontaminating the environment. This approach can reduce the risk of adverse impacts of toxic and hazardous chemicals though it is not the solution for all pollution problems. Bioremediation has great potential for destroying pollutants particularly creosote contaminated soils and is being used for saving the environment.

Environmental Modification for Bioremediation

Environmental factors such as unfavourable temperature, p^H, lack of oxygen, moisture and nutrients affect the bioremediation of hazardous chemical wastes. Generally in majority of cases an inoculation with a specific microorganism is neither necessary nor useful. The indigenous microflora allows enrichment of the appropriate microbe under suitable environmental conditions. Inoculation with preadapted microbial cultures may hasten the biodegradative cleanup. However, inoculation should always be combined with good growth conditions in the polluted environment such as suitable growth temperature, p^H, adequate water potential and oxygen for aerobic processes.

Microbial Seeding for Bioremediation

Microbial pollution control products have a potential market all over the world worth more than 200 million dollars. Inoculations with hydrogen degrading bacteria significantly improve the performance of petrochemical waste–treatment system. Several examples of pollution abatemest by specific microorganisms are known including removal of tertiary butylalcohol from a waste stream and formaldehyde from large accidental spill.

Microbial seeding has been proposed to clean up the pollutants from environment. However, the process is dependent on biodegradative capacity of microorganisms. Microbial seeding augments the metabolic capabilities of indigenous microbial populations. Seeding with microbes is known as *Bioaugmentation* which is particularly effective for the bioremediation of xenobiotic compounds. It is defined as the "*introduction of microorganisms into the natural environment for the purpose of increasing the rate of biodegradation of pollutants*". Commercial microbial mixtures are in use for degrading oil pollutants. *Pseudomonas* species are often used for the bioremediation of a wide range of hydrocarbon pollutants. *Burkholderia (Pseudomonas) sepacia* strain is more effective in bioaugmentation in comparison to conventional bioremediation techniques for groundwater. *Burkholderia* cultures remove twice the amount of trichloro ethylene (TCE) from groundwater.

The white rot fungus, *Phanerochaete chysosporium*, is a good candidate for bioremediation of many more complex pollutants as it produces laccases and peroxidases. These enzymes are able to attack a diverse range of compounds including DDT (Dichloro diphenyl trichloroethane), TNT (Trinitro – tolueue), Benzopyrene and plastics (Polyethylene).

Recombinant DNA technology provided a useful tool to genetically modify bacteria with specific biodegradative metabolic capabilities, which are potentially useful for bioremediation. The uses of bioremediation are enormous. The leakages of gasoline from underground storage tanks into soils and groundwaters are being treated by bioremediation. Marine oil spills are also effectively treated by bioremediation involving either intrinsic bioremediation or by the application of fertilizer such as oleophilic Inipol EAP 22. Successful bioremediation of petroleum spills produces harmless CO_2 and water.

Various air pollutants have also been treated by bioremediation. Biofilters, biotrickling filters and bioscrubbers are used to remove odour emanating compounds from air near farms, sewage treatment plants and various industrial operations. Bioreactors have biofilms of specific microbes that can also remove other toxic compounds such as toluene, hydrogen sulfide and chloro and nitrobenzenes.

Bioleaching

Recovery of oil and metals from low grade deposits using current recovery techniques is uneconomical. Ores with low metal content are not suitable for direct recovery by smelting. Microorganisms provide an alternative due to their metabolic capabilities to improve the recovery of metals from the environment. Sulfur-oxidizing Thiobacilli bacteria are commercially used in bioleaching operations for the recovery of Copper and Uranium. Recovery of metals or fuels by microorganisms is known as *bioleaching* or '*metal leaching*'. Microbial recovery of metals is often referred as '*Microbial mining*'.

Copper and Uranium are mined on a commercial scale from the low-grade ores using *Thiobacillus ferrooxidans*. Several other metals like nickel, cobalt, tin, zinc, cadmium, molybdenum, lead, arsenic and selenium may also be recovered using the thiobacilli. The bacterium derives energy through the oxidation of either a reduced sulfur compound or ferrous iron. Direct oxidation of metal sulfide and/or indirect oxidation of ferrous iron content of the ore to ferric iron which in turn oxidize the metal chemically for recovery by leaching.

A veracity of factors such as the chemical form of the element within the ore and the size of the mineral particles affect the efficacy of metal leaching. The surface area may be increased by crushing or grinding to enhance the rate of bioleaching. Ecological factors like temperature, p^H and availability of other nutrients influence the activities of the most important leaching bacterium, *T.ferrooxidans*. The obligately thermophilic bacteria *Sulfolobus* oxidize ferrous iron and *Sulfolobus* has been employed in the bioleaching of molebdenite (molybdenum sulflide). Besides these bacteria, other microorganisms in mixed populations may be potential candidates for use in bioleaching processes.

8.6 Biosensors

With the advent of biotechnology as a growth engine of modern day economy, the prospect of developing sophisticated but simple and user friendly diagnostic instrumentation seemed not only possible, but also a sound entrepreneurial proposition. Earlier in the arena of health care instruments, for instance, diagnostic tools nearer to the patient were limited to very simply equipment like stethoscope or thermometer. Any new biochemical test has to be performed it has to be done in sophisticated laboratories, often employing expensive, laborious and above all time consuming systems. However, the demand of market is to have a simple but accurate on the spot testing with instant result so that the diagnosis can be done speedily to save the delay in the treatment. Current developments in biosciences, particularly molecular genetics and biochemistry and boom in biotechnology based industry is rapidly enabling this demand to succeed. The fusion of knowledge between biosciences and microelectronics is quite revolutionizing the face of rapid testing and analysis in every sphere of human activity.

Metabolism in all life forms is essentially a network of biochemical reactions mediated by gene controlled enzymatic reactions which universally occur and can be induced in every organism, whether man, animal, plants, bacteria or viruses. The condition of organisms can be assessed by measuring, the extent and rate of biochemical reactions involving metabolism. It is now possible to design and develop biosensor to measure and quantify biochemical changes. Biosensors are the end products of synergy between biochemistry, membrane technology and micro-electronics, which permit the signal produced by biochemical reactions that are measured, quantified and recorded digitally. Historically, L. C. Clark (Jr) is credited with the idea of developing the biosensors conceptually and hence he may be considered as a father of biosensor concept.

Some of the technology that went into the design and development of biosensor has been available for the last three decades and prototypes have been developed for several applications. But these were limited to the laboratories where they originated. Since the biosensors are integrated systems with a high degree of sensitivity, stability and specificity of performance, these will be able to detect and measure biochemical reaction and offers a wide range of potential applications in human and animal diagnostics, industrial process control, pollution monitoring and detection of microbial contamination and the presence of toxic gases. The commercial application of biosensors is going to be widely employed in 21st century because of biotechnology boom and also due to availability of cheaper biological materials for design and construction of biosensors.

What is a Biosensor ?

It is necessary to define clearly what is a biosensor as opposed to other instruments used as biological probes.

The term biosensor has been variously employed to indicate a number of devices either used to monitor living systems or incorporating biotic element. Recently an international committee has surveyed literature that has one time or other used the term biosensor to describe a thermometer, a mass spectrophotometer, daphnia in pond water, electrophysiology equipment, chemical labels for imaging and ion specific selective electrode etc. A consensus has however emerged that the term biosensor should be reserved for use in its modern context of a sensor incorporating a biological element such as an enzyme, antibody, nucleic acid, microorganism or cell. This has been widely accepted and consequently biosensor is defined as '*a compact analytical device incorporating a biological or biologically derived sensing element either integrated within or intimately associated with a physicochemical transducer that can convert a biochemical signal into a quantifiable electrical signal proportional to a single analyse or related group of analyses*'.

The schematic diagram presented in Fig. 8.7 summarizes the essential concept behind the design of biosensor as defined above. To provide a brief outline of the principle involved is as follows : The targeted substrate (S) to be analysed along with coreactants if any diffuse through the outer safety membrane which also serves the function of selectively removing interfering moieties like proteins, other undesirable molecules and cells in biological sample. Reactions that take place with biological entities such as enzyme, antibody or cell produces the product (P) to be detected. In some instances the product is made to pass through another (second) membrane that is selectively permeable, and is detected by the transducer (T). The signal emanated from the transducer is processed by a signal processing electronic gadget (DP) and converts into a digital display (DR) that can be read out and recorded by another device (R).

Different biologically sensitive components *viz*., enzyme, multienzyme system, antibody, antigen, whole cell or organelle of any origin are employed in biosensors which are approximately immobilized to the transducer required. One problem with biological components, even though permit requisite specificity and sensitivity, they are unstable and are likely to create interferences from unexpected angles. The recent developments in membrane technology, in novel formulations and varied design combinations are enabling sophisticated analytical possibilities inspite of the presence of interfering substances. Furthermore biosensors made in such a way that they are self contained analytical system designed for a specific purpose. The miniaturization technology and the microelectronics utilized in such a way the chemical signal interpreted instantaneously providing the end result definitively (diagnosis) unlike other standard analytical equipments.

Thus the key to biosensor design and construction is type of membrane, the process of immobilization, and the kind of molecules immobilized. In other words it is a type of probe in which biological component *viz*., enzyme, antigen antibody or nucleic acid etc., are immobilized in semipermeable membrane which interact with the analyse and in which this reaction is detected by an electronic component and translated into a quantifiable electronic signal.

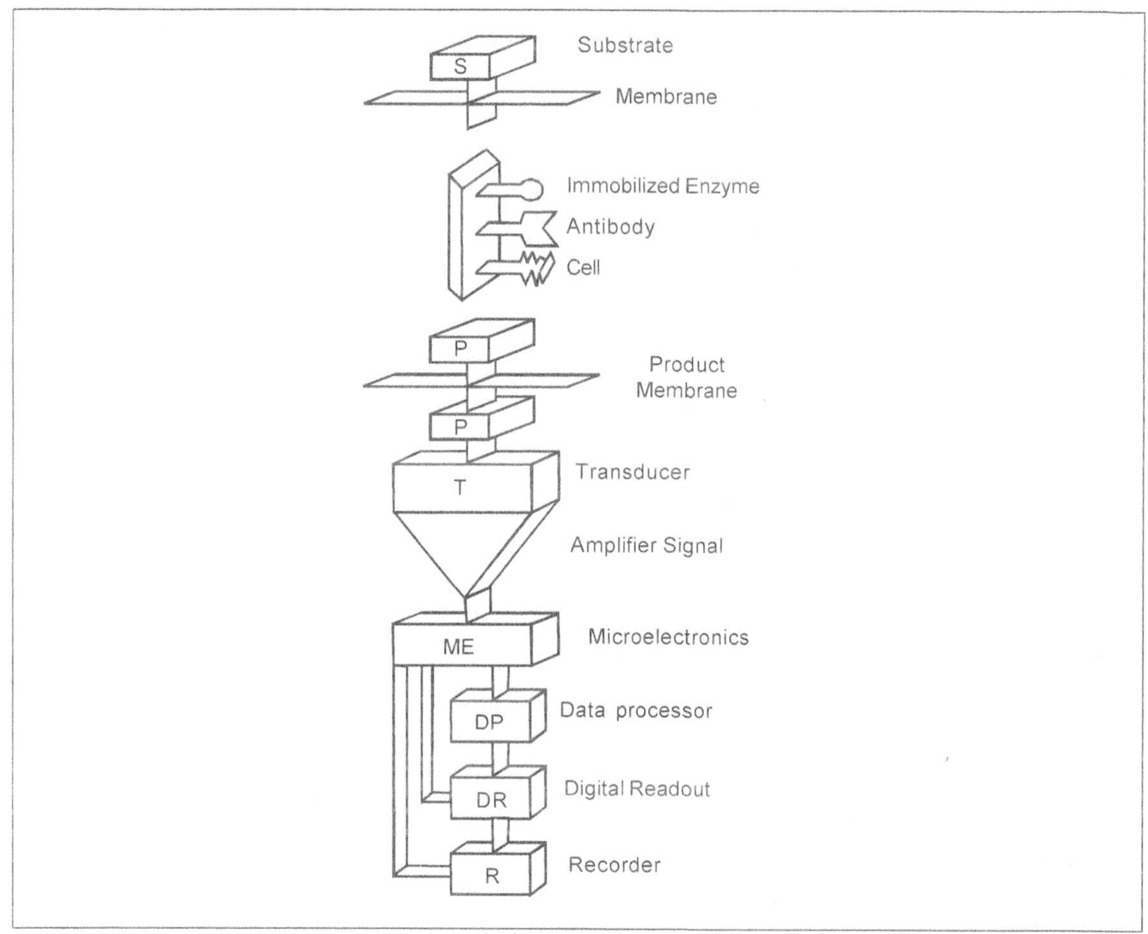

Fig. 8.7 Schematic diagram of a conceptual biosensor

The term immobilization refers to the binding or incorporation of biomolecules to different natural and man made materials, so that such immobilized system provides a catalytic surface to react with analyses. The last two decades have seen a spurt in the technology of immobilization and diverse protocols like covalent attachment and physical entrapment etc., has been developed. One of the most apt methodology relevant to the development of biosensor is entrapment within the pores of asymmetric membranes, giving a very thin but highly active layers of biomaterial.

Membrane technology has been developing rapidly for the last one decade or so and is employed intensively resulting in diverse large scale commercial activity in their syntheses and generated several kinds of materials that can be used in design and biosensor construction. One of the earliest commercial examples relevant to biosensor construction was the entrapment of glucose oxidase in cellulose acetate film.

Many kinds of membranes are used in biosensor design and construction from different sources, including those material developed for commercial purposes for microfiltration, ultrafiltration and haemodialysis. Besides different polymer meshes, and gels can be coated *in situ* over the sensing elements in conjuction with standard membranes. Membranes have to perform several critical functions to make the performance of biosensor efficient. Membranes have to act as a screening interface for the interferant in filtering out cell debris from the sample and prevent extraneous material to come in contact with transducer. Otherwise it gets clogged and modifies its sensitivity. Another most important function of a membrane is to limit the rate of diffusion of an analyse thus preventing a '*swamping effect*', of a high analyses concentration; and thus permitting the linear response to concentration. This is the most critical factor when biosensors have to quantify the analyses over a wide range of unknown concentrations. Some of the examples of membrane types employed for making biosensors are given in Table 8.3. Yellow Springs Instrument Company (USA) is the only company currently marketing, quite successfully, widest range of enzyme containing membranes for the determination of glucose, ethanol, lactic acid etc. A number of Japanese companies are also targeting this area for commercial exploitation of making biomembranes for the biosensor. The development of biosensors may well stimulate commercialization of a greater variety of membrane types. The Japanese Ministry of Internal Trade and Industry (MITI) has recognized the membrane production as a critical area in its drive to support commercial biosensor development and has set up a Biomembrane project at Ibaraki to develop membranes for biosensors.

Table 8.3 *Membrane types used in Biosensor design and construction*

Kind of membrane	Constitutents and composition characteristics	Commercial feasibility
Multi layer Heterogeneous	Multi layered structures composited each layer having a different function often with immobilized biological material.	Difficult to produce commercially, quality is not certain Yellow Springs membranes consists of immobilized oxydase enzymes.
Microfilteration	Developed for accurate and rapid aqueous filteration with careful control of pore size. Can have, straight, cylindrical pores or tortuous infrastructure	Polycarbonate membrane polyamides or Nylons ceramic ones are being developed
Material of Biological origin	Collagen, cellulose and gelatin have prospects, because of good porosity and have a property of immobilizing biological materials	Used for filteration of blood in haemodialysis
Hydrophilic structures & Hydrogels	Synthetic polymers with high water regain and entrapment	Some utilized for analytical purposes (e.g., polyacryalamide). others used in various clinical analysis.
Hydrophobic	impermeable to large molecular weight water soluble substances but lets ions and gases to passthrough.	Used in ion selective electrodes.

Table 8.3 *Contd...*

Kind of membrane	Constitutents and composition characteristics	Commercial feasibility
Active transport films	Polymer films either stop or help transport of molecules Conducting polymers for electrochemical transport	Polyvinyl chloride semi conductive polymers ranging from 10^5 - 10^9s/cm like doped polyacetylene
Biocompatible	Needed for implant or *in vivo* sensors to prevent premature rejection	Quite a variety of natural materials are available but synthetic ones such as acronitile sodium methylallyl suphonate are in use.
Photodiodes, photo sensitive films		Electrochemical-light absorption combination, luminiscence is measured
Piezoelectric acoustic crystals, surface wave sensors (SAW)		Changes in mass on surface of crystal gives variation in vibrational frequency
Semiconductor surfaces		Electron acceptor materials redox changes monitored,
Thermistor devices		Very small changes in temperature measured

Transducers

Transducers are devices that converts one form of energy to another. In biosensors a transducer converts the chemical energy into electrical energy. Sometimes intermediary substances are involved. Some examples of modern day transducer combinations that are utilized in biosensor design and construction is given in Table 8.4.

Table 8.4 *Transducers used in biosensors*

Transducer type	Response output of interaction
Conductimetric, electrochemical electrodes	using diverse electrode combinations differences in potential and conductivity is measured
Field effect transistor (FETs)	Miniaturised modified semiconductors used for microelectronics quantifies potentiometric change at gate
Gas sensing electrodes	Gases (like CO_2 and NH_3) are detected - potentiometer response
Ion-selective electrodes	Measures common ions present in biological material
Hall devices	Magnetic strength changes are detected
Fibre optic devices/optothermal (photoacoustic sensors)	Generation of absorption, and reflectance changes in wavelength of light is measured.
O_2 electrode, ampherometric electrode Photodiodes, photo sensitive films	O_2 consumption and evolution H_2O_2 is measured

There have been a considerable diversification in biosensor construction and design. Some of them have shown promise in Laboratory, but have not been commercialized (Table 8.5). However, with the advances in the construction of sophisticated membranes and the increasing supply of microelectronic gadgets like '*transputers*', some of the transducers currently not used, may find place in future biosensors.

Biosensors Construction based on Bioaffinity Principles

In addition to the enzyme biosensors novel kind of biosensors named '*bioaffinity sensors*' have been developed. Bioaffinity principle is employed in the construction of these biosensors. A weak binding of a labelled receptor to determinant analogue immobilized on to a transducer surface. Upon addition of the sample the receptor is displaced by forming tightly bound complex. Increase in determinant (sample) concentration is measured by reduction in signal coming from the labelled receptor. Biotin-avidin system in conjunction with an O_2 electrode is used in making a biosensor. Biotin concentration of 10^{-5}–10^{-8} g/ml can be measured in 60 seconds.

Other type of biosensors based on the bioaffinity principles can be designed and constructed using biological substances capable of molecular recognition :

Antibodies - antigens like drugs and protein, Antigens - for antibodies, like autoantibodies and AIDS antibodies, Lectins - for saccharides, Hormone receptors - for hormones, Drug receptor - for drugs and active drug metabolites, Nucleic acids (DNA/RNA) - as gene probes for inherited diseases, lineages, finger printing, etc. Examples of Enzyme - Immune assay biosensors is given in Table 8.5.

Table 8.5 *Biosensor based on bioaffinity principle for immune assay*

Analyte/determinant	Enzyme conjugate label	Transducer type	Sensitivity
Harmones	(a) Catalase	Ampherometric (O_2)	0.02-100 Iu ml^{-1}
human chorionic gonadotrophin (hCG)	(b) glucose oxidase	Ampherometric-Redox (Ferrocene)	0.15 Iu ml^{-1}
Oestradiol - 17β	Horse raddish peroxidase	Electrochemical	5.7×10^{-13} to 9×10^{-9} mol L^{-1}
Drugs			
Digoxin	Alkaline phosphatase	Electrochemical (phenol)	50-10^{-12}g
Theophylline	Catalase	O_2 electrode	5×10^{-9}g ml
Immunoglobulins	Catalase	Ampherometric	$10^{-3}-10^{-6}$g ml
I_gG	Alkaline phosphate	Electrochemical	$10^{-11}-5\times10^{-9}$g ml
Cancer Diagnosis *α - Fetoprotein*	*Catalase*	*Ampherometric*	$10^{-11}-10^{-8}$*g ml*$^{-1}$

8.7 Bioindicators

Over the past few decades, scientists have discovered many examples of life forms that can signal us the extent of environmental degradation and pollution, including water quality. The term bioindicator refers to the diverse group of biological organisms, species and their assemblage, or communities, whose presence and/or abundance is indicative of a particular set of environmental status. It includes assemblage of species that show outstanding sensitivity or a strong correlation with a particular natural factors (stressors) causing environmental degradation. Several organisms, like fungi, algae, lichens, protozoans, plants and animal species have been discovered having the ability to detect and indicate the erosion of environment endangering the ecological balance and niche necessary for the life form to survive and to coexist and build a web of life without which the survival of humanity will be on the edge of mass extinction.

In recent times several biotechnological companies exploiting the basic knowledge of biosciences and ventured into developing commercial technologies using microbes or other life forms or tissues. Several processing designs using genetically engineered organisms have been developed not only for the detection and monitoring but also for remediation purposes (bioremediation).

Factors Affecting the Quality of Environment

Several factors, anthropogenic (man made) as well as natural, degrade the environment and disturb the delicately poised ecosystem of the environment. Environmental pollution is all pervading: surface soils, water and air. Several matrices (indices) can be developed to get the indication of degradation by using a range of organisms with a varying degree of sensitivity. Some of stressors are listed below :

1. *Hydrological stressors*
   Changes in the levels of surface water or water tables (ground water) and quality etc. constitute hydrological stressors that affect the 'life forms'. Periodic drought, floods, drainage, excessive withdrawl of groundwater and global climatic changes will affect the life forms and the extent of damage could be indicated by the indicator species.
2. *Vegetative cover conditions*
   Changes in aerial cover and density of vascular plants also can act as stressor and can be used to study the degradation of environment.
3. *Salinity*
   Total dissolved salts in the soil and water can also act as stressor, which could be due to natural causes or anthropogenic.
4. *Sedimentation and turbidity*
   Water bodies and the quality of water can also act as one of the stressors which could be excerting perturbation of the life forms existence.
5. *Excessive Nutrient loads and Anoxia*
   The water quality for aquatic forms is a vital for maintaining their ecological balance. Any increase in the nutrient loads lead to anoxia condition thereby acting as a stressor and endanger the life of organisms.
6. *Pesticides, herbicides and metal contamination*
   These affect the organism critically and act as stressors for the various life forms.

Assaying of Bioindicators

Several protocols have been developed to assay the utility of organisms as bioindicators. The response of group of any given species to various environmental factors (stressors) and the sensitivity they display to various matrices can be quantified to recognise the useful bioindicators to a given situation. The general outline of assay systems depicted in the flow diagram given below (Fig. 8.8).

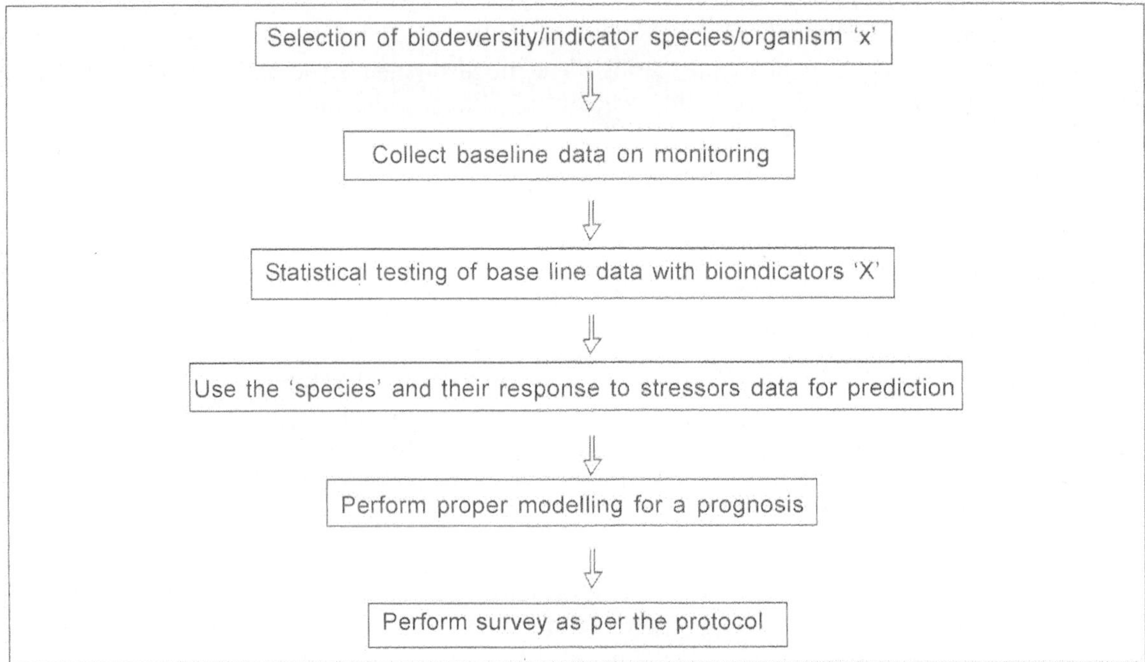

Fig. 8.8 General outline for bioindicator species identification assay and monitoring the environmental degradation

Microbes as bioindicators

Different microbial species have been identified which are very sensitive to assay the degradation of environment. Microbial diversity, structure and dynamics of microbial population serve for this purpose. Major concern is to use the microbial systems for assaying the contamination of water and defining microbial risk reduction in food, feed, animal manure etc., The microbial systems are also useful as bioindicators for the detection of heavy metals and pesticide pollution as well as desertification. Microbial activity and their abundance are the criterion. Several microbial species like *E. coli, Salmonella Cryptosporidium parrum*, algae, fungi, lichens, protozoa etc., have been used as bioindicators. The delicate symbiotic nature of lichens is one of the reasons for their sensitivity and attitude as bioindicators of environmental parameters and air pollution. Folicolous lichens have been widely used for assaying the air pollution.

A Partial list of utility of different organisms based on their response to stressors is given in Table 8.6.

Table 8.6 *Utility of various organisms based on their response to stresses*

Stressors	Possible indicator	Evaluation
Microbial Species		
Hydrological condition	1. Species composition	Fair
	2. Richness	Fair
	3. Density and biomass	Poor
	4. Decomposition	Fair
	5. Denitrification	Fair
Changes in vegetative cover conditions	1. Species composition	Good
	2. Richness	Fair
	3. Density, biomass	Poor
	4. Decomposition	Poor
Salinity	1. Species composition	Good
	2. Richness	Good
	3. Density, biomass	Fair
	4. Decomposition	Poor
Sedimentation and turbidity	1. Species composition	Fair
	2. Richness	Fair
	3. Density composition	Fair
	4. Decomposition	Poor
Excessive nutrition and anoxia	1. Species composition	Good
	2. Richness	Poor
	3. Density and biomass	Good
	4. Decomposition	Poor
	5. Denitrification	Fair
Herbicides	1. Species composition	Poor
	2. Richness	Poor
	3. Density & biomass	Poor
	4. Decomposition	Poor
	5. Denitrification	Poor
Insecticides	1. Species composition	Poor
	2. Richness	Poor
	3. Density & biomass	Poor
	4. Decomposition	Poor
	5. Denitrification	Poor
Plants		
Hydrological condition	1. Species composition	Good
	2. Community zone location	Good
	3. Richness (mature plants)	Fair
	4. Biomass and cover ratio	Good
	5. Seed density	Fair
	6. Germination rate	Poor
Changes in vegetative cover condition,	1. Species composition	Good
	2. Community zone location	Good

Table 8.6 *Contd...*

Stressors	Possible indicator	Evaluation
	3. Richness	Poor
	4. Biomass and cover ratio	Good
	5. Seed density	Fair
Salinity	1. Species composition	Good
	2. Community zone location	Good
	3. Richness	Poor
	4. Biomass and cover ratio	Good
	5. Seed density	Fair
Sedimentation & Turbidity	1. Species composition	Good
	2. Community zone location	Poor
	3. Richness (mature plants)	Fair
	4. Biomass cover ratio	Poor
Excessive nutrient loads and anoxia	1. Species composition	Good
	2. Community zone location	Poor
	3. Richness and community	Fair
	4. Biomass and Cover ratio	Poor
Herbicides	1. Species composition	Fair
	2. Community zone location	Poor
	3. Richness (mature plants)	Poor
	4. Biomass and Cover ratio	Poor
Insecticides	1. Species composition	Fair
	2. Community zone location	Poor
	3. Richness (mature plants)	Poor
	4. Biomass and Cover ratio	Poor
Heavy metals	1. Species composition	Fair
	2. Community zone location	Poor
	3. Richness (mature plants)	Poor
	4. Biomass and Cover ratio	Poor
Birds **(prairie wetland)**		
Hydrological condition	1. Species composition	Fair
	2. Single species indication	Poor
	3. Richness	Fair
	4. Density, biomass	Good
	5. Reproductive success	Good
Changes in vegetative cover condition	1. Species composition	Good
	2. Single species indication	Good
	3. Richness	Fair
	4. Density, biomass	Fair
	5. Reproductive success	Good
Salinity	1. Species composition	Good
	2. Single species indication	Good
	3. Richness	Fair
	4. Density, biomass	Fair
	5. Reproductive success	Good

Table 8.6 *Contd...*

Stressors	Possible indicator	Evaluation
Sedimentation and turbidity	1. Species composition	Fair
	2. Single species indication	Fair
	3. Richness	Fair
	4. Density, biomass	Poor
	5. Reproductive success	Poor
Excessive nutrient load and anoxia	1. Species composition	Poor
	2. Single species indication	Poor
	3. Richness	Fair
	4. Density, biomass	Poor
	5. Reproductive success	Poor
Herbicides	1. Species composition	Fair
	2. Single species indication	Poor
	3. Richness	Poor
	4. Density, biomass	Poor
	5. Reproductive success	Poor
Insecticides	1. Species composition	Fair
	2. Single species indication	Poor
	3. Richness	Poor
	4. Density, biomass	Poor
	5. Reproductive success	Good
Heavy metals	1. Single composition	Fair
	2. Community zone location	Poor
	3. Richness (mature plants)	Fair
	4. Biomass and cover ratio	Poor

Recently microorganisms as bioindicators for assaying DNA damage has been developed to test the contaminated water. Microorganisms exposed to the contaminated water can indicate whether such pollution is capable of causing genetotrophic by performing comet assay. The damaged DNA due to the effect of contaminant or pollutants forms a 'comet like' tail upon electrophoresis, which could be quantified and the samples can be tested using the sensitive microorganisms as a bioindication of pollution.

8.8 Microbial Resources for Food and Feed Additives

Fungi are one of the potential sources for biotechnology having a long history of utility in traditional industrial processes. The cultivation of mushrooms, use of yeast in alcoholic beverages, penicillia in cheese production, soy sauce through Koji process and other food products are some of the commercially processed practices through ages. The production of citric acid by *Penicillium* and *Aspergillus* is a good example of exploiting fungi for chemical products under aerobic conditions.

Fungal enzymes are used in making different foods and drinks. Various species of *Aspergillus* produce enzymes like α- amylase and pectinase which are useful in bread making and fruit processing respectively. Thus fungi can be utilized biotechnologically in various ways i.e., (a) as a biomass, (b) as producers of enzymes and (c) as producers of primary and secondary metabolites such as organic acids etc.

Industrial products obtained from fungi have been recognized into six categories, such as organic acids, pharmaceuticals, enzymes, pesticides (biocontrol agents), polysaccharides and biomass. Fungal biomass of bakers yeast constitutes 2 % by weight of bread, while the whole of the biomass is utilized in case of mushrooms.

Mushrooms are the fruit bodies of fleshy fungi mainly belonging to the group *Basidiomycotina* and *Ascomycotina*. The fruit bodies are either epigeous growing above the ground (*Agaricus* spp.) or hypogeous forming fruit bodies underground (Tubers and Truffles). Mushrooms are considered as wonder plants of medicinal and food value that give fruits without any leaves, buds and flowers. There are about 2000 species of mushrooms which are edible, and eighty of them were experimentally grown. Among them, cultivation of only 40 species found to be economical. Over a dozen mushrooms are cultivated on commercial scale in more than, hundred countries. Chinese were first to cultivate black ear mushroom (*Auricularia* sp.) about 1000 years ago. Later on shiitake and paddy straw mushrooms were cultivated.

Among the cultivated mushrooms white button mushroom (*Agaricus bisporus*), Shiitake Mushroom (*Lentinus edodes*), Oyster mushroom (*Pleurotus* spp.) and paddy straw mushroom (*Volvariella volvacea*) are some of the important and most popular fleshy fungi produced on a large

Table 8.7 *Nutrients available in different types of Mushrooms*

Mushroom	Moisture content	Protein	Fats	Carbohydrates	Fibre	Energy
Agaricus biosporus	78.3 – 90.5	23.9 – 34.8	1.7–8.0	51.3–62.5	8.0–10.4	328–368
Agaricus campestris	89.7	33.2	1.9	56.9	8.1	354
Auricularia sp.	89.1	4.2	8.3	82.8	19.8	351
Boletus edulis	87.3	29.7	3.1	59.7	8.0	362
Flamulina velutipes	89.2	17.6	1.9	73.1	3.3	378
Lentinus edodes	90.0 – 91.8	13.4 – 17.5	4.9 – 8.0	67.5 – 78.0	7.3 – 8.0	387 – 392
Pleurotus florida	91.5	27.0	16	58.0	11.5	265
Pleurotus ostreatus	73.7 – 90.8	10.5 – 30.4	1.6 – 2.2	57.6 – 81.7	7.5 – 8.7	345 – 367
Pleurotus sajar-caju	90.1	26.5	2.0	50.7	13.3	300
Volvariella diflacium	90.4	28.5	2.6	57.4	17.4	276
Volvariella volvacea	89.1	25.9	2.4	-	9.3	276

scale. Morels (*Morchella* spp.) are the most prized mushrooms collected as wild growth from the coniferous forests of Himachal Pradesh, Jammu and Kashmir and Uttar Pradesh. White button mushroom is leading world mushroom production sharing major part (56 %) followed by shiitake (14.4%) and paddy straw (8%) mushrooms. Mushrooms are preferred and priced commodity due to its nutritive value, characteristic aroma and flavour.

Poverty and malnutrition are the major problems confronting the developing world including India. Pulses are the main source of proteins in Indian diet. The per capita consumption of pulses is on decline since independence. The average consumption of pulses, in 1985 is 37g as against the prescribed 80g of pulses by Food and Agriculture Organization. A balanced diet should be able to provide the right kind of proteins, vitamins and minerals. Mushrooms provide a right supplement to the diet in augmenting proteins, carbohydrates, minerals and vitamins (Table 8.7). The amount of proteins in mushrooms (19-35 %.) is much higher than rice (7.3 %) and wheat and it is double the amount of proteins available in many vegetables.

Mushroom proteins contain various amino acids that are essential for growth and development in human beings. Essential amino acids like tryptophan and methionine are present in mushrooms which are absent in vegetable proteins. Lysine, threonine and phenyl- alanine are available in plenty in many types of mushrooms like *Lentinus*, *Pleurotus* and *Volvariella* (Table 8.8).

Mushrooms are an excellent source of vitamins, riboflavin, nicotinic acid or niacin and pantothenic acid. Other vitamins such as ascorbic acid, folic acid and thiamin are also present in appreciable quantity (Table 8.8).

Table 8.8 *Vitamin content of some edible mushrooms (mg/100 g dry wt)*

Species	Thiamin	Riboflavin	Niacin	Ascorbic acid
Agaricus bisporus (White button Mushroom)	1.1	5.0	55.7	81.9
Pleurotus ostreatus (Oyster Mushroom)	4.8	4.7	108.0	–
Volvariella volvacea (Paddy straw Mushroom)	1.2	3.3	92.0	20.0

Carbohydrate and fat content of edible mushrooms is remarkably low. The absence of starch makes mushroom an ideal food for diabetic patients. Mushrooms contain unsaturated fatty acids like linoleic acid in higher amounts along with minerals such as potassium, phosphorus, sodium, calcium and magnesium.

Mushroom Cultivation

India with its diversified agroclimate and abundant agri-wastes provides good scope for the cultivation of different mushroom varieties. Mushrooms can be grown on substrates consisting of various agricultural wastes and compost. This helps in the recycling of waste material besides providing employment to men and women. Mushrooms can be cultivated through extensive cultivation by utilizing the naturally colonized substrates like wood and humus. In this method, mushrooms such as *Lentinus*, *Pleurotus*, *Keuneromyces*, *Tremella*, *Auricularia* and *Agrocybe* are cultivated extensively.

Intensive cultivation of mushrooms is an indoor activity requiring minimum space. White button mushrooms are produced either during favourable seasons or by providing air conditioning facilities to maintain required temperature (14 - 20 °C, optimum 16 ± 2 °C). Oyster mushrooms (*P. florida* and *P. sajar-caju*) are being successfully cultivated in many parts of India where favourable temperature (20 – 33 °C, optimum 25 ± 2 °C) prevails or by using coolers to maintain it. Paddy straw mushroom cultivation is feasible where high temperature (25 – 38 °C) is prevalent.

Oyster Mushroom Cultivation

The mushroom produces soft fleshy fruit bodies resembling oyster in different colours ranging from black to white. The pileus is tongue like, eccentric measuring 5.3 cm in diameter. They grow into different sizes depending on the type of substrates. The fruit body is smaller on wood substrates and larger on grass and other substrates. Different species of *Pleurotus* have been cultivated so far including *P. ostreatus*, *P. flabellatus*, *P. sajor-caju*, *P. florida*, *P. citrinopileatus*, and *P. sapindus*.

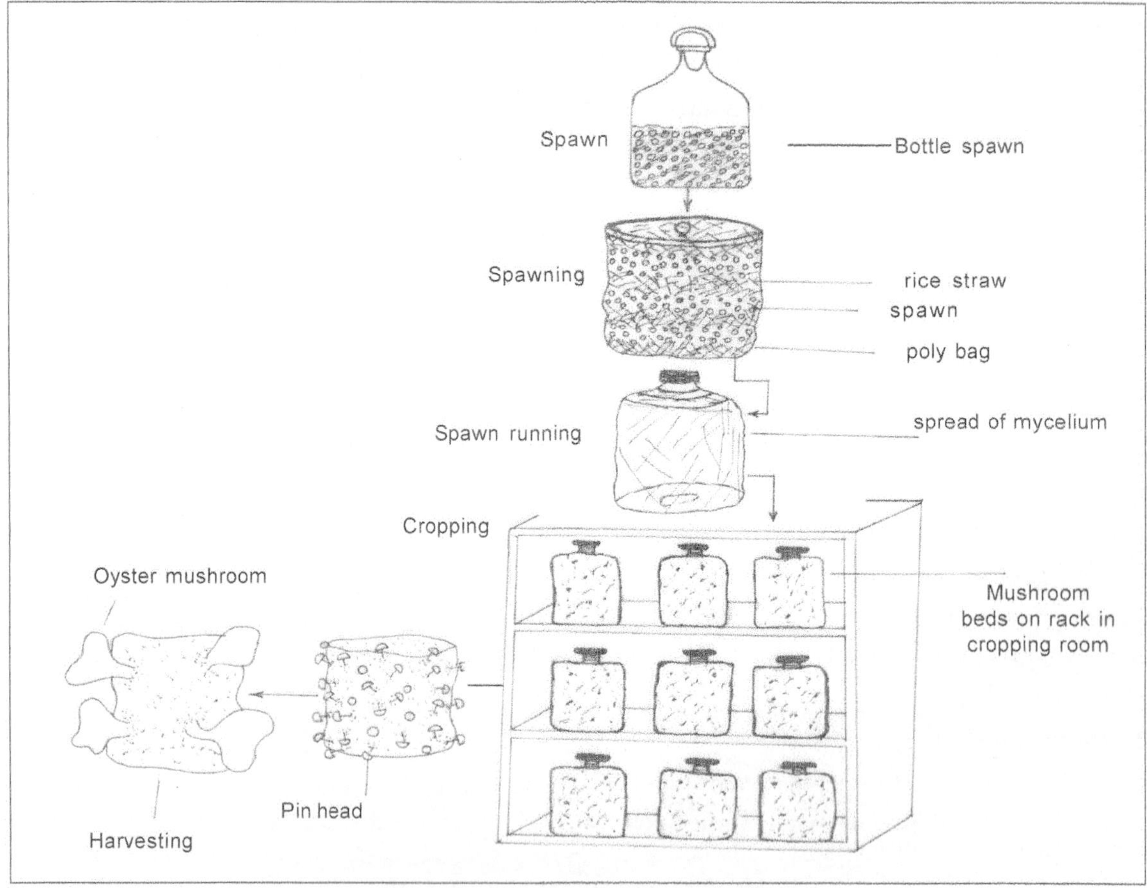

Fig. 8.9 Oyster mushroom cultivation (flow chart)

Cultivation of oyster mushroom is simple, economical and can easily be adapted even by small farmers. It is mandatory that aseptic condition have to be maintained throughout the growing period. The following steps are involved in the cultivation of mushrooms: 1. Spawn production, 2. Spawning and spawn running 3. Cropping and 4. Harvesting.

Spawn is the seed material of a mushroom fungus. It consists of growing mycelium. Good spawn is a prerequisite for a good crop of mushroom. Inoculum for spawn production is selected from healthy and efficient strain of the fungus. Method of oyster mushroom cultivation is outlined in the following diagram Fig. 8.9.

Mother Spawn

It is the first generation of the spawn developed from the mycelium or from a single spore of the fungus. Single spore or a hyphal tip of the mushroom is aseptically transferred on to a convenient medium (Potato dextrose agar) and incubated for about a week. A radial pure colony of the fungus develops and this is referred as the mother spawn. Mother spawn serves as an initial inoculum for large scale production of substrate based mushroom spawn.

Bottle Spawn

For production of spawn on a large scale the mother spawn is multiplied on grain based substrate either in bottles or in polypropelene sachets. For growing spawn, jowar, bajra or wheat grain may be used. Jowar and bajra are good for spawn preparation. However, other grains may also be used depending upon availability and convenience. Clean, half boiled jowar seeds are mixed with $CaCO_3$ or chalk powder to avoid excess moisture and to maintain a neutral p^H, before filling the glucose drip bottles or polypropelene bags. These bottles are sterilized and inoculated with the mother spawn. Spawn or spreading white mycelium over the grains is formed within two weeks of incubation. This can be used to spawn or inoculate the crop beds.

Mushroom beds are prepared by chopping the fresh paddy straw into 1-2 cm length pieces that are soaked over night. The paddy straw was treated with boiled water at 65 °C for half an hour and allowed to cool down. After draining off the excess water the straw is filled into poly bags while introducing the spawn simultaneously.

Spawning

Spawning is the introduction of mushroom seed material into the straw beds. The spawn is placed in layers for every 5 cm of the bed height. After filling the bed, small holes are made to allow aeration and transferred to spawn running room for incubation.

Spawn Running

Spawn running refers to the spreading of mycelium throughout the straw in the bed. After 2-3 weeks of incubation, mycelium spreads throughout the bed. Spawn running requires dark, well ventilated room and temperature between 25-27 °C.

Cropping

At the completion of spawn running, the polythene cover is aseptically removed and transferred on to the racks of the cropping room. Humidity should be about 85-90% in the room and it may be maintained by using the humidifier or through spraying water at regular intervals. Pin heads of mushrooms start emerging from the beds within four days and develop into a well grown oyster like body. Light is essential for the development of fruit bodies. Sporophores (fruit bodies) develop well at lower temperatures compared to spawn running. However, *P. ostreatus* can withstand higher temperature and produce sporophores between 25-30 °C.

Harvesting and Preservation

Oyster mushrooms are harvested once the fruit bodies are fully developed. As mushrooms appear in flushes, the straw is removed up to 1 cm depth after removal of the fruit bodies followed by water spray twice a day. Second crop comes within 7-10 days and 3-4 crops can be harvested from each bed in this manner.

The harvested mushrooms may be either directly used or preserved in Brine solution (18-20%) containing salts. The fleshy mushrooms may also be dried, preserved and used. However, marketing is the bottle neck in the cultivation of mushrooms as their shelf life is very low, with the availability of raw material and market demand, mushroom cultivation may become an attractive enterprise.

Microbial cells, in certain fermentations, are themselves the desired fermentation products which may directly serve as food and feed. Bakers' yeast, *Torula* food and feed yeast, mushrooms and algae are good examples of direct utilization of microbial biomass. Various yeasts are good sources of protein and B vitamins. Yeasts are commercially produced for animal feed and it may gain greater importance as human food to feed the underfed of the world. *Saccharomyces cerevisiae* grown on molasses may be used as a food or feed yeast. However, *Torula* yeast (*Candida utilis* or *Torulopsis utilis*) can utilize a wide variety of carbon and nitrogen compounds under simple fermentation conditions. A sulphite waste liquor medium in a continuous fermentation is usually employed for producing *Torula* food and feed yeast.

8.9 Microbial and Industrial Production of Enzymes

Enzymes are the biological catalysts for the biochemical reactions involved in microbial growth and respiration leading to the formation of fermentation products. Enzymes may be constitutive or adaptive. The cell produces the constitutive enzyme regardless of the substrate. Adaptive enzymes are produced by microbial cells in usable amounts only in response to a particular enzyme substrate in the medium. Microorganisms produce adaptive enzymes only when required to bring about degradation or change in the substrate to provide the nutrients, otherwise not available to the cell. Species of *Pseudomonas* are known for their outstanding ability to produce adaptive enzymes in response to the substrate availability in the media.

Enzymes are either exocellular or endocellular. An endocellular enzyme is produced within the cell and not released into the fermentation medium surrounding the cell. Extra cellular enzymes, on the other hand, are produced by the microbial cell and liberated into the fermentation medium. Most of the microbial enzymes are exocellular and degrade polymeric substances hydrolytically to enable the substances to pass through the cell wall. Amylases and proteases are good examples of extra cellular enzymes (Table 8.9).

Enzymes are proteins and each enzyme has its own optium p^H and temperature for activity. More than one enzyme can attack a specific substrate. Enzymes may also be specific to the particular substrate that they attack. Enzymes act relatively at low temperatures, i.e., below the denaturing temperatures of the enzyme protein. Enzyme reactions can be stopped at a desired moment simply by denaturing the enzyme either using the heat or acids. The protein nature of the enzymes is also advantageous in the easy recovery of the enzyme from fermentation culture broths. The enzymes can be precipitated with various alcohols and ammonium sulfate.

Table 8.9 *Industrially produced microbial enzymes and their applications*

Enzyme	Fungal source	Application	Type of activity
Acid-resistant amylase	*Bacillus subtilis* *Aspergillus niger*	Textiles, syrup, Glucose production Digestive aid	Starch liquifying amylasess
Amyloglucosidase	*A.niger, Rhizopus niveus*	Glucose production	Starch saccharifying amylase
Invertase	*Saccharomyces cerevisiae* *Aureobasidium pullulans*	Confectionaries, to prevent crystallization of sugar, chacolate	Starch saccharifying amylase
Pectinase	*A.oryzae* *A. flavus* *A. niger* *Coniothyrium diplodiella, Scle- -rotinia libertina*	Removal of pectin, increase yield & for clarifying juice, coffee concentration	Starch saccharifying amylases
Rennet	*Mucor pusillus*	Cheese manufacture	
Protease	*A. niger*	Feed, digestive aid	Microbial proteases
Protease	*A. oryzae* *A. niger* *Mucor pusillus* *M. michei*	Flavoring of sake, haze removal in sake	Microbial proteases
Protease	*B. subtilis, Strep-tomyces griseus*		
Glucose oxidase	*A. niger, Penicill-ium amagasakiense*	For removal of oxygen or glucosefrom various foods, dried egg manufacture	
Glucose oxidase	*Penicillium chrysogenum*	For glucose determination	
Naringinase	*A. niger*	Removal of bitter taste from citrus juice	
Cellulase	*Trichoderma koningi* *A. niger*	Digestive aid	
Penicillinase	*B. subtilis* *B. cereus*	Removal of penicillin	

Since ancient times, crude fungal enzyme preparations are in use to bring about starch hydrolysis for brewing. Commercial use of fungal enzymes was started just before the 20th century while the use of bacterial enzymes was from the first world war. The ease in recovery of microbial enzymes provided them an extra advantage to complete with the enzymes from the plant and animal sources. Microbial enzymes are commercially produced from various fungi including yeasts and bacteria. Bacteria and fungi often produce similar enzymes which may vary in p^H and temperature optima. There is a great variation among fungi and bacteria in their ability to produce a specific enzyme. The ability may also vary within species and even in strains. Thus, genetic stability of microbial strains plays an important role in enzyme production demanding the maintenance and preservation of cultures. The important fungal enzymes produced by fermentation include amylases, proteases, pectinases, invertase, catalase, penicillinase, glucose-oxidase and others (Table 8.9) with wide range of applications.

Amylases

Fungi and bacteria produce α-amylases. Amylases hydrolyze 1-4 glucosidic linkages in polysaccharides such as starch and glycogen.

Amylases are commercially used for the preparation of sizing agents and removal of starch sizing from woven cloth. These enzymes are useful in the manufacture of corn and chacolate syrups, production of bread and brewing industry.

Fungal amylase is used to hydrolyze starch for yeast-alcohol production and it is referred as '*Amylo*' process. Fungal amylase is produced from strains of *Aspergillus oryzae* in a stationary wheat bran culture or from strains of *A. niger* in submerged aerated-agitated culture. The wheat bran stationary culture has been employed extensively for the enzyme production. However, *A. niger* growing in starch-salt medium has been commercialized in recent times.

Amylases produced by *Bacillus subtilis* and *B. diastaticus* have been produced on a commercial basis and employed when amylases from other sources hydrolyze starch relatively less. High yields of bacterial amylase by *B. subtilis* is obtained in stationary culture as well as in highly aerated submerged culture by a highly starchy medium.

Proteolytic Enzymes

Proteolytic enzymes are produced by both bacteria (*Bacillus, Pseudomonas, Clostridium, Proteus, Serratia*) and fungi mostly aspergilli such as *A. niger, A. oryzae, A. flavus* and *Penicillium roquefortii*. Commercial applications of proteolytic enzymes include bating of hides in the leather industry which provide finer grain and texture, greater pliability and better quality. Proteolytic enzymes are used in textile and silk industry. These are also employed in the tenderizing of meat and removing the direct spots in dry-cleaning industry. Fungal proteases have a wider p^H activity range compared to animal or bacterial proteases resulting in a wider range of uses.

Fungal proteases are commercially produced by utilizing *A. flavus, A. wentii, A. oryzae, Mucor delemar* and *Amylomyces rouxii*. The fungi are usually grown on wheat bran and the fungal proteolytic enzymes are released into the medium at the time of sporulation. Bacterial protease production utilizes *B. subtilis* strains under the fermentation conditions similar to amylases.

Pectinases

Pectinases are produced by various bacteria and fungi which facilitate the clarification of fruit juices by eliminating pectin and pectin like colloids from the fruit juice. Commercial pectinase production utilizes species of *Penicillium* or *Aspergillus*. Pectin and pectin containing substances in the medium induce and stimulate the enzyme production. Pectinases are excreted and as well as retained in part in the cells of the fungus, hence, enzyme has to be recovered from both sources. Pectinase is extracted from the dried and ground mycelium of the fungus with water.

Invertase

This enzyme mainly hydrolyzes sucrose to yield glucose and fructose and it is also known as sucrase or saccharase. Yeasts contain considerable amount of this enzyme but it differs from the enzyme produced by other fungi. Yeast produces fructosidase type of enzyme while other fungi produce glucosidase type of enzyme. This enzyme prevents sugar crystallization during the preparation of candies and ice creams. Invertase enzyme is also employed in yielding invert sugars for use as plasticizing agents in paper industry.

Invertase is an endoenzyme, liberates into the medium only after the autolysis of the cells. Therefore, enzyme recovery involves mechanical disruption of yeast cell walls or by autolysis induced by chloroform, ethyl acetate or toluene, and precipitated by the addition of alcohol. Invertase is commercially produced from bakers or brewers' yeast, *Saccharomyces cerevisiae*.

Glucose oxidase is produced by *A. niger* along with catalase in a submerged-aerated fermentation. This enzyme is useful in removing glucose from egg white and whole eggs before drying, which prevents browning and deterioration of the dried egg. *Endothia parasitica* is employed in the commercial production of rennet like enzyme that helps the curdling of milk for cheese production.

Antibiotics

Antibiotics are the secondary metabolites produced by bacteria, *Streptomyces*, *Nocardia* and fungi. An antibiotic may inhibit the growth of or kill one species or few species or many types of organisms. Waksman defined the term antibiotic as *an organic compound produced by one microorganism that inhibits or kills another microorganism at greater dilutions*.

The discovery of penicillin by Alexander Fleming (1929) has revolutionised the course of modern medicine and it is the first and the best antibiotic produced on a large scale. It acts against many Gram positive bacteria by interfering with the cell wall synthesis (Table 8.10).

Table 8.10 *Antibiotics produced by Fungi and their activity against microorganisms*

Antibiotic	Producing organism	Target organism	Site of action
Penicillins	*Penicillium chrysogenum*	Gram positive and Gam-negative bacteria	Wall synthesis
Cephalosporins	*Acremonium* spp.	Gram-positive bacteria	Wall synthesis
Griseofulvin	*Penicillium griseofulvum* *P. nigricans*	Most fungi	Nucleic acid
Fusidic acid	*Fusidium coccineum, Mucor ramannianus*	Gram-positive bacteria	Ribosomes
Fumagillin	*Aspergillus fumigatus*	Bacteriophage and amoebae	–

Different fungi are known to produce many kinds of antibiotics such as cephalosporins, griseofulvin and fusidic acid. Penicillins are of different kinds and are produced by various fungi. The penicillin observed first time by Fleming was Penicillin F. Today, commercial production of penicillins is obtained from the highly mutated strains *P. chrysogenum*. Initially penicillin was produced commercially in stationary mat culture. Later on deep tank aerated fermentation process was found more efficient from the commercial view point.

Organic acids

Various microorganisms like bacteria and fungi have the ability to convert carbohydrates into organic acids. Organic acids produced by fungi include citric, fumaric, itaconic, gluconic, kojic and gibberellin acids (Table 8.11).

Table 8.11 *Organic acids produced by Bacteria and Fungi*

Organic acid	Bacteria	Fungi
Citric acid		*Aspergillus niger* *Candida hypolytica*
Itaconic acid		*A. terreus*
Gluconic acid	*Acetobacter*	*A. niger*
Kojic acid		*A. oryzae*
D - Araboascorbic acid		*Penicilliun notatum*
Fumaric acid		*Rhizopus delamar*
α-ketoglutaric acid	*Pseudomonas*	*Candida hydrocarbofumarica*
L(+)Allocitric acid		*P. purpurogenum*
Gibberillic acid		*Gibberella fujikuroi*
Acetic acid	*Clostridium*	
Butyric acid	*Acetobacter*	
Lactic acid	*Lactobacillus* *Streptococcus*	

Citric acid

Citric acid being a component of many fruits has been commercially produced for many years from fruit processing by products. Fungal fermentation has replaced the fruit processing for citric acid. Citric acid is useful industrially in many ways.

1. As an acidulant in food (Jams, fruit drinks etc) and pharmaceutical industry
2. As a chelating and sequestering agent (Tanning animal skins)
3. In the production of carbonated beverages
4. Citrate and citrate esters as plasticizers

Citric acid accumulates during the controlled fermentative growth and is produced by fermentation technology either through surface-culture process or by submerged culture process. Many fungi are capable of producing citric acid such as *A. niger, A. clavatus, Penicillium luteum, P. citrinum, Paecilomyces divaricatum, Mucor piriformis, Ustuline vulgaris* and others. However, *A. niger* is the efficient, high yielding fungus with uniform biochemical properties. Yeasts also contribute in citric acid production (*Candida guillermondii*) in submerged culture process. *C. lipolytica* utilizes hydorcarbons as a raw material in citric acid fermentation.

Citric acid is produced on an industrial scale presently both by the submerged - aerated technique and by utilizing stationary pans or trays with a layer of medium.

Fumaric acid

Most microorganisms produce fumaric acid in small amounts as an intermediate of tricarboxylic acid cycle, but only few microbes accumulate it as an end product of metabolism. Members of mucorales such as *Rhizopus nigricans* produce this acid under aerobic condition.

Itaconic acid

It is obtained from *Aspergillus itaconicus* for the first time. It is an important organic acid used in the resin, detergent and plastic industry. Itaconic acid forms a material for artificial dentures and for certain synthetic fibres. Industrial production of itaconic acid is achieved by fermentation process (surface and sub-surface methods). *A. terreus* is the chief producer of itaconic acid.

Gluconic acid

It is used as a mild acidulant in metal processing, leather tanning and foods. Sodium gluconate is a good sequestering agent in preventing precipitation of lime soap scums on cleaned products. Calcium gluconate is a theraputant. *Aspergillus* species produce *gluconic acid in good yields in several* fungi including *Penicillium*. Industrial production of gluconic acid by *A. niger is carried out only* by submerged fermentation process.

Kojic acid

Kojic acid is produced by aspergilli belonging to *Aspergillus flavus-oryzae* and *A. tamarii* groups in surface culture. High yields of this acid is achieved on a medium containing ammonium nitrate, salts and glucose, sucrose or xylose.

Gibberellins acid

The gibberellins are plant hormones comprising gibberillic acid as the most active form promoting plant growth by both cell enlargement and cell division. Gibberellins are produced by the fungus, *Gibberella fujikuroi*, a perfect stage of the rice pathogen *Fusarium moniliforme*. Gibberellic acid is produced in aerated submerged fermentation culture at 25C for a period of 2 to 3 days.

Ethanol

The fermentation product of yeasts include ethanol, glycerol, acetic acid, higher alcohols and acids. Brewing is a complex fermentation process producing various malt beverages including beer, ale, porter, stout and malt tonics. Brewing utilizes both bottom and top yeast strains. *Saccharomyces cerevisiae*, a top yeast, raises to the surface during fermentation while a bottom yeast, *S. carlsbergensis* settle to the bottom. Top yeasts also include distillers, 'bakers' and wine yeasts and are useful in the production of ale in open tanks. Bottom yeasts are employed in beer fermentation.

Ever since the beginnings of 20th century, the production of industrial alcohol, ethanol, became commercially feasible. Ethanol is extensively used as a solvent and also mixed with motor fuels such as gasoline. Several fungi are known to produce ethanol from various raw carbohydrate substrates. Selected strains of *S. cerevisiae* utilizes starch and sugar raw materials and produce industrial alcohol.

Candida pseudotropicalis on the other hand, converts lactose of whey, into ethanol after the removal of proteins. *Candida utilis* is able to ferment pentoses and it is employed in the production of alcohol from sulphite-waste liquor although *S. cerevisiae* also can utilize this substrate.

An efficient strain for the fermentation must grow rapidly, and tolerate high concentrations of sugar, must be able to produce large amounts of alcohol and also be relatively resistant to this alcohol.

Molasses is the principal media for the commercial production of industrial alcohol in addition to sulphite waste liquor, corn, rye and barley grains, whey, potatoes and wood wastes. Ethanol production is carried out in very large fermentors provided with 3 to 10 per cent inoculum temperature between 21-27 °C. The formation lasts approximately 2 to 3 days depending upon the substrate utilized and on the temperature of incubation. Ethanol is separated from the fermentation broth by successive distillation.

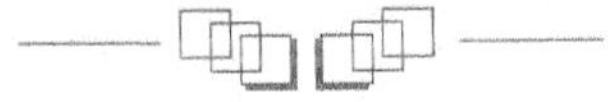

Essay Type Questions

1. What are biofertilizers and describe production methods of BGA biofertilizers ?
2. Describe mycorrhizae and add a note on their significance as biofertilizers.
3. Write about isolation and production of *Rhizobium* inoculate.
4. What are biofuels and write about various types of biofuels ?
5. Describe the structure, biosynthesis and polymer properties of PHB.
6. Describe the role of mcroorganisms in waste treatment.
7. Describe the application of biosensors.
8. Give a detailed account of industrially produced microbial enzymes and their applications.

Short Answer Type Questions

1. BGA biofertilizers
2. VAM fungi
3. Biofuels
4. Single cell proteins
5. PHB polymer
6. Bioremediation
7. Bioleaching
8. Biosensors
9. Transducers used in biosensors
10. Bioindicators
11. Fungal enzymes
12. Antibiotics
13. Organic acids

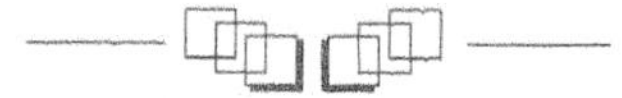

9 Plant Biotechnology

9.1 Introduction

The word biotechnology coined by Karl Ereky (1919), was originally defined to indicate "all lines of work by which products are produced from raw materials with the aid of living organisms". In fact the prefix - 'bio' - means 'life' and technology is defined as a branch of knowledge dealing with the application of science to economic ends. The plant scientists are constantly facing the stupendous task of increasing the yields of plants for food as well as for fiber and continue to pioneer the development and exploitation of both plant and animal biotechnologies. Historically, domestication of plants and animals might be the starting point of, one of the earliest and most vital biotechnologies, that transformed wild man, a hunter gatherer, to a civilized and cultured man, thus enabling the transformation of nomadic societies to more stable enduring and civilized human societies forever. Now that human being could stay put in one place and be reasonably assured of a higher quality nutrition at less expense of human capital (energy), he had more time to think, fantasize and envision Julian Huxley's (1946) Brave new world.

Today, the word '*Biotechnology*' conjures up visions of Genetic Engineering (Genetically modified crops or organisms) and cloned animals. Plant and animal biotechnology are receiving increasing attention, because, these are sources of innovation for the agriculture, horticulture, food, animal breeding, chemical and pharmaceutical industries. Basically the plant biotechnology provides the ways and means of translating the knowledge of fundamental process of growth, development and reproduction to modify the plant metabolism to produce new products and varieties capable of enhanced productivity. Over the next ten years, it is likely that genomes of many crop plants will be completely sequenced and the functions of a large number of genes resulting plant development will be analysed. Genes conditioning resistance to biotic (fungi, bacteria and virus) and abiotic stress (drought, salinity, alkalinity, and extreme temperatures) will be identified, isolated and cloned for exploitation in crop improvement. Seed with commercially important fatty acid profiles (and content), proteins, vitamins, mineral contents and starch will be designed and produced. Such genetically modified crops (transgenics) will become part of the future agriculture. With the developments in plant biotechnology a range of new options to both consumer as well as industries will be available, leading ultimately to eco-friendly agriculture. This kind of twist in plant biotechnology research is an inevitable necessity, because, providing food to a current yet exploding population of almost more than 6 billion people could not just be done using antiquated technologies of 1940 - 50s. Furthermore, the population at the current growth rate is going to be 8 billion by 2020 and 10 billions by 2030. Even if the population growth comes to a grinding halt at this moment, the agricultural scientists have to double the world agricultural production to simply meet food and feeding needs of current population in the year 2030. To avoid economic crises presently occurring in Afro-Asian countries, it is quite essential that method of food production from plant and animal sources, not only be refined and accelerated, but also new approaches be developed that will permit us to meet global demands for plant and animal products more efficiently without endangering future. Feeding the world in the 21st century is a gigantic challenge that plant scientists face today.

9.2 Economic Potential of Plants

Plants are of vital importance to the survival and sustenance of life on this space-ship, Earth, because they harness the solar energy of the sun by photosynthetic process and synthesize a variety of life molecules including carbohydrates, proteins, fats etc. and make them available to the animals and humans in the form of food. Plants absorb carbondioxide, air, water, minerals and generate oxygen using sun's energy which is so vital for the life besides maintaining the balance of gases in the atmosphere. The process of production of building blocks of macromolecules, both informational (proteins and nucleic acids) and non-informational (phospholipids and polysaccharides) by a plant cell proceeds through intermediates constitutes the primary metabolism and the resultant products are called primary metabolites Fig. 9.1. Plants have been used in diverse ways by humans and had far-reaching impact on human history for several millenniums and still holds immense economic potential.

Plants Provide Food and Feed

The bulk of the human food, either directly or indirectly come from plants. Numerous plant species were collected and eaten by the primitive men and agriculture has been discovered, developed and cultivation of selected and improved species of plants for food, feed and fibre was accomplished. Currently bulk of world food (about 90%) is derived from a mere twenty species or so. The cereals, wheat, rice and maize contribute to about half of the world's food needs. Besides the afore mentioned three grass species, other crop plants like sorghum, minor millets like, bajra, *Setaria, Elusine* etc, provide carbohydrates (starch) in addition to tubers of potato and roots of yam and cassava. The principal crops cultivated in India along with their annual production data are given in Table 9.1. Some of the plants also serve as vegetables, edible parts being leaves, stems, flowers,

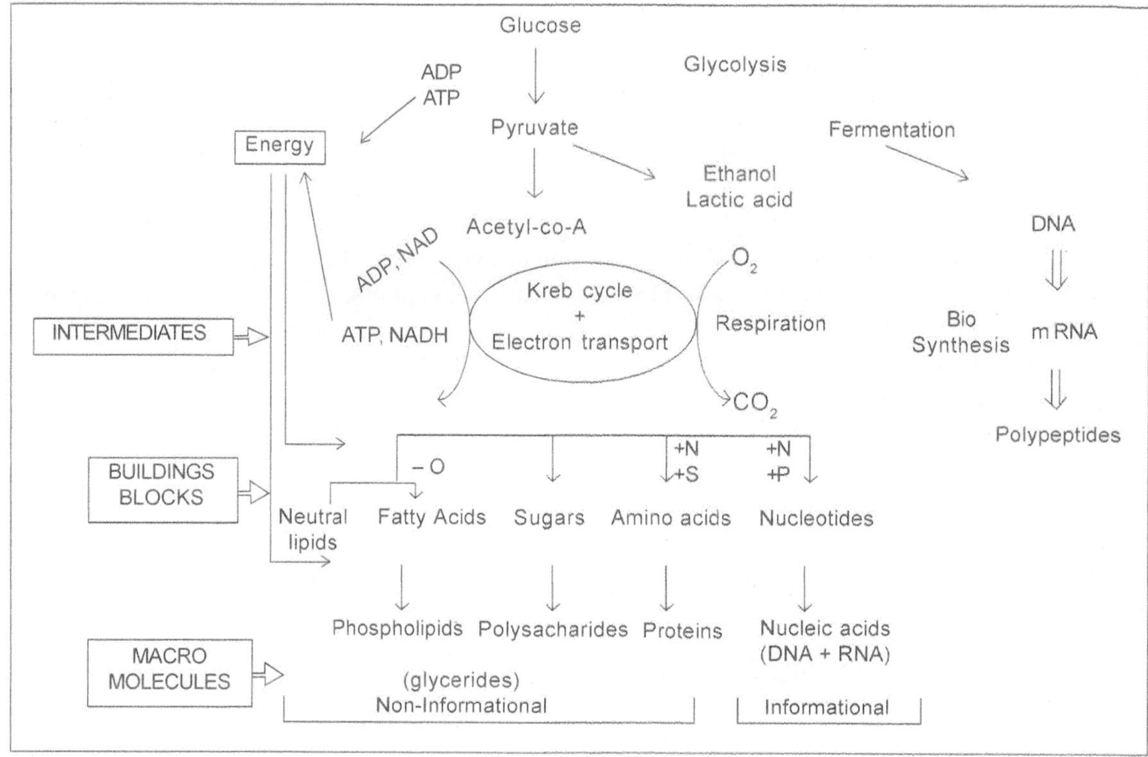

Fig. 9.1 Major metabolic activities of plant cells resulting into primary metabolites

with less starch but rich in minerals, vitamins and soluble and insoluble fibre that provide roughage vital for maintaining health; most important in this group are *Brassicas*. There are yet another group of plant species called pulses (Legumes) that are rich sources of protein (soybean, pigeonpea, chickpea etc.) and oilseed crops like groundnut, sunflower and mustard, soybeans, safflower. Sesame, provide fats necessary for the diet as well as for industry for the manufacturing of soaps, varnishes and paints etc.

A wide range of trees and herbaceous plant species provide fruits rich in sugars (fructose, glucose, sucrose, mannose, dextrose etc.,) not only useful as a source of energy, but also as sweeteners in the confectionery industry. The conventional plant breeding methods have generated mind-boggling array of improved varieties with high yield levels in many diverse crop species including plantation and fruit crops like banana, orange, apples etc. A wide range of plant families mostly from tropics are sources of prized spices like black pepper, turmeric, ginger, cinnamon, cloves and Capsicum, export of which brings millions of dollars to the country's exchequer.

Plants have other uses besides providing humanity with food, beverages and drink. There are several plant species like hemp, flax, jute, manila hemp and sisal that provide fabrics and ropes. Cotton, a seed appendage of the genus *Gossypium* (Malvaceae) has been the pride of India since time immemorial. Another fibre yielding tree kaphok, *Ceiba pentandra* (Bombaceae) produces cotton like fibre as a seed appendage which is used as light stuffing for mattresses and life jackets.

Table 9.1 *Important cereal, pulse and fibre yielding crops of India and their productivity*

Name of the crop	Botanical Name	Production data during period					
		1950 - 1951			1999 - 2000		
		A	P	Y	A	P	Y
A. Cereals							
1. Rice	*Oryza sativa*	308.16	205.76	668.00	449.72	894.72	1,990.00
2. Wheat	*Triticum aestivum*	97.40	64.62	663.00	274.34	755.74	2,755.00
3. Jowar	*Sorghum vulgare*	155.70	54.95	355.00	103.98	88.63	852.00
4. Bajra	*Pennisetum typhoideum*	90.23	25.95	288.00	88.55	56.57	639.00
5. Maize	*Zea mays*	31.59	17.29	54.70	64.27	114.73	1785.00
B. Pulses							
6. Gram	*Cicer arietinum*	75.70	36.51	4820.00	63.05	50.82	806.00
C. Fiber crops							
7. Cotton	Gossypium spp.	58.82	30.44*	88.00	87.59	116.44	226.00
8. Jute	*Corchorus capsularis*	5.71	33.09**	1043.00	85.00	94.17	1,995.00
9. Mesta	*Cannabis sativa*	-	-	-	1.86	11.13	1078.00
D. Industrial crop							
Sugarcane	*Saccharum officinarum*	17.07	570.51	33,422.00	42.25	2990.36	70,780.00

A - Area cultivated in lakh hectares, P - Production in lakh tones, Y - Yield in Kg per hectare,
* production in lakh bales of 170 and 180 Kg each, in cotton and jute respectively.

Plants as a Source of Industrial Material

Many plant species produce latex, gums and resins which are of paramount importance for industries. The latex of Brazilian tree, *Hevea brasiliensis* and other related species is used in the manufacture of rubber which has revolutionized the automobile industry in providing tyres and tubes. Rubber production was exclusive precinct of Brazil for a long time but other countries like Malaysia started exploiting their rubber plantation. New technology has been developed for the trapping of latex from the tree, without inflicting permanent damage to trees, and this proved to be a boon ecologically since soil stabilization as well as maintaining the forest cover is accomplished. Many of the plants also produce resins that are useful in making varnish and lacquer. These constitute important base for paint industry including those manufacturing glues, cosmetics and organic solvents like turpentine. Plants are also the source for many dyes and pigments. Dyes are extracted from leaves, roots, flowers, fruits or seeds. The traditional dyes include henna from the leaves of Henna (*Lawsonia inervis*), and indigo from Indigo (*Indigofera tinctoria*) plants.

Plants as a Source for Wood Industry

Plants are also invaluable source material for the construction of buildings as well as furniture industry. Wood is made of elongated hollow cells of xylem tissue thickened with lignin and suberin making it much stronger and durable against pest attack. The ability of some wood to be smoothed and stained has made them an important medium for sculpturing and other ornamentation including musical instruments. Violins are made from mapple and pine or spruce tree wood while Clarinets are made

Table 9.2 *The Productivity of Important Oil Seed Crops in India*

Crop	1989-91			1994			1995			1996		
	A*	Y*	P*	A	Y	P	A	Y	P	A	Y	P
Groundnut	85.6 (202.6)	884 (11.47)	75.7 (232.5)	78.7 (218.9)	1049 (1313)	82.5 (287.5)	83 (223.6)	855 (1251)	71 (279.7)	83 (224.1)	988 (1286)	82 (288.2)
Rape seed	51.9 (182.4)	880 (1368)	45.8 (250.0)	62.9 (228.0)	847 (1311)	53.3 (299.1)	62.3 (241.2)	944 (1432)	58.8 (345.4)	62.3 (218.7)	979 (1386)	61.0 (303.2)
Soybean	26.6 (564.3)	886 (1883)	23 (1062.5)	39.9 (616.8)	918 (2207)	36.7 (1361.7)	50.0 (621.4)	890 (2031)	44.5 (1262.3)	500 (634.3)	920 (2081.9)	46 (1320.3)
Sunflower	16.3 (163.6)	562 (1365)	9.3 (223.0)	27. (191.1)	446 (1145)	12 (218.8)	27.5 (216.4)	462 (1206)	12.7 (260.9)	27.5 (213.7)	473 (1171)	13 (250.2)
Sesamum	25.1 (64.0)	304 (351)	7.6 (22.4)	24.7 (71.6)	340 (366)	8.4 (26.2)	27.0 (79.2)	462 (340)	12.7 (27.0)	27.5 (79.3)	473 (344)	8.4 (27.3)
Safflower	8.3 (12.1)	504 (630)	4.2 (7.6)	8.0 (10.8)	733 (855)	5.9 (9.2)	7.6 (11.0)	548 (752)	4.2 (8.2)	7.6 (10.8)	589 (769)	4.5 (8.3)
Linseed	11.4 (38.5)	298 (684)	3.4 (26.2)	9.4 (29.1)	353 (870)	3.3 (25.4)	11.2 (32.8)	290 (80.6)	3.3 (26.4)	11.2 (30.4)	303 (796)	3.4 (24.2)
Niger	6.6	238	1.9	-	-	1.9	-	-	1.9	-	-	1.9

* World's production is given in parenthesis; A* = Average (Lakh/Ha); Y* = Yield (Kg/Ha); *P = Productivity (Lakh metric tonnes)

Source : (1) FAO QBS, Vol. 9 No. 3/4, 1996

(2) CMIE (P) Ltd, July, 1996.

from African black wood. Teak wood (*Tectona grandis*) and rose wood are the highly priced ones used in making furniture. Paper industry thrives on the soft wood from conifers, Eucalyptus, bamboo which are pulped, glued, pressed and bleached before made into paper. Wood pulp is also used in making fine clothing material after chemically modifying the cellulose. Wood has been used as an important alternative source of energy fuels directly or indirectly as fossil fuels, coal and lignite.

Plant fibre industry mainly thrives on few plant species like flax (*Linum spp.*), hemp (*Cannabis sp.* and jute (*Corchorus sp.*). Plant fibers are usually nothing but sclerenchymatous cells and some times consists of collenchyma or conducting cells.

Plants as a Source of Oils and Fat Industry

The world's oil industry is essentially based on plant origin. According to a conservative estimate, the global production of edible is more than 84 million metric tons of which 63 million metric tons are vegetable oils and fats. One of the important industrial sectors that is undergoing revolutionary changes world wide is that of vegetable oils and fats sector. The important edible oil seed crops cultivated in India are groundnut, rape seed, mustard, soybean, oil-palm, sunflower, linseed, safflower, sesame and niger (Table.9.2). The oil seed crop not only meet the needs of edible oils for indigenous population, but also used for exploiting their high quality protein meal product, which is having a high marketable value mainly as a raw material for animal feed and confectionery industries.

Plants and Religion

Use of plants in religious ceremonies is known since ancient times. The culture of worshiping plants has lead to the preservation of plants at certain specified places called sacred grooves in almost all ancient civilization all over the world. Sacred banyan trees in India and Babox (*Adansonia digitata*) tree in Africa are good examples of peoples reverence to plants. Ginkgo (*Ginkgo biloba*), a lone member of the Ginkgoales is saved from the extinction because of its association with religious ceremonies; besides seed of *Ginkgo* are edible and are consumed in Japan and China.

Plants and flowers are intimately associated with festivals and marriages. The olive branch is a global symbol of peace. The association of plants to religious ceremonies and festivities can be traced in different civilization to various paintings and writings. Plants are also grown purely for aesthetic purpose. Many orchids are prized for their exotic blooms and have a great economic potential for export. The whole cut-flower industry is centred around beautiful flowering plants.

9.3 Biotechnology and Secondary Metabolites

Plant derived chemical compounds may be classified as either primary or secondary metabolites. Generally, macromolecules that perform structural and/or catalytic functions (Enzymes-proteins) and informational biomolecules (DNA, RNA) are excluded from this classification. Primary metabolites are the substances that are ubiquitous in their distribution occurring in one form or other universally in all organisms. In plants such organic compounds are concentrated in seed (e.g. seed oils and fats) and vegetative storage organs (e.g. starch and sucrose in roots rhizomes and tubers) are essential for growth, physiological functions, and development, since their effective role in primary cell metabolism. Since the dawn of history man has been exploiting the primary metabolism for use as foodstuffs and raw materials from plants. As a rule the primary metabolites of plant origin obtained for commercial purposes are low value-high volume commodity chemicals. These are basically used as foods, food additives and industrial raw materials. These include products such as vegetable oils, fatty acids used for manufacturing soaps, detergents etc., and carbohydrates like table sugar (sucrose), starch, pectin, gums, cotton (cellulose). These usually fetch 2 - 5 dollars per kilogram and most of them are readily available in the grocers shop as commodities even. There are certain exceptions to this general rule. For instance, myosital is synthesized naturally by plants in trace quantities and hence difficult to extract and purify. There is another primary metabolite, β-carotene, even though accumulates in large quantities by plants like carrots and sweet potatoes it is highly labile and susceptible to oxidation and photodegrades on exposure to air. Hence the isolation and purification of these kinds of primary metabolites is highly expensive and tedious. That is the reason β-carotene is produced synthetically on commercial scale. Primary metabolites some times used as intermediates in the manufacture of high value semi-synthetic pharmaceuticals. One such instance is the production of steroid hormones from stigmasterol derived from soybean.

Secondary metabolites even though biogenetically derived from primary metabolites, differ from the latter in that these have limited distribution in the plant kingdom, often restricted to specific taxonomic groups such as species, genus, family or closely related taxa. Due to convergent evolution, the secondary metabolites that are not too widely separated from the primary metabolic pathways may arise in unrelated taxonomic groups as is the case with nicotine and related alkaloids such as anabasine. Another feature of secondary metabolites is, that these are the substances that do not have obvious roles in plants mainstream metabolism in that they are non-nutritive and are not essential for growth and development unlike primary metabolites. However, the secondary metabolites may play significant ecological role in dealing with the adaptation to the stressful biotic and abiotic environment and hence very vital for their survival. In summary the secondary metabolites in plants serve basically to combat diseases, to help in weed agrossiveness, to discourage herbivores, besides attracting ollinators (entomophily). By definition the secondary metabolites represent an assemblage of a wide array of bio-active compounds, which might prove highly useful in the production of plant based drugs, pharmaceuticals, aromatic volatile oils, insecticides and pesticides, food, as food additives etc., and hence constitute targeted area of plant biotechnology industry.

Certain examples of economically important plant derived pharmaceuticals and intermediates that are still obtained commercially *in vivo* is given in Table 9.3. Major groups of secondary metabolites and their biogenetic pathway is presented in Fig. 9.2.

Phytochemicals as Pharmaceuticals Products

From times immemorial plants have been used in many civilizations as a principal sources of medicines for human health as well as for protecting the health of domesticated animals and plants. Many a modern medicines are derived from plants as extracts or synthesized chemically in laboratories as a copy of phyto-chemicals. Plants synthesize innumerable bioactive secondary metabolites and accumulate in leaves, bark, roots, stems, seeds, fruits and flowers. Hence, the medicinal plants constitute one of the segments of world's economy and basis for alternative medicare systems *viz*, Ayurveda, Siddha, Unani and Homeopathy. As per the World Health Organization's survey (WHO) about 6 billion people (80% of the world's population) rely on plants for their primary health-care, including, animals and agriculture crops. Pharmaceuticals based on synthetic chemicals used in western system of medicine, allopathy are not only quite expensive but are inaccessible to under developed countries and are unsuitable because of undesirable side effects. In China, for example traditional medicine is mainly based on some 5000 plant species and employed to treat 40% of urban patients and 90% patients in rural areas. More than 7×10^6 tonnes of plant material was used for medicine in China, 80% of which is collected from the wild plants.

Table 9.3 *Widely used Drugs of plant origin in modern medicine (Allopathy)*

Drug	Plant derived from	Ailment Treated
Quinine	*Cinchona ledgeriana*	Malaria
Reserpine	*Rauvolfia serpentina*	Hypertension, as a tranquillizer for snake bite
L - Dopa	*Mucuna deeringiana*	Parkinson disease
Ephedrine	*Ephedra sinica*	decongestant
Digitalin and digoxin	*Digitalis purpurea*	heart drug
Eucatptol	*Eucalyptus spp.*	antiseptic throat soars and infection, bronchitis, cough, cold and asthma
Aspirin	*Salix alba, Filipendula ulmaria*	Analgesic antipyretic, heart diseases.
Picrotoxin	*Anamira cocculis*	Nervous system stimulant especially in case of barbiturate poisoning
Vinblastine, Vincristine	*Catharanthus roseus*	Cancer.

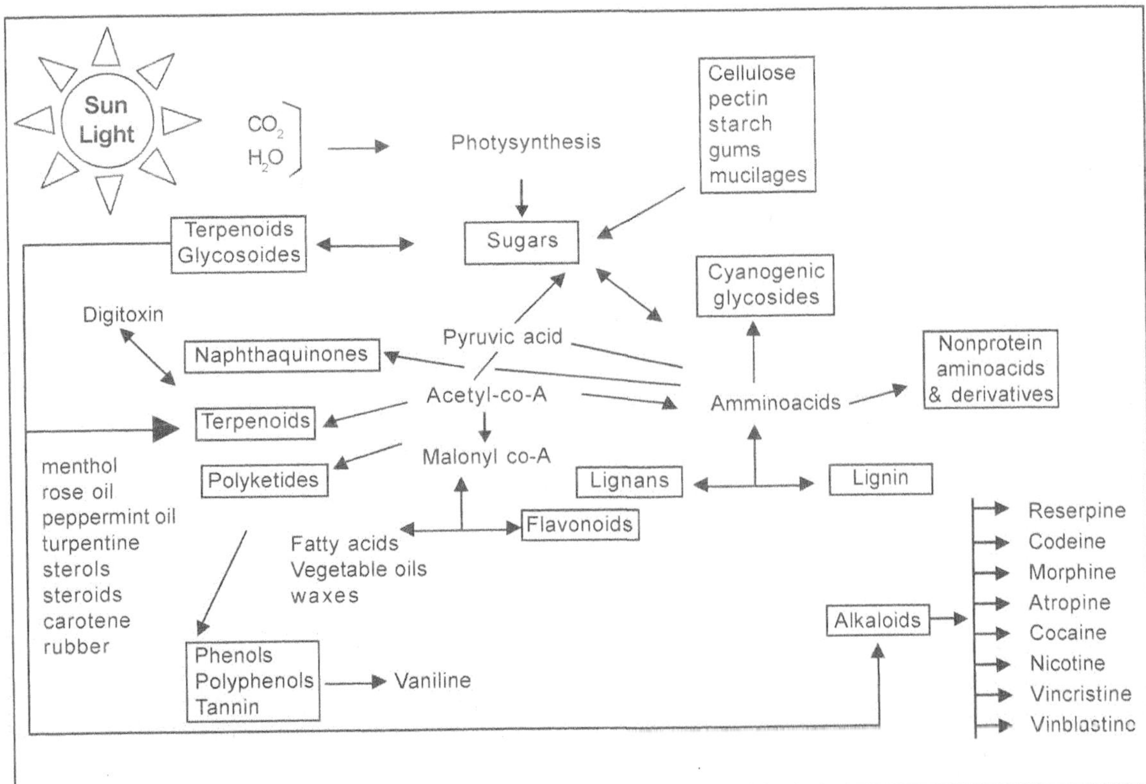

Fig. 9.2 Biogenetic pathway of some commercially important phytochemicals (secondary metabolities)

In India, traditional health care systems of medicine viz, Ayurveda, Unani, Siddha, is practised along with, Allophathy, the modern western health care system. There are 4×10^5 registered medicinal practitioners in India as compared to 3.32×10^5 registered Allopathic doctors.

In developed countries the use of medicinal plants for the management of diseases has declined considerably. However, plants have contributed more than 7,000 different phytochemicals, being used today as cardiovascular drugs, laxatives, anti-cancer agents, hormones, contraceptives, diuretics, antibiotics, decongestants, analgesics, anaesthetics, ulcer treatments, antiparasitic compounds. All over the world extensive pharmaceutical screening of phytochemicals are being made to discover the new chemical compounds possessing therapeutic properties that may offer new treatments for many recalcitrant and dreadful diseases.

In recent time there has been remarkable revival in the use of herbal medicines in many developed countries in the west due to increasing interest in natural products. Some of the plants that contributed life saving valuable drugs for modern medicine (Allopathy) is given in Table 9.3.

Ayurveda and traditional Chinese medicine are among the most popular health therapies. In Chinese medicine, herbs, especially tonics like reshi made up of a bracket fungus, *Ganoderma* were regarded as medicine that offers long life and good fortune. Indian traditional medicine is even older than traditional Chinese medicine. It has its roots in health and hygiene theories of early Dravidian civilization of Southern India around 5000 BC.

The *Rig Veda* is the earliest written text date at around 2500 BC and also in the works of Charaka and Susruta. Use of herbals in medicine was also recorded in *Ayurveda* which is part of *Atharva veda* and also in Egyptian, Greek and European literature. The earliest Indian medicinal plant, Soma is considered to be *Ephedra pachyclade*, was used by Indo -Aryans for health and immortality.

Plants contain a wide range of different, biologically active substances with diverse uses. Practically almost in every form of medicine such as the Ayurvedic, the Unani, the Sidha, the Allopathic and the Homeopathic medicinal herbs are used invariably. Table 9.4 provides a list of some of the important medicinal plants and their uses. Drugs of medicinal value have been extracted from plants such as Morphine, Cocaine, Strychnine, Digitalin, Ephedrine, Rauvolfine, Atropine, quinine etc. Various plant parts are of use and some times the whole plant may be used in medicine in different formulations

Table 9.4 *Medicinal plants used in alternative system of medicine (Ayurveda, Siddha Unani and Homeopathy)*

Name of the species (common name)	**Organoleptic part (OLP) of the plant**	**Used as**	**Ailment treated**	**Therapeutic Action**
Achillea millefolium (yarrow)	Leaves, flowers	Infusions Tinctures Massagerubs Essential oil	Colds, Hay fever catarrah	Astringent diaphoretic Digestive stimulant anti inflammatory anti allergenic anti spasmodic

Contd.. Table 9.4

Name of the species (common name)	Organoleptic part (OLP) of the plant	Used as	Ailment treated	Therapeutic Action
Agrimonia eupatoria (Agrimony)	Aerial parts/ leaves	Infusions Tinctures	Diarrhoea	Diuretic Astringent Tissue healer To stop bleeding
Alchemilla xanthoclora (Lady's mantle)	Aerial parts, leaves	Infusions Tirctures Ointments	Diarrhoea, sore throat, Dermatitis Menstrual regulator	Astringent Menstrual regulator digestive tonic autinflammatory wound healer
Allium sativum (Garlic)	Clove Oil	Capsules/ tablets ointments syrup	Colds, catarrah	Antibiotic expectorant diaphoretic hypotensive hypo glycoemic antihistamine etc.
Aloe vera (Aloe)	Sap, leaves	Ointments Tinctures Capsules Tonics	Eczema Thrush piles & fissures, stomach ache	Purgative, bile stimulant wound healer Antifungal helminth sedative etc.
Angelica polymorpha var. *sinensis* (Dang gui Chinese angelica)	Rhizome	Decoctions capsules Tinctures Tonics	Anaemia, Menstrual problems	Laxative, Blood Tonic circulatory stimulant antie spasmodic
Apium graveolens (celery)	Seeds, essential oil	Juices, Infusions, Tinctures powders	Arthritis	antirheumatic sedative, urinary antiseptic antispasmodic divretic
Aconitum ferox (Vatsanabha)	Tuber	Mrityunjayaras Hinguleswar ras Anand bhairav	Asthma diabetes dyspepsia, Leprosy paralysis Rheumatism Typhoid	Diaphoretic, divretic expectorant febrifuge, narcotic rervinetonic
Adhatoda vasica (vasaka)	Leaves	Decoctions powder	Asthma, chromic Bronchitis	Expectorant, antispasmodic
Atropa belladonna (Belladonna)	Roots, leaves		Whooping cough Night sweats, spermatorrhoea	Diuretic, sedative mydriatic, antispasmodic, anodyne
Azadirachta indica	Bark, leaves	Creams	Fever, skin disease boils	Astringent, antiperiodic
Butea monosperma	Gum, seeds bark		Diarrhoea, skin diseases	Antihelmintic
Cinchona callisaya	Bark	Tablets gargles	Malaria, dyspepsia gastric catarrh, adynamia hooping, cough pneumonia, Acute rheumatism, Acute tonsillitis	Antiperiodic
Cinnamomum camphora (camphor)	Whole plant		Diarrhoea, colds Inflammations bruises, sprains	Sedative, antispasmodic diapholetic, anthelmiatic
Datura alba	Whole plant leaves, seeds,	Tinctures Lotions	Rheumatic swellings diarrhoea skin disease Dogbite Elephantiasis ear ache, asthama	Anasthetic

Contd.. Table 9.4

Name of the species (common name)	Organoleptic part (OLP) of the plant	Ailment treated	Used as	Therapeutic Action
Digitalis lanata	Leaves	Tonic	Heart disease	
Arnica montana (Arnica)	Flowers Rhizome	Tissue repair	Ointments, creams homeopathic tablets, and extracts	Antiinflammatary healing circulatory stimulant
Calendula officinalis (pot marigold)	Flowers esse-oil	Dry skin, eczema	Infusions, timctures, infused oils, creams, ointments, mouth washes	Astringent, antiseptic, antispasmodic menstrual regulator
Capsella bursa - pastoris (Shepherd's purse)	Leaves aerial parts	Haemorrage	Infusions, tinctures compresses	Astringent uterine relaxant stypic, urinary antiseptic
Capsicum frutescens (chilli)	Fruit	Warming stimulant to combat chills	Ointments, Tinctures gargles, capsules, tablet, infusions, massage oils	Circulatory stimulant, Tonic, antispasmodic, analgesic
Centella asiatica (Brahmi)	Aerial parts	Remedy for failing memeory & old age problems (senile dementia)	Infusions Tinctures powders poultices	Tonic antirheumatic, peripheral/vasodilator laxative, sedative
Chionanthus (Fringe Tree)	Root bark	Jaundice, gallstones hepatitis	Decoctions, Tinctures, capsules	Liver stimulant, laxative, diuretic, tonic
Phyllanthus amana	Whole plant	Jaundice		
Citrus limon *Lemon*	Fruit, essential oil	Colds, flu	juice syrups, massage oils	antiinflammatory, antibacteria, antivtral, antihistamine, antioxidant
Commiphora myrrha (Myrrh)	Oleo-gum resin	Tonsillitis, mouth ulsers gumdisease	Capsules, Tinctures gargles & mouth washes, powder	Antispasmodic antiseptic, stimulant antiinflammatory, astringent, expectorant
Dioscorea villosa (wild yam)	Root and Rhizome	Asthama, gastritis cramps, gall gladder problems	Decoctions Tinctures	Antispasmodic Relaxant for smooth muscles anti inflammatory, bile stimulant
Echinacea spp. (purple cone flower)	Root and Aerial parts	Infections	Capsules / tablets Decoction, ointments creams, infusions	Antibiotic, immune stimulant, anti allergenic, antiinflammatory, diaphoretic wound healer
Equisetum arvense (Horse tail)	Aerial parts juice	Urinary tract problems, prostate disorders, chronic bronchitis	Dicoctions, tinctures juices, capsules	Astringent styptic diuretic, anti inflammatory, tissue healer
Eucalyptus globulus	Leaves essential oil	Muscle and joint problem	Steam inhalations chest rubs, infusions, tinctures, capsules/ tablets	Antiseptic, antiviral, fungal, antispasmodic, hypoglycaemic, febrifuge
Ephedra gerardiana	root, aerial parts	Rheumatism, syphilis	Decoction	antipyretic, divretic, diaphoretic
Foeniculum vulgare (Fennel)	Seeds aerial parts root essential oil	Indigestion flatulence gum	Infusions, decoctions, tinctures, gargles, mouthwashes	Carminative culinary stilumalnt antiflammatory diuretic

Contd.. Table 9.4

Name of the species (common name)	Organoleptic part (OLP)	Ailment treated	Used as	Therapeutic Action
Ginkgo biloba (Ginkgo)	Leaves, seeds	Asthma, urinary & blood disorders	Tablets/capsules tinctures, infusion, decoctions	Vasodilator antiinflammatory astringent (seeds), antifungal and antibacterial (seeds)
Ganoderma lucidum (Reishi)	Fruiting body	AIDS, chronic fatigue syndrome	Tinctures, powders, capsules	Hypotensive, hypocalcaemic, antiviral immune stimulant, expectorant, antihistamine
Lavendula angustifolia	Flowers essential oil	Head ache, digestive upsets, nervous problems	Infusions, tinctures, massage rubs	Antiseptic, analgesic, bacterial antidepressant relaxant, antispasmodic culinary stimulant, bile stimulant
Ocimum basilicum (Basil)	Leaves essential oil	Digestive problems nerve tonic	Infusions, decoctions juices, syrups, fresh leaves inhalation	Antidepressant antiseptic carminative febrifuge, expectorant, stimulates adrenal carex
Panax ginseng (Ginseng)	Roots	Aphrodisiac tonic	Tablets/capsules Tonic, decoction	stimulant, hypoglycaemic, immune stimulant, reduces cholesterol levels
Rosa spp. (roses)	Petal, rose essential oil	Skin and emotional problems, liver and menstr-	Infusions, tinctures syrups, decoctions, creams lotions, massage rubs, gargles and mouth washes	Anti depressant, antispasmodic astringent sedative, digestive stimulant, expectorant antiviral, anti bacterial, antiseptic
Santalum alba (sandal wood)	Wood, essential oil	Urinary problems nervous disorders Gonorrhoea, skin disease	Massage rubs pastes,	Antiseptic, antispasmodic, antibacterial, diuretic
Solidago vigaurea (Golden rod)	Leaves flowers	Kidney and bladder problems coughs, tonsillitis soar throat	Infusions, Tinctures compresses, gargles	Antiinflammatory antiseptic, diuretic, diaphoretic
Syzygium aromaticum (Clove)	Flower buds essential oil	Nausea and digestive upsets, toothache	Massage lotions tinctures, infusions	antiseptic, carminative, warming stimulant, antiparasitic
Strychnos nuxvomica (Nux-vomica)	Seeds	Paralysis		Stimulates central nervous system
Rauvolfia serpentina (sarpagandh)	roots	Schizophrenia Insomnia, violent insanity,m hyper tension		

9.4 Tissue Culture for Mass Propagation of Plants

The term plant tissue culture denote conglomeration of techniques for aseptic culturing of callus (loose unorganized mass of cells), tissues (such as apical meristems), organs (buds), embryos, protoplasts and plantlets for rapid and mass propagation. Because these are grown in a controlled artificial conditions on sterile but defined culture media, these are often referred to as *in vitro* growth or propagation, as opposed to the plant propagation carried out through the normal developmental cycle in nature, either through sexual or asexual mode. Such mode of production is called *in vivo* propagation.

Tissue culture techniques are unique in that these offer several advantages over the conventional methods mainly because of ease with which one can select and isolate new traits at the cellular level in the confines of a laboratory instead of whole plant level in the field. Besides, this affords a rapid accelerated breeding and improvement of crop varieties in that, the cellular screening for traits require weeks or months, whereas, screening whole plants typically requires entire growing season involving several generations. Another related advantage is that the space required for cell screening is substantially less than that required for field screening of characters. Yet another substantive advantage in this method is that plants derived from this are more uniform, because of the fact that those are obtained from single cells in laboratory than through sexual reproduction and also free from pathogens, either bacteria, virus or fungi. On negative side, unlike recombinant method of breeding associated with sexual reproduction, where there is a generation of diverse variation, in this method there is no possibility of generating a wide range of variation.

Pros and Cons of Propagation of Plants through Tissue Culture

Methods followed for the propagation of plants *in vitro* are mainly extension of those evolved for the conventional asexual propagation. Nonetheless, the *in vitro* techniques have the following advantages over conventional methods :

- Since cultures are started with very small pieces of plants (explants) and subsequently small shoots are propagated (micropropagation), only a small amount of space is required to maintain and multiply large number of plants.
- Since the propagation is carried out in sterile conditions, free from any pathogens, there is no loss of plants due to diseases and the plantlets produced will be devoid of any pathogen (bacteria, fungi or virus)
- By manipulating the techniques, it is possible to produce virus free plants and certified virus-free plants can be mass propagated for commercial purposes
- Since there is a flexibility of adjustment of factors responsible for regeneration viz., levels of growth regulators, light, temperature, the rate of propagation is enormous when compared to the conventional (*in vivo*) macropropagation.
- A large plant populations can be produced for a unit time. This feature of plant tissue culture is desirable for it enables newly developed varieties to be made available for cultivation very rapidly. In fact a large number of plants can be mass produced in a shortest time.

- The valuable plants which are recalcitrant, slow and difficult on the border line of impossibility can be propagated through tissue culture.
- Production of plants through tissue culture is independent of seasonal influences and the propagation can be continued all the year round.
- Enables the plant material to be stored for a long period (cryopreservation).
- Very small place (glass house) is required for propagation and maintenance of the stock.
- Since plants need no care between subcultures and hence there is no labour or materials requirement for watering, weeding, spraying etc., in this method.

The main disadvantages of *in vitro* methods are that sophisticated skills are needed for their successful operation, a custom-made specialised and costly production facility is quite essential, fairly specific protocols are needed for each species to get optimum response and the cost of propagules is relatively high. Further more, as a consequence of using artificial *in vitro* conditions for the growth, certain adaptations are necessary before their transfer to field conditions. Plant population obtained through tissue culture are initially small even though they may be mass produced. Since the small explants are cultured in a medium containing a suitable carbon source (sucrose), the plants initially obtained are not autotrophic and hence have to go through a transitional period before they are capable of independent growth. Furthermore, as they are raised under artificial condition within a containers with a high relative humidity, and are not self sufficient photosynthetically, the regenerated plantlets are more susceptible to water stress in an external environment and hence may have to be hardened in an environ of slowly decreasing humidity and increasing light intensity. Finally, there is a danger of producing genetically aberrant plants in plant populations obtained through tissue culture method.

Micropropagation

The methods that are theoretically feasible for the *in vitro* propagation of plants are illustrated in Fig. 9.3. They are essentially follow any one or more pathways presented below :

1. Multiplication of shoots from axillary buds
2. Induction of adventitious shoots, and or adventitious somatic embryos either (a) directly on explants removed from the mother plant or (b) indirectly on callus tissues or in suspension cultures that have been obtained by the proliferation of cells from explants or on the semiorganised callus tissues or propagation bodies such as protocorm or pseudobulbils that can be obtained from explants of organs.

Presently, most micropropagated plants in practice are produced by the multiplication of shoots from axillary buds and in some species by the induction of adventitious shoots and/or adventitious somatic embryos with subsequent regeneration of plantlets. Often shoots and plantlets do not originate in a tissue culture by a single method.

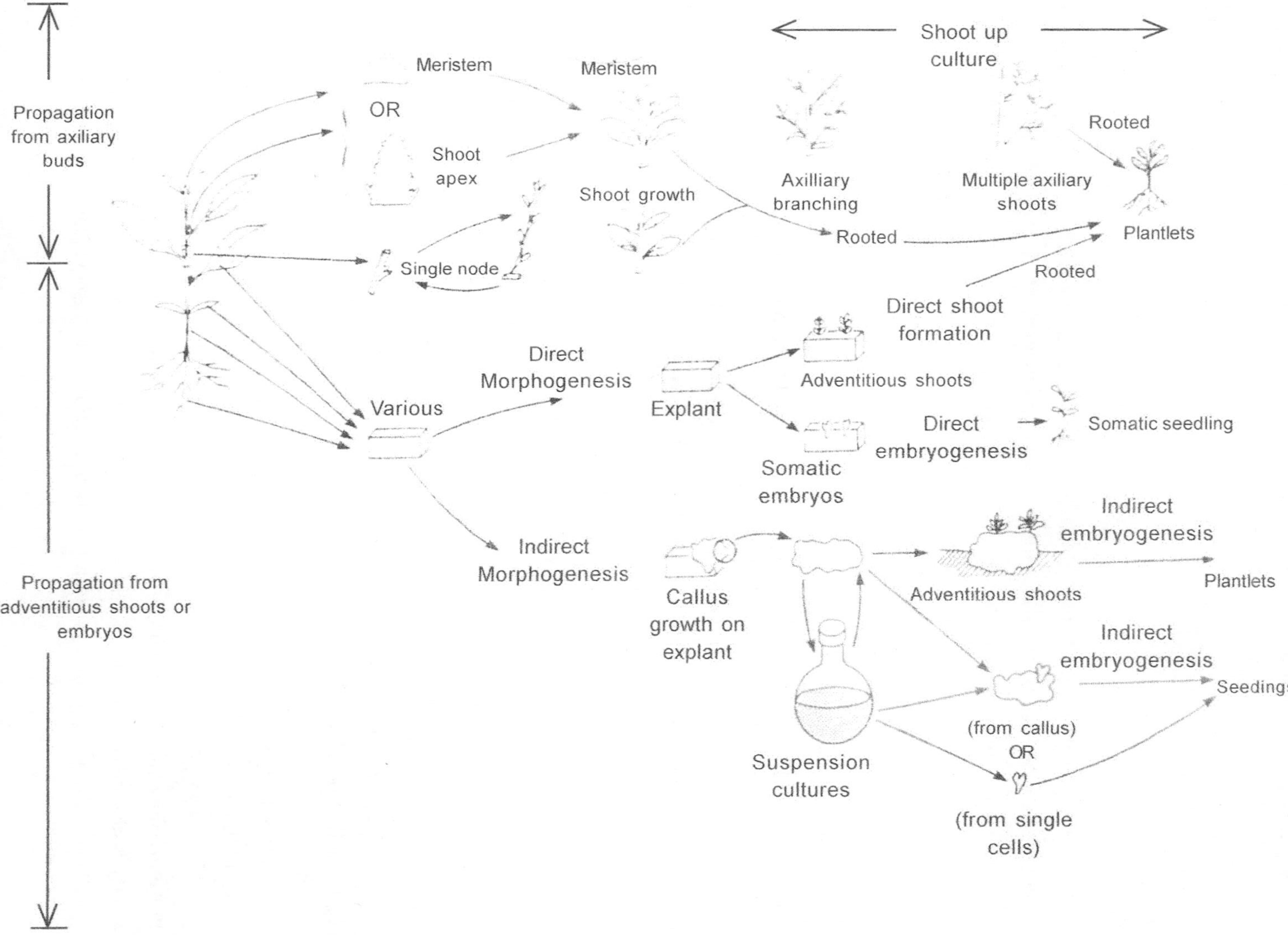

Fig. 9.3 The theoretically feasible micropropagation methods

Stages of Micropropagation

Three steps in the *in vitro* multiplication of plants have been defined by Murashige (1974). These has been widely adopted by the commercial tissue culture as well as research laboratories all over the world, since these describe not only protocols in the micropropagation process but also outlines the end points at which the cultural environment needs to be altered. Some researchers have opined that the treatment and preparation of stock plants should be considered as a distinctly numbered stage. De bergh and Maene (1981) proposed that the preparative stage should be called as stage zero (0). A fourth stage representing the transfer of regenerated plant into the external environment has also been recognised. A general description of stages 0 - II is given below. The manner in which stage, I - III might be employed to different types of micropropagation is given in Table 9.5.

Table 9.5 *Different Stages and Methods of Micropropagation of Plants*

Mode of propagation	STAGE/OPERATIONS		
	I. Culture initiation	**II. Multiplication of propagules**	**III. Transfer of plants to soil**
A. Shoot tip culture	Transfer of aseptic shoot tips/lateral buds to media (solid, liquid), and the shoot growth.	Induction of multiple shoot formation growth of shoots to a sufficient length for isolation to become starting point of another cycle of stage II plants or for passage to stage III.	Elongation of buds formed in stage II to uniform shoots. Root induction *in vitro* or externally.
(i) Shoots from floral meristems	Isolation of explants of compound floral meristems aseptically	Induction of multiple meristems to produce vegetative shoots, then as shoot tip culture.	As in the case of shoot culture (A)
(ii) multiple shoots from seed	Germination of aseptic seeds on a medium rich of in cytokinin	Induction and proliferation multiple shoots, subculturing of shoots.	As in the case of shoot tip culture (A) (Stage III)
B. Meristem culture (for virus free plants)	Transfer of tiny shoot tips (0.2 – 0.5mm) to culture. In case of heat treated plants the shoot tip length could be longer (1 - 2 mm)	Allow the shoot explants to grow to a length about 10mm then proceed to shoot tip culture. Alternatively, shoot multiplication is omitted and shoots are transferred to stage III.	As in the case of shoot tip culture (A)
C. Single Node Culture (*in vitro* layering)	As in the case of Step I of (A) Shoot tip culture but growth is prolonged till clear internodes are formed.	Induction of auxiliary buds at each internode to form a single shoot can be repeated indefinitely.	As in the case of (A) shoot tip culture
D. Direct shoot regeneration from explants	Establishing leaf, stem explants from mother plant.	The direct shoot induction on the explant without callus formation. Shoots formed are separated for the onset of another cycle of shoot tip culture.	As in the case of shoot tip culture (A)

Contd... Table 9.5

Mode of propagation	STAGE/OPERATIONS		
	I. Culture initiation	II. Multiplication of propagules	III. Transfer of plants to soil
E. Direct Embryo-genesis	Establishing suitable embryogenic tissue explants previously formed somatice embryos.	Direct embryogenesis on the explant with out callus formation	Growth of the induced somatic embryos into plants and subsequent transfer to field
F. Indirect embryo genesis from embryo gentic callus or suspension cultures	Initiation and isolation of embryogenic callus or obtaining embryo genic callus or by *donovo* induction	Subculturing of the embryo genic callus or suspension cultures and subsequent transfer to a medium conducive to embryo development	Growth of the embryo into "seedling"
G. Indirect shoot regeneration from morpho genetic callus	Induction and isolation of callus with superfecial shoot meristems.	Repeated subculture of callus explants with subsequent transfer to a shoot induction media till it attains a growth of 10mm in length.	Individual shoots are grown and rooted.
H. Storage organ formation	Isolation and culturing of tissue/organ with a potential to form storage organs	Induction of storage organ formation splitting them to start another cycle of stage III culture	Growing shoots plantlets formed from storage organ and transer to soil

Stage 0 - Selection and Preparation of Mother Plant

Before the starting of micropropagation, caution should be exercised for the selection of a mother plant typifying the variety. Furthermore, care should be taken that such selected mother plant should be free from diseases. Some times it may be necessary to pretreat the mother plant in some way to make it amenable for the *in vitro* culture successfully, steps to reduce contamination constitute a part of this stage in commercial micropropagation programme. Growth, morphogenesis, and rates of propagation can be improved by providing apt environmental condition and chemical pre-treatment of mother plants.

Stage I - Establishment of an Aseptic Culture

The first step in the micropropagation process is to establish an aseptic culture of the selected '*mother*' plant material. Success of this is assessed by two parameters : Firstly, the explants should be safely transferred to the cultural environment and secondly, that the explants should respond exhibiting apt reaction such as growth of a shoot tip or formation of callus on a stem piece. Generally a batch representing certain number is transferred simultaneously.

Stage I manipulations are regarded as successful if adequate frequency of them have survived without any contamination and are showing growth.

Stage II - The Propagule Production

The multiplication of organs and structures that are capable of giving rise to plantlets constitute the object of this phase. As shown in Fig. 9.3 these may be newly formed axillary or adventitious shoots, embryoids or miniature storage or propagative organs. Some times prior induction of meristematic zones from which adventitious organs are picked up is needed. Shoots produced in this step can be used as propagules for further culturing and multiplication to enhance their number.

Stage III : Conducive for Preparing the Conditions for Growth in Natural Environment

In this stage adequate steps are taken to grow plantlets individually so that these can syntheses their own ford by photosynthesis and survive without the supplementation of external carbon source. Some plantlets need special treatment. *In vitro* rooting of shoots is a prime necessity prior to their transfer to the soils. The propagules/buds formed in the previous stage are elongated to yield uniform shoots. These subsequently are used for the root induction *in vitro* or *extra vitrum*.

Stage IV : Transfer to the Field Condition Natural Environment (Extra vitrum)

The process of regenerated plant to the external environment (extra vitrum) constitute this phase. If not done properly, such transfer results in the loss of output of plants, because of two reasons : (i) Firstly, plantlets regenerated *in vitro* condition are often raised in high humidity and a low light intensity. Consequently there is a reduced formation of peculiar wax when compared to the plants raised in growth chambers, green houses in natural environment. Hence, tissue cultured plantlets tend to lose water quickly when transferred to external environment. Secondly, when plantlets are raised on *in vitro* on sucrose or any other carbohydrate containing media, they tend to become dependent on external media for their carbon source and donot press upon photosynthesis. A period of several days is needed to become fully autotrophic, capable of feeding themselves from elements in nature and synthesize their own food.

In practice the protocol involves the careful isolation of plantlets from stage III and washed with water gently to remove the adhered agar from the roots. The application of a thin film of anti-transpirant to leaves is desirable. Such treated plantlets are transferred to containers having a 'sterile' rooting medium (peat sand-compost) and allowed to remain for several days in an atmosphere of high humidity and reduced light intensity. The plants may require intermittent misting with water or the plants are placed inside a transparent polythene enclosure. In case of some plant species, the III stage is omitted and shoots from stage II rooted directly in high humidity, and allowed to hardened gradually to the external environment.

Presently many cash crops are mass propagated clonally under controlled conditions. The technology involves culturing of a large number of protoplasts obtained from a single plant in a bioreactor. Since the cells are diploid (somatic) containing the genetic information from both male and female parents, the plants raised from them are identical or clonal. The technique finds acceptance with high valued crops where uniformity is advantageous. Orchids, oil-palms, potatoes, sugar-cane obtained from clonal propagations are already commercialized by seed companies. The technique is highly useful where seed or planting material are likely to carry plant pathogenes. The bioreactor culturing protocols ensure the production of disease free material. Several species of disease-free economically important crop plants and plantation are now commercially sold.

The flow diagram depicting *in vitro* multiplication potatoes (Fig. 9.4) and *in vitro* micropropagation as well as tuberisation of potatoes is shown in Fig. 9.5.

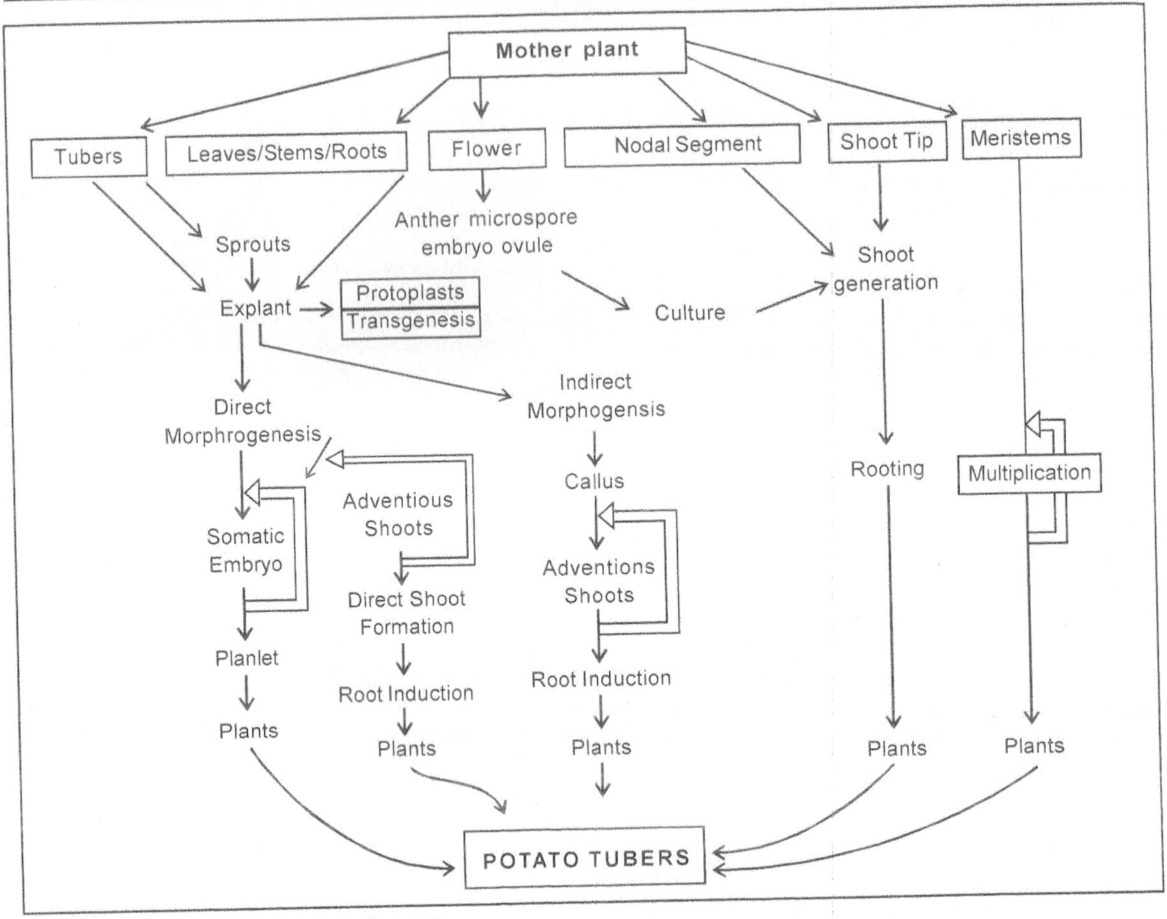

Mother plant
Tubers
Leaves/Stems/Roots
Flower
Nodal Segment
Shoot Tip
Meristems
Sprouts
Anther microspore embryo ovule
Shoot generation
Explant
Protoplasts
Transgenesis
Culture
Indirect Morphogensis
Direct Morphrogenesis
Rooting
Multiplication
Callus
Adventious Shoots
Somatic Embryo
Adventions Shoots
Direct Shoot Formation
Planlet
Root Induction
Root Induction
Plants
Plants
Plants
Plants
Plants
POTATO TUBERS

Fig. 9.4 *In Vitro* multiplication of potato

Stage 1. Propagation of nodal explants
Nodal Cuttings
In Vitro Plantlet
Stage 2 : Propagation in liquid culture using gyratory shaker
Rapid mass propagation by nodal explants
Liquid medium
Tuber induction by modifying medium
Change of environment
In Vitro tubers
Shoot culture
Stage 3 : Tuberisation

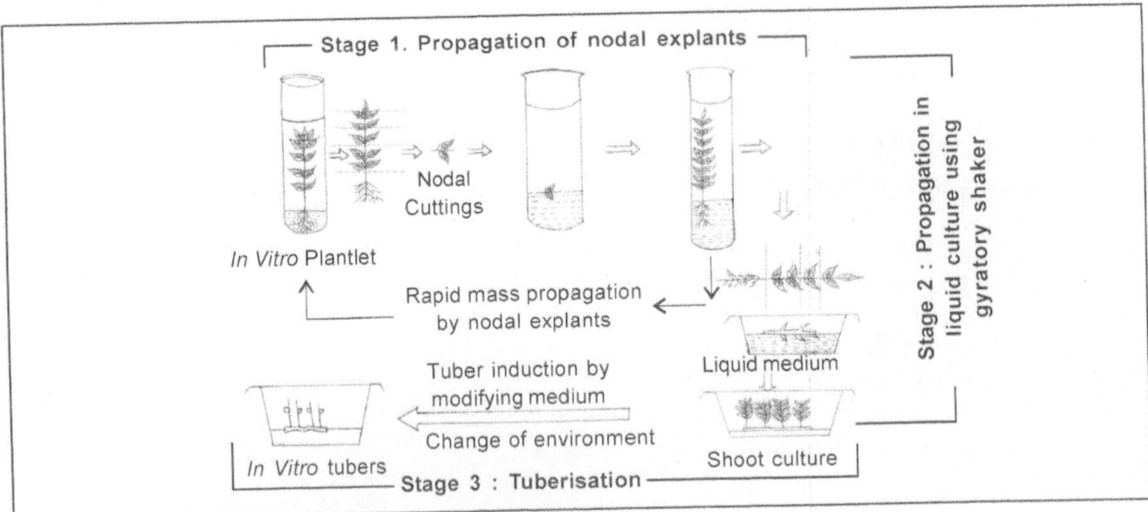

Fig. 9.5 *In Vitro* micropropagation of potato and tuberzation

9.5 Genetic Manipulation of Plants : Production of Transgenics

Traditionally the improvement of crop plants have been accomplished by crossing different plants possessing desirable traits and selecting the recombinants among the progeny depicting requisite combination of desirable traits, both agronomically as well as from the point of view of consumer satisfaction and commercial interest. Selection is possible because, in progeny generation of hybridized plants, the traits segregate as per the basic Mendelian tenets of biological inheritance. In fact the enormous developments made in crop yields including heralding of green revolution in agriculture has been made possible by conventional methods of plant genetic modifications. However, the conventional methods of plant breeding is limited by the fact that the recombination is possible only between conspecific individuals or at the best among the related species, since hybridization between distantly related species yields sterile plants or quite often no progeny at all. Another draw back of conventional crossing is that all genes, desirable as well as undesirable, from both parents get assorted in all sort of combinations and more often will result in plants with desirable traits get combined with undesirable genes. To eliminate and purify the lines, the breeder has to go through time consuming, often quite cumbersome back-crossing programme. The strategy of getting rid of undesirable traits by back cross programme often interferes with yield pre-potency gene constellations and retards the progress towards reconstructing the best variety, on which improvement is targeted. By the time one succeeds in achieving the convergence on the most widely accepted variety of a given region the time spent is so much that it becomes obsolete. Most often the progeny plants, along with desirable gene combinations will also show undesirable traits as well.

Despite the spectacular achievements of the classical crop improvement programme by hybridization and selection in segregating generation, this approach has its disadvantages. The major limitations in traditional plant breeding methodology are: barriers to interspecific gene transfer, reassortment of good and bad traits in the progeny and time factor involved in the back-crossing programme etc. However, some of these limitations can be removed by using the revolutionary methods developed in genetic engineering and plant tissue culture to obtain genetically modified crop varieties very rapidly. This modern plant breeding approach is called as molecular breeding.

Molecular plant breeding has advanced to such an extent that it is possible now to produce varieties by transferring any genes into them by breaking the barriers of species/genera/kingdom in shortest time to obtain transgenic crop plants. The present day developments in molecular biology and genetic engineering has opened up several possibilities as listed below :

- Isolation, characterization and amplification of genes encoding for traits *in vitro*.
- Addition of signal sequence for regulation and expression of isolated genes by the transcription and translation machinery of the host plant (transgenic).
- Introduction of modified genes into plant cells; and
- Regeneration of the transgenic plant cells into mature and fertile plants.

There are two main advantage of molecular plant breeding. Firstly it permits the addition of a new trait from whatever source (plants, animals and prokaryotes) to an existing most popular and valuable crop variety. Secondly, it accomplishes the genetic modifications in shortest possible time. However, this methodology is not without limitations. The current limitations of molecular or genetic modifications are: (i) not all plants can be transformed (ii) tissue culture methodologies are inadequate for many crop species. (iii) number of well characterized, isolated and identified gene and or genome sequences (genomics) is limited and (iv) the process of expression of introduced genes is not always understood and often these are silenced (gene silencing).

Gene Transfer Methods in Plants

General Concepts

Gene transfer technology supports both plant genetic engineering, where it is an imperative need and the plant sciences in general, where it is emerging as an essential tool for solving basic problems of physiology, biochemistry, developmental biology, molecular biology and genetics. The use of transgenic technology will enable the development of genetic based strategies for gene cloning, permitting the identification of connection between genotype and phenotype and will provide novel approaches for isolating genes useful for crop improvement as well as for applied research.

The revolutionary impetus for the transgenic plant technology has been provided by the development of innovative protocols in mammalian and yeast gene transfer systems coupled with the great advances made in reproductive biology in which *in vitro* fertilized ovules are implanted in a receptive uterus to develop into perfectly normal embryos. Transgenic animals are created through a process of micro-injecting the targeted gene (DNA) into the nuclei of embryonic cells at a very early stages of development. Although the first transgenic animal is chimeric for the transferred gene, when the germ line tissue arises from the transformed cell lineage, wholly transformed animals can be generated in the next generation. The development of transgenic technology in plants equivalent to what has been achieved in animal systems has been tardy and difficult because of vast diversity of plants and the very large number of different species. *In vitro* fertilisation techniques to manipulate the ovule, zygote and early embryo analogous to animals has not been accomplished in the case of plants so far. However, genetic transformation of plants involves the stable introduction of DNA sequences, usually into the nuclear genomes of the plant species using direct and indirect methods which have been standardized.

Target Plant Cells for in Vitro Genetic Modification

The successful production of transgenic plants depends upon the gene transfer protocols for its stable integration into the nuclear genome of cells capable of giving rise to a whole transformed plants. Genetic transformation without regeneration and vice versa is of no value. Theoretically there are several pathways for the recovery of transformed plants which shown in Table 9.6. Ideally any procedure that effects gene transfer without impairing the normal developmental process is important.

Table 9.6 *Paths for Recovering Transformed Plants*

Target cell/tissue/organ	Protocol of obtaining transformed plant
(a) Cell culture	Organogenesis/somatic embryogenesis via callus.
(b) Immature embryo/organ/ meristem cells	*In vitro* plant regeneration from transformed cell lineage following continued development of the embryo/organ.
(c) Cells in immature embryos/ shoots/flower meristems	Continued development of embryo/shoot/ flower meristems cells to obtain chimeric plant pollen from transformed cell lineages is employed to produce genetically modified seed via normal fertilization.
(d) Pollen	Direct production of transformed plants via fertilization with DNA - treated developing mature or germinating pollen.
(e) Zygote	Direct development of transformed plant.

Transformation Vectors

For transferring desirable gene(s) from whatever source to effect transformation, and subsequent insertion as well as stable integration of given DNA sequence (gene) into host plant needs a vector or a gene construct from which one would be able to screen and identify whether the gene is transferred. The vectors should have certain essential features. Firstly, the vectors should carry "*marker*" genes to enable the identification of transformed cells either by selection or screening. These genes depict dominance and are usually obtained from microorganisms, and placed under the control of a strong, constitutive eukaryotic promoters often of eukaryotic viral origin (Fig 9.6). A list of most popularly used marker genes of plant transformation vectors that permit easy selection of transformed cells given in Table 9.6.

Table 9.6 *List of Selectable Marker Genes Employed in Eukaryotic Vectors*

Gene	Encoding Enzyme	Agent (s) used for selection	Remarks
ble	Enzyme not known	Bleomycin	Antibiotics
dhfr	Dihydrofolate reductase	Methotrexate trimethoprim	Antibiotics
hpt	Hygromycin phosphotransferase	Hygromycin B	Antibiotics
npt II	Neomycin phosphotransferase	G 418, Kanamyc in and Neomycin	Antibiotics
als	mutant acetolactate synthase	Chlorsulfuron	Herbicides
aro A	5 - Enolpyruvylshikimate 3 - phosphate synthase	Glyphosate (Round up)	Herbicides
bar	Phosphino thricin acetyltransferase	Phosphinothrioin (Bialphos)	Herbicides

The markers that enable screening transformants are referred to as '*screenable markers*' or '*reporter*' genes, which are often developed from bacterial genes coding for easily assayed enzyme like chloramphenical acetyltransferase, β - glucuronidase, luciferase, nopaline synthase and octopine synthase. These are useful only when the comparable activities are not present in non transformed plant tissues. The list of commonly used 'reporter' genes along with their properties is given in Table 9.7.

Making Transgenic Plants Using Agrobacterium tumefaciens - Nature's (DNA) Vector

Since the turn of the century the soil bacterium *Agrobacterium tumefaciens* has been implicated in tumour (gall) formation at the site of injury caused in many dicotyledonous plants. This tumour induction is due to the occurrence of a large Ti (tumour inducing) plasmids normally present in virulent strains of *Agrobacterium* Fig. 9.6.

Table 9.7 *List of Commonly employed screenable marker (reporter) genes in plants*

Gene	Enzyme encoded	Substrates and assays
CAT	Chloromphenicol acetyl transferase	$\begin{bmatrix}14\\C\end{bmatrix}$ chloramphenicol and acetyl COA, TLC Separation of acetylated $\begin{bmatrix}14\\C\end{bmatrix}$ chlorampehicol by autoradiography
Lac Z	β - Galactosidase	As β- -glucuronidase, problem with background activity in certain species.
GVS	β - Glucoranidase	Range of substrates depending on assay. Colourimetric flourometric, and histochemical techniques available.
Lux	Luciferase / bacterial / insect	Decanal and $FMNH_2$ ATP and O_2 and luciferin Biolumiscence assays : quantitative tests on extracts or *in situ* tissue assays with activity detected by exposure of X-ray film
npt-II	Neomycinphosphoryl-transferase	Kanamycin and $\begin{bmatrix}32\\P\end{bmatrix}$ ATP *In Situ* assay on enzyme fractionated by non-denaturing PAGE; autoradiography to detect the enzyme. Quantitative dot binding assay on reaction products

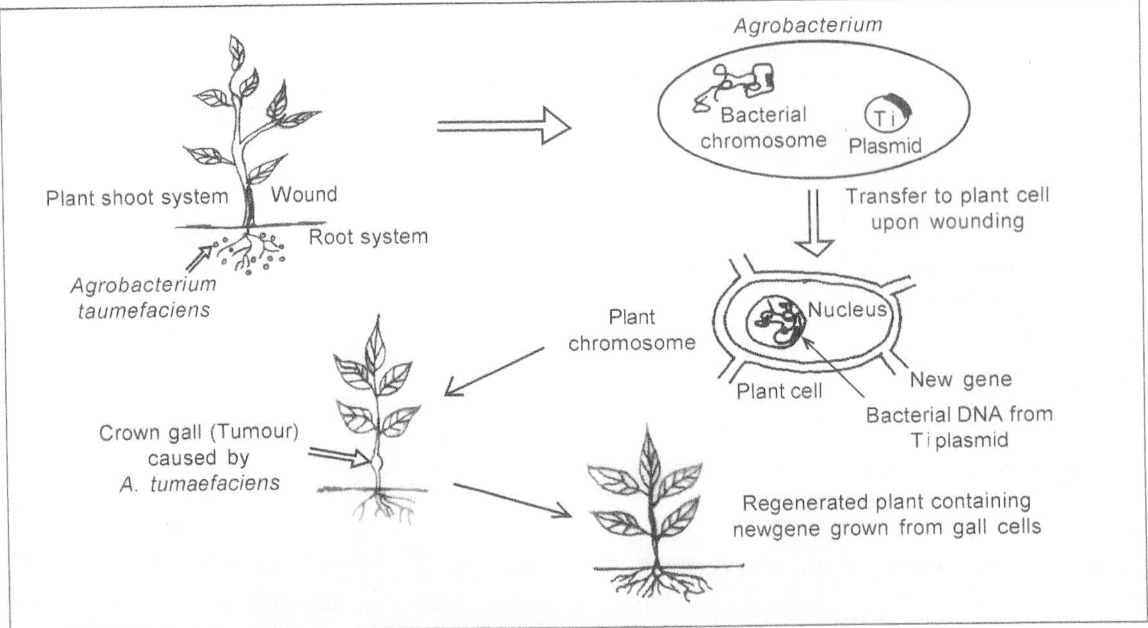

Fig. 9.6 *Agrobacterium* infection at wound sites of a dicot plant and formation of crown gall (tumour)

Agrobacterium Mediated Gene Transfer

Methods of transfer of DNA (genes) from one organism to other were known since early 1940s. Genetic recombination by transfer of genes or chromosome segments between bacteria and their plasmids, sex-factors, bacteriophages (viruses) or naked DNA through processes like conjugation, transduction and transformation were well understood prior to the heralding of the recombinant DNA (rDNA) technology. In animal biotechnology, viral medicated transfer, direct DNA uptake, and micro injection of purified DNA have been accomplished successfully and transgenic animals have been produced. In case of plants the most widely used, and well developed successful method of introducing alien DNA is by a unique system of bacterial mediated transfer. The crux of the problem is that even if one has cloned a desirable gene into a plasmid of *E. coli*, how does one gets it from the bacterial cell, breaking the barrier of the plant cell wall into the nucleus and permanently integrated into the plant genome ? Theoretically there could be two possibilities : (1) by integration of a targeted gene into a plasmid that is able to replicate in plant nuclei, or (2) by integration of targeted foreign gene into a plant chromosome which would permit the transferred gene to be passed on to the daughter cells through the normal segregation of chromosome that occurs during cell divisions (mitosis). The first approach is not feasible because plants cells do not have plasmid. The second option is possible for producing transgenics by transferring foreign DNA.

Cloning foreign gene (DNA) sequences into plants cells is not an easy task. Further more obtaining a stable integration of alien DNA sequence into plant chromosome is even more difficult. However, the plant molecular biologists have been able to harness the commonly occurring soil bacteria, *Agrobacterium tumefaciens* and *Agrobacterium rhizogenes* with their naturally occurring respective plasmids (Ti and Ri plasmids) to act as vectors for gene transfer. Several modifications of plasmids have been made to effectively transfer the genes to develop transgenic plants and have been successfully released to farmers for commercial purposes using the bacterium.

The tumours synthesize chemical substances called opines, that are derivatives of amino acids, which are not of normal occurrence in plants. These compounds are used by the *Agrobacterium* as a carbon and nitrogen source. Thus the bacterium is able to coax the plant to produce the compounds it can use for its own metabolism. This transformation is brought in the crown gall cells by the introduction of a small fragment of *Agrobacterium* DNA known as (transfer) T - DNA into genomic DNA of plant cell.

This segment of DNA (T-DNA) of bacteria contains the genetic information for the biosynthesis of opines as well as for tumour induction. The transfer of genetic information from bacteria to plants is unique in nature and has been discovered in late 1970s. These discoveries became basis for the development of efficient protocols for *Agrobacterium* mediated gene transfer by genetic transformation methodology.

The mechanism of crown gall formation or hairy root disease is identical in the sense that both of them are transformed by a piece of plasmid DNA originating from tumour inducing bacteria (*A. tumefaciens*) or from the hairy - root induction bacteria (*A. rhizogenes*). These two bacteria have large plasmids (about 100, 000 bp) consisting of three distinct sets of gene sequences for virulence, for transfer DNA and for opine synthesis.

Plant tumours (crown galls) are formed due to the expression of oncogenes (onc) present on T-DNA, the region of the T plasmid which is transferred to the plant cell upon infection. The T-DNA segment about 25,000 bp of DNA consisting of genetic information for three genes coding for isopentyl transferease (ipt), tryptophan mono oxygenase (iaa M) and indolacetamide hydrolase (iaaH). These code for enzymes for the synthesis of plant hormones *viz*, isopentenyl AMP (a cytokinin) and indolacetic acid (an auxin). Because of the ability to synthesis these hormone by transformed cells, they are able to grow in the absence of these hormones in the culture medium, while the normal cells require them (Fig. 9.7 - Fig. 9.9).

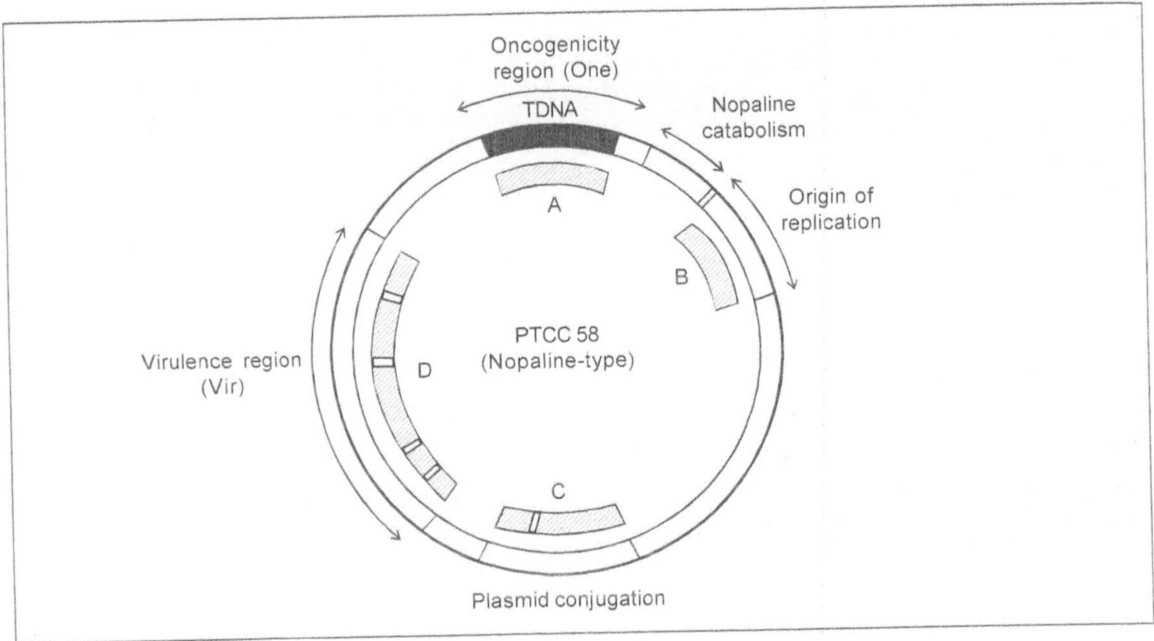

Fig. 9.7 General feature of Ti plasmid; T-DNA, region transferred to plant genome in crown gall cells, A, B, C and D region of homology common to all Ti plasmids

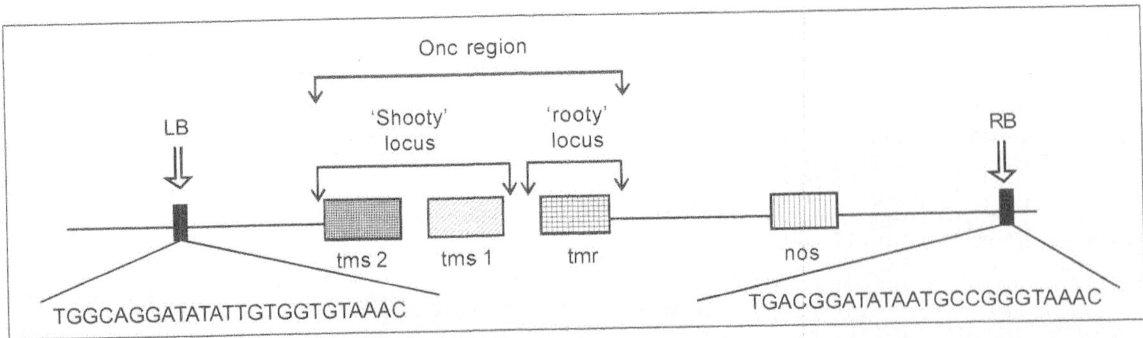

Fig. 9.8 Generalized structure of T - DNA region; turns 2 and turns 1 m mutants exhibiting attenuated oncogenicity and/or induced tumours with altered shooty morphology; tmr, tumour altered with 'roots' morphology

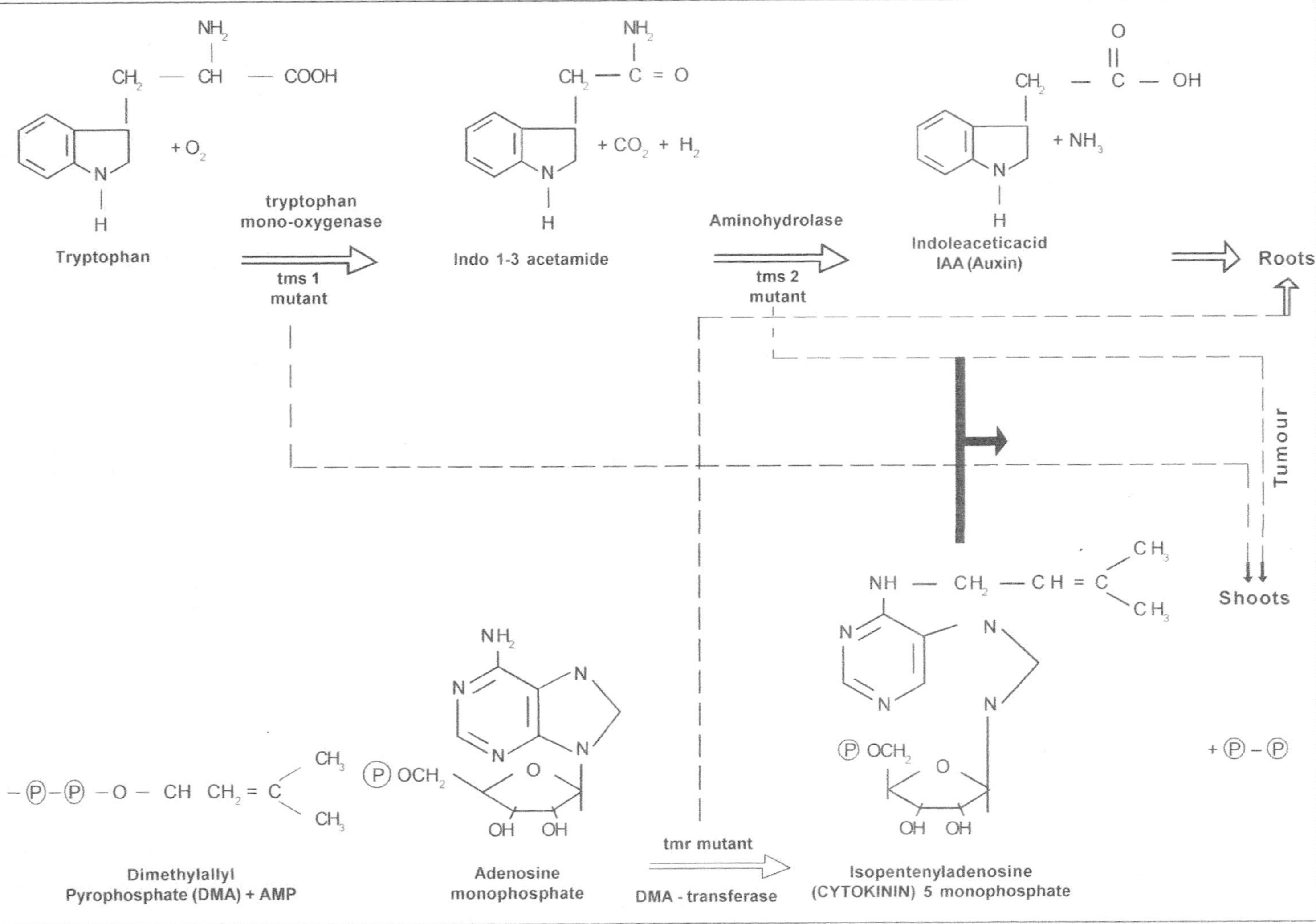

Fig. 9.9 Biosynthetic pathway for auxin and cytokinin encoded by the T-DNA in crown gall tissue and expression of tumour modification and / or tumour induction mutants

In the case of *A. rhizogenes*, the genes present on Ri plasmid of the bacteria are causative factors for the transformation of normal cells to hairy roots. Four root locus genes are prevalent and have been named as A, B, C and D. These genes probably increase the sensitivity of the transformed cells to plant hormone / auxin thus making hairy roots. The hairy roots can be regenerated into plants readily, hence, *A. rhizogenes* transfer system is used to study the gene expression of foreign genes in transgenic plants.

In addition to the *onc* genes, the T-DNA also has genes encoding enzymes involved in the production of tumour specific metabolites called opines like, octopine, nopaline, leucinopine, agropine and mannopine which are nothing but condensates of aminoacids and sugars. There are other substances like agrocinopines which are phosphorylated sugar derivatives. Chemical structure of main opines is given in Fig. 9.10.

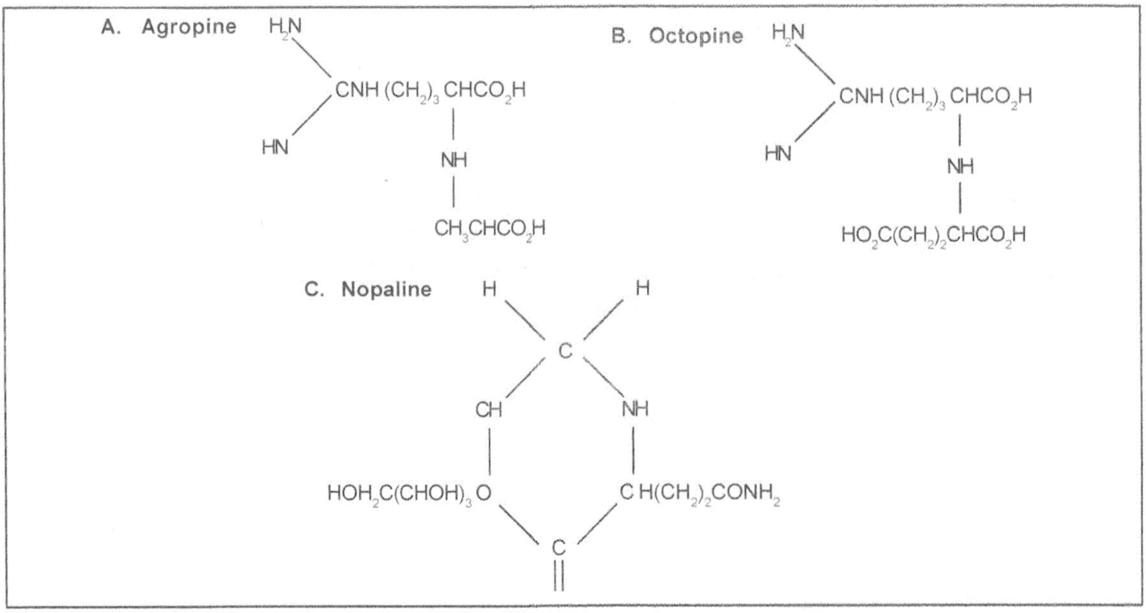

A. Agropine H_2N / HN — $CNH(CH_2)_3CHCO_2H$ — NH — CH_3CHCO_2H

B. Octopine H_2N / HN — $CNH(CH_2)_3CHCO_2H$ — NH — $HO_2C(CH_2)_2CHCO_2H$

C. Nopaline H, H, C, CH, NH, $HOH_2C(CHOH)_3O$, $CH(CH_2)_2CONH_2$, C

Fig. 9.10 Structures of three principal opines

The transformed plant cells excrete opines that are utilised by *Agrobacterium* because of enzymes for catabolism are present on the Ti and Ri plasmids. These two plasmids can be distinguished by their presence of catabolic/anabolic genes they carry. Ti plasmid is identified by the presence of enzyme that metabolise octopine, nopaline, and leucinopine while Ri plasmid by virtue of genes metabolizing agropine and mannopine. In nature, *Agrobacterium* transfers T-DNA segment into plant cells for the synthesis of opines which could be used by the bacterium for its metabolism. In other words these bacteria have evolved a mechanism by which the plant cells are transformed to manufacture substrates needed for their growth, development and survival.

Modus Operandi of the Transfer of DNA from Bacterial Plasmid to Plant Cells

In the induction of crown galls (tumours) in plants, in addition to '*onc*' genes several other genes present on Ti (R_1) plasmid as well as on the bacterial chromosomes have a role to play. The essential first step in this direction is to effect the transfer T-DNA from plasmids to plant cells. Two sets of genes, vir present on the Ti plasmid and present on bacterial chromosome (chv = the chromosomal virulene) are *sin que non* for the transfer of T-DNA segment from the plasmid to plant cells. The roles of these genes is illustrated in Fig. 9.11.

The Role of Chromosomal Virulence Genes (chv)

The chromosomal virulence genes (*chv*) of *Agrobacterium* has two genes *chv* A and *chv* B; *chv* B gene encodes for a protein necessary for the formation of a cyclic β - 1, 2 glucon, whereas *chv* A gene determines a transport protein located in the membrane that is necessary for the transport of β, 1-2 glucon into the interspace between the cell wall and plasma membrane (periplasm). Thus β, 1-2 glucon has a vital role to play in facilitating the attachment of bacteria to cell wall without which the T-DNA of plasmid origin cannot enter into the plant cell and subsequently integrated into its genome.

The Roles of Plasmid 'vir' Genes

There are many vir genes on the plasmids whose protein products are quite essential for the actual transfer of the T-DNA from bacterial plasmid to the plant genome. Their main role is cutting the T-DNA region and transport over the bacterial cell membrane into the plant cell across cytoplasm into the nucleus and finally integrate the T-DNA into plant genome (chromosome complement). There are as many as 22 *vir* genes are present, which are under the control of a set of seven operons designated as *vir* A to *vir* G. Each one of them some times code one or more than one protein. For instance vir B operon has eleven open reading frame sequences (i.e code for proteins), while vir A operon has only one. These operons and genes under their control are activated by phenolic compound that are often released at the woundings site of the plants. Acetosyringone or hydroxy acetosyringone are some of the common phenolics that help in *Agrobacterium* and plant interaction. Two proteins '*vir*' A and *vir* G, coded by two genes of the virulence region are activated by plant phenolics. The *vir* A proteins is situated in the membrance of bacteria and helps signal transduction and consequently sense the presence of phenolics that induce the operons for action. The resultant *vir* A protein might phosphorylate the *vir* G protein which is present in the cytoplasm of bacterial cell. The modified (phosphorylated) *vir* G proteins then bind to the promoter regions of other plasmid *vir* genes thus inducing them for their expression.

It is not clear about the role of other *vir* genes. However the *vir* B operon under its control has eleven open reading frames (ORFs) and hence may code for 11 different proteins. The predicted amino acid sequence present us with a clue that these proteins might be involved in the transportation function and are localized in the periplasm of the bacterial membrane. These might be quite essential for T-DNA transfer and might determine the pilus or a pore in the bacterial membrane that makes the transfer of T-DNA segment possible. The role of one of the *vir* C proteins and two of the *vir* D proteins have been shown to be involved in the actual transfer of T-DNA Fig. 9.11. There are also other accessary gene products of the plasmid *vir* region that may determine the host range of the *Agrobacterium*, beside chromosomal genes. In nature this is quite essential to select ideal host for the production of opines.

Molecular Mechanism of T-DNA Transfer

No other parts of the Ti or Ri plasmid of *Agrobacterium*, excepting T-DNA, gets integrated into the plant cell genome. It is really amazing that no physical linkage of T-DNA and plasmid is needed for

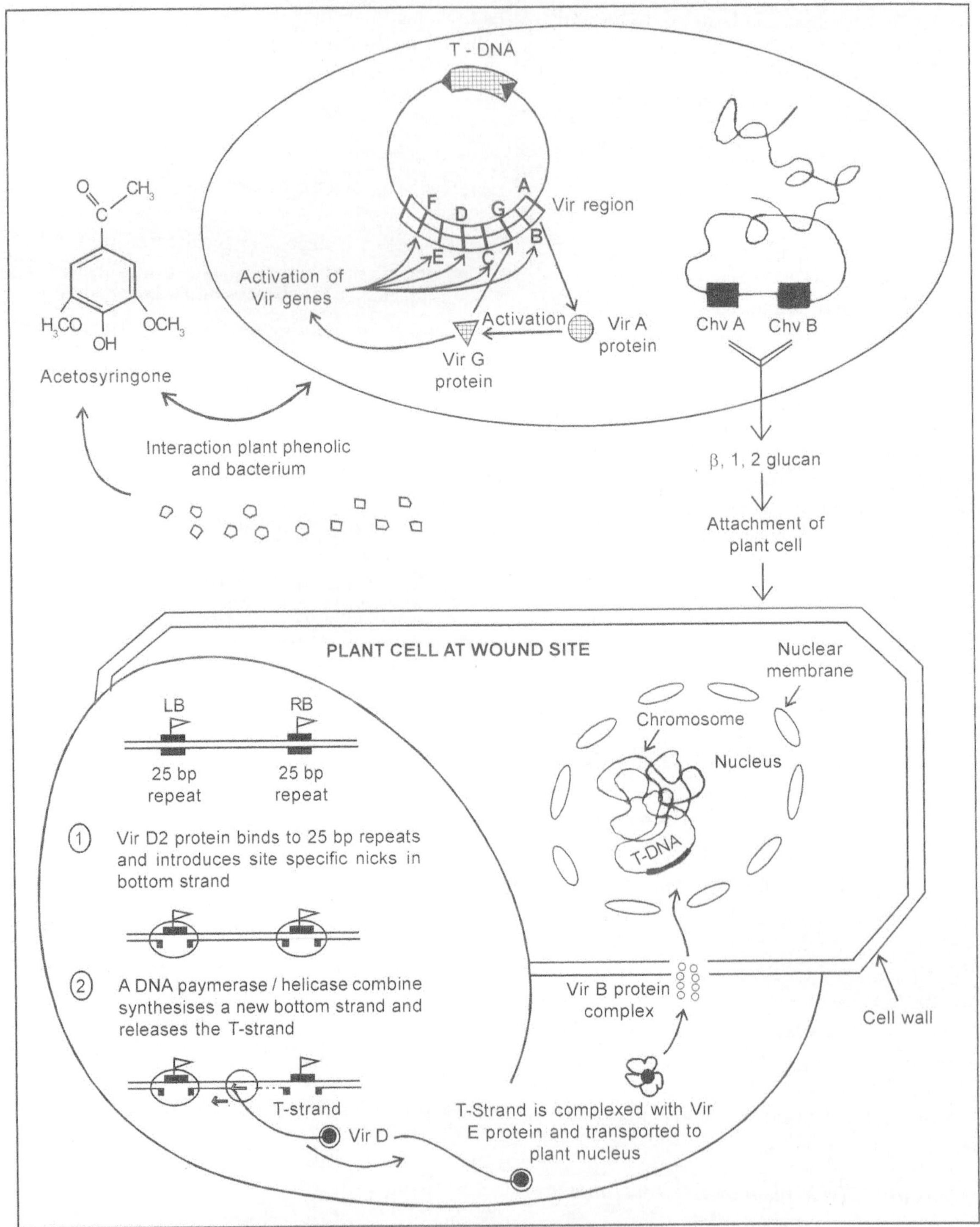

Fig. 9.11 The role of *vir genes* in the transposition of T-DNA from bacterium to plant genome

the T-DNA transfer to occur. Another bizzare feature is that none of the genes of T-DNA are involved in the transfer. Only the genes residing in the *vir* region of plasmid and chromosome (*chv*) have an exclusive role in the process of T-DNA transfer. Nonetheless, the 24 bp border sequences of T-DNA (LB and RB) are essential for the transfer and plays a vital role as recognition and signalling for the transfer apparatus to be built up. The essential steps as we understand today, in the molecular mechanism of transfer of T-DNA and its integration into the genomes of plant cells are : phenolic compounds, like acetosyringone, often produced due to injury to plant cell induces the *vir* regions to produce a battery of proteins; subsequently single stranded nicks are caused in the double strand of the border sequences of T-DNA region. The T-strand, now single stranded molecule, possessing the information of T-DNA can be found in the bacterial cell. The process of induction of nicks as well as the formation of single stranded DNA is mediated by the action of two proteins, *vir* D1 and *vir* D2 encoded by the *vir* D operon of the *vir* region of the Ti - plasmid. These two protein molecules conjointly act as endonucleases causing the cuts in the border sequences (LB and RB) of T-DNA at a specific sites very precisely. The resultant nicks might provide a signal as well as for starting the repair synthesis of new bottom strand of T-DNA in 5'-3' direction by and consequent release of the T-strand. The endogenons DNA polymerase / helicase. The release of T. strand is achieved by a second nick at 3' side. The process of transporting and introducing into the plant cell is catalysed by *vir* D2 protein which attaches to the 5' terminus of single strand T-DNA covalently and another protein *vir* E2 also binds the T-strand and coats it to make a long thin nucleoprotein filament, presumably confer protection to the DNA fragment against the DNA degrading enzymes present in the plant as well as the bacterial cells. *Chv* - A and *Chv* - B genes synthesize proteins which are involved in making a pore or pilus in the membrane to enable the transport of T-DNA nucleo-protein complex and helps in attachment of bacteria to plant cell.

Host range of Agrobacterium

In nature *Agrobacterium* infection is generally restricted to dicotyledonous plants. Plant species such as tobacco, tomato, potato, petunia, rape seed, cauliflower, lettuce, sunflower, chrysanthemum, rose, apple and many more have been experimentally used for the *Agrobacterium* mediated tumour induction (transformation). However, most of the monocotyledonous crop species like wheat, rice, maize and grasses are recalcitrant and do not respond favourably in the induction of tumour upon infection with the bacterium. One possible reason being that these plants may not possibly have enough right kind of phenolics for the induction of *vir* genes to permit the T-DNA transfer by the bacterium.

Perhaps there exists a basic difference between di and monocotyledonous plant species the way each category of plants respond to the wound healing process. In dicotyledons injury to plant tissue leads to the process of dedifferentiation by which the cells go into embryonic mode and forms the callus at the site of injury. These callus cells isolated from the injury site possess capacity for the regeneration into new plant (differentiation). Singularly it is this characteristic that is exploited in plant tissue culture. Monocotyledons generally lack the capacity for the formation of callus tissue (dedifferentiation) surrounding the wounded site. Hence, these are not ideally suitable for the plant regeneration. In tissue culture of monocotyledonous plants usually the immature or meristematic tissues that still retain the capacity to differentiate and go through cell division are used. Thus in the case of monocotyledons, even if the T-DNA integrates, the transformed cells cannot differentiate (regenerate) into a plant. This recalcitrance comes in the way of genetic manipulation of crop species of vital importance. However, these hinderances have been over come, and transgenics are being produced in many plants by modifying and developing the plasmid specifically suitable for mnocotyledonous crop species.

T_i *Plasmid Based Vector Systems for Gene Transfer to Plant Cells*

Since LB and RB repeat 24 bp sequences are needed, plant molecular biologists have developed useful vectors for transferring genes into plant cells (Fig. 9.12-13). These vector systems are of two kinds (i) In *Cis* system of plasmid design, the *vir* region and the genes to be transferred, which are inserted artificially into T-DNA are part of the same Ti plasmid. This kind of physical linkage in the same plasmid construct is called *cis* condition and hence this category of plasmids are called *cis* plasmids. In the case of (2) trans or binary vector system the genes to be transferred are cloned in between border sequences into a plasmid, which have competence to replicate in *E. coli* as well as in *Agrobacterium*. This facilitate to take up manipulations that can be handle easily in *E. coli*, which subsequently, with the genes to be transferred can be conjugated with the *Agrobacterium*. This *Agrobcaterium* contains a second plasmid containing *vir* region. In other words the alien genes to be transferred and the genes responsible for the transformation are not physically linked but located on two different plasmids (trans condition). Since, many of *onc*-genes responsible for tumour induction are removed and replaced by the gene construct that has to be transferred to plant cells, these are called *disarmed vectors*. Both these systems (cis and trans) have been successfully used in gene manipulation for the production of transgenics (Fig. 9.12 - 9.13(a)).

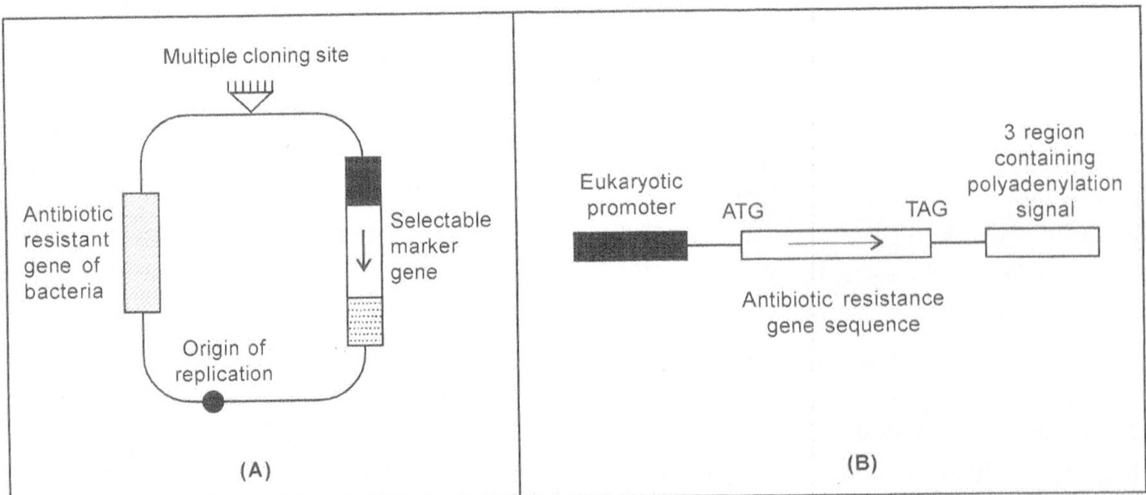

Fig. 9.12 (A) Generalized vector construct for the transformation of plant cells (B) Details of the organization of gene construct with promoter, selectable marker and termination signal sequence

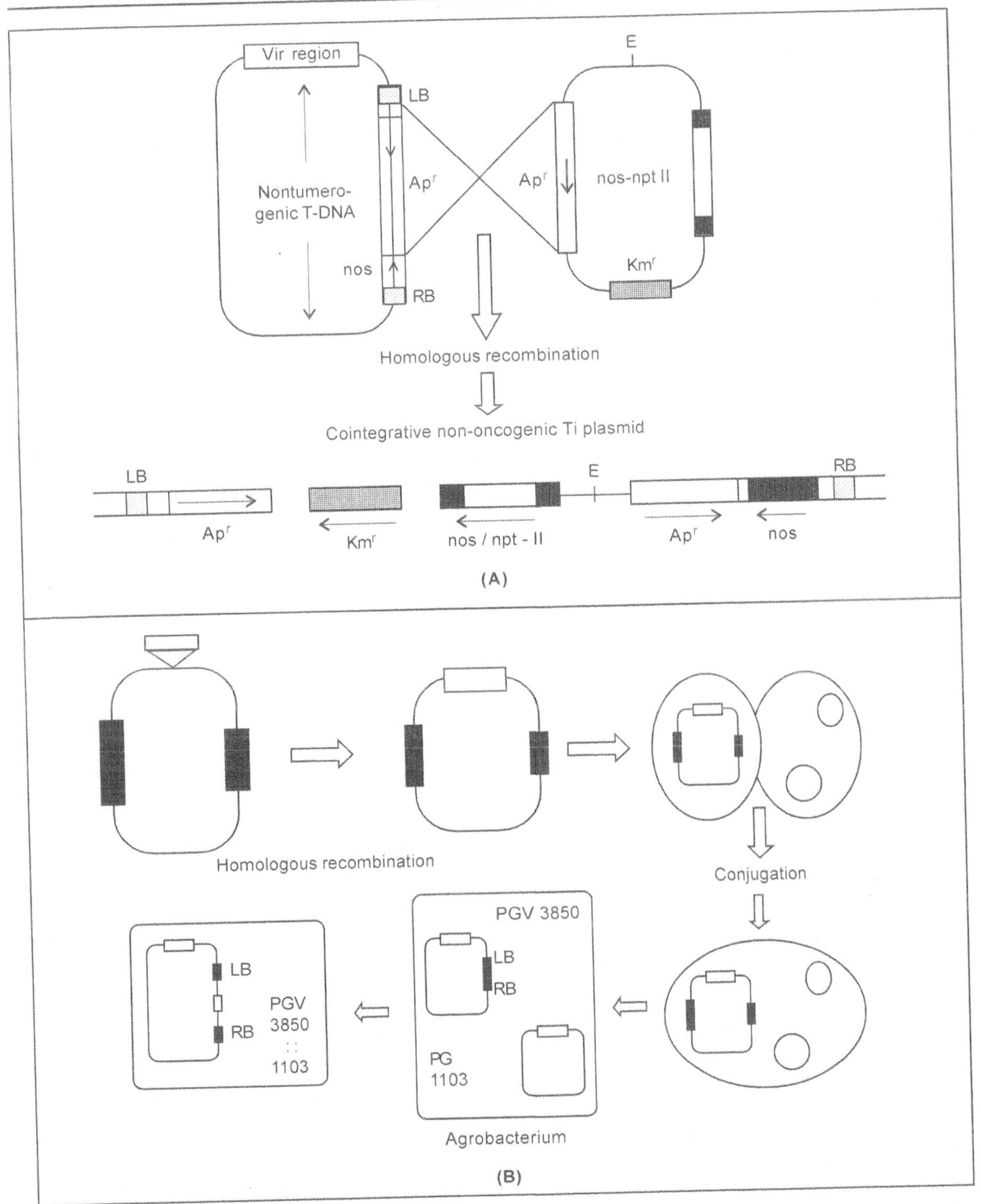

Fig. 9.13(a) (A) Cointegrate vector (non-oncogenic) with an intermediate vector (B) Preparation of a cointegrative vector for the transfer of a desired gene sequence

The gene construct (the artificial T-DNA) used in these vector systems comprises of minimally the two border sequences between which sandwitched are selectable marker gene, desired gene (the gene under transfer) and reporter gene. All *onc* genes involved in tumour formation including those responsible for opine synthesis are removed (disarmed) since these are unneccesary for the transfer of foreign gene. The plant cells transformed this way are normal and can be regenerated normally to produce a normal looking plants. The requirement for selectable marker gene is satisfied by the antibiotic resistant genes. Early in 1980s a gene of bacterial origin coding for an enzyme neomycin phosphotransferase (NPT) has been transferred to plant cells that inactivate the antibiotic knanamycin. Since the plant cells are susceptible to kanamycin and only those transformed and possess kanamycin resistant gene will survive, thus serving as a marker for selecting the transformed cells.

The requirement for recognition of signals for the expression of marker and reporter genes in plant cells has been accomplished by the use of chimeric gene consisting of promoter and terminator sequences of original opine synthesis derived from Ti plasmid T-DNA region and fused to NPT gene for synthesizing the enzyme in these chimeric genes. These engineered genes were active in plants conferring the antibiotic resistance, with which the transformed cells can be selected.

Importance of Promoters for the Programmed Expression of Alien Genes

Both for animal and plant gene expression certain sequences called as promoters at 5' end of the gene construct are quite essential. The expression of desired gene can be directed to limit to certain tissues like leaves or roots or tubers of flowers or seed. The promoter expression also can be regulated specifically by the stimulous of extraneous factors like light, heat frost, injury, pathogen attack etc. The isolation and characterization of specific promoter with requisite properties accomoplishes the possibility of expression of foreign genes limited to specific tissues or conditions or during specific developmental stages. This kind of regulated expression will be very attractive proposition and a boon to molecular breeders for a strategy, can be evolved to develop strains resistant to pathogens to express in leaves, but not in seed or fruit, so that the edible quality of fruits or seed is kept intact. Hence a great deal of research gone into the development of ideal of research gone into the development of ideal promoters rather than earlier developed nopaline synthase gene promoters. Several promoters such as cauliflower mosaic virus (CFMV), 35s RNA gene promoter (35s promoter), 2'-promoter derirved from the T-DNA of the Ti plasmid etc, have been developed. The 35s promoter is expressed in all tissues but not in all cell types of all tissues of transformed plants. The uniqueness of 2' promoter is that it is expressed normally in roots, but in case of injury to cells it also expresses in cells adjacent to the wounded site. This kind of promoter is aptly suitable for the expression of genes toxic to pathogens so that host (tansgenic) plants are protected.

Another promoter of this gene, derived from proteinase inhibitor II gene of potato is also ideally suitable for the expression in wounded cells/tissues, since this gene is silent in leaves but is systematically gets inducted throughout the plant upon injury to even a single leaf.

To protect the plants from the extremes of environment (abiotic) factors such as high temperature, drought, salinity, alkalinity, and frost; specific promoters have been designed. Heat shock promoters are one such category, that induces the expression of genes upon a heat shock. The heat shock proteins are generally confer protection to plants and animals against the damage to high temperature. The heat shock promoters specifically are able to induce the expression of genes under their control with in one hour upon heat shock.

Another class of promoters that are of light driven have been identified under whose tutelage a large number of genes are expressed in leaves inducted by the light. The 5' promoter sequences of nuclear genes involved in the coding for example the small subunit of the RUBISCO gene (ribulose 1,5 - biphosphate carboilase) or the gene necessary for encoding chlorphyll a/b protein for the harvesting of light etc., confer light regulated expression. These sequences are small segment of DNA about few hundred base pair length present up stream 5' region of the gene. Some of them have been depicted to express as light dependent enhancers of gene expression in the leaves and remain silent in roots. Chalcone synthase constitute yet another type of light regulated gene. Using its 5 upstream sequences, it has been possible to construct chimeric genes that would be specifically silent in transgenic plants but these would get expressed in plants exposed to UV - rays for 20 hours or so.

Besides several organ/species specific regulation of gene expression has also been developed for a number of genes. For instance, the patatin gene is solely expressed in potato tubers and the phaseolin gene only in seed of *Phaseolus vulgaris*. Chimeric genes consisting of the promoters of such genes have been developed that have strict organ specific expression of reporter genes. Further more, genes have been isolated that express only in specific parts of the flower, making flower specific expression of alien genes feasible. Thus the genetic engineering techniques enabled the isolation of regulating gene sequences that express specifically and it is possible to direct alien proteins to a predefined organ/tissue of plants or alternatively get expressed only under certain environmental conditions, biotic or abiotic.

Methods of Transformation Using of Agrobacterium

Based on the property of *Agrobacterium* infecting injured cells and transferring its DNA several efficient protocols have been developed to transfer isolated genes ligated disarmed T-DNA of plasmid into the nuclear genome of plant cells. These methods are solely based on the knowledge about the feasibility of regenerating certain plant tissues to plants. Somatic plants are totipotent in principle and new plants can be regenerated. Any explant (peace of leaf tissue), when incubated in the presence of plant growth hormone like auxin and cytokinin can regenerate to shoots which can be rooted and grown to mature, fertile plants. Since explants are wounded due to cutting procedure these are ideally suited for transformation. It was demonstrated as early as 1980 that simple incubation with *Agrobacterium* resulted into gall formation at wound site and therefore transfer of T-DNA took place. Subsequently even the engineered plasmids incorporated with alien gene into disarmed T-DNA were also transferred incubating with bacterium and the transformed plant cells tissue recovered using the antibiotic resistance property, which subsequentially regenerated into a plantlet. After co - cultivation the bacterias are killed with different antibiotic to enable the culturing of transformed cells aseptically. Any agronomically useful gene such as resistant to biotic and abiotic factors or any quality traits etc. can be linked to the antibiotic resistant gene and following the *in vitro* protocols of regenerating could be followed to get a produce a transgenic plants of desired type (Fig. 9.13 (b) - 9.15).

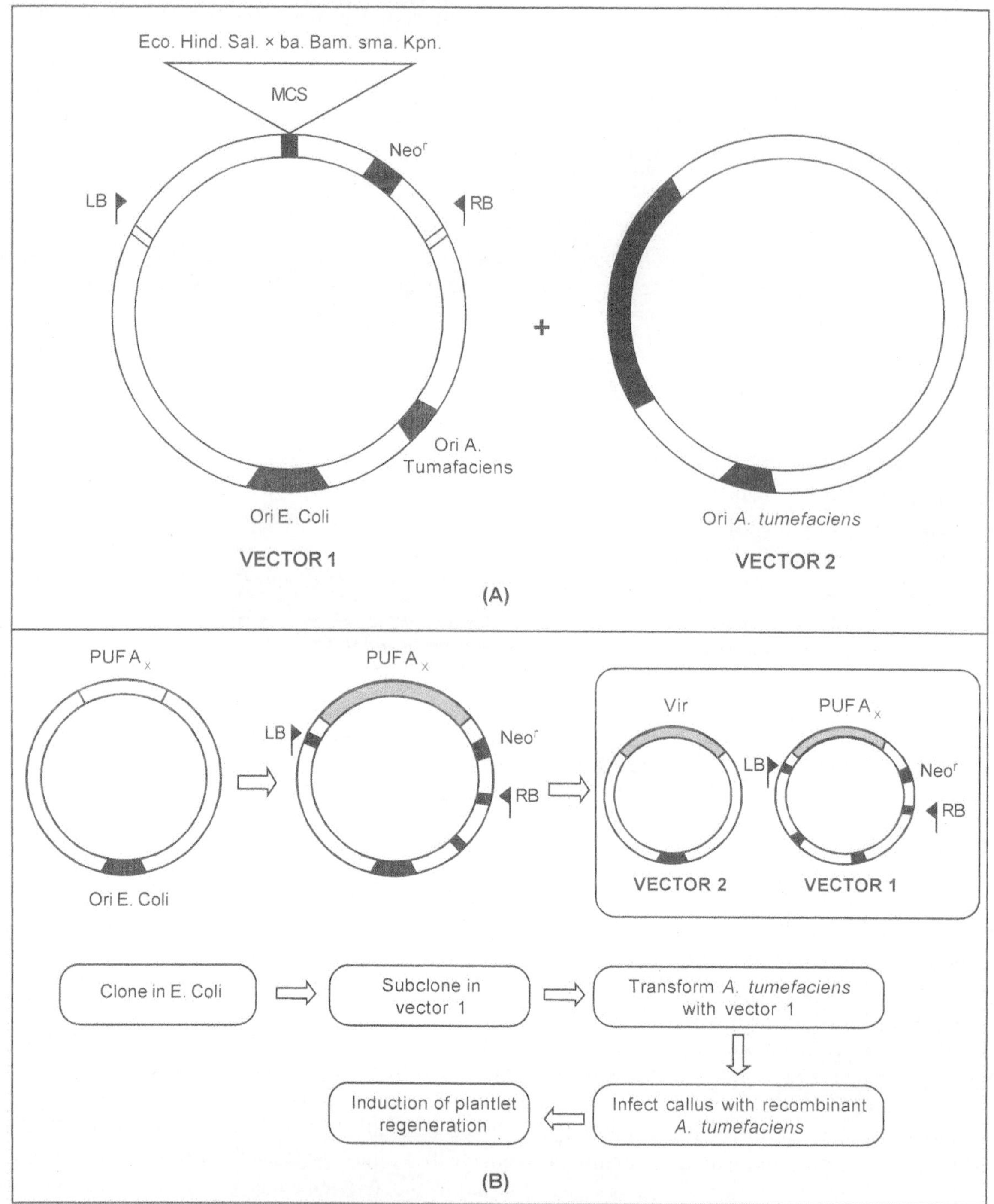

Fig. 9.13 (b) The binary vector system, derived from Ti plasmid B. The use of binary vector system of *A. tumefaciens* for plant transformation (MCS = Multiple cloning restriction endonuclease sites, LB = left border, RB = right border, NEO = neomycin resistance)

Direct Gene Transfer

The other methods of genetic transformation like gene transfer by PEG - directed DNA uptake and electroporation, micro and macro-injection of DNA, and biolistics have been developed which are ideally suited for monocotyledonous plants that are not susceptible to transformations by *Agrobacterium*. An outline of these protocols are shown in Fig. 9.14.

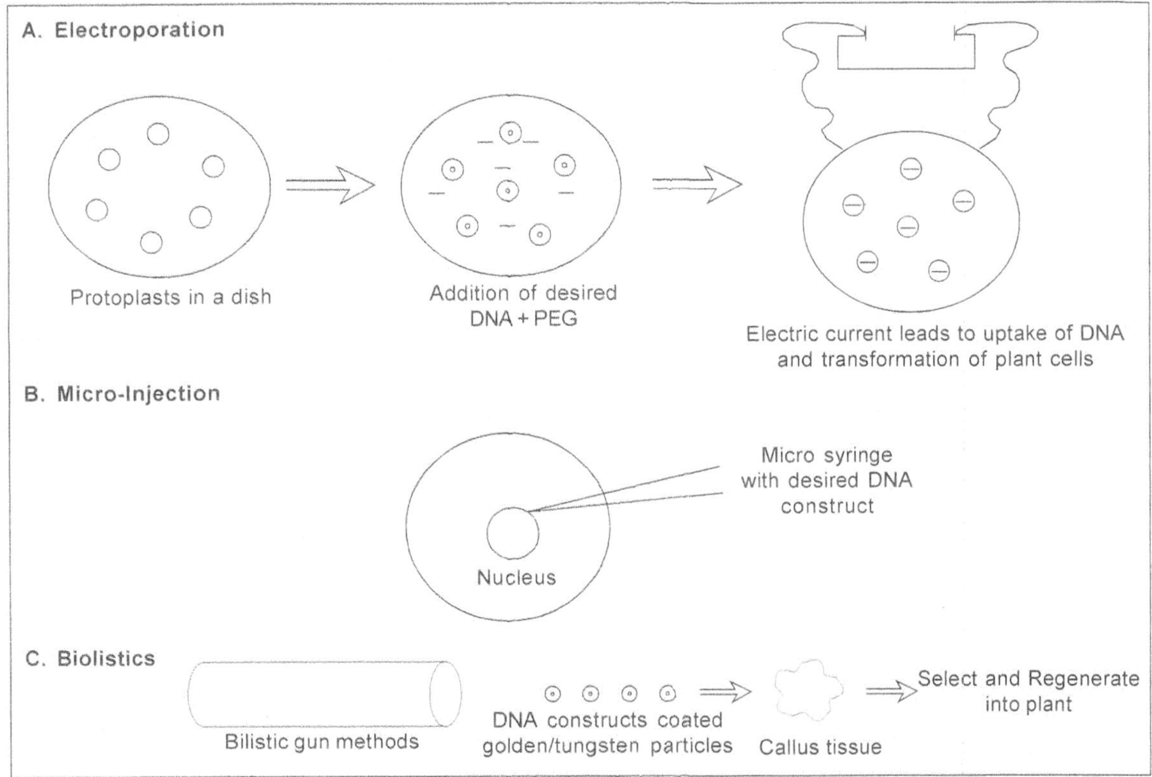

Fig. 9.14 Direct gene transfer methods alternative to the use of *Agrobacterium* to transform plants

9.6 Strategies Towards Food and Nutritional Security in 21st Century - GM Foods and Crops

An outstanding progress has been made in the last millennium both in developed and developing countries to challenge the poverty, hunger, disease, and ignorance to a certain extent. One of the greatest achievement has been the phenomenal increment in the productivity of crop plants consequent to research based agriculture, that has fed millions of people and has become the basis of economic transformation in many underdeveloped countries especially in the Indian subcontinent. This heralding of '*Green revolution*" has prevented dire prediction of famine and pestilence in Asia. However, much more has to be accomplished since poverty still haunts the people especially in developing countries limiting their access to food. Consequently, millions of people are malnourished. Increasing population puts a tremendous pressure on food security in future. As per the estimate of Norman Borlaug, a Noble laureate, to meet the food demands by 2025, average cereal yield must increase nearly 100% over the 1990 average. It is doubtful that conventional plant breeding approach will be able to accomplish this and continue to provide the food security to the people.

Plant biotechnology is receiving increased attention towards the goal of providing nutritional security in 21st century, because it is a source of innovation for the agriculture, horticulture, food, chemical and pharmaceutical industries. It provides the means to translate the knowledge of gene cloning technology and genomics to alter or genetically modify the fundamental process of plant reproduction, growth and development into new products and enhanced productivity. Thus the transgenic plants can contribute to future food through the revolutionary manipulation of genes in the form of production of genetically modifing food crops. The status of future demand for various food of India is presented in Table 9.9.

Table 9.9 *Status and future demands (in m. tons) of different food crops in India.*

Food	Status during		Demand by 2030	
	1960	2000	LIG	HIG
Caloried Food :				
Cereals	70.0	180.06	239.24	233.7
Potato	2.7	19.90	30.76	34.4
Protein Food :				
Pulses	12.0	16.6	24.12	26.3
Milk	20.4	76.93	130.50	152.0
Meat	-	5.34	10.12	14.0
Fish	<1.0	5.57	10.44	14.0
Egg	~0.6	1.88	3.57	5.0
Oil Food :				
Edible oil	2.3	8.15	11.87	12.0
Other Foods :				
Vegetables	-	89.39	150.82	193.0
Fruits	-	47.69	84.34	106.0

LIG - Low Income Group; **HIG** = High Income Group

Global Status of GM Plants

Biotechnology in the area of agriculture is moving very fast. More traits and more land is being planted with GM varieties of an ever expanding number of crops. Billions of US $ are being invested by companies in consolidation to ensure access to these rapidly growing market, while investing billions for R & D. Public Debate the GM Crops - Biotechnology is being commercialized very fast. In 1996, 1.7 million ha were planted with GM crops. This rose to 11 million ha in 1997 and about 28 million ha in 1998 (Table 9.10). In the first three years of commercialization (1996-98) eight countries have contributed to more than 15 fold increase in the area covered by transgenic crops globally. During 1997, the number of countries growing commercially transgenic crop were five viz., USA, Argentina, Canada, Australia, and Mexico. In 1998 this has increased to eight, when South Africa, Spain and France grew TG crops for the first time.

Table 9.10 *Global area of Transgenic Crops (in millions of Acres)*

Year	Area under cultivation (× 10^6 Acres)		Total area (× 10^6 Acres)
	Developing countries	Developed countries	Total
1996	-	1.7	1.7
1997	1.5	9.5	11.0
1998	4.4	23.4	27.8

The main transgenic crops which were commercialized include Soybean (52), Corn (30), Cotton (9), Canola (9%) and Potato (≃ %). Transgenic have been produced for herbicides tolerance (71%), insect resistance (28%), and quality traits (1%). Research, however is being done to evolve GM plants that are of high economic value such as cereals, fruits, floriculture and horticultural species Table 9.11.

Table 9.11 *Transgenic crops released and the characters incorporated*

Crops Species	No. of field releases	Introducted trait					
		Herbicide resistance	Insect resis-tance	Viral resis-tance	Improved product quality	Others	% of transgenic crops releases
Maize (*Zea mays*)	46	29	16	1	3	5	30.0
Soybean (*Glycine max*)	27	25	-	1	1	-	52.0
Cotton (*Gossypium*)	24	16	15	-	-	-	-
Tomato (*Lycopersicon*)	19	16	15	-	-	-	-
Potato (*Solanum tuberosum*)	13	-	1	16	-	-	≃ 1.0
Other species	30	-	-	-	-	-	17.0
Total	159	70	34	9	20	11	100.0

Some of the current applications of genetically modified organisms by transgenic technology include :

* Crop with improved agro-botanical traits like, less need for agrochemicals, thrive marginal and resistant to pests and diseases, give increased yields, (even today, an estimated 30% of the food grown is lost to pests).
* Crops with improved handling properties-delayed ripening and increased self life.
* Crops with new qualities-improved nutritional status such as enhanced protein or starch content added vitamins and improved flavours.
* Plants which yield entirely new products - pharmaceuticals, edible vaccines, oils, fuel, and other non-food products.

Developing countries also taken a step ahead of the 159 field releases, Argentina leads with 43, followed by Chile, Mexico, Puerto Rico, and Republic of South Africa 17-20 each. India is far behind with just two field releases.

The Present Global Scenario of Food and Nutrition

Not withstanding the impressive and unparallel growth in the productivity of crops that prevented the recurrence of famines and attendant starvation deaths in the third world countries, sizeable population still remain below the poverty line and do not have food and nutrition security. About one billion people in the world are under nourished. Every year 13-18 million people die of hunger and malnutrition. Sadly, of this about 6 million are kids below five years of age. About 150 million children are under weight. Even though, India boasts of surplus food grains, the situation is no different from the rest of the world. About 250 million population remain under nourished. About 70-80 per cent of urban and rural population consume less calories than recommended. Over 400-700 million, of urban and rural population will go hungry and will not have even two square meals a day. About 53% of the children (about 40%) of the world below four years age are malnourished and over 30% of new born children are of low birth weight. Hence, nutrition security calorie-malnutrition, one has to evolve strategies to produce more food to keep calorie-hunger, by the exploding population, at bay. Some of the biotechnological approaches are discussed below.

Status of Genetically Modified Plants in India

The 1st experiment on TG plants in the field was started in 1995, when *Brassica juncea* plants containing *Barnase, Barstar* and *Bar* genes were planted at gurgaon (Haryana) to assess extent of escape. Subsequently in several crops like mustard, cotton, tomato, modified by several Indian Institutes or Organizations claim that they have developed TG which are ready for green house testing. The Government of India, through DBT has been mainly responsible for the propagation of research and development. The DBT has spent about 6756 million rupees during 1986-97 on all aspects of biology development in the country of which about 440 millions went into the development of TG crops. The status of research and development of GM crops in India is given in Table 9.12.

Table 9.12 *Current Research and Development Status of Genetically Modified Crops in India*

Organisation	Crop Species for GM	Trans Gene	Objective
Central Tobacco Research Institute, Rajamundry (A.P)	*Nicotiana toabacum* (Tobacco)	*Bt* toxin gene *cry IA* (b) *Cry Ic*	* To generate plants resistance *H. armigera, S. litura* * One field trial completed Evaluation under progress
Bose Institute, Calcutta	*Oryza sativa* (Rice)	Bt toxin gene	* To develop resistance to lepidopteran pests * Green house testing in progress
Tamilnadu Agriculture University, Coimbatore	*Oryza sativa* (Rice)	Reporter genes like *hph orgus*	* To study the extent of transformation
University of Delhi, South Campus, New Delhi	*Brassica campestris* Rape seed	*Bar, Barnase, Barstar*	* To develop herbicide resistance * Transformation work completed and ready for greenhouse evaluation
	Oryza sativa (Rice)	hph and gus	* Transformation completed * Green house studies in progress

Contd.. Table 9.12

Organisation	Crop Species for GM	Trans Gene	Objective
National Botanical Research Institute, Lucknow (UP)	*Gossypium sp.* (Cotton)	*Bt* toxin gene	* To develop resistance to Lepidopteran pests: * Lab transformation is in progress
Indian Agricultural Research Institute (IARI), New Delhi	*Oryza sativa* (Rice)	Bt toxin gene	* To develop resistance to Lepidopteran pests * Pests transformation in progress
IARI, CPR Institute, Shimla	Potato	Bt toxin gene	* To develop resistance to Lepidopteran pests * Green house trial in progress
Pro Agro-PGS India Limited New Delhi	*Brassica*	*Barstar, Barnase, Bar*	* To develop herbicide resistance in hybrid cultivar
	Tomato	*CryI IA* (b)	* Five field trials have been conducted
	Brinjal	*Cry* IA (b)	* Further resistance to lepidopteran pests
	Cauliflower	*Barnase, Barstar, Bar*	* Glass house experiments are in progress
	Cabbage	*Cry* IH Cry IC	* Lepidopteran pests resistance
Mahyco, Mumbai	Cotton	*CryI* IA (c)	* Multilocation tests completed
Rallies, India, Ltd.	Chillies	Lectin gene snow drops from *Galanthus nivalis*	* Resistance to Lepidoteran coleopterin and homopteran pests transformation is in progress
	Bell pepper	Lectin gene snowdrops from *Galanthus nivalis*	* Resistance against lepid-opteran, coleopteran and homopteran pests, * Transformation in progress
	Tomato	Lectin gene snowdrops from *Galanthus nivalis*	* Resistance against lepidoptera, coleopteran and homopteran pests * Transformation in progress
Indian ARI, New Delhi	Brinjal, tomato, Cauliflower, mustard/rape seed	Bt toxin gene	* To impart resistance to lepidopteran pests * Transformation is completed * Greenhouse trial completed * One season field trial is completed
JNTU, New Delhi	Potato	High lysine gene from *Amaranthus* plant	* Transformation is completed * TG under evaluation

Two examples, one for the improvement of resistance to insect pests, another for the improvement of nutrition is presented below.

A. *Transgenics Resistant to insect pests.*

To stabilize the yields of crops, it is necessary to prevent insect pest attack. This is generally done by spraying the costly chemical pesticides, which adds not only additional expense on farmers but also increases the pesticide residue levels both in soils as well as in foods. One of the strategies is to introduce genes coding for insect-toxic protein into the crop plants and release these transgenics, which will automatically reduce the consumption of toxic chemical insecticides. Farmers would be able to save large amount of money through reduced use of pesticides. Added advantage is the decreased level of pesticide residues in the environment as well as foods.

The selection of insect-toxic protein is based on the three key criteria : 1. The gene must code for the toxic protein in the plants in enough quality so that the insects are killed, 2. The protein should be toxic to only insect pests, not other animals or humans who consume the plant product and 3. The toxicity should be highly species specific killing only the targeted insect pests not other useful insects. The *cry* protein produced by the *Bacillus thuringenesis* meets the cirteria specified above and transgenic crops are being developed, cloning this gene into crop varieties. Bt.cotton is one such example which is stirring controvery all over the world.

Manipulating Plant Metabolism to Improve Human Health

The nutritional health and well-being of human entirely rests with plant food either directly or indirectly when plants are eaten by animals. Plant foods provide nearly all essential vitamins and minerals and a number of health promoting phytochemicals. However, micronutrient quantities are often low in most of the staple crops. Research programmes are in progress to understand and manipulate synthesis of micronutrients to improve crop nutritional quality of crop plants. The plant genome sequencing projects are providing none insight and ways and means of identifying plant biosynthesis genes of nutritional importance. The interface of plant biochemistry, human nutrition and genomics is now referred as *Nutritional genomics*.

Human beings require a variety of well-balanced diet comprising a complex mixture of macro and micro nutrients to maintain optimal health. Macronutrients like carbohydrates, lipids, and proteins constitute bulk of the food stuff the humans consume primarily as energy source. Micronutrients are an array of organic and inorganic compounds needed essentially for maintaining all the biological functions efficiently functions and keep the body healthy. The most essential micronutrients in the human diet include thirteen vitamins and seventeen minerals needed at optimum level to alleviate nutritional disorders. Non-essential micronutrients (Table 9.13) as group of unique organic phytochemicals that are not strictly essential in the diet but when present at adequate levels promote good health.

Table 9.13 *Essential Micronutrients their Daily Requirement and Rate Upper Limits for the Human*

Nutrient Calcium	Maximum recommended directly allowance/day (RDA)	Safe upper intake limit
Minerals		
Calcium	1200 mg	2400 mg
Iron	15 mg	75 mg
Iodine	150 µg	1960 µg
Selenium	70 µg	910 µg
Water-soluble vitamins		
Vitamin C	60 mg	960 mg
Vitamin B6	2 mg	250 mg
Folate	200 µg	10,000 µg
Fat soluble vitamins		
Vitamin E	10 mg α TE (a Tocopherol equivalent)	1000 mg
Vitamin A	1 mg RE (Retinol equivalent)	5 mg or 100 mg (β-carotene)

Modifying the nutritional composition of crop plants is an urgent world wide health-care issue because basic nutritional requirements for the majority of the world's population is yet to be fulfilled. A large populace of developing countries subsist on simple diets primarily based on staple foods like cassava, rice, corn and wheat which are poor sources of macronutrients as well as essential micronutrients. The startling reality is that about 40% of world's population doesnot consume enough micro and macronutrients. For instance, the magnitude of micronutrient deficiencies as per one estimate is that about 250 million children are at risk for vitamin A deficiency and this deficiency causes irreversible blindness to about half million children annually; about two billion people are at risk for iron deficiency with children and women of reproductive age are particularly vulnerable and 1.5 billion people are at risk for iodine deficiency.

Molecular genetic and genomics approach in dissecting plant metabolism for the reconstruction of pathways of nutritional importance in plants.

Dissecting plant secondary metabolism and pathways of nutritional importance is one of the targetted area of plant biotechnology in developing food biotechnology industry, consequently, many plant secondary metabolites as well as other micronutrient molecular genetic approach in identifying the genes involved in the metabolism of nutritional importance is being studied intensively (Table 9.14). The development of 'Golden rice' by introducing the complete b-carotene (provitamin A) biosynthetic pathway into rice endosperm through genetic engineering by Ingo Potrykus is one of the most spectacular achievement of transgenic technology.

Table 9.14 *Certain Health promoting Phytochemicals and their sources*

Phytochemical group	Diseases alleviated	Example of active compound	Source
Carotenoids (> 700)	Protease, cancer of oesophagus and others, cardiovascular diseases macular degeneration	Lycopene, lutein	tomatoes, kale, spinach
Glucosinolates (> 100)	Cancer	Glucorraphanin	broccoli and broccoli sprouts.
Phytoestrogens (> 200)	Cardiovascular diseases osteoporosis, breast, prostate and colon cancer	Genistein, Daidzein	soybean tofy, soyproducts
Phenolics (> 4000)	Cardiovascular diseases, cancer	Resveratrol	Redwine redgrapes

Vitamin A-deficiency presents a grievous health problem in many countries living on rice as a major staple food. This is because rice in its polished state is free of provitamin-A (β-carotene). To overcome this problems Ingo Potrykus carried out work to reconstruct the provitamin A-biosynthesis into rice endosperm by gene cloning technology.

The carotenoid pathway for the synthesis of β-carotene (provitamin-A) from the precursor geranyl diphosphate requires three enzymes, phytoene synthase, phytoene desaturase and lycopene cyclase whose genes are not present in rice. Two genes ioding phytoene shynthase and lycopene - cyclase, present in Daffodils (*Narcissus pseudonarcissus*) were placed under the control of Gtl (glutelin) promoter and another gene required for codingphytoene desaturase was obtained from a bacterium (*Erwinia*) and was placed under 35S promoter were used to transform rice (TP309) using the *Agrobacterium* as a vector. The whole carotenoid biosynthetic pathway is induced in the transgenic rice and found to synthesize β-carotene (provitamin-A) in the endosperm of the rice, which give golden yellow hue to it hence the name '*Golden Rice*'. This accomplishment is a real boon to the most of the developing nations, where the vitamin-A deficiency is a scourge (Fig. 9.15).

9.7 Production of Antigens and Antibodies (plantibodies) in Transgenics

Plants have been a main source of most of the drugs used in the management of several human diseases until recent times. With the technological developments and innovations in the synthesis of organic molecules, there spawned a lucrative pharmaceutical industry all over the world. Recent progress in developing transgenic plant has once again renewed the interests of several scientists to look upon plants as an ideal system of 'bioreactors' or 'biofactories' for the production of immuno-therapeutical molecules. As per one vision, by 2020 a huge plant industry based on what is called *molecular* farming is going to emerge producing '*fully human*' monoclonal antibodies (Mabs) worth 100 billion US dollars.

Vaccines and *antibodies* play a pivotal role in health care and in the management of infectious diseases caused by a range of microorganisms such as bacteria and viruses, including cancer. In blood there are several mixtures of plasma proteins called as immunoglobulins which are of relatively low electrophoretic mobility, often given to boost immunity. Nonetheless, there have been two major

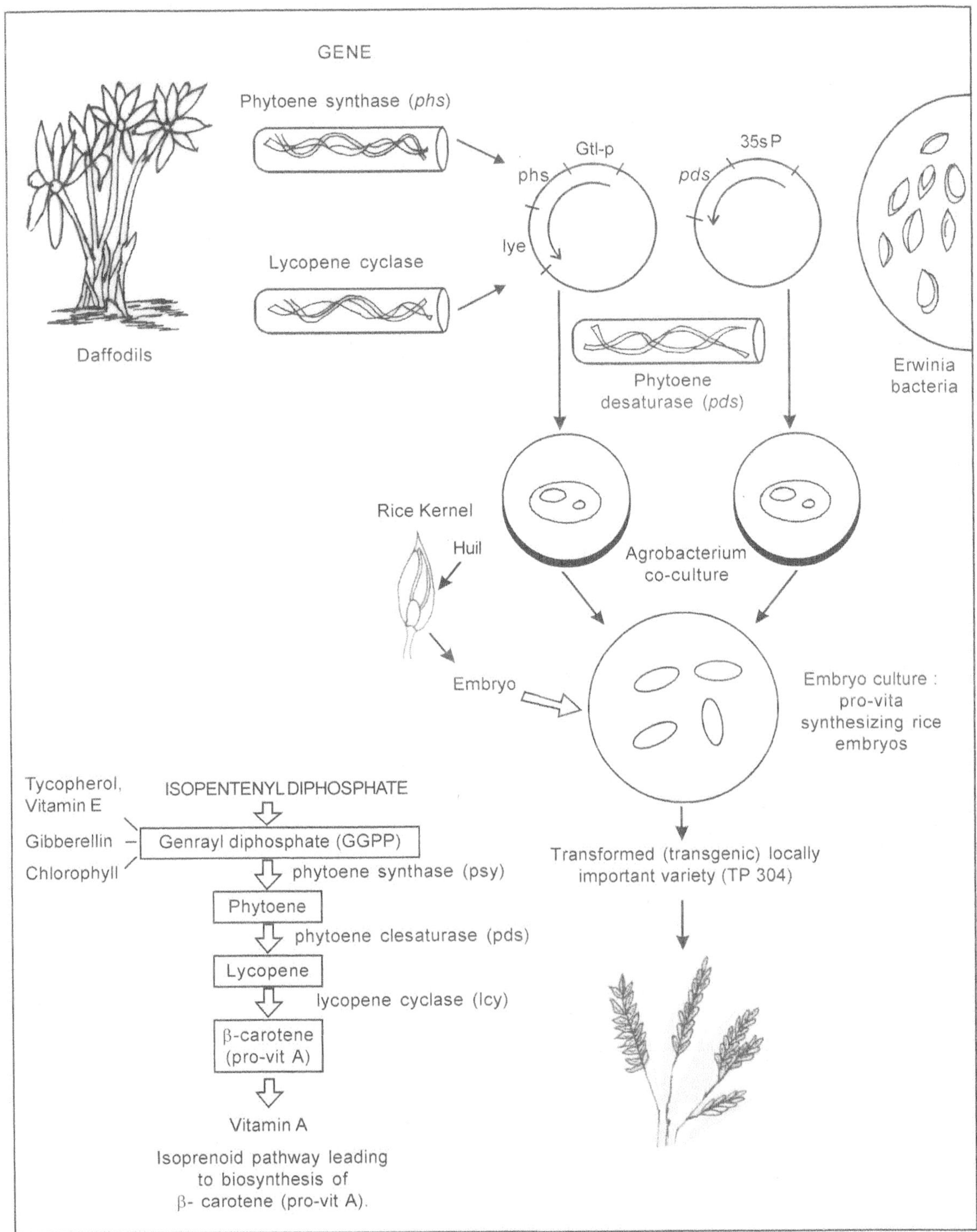

Fig. 9.15 Route map for the production of transgenic Golden rice

constraints in realization of the full potential of vaccines and antibodies *viz*., cost of production and maintaining the chain of their distribution. Expression of antigens as vaccines and of antibodies against antigens of pathogen in genetically modified crop plants is a cost effective strategy for the production of immuno therapeutical molecules. Other advantages of this approach is that the transgenic targetted molecules can be produced in the form of fruit or seed, so that these can be stored and transported from place to place even long distances, without any problem of damage or degradation. Further, a large amount of biomass can be easily produced in "*Pharmaceutical farms*" with relatively low inputs but can be able to realise high profits. Besides, several different range of chemicals can be created at any time by merely crossing transgenic plants capable of producing different products.

Plant bodies or plant antibodies are the human/animal anti bodies made by transgenic plants. Haitt *et al* (1989) attempted to produce antibodies in plant which could serve the purpose of passive immunization using the complementary DNA derived from a mouse hybridoma messenger RNA and transformed leaf segments of tobacco followed by their regeneration to become mature plants. These transgenics expressing single gamma or kappa immunoglobulins chains were crossed to obtain progeny in which both chains were expressed simultaneously. About 1.3% of total leaf protein was made up of a functional antibody Fig. 9.16.

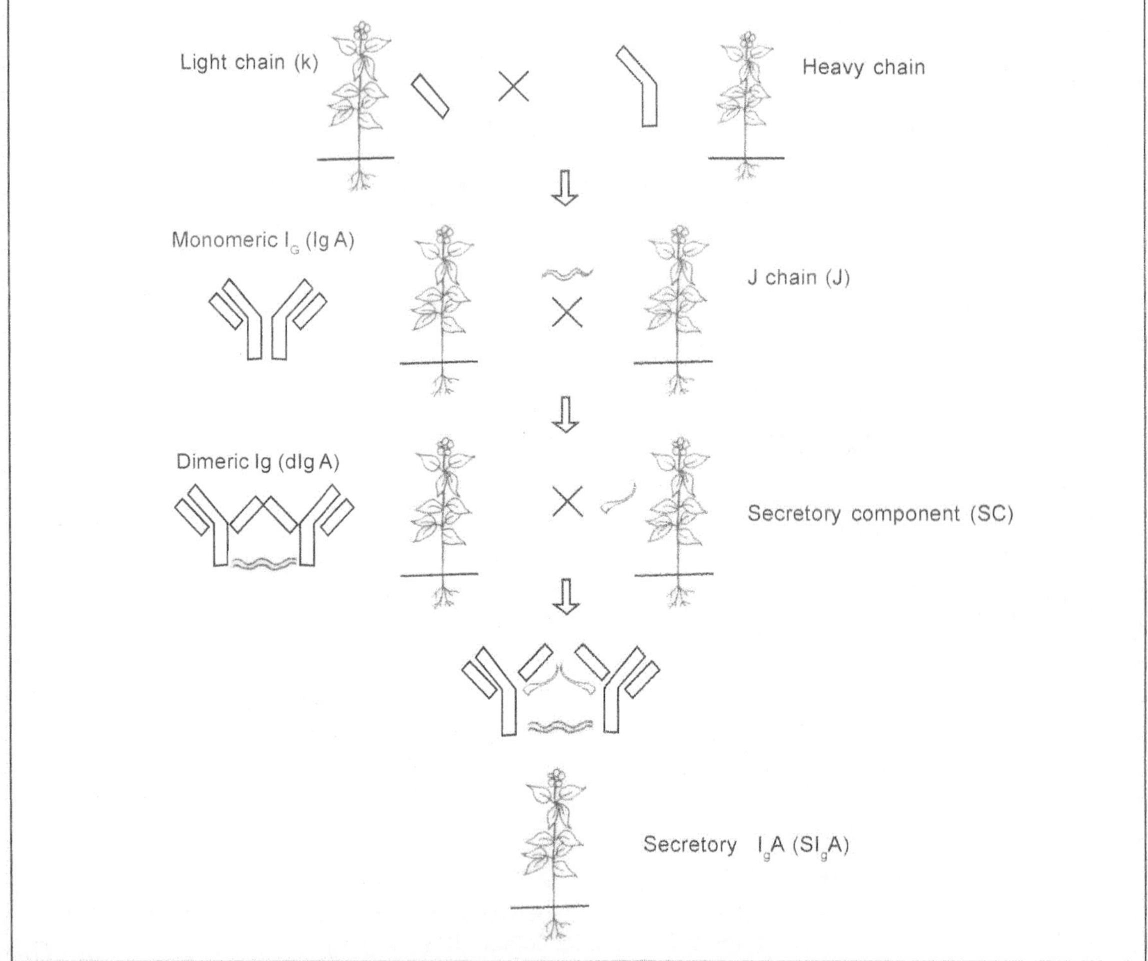

Fig. 9.16 Expression and Assembly of gamma (γ) immunoglobulins

A list of plantibodies produced by transgenic plants with the intent of treating and diagnosis the diseases is given in Table 9.15.

Table 9.15 *Therapeutic and Diagnostic Plantibodies Produced in Transgenic Plants*

Anti body Name or type	**Antigen Signal sequence**	**Plant species**	**Management of Disease**
Guy 13 IgA	Murine Ig G signal peptide	*N. tabacum*	Dental carries *Streptococcal*
(5–1 IgG)	Murine IgG signal pepties C5-1 (Ig G)	*Alfalfa*	Diagnostic anit – human Ig G
KDEL Sc FV	Murine IgG signal Peptide	*Wheat* and *Rice*	Cancer treatment Carcinoembryonic antigen
T84.66 Ig 6	TMV lender murine Ig G signal Peptide	*N. tabacum* transiently with *Agrobacterium* infiltration	Cancer treatment Cancer embryonic antigen
38c 13 (Sc FV)		*N. benthamiana* Rice α-amylae	β-cell lymphoma treatment idiotype-vaccine
C017–1A (lgG)	Murine lgG signal peptide; KDEL	*N. benthamiana*	*Onc* antigens of colon cancer, scal surface antigen
Anti–HSV–2(IgG)	Tobacco extensin signal	*Soybean*	Herpes simplex virus 2

9.8 Phytoremediation

The term phytoremediation refers to the use of plant species to clean up either soils or water degraded by pollution and contaminants. Scientists have discovered long time ago that certain plant species accumulate chemicals or elements that are present in their environs either naturally or unnaturally as pollutants/contaminants, as does happen in the surroundings, where there is an intenses industrial activity or mining. Most of the chemical compounds or elements are harmful to the flora and fauna, if certain critical level is exceeded. Yet, certain plant species thrive well on such contaminated soils and or water, since they have a capacity to absorb them without injury to themselves. Such plant species can be employed for the cleaning of environmental pollution. The process of employing appropriate plant species for the cleaning up of environmental pollution/contamination is termed as phytoremediation (phyto = plant; remediation = the process of removal).

Plants as Cleanup Devices of the Environment

The potential approach to cleaning the environment is known for a long time. Water bodies, when contain certain chemical compounds in abundance as a result of pollution, for example, phosphates, poplar trees have been used to clean the water bodies. The over abundance of phosphates results in the growth of algae and bacteria. Consequently the levels of dissolved oxygen in the water goes down the critical point and causes the several aquatic organisms to die. Many fish kills reported sporadically are result of oxygen depletion. By using the poplar tree roots to clean up and the phosphates, the oxygen level of aquatic environment or water quality is remedied.

Similarly, certain grasses and sorghums are used for remedial purposes. These plants lack nitrogen. However, when the nitrogen level is too high in the soil the other plants get '*burnt*' and do not survive. Nitrates in water are also dangerous to young animals and human beings. By using the grass and sorghums species, the nitrogen from water or soil can be reduced, so that other species of *flora* and *fauna* can thrive. The idea that plants can be used for the environmental remediation is not new.

The emeregence of biotechnology in recent times resulted in the development of low cost processing technology for the treatment of degraded water bodies, soils, mining sites for reclamation. This potential approach is drawn on centuries of experience of several scientists in cultivating crops and also findings. As early as 1960s, Timofeev-Resovsky (1962) and his group conducted extensive research on the use of plants for removing and cleaning the entire ecosystem from the radio-nucleotide contamination that took place accidentally. Since then, there have been several reports to show that aquatic plants such as water hyacinth, duck weed and water velvet can accumulate heavy metals (Pb, Cu, Cd, Hg, and Fe) from contaminated water. This capability is currently utilized for the removing heavy metals and organics from water bodies and wetlands.

Phytoextraction of Metals

The ability of plants to accumulate metals is not only used for the remediation but also being used for the extraction of heavy metals from the soils, sludges and sediments. Crop plants such as *Brassica* (mustard) have been used to extract heavy metals from soil and sediments and translocate those metals to the harvestable portions : stalks, leaves, and roots of the plants for extraction and recycle. The list of hyper accumulating plants is ever increasing (Table 9.16).

Table 9.16 *List of plants useful for phytoremediation of polluted waterbodies and contaminated soils*

Families / plants species	Useful for accumulation
Brassicaceae	Pb, Cd, Cr (vi), Nil
B. juncea	Zn and Cu
Euphorbiaceae	Nil (11% in latex)
Seberita accuminata	
Alyssum	Ni, Pb and Zn
Thalapsi caerulescence	Ni, Pb and Zn
Eichhornia crassipes	Pb, Cu, Cd, Fe, Hg
Hydrocotyle umbellata	Pb, Cu, Cd, Fe, Hg
Lemna minor	Pb, Cu, Cd, Fe, Hg
Azolla pinnata (water velvet)	Pb, Cu, Cd, Fe and Hg
Datura innoxia	Ba
Helianthus annus L.	U, Sr

Mechanism of Accumulation

No serious attempts have been made to explain the mechanism of hyper accumulation of heavy metals, as to why, certain plant species have developed the ability to accumulate heavy metals. It is common knowledge that most of the plants including crop plants accumulate metals in trace quantities. Most metal accumulating plant species known today were discovered in and around soils containing heavy metals including the mining sites. These plants are endemic to the soils contaminated with heavy metals suggesting that metal accumulation is associated with heavy metal resistance and limited to tropical areas. Very little is known about the significance of hyperaccumulation phenomenon of these plants. However, several hypotheses have been putforth to explain the reason why certain plant species endemic to contaminated soils hyperaccumulate heavy metals in such large quantities. Current hypothesis include (1) tolerance (2) metal from plants disposal (3) drought resistance (4) in advertent uptake, and (5) defense against herbivores or pathogens. Some experimental evidence however suggests that hyper accumulating of metals has evolved as a defence mechanism and protects such plants from herbivores.

Development of Technology for Economic Exploitation

Phytoremediation is a very new area of biotechnology which holds an immense potential for the future. Phytoremediation of heavy metal is designed to concentrate metals in plant tissues thus minimizing the amount of solid or liquid hazardous waste which needs to be treated and deposited at hazardous waste sites. As an ultimate goal, an economic method of reclaiming metals from plant residue should be developed. This will completely eliminate the need for costly off site disposal. Presently, methods for the further concentration of metal in plant tissue include : (a) Sun heat and air, drying, (b) ashing or inceneration taking precaution for environmental safety (c) composting, (d) pressing and compacting and (e) acid leaching.

In future, it is necessary to understand the diverse processes involved in heavy metals hyper accumulation by plants to harness their potential for the economic remediation and/or leaching. This requires a multidisciplinary approach embracing the diverse fields of biology, agricultural engineering, agronomy, soil science, microbiology and genetic engineering. In future genetically engineered plants will be grown for the extraction of gold, copper, and other heavy metals.

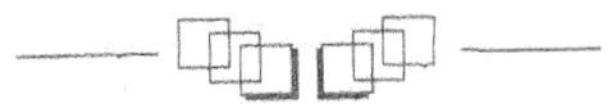

Essay Type Questions

1. Discuss economic potential of plants.
2. Narrate the role of phytochemicals as pharmaceutical.
3. Give an account of medicinal plants used in traditional system of medicine.
4. Discuss plants of medicinal value and their therapeutic use.
5. Describe gene transfer methods in plants.
6. Write about *Agrobacterium* mediated gene transfer in plants.
7. Describe modus operandi of the transfer of DNA from bacterial plasmid to plant cells.
8. Discuss Ti plasmid based vector systems for gene transfer to plant cells.
9. Describe alternative methods to *Agrobacterium* mediated gene transfer.
10. Describe plantibodies produced in transgenic plants.
11. What are GM crops and give an account of genetically modified crops in India?

Short Answer Type Questions

1. Plants as a source of industrial material.
2. Differentiate secondary metabolites from primary metabolites.
3. Mention some common plants used in medicine.
4. Mention selectable marker genes employed in eukaryotic vectors.
5. Ti plasmid
6. Role of plasmid '*vir*' genes.
7. Molecular mechanism of T-DNA transfer.
8. Electroporation
9. Plantibodies
10. Transgenic plants.

10 Animal Biotechnology

10.1 Introduction

Animals have become man's constant companion, ever since their domestication from time immemorial. All domestic animals have undergone drastic changes due to the process of artificial selection by humans under captive breeding with a restricted gene pool. Consequently, with the passage of time, and continued isolation from the wild gene pool, the tamed animals acquired startling traits which have been put to use by the man for his survival and sustenance. Some have changed in such a way that they started yielding increased quantity of better milk or wool than others. Some others have displayed rapid growth and gaining of body mass or developed high fecundity, some became more vigorous and some just looked bizzarrely different and arose curiosity. The classical animal breeders have taken into account natural variability of animals and mated them selectively *inter se* to develop innumerable breeds of domestic animals such as horse, dog, sheep, goat and cattle and other live-stock animals that we see all around us in the present day world.

Animal improvement remained no longer a matter of interest of indolent bourgeois society, but it became a matter of necessity for human survival. The incessent population explosion puts a tremendous pressure on biologists to explore the possibility of developing new technologies for the animal improvement under the conditions of shrinking resources, at the same time ensuring the quality of food, fiber, drugs and medicines for health care. Eventhough, the production of food from animals is not only expensive but also inefficient from the angle of per capita utilization of energy, nutrients and space to produce similar quantity and quality of biologically assimilable protein (Table 10.1), there is an imperative necessity to continue to do animal farming for food, hide, fiber, various by-products, sources of dietary supplements like necessary amino acids, drugs vitamins etc.

Table 10.1 *Comparative Protein Yield and Use of Resources by Plant and Animal Systems*

Source of protein	Protein (Kg h^{-1} y^{-1})
Plants	
Algae	30×10^3
Potato	0.8×10^3
Rice	0.6×10^3
Pea-nuts	0.45×10^3
Wheat	0.36×10^3
Animals	
Fish	1.0×10^3
Milk	0.12×10^3
Meat	0.08×10^3

The recently developed rDNA technology (recombinant DNA technology) provides us powerful tools that enable not only to solve the basic food and nutrition problems but also to develop and produce a range of health care products. The coming hundred years would be the era of animal biotechnology and all over the world the strategies for animal production would change radically as the biologists learn to use these tools.

"*The use of animals, tissues, organs, cells, subcellular organelles, and/or parts of those structure, as well as the molecules to effect physical or chemical changes needed to generate new products for research and commercialization*" constitute the animal biotechnology. Inspite of such broad perspective, animal biotechnology is considered to be synonymous to rDNA technology. However, in practice it includes not only genetic engineering or gene cloning technology, but also some of the older technologies along with the state of the art cutting edge technologies and related methods such as cell culture, monoclonal antibodies, bioprocess engineering and manipulation of reproduction. Thus, animal biotechnologists not only manipulate genomes of targetted animals, but also processes that exists in the organism but out of reach for the manipulation. Hitherto, the modification of such traits was either made fortuitously, through the microevolutionary forces or by the strategies of artificial selection practiced by the animals breeder. Animal biotechnology offers supplementation of selective breeding, helping to effect changes at the organism level by the manipulation of cells and genes within an organism. It creates a novel procedures (protocols) for the effective intervention of biological system and processes with potential for bringing about modification, which otherwise is not possible.

Major Areas in Animal Biotechnology

The cutting edge state of the art animal biotechnology includes the following major areas of interest :

- Manipulation of the processes of reproduction
- Cloning of macroorganisms

- Genetic engineering of microorganisms and molecules including cell engineering (hybridomas) to synthesize desired end products like vaccines, gene probes, monoclonal antibodies and growth products.

Animal Response to Manipulations

All organisms have intrinsic limitations in their production systems *viz*., mating system, genetic variability and maintenance process, which includes health as well as nutrition. Animals and plants differ basically from each other from the production point of view. Though, the nutritional and environmental stress and diseases are common factors to both of them, however, the animal production solely depends up on the health of pasteur/fodder of plants. The production strategies are interrelated. The plants convert inorganic nutrients from the environment to organic food for animals; animals inturn provide organic mannure to plants. Reproduction system is basically different in animals and plants. The plants

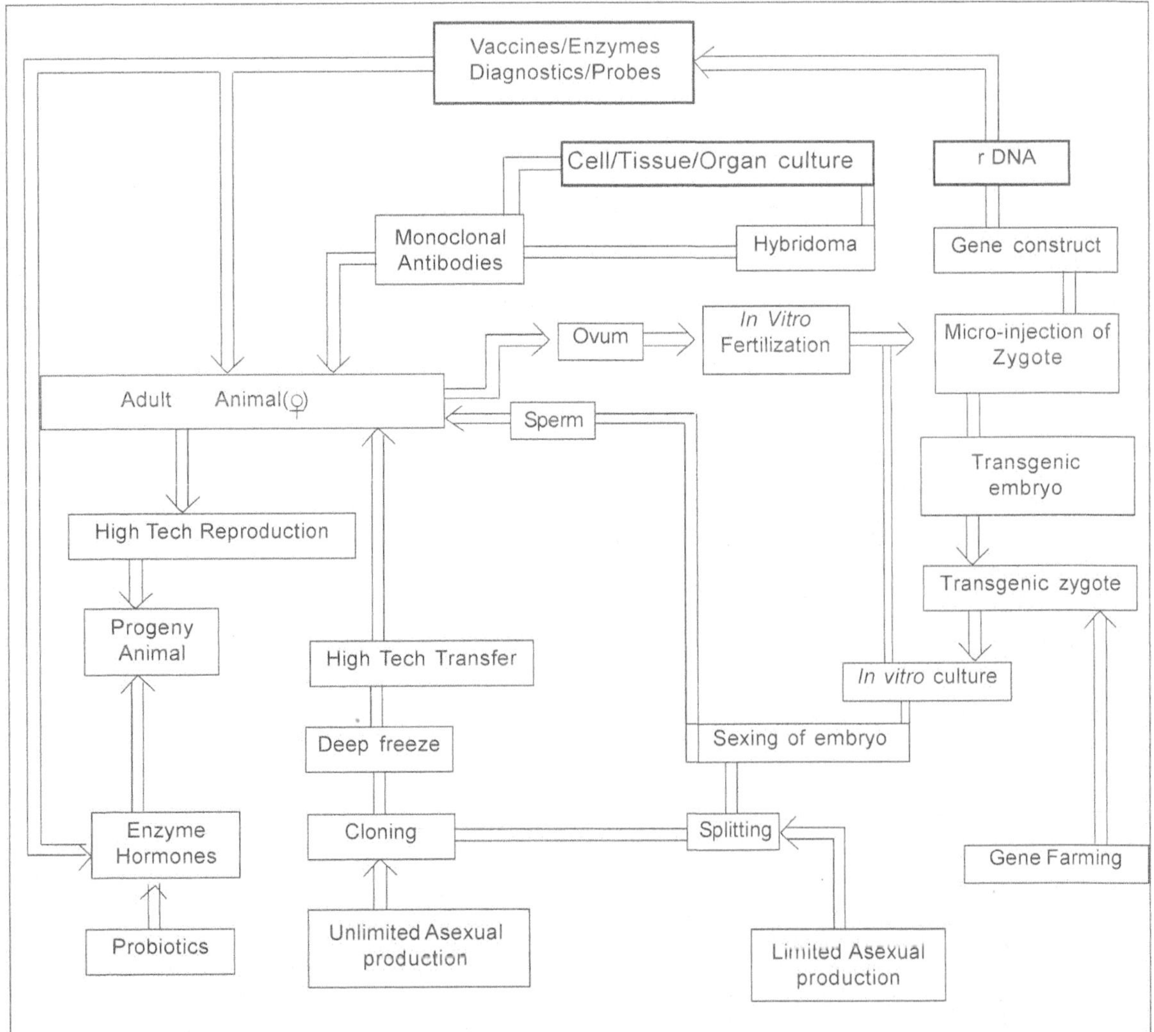

Fig. 10.1 Animal Biotechnology : A Sum up - A conspectus of high-tech innovations in Animal productivity

reproduce both sexually and asexually. When reproduction is through amphimixis (union of gametes) plants produce a large number of gametes, both male and female, whereas animals give rise to few. All live stock animals produce only one per egg each. Life cycle most of them are annual or pernniels Whereas in animals, it takes many years from birth to maturity and old age. The low, fertility of females in animals has profound impact on genetic manipulation strategies. The existing variability can be compounded and new genetic recombination could be generated very easily in plants, since several thousands of propagules (seed) can be produced from a single hybrid whereas to achieve the same it is very difficult in animals. Even quantitative trait loci (QTL) can be identified through molecular probes very easily in plants, but not so easy with the animals. When a desirable genetic variant is isolated, it can be multiplied rapidly in plants, but the same is very slow process in animals. The important goal of animal biotechnology is to make feasible rapid animal production as that of plants and find out ways and means of genetic manipulations at cellular, subcellular, tissue or organisms level for the possible commercialization (Fig. 10.1).

Manipulation of the Process of Reproduction

Manipulation of reproductive process is a *sin que non* for the successful exploitation of biotechnological potential of a vast array of gene and genome pools of domestic animals. Historically the manipulation of the reproductive processes has started in early 1930's with the development of procedures for artificial insemination (AI). When it was feasible to deep freeze the semen, the use of AI became wide spread in 1960-70s. With the concurrent developments in reproductive research on female reproductive process, the procedures were developed for the induction of *Oestrus cycle* and synchronisation. Initially this was done using steroid hormones, but subsequently the same was accomplished by using *prostaglandins*. The embryo transfer technology perfected in 1980s owes its success in the development of methods for super ovulation using gonadotrophic hormones. Ever since, the embryo transfer technology has became routine affair especially because of technological advances for the embryo preservation by freezing.

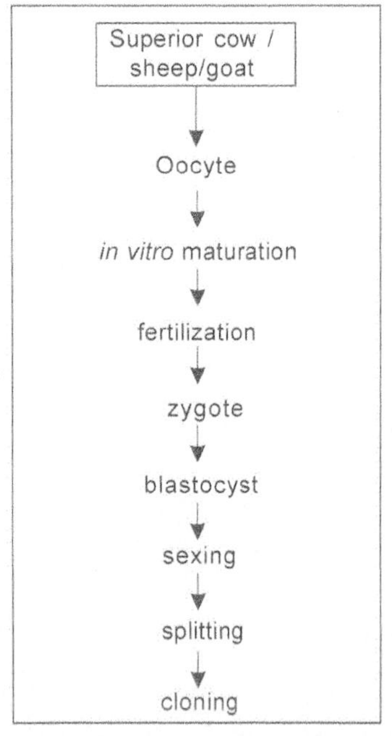

Fig. 10.2 Reproductive process

Artificial insemination and embryo transfer technique together offer immense possibilities of enhancing the rate of genetic progress in national breeding programmes. In recent times, with the development of more advanced techniques for manipulating reproduction as well as *in vitro* embryo manipulation and production, including the perfection of technologies for gene transfer to a whole animal for the possible production of transgenic live stock has a distinct possibility. Basically, the protocol includes the following : (i) removal of oocyte from a genetically superior cow (goat, pig, sheep etc., or theoretically any animal) during the *dioestrous* period, (ii) *in vitro* maturation and fertilization of the oocytes, (iii) culturing of zygote (fertilized oocytes) *in vitro* till blastocyst stage, (iv) sexing, splitting and (v) cloning of embryo. Micromanipulation techniques gave further phillip to the development of protocols for splitting and cloning of sheep and bovine embryos. Marketing of bovine embryo that originate from *in vitro* manipulations, cloned embryos and transgenic animals, becomes a reality and these procedures can be adapted for the industrial production of animals, with varied applications and utility, thus revolutionizing the animal productivity (Fig. 10.2).

10.2 Genetic Engineering and the Production of Transgenic Animals

The term *transgenic* or *genetically modified* animals means that these animals possessed some genetic material or sequence of DNA or gene added to their genome or had their DNA altered by some *in vitro* means. This added DNA sequence may be from another conspecific animal or may be from a different source altogether across the species/genera/kingdom barrier. The transgenic or genetically modified animals are basically used :

(i) to study and understand the normal physiology and development
(ii) to study the progression and regression of diseases of pathological origin
(iii) to manage, if possible to find the gene therapy for genetic disorders
(iv) to design and develop drugs for the treatment of diseases
(v) to produce biologically useful gene products especially therapeutics
(vi) to test the safety of vaccines and chemicals
(vii) to provide organs for transplantation and
(viii) to increase the quality and quantity of farm animal products.

When foreign DNA is introduced into a fertilized oocyte before the cleavage initiated, there is good probability that the introduced gene sequence/DNA will be incorporated into the host genome and inherited by all the cells of the embryo, both somatic as well as germinal. Consequently, after differentiation and development of adult animal, the alien DNA is transmitted to all progeny. Animals that have been manipulated in this way are referred to as transgenic or genetically modified animals. Introduction of DNA sequence into any later stage of differentiating embryo may give rise to a mosaic (or chimeric) animal. By chimera it is meant that the organism contain cells and tissues of more than one distinct genotype.

A. *Genetic Engineering : Cloning Strategies*

Genetic engineering process is something akin to the genetic recombination that naturally occurs during sexual reproduction, where the breakage and rejoining of DNA molecules of the chromosomes occurs, which is vital importance to all the organisms. The naturally occurring process, however, limits the recombination of genes within a species, because of strong reproductive isolating mechanisms. Whereas the *gene cloning* or *genetic engineering* offers unlimited opportunities for bringing about new combinations of genes across the species/kingdom barriers which does not takes place under natural conditions.

Genetic engineering has been defined as "*the formation of new combinations of heritable material by the insertion of nucleic acid molecules produced by whatever means outside the cell, into any virus, bacterial plasmid or other vector system so as to allow their incorporation into a host organism in which they do not naturally occur are capable of continued propagation*".

These techniques allow the joining (splicing) of DNA molecules of quite different origin and when combined with gene delivery techniques, including transformation etc., facilitate the introduction and integration of foreign DNA (gene) into other organism.

Thus DNA can be isolated routinely from plants, animals and microbes serving as the donors and can be fragmented into convenient sizes comprising of one or few gene blocks. Such fragments can then be joined to another fragment of DNA, (the vector) and then passed into the recipient (new host) cell to be integrated into its genome. The host can be then mass propagated to yield population of identical cells called *clones*. In this way an organism can be made to acquire novel genes with desirable properties that were impossible with the conventional methods of breeding and selection, including by the process of induced mutation.

The technological development in gene cloning methods is so perfected presently that, it is now feasible to modify any genome by transferring alien genes or a gene construct as an expression cassette or module so that the transgene gets integrated and expressed. Virtually there is no limit to the production of transgenic animals capable of producing an array of useful products that can be manufactured on industrial scale for the possible commercialization. The gene cloning technology is undoubtedly one of the most potential areas of biotechnology and enabler of development of several industrial processes for the production of simple as well as complex chemical compounds hitherto deemed impossible by conventional methods including microbial manipulation. Examples include the synthesis in microorganisms of specific animal proteins such as insulin, enhanced ranges of enzymes, hormones, antitumour and antiviral compounds such as interferon, fine chemicals or bulk chemicals such as ethanol, or creating ability to utilize complex substrates such as cellulose and lignin and to produce useful products from them.

The transgenic technology holds the potential to extend the range and power of each and every aspect of biotechnology. In the first instance, these techniques will be widely used to improve currently used microbial processes leading to the improved yields of specific fermentation products by stabilizing the existing cultures and elimination of undesirable side products. With the rDNA technology new microbial strains will be created with new and unique metabolic properties. The industrial microbiology based on fermentation processes will acquire a competitive edge over the petrochemical based industry in producing the whole range of new chemicals. For instance, chemicals like ethylene glycol used as

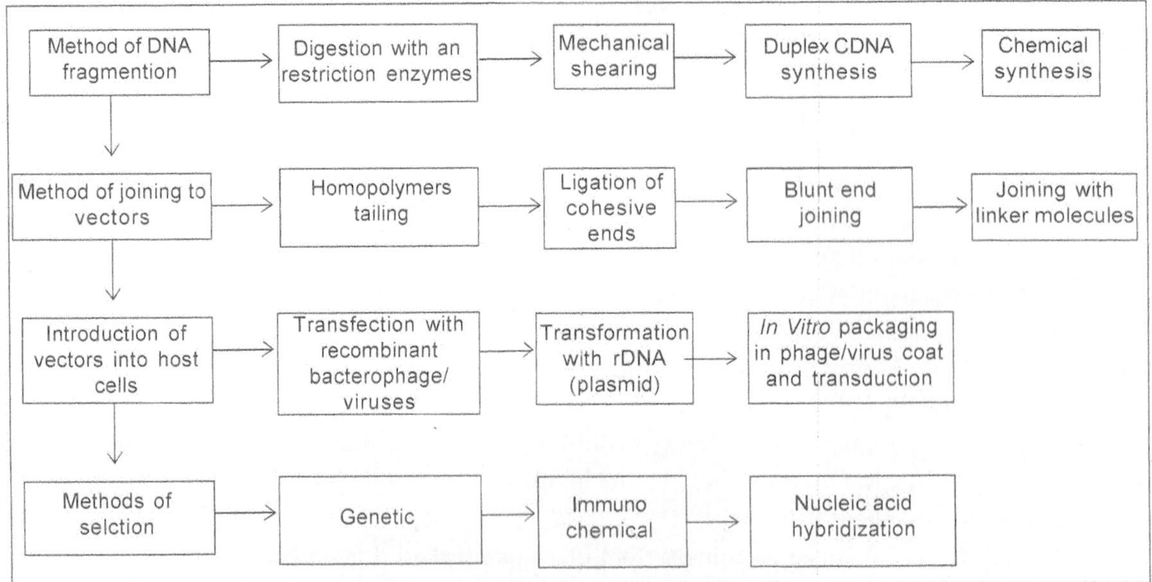

Fig. 10.3 Gene Cloning Strategies

anti freeze agent, ethylene oxide used in manufacturing polysters and as surfactants, and propylene glycol used in the plastic industry etc., which are being produced using the petrochemicals as feed stock will be replaced by the biotechnological process using rDNA microbes. In food industry, particularly in chese manufacture and baking, the traditional process will replace with genetically modified fungi and bacteria that will impart greater control and reproducibility of flavour, quality and texture.

On agricultural front, animal biotechnology is going to be revolutionized with the advances in transgenic animal production. The innovation will have a great impact on the animal productivity including growth rate, meat, milk and egg production. Gene farming and the genetic manipulations improving the fertility, reproduction, growth and animal cloning. Proper insight into gene technology that requires the comprehension of advanced molecular biology is also needed. A summary of strategies involved in genetic engineering is presented in flow diagram (Fig.10.3).

B. *Gene Transfer to a Whole Animal : The Production of Transgenic Live-Stock Animals through Micromanipulation*

Micromanipulation techniques have been used to produce transgenic mice, sheep, pigs and cows ever since T. P. Lin (1966) first outlined the technique of microinjection into mouse eggs. Gene constructs coding for a known characteristic can be injected into the pronucleus of a fertilized ovum. This technique permits the transfer of any gene from any source - across the living kingdom, no matter how remote the relationship is between donor and recipients. The ultimate aim of gene transfer to a whole animal is to get transgenic or genetically modified organism.

Microinjection is carried on using two micromanipulators. Zygotes maintained in a microdrop of embryo culture media are held individually at the polar body by suction applied by a microsyringe or picoinjector attached to a fine-polished microholding pipette. The injection pipette attached to microsyringe or automated injector is used for gene delivery and approximately 1-2 picoliters of the DNA solution consisting of 2000 copies of a gene construct is injected into the male pronucleus; successful injection is monitored visually by observation of the pronucleur volume expansion (approximately 25%). Microinjection is carried out under a magnification of 400X using an inverted microscope equipped with Marski or Hoffman modulation optics. The reason for the choosing male pronucleus in mouse is that it is larger than the female pronocleus and therefore it is easier to use for injection. Special equipment are now available which makes it possible to inject about 100 fertilized eggs per hour in the case of mouse. Unlike mouse, the manipulation of fertilized eggs of livestock animals is a cumbersome process. One reason is that, substantial number of these eggs do not have pronuclei. The second reason is that cytoplasm is opaque and not translucent like the mouse eggs. To overcome this problem, the differential interference contrast microscopy and centrifugation methods have been employed for visualising the pronuclei.

The microinjected eggs are subsequently transferred to a pseudopregnant recipient on the same day, or following incubation *in vitro* for 12–24 h, by which time the eggs have divided once. Pseudopregnancy results from an infertile mating and prolonging the luteal phase of the oestreous cycle to favour implantation of the blastocyst.

By following the procedure outlined above, an overall transgenic efficiency of about 1–10%, based on the total number of eggs employed for microinjection, has been achieved in the mouse system. The efficiency in the case of livestock animals is one magnitude less, mainly because of technical problem. The Fig. 10.4 illustrates the gene transfer method in mouse system.

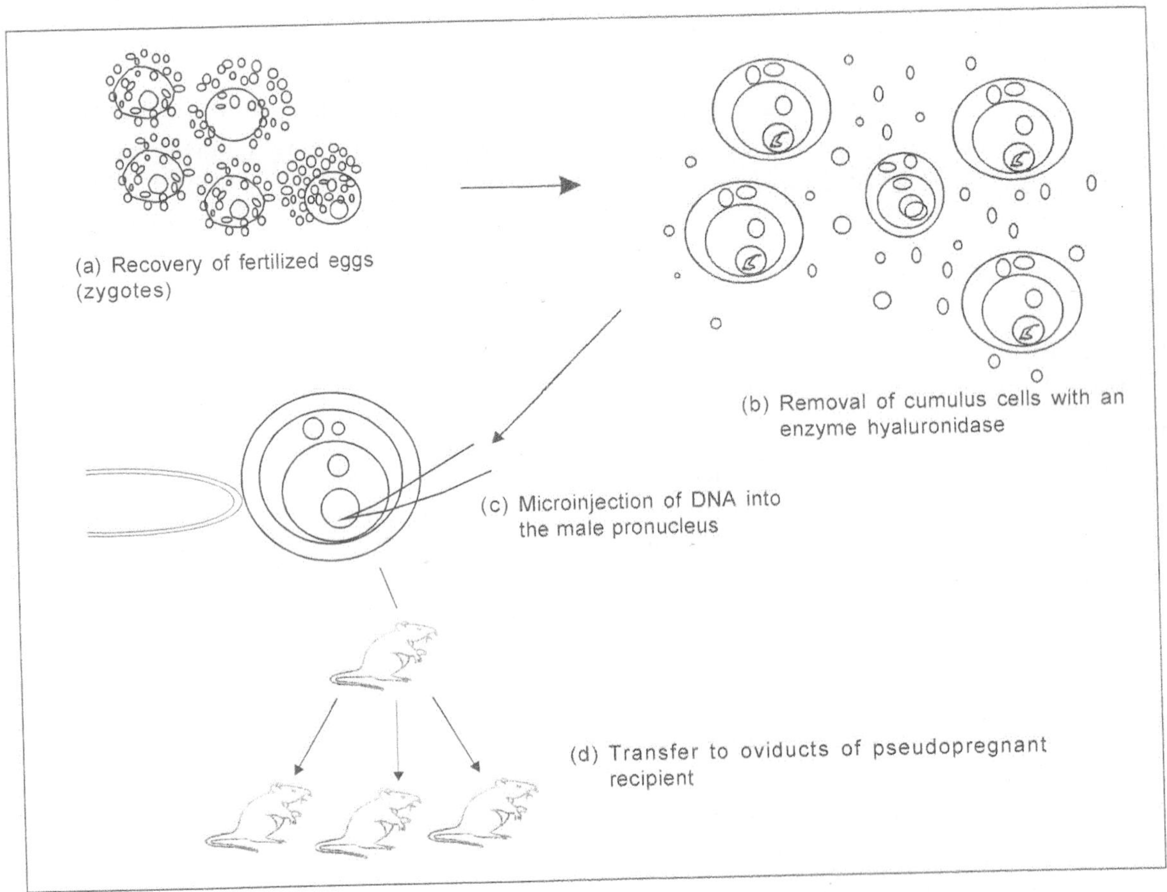

Analysis of off spring

* Southern blot – probe for gene integration
* Northern blot – probe for RNA to assess gene expression
* Western blot – direct assay for protein production (assess gene expression)

Fig. 10.4 Gene cloning strategy in mouse system by microinjection of foreign DNA into male pronucleus of zygote

(i) *Viral Mediated Gene Transfer*

A significantly efficient method of introducing foreign gene (DNA) into animal cells is the use of virus as a vector. Retroviral vectors have been increasingly used in gene transfer technology in animals. Retroviruses are single stranded RNA viruses upon transfection (entry into a host cell), the retroviruses introduce their genome (RNA) into eukaryote cell, and are multiplied by replication using the reverse transcriptase enzyme. The resulting single stranded DNA copy is made double stranded and subsequently integrated into the host chromosome as a single copy. Any gene that has been cloned into a retroviral vector can be transferred to eukaryotic cell in principle. The cellular RNA polymerse-II of host can transcribe the DNA to make viral RNAs which in turn are capable of infecting other cells.

No special equipment is necessary for gene transfer for viral mediated gene transfer. Cells of the mouse embryo can be infected with murine retrovirus either by co-culturing 4-8 cell embryos with virus-producer cells or by injecting the virus directly into the blastocyst. Virus producer cells are those that are already infected by the virus in which viruses are produced and released. Only when foreign DNA is integrated into embryonic cells which during development contribute to the germ line with the gene (DNA) be transmitted as a Mendelian traits since integration can takes place at different stages, sites and cells, the resultant off spring could be a chimera. Outbreeding of the offspring are necessary to obtain pure lines. The Fig. 10.5 illustrates the essential process of gene transfer using retroviruses as a vector.

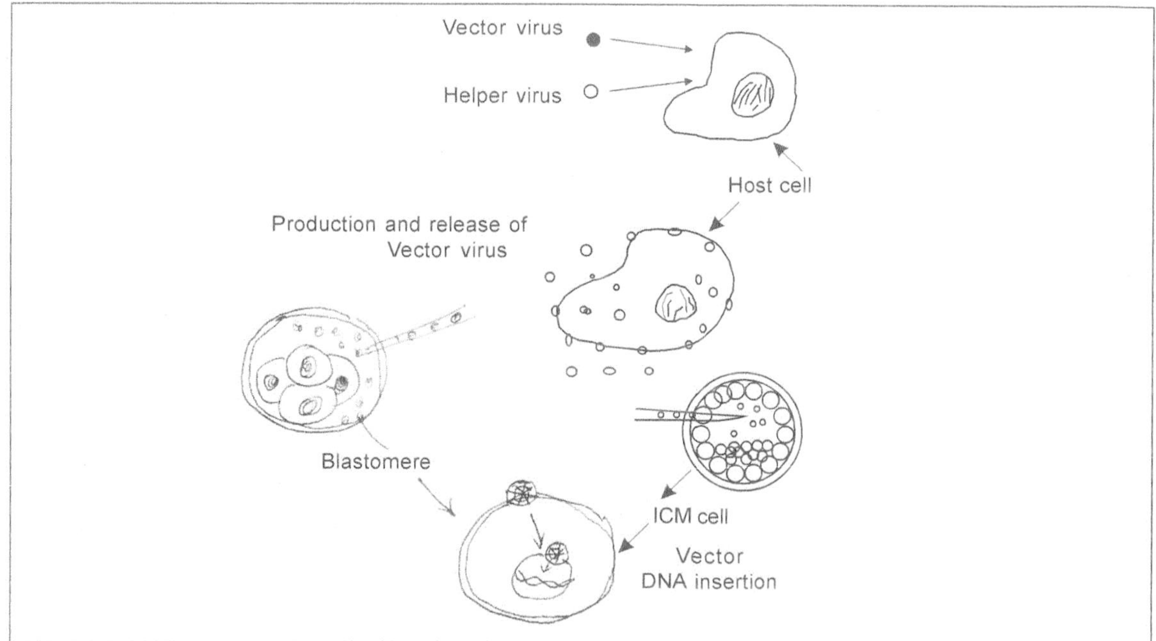

Fig. 10.5 Gene transfer through (Virus mediated) retrovirus vectors

(ii) *Electroporation Mediated Gene Transfer*

The exposure of a cell to an electric field results in the polarization of membrane lipo-proteins. Consequently, after a threshold level is reached, the membrane develops localized pores and the cell becomes permeable to exogenous molecules. The induced permeability is reversible provided the magnitude, duration and other characteristics of generated electrical potential do not go beyond a critical point, beyond which the damage to cell is permanent. The use of electric field to induce micropores in cell membrane has been termed electroporation (Zimmermann, 1982). The cell membrane upon exposure to a high impulse voltage electric impulse could be used profitably as a delivery system for gene transfer (DNA sequence) by keeping targeted cells suspended in media in a chamber across which potential is developed. Neumann (1982) for the first time developed the mouse lymphocyte cells deficient for thymidine kinase gene (TK^-) upon exposure to a high voltage DC electrical impulse in the presence of DNA with a foreign gene from herpes simplex virus acquired phenotype TK^+. Since then the electroporation has been routinely used for gene transfer in higher animals as well as plants. The Fig. 10.6 illustrates the process of electroporated gene transfer.

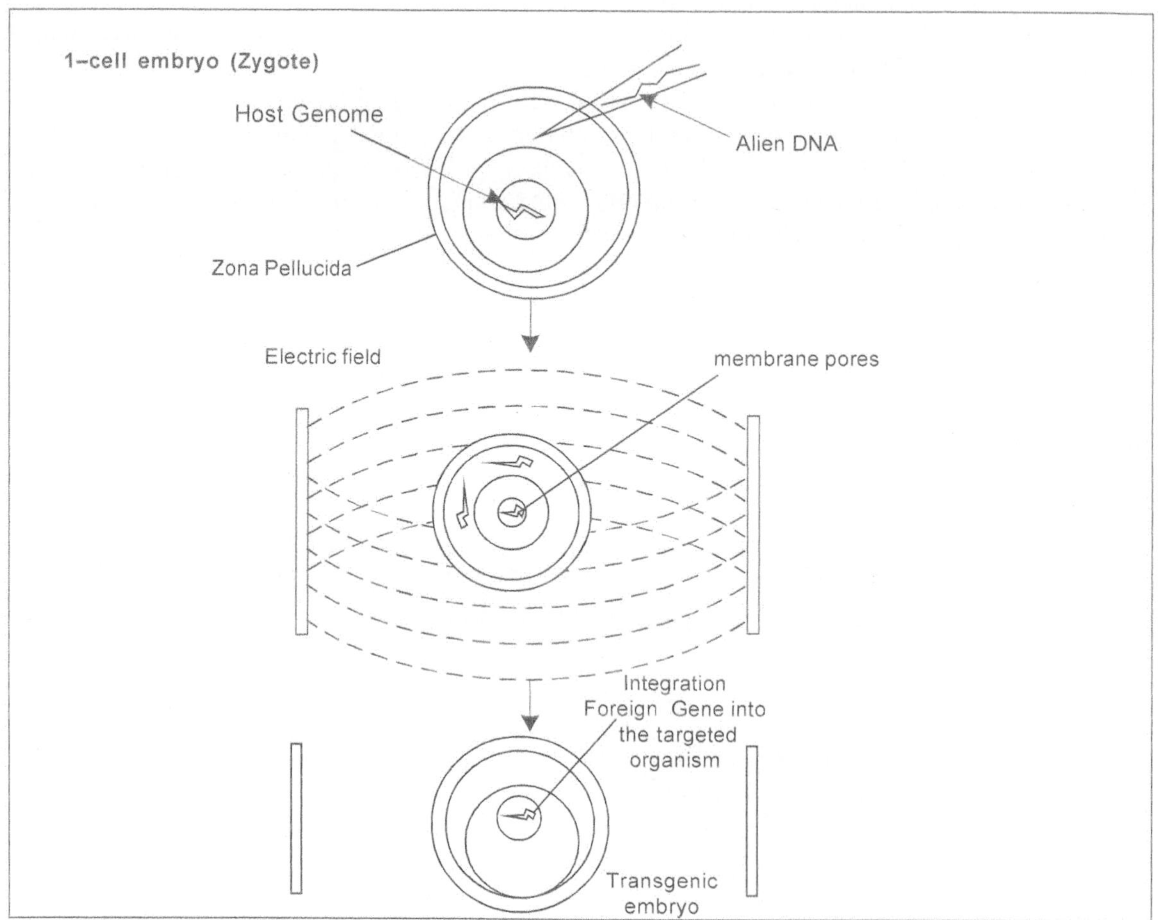

Fig. 10.6 Electroporated gene transfer

C. *Production of Transgenic Livestock and its Applications in Biotechnology*

The adoptation of the transgenic technology developed in mouse to livestock animals has not been as strait forward and successful as anticipated. The efficiency has been very poor. Several parameters have been shown to affect the efficiency of generating transgenics. Some of the factors include : DNA concentration, injection buffer, nature of injected DNA (linear or circular), the site of injection, nuclear or cytoplasm size and taper of pipettes used for micro injection, the timing of injection and regulation of transgenes. Futhermore, the business of transgenic production with higher animals is quite expensive, especially to purchase and maintain. The age of sexual maturity also varies from species to species. In the case of pigs it is 25–32 weeks, while in cattle the same is almost double 52 – 65 weeks, when compared to mouse (6 weeks). Hence, transgenesis of livestock animals is not only cumbersome and costly, but also time consuming. Despite, several genes have been tested on wide ranging higher animals, including cattle. The technology has developed to such an extent that it is feasible routinely to develop the transgenic animals.

Although it is a very expensive and sophisticated technology, with increasing knowledge of embryology and gene cloning on diverse group of higher animals, it will become a new and profitable area in biotechnology.

D. *Potential Applications of Transgenic Animals*

Some of the examples of the application of transgenic animals are :

* *The production of a large amount of specific protein (therapeutic) in farm animals.* For example, tissue plasminogen activator (tpa), blood-clotting factors VIII and IX, human serum albumin, subunit-vaccine development are feasible. In fact the transgenic livestock animals can be considered as a bioreactors. The commercial potential can be exemplified with an example. For instance the worlds requirement for factor IX, a key element in blood clotting cascade required to manage haemophilics is around few kilograms (1-2 Kg) of purified protein per year. The value per gram oscillates between US $ 25,000-150,000 i.e., a staggering market of 25-150 million dollors, which can be met with a small flock of lactating ews and cows, if only the factor IX gene expresses in the mammary gland of animals and produces the same level of protein as that of endogeneous milk protein.
* *Change in the quality and composition of milk.* That is to say that bovine milk is humanised and become lactose – deficient so that a special quality of milk can be marketed for the use of a large number of human population who are intolerant to lactose.
* *Introduction of disease resistant genes.* Many more applications are feasible. It requires capital investment for research and development. Further boost in the development of transgenic animals for commercial production of drugs and therapeutic gene products largely depends upon the technological advances in embryology, cell biology, physiology, biochemistry, genetics and molecular biology, when integrated successfully. The list of some of the possible products through transgenic animals is given in Table 10.2.

Table 10.2 *Some of the products of Animal cell Biotechnology*

Kind of product	Utility
Vaccines	
Poliomyletis Rabies Rubella Mumps	Prevention of viral infection
Interferons	
Interleukins IL 1, IL 2, etc.,	Prevention and treatment of infection Anticancer, immune modulation and potentiation
Growth factors	
GM – colony stimulating factor	Immune stimulator
Erythropoietin, Epidermal growth factor	Promotes the growth of blood cells Stimulates the growth of skin and tissue
Fibroblast growth factor	Promotes tissue growth
Blood Regulation proteins	
Auriculin	Hypertension control
Prourokinase	Dissolves blood clots in strokes and heart attacks
Tissue plasminogen activator	

(i) *Evaluation of Vaccine using Transgenic Animal*

Transgenic animals such as mice are being developed for use in evaluating the safety of vaccine prior to their administration to humans. For instance, the aptness of employing transgenic mice to test the safety of poliomyeletis virus vaccine is presently being investigated. If the use of transgenic mice for evaluating the efficacy of vaccines is found to be most efficient and dependable. Very soon the need of testing large animals, like primates will be dispensed with to test the safety of batches of the vaccines.

(ii) *Toxicological Testing for the Safety of Chemicals / Drugs*

Chemicals and drugs intended for treating the maladies especially inborn errors of metabolism have to be tested for their toxicological properties and side effects, if any, using a battery of tests, often employing several animals. This procedure is called as *toxicity* or *safety testing*. Cloning technology enabled the designing of animals that carry genes which makes them ultrasensitive than normal animals to toxic substances. This enables the evaluation of drugs and chemicals designed for therapy more rapidly and that too with few test animals. This saves lot of rigour and the expenses of the pharmaceutical companies that otherwise they have to put up with in the development of chemical molecules as drugs for alleviating human suffering.

(iii) *Organ Donors for the Transplantation*

Organ transplantation is the frontier area of medicine, where there is a growing need for the replacement of damage or dysfunctional vital organs like kidney, liver, heart valves etc. Recently transgenic pigs containing human genes in their cells have been developed with aim of providing organs for patients who require a transplant. Normally, an organ transplant from animal world will be rejected by a patients immune system. Immune system is a complex dynamic system involving protective cells, tissues and organs in the body that especially recognise and destroy alien organisms, which otherwise could harm the host organism. Immune system is also capable of recognizing the animal cells if transplanted to humans as 'foreign' and go into destruction mode. However, with the insertion of human gene that results into the presence of human protein on the surface of the animal cells, it is expected that the transplanted organ will be recognised as 'human' or accept as into graft by the immune system of the patient, and will not be destroyed by the cells that protect human bodies. There are two difficulties with this kind of approach. Firstly, just adding one human gene to the animal cells is not enough to block the action of other components of the very complex human immune system, which still could reject the grafted organ of animal origin. Secondly, there is a particular concern that such operations could help bacteria and virus endogeneous to animal may turn to human by changing their specificity and cause the spread of disease from animals to humans.

(iv) *Transgenics for Agricultural Production*

Several transgenic farm animals have been successfully produced including cows that produce more milk, sheep that produce more wool and fish which can grow bigger or can endure and survive colder temperatures than normal. But there is a growing concern by Animal Welfare Activists about the use of transgenic animals for agricultural production, since these animals suffer from various infirmities and hence their use is considered redundant and unnecessary. For instance, the transgenic pigs with the cows growth hormone gene were found to fatten pigs faster and yield increased quantity of meat, but nonetheless were found to suffer from arthritis, ulcers, kidney diseases and fecundity related problems.

(v) *Moral Concerns and Welfare Problems Associated with the Production of Transgenic Animals*

Many opponents of transgenic technology and animal activists object to transgenesis because the production of transgenic animals can severely affect their welfare. Since some of the protocols for the production of transgenic animals give unpredictable results consequent to the insertion of the foreign gene randomly in wrong places in the genomes of their cells. Consequently the transgenic animals suffer from congenital diseases, deformities and organ failure leading to the suffering of the animal. Furthermore only a small percentage of embryo used in the study express the foreign DNA and not all of them will survive. Surviving animals that do not express the transgene are quite often killed. It is a cruelty. The transgenic animals used as a specific disease model for humans are designed to develop diseases as close to as that of humans, and consequently suffer pain and distress unnecessarily. Animal activist object to this kind of research on moral and ethical grounds. Besides, there might be far reaching consequences from the evolution point of view.

10.3 Animal Cell, Tissue and Organ Culture

Introduction

The animal cell, tissue and organ culture has been a difficult area until recently because of poorly characterized techniques and cell lines. The revolutionary developments in biotechnology initiated significant changes in animal cell culture technology for the past few decades. The discovery of biomolecules with immense potential for therapy and diagnostics has fuelled the commercial motivation for the rapid development of cell culture technology. The advanced analytical techniques and methods of genetic modifications at the cellular level with parallel developments in understanding molecular and cell biology at all levels of complexity of organization have provided new research tools to help and bring the benefits of the new discoveries to the human welfare as a distinct technology.

Animal Cell Culture Techniques

The '*architectural design*' of animal cells in form, structure and function has been evolved by millions of years evolutionary process. Unlike the microbial cell system, like bacteria or yeast, the work horses of industrial biotechnology, which are relatively simple and well equipped to survive, grow rapidly and reproduce in a range of diverse environs, the animal cells are complex, recalcitrant and hence have to be coaxed to respond to *in vitro* cultural condition.

Typically, the animal cells have a fragile membrane, an intricate network of circulatory system for providing nutrients as well as to drain the waste products with a natural environment that features a high cell density up to 10^9 cells per cm^3 of tissue. Besides high degree of control of p^H, temperature etc., regulation by cascade of hormones and growth factors are needed. Any individual cells are typically differentiated to perform a specific function designated in the tissue or organ. This process of differentiation commits the cell in the direction of a specific course and annul the most of the available genetic information. From this point of view, there is no such thing as *typical animal cell.* Any cell line cultured and maintained *in vitro* should be scrupulously characterized with respect to its original source tissue and the manipulations that have been made to bring to the current status of pure cell lines.

Method of Development of Pure Cell Cultures and Cell Lines

The term tissue culture broadly refers to *in vitro* culturing of organs, tissue explants or dispersed cells in a specific medium that promotes their growth and maintenance. Accordingly it is divided into

(a) organ culture and (b) cell culture mainly differentiated based on the presence or absence of tissue organization.

Historically the onset of animal cell and tissue culture goes back to the work of Arnold (1880), who demonstrated that the white blood corpuscles (WBC, leucocytes) can divide out side the body (*in vitro*). Subsequently, Jolly (1907) studied the animal tissue cultures grown in serum, lymph and ascite fluids. However, the first documented endeavour to grow animal cells *in vitro* is credited to Ross Harrison (1907), who cultured the frog embryonic nerve cells following the protocol of hanging drop technique, which later extended by others, who standardized methods for cultivation of a wide array of animal cells.

In organ culture, the entire embryonic organ or a part of a tissue (explant) derived from the organ is cultured *in vitro* in such a way that it retains its tissue organization and identity. Whereas, the cell cultures are obtained by either mechanical or enzymatic disintegration of tissues to yield individual cells and are maintained subsequently in mono–layer, or as cell suspensions in a defined media. Freshly isolated cell cultures are often referred to as primary cultures, which are highly heterogeneous, slow growing but the properties of cell types of their origin is retained and manifested.

Once the primary cell cultures are obtained, the same are subcultured to generate cell lines. Two types of cell lines are obtained (1) *finite cells lines*, characterized by their termination after a finite number of subculturing *in vitro* and do not survive in suspension cultures for several generations. These features limit their utility. On the other hand, (2) *continuous cell lines,* distinguished by their capacity to proliferate indefinitely and hence immortal. However these cell lines exhibit chromosomal instability in their genomes and some times fail to respond to hormone signals as well as for growth factors. As a thumb rule, the normal tissues provide finite cell lines and tumour cells give rise to immortal cell lines, though there are exceptions. For instance, MDCK dog kidney cell line, 3T3 fibroblast cell lines etc; are obtained from normal tissue, but can be cultured continuously. He la cells are of oncogenic origin.

Establishing and Handling Mammalian Cell Lines

Management of cell lines in terms of documentation, cataloguing and handling is critical for meeting regulatory prerequisites in establishing a reliable process for the possible use in biotechnology industry. Once an useful cell line is established, an aseptic sample of productive cell line should be preserved and maintained in vials or ampules in a liquid nitrogen to form a master seed lot from which one of them expanded and used to form a working lot. The frozen master seed lot should be replicated and maintained independently at two separate locations. Both the master and working seed lot should be tested periodically to ensure the integrity of manufacturing process. The pure cell cultures should be free from any sort of microbial contamination either aerobic or anaerobic including bacteria, fungi , mycoplasma and viruses. Cross contamination of cell lines is a major problem. Developing any standardized test for identification of cell lines is also necessary. Table 10.4 lists the procedures available for testing and characterizing mammalian pure cell lines. The flow diagram (Fig. 10.7) outlines how cultured cells from somatic tissue is obtained.

Laboratory-Scale Culture of Animal Cells

Several methods for culturing of animal cell lines are available at laboratory scale from seed lot once a desirable line gets isolated and established. One of the earliest methods, the researchers of animal cell

Table 10.4 *Methods for analysis and characterization of cell lines*

A. *Microbial contamination tests*

Type of contamination	Method	Observation
Bacteria/Fungi	1. Microscopy	Presence of cell colonies
	2. Nutrient *broth* culture	Turbidity in broth to test fungal/ bacterial contamination
Mycoplasma	1. Nutrient broth	Mycoplasma morphology, +ve
	2. Immunofluorescence	immune reaction to antibodies
	3. DNA testing	Southern blotting (hybridization to mycoplasma DNA)
Viral		
MAB testing	1. Antibody production test	Detection of antibody production
Cytopathogenicity	2. Co-culture with sensitives cells	Pathological effect
Retrovirus	3. Enzyme assay	Detection of reverse transcriptase
B. *Cell line characterization tests*		
Type of test	**Method**	**Observation**
Karyotyping	Standard metaphase spread of chromosomes post colchicine treatment of cultures.	Establish, distinguish karyotype for referrence
Isozyme Analysis	Electrophoresis	Zymogram profile
Antibody characterization	Immunological technique	Distinct antigen antibody reactions
Nucleic acid hybridization	Southern blotting	Distinct pattern of Hybrid bands
Immunofluorescent detection of cell surface antigens	Immunofluorescent techniques	Characterization of surface antigens

culture got established was direct injection of cancerous cells in to a susceptible host animal to isolate different tumour lines. This protocol even now has a commercial importance for the production of antibodies from hybridomas. The methodology is simple, elegant and efficient in yielding significant quantities of specific antibodies.

In vivo antibody production : It starts with the injection of hybridoma cells into the peritoneum of primed nude mouse. After 4-6 weeks *aliqots* of ascites fluid (about 10 ml) can be drawn at frequent intervals regularly, which contains a concentration of antibodies (immunoglobulins) in the range of several mg ml^{-1}. However, when large quantities of product for purposes of commercial necessitated this, *in vivo* approach is not suitable and economically feasible. Besides this methodology is severely limited with handicaps like, low reproducibility and the presence of other protein of host animal.

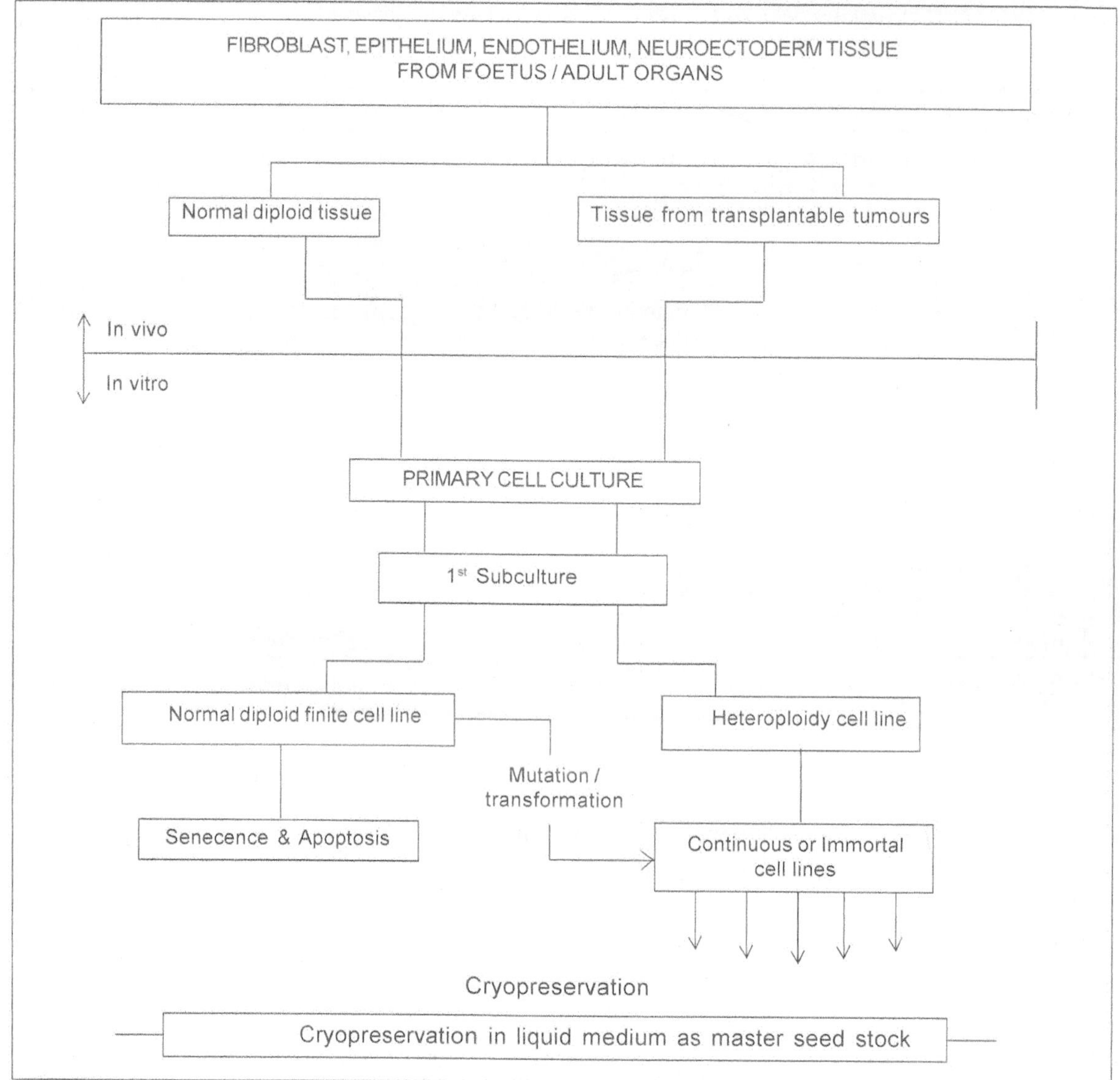

Fig. 10.7 Method for obtaining pure cultured cell lines from somatic tissues

Suspension Cultures

The success of in vitro cell culture technology for economic purposes depends on the kinds of cells to be cultured. There are two modes of culturing the cells; based on, the cell type suspension and Ancorage dependent cells. The brief outline of method of culturing is presented below :

Once the suitable well defined media that support cell growth optimally conditions for physical attributes are established and, the culturing procedure commences either for laboratory scale/or commercial scale. Accordingly suitable reactors are designed. At laboratory-scale of culturing, different types of reactors are employed. Essentially, the suspension cell cultures are made in batches, (hence the name '*batch culture*'), in suitable culture containers in a semi-continuous mode. By repeating the culturing in batches at regular interval with replacing bulk of the media and maintaining physical conditions. This procedure is similar for the suspension cell culture at laboratory scale or industrial scale.

The most important factor that has bearing on the success of suspension cell cultures is that the cultures should be free from any stress as far as possible. The stress can result into the selection of sub population of cell lines with low potential for the targetted product, beside several other undesirable attributes. The factors that contribute to stress are : non-availability of enough oxygen, nutrients, accumulation of toxic metabolic products, change of p^H, stationary or dilution below critical level, *non-stirring* etc. Cultures in closed flask, which has little scope for agitation generally depict gradual reduction in growth rate of cell and ultimately lead to the cessation of cell-growth. This phenomenon is called as stationary phase. As the cell population nears 1.5×10^6 cell ml concentration in the media, the stationary phase is attained. One of the strategy employed in semicontinous culture to go smoothly is to replace the medium at regular interval well before the critical concentration of cells in media (about 80% of the maximum concentration) is reached (1.5×10^6 cell ml^{-1}).

T – flask

When there is a need for a small scale culturing (10 – 50 mL) the obvious choice is to culture in *T – flasks* (Tissue culture flasks) which are nothing but a rectangular container made of prism glass with a raised rack to keep away the liquid media from the cap. Typically, when sodium biocarbonate buffer is used, it is possible to maintain the P^H within a tolerable limits employing 5% carbon dioxide in the environs. In order to facilitate gaseous exchange through the gas-liquid interface, the volume of the media in the vessel should not be filled beyond 20 % of the total volume of the cultured flask.

Spinner flask

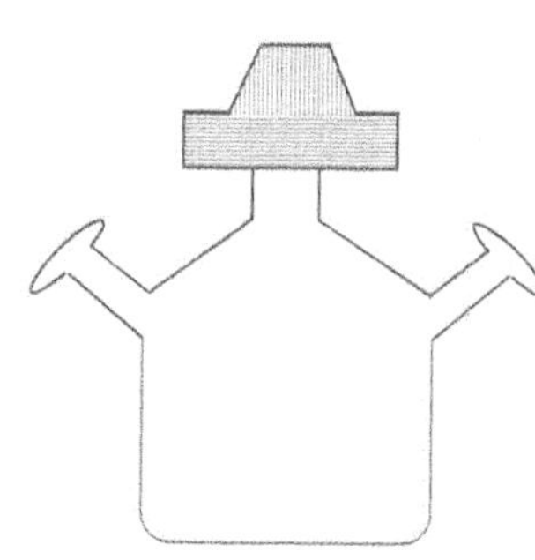

When larger volumes of cell culturing is needed, the agitation of cultures is necessary; the cell concentration as well as productivity of targetted compounds can be enhanced by such provision. To achieve this, *spinner flasks* are employed. These are cylindrical glass vessels having a suspended magnetic stirrer (bar) on a swivel and side arms for addition or withdrawal of inoculum or medium components. Generally about 50% of the flask volume is filled with the culture medium. The spinner flask may serve for routine cell culture needs even for large volumes upto 10 litres. Beyond this scale of volume aeration through the liquid surface hampers the cell growth.

Culturing of Anchorage - Dependent Cell Lines

The culturing of anchorage - dependent cells is a bit delicate process needing conditions that increase the surface area for the cell attachment. These cell lines can be initiated and be maintained in T-flasks routinely. However, the spinner flasks are not employed in this case, since such contraptions do not help to provide enough surface area for cell attachment. The maintenance of anchorage-dependent cells need altogether new techniques as compared to suspension cell culturing. Besides, fastidious care, density and physical condition are needed.

Attachment

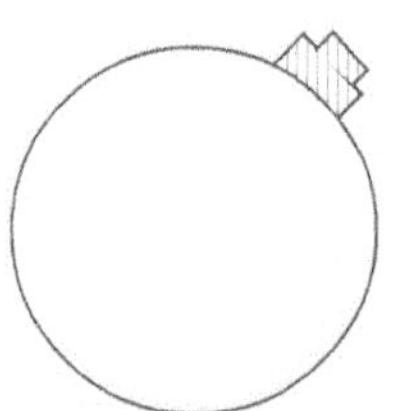

When cells are provided with a compatible surface the same will attach, spread physically and assume into a flattened conformation. Subsequently they commence to reproduce by cell division and permeate along the surface available. The process of attachment depends on physical and biological factors like, the nature of cell line and their ability to produce their own attachment factor, density of cells, the history of the culture and the presence of attachment factors in the medium, mixing conditions and the surface characteristics etc. The widely used material in the culturing of anchorage-dependent cells are, glass, polycarbonate and treated polysterone etc. As cells near the confluent growth, the growth rate will either stop or come down to a bare minimum. Many types of cells continue to grow even after reaching the confluence growth stage forming several layers or leak cells into the medium.

Spreading

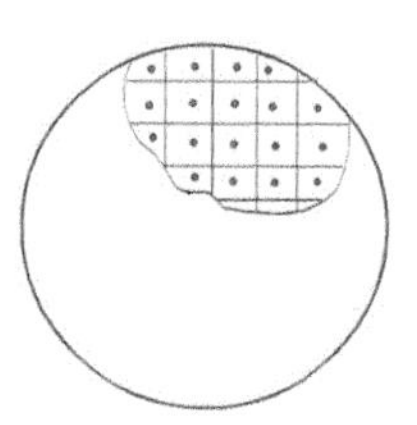

Subculturing of these lines necessitates removal of an aliquot from the surface and is placed on the fresh media surface for attachment after a proper dilution. To detach cells from the surface a solution of trypsin or EDTA is used. Since trypsin can damage the cells, such an exposure is generally kept for the minimum necessary for detachment of cells. To obtain larger culture volumes of anchor dependent cell lines, the Roller bottles are employed.

Confluent monolayer

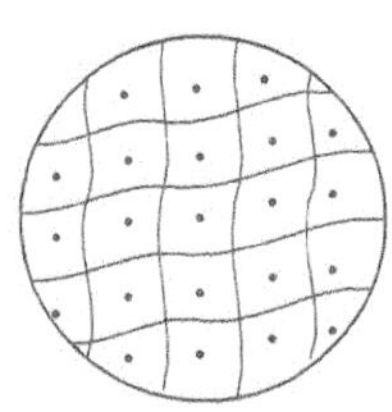

Roller culture bottles are made of plastic, generally filled to about 20% (1/5th volume) with culture fluid. The culture bottles after inoculation are paced or roller apparatus alternately submerging the cells and exposing them to co_2 enriched air. The routinely used Roller bottles the size of 850 cm^2 capable of holding 500 m*l*. For the commercial scale of production, these are unwieldy as there is great requirement for material. However, the use of microcarriers in the culture flasks will enable to expand the volume of anchor dependent cell cultures. However, for commercial large scale production reactor technologies are employed.

10.4 Stem Cell Research and their Potential Applications

Stem cells are special types of animal cells that can be isolated from inner mass of cells or stem of embryo. Hence the name stem cells. These cells initially show no tendency to differentiate into a specialized tissues but are initiators of the process of differentiation subsequently upon receiving the right kind of signal for gene expression during growth, development and differentiation. Until such phase, these develop and multiply without differentiation. However, these are pluripotent in the sense that once the right signal is transduced, cells commit themselves for differentiation and change their morphology, structure and function to develop into a specialized tissues and organs. For instance, muscle cells (myoblasts) differentiate to produce the long fibrous muscle tissue. The cells of the eye lense epithelium undergo morphological changes from a compact globular shape into long fibrous entities that make the eye lense. A stem cell has the capacity to divide without limit and give rise to progeny which can be terminally differentiated. The stem cell development does not follow one-way path differentiation. When it divides, the daughter cell has an option of remaining as stem cell like its progenitor cell or to follow one way route of differentiation depending upon the right kind of cell signal these receive and commit towards the terminal differentiation (Fig. 10.8). Hence, the stem cells are called pluripotent (analogous to totipontency in plants) and can give rise to diverse cells and tissues and organs in the embryo.

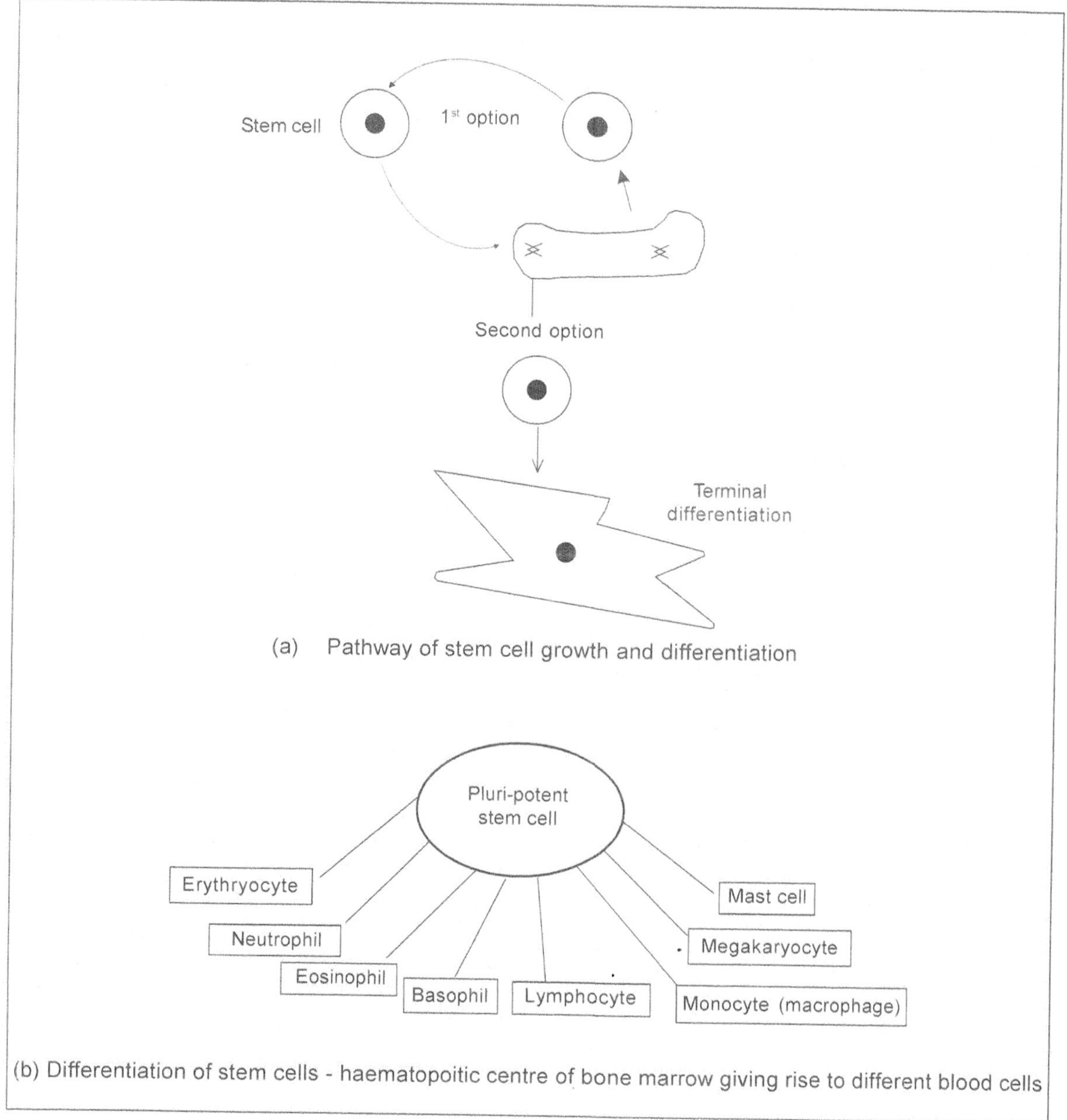

Fig. 10.8 Stem cells, their two-way pathway and pluripotency

Application of Stem Cell Research

Embryonic stem cells are an ideal route for the introduction of foreign genes into a whole animal to produce transgenics. For transfection (introduction of exogenous donor DNA) into embryonic stem cells, both the microinjection and viral mediated gene transfer can be used. Following transfection of ES cells, appropriate *in vitro* selection methods can be deployed to isolate the transformed (transgenic) cells expressing the inserted DNA, which subsequently can be utilized for transfer to the early embryo. The embryonic stem cells are in active stage of research and development, especially for the production of clones of transgenic animals/cattle that are genetically modified towards the desired direction. Recent developments in cloning of cattle (sheep, goat, cows, etc.) is unique in that the transfected embryonic stem cells were fused with an enucleated mature oocyte and reintroduced into pseudopregnant females so that the genetically modified animals are produced with a felicity.

Another important application embryonic stem cell system is targeted insertion of transgene in place of endogenous gene by a process what is known as *Homologous recombination.* Homologous recombination is a rare occurrence. Such recombinants can be identified by a battery of sophisticated sensitive tests against the noise of random insertions. Nonetheless the potential of this systems is enormous in principle, any kind of transgenic animal with a desirable gene added can be generated. Genes could be mutated in directed way, indigenous gene could be made a null gene and be replaced by a mutant version or by an analogus gene from a different species, and defective genes can be repaired at will.

Some of the applications in health care sector is presented in Table 10.5. The potential is so enormous that in future all worn out tissue, organs or cells can be replaced with new ones. A revolutionary bio-tissue/organ trading industry is possible due to new development in biotechnology. Stem cell research which was banned in many countries in Europe is now revoking by legislation on the stem cell research.

Table 10.5 *Possible use of tissues derived from stem cells for health care*

Cell type differentiated from the stem cell	Target disease for the treatment
Nerve cells	Stroke, Parkinsons's disease, Alzeimers disease, spinal cord injury, multiple sclerosis
Heart muscle cells	Congestive heart failure, heart attack
Insulin producing cells (Islets of Longherhan)	Diabetes
Cartillage cells	Osteoarthritis
Blood cells	Cancer, immunodeficiency, inherited blood diseases (Hemaglobinopathies, Leukaemia)
Liver cells	Cirrhosis of liver, Hepatitis
Skin cells	Wound healing, burns
Bone cells	Osteoporosis
Eye cells	Macular degeneration
Skeletal muscle cells	Muscular dystrophy

Essay Type Questions

1. Discuss about the applications of animal biotechnology to the welfare of mankind.
2. What is genetic engineering? Explain the gene cloning strategies involved in it.
3. Illustrate the mechanisms of production of transgenic animal. Add a note on importance of transgenic animals to biotechnology.
4. Give a critical account on animal cell, tissue and organ culture.
5. What are stem cells? Write an account on scope, potential and application of stem cells to biotechnology.

Short Answer Type Questions

1. Write short note (about 300 words) on the following:
   (i) Animals as resource for biotechnology
   (ii) Genetic engineering
   (iii) Transgenic animals
   (iv) Cloning strategies
   (v) Gene transfer techniques
   (vi) Animal cells culture techniques
   (vii) Cell lines and their application
   (viii) Stem cells and their potential

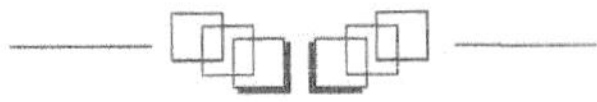

11 Molecular Medicine and Medical Biotechnology

11.1 Clinical Genetics : Genetic Counselling

With the advances in Genetics and Molecular biology, especially in the areas of gene cloning and subsequent developments in genomics and proteomics, our understanding of the role of genes in pathological processes has improved to a great extent, so much so, that it is not only possible to identify genetic disorders more precisely, but also it is possible to provide an accurate genetic counselling and prenatal diagnosis for genetic diseases and offer ultimately gene therapy to improve the health and alleviate suffering of many people afflicted with such conditions. Some of the areas where the recent developments of molecular genetics have impacted genetic-medicine are given below :

* **Medical genetics** will emerge as a new discipline of medicine exclusively dealing with the mode of inheritance, diagnosis and treatment of genetic diseases which may be due to gene mutation, chromosomal mutation and multifactorial causes. Two more aspects that follow suite are : genetic counselling and screening.
* **Clinical genetics** is the realm of application of genetics to clinical problems to manage the individual families and patients suffering from genetic diseases, including gene therapy.
* **Dysmorphology** is the area that studies and analyses abnormalities of development especially morphological.

The management of patients with genetic disorders involves the process of diagnosis and treatment which are basically the same as applied in other super specialities. However, what is unique to medical genetics is, it concentrates on prevention or avoidance of disease. Main emphasis of this approach is to focus on presymptomatic diagnosis, genetic counselling, screening for genetic disorders and prenatal diagnosis.

A. *An approach to clinical genetics*

To deal with innumerable genetic disorders, and malformation that float in population it is essential that one need to have a thorough knowledge about anatomy and physiology of normal condition so that the clinical presentation of the same can be done with a high degree of adroitness and maturity. In other words, a precise diagnosis is a must. This varies according to the severity of condition, in terms of whether there is an abnormal fetus, still born infant, dysmorphic infant or dysmorphic adult or an adult with a suspected genetic abnormality. Each condition necessitates gathering of information as much as possible, and analyse the data to formulate a hypothesis regarding the diagnosis and estimate the most likely recurrent risk. It is a difficult task as the family may not be aware of the seriousness of the problem. Social ostracism or stigma attached to being affected with dementing disorder may be the cause of not forthcoming with information, but nonetheless becomes a delicate problem to handle at the time of genetic counselling. Despite problems, all data to the extent of possibility in term of accuracy and completeness, must be recorded. Family history plays a key role in clinical genetics. If properly recorded in terms of relevance to the current genetic illness, period of onset, information about similar affliction among first degree and other informative relatives, racial and ethnic back ground, finally inquiry about consanguinity will be an asset for genetic counselling. There are three reasons at least for the necessity of family history. Firstly, it is one of the most useful and accessible information tools that provides definitive aid for correct diagnosis. For instance, if a person comes to a genetic clinic with a pain in chest, typical of unstable angina pectoris, with the clinical information about the existence of coronary artery disease in family, one might suspect it to be related familial hypercholestremia and diagnose very fast by necessary tests and suggest appropriate therapy and prevent fatal heart attack later. Similarly, a lump in a breast of a patient, if information is available about the history of breast cancer in the family, can help the patient very early to prevent suffering from metastasis and full blown cancer later. The second reason for family history is, it provides a means of accurate prognosis. The third and most importance utility of family history is for the presymptomatic diagnosis of genetic disorder and the prevention of full blown clinical disease later.

B. *Construction of the Pedigree*

The meticulous recording of family history of *proband* in terms of similar occurrence in different filial generation constitutes the pedigree. The starting point is data gathering about genetic diseases despite the availability of sophisticated molecular analysis. A pedigree is a graphic representation of the occurrence/non-occurrence of the genetic disease in different generations of a family with various symbols used for representing relationships and for clinical findings. Pedigree analysis will provide an insight into mode of inheritance, differential expression in sex etc., which is a crtical information for *genetic counselling*. The typical pedigree of autosomal dominant, recessive and sex-linked inheritance along with a variety of symbols used is shown in Fig. 11.1.

C. *Genetic Counselling*

Genetic counselling is a communication process indicating individuals or families having a genetic disease or at risk of having such a disease. With the recording of accurate information about their condition and to provide information that would allow couples at risk to make informed decision about having children in future. The following definition of genetic counselling was adopted by the adhoc committee on genetic counselling that was chaired by C. J. Epstein as suggested by the American Society of Human Genetics (1975).

"Genetic counselling is a communication process which deals with the human problems associated with the occurrence or the risk of occurrence, of a genetic disorder in a family. This process involves an attempt by one or more appropriately trained person to help the individual or family to : (1) comprehend the medical facts including the diagnosis, probable course of the disorder and the available management, (2) to appreciate the way heredity contributes to the disorder and the risk of occurrence in specified relatives, (3) understand the alternatives for dealing with risk of occurrence, (4) choose the course of action which seems to them appropriate in view of their risk, their family goals, and their ethical and religious standards, and to act in accordance with that decision, and (5) to make the best possible adjustment to the disorder in an affected family member and / or to the risk of recurrence of that disorder". (Am J Hum Genet 27 : 240 - 24 - 2, 1975).

(i) *Situations when genetic counselling is needed*

The need for genetic counselling is generally indicated in following cases :

- When pregnant mother belongs to advanced age group i.e., beyond 35 years or older
- A family with the history of known genetic disorder or suspected hereditary condition
- Single and multiple birth defects or malformations in fetus or child
- Mental retardation that cannot be explained
- Recurrent spontaneous abortions
- Teratogen exposure either known or unknown
- Consanguinity

(ii) *Constitution of the Team for Genetic Counselling*

The medical specialists trained in pediatrics with additional training as a clinical genetics is ideally suited for being in the team of genetic counsellor. The general training in genetics also will be useful. The role of clinical/medical geneticist is to establish the diagnosis of the disorder accurately and to counsel the patients about the medical implications of the disorder clearly. Often any other person with a nursing background, social work and education with a master's degree in genetics also can act as a

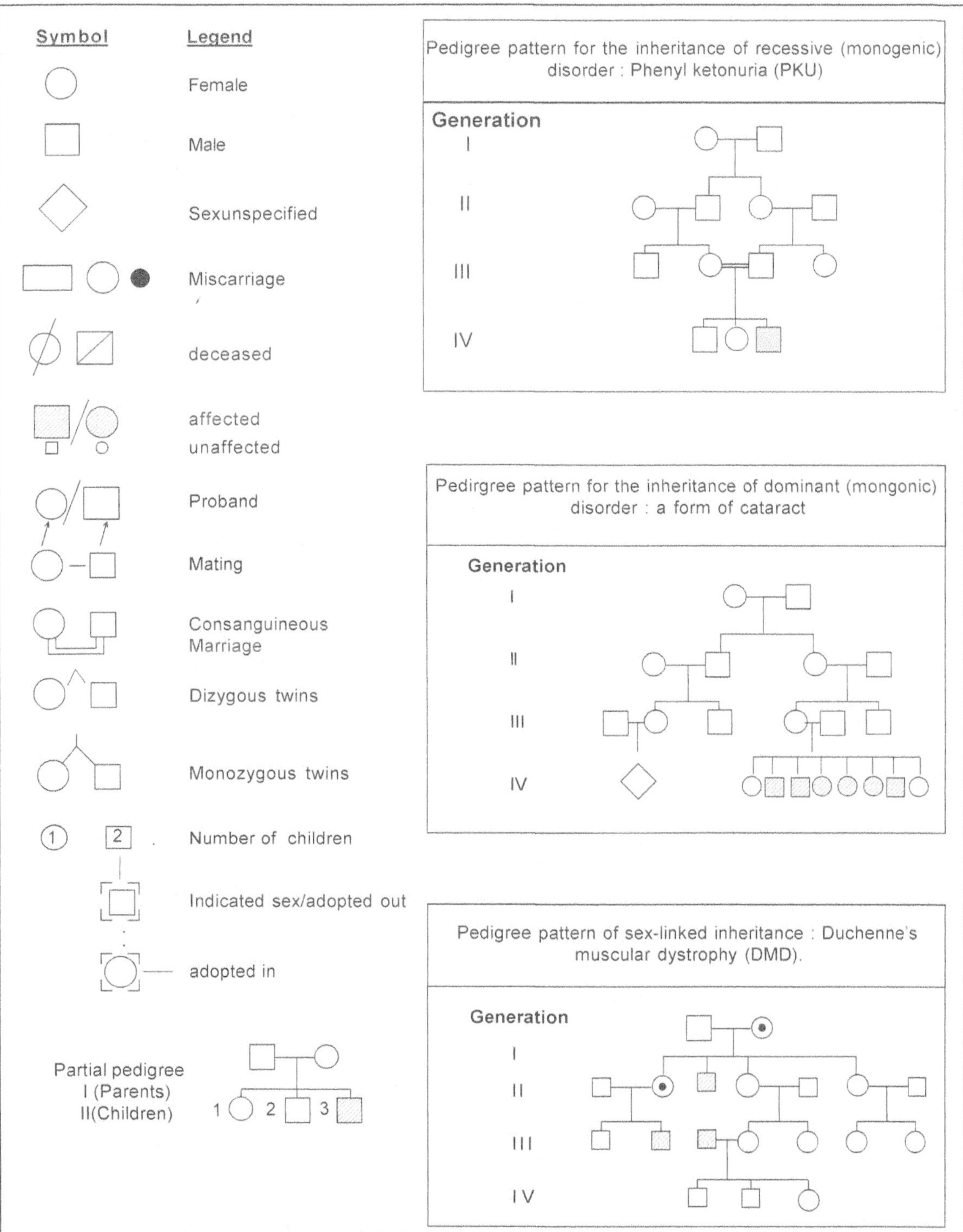

Fig. 11.1 Pedigree pattern of inheritance in some disorders

counsellor. The role of genetic counsellor is to be involved in the process of counselling besides following up of the family and give moral support. One more member can be included additionally in the team, who could be a medical specialist being consulted for specific investigation and opinion, or a social worker/religious support person parent for ongoing family support.

Genetic counselling facilities are often provided by the teaching university-based testing care hospitals or by a component of a specialised clinic, or as a part of screening programme and / or out reach programme.

(iii) *Genetic Counselling Goals*

Following are the goals of genetic counselling as outlined by American Society of Human Genetics (1975) :

(a) ***Comprehending the medical facts*** - As soon as the diagnosis is made accurately, the family should be informed about the case, prognosis, apt investigation (s) and on going care about the condition.

(b) ***Understanding the pattern of inheritance and the possibility of recurrence risk.*** The family should be clearly explained and informed about the mode of inheritance and the calculated risk (at least empirical), based on the knowledge of Mendelian genetics or risks observed in population. The origin of risk may pertains to a specific diagnosis, say for example *cystic fibrosis,* the recurrence risk after one child is 25%. If the diagnosed condition is non-specific one (as in the case of non-specific mental retardation) where the cause is unknown the recurrence risk calculation is only empirical.

It is better to explain the binary terms i.e., the condition may or may not occur again, but should be told unambiguous about losses and gains leaving the decision to couple whether to go for another pregnancy. The numerical risk in terms of percentage should be given to them.

(c) ***Understanding the options when confronted with a discrete recurrence risk.*** The couples should be informed about the other mode of reproductive options such as, methods of birth control, adoption, artificial insemination by the donor sperm or surrogate mother and use of donated ova, prenatal diagnosis with or with abortion and unmonitored pregnancy.

(d) ***Choice of the course of the action.*** The couple should be advised to choose the best course for themselves based on the recurrence risk data, and the perceived social economic and psychological burden on the family, keeping in view the goals of family and their religious and ethical standards.

(e) ***Helping the family to adjust to the condition.*** The role of the counsellor is to assess the family's reactions, followup with a panel of medical genetic team or psychologist to help family in adjusting to the condition.

(iv) *The Process of Counselling*

To accomplish the goals of genetic counselling, it is essential that the family is in a position to comprehend the information given and convert the same into a part of their decision making. The role of the genetic counseling team becomes very important at this point and these people will be in a better position to interact with the family most effectively in following ways :

(a) Reconing the psychological and emotional burdens associated with the diagnosis like guilt, masked or manifest, realistic or unrealistic shame etc.

(b) Taking into account all the factors that influence counselling, such as, diversity of ethnic background, socio-economic status etc.,

(c) The advise given should be non-directive giving full disclosure of option without being paternalistic or any bias for the preferred action. However, other family examples can be given.

(d) Tackling the problem of feelings of shame and guilt.

Since the counselor may become a figure of authority during the period of investigation and diagnosis, he should be in a position to relieve the sense of feeling of guilt and shame by the parent for the genetic disorder. The family should be advised and told that the particular genetic disorders is not their making and hence is not their fault so that subsequently the family member may remember this and get comforted. One should address the feeling directly and normalize them by comparing other family members with similar condition.

The parent's sense of anxiety and responsibility for the child's problem may be told in different way by saying that "*it is normal to feel that responsible and be anxious worrying about child's care*" while they are not cause of the problem itself. To make feel better and positive, it is advisable to make them clear that which aspects of the child's condition can be controlled and helped as well as inform which aspect is beyond their (human) control.

It is also necessary to help the family to identify its preferred goals by asking the family in the initial stages itself, its expectation of the counselling session and set a clear agenda, which might permit the counsellors to address other related issues. One of the early goals is to understand medical implication and later to decide whether to go for another pregnancy or not. Given the necessary information and option, the family is helped by the counseling process. The counselor biases should not be manifest in any form. Sometimes, family might make a choice contrary to the preferred choice indicated by the counsellor. In that case the counsellor has to test different options available in any particular order but in a non-judgemental way leaving the final choice to the family to choose from different options.

(iv) *Problems faced in counseling*

Genetic counselling is often not an easy process. In the course of counselling certain factors might make it difficult to arrive at the accurate diagnosis. For example, genetic heterogeneity often complicates the diagnosis. Similar genetic disorders due to mutation at different loci or involving different alleles may have different pattern of inheritance. For instance, *retinitis pigmentosa* might be inherited either as an autosomal recessive or dominant or sometimes as a sex linked inheritance. It becomes difficult to arrive at estimating the risk. Sometimes the disorder could be due to phenocopies mimicing the condition in certain environment which will not recur. But to arrive at the accurate diagnosis of the condition whether it is due to phenocopy or true gene mutation, is difficult to assess based on the meagre data. For example microcephaly often encountered in pregnancies might be of genetic or non-genetic origin. The sporadic occurrence of a genetic disorder presents a different problem. The pedigree analysis will not be helpful, nonetheless the genetic counsellor must depend on accurate diagnosis. The advent of molecular diagnosis has been helpful as in the case of Duchennes muscular distrophy due to deletion.

The genetic counsellor might confront with the ethical and moral problems if there is an inadvertent discovery of non-paternity, which invalidates the risk assessed in a counselling process.

11.2 Prenatal Diagnosis of Hereditary Diseases

The term prenatal diagnosis refers to the screening and identification of any genetic disorders or deformities in the embryo before the birth (pre = before, natal = birth). It is a rapidly developing field encompassing both screening and definitive testing with a commercial potential. The primary objective of prenatal diagnosis is to offer prospective parents the assurance of having normal children when the risk of having children with genetic diseases or malformation is especially high. Prenatal diagnosis allows one to convert a probability statement about the risk of genetic disorder or disease to a certainty.

Prenatal diagnosis represents a paradigm shift in clinical medicine with the biotechnological developments to treat and diagnose several genetic and metabolic disorders. The development of various diagnostic procedures resulted from advances in obstetrics and molecular biological discoveries, including human cell culture and cytogenetics etc., during 1960-1980. In Europe and America consequent to the planned birth control and reduction in family size, with an emphasis on a positive outcome of each pregnancy, the prenatal diagnosis had become an imperative need. Furthermore, physicians have also recognized the necessity to ascertain the genetic and environmental risks associated with a pregnancy and the need to be aware of available parental diagnostic services.

Prior to offering prenatal diagnostic technique, including rDNA and other molecules as a service, several criteria has to be met. The criteria are : (i) safety (ii) accuracy (iii) identification of risk population (iv) quality control (v) limitation of diagnostics. The prenatal diagnostic techniques can be categorized into two broad categories : non-molecular and molecular diagnostics.

A. *Non-molecular Diagnostic Techniques* (Fig. 11.2)

A brief description of non-molecular diagnostic techniques is given below :

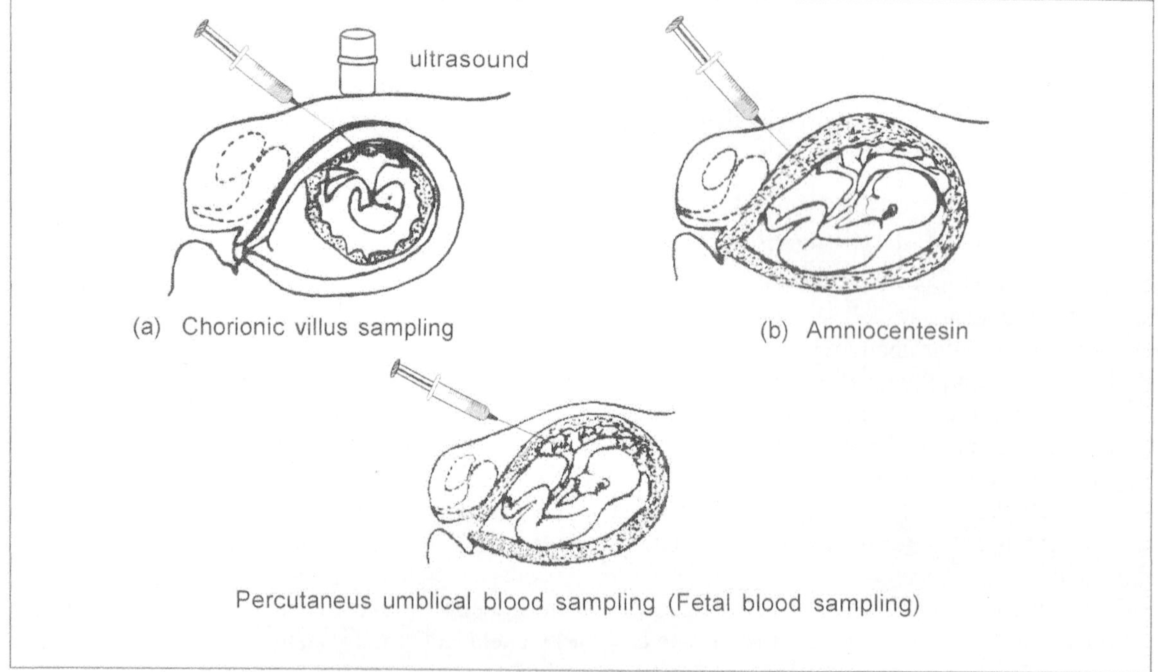

Fig. 11.2 Some of the non-molecular prenatal dignostic techniques.

1. ***Ultrasound and Doppler analysis :*** Ever since the first use of ultrasound in 1972 for prenatal diagnosis of anencephaly, many fetal structural abnormalities have been diagnosed using ultrasound technology. Innumerable uses of ultrasound in pregnancy are in vogue. Doppler analysis involving the generation of flow velocity wave form with a continuous wave is used to evaluate the flow of umbilical/placental/fetal blood circulation to detect fetal cardiac pathology. In normal pregnancies with no growth retarded fetus the blood circulation in cord/placenta will have low resistance with high flow rate, whereas in case of growth retarded abnormal fetus, placental blood flow rate may be abnormal with high resistance.
2. ***Chronic Villus Sampling (CVS) :*** It is a first trimester method of prenatal diagnosis involving the transcervical or trans-abdominal biopsy of the chorionic villi (developing placenta) with the help of a flexible catheter or aseptic needle, first employed in 1960s. The CVS was not used widely until the ultrasound technology was available. The chorion differentiates into chorion frondosum (placental site) and the chorion laeve (membranes). The first one, the frondosum is actively dividing mitotically. The villi originates from three different cell lineages, viz., *Polar tropectoderm, extraembyronic mesoderm* and *primitive embryonic streak*. Both the extraembryonic and the interior cell mass, but only a small number of cells (3-8), become progenitors of the embryo. CVS can any time during 9-12 weeks of pregnancy. The sample tissue can be analysed directly or cultured for both for molecular, biochemical and cytogenetic analysis (karyotype). The results generally take two weeks time.
3. ***Amnioscentesis :*** This procedure involves removal of 15-20 ml of amniotic fluid with a fine gauge spinal needles. The amniotic fluid which bathes the developing fetus during the early part of gestation period is believed to represent a transudate and its composition reflects that of fetal extracellular fluid. Within this fluid one finds cells sloughed off periodically from fetal skin and from epithelium urinary tract. As the pregnancy advances the amniotic fluid volume increases, major contributor to the fluid is fetal urine. Amniocentesis serves the useful purpose when done early period of pregnancy, generally during 14-17 weeks after the conception. The indications for amniocentesis are : neural tube defect/abnormal maternal serum α-fetoprotein, inborn errors of metabolism and disease diagnosis by DNA analyses.
4. ***Fetal blood sampling :*** Percutaneous umbilical blood sampling (PUBS) was first developed to diagnose disorders not expressed in cells sampled by amniocentesis or CVS. This technique attained wide popularity as a faster way of getting genetic information on the fetus as ultrasound resolution improved. Essentially, the technique consists of percutaneous puncturing of the umbilical cord near the placental insertion with a spinal needle under ultrasound guidance. The procedure can be performed from 14 weeks after conception to near term of the pregnancy. Usually 0.5-1.00 cc of fetal blood is removed, and the sample is immediately analysed to make sure that the sample is fetal origin. Results are usually made available within 48 hours.

B. *Applications of Molecular Techniques for Prenatal Diagnosis*

The expression of most of the genes is highly tissue specific. Hence, the prenatal diagnosis of hereditary diseases using biochemical approach is often limited to those genes which express in amniocytes. Deficiency of gene products that express tissue specifically, in a highly specialized and differentiated tissue cannot be diagnosed. For instance, the deficiency of ornithin carboxylase that expresses in highly specialized liver tissue cannot be detected through biochemical techniques. However diagnostics, can serve the function of detecting genetic diseases in any cell, since all the somatic cells are essentially the same and will have the sequence for all the genes. Hence, it is possible to detect any defects in genes (mutation) by analyzing DNA using molecular techniques.

Recombinant DNA technology has revolutionized the diagnosis and treatment of genetic diseases. Many of these diseases have their origin due to mutation resulting in either in the complete absence of a catalytic protein or in a defective protein that does not do its functions. The molecular biological techniques, especially, r DNA technology has immensely helped precised diagnosis and localization of the defects causing the hereditary disorders. Consequently, most of the diseases can be diagnosed prenatally in the fetus itself at an early stages of gestation period. Besides, the hither to undetectable heterozygotes (carrier state) of some of the genetic abnormalities can now be identified. As a result prospective parents have an option of making an informed choice about whether to continue with the pregnancy or terminate it. Prenatal diagnosis by DNA analysis for selected genetic diseases, under the control of single genes amenable to prenatal diagnosis is presented in **Table 11.1**.

Table 11.1 *Molecular Biological Techniques for Prenatal Diagnosis of Genetic Disorders*

Genetic Disease	Diagnosis by Molecular Technique
Autosomal dominant	
Adult polycystic kidney disease Huntigton's disease Neurofibromatosis - 1	1. Restriction endonucleuse analysis to localize point mutations that alter restriction sites
Autosomal recessive	
Sickle cell anaemia	2. Allelic specific oligonucleotides to detect known point mutation
Thalassemia - α and β Cystic fibrosin Phenyl ketonuria	3. Southern blot analysis to detect deletion or other major structural rearrangements
α_1 - Antitrypsin deficiency Tay - Sach's disease	4. Allelic-linked RFLPs to detect unknown mutation to be in linkages disequilibrium with specific RFIP hyplotypes
X-linked recessive Haemophilia A & B Duchenne muscular dystrophy ornithin transcarbamylase deficiency	5. Locus linked RFLPs to detect unknown mutations whose malposition is known
Infectious diseases	
AIDS	6. PCR analysis
Malaria	7. DNA sequencing
Metabolic disorder Diabetes	8. Electrophoresis of single stranded or denatured DNA

Tools for DNA Analysis

Recent advances in molecular biology, especially our knowledge about eukaryotic gene structure, function and regulation, including human genome has led to dissect and diagnose precisely the genetic diseases at the level of malfunctional gene itself. Several protocols developed for the analysis and expression of rDNA have provided important applications to clinical medicine. Some of them are presented below :

(i) *Southern blot*

This technique essentially couples gel electrophoresis with the provision for the use of specific probes. An out line of southern blot analysis is diagrammatically presented in Fig. 11.3.

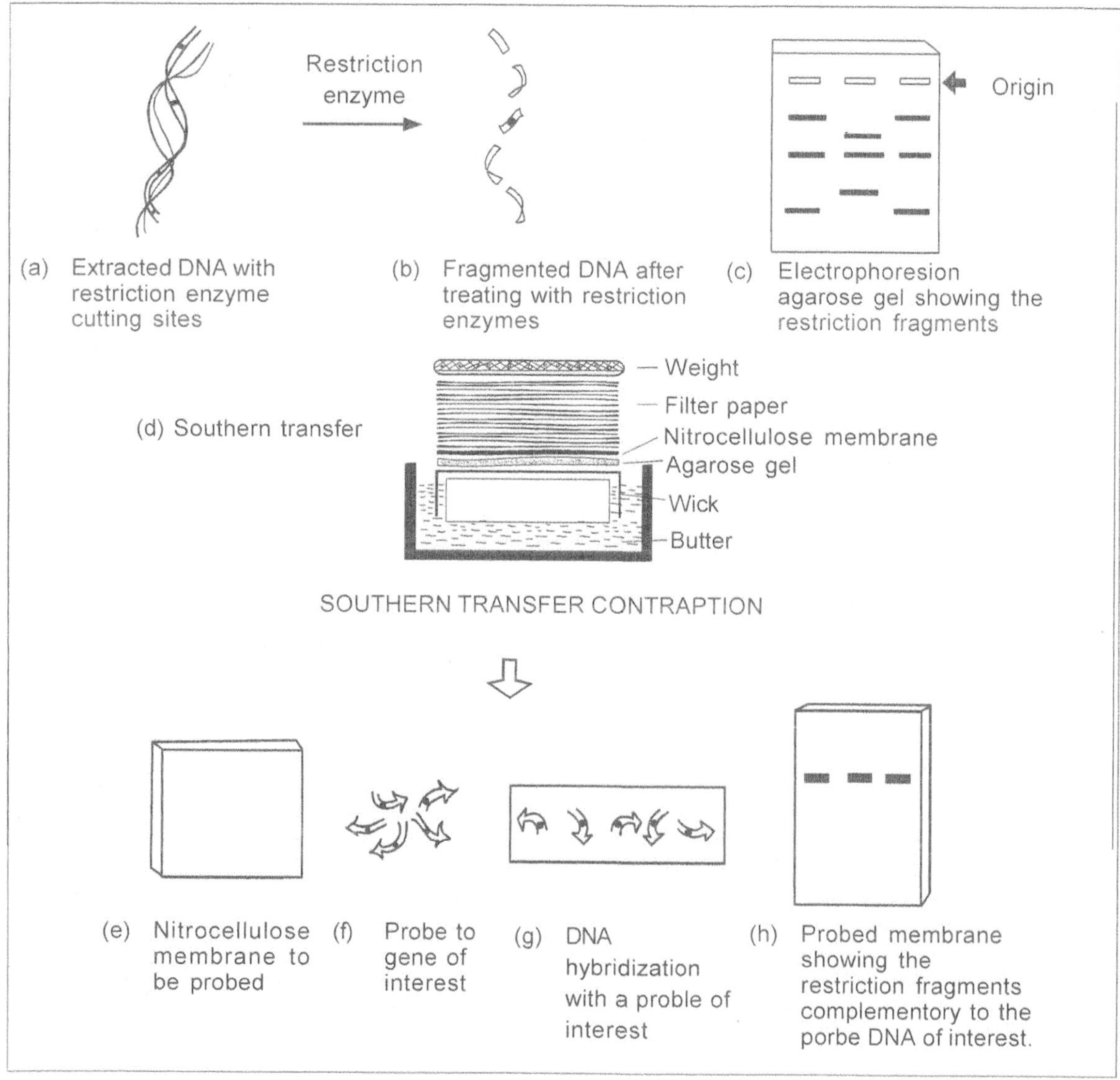

Fig. 11.3 The outline of protocol involved in the southern blot DNA hybridization.

In southern blot DNA hybridizations, the DNA molecule of interest is first digested after extracting from a donor (source) with appropriate restriction enzyme and the fragments are separated by agarose gel electrophores in (a - c), the separated fragments are then transferred from agarose gel on to a nylon membrane (nitro-cellulose) by a capillary action in southern apparatus (d), which subsequently probed and hybridized to detect the restriction fragment with the desired targetted candidate gene (e - h)

There are three stages in this method : 1. Approximately 5-10 mg of DNA obtained from white blood cells (WBC) is treated with a restriction enzyme that makes sequence specific nicks (cuts). The resultant DNA is separated by gel electrophoresis. 2. After electophoresis, the DNA is denatured to separate the strands and is transferred to a membrane by placing the membrane between the nitrocellulose gel and a stack of paper towels. This lets the DNA to move out of the gel on to membrane, 3. Labled DNA or RNA probe is added, which identifies and hybridizes with the complementary sequence in DNA, thus allowing the detection of the specific fragment of interest.

Applications

The southern blot analysis is a very common strategy employed in the diagnosis of genetic diseases. The principle of molecular hybridization is exploited in this methodology, i.e., the sequence of probe employed recognizes its complementary sequence in targeted DNA. Many of the disease that have a genetic basis in terms of mutation (deletion, duplication, base substitution, addition or any nucleotide polymorphism at the extreme, gross structural change etc.,) can be detected by DNA analysis (southern blot). Infact in the year 1976 thalassaemia, involving the deletion of globin genes (α or β) became first genetic disease isolated (haemoglobinopathies) to be successfully diagnosed by prenatal DNA analysis. Diagnosis can be achieved in the case of many haemoglobinopathies, by probing whether the DNA from cells obtained by amniocentesis has functional globin genes or not. Probes can be developed depending on the type of abnormality (genetic disorder) one is targeting to detect eventhough the development of southern blot analysis by Edwin southern of the university of Edinburgh, UK, (the technique is named after him), has very widely used and greatly simplified prenatal diagnosis of several genetic disorders, it has certain limitations. Significant amounts of DNA are needed for the analysis. The technique is labour intensive and may need 1-2 weeks to get the results. Exposure to radioactive material used in probes constitute health hazard and is a limitation. However, new technique of the use of non-radiolabeled probes has been developed in recent times. Yuet Wai Kan and his associates at the university of California, Sanfrancisco used a hybridization method in which radio labeled copy of the probe (α-globin gene sequence) for detecting the genes in subjects DNA. When the radioactive probe and the proband's DNA are mixed in a solution, the probe will adhere to the globin gene. Under controlled condition, the degree of binding to the subjects DNA is related to the number of copies of the globin gene that are present. Thus, it can indicate whether the individual under investigation has the homozygous form of thalassemia (α or β). Although the technique of hybridization is feasible, it is a difficult technique and requires a large amount of DNA that can be only obtained by culturing the amniocytes, which takes 3-4 weeks.

(ii) *Polymerase Chain Reaction (PCR)*

The PCR method can help greatly in increasing the quantity of a specific DNA sequence from as little as one molecule of DNA/RNA, consequently it is now possible to detect mutations (in restriction sites) affecting genetic hygeine of humans by gel electrophoresis of PCR products. Simply by hybridization of allelic-specific oligonucleotides to PCR amplified products will enable one to detect even single base differences between normal and disease causing allele.

The PCR methodology is a powerful technique which helps in selective and rapid amplification of targeted DNA or/CDNA (RNA). This method for amplifying DNA sequences was developed by

K.B. Mullis (1986) and received Nobel Prize for this discovery. Essentially the technique is based on the enzymatic amplification of DNA fragments that is flanked by two stretches of nucleotide primers, which hybridize to opposite strand of the sequence being investigated (Fig. 11.4). The technique makes use of the property of two strands of DNA which separate when heated. The separated strands act as template for primers sequence consequently the binding to the segment between primers is amplified in the presence of heat resistant DNA polymerase (*Taq* polymerase) and deoxy ribo nucleotide. Further elevated temperatures inhibit the reaction, which resumes when the temperature is decreased thus, the reaction takes place cyclically and the quantity of amplified DNA increases rapidly with each cycle of PCR amplification reaction.

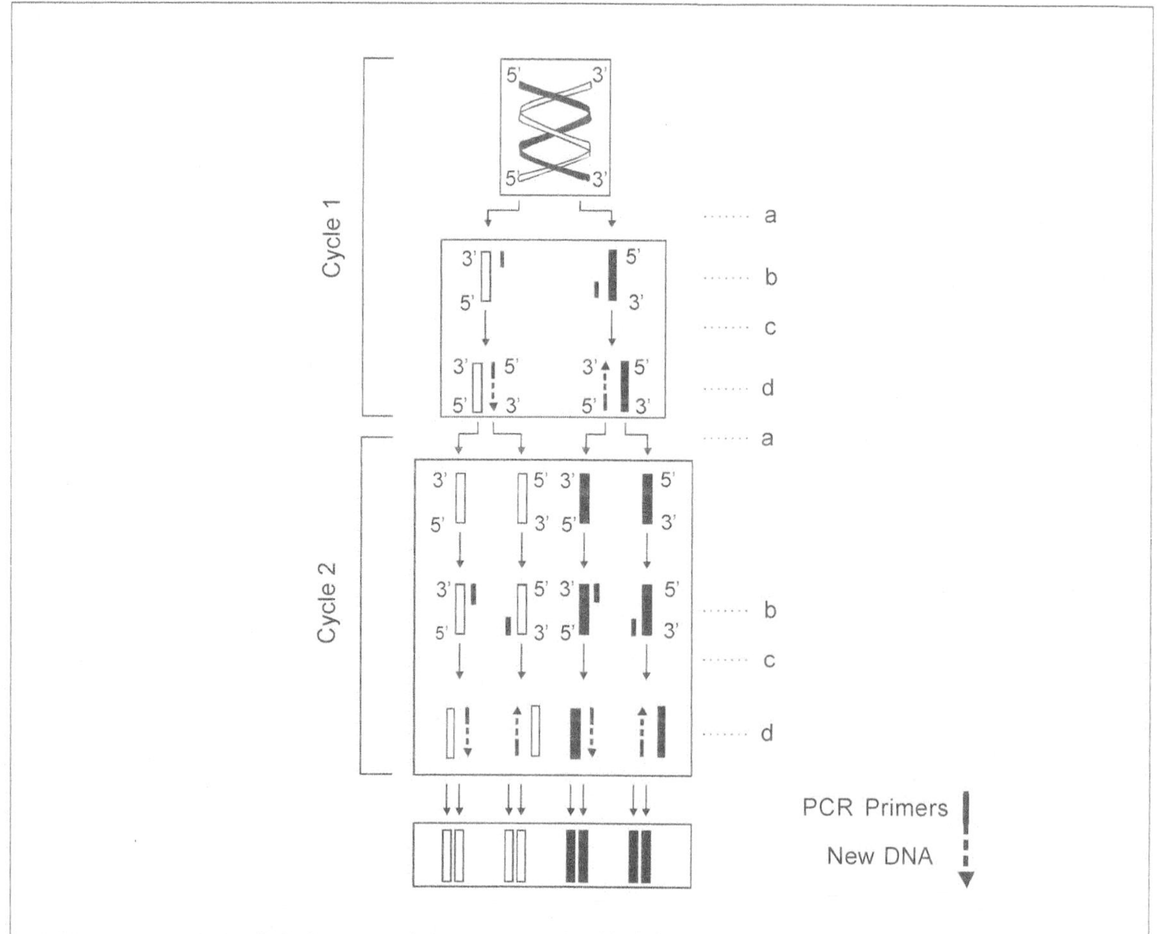

Fig. 11.4 Polymerase chain reaction (PCR) –
Amplification of a DNA fragment flanked by two primers that are complimentary to opposite strands of the sequence being targetted (a) Heat denaturation separates the strands of the target DNA, (b) Primers are added in excess and hybridized to complementary fragments, (c) Deoxynucleotides and Taq polymerase are added while the temperature increases, (d) The primer is extended in the 3' direction as new DNA is extended in the 5' direction.

There are certain advantages as well as limitation of this technique in the disease diagnosis processes. One of the major advantages of this technique is that only a very small amount of DNA, as little quantity as 1 μg or even less, or even only one single cell is enough for amplification. Besides, the results can be achieved in less than 24hours. The limitation of this technology is the imperative need to know the sequence information of the target gene to synthesize the oligonucleotide primers for PCR.

(iii) *Sequencing of DNA*

Most of the genetic disorders involve the alteration in sequence of nucleotides in DNA. When the normal sequence is compared to the diseased one, there will be alteration in the sequence of always DNA. Hence one of the most precise ways to characterize the genes involved in genetic disorders is to isolate the gene (DNA segment) and to get a precise sequence and compare with the normal gene (Fig. 11.5).

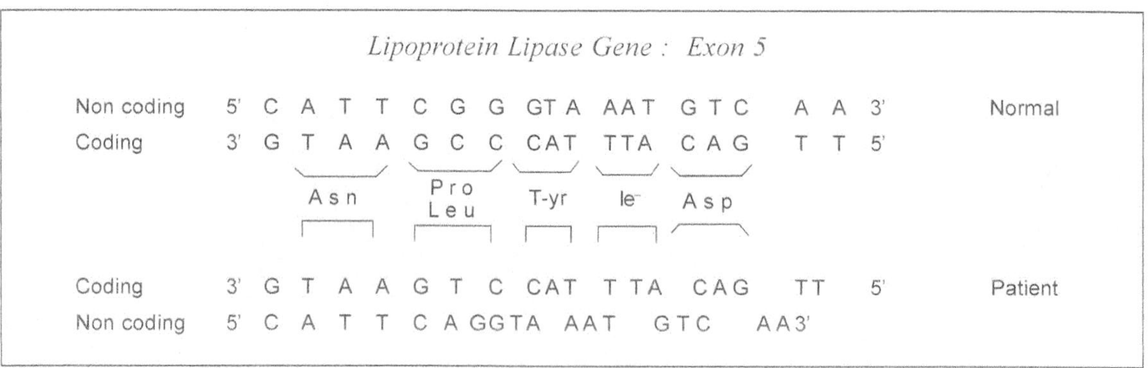

Fig. 11.5 Sequence difference in patient and normal individual for Lipoprotein lipase gene

Two methods are available presently for determining DNA sequence; Sanger method and Maxam-Gilbert method; both the methods rely on separating fragments whose lengths vary by one nucleotide. The established methods for analysis has now been automated and instruments have come into the market which enable the rapid sequencing of given DNA by a machine. DNA sequencing methods have revolutionized the whole process of diagnosis of genetic disorders. The products of PCR reaction can be sequenced very rapidly. Major advantage of this protocol is that it is possible to assess directly whether the gene of a patient has any sequence changes as compared to the normal. With development of micro-array chip technology the screening for genetic abnormalities can be done in future routinely.

11.3 Gene Therapy

Early in the evolution of gene cloning technology, it was perceived that the ability to recombine and manipulate genes (DNA) will have a great potential for human gene therapy for treating hereditary diseases. Much of the initial excitement was mired in controversy concerning the recombinant DNA technology, whether it should be applied for the manipulations in humans. Research in gene therapy as a biomedical discipline commenced in right earnest in the mid 1980s. The gene therapy as a discipline was originally conceived as the ultimate treatment for inherited metabolic disorders. Several attempts at gene therapy in humans were made as early as 1979. These early experiments were not only unsuccessful but fuelled the controversy about the moral and ethical propriety of genetic manipulations in humans and the means by which such experimentation should be conducted and be regulated.

With the advancement in rDNA technology and attendant refinements in mode of gene delivery as well as developments in embryology including cloning technology during the past few years, the concept of gene therapy for human diseases has become attractive proposition and has been widely accepted by the medical fraternity, scientists and public policy makers. Even before the clinical demonstration about efficiency and repeatability of gene therapy, in humans the concept that the inherited diseases can be corrected at genetic level, gained momentum and support and become a driving force in medicine. As a sequel, over the past a score of years the gene therapy research has progressed from laboratory to the clinic more quickly than any of the investigators even anticipated. More than one hundred clinical applications of gene transfer into human patients for both therapeutic and cell marking purposes have now been approved in USA and in a number of East European countries.

Basis for Hereditary Diseases

The molecular basis to human diseases, centers round on the premise that discrete aspects of metabolism can be traced to the gene function and their regulation. This molecular approach to understand inborn errors of metabolism or genetic disorders had its origin in the one gene-one enzyme hypothesis, first enumerated by Archibald Garrod (1908), which subsequently proved by Beadle and Tatum (1945). Garrod recognized that specific gene products have essentially unique functions and that any deficiency of a given gene product would cause a specific pathology due to failure to perform that function in a metabolic pathway. The most obvious example of the role of individual gene functions is the effect of hormones in regulating metabolism and growth. Many of genetic diseases are being identified which results from the absence of functional protein and / or enzymes. In recent years the whole gamut of growth, development, ageing process has been dissected and was established that every aspect of metabolism is under the control of gene function including immune system, neurological development, and apoptosis (cell death). Increasingly, even the infection disorders like acquired immune disease syndrome AIDS, mental retardation, behavioral disorders, as well as cancer are being understood as maladies due to malfunctioning of genes leading to impairment of molecular events.

What is Gene Therapy ?

Since diseases process are the end result of interaction between myriads of molecules and genetic interactions, it should be plausible to intervene in these processes by a requisite genetic manipulation that restores normal production of gene products so that cell, tissues and body is restored to health.

This is the essence of gene therapy. The term '*gene therapy*' may be defined as the "*introduction of normal genes into somatic cells to provide gene functions that are deficient because of genetic or acquired (infections) disease*" (Ledeley, 1987). Initially, research was focused on developing ways and means for treating diseases where it is possible to reconstitute the functional gene, hence product, to replace with the defective one. The technology was developed for treating diseases of the blood cells like, thalassemia, which results from defective globin gene, that could be treated by introducing a normal globin gene into red blood cell precusor (hematopoietic stem cells). Similarly, phenyl ketonuria, which results from the deficiency of hepatic enzyme, phenylalanine hydroxylase, could be treated by introducing a normal phenylalanine hydroxylase gene into liver cells.

As the technological advancement for *in vivo* and *in vitro* gene therapy progressed, it became obvious that the same technology can be deployed to deliver genetic material to suppress undesirable activities *in vivo* consequent to infectious and other non-genetic diseases. Genetic manipulations could be used to make cells resistant to destruction due to auto-immune diseases, parasitic diseases, drugs or toxins, etc. Besides the technology could be used to for the production of specific antibodies or restore functions which have been lost. Gene therapy has been broadened to encompass the treatment of acquired diseases such as cancer, AIDS etc, that involve genetic alteration of cells.

Gene therapy promises to be a most fertile area of scientific and clinical research for 21st century, especially in the light of discoveries made due to genomics and proteomics. It is undoubtedly will be a major clinical practice. There is going to be a paradigm shift in the whole of medicine.

Basis for Gene Therapy and Approaches

The success of gene therapy depends on the ability to accomplish gene transfer. Thus the gene transfer is the basis for gene therapy. Gene transfer is nothing but a mode of introducing genetic material into a targetted human cell. There are several strategies for gene transfer to specific cells. Some of the methods are : DNA mediated gene transfer through transfection (retroviral infection), calcium phosphate transfection, protoplast fusion, lipofection, electroporation, microinjection and microprojectiles (biolistics).

There are basically four approaches to gene therapy : (i) addition of normal functioning gene to replace the function of mutant gene (ii) replacement of mutant gene sequences with normal gene sequences, (iii) establishing alternative pathways to overcome mutant functions and, (iv) altering regulation of genes, normal or mutant.

One of the vital issues in gene therapy is whether genes to be introduced for correction of disorder go into germ line or somatic cells. Accordingly, the therapy is refered to as germ line therapy or somatic gene therapy. The distinction primarily rests upon where the genetic alteration is made. In principle when *germ line gene therapy* is achieved the recombinant gene might become part of the human germ line and be passed on to offsprings, while in *somatic gene therapy*, the recombinant gene will serve only to alter the pathological phenotype limited to patients life time at most thus its effect is transient. For both technical and ethical reasons, the somatic gene therapy is preferred.

Many organ or tissue system have become legitimate targets for gene delivery. These include post mitotic tissues such as hepatocytes, myoblasts, fibroblasts etc., In the beginning research centered on *ex vivo* gene therapy. This involves the excision of tissue (such as hepatocytic) from the patient and cultured *in vitro* and transduced with therapeutic gene construct and finally make autologous transplantation of the genetically modified cells. While this protocol has been successful in laboratory animals, it entails a complicated clinical procedure that would have only limited applications in treating a large number of people. As a result of recent developments, another approach viz., *in vivo* gene therapy has come into focus. In this approach the therapeutic gene constructs are delivered directly into the target organs. When perfected this approach is going to be highly potential to provide genes as medicine for the effective treatment of variety of genetic disorders.

Candidate Diseases for Somatic Gene Therapy

Some of the diseases which could be candidates for gene therapy are presented in Table.11.2. These are models being used to develop new technologies that will be applicable to a variety of genetic diseases.

Table 11.2 *Gene Therapy - Different Disease Models as a Candidates for Somatic Gene Therapy*

Disease	Gene involved	Whether successfully cloned	Target cells / tissue organs	Remarks
Immune Deficiencies				
Severely combine immune deficiency disease (SCID)-Typ1	Adenosine deaminase (ADA)	Yes	Bone marrow/ Fibroblast	Successful, the first instance of gene therapy
SCID -Type 2	Purine nucleotide phos-phorylase (PNP)	Yes	Bone marrow/ Fibroblst	
Agammaglobubinemia	Gamma globunin	Not yet	Bone marrow	
Chronic granulomtous disease		Yes	Bone marrow	
Hemoglobinopath ies				
Thalassemia α and β	α/β globin	Yes	Bone marrow	High level, regulated
Sickle cell	β-globin	Yes	Bone marrow	Expression
Hereditary Anemias				
Hereditary hemolytic	Spectrin	Yes	Bone marrow	
Organic acidemias	band 4.1	Yes	Bone marrow	
Methylmalonic acidemia	methylmalonyl CoA mutase	Yes	Liver, Fibroblast, bone marrow	
Propionic acidemia	propionyl CoA carboxylase	Yes	Liver, Fibroblast bone marrow	
Short chain acyl CoA dehydrogenase deficiency	acyl CoA dehydrogenase	Yes	Liver	
Aminoacidopathies				
Phenylketonuria	Phenylalanine hydroxylase	Yes	Liver	Requires biopterin

Maple syrup urine disease		No	Liver	
Homocystinuria	Cystathionine β-synthase	Yes	Liver, bone marrow Fibroblast	
Tyrosinemia	-	No	Liver	
Urea cycle defects				
Ornithin transcarbamylase deficiency citrullinemia	Ornithintranscarbomylase	Yes	Liver	Requires carbamyl phosphate
Arginosuccinate synthesis				
Carbomyl phosphate Synthetase deficiency	Carbomyl phosphate synthatase	Yes	Liver	
Argininemia	arginase	Yes	Liver, bone marrow, Fibroblast	
Endocrinopathies				
Hypopituitary	TSH, ACTH, GH	Yes	Fibroblast, bone marrow	Regulation/ hornome therapy
Insulin dependent diabetes mellitus (IDDM)	Insulin	Yes	Fibroblast, bone marrow islets of langerhans	Regulation/ hormone therapy
Hypoparathyroid	Thyroid hormone	Yes	Fibroblast, bone marrow	Regulations / hormone therapy
Circulating proteins				
Hemophilia	Factor VIII Factor IX	Yes Yes	Liver, fibroblast, bone marrow, vascular endothelium pancreas (islet of langerhans)	Inhibitors factor therapy
α - Antitrypsin deficiency (pulmonary)	α-Antitrypsin	Yes	Liver, pulmonary epithelium, fibroblast, bone marrow	Late malignancies
Transcobalamine II deficiency	Transcobalamine	No	Intestinal endothelium, liver	
Other Metabolic disorders				
Galactosemia	gal-1-p-uridyl transferase	Yes	Liver	
Glycogen storage disease	-	-	Liver, muscle, central nervous system	Storage in many tissues
Various mucopolysacharidoses	-	-	Central nervous system	
Tay Sachs disease	Hexasaminidase	Yes	Central nervous system	
Lesch Nyhan syndrome	(HPRT)	Yes	Central nervous system	

Gaucher's disease	Glucocerebrosidase	Yes	Bone marrow, fibroblast	
Crigler Najar	VDP-glucoronyl transferase	-	Liver	
Miscellaneous diseases				
Cystic fibrosins		Yes	Pulmonary epitehlium	
Hypercholestorolemia	LDL-receptor	Yes	Liver, firbroblast, vascular endothelium	Heterozygote phenotype
Duchenne's dystrophy	Distrophin	Yes	Muscle	Very large gene
Acquired diseases				
AIDS				Experimental
Cancer				stage
Neurological disorder				

Example of Successful Human Gene Therapy

As a result of developments in the technology necessary for the transfer of '*alien*' genes, the prospects for treating and findings a possible cure for genetic disorders has improved enormously. The first gene therapy to be accomplished is the work spearheaded by French Anderson at the University of Southern California on nine years old patient, Ashanti De Silva, a Ceylonese girl in 1995. Anderson's group was focussing for the past several years to develop a strategy to treat the human inherited disorder, Adenosine Deaminase Deficiency (ADD) by gene manipulation. Adenosine Deaminase (ADA) is an enzyme that plays a vital role in developing human immune system. Patients suffering from ADD carry a malfunctional genes for ADA. Consequently such individuals have Severe Combined Immune Deficiency (SCID), rendering them unable to fight the routine opportunistic infections. ADD is primarily a disease that occurs early in childhood, rarely such children survive beyond two years of age.

The method initiated by Dr. Anderson's group comprises of five steps : (i) isolation of defective T-cells from the white blood cell fraction of the blood after drawing the blood, (ii) introduction of a normal functional adenosine deaminase gene into the genome of the defective T-cells with the help of genetically engineered retrovirus containing a normal ADA gene, (iii) selection of T-cells that have integrated retrovirus expressing the normal activity of ADA. (iv) culturing and proliferating the genetically modified T-cells to increase the number and finally (v) the introduction of normal T-cells with restored functioning of ADA back into the patient by transfusion (Fig. 11.6).

After performing such a protocol it was seen that the patient recovered dramatically, and gradually the levels of ADA improved and all the symptoms of disease vanished.

Future Prospects of Gene Therapy

Other human genetic disorders including infectious diseases like AIDS, likely to be future targets of gene therapy. Some of the disease holding promise to be treated with gene therapy include diabetes, haemoglobinopathies, aminoacidopathies, cystic fibrosis, certain cancers of the brain, breast, lungs, kidney and liver (Table 11.2). The future seems to be promising for the application of modern genetic techniques to the treatment of human disease. In fact there is going to be a pardigm shift in medical science.

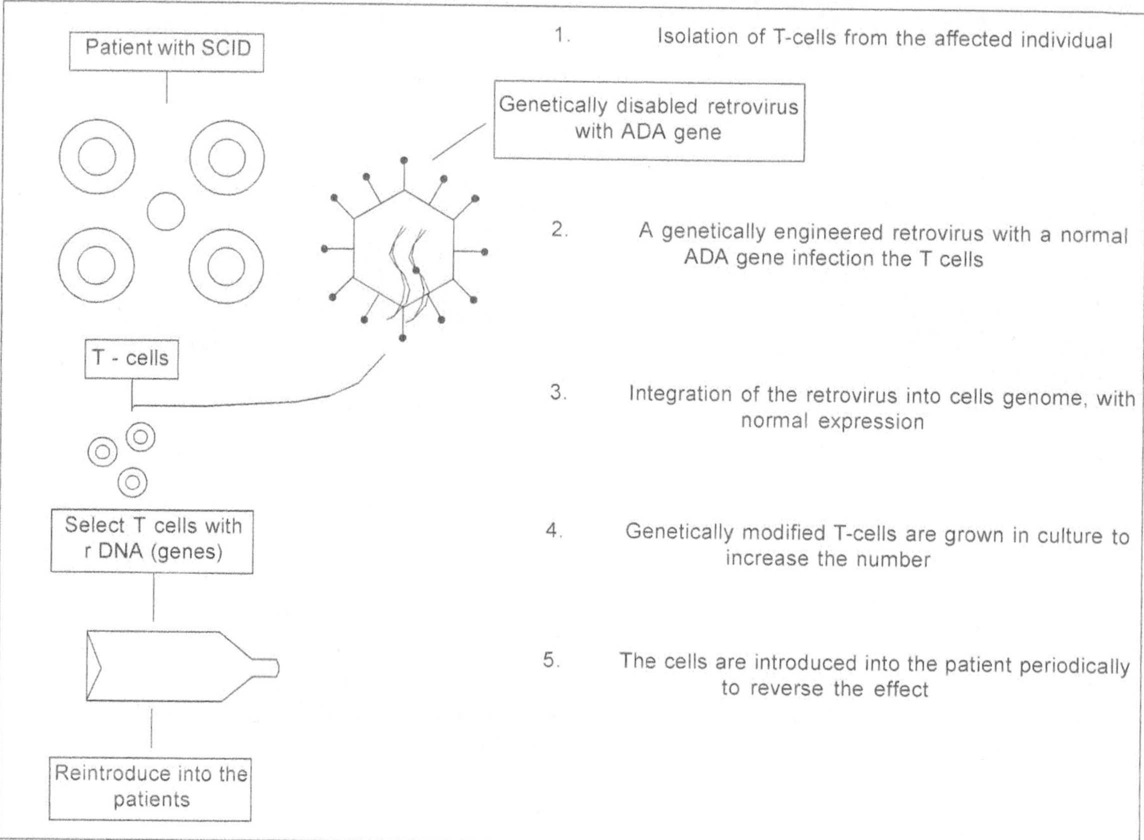

Fig. 11.6 Gene Therapy protocol for treating SCID

11.4 Development of Vaccines by DNA Technology

The medical sciences are being revolutionized by biotechnology. There is going to be several paradigm shifts in the whole gamut of clinical medicine. A major focal point is the production of large amounts of therapeutic proteins with important utility in four broad areas of management of human health : (i) prophylaxis, to prevent diseases or infection, (ii) therapy, to treat effectively clinically manifested diseases, (iii) diagnosis, to detect the latent infection, progression and regression of diseases, and (iv) discovery, to evolve, design and develop new therapies. Furthermore one of the important areas of prophylaxis include the development of vaccines for various infectious diseases, both tropical as well as temperate regions of the world, including pandemic diseases like AIDS.

The advances in the knowledge of the immune responses and molecular biology, especially the cloning technology of the last two decades have enabled the researchers the identification of a large number of infectious agents and proteins of immunological interest and their expression in different vectors for amplification. The use of cloned genes (recombinant-derived genes) and hormone are refferred to as gene products or biological products to distinguish those manufactured by pharmaceutical industry by chemical synthetic process and marketed as drugs.

Rationale Behind the Use of rDNA Technology to Produce Proteins

There are atleast four reason for following the strategy of rDNA technology to produce clinically useful proteins : 1. *Absolute supply* is ensured by the rDNA technology permitting wide spread use of any protein if need be. The supply of the natural proteins are too limited to meet the growing demand and also might not be pure. 2. *Relative supply* – the supply of natural proteins are available but it is cumbersome to exploit and or may become limited in the future as demand expands. One of the prominent examples is hepatitis vaccines. Prior to the advent of rDNA technology, the source for the production of hepatitis B vaccines was plasma of chronic patients suffering from hepatitis. The plasma protein based vaccine was available as early as 1980s. However, it was soon realized that the supply of plasma would probably become limited in future. Therefore the attention was directed towards rDNA based production of the antigenic protein. 3. *Safety* – The source material for the natural protein is proven or suspected of being contaminated with a pathogenic agents whose removal or inactivation cannot be assured. For instance, there have been several tragic incidents involving the use of naturally derived proteins of human origin. Specially threat of AIDS has affected human derived biological products. Hence, it is safe to follow the route of rDNA technology for the mass production of therapeutic proteins. 4. *Specificity* – Residual source may cause unwanted background, especially in a diagnostic test. One of the most important issue in diagnostic testing is specificity. If a diagnostic protein employed for the detection of specific antibodies is not pure and homogeneous, it will obsecure the detection of antibodies to diagnose the disease. Hence the rDNA approach is ideal for developing diagnostic proteins.

Recombinant DNA Methods for the Development of New Generation Vaccines

Vaccines are proteinaceous susbstances with antigenic property capable of mounting immune response by producing antibodies to neutralize the pathogens and ultimately destroy them. Some of the interaction that takes place during the generation of an immune response is presented in Fig. 11.7. There have been several advances in the modern biosciences which impact the four generations of advances in the development of vaccines. Functionally, vaccines can be divided into two categories : 1. *Live vaccines* and 2. *Inactivated vaccines*. Live vaccines are defined by the ability of the vaccine strain, for example, attenuated virus capable of replication within the human host. Non-live or killed or inactivated vaccines are incapable of replication or infection in the host.

General Characteristics of the Vaccines

The following are the broad features of the two categories of vaccines.

Live vaccines are : 1. Attenuated with respect to pathogenicity, 2. Elicit both cell-mediated and humoral immunity, 3. Provide long enduring protection than the inactivated vaccines 4. There is a tendency to reactogenicity, and 5. Capable of reversion to a more pathogenic form. On the other hand, the killed vaccines are 1. non-replicating, 2. non-infectious, 3. have reduced reactogenicity when compared to live vaccines, 4. need booster inoculation to optimise their effect and 5. are highly purified and characterised, physically and chemically.

First generation vaccines are produced by classical approaches. Several strategies have been employed in the production of live vaccine for viral infection. Such as (a) modification by passage in cell cultures (b) variant virus from other species (c) selection of temperature sensitive mutants and using hybrid genomes. In the case of killed vaccine, the developmental strategy in classical approach has been by (a) killing whole pathogens (b) using toxides from pathogens (c) purified surface components and (d) by conjugated surface components.

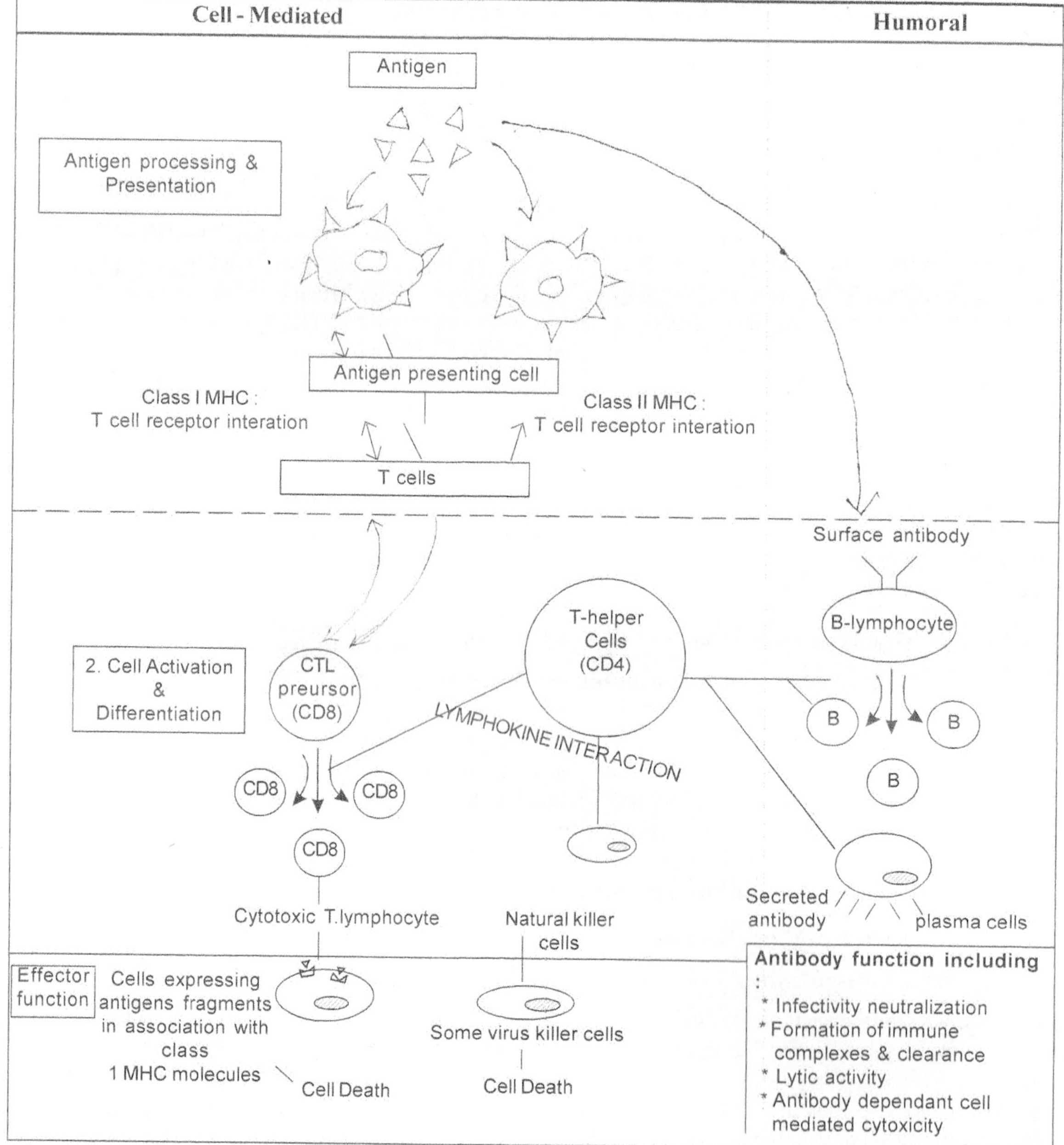

Fig. 11.7 Some of the interactions that occur during the generation of an immune response

New Generation Vaccines

The basis of developing new generation vaccines is to identify the protein components of infectious agents that are capable of inducing an immune response in the host organism in a similar way to that produced by the whole (live) agent. Furthermore, the candidate proteins are identified in such a way that these have two more attributes viz., have no role in replication are not related with the process of virulence.

Using the rDNA technology, the genes coding for these proteins are selected, cloned and expressed using suitable vectors such as *E.coli* (bacterial), yeast (Fungus) and Baculovirus (Virus). Such proteins used as vaccines are called *Subunit Vaccines*. Another approach is to synthesize proteins if the sequence of genes coding for the antigenic determinant is known. Such vaccines are called *Synthetic Vaccines*. Live vaccines can also be produced by attenuation of pathogen using modern techniques in molecular biology either by modification or deletion of key genes of pathogens. Yet another approach is producing vaccines using recombination pathogens (viruses) carrying alien genes capable of mounting immune reaction. Other interesting aspects of rDNA vaccines are, when the designing of new vaccines is feasible for given infection there is possibility adding new protein information so that immunologically interesting proteins sequence of other antigens capable of increasing stimulation of B and T-cell types and even cytokine release. The outline of the types and production strategy of modern generation vaccines is presented diagrammatically in Fig. 11.8.

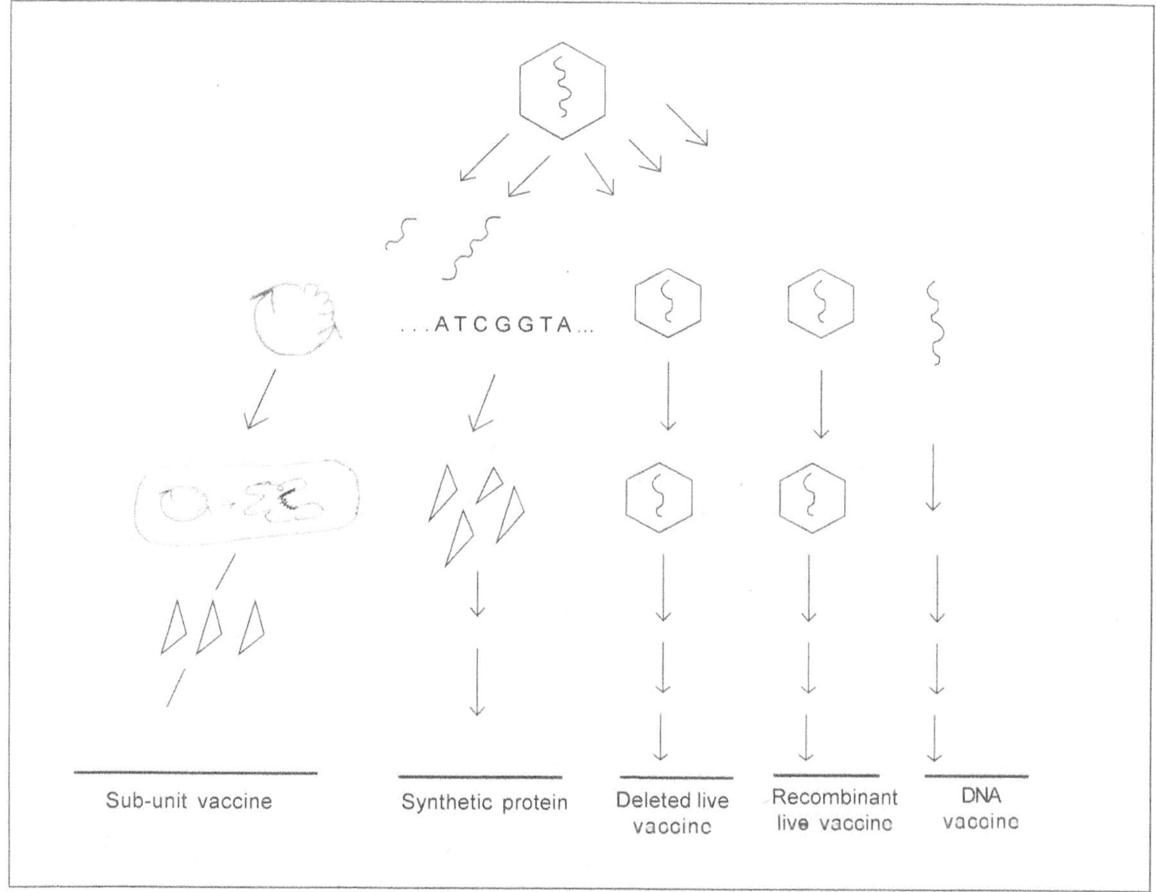

Fig. 11.8 Types and Strategy of the designing and production of New generation Vaccines

11.5 Chemicals and Drugs from Animals

Simple protozoan to highly evolved mammal synthesize a variety of chemicals to maintain their life function. These may include simple amino acids, sugars, proteins, fatty acids as well as complex glycosides, derived lipids, glycoprotein, hormones, enzymes, vitamins etc. A wide variety of animal species also produce naturally toxic substances for securing food as well as defending themselves. Snake poisons, bee venomes etc., come under this category. Animals especially those that are sessile and stay put at one plane (sponges, tunicates etc) produce chemicals to discourage predators. These array of chemicals from animals are used as tools of biomedical research and also show potential to be useful as drugs. These chemicals kill bacteria, viruses and fungi. The structural and functional components (bioactive molecules), venoms and toxins from animals are sources for chemicals and drugs. A drug is a pure substance intend to mitigate, treat, cure or prevent a disease in humans or other animals. Thus, the animals become a magnificent source of biochemicals-structural and functional constituents and / or secretions in the form of venoms and toxins-for nutritional, medicinal, pharmaceutical and industrial applications. Recently with advances in genetic engineering and biotechnological processes, the animals serve as potential source for mass production of designed chemicals. The animals are being used for drug production (pharming) with genetically modified and cloned animal species or mammalian cell culture (e.g., erythropoietin, blood coagulation factor VIII).

A. *Constituents and Derivative Bioactive Molecules of Animal Origin*

The basic metabolic constituents such as amino acids, proteins, fatty acids, derived lipids, vitamins, enzymes, hormones etc., of animals can be source of production of chemicals with myriads of clinical and biological applications. Many chemical mediators in animal systems like human growth hormone, histamine, leukotrienes, prostaglandins, immunogloulins, antibodies, insulin, estrogen, progesterons, and vitamins are quite useful as drugs for therapeutical purposes. These bioactive molecules have important role in correcting the metabolic disorders. The chemicals derived from coral reefs have potential to treat leukemia and Hodgkin's diseases. The *cortisone* obtained from adrenal gland of animals is used for curing rhumatoid arthritis. The *Heparin*, used as anticoagulent is prepared from liver and lung of animals. The digestive protein like *pepsin* is obtained from animal stomach extract and used in correcting the digestive disorders. The *B-complex* vitamins are obtained from animal sources. The '*aphrodisiacs*, from animal sources (like horn of rhinoceros) are used in allopathy and in ayurveda system of medicine. The dried skin of toad (with traces of adrenalin) was used to treat toothache and bleeding. *Musk* is obtained from musk deer. The *Gelatin* which has wide application is extracted from horns, hoofs, skin, tendons and bones of animals. *Squalene*, a constiuent of fish oils is used as mordent in dyeing industry. *Lecithine* from shark liver oil is widely used in chocolate industry.

The *vitamin A and D* are obtained from liver of cod fish. *Insulin* is extracted from the pancreas of bovine or porcine. *Immunoglobulin* G is prepared by injecting antigen into animal body. For example, horse globulins containing antitetanus and anitidiptheria toxins have been obtained and used prophylactically. Human normal *immunoglobulin* is prepared from pools of donations of human plasma containing antibody to measles, mumps hepatitis A and other viruses. Similarly the antirabies vaccine is also obtained. Human menopausal *gonadotrophins* is isolated from the urine of post menopausal women and contains a mixture of follicle stimulating hormone (FSH) and (LH). Human chorion gonadotrophin (HCG) is produced by the placenta and can be isolated and purified from urine of pregnant women. The specific *monoclonal antibodies* and other bimolecules produced from animals are used in diagnosis as well as therapeutics.

B. *Animal Toxins*

Some speices of animals produce toxins. Toxins are special kind of chemical molecules produced by organisms which act as potent poisons to competitors/predators of host. Some are passively venomous others are actively venomous. The chemistry of animal toxins extends from neurotoxic, cardiotoxic peptides and proteins, enzymes to many small molecules such as biogenic amines, glycosides, terpenes etc. The venoms are sometimes mixtures of both proteins and other small molecules. For example, bee venom contains a biogenic amine (histamine), three peptides (apamine, Melittin, mast cell degranulating peptide) and two enzymes (phospholipase A, Hyaluronidase). The toxicants of other insects also include formic acid, benzoquinine or terpenes.

Poisonous animals are widely distributed throughout the world and these range from tiny protozoan species to higher animals including mammals. For example, dianoflagellates (protozoan) release poisons called *saxitoxins* which cause burning sensation, numbness and hyper salivation. The jellyfishes (coelenterates) are poisonous. Some arthropods secrete chemicals that induce vomiting, convulsions, coma and death. Sharks, rays, puffer fish, some amphibians and reptilies are also poisonous. Snake venoms have been studied extensively. They are small peptides comprising 60-70 amino acids and are cardiotoxic or neurotoxic. Their effects are usually accentuated by proteolytic enzymes like phospholipases, peptidases, proteases present in venom.

Zootoxins, produced by animal species, can be divided into three categories, viz., *oral poisons*, *parental* and *crino-toxins*. Oral poisons are those that are poisonous when eaten up. The parental poisons are produced by a specialized gland/organ and administered by means of a venom apparatus. The crinotoxins are also produced by specialized poison glands but they are merely released into the environment (Table 11.3).

Table 11.3 *Chemicals and Toxins produced by different Animals.*

A. *Oral Poisons from animals*

A. Animal	Species	Type of Toxin	Toxic Effect
Dinoflagellate	Gnoyaulax catenella	Saxitoxin	Burning sensation and numbness
Sea cucumber	Holothuria	Holothurin	Fatal
Mollusc	Neptunea	Tetramine	Nausea, vomiting, photophobia
Japanese dosinea	Dosinia japonica	Venerupin	Nervousness, bleeding
Moray eel	Gymnothorax javanicus	Ciguatoxin	Abdominal cramps, skin rash
Castor oil fish	Ruvettus pretiosus	Oleic acid	Painless diarrhoea
Skip-jack tuna	Thumnus thynnus	Saurine	Throbbing of large blood vessels
Deadly death puffer	Arothron hispidus	Tetrodotoxin	Floating sensation, hypersalivation
California newt	Taricha torosa	Tarichatoxin	Effects are unknown
Hawksbill turtle	Eretmochelys imbricate	Chelonotoxin	Hypersalivation

Table 11.3 Contd...

B. *Parental Poisons of Animal Origin*

B. Animal	Species	Toxin	Toxic Effect
Coral	Millepora	Unknown	Skin irritation, sting sensation
Portuguese Man or war	Physalia sp.	Tetramine	Throbbing or burning sensation
Sea wasp	Chironex sp.	Cardiotoxin	Respiratory distress, extremely painful
Sea anemone	Actina equina	Unknown	Itching, swelling and redness
Conus shell	Conus asteriosus	Ammonia	Cyanosis, blurring of vision
Spotted octopus	Octopus sp.	Cephatoxin	Neuromuscular pain
Kissing bug	Triatoma sp.	Unknown	Palpitation
Honey bee	Apis sp.	Neurotoxin Phospholipase A Histamine	Redness, pain, Inflammation, local itching
Wasp	Polistes sp.	Neurotoxin	Redness, pain, Inflammation, local itching
Velvet ant	Dasy mutilla	Neurotoxin	Redness, pain, Inflammation, local itching
Fire ant	Solenopsis	Neurotoxin	Very painful, burning sensation
Tarantula	Dugesiella and Lycosa sp.	Neurotoxin	Highly painful
Scorpions	Centruroides Tityus and Leiurus	Neurotoxin, cardiotoxin hemolytic	Sweating, restleessness
Seaurchin	Toxopneustes	Nature of poison unknown	Intense radiating pain
Gila Monster	Heloderma suspectum	Neurotoxin	Cardiac failure
Indian cobra	Naja naja	Carboxy peptidase cholinesterase	Pain radiating from site of bite, convulsioons, death
King cobra	Ophiophagus hannah	Cholinesterase oxidase	Hemorrhages from nose and mouth

C. *Certain Chemicals released by Crinotoxic Animals*

Animal	Species	Type of Toxin	Toxic Effect
Sponges	Microciona prolifora	Unknown	Redness, Stiffness of the finger
Flatworm	Leptoplana	Unknown	Cardiac arrest
Blister Beetles	Cantharis vesicatorea	Cantharidin	Gastroenteritis kidney damage
Limprey	Petromyzan marinus	Unknown	Ingestion may cause diarrhoea
Paufie Sea Bass	Grammistes sex lineatus	Neurotoxin	Effects on man are unknown
Toad	Bufo species	Bufotoxin	Skin irritant
Tree frogs	Hyla and Phyllobates	Batrachotoxin	Severe inflammatory reaction

Several pharmacologically active substances are released during the inflammatory process following envonomation. These include histamine, serotonin, kinin, prostaglandins, phosphotides, and leukotriences and cause clinical symptoms and physiological derangement in organisms. Some of the examples of effects of toxins are, tetrodotoxin blocks sodium, sea anemone toxin block potassium channel, bungarotoxin blocks acetylcholin release, phospholipase A of cobra venom has cardiovascular effect, neurotoxins cause respiratory paralysis, and blocks acetylcholine receptors. Some of venoms or toxins have direct cardiotoxicity, cornonary vasospasm, hyperkalemia, and pulmonary edema, myonecrosis and vasculotoxicity. The effects of spider's bite are pulmonary edema, acute metabolic acidosis, intracranial hypertension, hypotheria and local skin lesion. Scorpion toxins act on calcium and sodium binding channels and blocks neuronal transmission. Basically, the animal toxins properties depend on amino acid residue present. The modifications in amino acid position that have occured in the catalytic region have resulted in various structural and functional diversifications in toxins, due to which different biologically active toxins have evolved in animal kingdom.

The knowledge of bioactive chemical molecules and toxins that occur in animals coupled with latest advances in molecular biology and gene cloning technique will enable to produce transgenic animals capable of synthesizing them. There is an immense potential or industrial production of chemicals for industrial and medical application from animals in human welfare. The knowldege of bioactive chemical molecules that are prevalent in animal kingdom coupled with advances in molecular cloning enable their exploitation for the production of animal and health care products.

C. *Recombinant DNA Technolgoy and the Production of Animal Based Chemicals, Drugs and Foods*

Animals that have had their genomes modified *in vitro* to add an extra desired characteristics from another source using gene cloning technology are called transgenics. Since 1990, transgenic animals like rats, rabbits, pigs, sheep, cow and fish are being produced, routinely even though 95% of all existing transgenic animals are mice. Transgenic animals can be used to have a deeper insight into physiology of growth and development, to study the progression of diseases caused by the pathogens and in the development of treatments, to produce useful gene products, to test the safety of vaccines and chemicals, to provide donor organs that could be used for transplantation and importantly increase the quantity and quality of foods and other products from animals.

D. *Production of Chemicals and Drugs for Therapy : Animal Pharming*

Chemical molecules and Medicines (drugs) are required to treat (certain) human as well as animal diseases Many biological products (gene products) like hormones, proteins, enzymes, toxins, small peptides etc., possess innumerable therapeutic properties and these can be created by the introduction of the segments of DNA which codes for a particular product. For instance, large amounts of a human protein, α-1 antitrypsin are used to treat life-threatening condition called emphysema. Transgenic sheep have been developed that make the protein in their milk in larger quantities than could be produced by cell culture protocols. The first successful products of the genetic engineering process were protein drugs like insulin and growth hormone. These drugs do not have to be produced by mammals to be active in them. An inexpensive easy to grow culture of genetically engineered *E.coli* can manufacture these protein drugs. An outline of cloning stratagies for the production of insulin in *E.coli* is given in Fig. 11.9. Other human drugs, such as tpA for blood clots, erythropoietin for

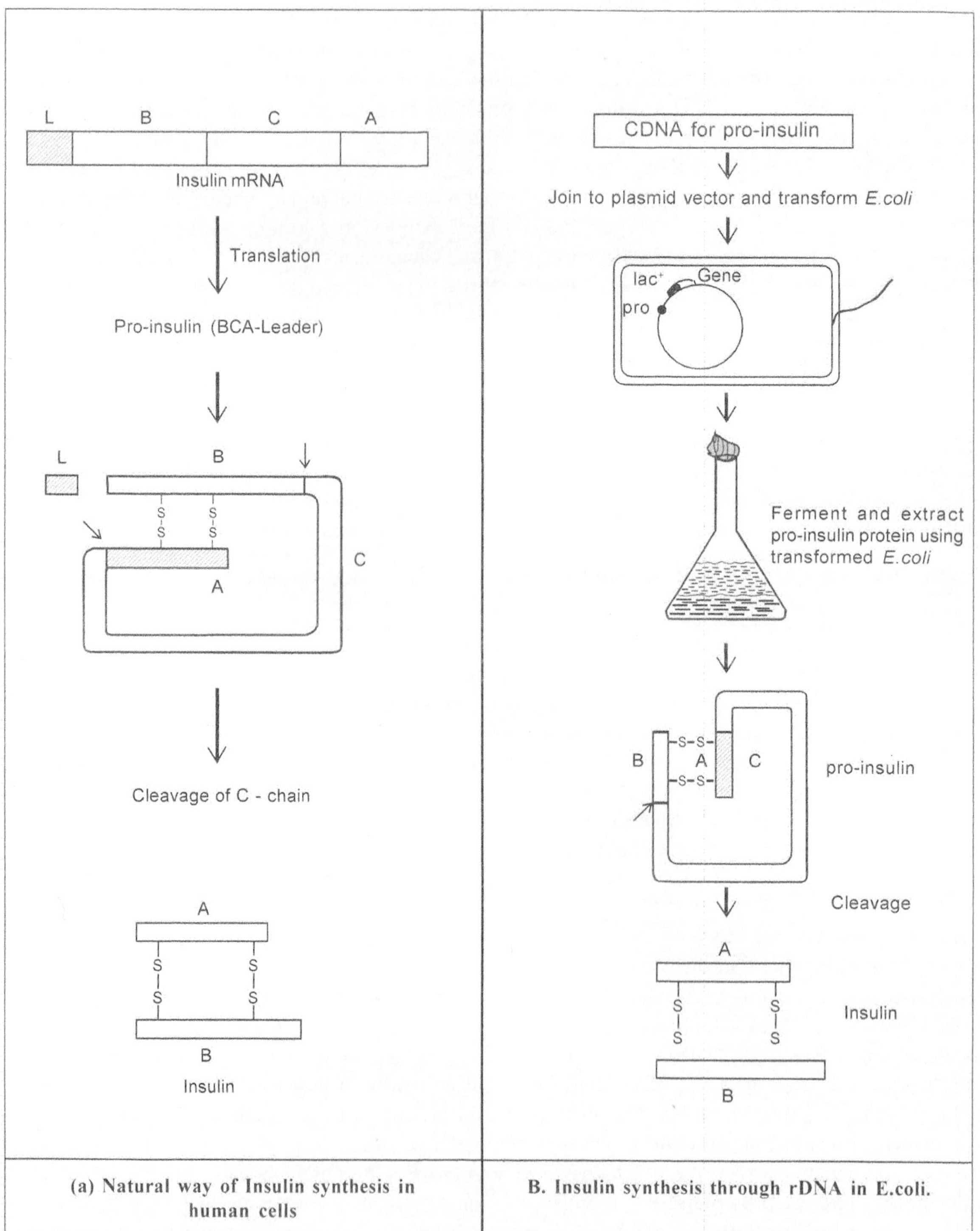

(a) Natural way of Insulin synthesis in human cells

B. Insulin synthesis through rDNA in E.coli.

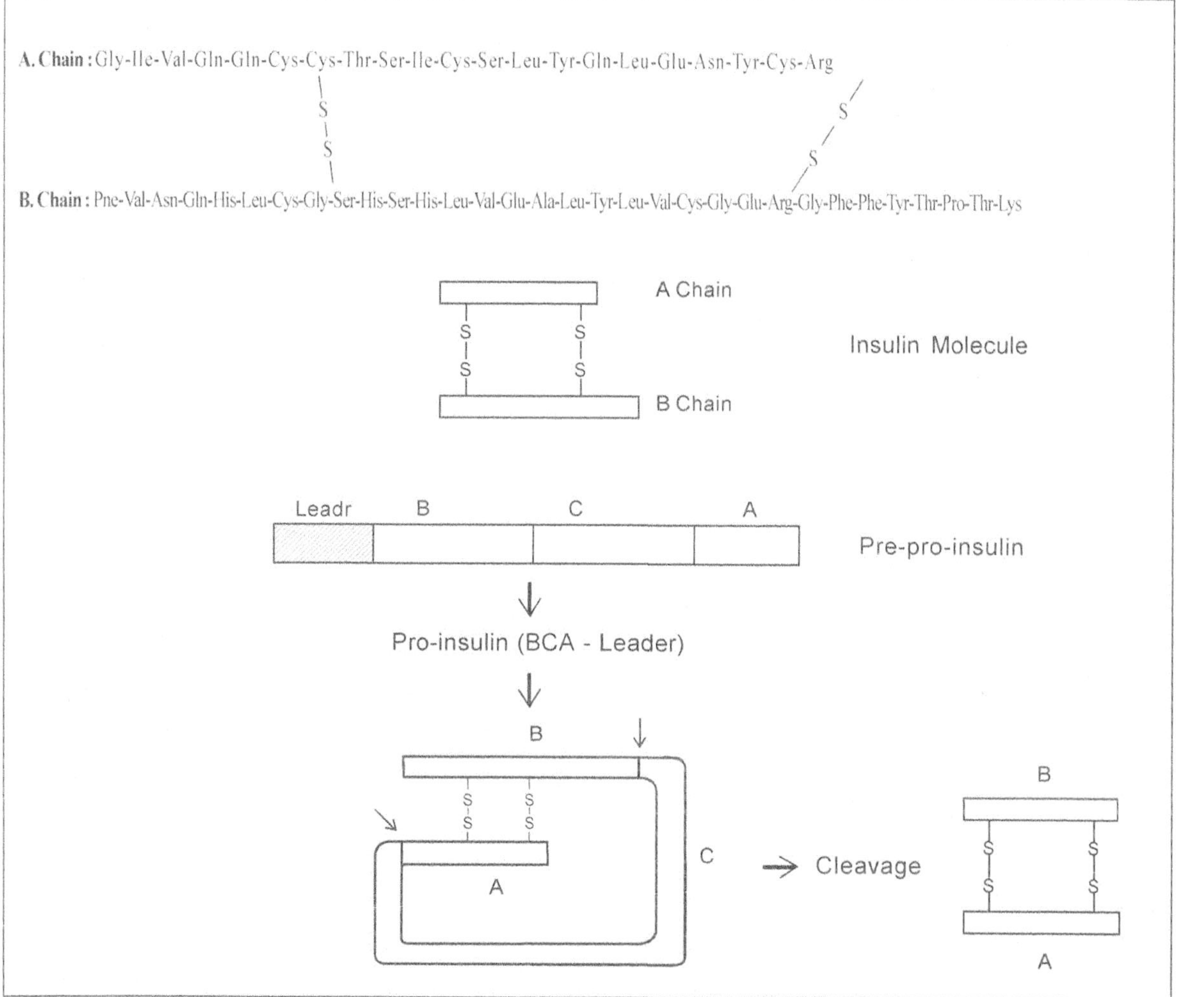

(c) Sequence of aminoacids and structure of insulin molecule

Species	A Chain 8	A Chain 10	B Chain 30
Bovine	Ala	Val	Ala
Procine	Thr	Ile	Ala
Human	Thr	Ile	Thr

(d) Differences in aminoacid sequence at positions

Fig. 11.9 The synthesis of insulin (a) in human cells natural way, (b) Synthesis of insulin by rDNA technology (c) sequence of amino acids in insulin and (d) difference in aminoacid position in insulin in certain mammals.

anaemia, blood clotting factors VIII and IX for haemophilia, require modifications that only cells of higher organisms like mammalian systems can provide. The high cost of maintaining animal cell cultures that produce only minute quantities of the drugs have been a bottle neck to the commercial development of recombinant cell culture method.

By the rDNA technology, the gene for a protein drug of interest can be transferred into another organism for production. Which animal system has to be employed for production is a technical and economic decision. For certain other protein drugs that require either complex modifications or are needed in large quantity, the production of transgenic animals seems to be inevitable and most efficient alternative. The farm animal becomes a production facility with several advantages : it is reproducible, has flexible production ability through the number of animals bred, and maintains its own energy (fuel) supply. Most attractive aspect of the strategy of animal drug production is that the drug is delivered from the animal in a very convenient form of package such as milk.

Essentially the transgenic animal for pharmaceutical production should satisfy two conditions. Firstly, the animals should be able to produce the desired drug / chemical molecule at high levels without endangering its own health. Secondly, the transgenic animal should have ability to pass on the gene to subsequent generation and be able to express the production capacity at the same high levels among it offsprings. The current strategy is to couple the DNA segment representing the sequence coding for the protein drug with a promoter and signal sequences, directing the synthesis, in the mammary gland. The 'transgene' while presenting in every cell of the animal, expresses only in the mammary gland in such a way that the protein drug is made in the milk. The question of the transgenic animal being harmed or endangered by the foreign gene does not arise at all, since the mammary gland and milk are essentially extraneous to the main life support system of the animal. After the gene sequence (DNA) for the protein drug has been joined with the mammary, directing signal, the same is injected into fertilized cow, sheep, goat, or mouse embryos with the help of a micromanipulator and microscope. The micro injected embryos are subsequently implanted into a surrogate mothers they survive and are borne normally.

Drugs currently made by the transgenic approach are listed in the Table 11.4.

Table 11.4 *Protein Drugs currently being made in Transgenic animals*

Drug Description	**Used for the treatment of**	**Transgenic animal**	**Values $ per animal per year**
Alpha-1-antitrypsin (AAT)	inherited deficiency leads to emphysema	Sheep	15,000/-
Tissue plasminogen activator (t pA)	treatment for blood clots	Goat	75,000/-
Factor VIII	clotting factors,	Sheep	37,000/-
Factor IX	for treatment of haemophilia	Sheep	20,000/-
Haemoglobin	blood substitute for human transfusion	Pig	3,000/-
Lactogerrin	infant formula food additive	Cow	20,000/-
Cystic fibrosis trans-membrane conductance regulator (CFTR)	Treatment of cystic fibrosis (CF)	Sheep Mouse	75,000/-
Human protein C	anticoagulant, treatment for blood clots	Pig	1,00,000/-

Essay Type Questions

1. Discuss about the prospects of genetic counselling.
2. Describe the techniques involved in prenatal diagnosis of heriditary diseases.
3. What do you understand by gene therapy? Give an overview of the disease in which the gene therapy can be applied.
4. What are vaccines ? Explain the role of DNA technology in generation of vaccines.
5. Write an account on chemicals and drugs from animals as a resources.
6. Explain how the recombinant DNA technology useful in production of chemicals and drugs from animals.

Short Answer Type Questions

1. Write short notes (about 200 words) on the following :
   (i) Techniques of prenatal diagnosis
   (ii) Gene Therapy
   (iii) Vaccines and new generation vaccines
   (iv) Bioactive molecules and biotechnology
   (v) Animal toxins and drugs.

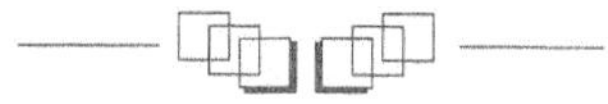

General Bibliography

- **A Text Book of Fishery Science** and **Indian Fisheries**, C. B. L Srivastava, Kitab Mahal, Allahabad, 1999.
- **A Text Book of Microbiology,** R. C. Dubery and D. K. Maheshwari, S. Chand and Co. Ltd., 1999.
- **An Introduction to Gymnosperms**, Trivedi, B. S. and Singh D. K., Sashidhar Malavlya Prakashan, Lucknow, 1965.
- **Animal Biology**, Richard D. Jurd, Bios Scientific Publishers, Oxford, 1998.
- **Animal Parasitology,** J. D. Smith, Combridge University Press, 1994.
- **Aquaculture - principles and practices,** Pillay TVR, Oxford Publications, London, 1996.
- **Biological Science**, N.P.O Green, G. W Stort, D. J. Taylor, Soper, Cambridge University Press, Cambridge, 1996.
- **Biotechnological Innovations in Animal Productivity**, Biotol, Butterworth-Heinemann, 1992
- **Biotechnological Innovations in crop improvement**, Biotol, Butterworth-Heinemann, 1991.
- **Biotechnology - The Science and Business**, V. Moses and R. E. Cape, Harwood Academic publishers, 1994.
- **Cryptogamic Botany - Bryophyta and Pteridophyta** Vol. II. Smith, G. M. Tata McGraw-Hill Publishing Co. Ltd., 1988.

- **Economic Botany of Crop Plants** - AVSS Sambamurthy and N. S. Subrahmanyam, Asiatech Publishers Inc, New Delhi 2000.
- **Fish and Fishery of India**, V. G. Jhingran, Hindustan Publishing Co, 1984.
- **Gene Cloning** by T. A. Brown, Chapman And Hall, 1990
- **General Parasitology** Thomas C. Cheng, Academic Press Inc, London, 1986.
- **Genetic Transformation in plants** by R. Walden, Open University Press, Miton Keynes, 1988.
- **Genetics** by Jan M. Friedman, Fred J. Dill, Michael R. Hayden, Barbara C. MC gillivray, Harwal publishing company, Malvern, Pennsylvania, 1992.
- **Hand Book of Medical Parasitology**, Viqar Zaman and Loh ah Keong, Keang Publishing Pvt. Ltd, Singapore, 1989.
- **Immunology**, I. Roitt, J. Brostoff and D. Male, Times international publishers Ltd. London, 1998.
- **Introduction to Microbiology** - A. S. Rao, Prentice-Hall of India Pvt. Ltd., 1997.
- **Introduction to Modern Mycology** - J. W. Deacon, ELBS/Blackwell Scientific publications, 1988.
- **Introductory Phycology**, (2nd Edition) Kumar, H.D.Affiliated East West Press Pvt. Ltd., New Delhi, 1999
- **Modern Text Book of Zoology - Invertebrates** R. L. Kotpal, Rostogi Publications, Meerut, 2002.
- **Plant Biology-Instant Notes** - Lack, A. J. and D. E. Evans Viva Books Pvt. Ltd., 2001.
- **Plant Genetic Engineering**, Don Grierson (ed), Blackie, Chapman and Hall, 1991.
- **Plant propagation by Tissue culture - Hand Book** and Directory of Commercial Laboratories, Edwin F. George and Paul D. Sherrington, Exegetics Ltd, 1984.
- **Principles of Medical Genetics** by Thomas D. Gelehrter and Fracis S. Collins, Williams and Wilkins, 1990.
- **Principles of Systematic Zoology**, Ernst Mayr and Peter D. Ashlock, MC Graw - Hill Inc, 1991.
- **Studies in gymnospermous Plants - Cycas**, Pant, D.D. and Mehra, B. Central Book Depot, Allahabad, 1962.
- **Taxonomy of Angiosperms**, (2nd edition) V. Singh and D. K. Jain, Rastogi Publications, Meerut, 1999.
- **The Embryology of Angiosperms** (4th edition), Bhojwani, SS and Bhatnagar SP. Vikas Publishing House Pvt. Ltd, New Delhi, 1999.
- **The Wealth of India**, Raw materials Vol IV, Fish and Fisheries, CSIR, New Delhi, 1976.

Index

D

E

F

G

H

I

K

L

M

N

O

P

R

S

T